T0245237

CAMBRIDGE LIBRARY COLLECTION

Books of enduring scholarly value

Technology

The focus of this series is engineering, broadly construed. It covers technological innovation from a range of periods and cultures, but centres on the technological achievements of the industrial era in the West, particularly in the nineteenth century, as understood by their contemporaries. Infrastructure is one major focus, covering the building of railways and canals, bridges and tunnels, land drainage, the laying of submarine cables, and the construction of docks and lighthouses. Other key topics include developments in industrial and manufacturing fields such as mining technology, the production of iron and steel, the use of steam power, and chemical processes such as photography and textile dyes.

Electrical Papers

A self-taught authority on electromagnetic theory, telegraphy and telephony, Oliver Heaviside (1850–1925) dedicated his adult life to the improvement of electrical technologies. Inspired by James Clerk Maxwell's field theory, he spent the 1880s presenting his ideas as a regular contributor to the weekly journal, *The Electrician*. The publication of *Electrical Papers*, a year after his election to the Royal Society in 1891, established his fame beyond the scientific community. An eccentric figure with an impish sense of humour, Heaviside's accessible style enabled him to educate an entire generation in the importance and application of electricity. In so doing he helped to establish that very British phenomenon, the garden-shed inventor. Illustrated with practical examples, the subjects covered in Volume 1 include voltaic constants, duplex telegraphy, microphones and electromagnets.

Cambridge University Press has long been a pioneer in the reissuing of out-of-print titles from its own backlist, producing digital reprints of books that are still sought after by scholars and students but could not be reprinted economically using traditional technology. The Cambridge Library Collection extends this activity to a wider range of books which are still of importance to researchers and professionals, either for the source material they contain, or as landmarks in the history of their academic discipline.

Drawing from the world-renowned collections in the Cambridge University Library, and guided by the advice of experts in each subject area, Cambridge University Press is using state-of-the-art scanning machines in its own Printing House to capture the content of each book selected for inclusion. The files are processed to give a consistently clear, crisp image, and the books finished to the high quality standard for which the Press is recognised around the world. The latest print-on-demand technology ensures that the books will remain available indefinitely, and that orders for single or multiple copies can quickly be supplied.

The Cambridge Library Collection will bring back to life books of enduring scholarly value (including out-of-copyright works originally issued by other publishers) across a wide range of disciplines in the humanities and social sciences and in science and technology.

Electrical Papers

VOLUME 1

OLIVER HEAVISIDE

CAMBRIDGE UNIVERSITY PRESS

Cambridge, New York, Melbourne, Madrid, Cape Town,
Singapore, São Paolo, Delhi, Tokyo, Mexico City

Published in the United States of America by Cambridge University Press, New York

www.cambridge.org
Information on this title: www.cambridge.org/9781108028561

This edition first published 1892
This digitally printed version 2011

ISBN 978-1-108-02856-1 Paperback

ELECTRICAL PAPERS.

VOL. I.

ELECTRICAL PAPERS

BY

OLIVER HEAVISIDE

IN TWO VOLUMES

VOL. I.

London
MACMILLAN AND CO.
AND NEW YORK
1892

PREFACE.

THIS Reprint of my Electrical Papers comes about by the union of a variety of reasons and circumstances.

First, there was a demand for certain of my papers, especially for a set relating to Electromagnetic Waves. Although I distributed 49 copies in a collected form, I was asked for more, and also received assurances that a republication of my papers in general would be useful. But this demand was too small to lead to an immediate supply.

Secondly, however, at the beginning of 1891 it was proposed to me by the publisher of *The Electrician* that my articles on "Electromagnetic Theory," then commencing and now continuing in that journal, should be brought out later in book form. This was satisfactory so far as it went, but it brought the question of a reprint of the earlier papers to a crisis. For, as the later work grows out of the earlier, it seemed an absurdity to leave the earlier work behind.

Thirdly, the experimental work of Hughes in 1886, furnishing the first evidence (in the sense ordinarily understood, though other evidence was convincing to a logical mind) of the truth of the theory of surface conduction along wires under certain circumstances, first advanced by me a year previously; followed in 1887-8 by the experimental work of Hertz and Lodge on electrical vibrations and electromagnetic waves, still further confirming the above, and also broadly confirming the truth of the theory of the propagation of disturbances along wires I had worked out on the basis of Maxwell's doctrine of the ether in its electromagnetic aspect, and the correctness of Fitzgerald's ideas concerning electrical radiation, and of the nature of the energy-flux developed by Poynting and myself from Maxwell's theory, were the means of stirring up an amount of interest in this theory that was quite wonderful to witness. That electrical disturbances were propagated in time through a medium was raised from a highly probable speculation to an established fact. A careful study by electrical physicists of Maxwell's development of Faraday's ideas became imperative, especially on the Continent, where Maxwell's work had hitherto met with a singular want of appreciation, arising, I believe, mainly from misconception of his theory of electrical displacement. This misconception, I think, exists even now, since some writers apply to Maxwell's theory ideas and processes which seem to me to be thoroughly antagonistic to his views. But even in England the theory had been much neglected. For one thing, much attention was being devoted to the dynamo. Then again, the form in which Maxwell presented his theory did not, I think, display its merits in a manner they deserve, and suited for legitimate development. Moreover, the contrast between the old notions of electricity and Maxwell's was so great that mere natural conservatism stood in the way. A stimulus was wanted in favour of a theory so ill-understood and (apparently) so far removed from actual observation. But the experimental stimulus having come, the result has been a flood of other experimental work, mostly tending to confirm the general theory. A work, therefore, like the present, which is, in the main, devoted to the elucidation and extension of Maxwell's theory, and of the mathematical methods suited to it, should have a legitimate place amongst others. Though it was nearly all done before the electrical "boom" began, it may not be out of date, and may perhaps be, in some respects, ahead.

Fourthly, it had been represented to me that I should rather boil the matter down to a connected treatise than republish in the form of detached papers. But a careful examination and consideration of the material showed that it already possessed, on the whole, sufficient continuity of subject-matter and treatment, and even regularity of notation, to justify its presentation in the original form. For, instead of being, like most scientific reprints, a collection of short papers on various subjects, having little coherence from the treatise point of view, my material was all upon one subject (though with many branches), and consisted mostly of long articles, professedly written in a connected manner, with uniformity of ideas and notation. And there was so much comparatively elementary matter (especially in what has made the first volume) that the work might be regarded not merely as a collection of papers for reference purposes, but also as an educational work for students of theoretical electricity.

As regards the question, "Will it pay?" little need be said. For, fifthly, however absurd it may seem, I do in all seriousness hereby declare that I am animated mainly by philanthropic motives. I desire to do good to my fellow-creatures, even to the *Cui bonos*.

Having thus justified the existence of this reprint, it remains for me to indicate the general nature of the contents, and, in doing so, I will imagine myself (usually) to be addressing an intelligent and earnest student, who means business. The first twelve articles, pp. 1 to 46, are on matters dealing mainly with telegraphy, and are but loosely connected. But a sort of continuity then begins, for the next eight articles, up to p. 179, deal mainly with the theory of the propagation of variations of current along wires, beginning with applications of the simple electrostatic theory of Sir W. Thomson (1855) to cables under different circumstances (terminal resistances, condensers, etc., intermediate leakage, etc.), and followed by extensions to include self-induction, or the influence of the inertia of the magnetic medium, and the mutual influence, both electrostatic and magnetic, of parallel wires. The last of this set, Art. XX., has not been printed before. It is, however, in its right place, having been written in 1882 as a sequel to the papers preceding it. It may be found useful to those who are interested in the subject as an intermediate between the papers of this set and the later series in the second volume, wherein the subject is treated from a more comprehensive point of view, viz., Maxwell's theory of the ether as a dielectric. There is no conflict. The later investigations are generalizations of the earlier, or the earlier are specializations of the later; and I can recommend the earnest student to read the earlier set first, before proceeding to the more advanced treatment in the later set.

We next come to a series of papers published in *The Electrician* between the autumn of 1882 and the autumn of 1887, when under the editorship of Mr. C. H. W. Biggs, to whom I desire to express my obligations for the opportunity he gave me of exercising my philanthropic inclinations, in the face, as I afterwards learnt, of considerable opposition. These papers extend over about 500 pages, mostly in this, partly in the second volume, and are usually long articles, with continuity. They relate to electrical theory in general. Beginning with the abstract relations of the electrical quantities, and the mathematics of the subject in vector form (of an elementary kind), including a general theory of potentials and connected quantities expressed in the rational units I introduced, we pass on to the consideration of the energy of the electric and magnetic fields, and the transformations concerned in the phenomenon of the electric currents including an account of Sir W. Thomson's theory of thermo-electricity. Next comes a pretty full study of the theory of the propagation of induction and electric current in round cores, to which I was led by my experiments with induction balances, in the endeavour to explain certain phenomena observed. The analogy with the motion of a viscous liquid is also introduced and developed.

Lastly we come (1885) to a more comprehensive treatment of electromagnetism, based upon Maxwell's theory, in "Electromagnetic Induction and its Propagation," of which the first half is in this volume. I here introduce a new method of treating the subject (to which I was led by considering the flux of energy), which may perhaps be appropriately termed the Duplex method, since its main characteristic is the exhibition of the electric, magnetic, and electromagnetic equations in a duplex form, symmetrical with respect to the electric and magnetic sides, introducing a new form of fundamental equation connecting magnetic current with electric force, as a companion to Maxwell's well-known equation connecting magnetic force and electric current. The duplex method is eminently suited for displaying Maxwell's theory, and brings to light many useful relations which were formerly hidden from view by the intervention of the vector-potential and its parasites. There is considerable difficulty in treating electromagnetism by means of Maxwell's equations of propagation in terms of these quantities, as presented in his treatise. The difficulty is greatly increased, if not rendered practically insuperable, when we pass to more advanced cases involving heterogeneity and eolotropy and motion of the medium supporting the fluxes. Here the duplex method furnishes what is wanted in general investigations, and is the basis of "Electromagnetic Induction" and of the whole of the second volume. The electric and magnetic forces (or fluxes) and their variations are the immediate objects of attention in the duplex method, whilst potentials are treated as auxiliary quantities which do not possess physical significance as regards the actual state of the medium, though they may be useful for calculating purposes.

Towards the end of this volume the electric and magnetic stresses are considered. The treatment was interrupted, but a later paper, "On the Forces, Stresses, etc.," in the second volume contains what was to have been its continuation, and developments thereof. The reason of the break was that the interest excited by Professor Hughes's 1886 experiments made it desirable that I should at once publish other matter long in hand, namely, developments of the views relating to the functions of wires and of the dielectric surrounding them, explained in Section II. of "Electromagnetic Induction." These developments are contained in the second half of that article (Art. XXXV., vol. II.) and in the article "On the Self-Induction of Wires" (Art. XL., vol. II.), published in the *Philosophical Magazine* in 1886-7. The reader is recommended to read the former first, as it is much more elementary than the latter, which contains mathematical developments and examinations unsuited to *The Electrician.* The subject is the diffusion of electrical waves into wires from their boundaries and the propagation of waves along the wires through the insulator surrounding them, supplying the wires themselves with the energy they absorb. Also the self-induction of various arrangements of apparatus, and the theory of induction balances.

But in the year 1887 I came, for a time, to a dead stop, exactly when I came to making practical applications in detail of my theory, with novel conclusions of considerable practical significance relating to long-distance telephony (previously partly published), in opposition to the views at that time officially advocated. On the official side the electrostatic theory was upheld, with full application of the retardation law of the inverse-squares to telephony; inertia being regarded as a disturbing factor, assumed to be of a harmful nature, but argued to be quite negligible in long copper-circuits, because telephony through such circuits of low resistance was so successful. On the other side was my theory asserting that owing to the rapidity of telephonic changes of current inertia was not negligible, that it was often important, and sometimes, as in the case of wires of low resistance, even a dominating factor. Furthermore, that it was not harmful, but was, on the contrary, beneficial in its effects, which was, in fact, the very reason why long-distance telephony was successful. Then, as regards the measure of the inductance, it was asserted on the official side that the inductance per centim. of a copper suspended circuit was (in electromagnetic units) only a minute fraction of unity; whilst on the other side it was declared to be some hundreds of times as big, say from 10 to 20 per centim. of circuit. Here was the most complete possible antagonism between my views and official views, both in principle and in detail, and a careful consideration and discussion of the matter was desirable. Yet I found it next to impossible to ventilate the matter. First of all, I was prevented by circumstances which need not be mentioned from bringing the matter before the S. T. E. and E. in the spring of 1887 (Art. XLI., vol. II.). Next, a little later, the editor of the *Philosophical Magazine* could no longer afford space for the continuation of my article on "The Self-Induction of Wires," Part VIII., dealing with the non-distortional circuit and telephony (p. 307, vol. II.). Thirdly, after a partial exposition in Sections XL. to XLVI. of "Electromagnetic Induction," a change of editor occurred, and the new editor asked me to discontinue. He politely informed me that although he had made particular enquiries amongst students who would be likely to read my papers, to find if anyone did so, he had been unable to discover a single one. Fourthly, he returned a short article (Art. XXXVIII., vol. II.) on the same subject of long-distance telephony, which pointed out official errors in detail, and directed attention to the contrary results indicated by my theory, this paper having been in official hands. And lastly, three other journals declined the same, for reasons best known to themselves.

Perhaps it was thought that official views were so much more likely to be right that it was safe to decline the discussion of novel views in such striking opposition thereto. There seemed also to be an idea that official views, in virtue of their official nature, should not be controverted or criticized. But there seems something wrong here, as the above facts, and the later evidence in support of my views, have shown. For what other object have scientific men than to get at the truth, and how is it to be done without free discussion?

The student is particularly recommended to read the articles referred to, not merely on account of the telephonic application, but because of the simplicity of treatment which the distortionless circuit allows, and as a preliminary to the study of Electromagnetic Waves, to which it supplies a royal road. The action of leakage in promoting quick signalling is treated of in the early set in this volume; now the inductance of the circuit has also a beneficial effect; and the two together conspire to annihilate the distortion which the resistance of the circuit produces. The same occurs (approximately) without the leakage, by the action of self-induction, if the frequency of alternation be sufficiently rapid, and the wires of not too great resistance.

Now in the theory of electromagnetic waves there is a similar property, which throws considerable light upon the subject of waves in general. I had introduced, in 1885, for purposes of symmetry, the fictitious quality of magnetic conductivity. When its effects upon the propagation of waves in a real conducting dielectric are enquired into, it is found to act contrary to the real conductivity, so that the distortion due to the latter can be entirely removed by having duplex conductivity. How this strange result comes to pass may be readily understood in detail by studying the theory of the distortionless circuit, in which the leakage conductance and the resistance of the circuit act oppositely in respect to distortion.

The remainder of the second volume consists of investigations growing out of "Electromagnetic Induction," viz., the set relating to electromagnetic waves; the electromagnetic wave-surface; propagation in a uniform conducting (duplex) dielectric, with the application to plane waves, either free or along straight wires; the connected theory of convection currents; the theory of resistance and conductance operators; with a few miscellaneous papers concerning propagation in moving media; finishing with an article discussing the forces and stresses concerned in the electromagnetic field.

Acting under advice, I have not carried out my original design to make large additions. Limitations of space prevented this, and I have confined myself to an occasional small addition or footnote. These are put in square brackets, all such signs in the original papers being cancelled. For the rest, I have corrected misprints and obvious slips, and have made verbal improvements and omitted occasional redundant matter. The scientific reader may therefore refer to this work as to the original papers. Their dates, etc., are given at the commencement of the articles.

I have introduced uniformity in the notation connected with vectors, though there was little change to be made except to put all vectors into Clarendon black type, as in some of the later of the original papers. The vector-algebra, I should mention, is of a rudimentary kind, and has nothing to do with quaternions; first, only addition and the scalar product are used, whilst later on the vector product is introduced and freely employed.

On the vexed question of vectors, the conclusions to which I have gradually settled down are as follows:—The notorious difficulty of understanding and working Quaternions will always be a bar to their serious practical use by any but mathematical experts. But, on the other hand, a vector algebra and analysis of a simple kind, independent of the quaternion, and readily understandable and workable, can with great advantage take the place of much of the usual cumbrous Cartesian investigations, and be made generally useful in all physical mathematics concerning vectors, and be employed, comparatively speaking, by the multitude. It should obviously be harmonized with the Cartesian mathematics. The quaternionic system is defective in this respect;

in its very nature it cannot be thus harmonized. The system I recommend is fully explained in "Electromagnetic Theory," chapter III. (*The Electrician*, Nov. 13, 1891, and after). The numerous letter prefixes of the quaternionic system, which greatly contribute to the difficulty of reading quaternionic investigations, are abolished, retaining only the symbol V before a vector product. Another difficulty is in the scalar product of Quaternions being always the negative of the quantity practically concerned. Yet another is the unreal nature of quaternionic formulæ. The terms do not stand for physical quantities. Again, in most physical mathematics, the quaternion does not even present itself for consideration, or, at any rate, may be readily dispensed with. Lastly, the establishment of vector-algebra on a quaternionic basis is very hard to understand, as chapter II. of Professor Tait's treatise shows. These troubles are obviated by the method I follow, basing the whole upon the definition of a vector, and of the scalar and the vector product of a pair of vectors. The notation is harmonized with Cartesians and transition is readily made. We may, indeed, regard a vector investigation, from this point of view, as a systematically abbreviated Cartesian investigation, and the latter as the full expansion of the former. And, considering that the bulk of special investigations are necessarily scalar, it seems to me that we should keep in touch with them as far as possible, and not try to abolish the Cartesian method, but make it a useful auxiliary to the vector method. That quaternionic experts may do valuable work is undoubted, but how can the bulk of mathematicians possibly understand it?

Lastly, on the question of units, it is not, I think, generally understood that the ordinary electrical units involve an absurdity similar to what would be introduced into the metric system of common units were we to define the unit area to be the area of a circle of unit diameter. A rational system of units founded upon a rational definition of a pole (electric or magnetic), associating the *unit* pole with *one* line of the corresponding force or flux instead of with 4π, was employed by me in some of the earlier papers (1882-3), but was not carried out further because I believed that a reform of the electrical units was impracticable. Now, I had commenced "Electromagnetic Theory" in January, 1891, with rational units merely to exhibit the theory in a fitting manner, intending to transform later to the common units. But I came afterwards to the definite conclusion that a thorough reform of the electrical units is practicable and perhaps indeed inevitable, and shall therefore continue the use of the rational units. But this decision was only arrived at after a considerable portion of this volume was in type. I have, therefore, not altered to rational units throughout, as I should have preferred; though, on the other hand, the long article LII. at the end of the second volume remains as it was written, in rational units. But we are, in the opinion of competent judges, within a measurable distance of a reform of the ordinary heterogeneous British units, by adoption of the metric system. I hope and believe that the smaller reform I advocate will be determined upon by electricians.

PAIGNTON, DEVON, June 16, 1892.

CONTENTS OF VOL. I.

CORRECTIONS.

p. 44, 8th line from end, *for* $\left.\frac{2h^2}{r}\right)$ *read* $\left.\frac{2h}{r}\right)^2$.

p. 99, last formula, *for* L_2 *read* L, and *for* R_2 *read* R.

p. 267, 27th line, *for* $\Sigma_2\mathbf{B}$ *read* $\Sigma\,\mathbf{B}_2$.

p. 415, *for* § 39 *read* § 40.

p. 555, equation (39a), for H_0 read $\mathbf{H}_0$

ELECTRICAL PAPERS.

I.—COMPARING ELECTROMOTIVE FORCES.

[*English Mechanic*, July 5th, 1872, p. 411.]

THE following null arrangement for comparing electromotive forces is, as far as I am aware, original:—Join up the two batteries E_1 and E_2 with a galvanometer, as in the diagram, so that their currents go through it in opposite directions. Also insert resistances R and r. Let x and y be the unknown resistances of the batteries, and i_1, i_2, i_3, the currents in the three branches. Then we have

$$i_1 - i_2 \pm i_3 = 0,$$
$$(R+x)i_1 + gi_3 = E_1,$$
$$(r+y)i_2 + gi_3 = E_2.$$

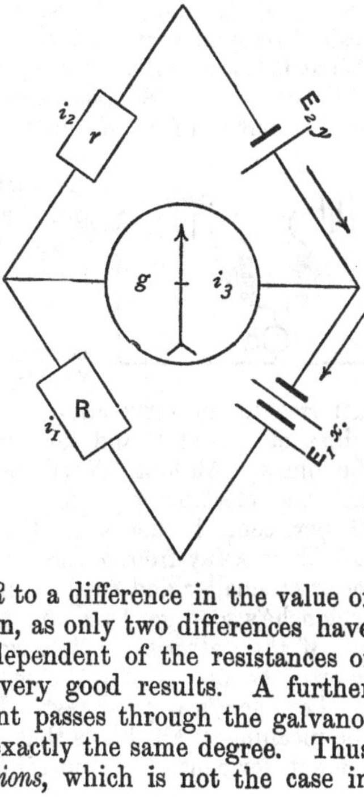

Now, by altering the resistance R, bring the needle to zero. Then $i_3 = 0$, and $i_1 = i_2$, therefore

$$\frac{E_1}{E_2} = \frac{R+x}{r+y}.$$

Here we have the unknown resistances, x and y, in our result; but by taking another value of R, say R', and finding the corresponding value of r, say r', we get the simple result

$$\frac{E_1}{E_2} = \frac{R-R'}{r-r'}, \text{ or } \frac{\Delta R}{\Delta r},$$

the ratio of a difference in the value of R to a difference in the value of r. This method, involving no calculation, as only two differences have to be observed, and being perfectly independent of the resistances of the batteries and galvanometer, gives very good results. A further advantage is that, as $i_1 = i_2$ and no current passes through the galvanometer, each battery is being worked to exactly the same degree. Thus they are compared under *similar conditions*, which is not the case in Poggendorff's and other methods.

II.—VOLTAIC CONSTANTS.

[*Telegraphic Journal*, May 15th, 1873, p. 146.]

THIS journal for April 15th contains an article on a "New Method of Determining Voltaic Constants." It is new, inasmuch as it is not to be found in any electrical books, as far as I am aware, but it is not entirely new. The method to which the diagrams 6, 7, and 8 refer was devised by me about three years ago, and it will be found in the *English Mechanic* for July 5, 1872, p. 411. A description and proof will be found there. I arrived at it nearly as M. Emile Lacoine has, by considering the potentials of the different points of a circuit containing two electromotive forces of the same sign. Perhaps a few remarks as to the value of this method may not be unacceptable. It gives very different results from Poggendorff's, and with reason. Poggendorff's method—in which the battery having the lesser electromotive force is not allowed to work—especially as improved on by Latimer Clark, most certainly is an exceedingly accurate way of comparing the electromotive forces of elements when not in action, which may be then very well called their *potentials*; but it is a notorious fact that these potentials fall more or less, generally more, when the batteries are called upon to make themselves useful. The new method in question compares the *working* electromotive forces of batteries when in action through any desired resistance, and can on that account be of some value in practice. (What is the use of a battery having a very great potential if it is only while sleeping?) Suppose we compare a number of Daniell's with an equal number of Leclanché's by the new method. Referring to the figure, let the right-hand battery be the Leclanché's with the big and lazy potentials, and the left the Daniell's with the smaller but more industrious potentials. Then

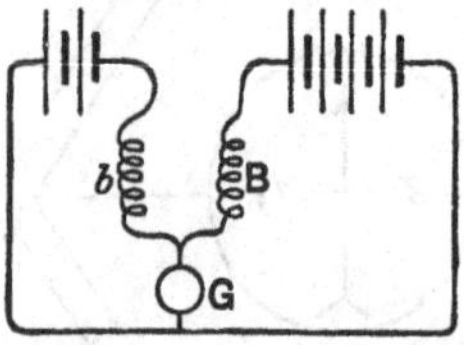

$$\frac{L}{D}=\frac{\Delta B}{\Delta b}$$

expresses their relative electromotive forces. Now we may watch the behaviours of these batteries in an instructive manner by commencing with very high values of B and b, and for convenience we may make Δb constantly 100 ohms. At first ΔB will be found much higher, say 150, showing that the electromotive force of the Leclanché's is *at that moment* 50 per cent. higher than the Daniell's; but by constantly taking 100 ohms away from b, the corresponding difference in B, namely ΔB, becomes smaller and smaller, and if we go on for a little while (for the Leclanché's soon get tired) ΔB will become actually less than 100 ohms, and, if the batteries be left working, may fall much lower. We may reverse the process, but ΔB will not become 150 again unless we give the Leclanché's a good rest. No two series of trials agree, however. The meaning of all this is that the electromotive force of the Leclanché element, for continuous working, is anything between nothing and 1·2 or 1·3 times that of Daniell's. I have even seen the current of a Leclanché cell reverse itself after a few hours' hard work, but it partially recovered after a rest.

III.—ON THE BEST ARRANGEMENT OF WHEATSTONE'S BRIDGE FOR MEASURING A GIVEN RESISTANCE WITH A GIVEN GALVANOMETER AND BATTERY.

[*Phil. Mag.*, Feb. 1873, S. 4, vol. 45.]

In the figure, a, b, c, and d are the four sides of the electrical arrangement known as Wheatstone's bridge or balance, e the galvanometer, and f the battery branch. Throughout this paper d is supposed to be the resistance to be measured, and e and f both known. The problem is to find what resistances should be given to the sides a, b, and c (which we are able to vary), so that the galvanometer may be affected the most by any slight departure from the balance which occurs

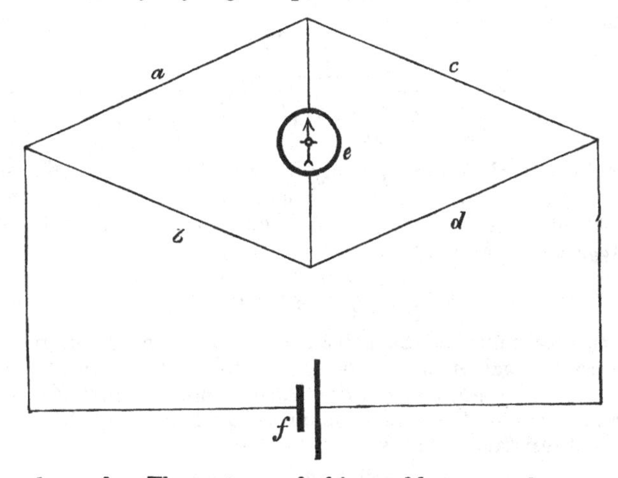

when $a:b=c:d$. The nature of this problem may be more easily understood from the following considerations:—

1. If b, c, d, e, and f are given, then there is only *one* value of a which will produce a balance, viz., $a=\frac{bc}{d}$.

2. But if c, d, e, and f are given, but not b, then there is an infinite number of pairs of values of a and b which will produce a balance by satisfying the relation $a:b=c:d$; and one particular pair will constitute the best arrangement, by which is meant that the galvanometer will be most sensitive to any slight departure from the equality of $\frac{a}{b}$ and $\frac{c}{d}$ when those particular values of a and b are used.

3. And if only d, e, and f are given, then for any value we give to c there is a pair of values of a and b which constitutes the best arrangement for that value of c; and there will be a particular value of c which, with the corresponding values of a and b, will be the best arrangement for the given values of d, e, and f.

In order to find what functions a, b, and c must be of d, e, and f to constitute the best arrangement, it will be first necessary to find the best values of a and b when c, d, e, and f are given. This I now proceed to do.

It is well known, and may be easily proved by Kirchhoff's laws, that the current passing through the galvanometer is represented by

$$u=\frac{E\times(a+b+c+d)(ad-bc)}{\{(a+b)(c+d)+(a+b+c+d)e\}\{(a+c)(b+d)+(a+b+c+d)f\}}, \quad (1)$$

in which E is the electromotive force of the battery. $(ad-bc)$ may be positive, negative, or nothing, in which last case $u=0$, and a balance is obtained, no current passing through the galvanometer.

Dividing both numerator and denominator of (1) by

$$(a+b+c+d)^2,$$

it becomes

$$u=E\times\frac{\dfrac{ad-bc}{a+b+c+d}}{\left\{\dfrac{(a+b)(c+d)}{a+b+c+d}+e\right\}\left\{\dfrac{(a+c)(b+d)}{a+b+c+d}+f\right\}}; \quad (2)$$

from the form of which it may easily be seen that the best value of the resistance of the galvanometer e, when a balance is obtained and the other resistances are fixed, is, as Schwendler has shown in the *Philosophical Magazine* for May, 1866,

$$e=\frac{(a+b)(c+d)}{(a+b+c+d)}=b\cdot\frac{c+d}{b+d}; \quad (3)$$

that is, the resistance of the galvanometer should equal the resistance external to the galvanometer, being the joint resistance of the two parallel branches $(a+b)$ and $(c+d)$. Also it may be proved that the best arrangement of the battery is obtained when its resistance equals the external resistance, that is,

$$f=\frac{(a+c)(b+d)}{a+b+c+d}=c\cdot\frac{b+d}{c+d}, \quad (4)$$

the joint resistance of the two parallel branches $(a+c)$ and $(b+d)$.

(In passing, I may notice that Schwendler, in the paper above referred to, and also in a later one in the *Philosophical Magazine* for January, 1867, has assumed it to be necessary for the battery resistance to be very small, in order that the relation exhibited in equation (3) may be satisfied. This appears to me to be totally unnecessary; for the resistance external to the galvanometer when a balance is obtained is quite independent of f, the battery resistance. In fact, the proper resistance for the battery when it is to be most advantageously used is given by equation (4).)

As in the present paper we are only concerned with such values of a, b, c, and d as produce a balance, or nearly so, one of these four resistances may be eliminated at once. Let it be a. Then

$$\frac{(a+b)(c+d)}{a+b+c+d}=b\cdot\frac{c+d}{b+d},$$

$$\frac{(a+c)(b+d)}{a+b+c+d}=c\cdot\frac{b+d}{c+d},$$

and

$$a+b+c+d=\frac{(b+d)(c+d)}{d}.$$

Substituting these in equation (2), we get

$$u=E\times\frac{\dfrac{(ad-bc)d}{(b+d)(c+d)}}{\left\{b\dfrac{c+d}{b+d}+e\right\}\cdot\left\{e\dfrac{b+d}{c+d}+f\right\}}$$

$$=Ed\times\frac{ad-bc}{(bc+ef)(b+d)(c+d)+ce(b+d)^2+bf(c+d)^2}.\;\ldots\ldots\;(5)$$

Now c, d, e, and f being fixed, and b the variable, we have to make u a maximum. As Ed is constant, it may be dismissed. As to the numerator $(ad-bc)$, it vanishes when at a balance; but of course such a thing as an exact balance is unattainable. Let $d\pm\Delta$ be the real value of the resistance we are measuring, d being the calculated value $\frac{bc}{a}$, and Δ a small difference, then

$$a(d\pm\Delta)-bc=\pm a\Delta.$$

Therefore the numerator varies as a or as b, since in the present case a and b vary together. Hence we may write b for $(ad-bc)$. Thus

$$u=\frac{b}{(bc+ef)(b+d)(c+d)+ce(b+d)^2+bf(c+d)^2}.$$

By differentiation and putting $\frac{du}{db}=0$, we obtain

$$(bc+ef)(b+d)(c+d)+ce(b+d)^2+bf(c+d)^2$$
$$=bc(b+d)(c+d)+b(bc+ef)(c+d)+2bce(b+d)+bf(c+d)^2;$$

therefore

$$ef(b+d)(c+d)+ce(b+d)^2=b(bc+ef)(c+d)+2bce(b+d),$$
$$def(c+d)+ce(b+d)^2=b^2c(c+d)+2bce(b+d),$$
$$b^2c(c+d+e)=de(cd+df+fc),$$

which gives the relation sought,

$$b=\sqrt{\frac{d}{c}\cdot\frac{cd+df+fc}{c+d+e}\cdot e};\;\ldots\ldots\ldots\ldots\;(6)$$

and as $a=\frac{bc}{d}$, therefore

$$a=\sqrt{\frac{c}{d}\cdot\frac{cd+df+fc}{c+d+e}\cdot e}.\;\ldots\ldots\ldots\ldots\;(7)$$

These values of a and b will be found to make $\frac{d^2u}{db^2}$ negative; therefore they give the most sensitive arrangement for the fixed values of c, d, e, and f.

If b vary from nothing upwards, it will be found that u rapidly increases up to its maximum value and then slowly decreases, from which it may be concluded that it is better to use too large values of a and b than too small.

In case $c=d$, formulæ (6) and (7) become

$$a=b=\sqrt{ce\frac{c+2f}{2c+e}}. \qquad (8)$$

As a numerical example of these formulæ, suppose the resistance to be measured $d=1{,}000$ ohms, the galvanometer $e=500$ ohms, the battery resistance $f=100$ ohms, and we make $c=1{,}000$ ohms; then the best values for a and b will be found to be $\sqrt{240{,}000}=100\sqrt{24}$, or nearly 500 ohms.

Having thus determined the relations of a and b to c, d, e, and f, the latter resistances being fixed, we now proceed to the second part of the problem, to determine the best values of a, b, and c when only d, e, and f are given. This is the case which occurs so often in practice, when we have a battery, a galvanometer, and a resistance to be measured, and three sides of a bridge to which we may give any values we choose (within certain limits).

Insert the values of a and b, as given in equations (6) and (7), in equation (5); then, after some reductions, we obtain

$$u=\frac{ad-bc}{2de(cd+df+fc)+\{(c+d+e)(cd+df+fc)+cde\}\sqrt{\frac{d}{c}\cdot\frac{cd+df+fc}{c+d+e}e}}.$$

We must now consider c the independent variable, a and b being dependent variables. $(ad-bc)$ still varies as a. It does not, however, vary as b, but as the product bc or ad, since d is constant. Therefore we may put the known value of bc in the numerator instead of $(ad-bc)$. Thus

$$u=\frac{\sqrt{cde\frac{cd+df+fc}{c+d+e}}}{2de(cd+df+fc)+\{(c+d+e)(cd+df+fc)+cde\}\sqrt{\frac{d}{c}\cdot\frac{cd+df+fc}{c+d+e}e}}.$$

Multiply numerator and denominator by $\sqrt{\frac{c}{de}\cdot\frac{c+d+e}{cd+df+fc}}$, and we have

$$u=\frac{c}{2\sqrt{cde(c+d+e)(cd+df+fc)}+(c+d+e)(cd+df+fc)+cde},$$

which has to be made a maximum. Differentiating and putting $\frac{du}{dc}=0$,

$$2\sqrt{cde(c+d+e)(cd+df+fc)}+(c+d+e)(cd+df+fc)+cde$$

$$=\frac{c}{\sqrt{cde(cd+df+fc)(c+d+e)}}$$
$$\times\{cde(cd+df+fc)+cde(c+d+e)(d+f)+de(c+d+e)(cd+df+fc)\}$$
$$+c(c+d+e)(d+f)+c(cd+df+fc)+cde.$$

Therefore

$$2\sqrt{cde(c+d+e)(cd+df+fc)}+df(d+e)-c^2(d+f)$$
$$=\frac{cde\{c(cd+df+fc)+c(c+d+e)(d+f)+(c+d+e)(cd+df+fc)\}}{\sqrt{cde(cd+df+fc)(c+d+e)}}.$$

Multiplying both sides of this equation by the denominator on the right hand side and reducing, we get

$$\{df(d+e)-c^2(d+f)\}\sqrt{cde(c+d+e)(cd+df+fc)}$$
$$=cde\{c^2(d+f)-df(d+e)\},$$

which is satisfied by

$$df(d+e)-c^2(d+f)=0,$$

which gives the required relation,

$$c=\sqrt{df\frac{d+e}{d+f}};\;\ldots\ldots\ldots\ldots\ldots\ldots\ldots\ldots \quad (9)$$

that is, c equals the square root of the product of the joint resistance of the battery and the resistance to be measured, into the sum of the resistance of the galvanometer and the resistance to be measured. Inserting this value of c in (6) and (7), we find the values of a and b to be

$$a=\sqrt{ef},\;\ldots\ldots\ldots\ldots\ldots\ldots\ldots\ldots \quad (10)$$

$$b=\sqrt{de\frac{d+f}{d+e}}.\;\ldots\ldots\ldots\ldots\ldots\ldots\ldots\ldots \quad (11)$$

In using the Wheatstone's bridge for measuring very high resistances, as, for instance, the insulation resistances of (good) telegraph lines, the battery resistance is usually very small in comparison with that of the line; hence $\frac{df}{d+f}$ will be very little different from f. When this is the case, formula (9) becomes

$$c=\sqrt{f(d+e)}.$$

If also the galvanometer resistance is small compared with the resistance to be measured, then these equations are sufficient for the determination of b and c,

$$b=\sqrt{de},$$
$$c=\sqrt{df}.$$

As a numerical example of these formulæ, suppose $f=100$ ohms, $e=1000$ ohms, and d is known to be about 1,000,000 ohms. Then by (10),

$$a=\sqrt{100{,}000}=316 \text{ ohms}.$$

By (9), $$c=\sqrt{\frac{10^6+10^2}{10^6+10^2}}(10^6+10^3)=10{,}004 \text{ ohms.}$$

$$b=\frac{ad}{c}=31{,}608 \text{ ohms.}$$

These values of a, b, and c will be the best. The more convenient arrangement,

$$a=300, \qquad c=10{,}000,$$
$$b=30{,}000, \qquad d=1{,}000{,}000,$$

would be very nearly the best.

It appears to me that the formulæ (9), (10), and (11), or those following, will be found of considerable practical value. If the same battery and galvanometer be always used, the side a of the bridge will be a constant resistance, and a table of the nearest convenient values of b and c could be easily calculated for different values of d. Formula (3), which is Schwendler's, can evidently have only a very limited application, as, for instance, to the construction of galvanometers for particular purposes. Formula (4) could be sometimes used; but it is a troublesome thing to make combinations of cells for "quantity" or "intensity" besides spoiling them if they are not all precisely similar.

In conclusion, if, to measure a certain resistance, the best resistances for the galvanometer, battery, and the three sides a, b, and c were required, then we should have to make $a=b=c=d=e=f$, which can be proved by combining equations (3), (4), (6), and (10). This, however, is more curious than useful.

IV.—SENSITIVENESS OF WHEATSTONE'S BRIDGE.

[*The Electrician*, February 15th, 1879, p. 147.]

SOME difference of opinion prevails amongst electricians as to what constitutes the most sensitive arrangement of Wheatstone's Bridge for comparing electrical resistances. Now, were Wheatstone's Bridge little used, this would be of no importance; but as it has, on the other hand, most extensive employment, it is certainly desirable that the matter should be thoroughly threshed out. When it is considered that Wheatstone's Bridge is by no means a complicated electrical arrangement, and that the laws regulating the currents in the different branches, and the proportions in which they are divided when division takes place, are extremely simple, and their accuracy as well established as that of the law of gravitation, the wonder is that there should be any doubt respecting a question which can be brought under mathematical reasoning without any hypothetical assumptions whatever. In the following is given an outline of the subject, omitting all the algebraical work.

Wheatstone's Bridge consists of six conductors, A, B, C, X, F, and G, uniting four points. Let the small letters a, b, c, x, f, and g stand for their respective resistances. The property which gives so much value to the bridge in comparing resistances is that, if the following proportion holds, viz., $a:b::c:x$, then an electromotive force in F causes no permanent current in G, or an electromotive force in G causes no current in F. Thus, when there is an electromotive force in F and no current resulting therefrom in G, we know that $a:b::c:x$; and, therefore, if any three of these four quantities are known, the fourth can be found by "rule of three." Or, if $c:g::f:b$, then an electromotive force in A causes no current in X, and *vice versa*. Or, if $a:g::f:x$, an electromotive force in B causes no current in C, and *vice versa*. It is only necessary to consider the first case, viz., $a:b::c:x$, making F and G conjugate.

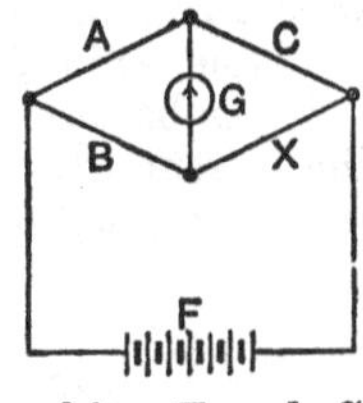

Suppose X is a conductor whose resistance is to be determined, and that A, B, C are conductors whose resistances are known and adjustable. Also, F to be a battery and G a galvanometer. Then, to find x, we simply alter a, b, and c, or any of them, till the current in G is rendered inappreciable, and then we know that $x = \frac{bc}{a}$. Now, there is, theoretically, a doubly-infinite number of ways of doing this—of getting a balance. For we have four quantities bound by two relations, viz., x constant and $a:b::c:x$. Therefore any two of the three conductors A, B, C may have any resistances we please for a balance.

As every one who has used the bridge knows, some arrangements—that is to say, some particular balances—are more sensitive than others. For instance, if a and b are extremely small, there is so little difference of potential between the terminals of the galvanometer that a want of sensitiveness arises that way. And if a and b are enormously large, the current from the battery is rendered very small, with a consequent want of sensitiveness. Now, if an operator can, by trial, easily find a sufficiently sensitive balance, and is satisfied with its sensitiveness, there is an end of the matter, and it is not worth while hunting after greater sensitiveness. But as it may happen that the greatest sensitiveness is desirable, or an approximation thereto, it becomes necessary to answer the question, What is the most sensitive balance? Unfortunately this cannot be done without algebra, for there is probably no man living who could find it out in his head unassisted by symbols.

It becomes, in the first place, necessary to give a precise meaning to the word sensitiveness. If as nearly perfect a balance as possible be obtained, and then any one of a, b, c, x be altered by a given small fraction of itself—as, for instance, x changed to $x(1+\Delta)$, where Δ is a small fraction—then a current will appear in G of the same strength whichever one of the four be altered, though of opposite direction for b and c as compared with a and x. Obviously one arrangement will be more sensitive than another if in the first the change of x to $x(1+\Delta)$, or corresponding changes in a, b, or c, causes a greater current through the galvanometer than in the second. And the importance of an error is to be reckoned by the ratio it bears to the quantity measured; thus

$\frac{\Delta x}{x} = \Delta$. Whence the balance of greatest sensitiveness is that one in which a given small change from x to $x(1+\Delta)$ causes the greatest current through the galvanometer. For the greater this current the nearer can approximation to accuracy be made by adjustment, and if it is inappreciable, as in a coarse balance, no further accuracy can be reached.

It can be easily shown, for it is a simple consequence of the laws of the current, that if E is the electromotive force of the battery and Γ the current through the galvanometer, that

$$\Gamma = \frac{E\frac{ax-bc}{s}}{\left\{\frac{(a+b)(c+x)}{s}+g\right\}\left\{\frac{(a+c)(b+x)}{s}+f\right\}-\left(\frac{ax-bc}{s}\right)^2},$$

where s signifies $a+b+c+x$. This is irrespective of a balance. Now let $a:b::c:y$, and $x=y(1+\Delta)$, and eliminate a. Then we shall have

$$\Gamma = \frac{E\frac{bc\Delta y}{s'}}{\left\{\frac{b(c+y)(c+x)}{s'}+g\right\}\left\{\frac{c(b+y)(b+x)}{s'}+f\right\}-\left(\frac{bc\Delta y}{s'}\right)^2},$$

where $s'=b(c+y)+y(c+x)$. When $\Delta=0$, we have $y=x$, $\Gamma=0$, and the perfect balance $a:b::c:x$.

If now we give different values to b and c, say b' and c', then $\frac{\Gamma}{\Gamma'}$, where Γ' is the new current through the galvanometer, is the ratio of the sensitiveness of the first arrangement to the second. The expression for Γ' is the same as that for Γ with b changed into b' and c into c'. That $\frac{\Gamma}{\Gamma'}$ is the ratio of sensitiveness of the two arrangements will be true in the limit when $\Delta=0$, *i.e.*, with an exact balance in each case. Therefore, dividing Γ by Γ', and *afterwards* making $y=x$ and $\Delta=0$, we shall find

$$\frac{\Gamma}{\Gamma'} = \frac{\dfrac{bc}{\{b(c+x)+g(b+x)\}\{c(b+x)+f(c+x)\}}}{\dfrac{b'c'}{\{b'(c'+x)+g(b'+x)\}\{c'(b'+x)+f(c'+x)\}}}.$$

It follows from this that the *sensitiveness of any balance* whatever is proportional to

$$\frac{bc}{\{b(c+x)+g(b+x)\}\{c(b+x)+f(c+x)\}}.$$

The most sensitive balance can now be easily found by the ordinary rules of the differential calculus, treating b and c as independent variables. The result is

$$a=\sqrt{fg},\quad b=\sqrt{\frac{gx}{g+x}(g+f)},\quad c=\sqrt{\frac{fx}{f+x}(g+x)}. \quad\text{.........}\quad \text{(A)}$$

The rule, therefore, is, for the most sensitive balance with a given galvanometer and battery. Make $a = \sqrt{fg}$. This is the same for all resistances to be measured. Find x approximately. Then make

$$b = \sqrt{\frac{gx}{g+x}(g+f)},$$

or as near as may be convenient, and get the nearest possible balance by adjusting c. Then $x = \frac{bc}{a}$.

Having found the most sensitive balance, we are able to estimate the relative sensitiveness of any other balance, using the same battery and galvanometer. If ρ is the ratio of the sensitiveness of any balance $a : b :: c : x$ to the most sensitive balance possible (this may be called the ratio or coefficient of sensitiveness), then

$$\rho = \frac{\{\sqrt{gx} + \sqrt{fx} + \sqrt{(f+x)(g+x)}\}^2}{\left(c+g+x+\frac{gx}{b}\right)\left(b+f+x+\frac{fx}{c}\right)}.$$

ρ is a quantity between 0 and 1, being 1 for the most sensitive balance.

A few numerical examples will serve to show the variation of the sensitiveness in different balances. Suppose $x = 4{,}000$, $g = 1{,}000$, $f = 90$ ohms. Then, in the following balances—

	a	b	c	x		
1.	1 :	1 ::	4,000 :	4,000,	we have	$\rho =$ ·003
2.	10 :	10 ::	4,000 :	4,000,	,,	$\rho =$ ·029
3.	100 :	100 ::	4,000 :	4,000,	,,	$\rho =$ ·24
4.	1,000 :	1,000 ::	4,000 :	4,000,	,,	$\rho =$ ·75
5.	10,000 :	10,000 ::	4,000 :	4,000,	,,	$\rho =$ ·38
6.	1 :	10 ::	400 :	4,000,	,,	$\rho =$ ·025
7.	10 :	100 ::	400 :	4,000,	,,	$\rho =$ ·219
8.	100 :	1,000 ::	400 :	4,000,	,,	$\rho =$ ·90
9.	1,000 :	10,000 ::	400 :	4,000,	,,	$\rho =$ ·58
10.	10 :	1 ::	40,000 :	4,000,	,,	$\rho =$ ·0024
11.	100 :	10 ::	40,000 :	4,000,	,,	$\rho =$ ·027
12.	1,000 :	100 ::	40,000 :	4,000,	,,	$\rho =$ ·14
13.	10,000 :	1,000 ::	40,000 :	4,000,	,,	$\rho =$ ·203
14.	100 :	1 ::	400,000 :	4,000,	,,	$\rho =$ ·0029
15.	1,000 :	10 ::	400,000 :	4,000,	,,	$\rho =$ ·015
16.	10,000 :	100 ::	400,000 :	4,000,	,,	$\rho =$ ·027
17.	300 :	1,800 ::	$666\frac{2}{3}$:	4,000,	,,	$\rho = 1$ nearly.

In the first five we have equality between a and b, and the sensitiveness increases rapidly, reaches a maximum, and afterwards diminishes less rapidly. In the next four, $a : b :: 1 : 10$, and the sensitiveness is greater and behaves similarly. There is a falling off in sensitiveness in the following four, where $a : b :: 10 : 1$, and a great falling off in the next three, where $a : b :: 100 : 1$. Finally, No. 17 is nearly the most sensitive balance, and $\rho = 1$ nearly.

It does not follow that the most sensitive balance is also the most convenient. This is obviously the case when the resistances of the conductors A, B, C giving greatest sensitiveness are out of reach. Also, if balance be obtained by adjusting the resistance of C, and the least change that can be made in C is one ohm, it may be convenient to make the ratio of $a:b$ and $c:x=10$. For, although there may be a loss of sensitiveness consequent on using this ratio, yet, since a change of one ohm in x then corresponds to a change of 10 ohms in c, x may be measured to a tenth of an ohm without calculation. But that any greater accuracy is attained by using the ratio 10 : 1 if the resistance of the conductor C can be altered by tenths or hundredths of an ohm, as by employing a divided wire with a sliding contact, or a duplicate arc, or other means, is a delusion; for the nearest approach to accuracy must finally depend on the least current recognisable in the galvanometer, and therefore on the sensitiveness.

I have given what I believe to be the most general way of estimating the sensitiveness of a balance with a certain galvanometer and battery. When special conditions are introduced, special results must follow. Thus, for example, if the sum of a and c is maintained constant, and also the sum of b and x, the most sensitive arrangement subject to these conditions is $a=c$ and $b=x$. And if the sum of a and b is kept constant, and also the sum of c and x, then the most sensitive balance is when $a=b$ and $c=x$. Further, if the sum of a, b, c, and x is kept constant, then we shall have $a=b=c=x$.

Also, if a, b, c, and x are separately kept constant, having any values producing an approximate balance, changing the places of the battery and galvanometer may produce benefit. The current in the galvanometer is increased or decreased by changing the positions of F and G according as $(g-f)(a-x)(b-c)$ is + or −. From this the rule follows for the best position for the galvanometer. Of the two, F and G, place the one of greater resistance so as to connect the junction of the two greatest consecutive resistances (out of A, B, C, X) with the junction of the two least. (By consecutive is meant that two resistances are next each other in the bridge. The two greatest resistances may not be consecutive.) This rule is always complied with when the most sensitive balance, $a:b::c:x$, is used.

As to the galvanometer and battery resistances. If the space in a galvanometer to be filled with wire is constant, then it is well known that the size of the wire should be such as to make its resistance equal to the external resistance to get the greatest magnetic force. And, to produce the greatest current externally, a battery consisting of a number of similar cells should be arranged when possible in so many rows of so many cells each, as to make the internal and external resistances equal. Applying these principles to the balance, we have

$$g=\frac{(a+b)(c+x)}{a+b+c+x}, \quad f=\frac{(a+c)(b+x)}{a+b+c+x}.$$

If, with these values of the battery and galvanometer resistances, we also employ the most sensitive balance, we shall find that the resistances of all the conductors should be equal, or

$$a=b=c=f=g=x.$$

This is absolutely the most sensitive balance if f and g can be varied in the manner stated, which is obviously impracticable in the vast majority of cases.

V.—ON AN ADVANTAGEOUS METHOD OF USING THE DIFFERENTIAL GALVANOMETER FOR MEASURING SMALL RESISTANCES.

[*Phil. Mag.*, April 1873, S. 4, vol. 45.]

In the usual method of measuring resistances with the differential galvanometer, the current from the battery is divided between the two coils, having opposite effects on the needle within them, so that, if the currents in both the coils are equal, the needle is unaffected. The introduction of resistance in the circuit of one coil will not affect the balance, provided an equal resistance is introduced in the circuit of the other coil. Hence, if on one side we place a rheostat, and on the other an unknown resistance, the latter may be determined by varying the resistance of the rheostat until a balance is obtained.

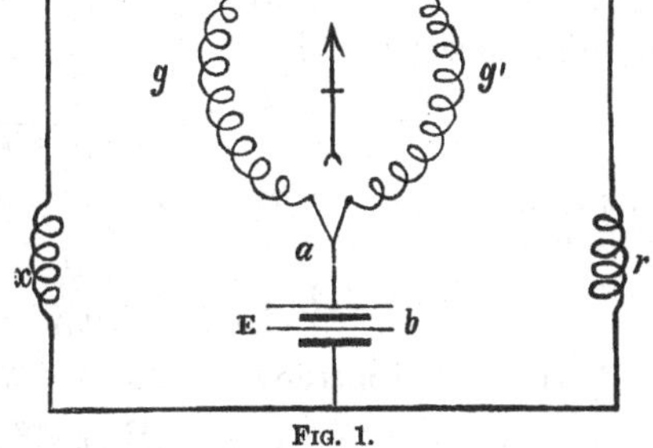

Fig. 1.

Fig. 1 is a representation of this arrangement. The current from the battery, having a resistance b and electromotive force E, divides at the point a between the coil g and resistance x, and the coil g' and resistance r. When $r=x$, the needle is unaffected.

By the following method of using the differential galvanometer, a much greater accuracy is obtained when the unknown resistance whose value has to be determined is small.

Instead of dividing the current from the battery E between the two coils, I join up the coils, so that the same current passes through both of them, and by reversing one of the coils, g', prevent the current from influencing the needle (see Fig. 2). The rheostat r is connected to the two ends of one coil, and the resistance to be measured, x, to the two ends of the other. It

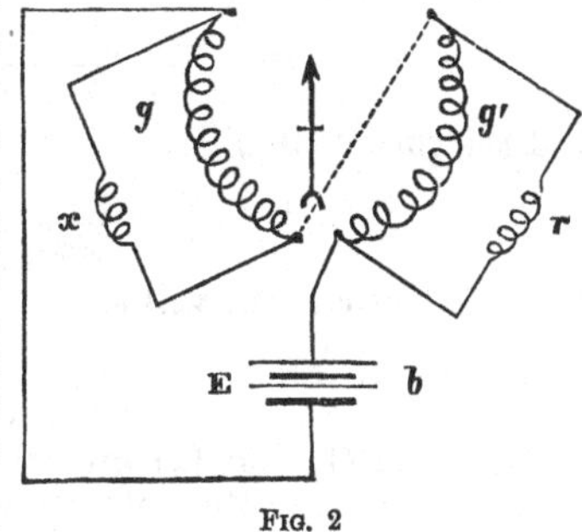

Fig. 2

will easily be seen, without further explanation, that when $r=x$, the currents in g and g' are equal; but should r not equal x, there will be a greater current in one coil than in the other, and the needle will move in obedience to the difference of these currents. It then only remains for me to show what, and under what circumstances, advantages are obtained by this method. To do so, we have only to compare the expressions for the difference-currents in the two methods.

By the first method the resistance external to the battery is

$$\frac{(x+g)(r+g)}{x+r+2g};$$

therefore the current from the battery is

$$B = \frac{E}{b + \frac{(x+g)(r+g)}{x+r+2g}} = \frac{E(x+r+2g)}{b(x+r+2g)+(x+g)(r+g)}.$$

This current divides between the two paths $x+g$ and $r+g$ in inverse proportion to their resistances; therefore the current in g is

$$G = B \times \frac{r+g}{x+r+2g} = \frac{E(r+g)}{b(x+r+2g)+(x+g)(r+g)},$$

and the current in g' is

$$G' = B \times \frac{x+g}{x+r+2g} = \frac{E(x+g)}{b(x+r+2g)+(x+g)(r+g)}$$

The effective current (that influencing the needle) will be the difference of G and G', say

$$D_1 = \frac{E(r-x)}{b(x+r+2g)+(x+g)(r+g)}. \quad \text{.................. (1)}$$

By the second method, the resistance external to the battery is

$$\frac{xg}{x+g} + \frac{rg}{r+g};$$

therefore the current from the battery is

$$B = \frac{E}{b + \frac{xg}{x+g} + \frac{rg}{r+g}} = \frac{E(x+g)(r+g)}{b(x+g)(r+g)+gx(g+r)+gr(g+x)}.$$

The current in g is

$$G = B \times \frac{x}{x+g} = \frac{Ex(g+r)}{b(x+g)(r+g)+gx(r+g)+gr(x+g)},$$

and the current in g'

$$G' = B \times \frac{r}{r+g} = \frac{Er(g+x)}{b(x+g)(r+g)+gx(r+g)+gr(x+g)}.$$

The effective current will therefore be

$$D_2 = G - G' = \frac{Eg(x-r)}{b(x+g)(r+g)+gx(r+g)+gr(x+g)}. \quad \text{........ (2)}$$

Equations (1) and (2) give the effective current in each case; and we may ascertain the relative sensitiveness of the two methods by comparing D_1 and D_2.

$$\frac{D_2}{D_1}=\frac{b(x+r+2g)+(x+g)(r+g)}{b(x+g)(r+g)+gx(r+g)+gr(x+g)}\times g;$$

and in the limit, when $x=r$,

$$\frac{D_2}{D_1}=g\times\frac{2b+r+g}{b(r+g)+2gr}.$$

When $r=g$, $\frac{D_2}{D_1}=1$, showing that the two methods are equally sensitive for that value of r or x which equals the resistance of one coil of the galvanometer. When r is greater than g, the ordinary method is to be preferred, for $\frac{D_2}{D_1}$ is then less than unity. It can, however, never be less than $\frac{g}{b+2g}$, which happens when r is infinite.

But for values of r less than g, $\frac{D_2}{D_1}$ is greater than unity, and increases rapidly as r is reduced, until in the limit, when $r=0$, $\frac{D_2}{D_1}=2+\frac{g}{b}$.

This proves that when the resistance to be measured is smaller than that of the galvanometer-coil, the second method is much to be preferred. For instance, let the battery have a resistance of 10 ohms, the galvanometer (each coil) 500 ohms, and $r=10$ ohms, then the second method is 17 times as sensitive as the first; and if r were 1 ohm, the second method would be 43 times as sensitive.

In fact, if, after getting as true a zero as possible by the ordinary method, the connections be altered to the second arrangement, the slight inequality between r and x, which was inappreciable by the ordinary method, will be at once rendered evident by a large deflection of the needle.

VI.—ON THE DIFFERENTIAL GALVANOMETER.

[*Phil. Mag.*, Dec. 1873, S. 4, vol. 46.]

THE great similarity between the systems of resistance measuring by means of the differential galvanometer and Wheatstone's Bridge, the latter having probably been suggested by the former, must have struck everyone who has had anything to do with them. In each case do we make one resistance a fourth proportional to three others, and, knowing the three, deduce the fourth. As in the bridge, for every resistance to be measured there is a certain arrangement of the three other sides which gives the most sensitive balance; so with the differential galvanometer, there must be a best arrangement for any particular case, which is the object of this paper to point out.

The expression for the strength of the current through the galvanometer in Wheatstone's Bridge is

$$E = \frac{v \cdot \dfrac{ad-bc}{a+b+c+d}}{\left\{\dfrac{(a+b)(c+d)}{a+b+c+d}+e\right\} \cdot \left\{\dfrac{(a+c)(b+d)}{a+b+c+d}+f\right\}}, \qquad (1)$$

(where v is the electromotive force of the battery, E the current through e, and the resistances as in the diagram) when at a balance, and therefore $ad-bc$ a vanishing quantity.

To deduce from this the expression for the force acting on the needle in the differential-galvanometer arrangement, we must make e infinite. Therefore, multiplying (1) by e and making $e=\infty$, we obtain

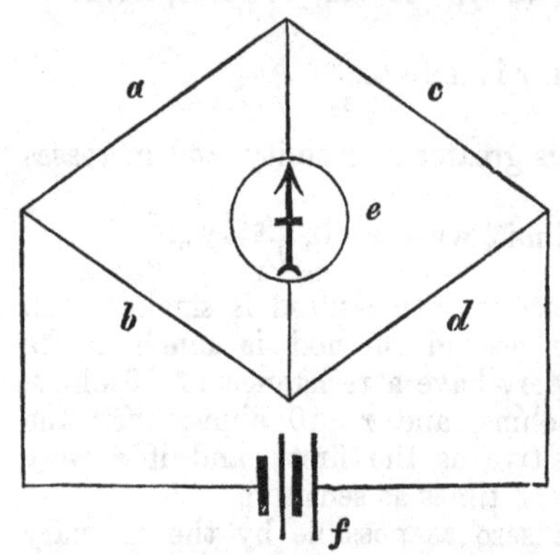

$$Aa - Bb = \frac{v \cdot \dfrac{ad-bc}{a+b+c+d}}{\dfrac{(a+c)(b+d)}{a+b+c+d}+f} \qquad (2)$$

for the differential galvanometer. [Here A and B mean the currents in a and b respectively.] This can, of course, be obtained independently of any consideration of Wheatstone's Bridge, but makes it evident that the best arrangement of the differential galvanometer with a given battery may be derived from the Wheatstone's Bridge formulae by making e infinite in them.

These formulae are as follows (*Phil. Mag.*, February, 1873) when c, d, e, and f are fixed :—

$$\left.\begin{aligned} a &= \sqrt{\frac{c}{d} \cdot \frac{cd+df+fc}{c+d+e} \cdot e}, \\ b &= \sqrt{\frac{d}{c} \cdot \frac{cd+df+fc}{c+d+e} \cdot e}, \end{aligned}\right\} \qquad (3)$$

and if c be not arbitrarily fixed,

$$\left.\begin{aligned} a &= \sqrt{ef}, \\ b &= \sqrt{de \cdot \frac{d+f}{d+e}}, \\ c &= \sqrt{df \cdot \frac{d+e}{d+f}}, \end{aligned}\right\} \qquad (4)$$

which is the most sensitive arrangement possible with a given battery and galvanometer. Finally, if the best resistances for the galvanometer and battery are also to be employed we must make every branch of the bridge of the same resistance, viz. that of d, the resistance to be measured.

Making $e=\infty$ in (3) and (4) they become

$$\left.\begin{aligned} a&=\sqrt{\frac{c}{d}(cd+df+fc)},\\ b&=\sqrt{\frac{d}{c}(cd+df+fc)}\,; \end{aligned}\right\}\quad\ldots\ldots\ldots\ldots\ldots\ldots\ldots\ldots \quad (5)$$

and

$$\left.\begin{aligned} a&=\infty\,,\\ b&=\sqrt{d(d+f)},\\ c&=\infty\,, \end{aligned}\right\}\quad\ldots\ldots\ldots\ldots\ldots\ldots\ldots\ldots \quad (6)$$

which may also be obtained by differentiation from (2). These formulae (5) and (6) would not be of any particular use if we had no means of varying at will the resistance of the coils a and b. This cannot be accomplished directly without considerable complication, but by means of shunts the same end may be reached. Thus, using (5), if the coils of our galvanometer have resistances greater than those best suited for the particular resistance to be measured, by means of shunts we may reduce their resistances to the required extent; and here a remarkable peculiarity presents itself. In general when a galvanometer is shunted its resistance and sensibility are reduced in the same proportion. Not so with the differential galvanometer, for its sensitiveness will be increased or reduced by shunts according as the normal resistance of its coils are greater or less than their best values in the particular case under consideration. The accuracy of (5) may be easily verified experimentally.

Proceeding to examine (6) we meet practical impossibilities, and can only carry it out by making the resistance of the coil a as great as possible by not shunting it. But as the values of b and c in (6) correspond to $a=\infty$, it will be necessary to find their values when $a=a$. Therefore, regarding both a and d as constant, and b and c variable, subject to the condition $ad-bc=0$, we shall find by differentiating (3) that

$$\left.\begin{aligned} c&=\sqrt{ad\,.\,\frac{a+f}{d+f}},\\ b&=\sqrt{ad\,.\,\frac{d+f}{a+f}}, \end{aligned}\right\}\quad\ldots\ldots\ldots\ldots\ldots\ldots\ldots\ldots \quad (7)$$

gives the most sensitive arrangement for measuring a resistance d with a battery whose resistance is f.

It is an evident conclusion that differential galvanometers intended for measuring resistances comprised within wide limits, both high and low, should have coils of long fine wire, having necessarily a high resistance; for a galvanometer with short and thick wire is only suitable for measuring small resistances, whereas if it have coils of fine wire it is suitable for both high and low resistances—for the latter by shunting.

VII.—ON DUPLEX TELEGRAPHY. (PART I.)

[*Phil. Mag.*, June 1873, S. 4, vol. 45.]

DUPLEX telegraphy, the art of telegraphing simultaneously in opposite directions on the same wire, which was first performed by Dr. Gintl in 1853, and subsequently engaged the attention of so many inventors, until lately seemed never likely to be carried out in practice to any extent. According to the very practical author of *Practical Telegraphy*, "this system has not been found of practical advantage"; and if we may believe another writer, the systems he describes "must be looked upon as little more than feats of intellectual gymnastics—very beautiful in their way, but quite useless in a practical point of view." However, notwithstanding these unfavourable reports as to the practicability of duplex telegraphy, the experience of the last year has negatived them in a striking manner, and made the so-called "feats" very common-place affairs. Circuits worked on a duplex system are now established in various parts of the United Kingdom—not to mention the United States, where the resurrection of these defunct schemes took place—and continue to give every satisfaction. There seems little reason to doubt that this system will eventually be extended to all circuits of not too great a length, between the terminal points of which there is more than sufficient traffic for a single wire worked in the ordinary manner—that is to say, only one station working at a time.

I propose in this paper to give a short account of the theory of duplex telegraphy by the principal methods, and to describe two other methods, which are, I believe, entirely original.

To begin at the beginning. Prior to 1853, it is said to have been the current belief of those best qualified to judge, that to send two messages in opposite directions at the same time on a single wire was an impossibility; for it was argued that the two messages, meeting, would get mixed up and neutralize each other more or less, leaving only a few stray dots and dashes as survivors (after the manner of the Kilkenny cats, who devoured one another and left only their tails behind). However, Dr. Gintl effectually silenced this powerful argument by going and doing it.

In order to be able to receive messages from another station, it is necessary for the receiving instrument to be in circuit with the line; and in order to send to another station, the battery must be in circuit. Hence, in order to receive and send at the same time, both the sending and receiving apparatus must be in circuit together. This can be arranged by making one continuous circuit between the two earths, and including the line and all the apparatus at each station. But if nothing further were done, the receiving instruments would be worked both by the received and sent currents; and if both stations worked at once, inextricable confusion would be the only result. Now, evidently, if the effect of the sent currents on the sending-station's instrument can be neutralized, the "feat" is accomplished. There are many ways of doing this. Dr. Gintl surmounted the difficulty in what was, to say the least, a very ingenious manner, although from a modern point of view, it was decidedly clumsy. He made his key, while being depressed to send a current to the line through his own relay, at the same time close a local circuit, including a coil of wire outside the principal coils of the relay, in such a manner that the current in this local circuit (which contained an independent battery) circulated round the cores of the electromagnets in the opposite direction to the current going out to the line; and by placing a rheostat in this local circuit he was able to vary the strength of the local current, so that the effect of the outgoing current on the relay was exactly neutralized. The relay then responded only to currents coming from the opposite station, which, of course, passed through the inner coils alone. Did both stations depress their keys simultaneously, the current in the batteries, inner coils, and the line was that due to both batteries; but in each relay as much of this current as was due to the corresponding battery was neutralized by the local current. The line-current might even be nothing, which would happen if each station had equal batteries and the same poles to earth. Then the relays would be worked entirely by the local current.

But local circuits are nuisances, and it is not to be wondered at that this method of Gintl's never came into practical use. But the possibility of the "feat" having been once demonstrated, it was not long before another and much superior method was introduced. It was discovered about the same time in 1854 by Frischen and Siemens-Halske, and may be called the *Differential method.* It is represented in its simplest form in Fig. 1. The relay at each station is wound with

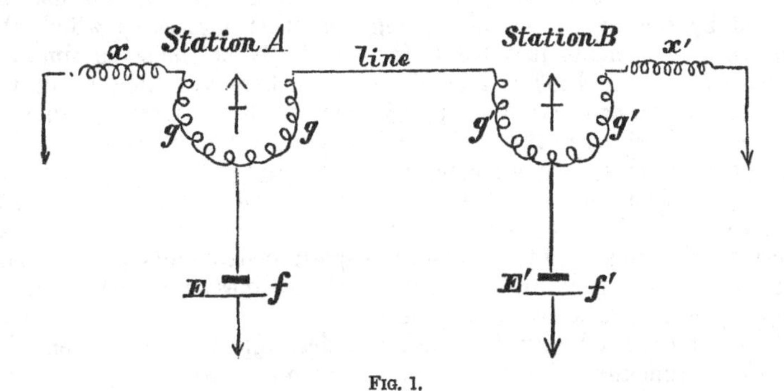

Fig. 1.

two coils of equal length; and the connections with the battery and the line are made in the same manner as if each station were taking a test of the resistance of the line with the differential galvanometer. The resistance x, then, at station A equals the whole resistance outside station A; and x' at station B equals the whole resistance outside station B. Then, when the battery E is in circuit, as its current divides equally between the two coils of the instrument $g\,g$, the latter is unaffected; but that half of the current which passes to the line necessarily influences the instrument $g'\,g'$ at the other station, since the whole of it passes through one coil, and then divides between the other coil and the battery E'. Thus each station does not work its own relay, but only that of the opposite station, and the conditions of duplex working are satisfied.

It is upon this system that nearly the whole of the existing methods of duplex telegraphy are founded. As the object is to prevent outgoing currents from working the sending station's instrument, it is plain that there may be many modifications having for object the easier production of balances under different circumstances—as by varying the distance of one or both coils from the armature instead of altering the resistance x (Fig. 1). There are also a few small points to be attended to before this system can be considered perfect. First, it is necessary for the external resistance to be as constant as possible, in order that the currents sent by a station (say A) may never affect its own instrument. But this external resistance includes B's apparatus; and B's battery is sometimes in and sometimes out of circuit. A variation in the external resistance will therefore be caused unless the transmitting apparatus is so arranged that a resistance equal to that of the battery is substituted for it when the latter is not in circuit. Again, there should be no interval of time during which neither the battery nor this equivalent resistance is in circuit. These things can generally be arranged with little difficulty. Thus, taking the case of the simplest transmitting instrument (the common Morse key), consisting of merely a lever with a front and back contact, the equivalent resistance may be connected with the back, and the battery-pole with the front contact; and the interval of disconnection may be avoided by the use of suitable springs, or other means, by which the front contact is made just before (or practically at the same time as) the back contact is broken, and *vice versâ*. There will then be only a very much smaller interval of time during which the received currents can pass both through the battery and its equivalent resistance. The application of this to more complicated instruments (as, for instance, Wheatstone's automatic transmitter) is not at first sight so evident; but I have done it in a very simple manner, which it is unnecessary to describe. On long lines, or with high-speed instruments, an attention to these *minutiæ* is desirable; but on short lines and with common Morse apparatus they are superfluous.

It is not essential, though sometimes desirable, to use differentially wound instruments. Most telegraph instruments are constructed with two separate coils of wire, each on its own core. By connecting the battery to the wire joining these coils, we have a differential arrangement, and frequently all that is needed. In fact, if the armature is polarized, as in most relays, the result is the same as if they were differentially wound. With an unpolarized Morse direct-writer, however, the effect of the out-going currents would not be completely neutralized. This is of little consequence, as the spring which draws the armature from the electromagnets may have a tension given it that only the received currents can overcome. The rheostats x and x' (Fig. 1) may even be dispensed with and a direct earth-connection substituted, provided the external resistance be not too great.

Quite recently another system has been brought forward, undeniably the most perfect, which may be called the *Bridge duplex*, its principle being that of Wheatstone's bridge. To whom the idea first occurred of using this arrangement for duplex telegraphy is unknown to me. It has been claimed by Mr. Eden, of Edinburgh; but it has been patented by Mr. Stearns, of Boston, U.S., who also patents a number of plans, all depending on the differential system before described.

The arrangement for the Bridge duplex is shown theoretically in Fig. 2. a, b, c and a', b', c' are resistances, g and g' the receiving instruments, and f and f' the batteries. By the well-known law of the

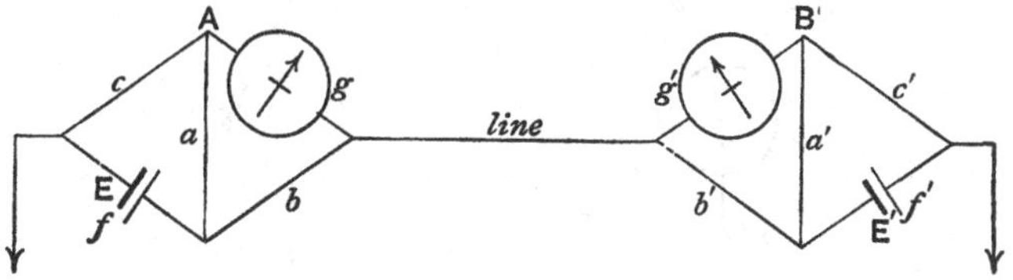

FIG. 2.

balance, when $a : b = c : d$, where d is the whole external resistance between station A and the earth at B, the electromotive force E will cause no current in g; and similarly for station B. The circumstance that the out-going currents do not pass through the receiving instruments is very important, as it allows any description of existing instruments to be used, and without any alteration. As in the Differential plan, it is not always indispensable to adhere rigidly to the conditions which give theoretical perfection.

Although the signals sent by station A are only received at B, and *vice versâ*, and it is convenient to assume that the currents producing these signals actually come from the opposite station, yet it does not always happen that such is the fact. To take an extreme case. Let all the apparatus at each station, and likewise the batteries, be exactly alike, and the line of uniform insulation. Now let A depress his key. The galvanometer at B will be deflected, but not A's. Let now B depress *his* key. No change will be produced in the deflection of B's galvanometer; but A's will be deflected. But if A and B have both the same pole of their batteries to line, there will be no current in the line, which, by Bosscha's first corollary, may be removed without producing any alteration in the currents in the remaining circuits. It is thus evident that A and B are both working *their own* instruments; and this supplies us with a very easy way of calculating the strength of the received signals—which would otherwise be very complicated. We have only to consider the current produced in $(g + b)$ by E having resistance f, with an external resistance c, $(g+b)$ being shunted by a resistance a, and we find the strength of the signal to be

$$G = \frac{Ea}{(f+c)(a+b+g)+a(b+g)}.$$

I will now describe two original methods of duplex working, which though perhaps not quite so easily put into practice as the foregoing, may be interesting from the theoretical point of view.

As in the Bridge arrangement, the out-going currents do not pass through the receiving instruments at all. In the first of these plans, as shown in Fig. 3, the receiving instrument at each station is connected

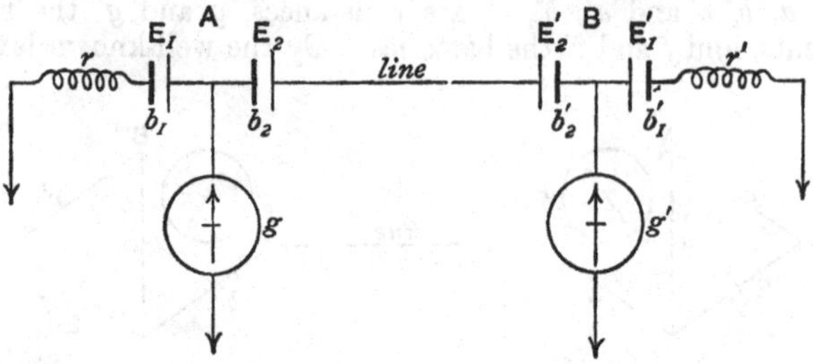

FIG. 3.

between the middle of two batteries and the earth. r and r' are rheostats. The condition that the batteries E_1 and E_2 at station A cause no current in g is that

$$E_1 : E_2 = r + b_1 : b_2 + d,$$

d being the exterior resistance as before; and similarly for the other station. The strength of current sent to line is $\dfrac{E_1}{r+b_1}$ or $\dfrac{E_2}{b_2+d}$. In the second plan, shown in Fig. 4, one of the batteries at each station

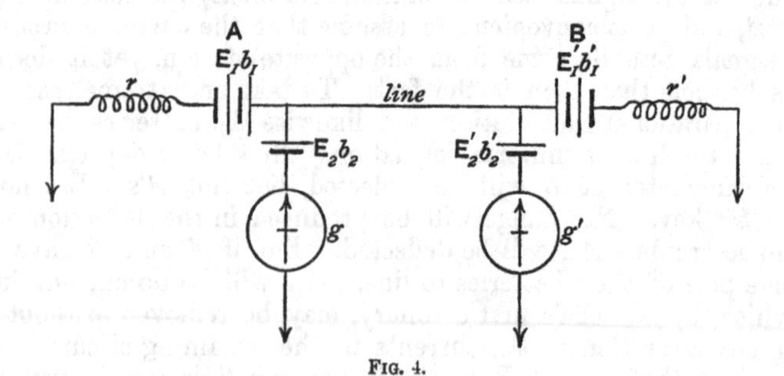

FIG. 4.

$(E_2$ and $E'_2)$ is placed in the same branch as the receiving instrument, and both E_1 and E_2 tend to send the same current to the line. When

$E_1 : E_2 = 1 + \frac{r+b_1}{d}$ (d having the same meaning as before), the electromotive force E_2 is neutralized and there is no current in g. The line-current is $\frac{E_1}{r+b_1+d}$.

I have adapted these plans to the direct-writing Morse by using an ordinary reversing key (which is nothing more than two keys insulated from each other and worked by the same lever), in order to put the two batteries E_1 and E_2 simultaneously in or out of circuit. I also found it necessary for there to be no interval of disconnection, but did not find it necessary to introduce equivalent resistances when they were out of circuit, though theoretically this should be done, and the same key could be made to do it.

The last plan will be easily recognized to be based on the method of comparing electromotive forces known as Poggendorff's compensation, in which the battery having the lesser electromotive force is not allowed to act. The other plan (Fig. 3) is also an adaptation of a method of comparing the *working* electromotive forces of batteries, which I devised three years ago, and subsequently published in the *English Mechanic*, for July 5, 1872, No. 380, p. 411. I mention this because it is claimed by Emile Lacoine as a "new method of determining voltaic constants," in the *Journal Télégraphique*, vol. 2, No. 13. See also the *Telegraphic Journal* for April this year.

The greatest drawback to duplex working (and this is common to all known systems) is the changeability in resistance of the line-wire itself, caused by defective insulation, variations of temperature, etc.; and in such a wet and changeable climate as ours, this fixes a limit to the length of line on which a duplex system can be worked with advantage, making it less than can be worked through in the ordinary manner. On short lines the resistance never varies much in any weather (unless actual faults occur), and it is not necessary to vary the balancing resistances. But on long lines this variation is sometimes very considerable; and it is questionable whether, in the present state of telegraphy, a long circuit in this country, as from Glasgow to London, could be profitably worked in wet weather. But the variations in the resistance of submarine cables (having no land-lines attached) are so very much less, that it seems probable, *à priori*, duplex telegraphy would be successful with them. Of course their electrostatic capacity must be balanced by condensers. It could also be applied to the system by which some long cables are worked, where there is no metallic circuit through the receiving instrument, which is placed between a condenser and the cable.

Those systems where the outgoing currents do not pass through the receiving instruments have a peculiar and perhaps what will some day (when telegraphy, now in its infancy, has arrived at years of discretion) be considered an important advantage over the Differential system. It is theoretically possible to send any number of messages whatever simultaneously in one and the same direction on a single wire. Now by combination with a "null" duplex system it becomes obviously possible to send any number of messages in the other direction while the opposite correspondences are going on, and without interference. Thus the working capacities of telegraphic circuits may be increased indefinitely by suitable arrangements. Practically, however, it would seem that a limit would soon be reached, from the rapidly increasing complication of adjustments required. Besides, to keep them going, the telegraph clerks must themselves be electricians of a rather higher order than at present. Nevertheless from experiments I have made, I find it is not at all a difficult matter to carry on *four* correspondences at the same time, namely, two in each direction; and if we may suppose the growth of telegraphy will be as rapid in the future as it has been in the past, it seems not improbable that *multiplex-telegraphy* will become an established fact.

In a following paper I intend giving the formulæ necessary for calculating the proper proportions of the resistances, etc., to suit different lines and apparatus, so that the greatest possible amount of current may be driven through the receiving instruments, where alone it is of practical service.

VIII.—ON DUPLEX TELEGRAPHY. (PART II.)

[*Phil. Mag.*, Jan. 1876, S. 5, vol. I.]

THE BRIDGE SYSTEM.

A THEORETICAL diagram of the arrangement of the conductors for duplex telegraphy by the Bridge system is given in Fig. 1. g and g' are the receiving apparatus, which may be of any kind, f and f' the

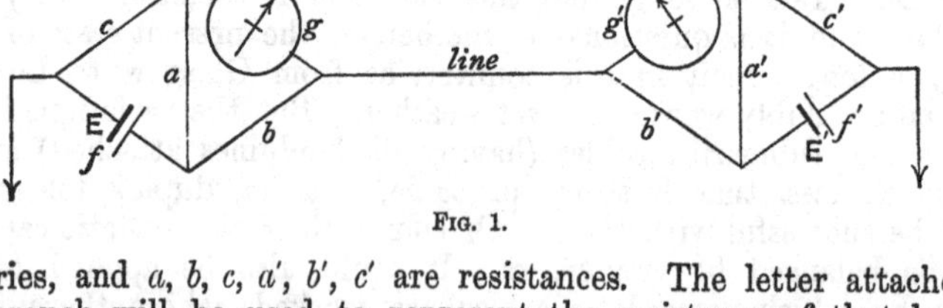

FIG. 1.

batteries, and a, b, c, a', b', c' are resistances. The letter attached to any branch will be used to represent the resistance of that branch. The two branches f and f' are preserved of constant resistance by mechanical means, whether the batteries are in or out of circuit. The object of the above arrangement of conductors is to enable two stations, A and B, to signal each other at the same time through a single line without mutual interference; and this is accomplished by adjusting the six resistances, a, b, c, a', b', c', so that

$$a : b :: c : x$$

and

$$a' : b' :: c' : x',$$

where x is the resistance outside station A, *i.e.* the resistance of the line *plus* the resistance of station B's apparatus, and x' is the resistance outside station B. When the above proportions hold good, f and g are conjugate, likewise f' and g'; hence each station working alone produces no current in its own receiving instrument; and when they are both signalling simultaneously, the currents in the receiving instruments are the algebraical sums of the currents which would be separately produced; thus station A gets only B's signals, and station B receives A's signals alone. It is obvious without any mathematical demonstration that the above is true whatever may be the direction and strength of the currents, provided only that the two conditions $ax = bc$, $a'x' = b'c'$ are satisfied, and that these are the sole necessary conditions for duplex telegraphy when the line has so little electrostatic capacity that the transient currents due to that cause are inappreciable. When the capacity of the line cannot be neglected, it may be perfectly balanced by distributing artificial capacity along the resistance c with the same uniformity it has along the line; and this may be approximated to by subdividing the resistance and required capacity as much as possible. If l is the resistance and l_1 the capacity of the line, and c_1 the required capacity, then the condition of balance as regards capacity is $cc_1 = ll_1$.

Since each station has three adjustable resistances, a, b, and c, and they are connected by the single relation $ax = bc$, it follows that any two of them may be taken arbitrarily and balance made with the third. Thus we may take b and c as independent variables, and eliminate a. The question then arises, in what respects an arrangement in which the resistances b and c have particular values differs from another in which b and c have other values. There are three principal differences: first, the received currents will be in general different; next, the balances will be of different degrees of sensitiveness, so as to be more or less affected by changes in the resistance of the line; and, lastly, different amounts of artificial capacity will be required to produce a balance with respect to the capacity of the line. Since a duplex apparatus is generally set up for permanent use, it is clearly of the first importance to obtain the maximum current with a given receiving instrument and battery. On cables this is quite a minor consideration, on account of the great delicacy of the instruments employed. But duplex telegraphy has not hitherto been very successful on cables; whereas on land lines, where such delicate instruments would be quite out of place, and much larger batteries are employed, the cost of the current is considerable, and it is desirable to get as much out of it as possible. I shall therefore in the first place endeavour to discover what the actual magnitudes of the resistances a, b, and c should be to render the received current a maximum,—and when that is done, consider the sensitiveness of the resulting arrangement, or its liability to disturbance.

In order to avoid useless complication, I shall suppose that the line is perfectly insulated, and that the receiving instrument at each station has the same resistance, and likewise the battery at each station, *i.e.* $f=f'$ and $g=g'$. This is very nearly fulfilled in practice; for the same description of instrument and battery is generally used at each end of a line. Then symmetry tells us that the resistances a, b, c should be the same at each station; or rather there is no reason why they should be different; and besides, if we do not make use of this simplification, the problem will become almost intractable.

Let E be the electromotive force of each battery, and G the current each station receives from the other through its receiving instrument g, and let both A and B send the same current to line. Then, from the identity of the arrangement at each end, there will be no current in the line, which may be removed without influencing the currents in the other conductors. Thus we find

$$G=\frac{E}{f+c+\dfrac{a(b+g)}{a+b+g}}\times\frac{a}{a+b+g},$$

or

$$G=\frac{Ea}{(f+c)(b+g)+a(f+c+b+g)}. \quad\text{.................. (1)}$$

Now this expression for the strength of the received current contains the constants E, f, and g, and the variables a, b, and c. The last two are independent; but the first is a function of all the resistances; for $a=\dfrac{bc}{x}$, and

$$x=l+\frac{gbx(c+f)+cfx(g+b)+bc(g+c)(b+f)}{(g+b)(c+f)x+bc(g+b+c+f)}. \quad\text{........... (2)}$$

This gives a quadratic equation for the determination of x, which, however, it is unnecessary to effect. By eliminating $a=\dfrac{bc}{x}$ from (1), we have

$$G=\frac{Ebc}{(f+c)(b+g)x+bc(f+c+b+g)}. \quad\text{.................. (3)}$$

We have to make G a maximum with respect to b and c; and therefore we must have $\dfrac{dG}{db}=0$ and $\dfrac{dG}{dc}=0$. Thus we have the following conditions:—

$$\left.\begin{aligned} gx(f+c)-b^2c&=b(f+c)(b+g)\frac{dx}{db}, \\ fx(b+g)-bc^2&=c(f+c)(b+g)\frac{dx}{dc}. \end{aligned}\right\} \quad\text{.................. (4)}$$

The only difficulty now lies with the complex function x. It would be most natural to obtain x, $\dfrac{dx}{db}$, and $\dfrac{dx}{dc}$ as functions of b, c, f, g, and l, and then find the values of b and c, in terms of the constants f, g, and l, which make G a maximum. But it will be found impossible to obtain an explicit solution in this manner, owing to the high degree of the final equations. However, a simple solution may be obtained in terms of f, g, and x, the external resistance. Thus, differentiating (2) with respect to b and then solving for $\frac{dx}{db}$, we obtain

$$\frac{dx}{db}=\frac{\{gx(c+f)+bc(c+g)\}^2+cx(bc-fg)^2}{\{x(g+b)(c+f)+bc(g+b+c+f)\}^2+bc(bc-fg)^2};\quad\text{(5)}$$

and since the right-hand side of (2) is unaltered by interchanging f and g, and b and c, we can find $\frac{dx}{dc}$ by making these changes in (5). Thus

$$\frac{dx}{dc}=\frac{\{fx(g+b)+bc(b+f)\}^2+bx(bc-fg)^2}{\{x(g+b)(c+f)+bc(g+b+c+f)\}^2+bc(bc-fg)^2}.\quad\text{(6)}$$

Equations (4), (5), and (6) must now be manipulated to obtain b and c in terms of f, g, and x. After going through the usual algebraical drudgery, which it is unnecessary to give here, we obtain

$$\left.\begin{aligned} b=\sqrt{\frac{xg}{x+g}(x+f)},\quad c=\sqrt{\frac{xf}{x+f}(x+g)},\\ \text{whence}\qquad a=\sqrt{fg}.\end{aligned}\right\}\quad\text{(7)}$$

It will be found that these values of a, b, and c make the received current a maximum with any given battery, receiving instrument, and line. The strength of the received current is then

$$G=\frac{E}{x+f+g+\sqrt{fg}+(\sqrt{xf}+\sqrt{xg})\left(\sqrt{\frac{x+f}{x+g}}+\sqrt{\frac{x+g}{x+f}}\right)},\quad\text{(8)}$$

or, which is the same,

$$G=\frac{E}{a+b+c+f+g+x+b\sqrt{\frac{f}{g}}+c\sqrt{\frac{g}{f}}},$$

It will be observed that in the above solution (7) one of the resistances (a) is independent of x, and is the same for all lines with the same receiving instrument and battery. Since balance is always obtained in practice by adjusting one of the resistances, say c, it follows that only the other two, a and b, need be calculated. This is readily done for a; but for b it is more difficult, since x cannot be simply expressed in terms of f, g, and l, the equation to determine it being of the sixth degree, viz.

$$l=\frac{(x^2-fg)\{2\sqrt{xf}+2\sqrt{xg}+\sqrt{(x+f)(x+g)}\}}{(f+g+x+\sqrt{fg})\sqrt{(x+f)(x+g)}+(2x+f+g)(\sqrt{xf}+\sqrt{xg})}.$$

This theoretical difficulty, however, is of no practical importance, since the value of x can be determined as closely as necessary in the very act of adjusting the instruments for duplex working for the first time. For long lines, x may be considered equal to $l+\sqrt{fg}$. It is actually rather greater.

The next thing to be considered is to what extent this arrangement is liable to disturbance from variations in the resistance of the line. Now, as I have shown in a former paper (*Phil. Mag.*, Feb. 1873), the values of a, b, and c, which make the most sensitive balance for measuring a resistance x with a battery of resistance f and galvanometer of resistance g, are precisely the same as those given in equations (7) above as giving the maximum current in duplex working. We are thus at once led to the conclusion that the arrangement of Wheatstone's bridge for duplex telegraphy which gives the maximum received current at both stations, is also the arrangement which is most easily disturbed by variations in the resistance of the line. We may show this otherwise. When $a:b::c:x$, station A, when sending alone, produces no current in his instrument. Let now the external resistance x be changed to x', then A sending alone will produce a current C_1 in his instrument, of strength

$$C_1 = \frac{\dfrac{Ebc(x'-x)}{b(c+x)+x(c+x')}}{\left\{\dfrac{b(c+x)(c+x')}{b(c+x)+x(c+x')}+g\right\}\left\{\dfrac{c(b+x)(b+x')}{b(c+x)+x(c+x')}+f\right\}-\left\{\dfrac{bc(x'-x)}{b(c+x)+x(c+x')}\right\}^2}.$$

If another arrangement be made in which b and c are altered to b' and c', and the current be now C_2, then C_2 may be found by changing b into b' and c into c' in the expression for C_1; and the ratio $\dfrac{C_1}{C_2}$ when $x'-x$ is small will express the relative sensitiveness of the two arrangements. In the limit, when $x'-x=0$, we have

$$\frac{C_1}{C_2} = \frac{\dfrac{bc}{\{b(c+x)+g(b+x)\}\{c(b+x)+f(c+x)\}}}{\dfrac{b'c'}{\{b'(c'+x)+g(b'+x)\}\{c'(b'+x)+f(c'+x)\}}};$$

from which we see that the sensitiveness of any exact balance to disturbances in the resistance of the line, either in duplex working or in testing, is proportional to

$$\frac{bc}{\{b(c+x)+g(b+x)\}\{c(b+x)+f(c+x)\}};$$

and this expression is a maximum when b and c have the values given in (7) above.

Hence it is perfectly hopeless to find any arrangement of Wheatstone's bridge for duplex telegraphy which shall give the maximum received current at both stations and at the same time be least liable to disturbance. Generally speaking, the more sensitive the balance the stronger the received current.

Since x, the resistance external to one station, includes the resistance at the other station, any alteration of adjustment at one station will theoretically cause a disturbance in the other station's balance; and it is true that an infinite series of successive adjustments must be made by each station to reobtain an exact balance whenever balance is disturbed. But these alterations are so excessively small that practically they have no existence. By making $bc=fg$ and adjusting solely by the resistance a, each station's balance becomes independent of the other's; but this is introducing a greater difficulty to avoid a lesser and inappreciable one, since to keep $bc=fg$ frequent measurements would have to be made of f, the battery-resistance, a variable quantity; and besides, such an arrangement would not give the maximum current, as is evident from equations (7).

The above investigations apply to any instrument, battery, and line, and therefore admit of immediate practical application in any particular case. There are, however, two principles frequently made use of by theoretical writers on electric circuits:—first, that if the space to be filled with wire in a galvanometer or relay is fixed, the greatest strength of signal is obtained when the wire is of such a size that its resistance equals the external resistance; and next, that if the quantity of metallic surface of a battery is fixed, and also the distance between the plates in each cell, to obtain the maximum current the cells should be of such a size that the total resistance of the battery equals the external resistance. These principles do not often admit of practical application in telegraphy; but we may just see to what they lead us when we apply them to duplex working with the bridge. We shall have the following equations to determine f and g:—

$$x^2-3g(x+f)+2\sqrt{fg}(x+g)=0,$$
$$x^2-3f(x+g)+2\sqrt{fg}(x+f)=0.$$

Either of these equations by itself can be made use of to determine g when f is constant, or f when g is constant. When they are combined, we have

$$a=b=c=f=g=x.$$

Now, although this result can be applied to the construction of instruments for testing purposes where x is constant, there is one insuperable difficulty that prevents its use in duplex working; and that is, x becomes infinite. We can only conclude that the finer the wire of the relay and the greater the number of convolutions, the smaller the cells are made, and the greater their number, the greater will be the strength of the signals—a fact which might be safely predicted without mathematical examination.

A comparison of the strength of the received current in ordinary simplex working and duplex working with the bridge will be interesting. In simplex working the received current is

$$\frac{E}{f+g+l};\qquad\text{(9)}$$

and in duplex working, when the arrangement is such as to give the maximum current, its value is given in equation (8) above as

$$G=\frac{E}{x+f+g+\sqrt{fg}+(\sqrt{xf}+\sqrt{xg})\left(\sqrt{\frac{x+f}{x+g}}+\sqrt{\frac{x+g}{x+f}}\right)},$$

where x is the external resistance, rather greater than $l+\sqrt{fg}$. When

l is very great compared with f and g, these expressions (8) and (9) are nearly equal, the duplex current being a little less than the other. (In the extreme case $f=0$, $g=0$, they are identical.) Numerical comparison, taking the most general values of f and g occurring in practice, will show that the duplex current is about one half or one third the strength of the current obtained when the same instruments and batteries are used for simplex working; so that in general more than double the electromotive force will be required to obtain signals of the same strength in both cases.

The third principal difference between one arrangement of the bridge and another, viz. that different amounts of artificial capacity are required, is of some importance as regards cables. Condensers of large capacity are such cumbrous and expensive affairs that the smaller the artificial capacity can be conveniently made the better. Now c_1, the required capacity, equals $\frac{l}{c}l_1$, where l_1 is the capacity of the line; consequently, to make c_1 as small as possible, c must be as large as possible; and this will occasion a great loss of working current. This, however, will be of little consequence with the sensitive instruments used on cables, if ever duplex telegraphy is successful on them.

THE DIFFERENTIAL SYSTEM.

Any instrument may be used in the Bridge system without alterations being made; but in the Differential system the coils must be differentially wound, or some equivalent device employed, so that two currents, one passing through each coil, may annul each other's action on the magnet or cores within them. On the other hand, only one balancing resistance is required instead of the three in the Bridge system. The following diagram (Fig. 2) is a theoretical view of the Differential

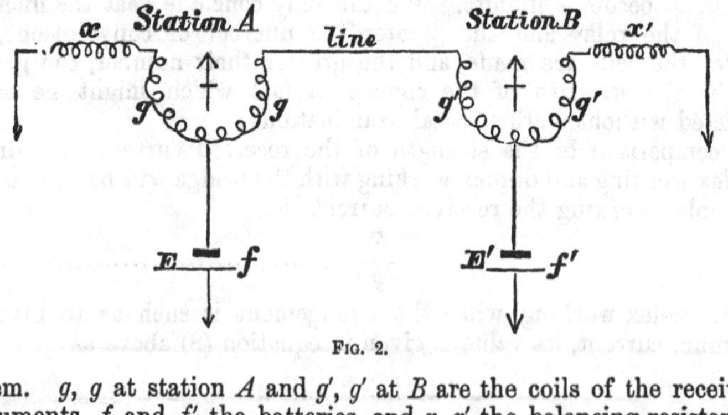

FIG. 2.

system. g, g at station A and g', g' at B are the coils of the receiving instruments, f and f' the batteries, and x, x' the balancing resistances. We shall suppose, as is usually the case, that $f=f'$ and $g=g'$. To find the strength of the signals each station receives from the other, let both send the same current to line; then, from the identity of the arrangement at each end, there will be no current in the line, in the right-hand coil at station A, and in the left-hand coil at B. Therefore, if S is the strength of the signal,

$$S=\frac{Em}{f+g+x}, \qquad (10)$$

where E is the electromotive force of the battery, m the strength of signal produced by the unit current circulating through a single coil of the receiving instrument, and x the external resistance. The value of x is

$$x=l+g+\frac{f(g+x)}{f+g+x}, \qquad (11)$$

or

$$x=\tfrac{1}{2}\{l+\sqrt{(l+2g)(l+2g+4f)}\}.$$

In simplex working with the same instruments and batteries the strength of the signals is

$$\frac{2Em}{f+2g+l}, \qquad (12)$$

when the current passes through both coils in succession. When f, the battery resistance, is very small, we see from (10) and (12) that the strength of the signals in duplex working is nearly one half their strength in simplex working with the same instruments, since x is a little greater than $l+g$.

Since there is only one balancing resistance at each station in this system, there is only one arrangement possible with a given receiving instrument and battery, leaving out minor details. We may, however, inquire what the resistance of the instrument should be to obtain the strongest signals on the supposition that the space to be filled with wire is fixed. In such case m will vary as the square root of g, and

$$S \propto \frac{E\sqrt{g}}{f+g+x}.$$

Therefore for S to be a maximum we must have

$$f+g+x=2g\left(1+\frac{dx}{dg}\right);$$

and we find from (11)

$$\frac{dx}{dg}=\frac{(f+g+x)^2+f^2}{(f+g+x)^2-f^2};$$

therefore

$$g=\tfrac{1}{3}\{-(x+f)+\sqrt{(f+2x)^2+4fx}\} \qquad (13)$$

is the best resistance for each coil of the receiving instrument. When $f=0$, $g=\frac{x}{3}$. Now (13) is identical with Weber's formula for the resistance of each coil of a differential galvanometer to obtain the maximum sensitiveness at a balance; thus again we see, just as in the Bridge system, the arrangement in which both stations get the strongest signals is also the most sensitive balance, and most liable to disturbance from variations in the external resistance.

Weber's formula (13) admits of considerable simplification if we arrange the battery so as to obtain the maximum current by making

$$f = \frac{g+x}{2}.$$

We then have $$g = \frac{x}{2}, \quad f = \frac{3x}{4}$$

as the best resistances for each coil and the battery. Now although this admits of application in testing with the differential galvanometer (when x is constant), yet it cannot be applied to duplex working, since x becomes infinite, which may receive an interpretation similar to that in the corresponding case of Wheatstone's bridge.

It is an interesting practical question whether with a given instrument and battery it is possible to obtain stronger signals by the Bridge than by the Differential system. To make the comparison fairly, in the former case the arrangement must give the maximum current. In the Differential system the strength of the signal is, by (10),

$$\frac{Em}{f+g+x},$$

and in the Bridge system

$$\frac{2Em}{f+x+2g+\sqrt{2fg}+(\sqrt{xf}+\sqrt{2xg})\left(\sqrt{\frac{x+f}{x+2g}}+\sqrt{\frac{x+2g}{x+f}}\right)}, \quad \ldots\ldots (14)$$

which is obtained from (8) by changing g into $2g$ and multiplying by $2m$. In (10) and (14) x has not the same signification; but the difference is not great. Effective comparison can easily be made numerically in any particular case. As general results, we may say that when f and g are very small in comparison with l, the advantage is in favour of the Bridge system; but when f and g are taken larger, the advantage becomes rapidly in favour of the Differential system. It may also be observed that in the latter the strength of signal is always less than one half the strength when the same instruments are used for simplex working, whereas in the former system the strength of the signal may be, but generally is not, more than one half.

If the practical success of duplex telegraphy were dependent on the continuous maintenance of an exact balance at each station, then would duplex telegraphy exist only on paper. The variations, sometimes large and rapid, which are always taking place in the resistance and insulation of overland wires would necessitate such frequent changes of the balancing resistances as to render efficient working the exception rather than the rule. But it is found practically that, instead of an exact balance being always required, the signalling can be continued for great lengths of time without any change of adjustment; and, moreover, the balancing resistances may sometimes be altered very considerably without actually interrupting the signalling. The actual received current may be considered as the algebraical sum of two parts—one the proper received current, the other an interfering current produced by inexact balance. In the double-current Morse system in common use in England the marks are made by one current (say, positive), and the spaces by the negative current. If C is the strength of the received current, then the whole range of the current is $2C$. In the single-current Morse system employed on the Continent and elsewhere there is no current during the spaces; hence the range of the current in the receiving instrument is only C. Therefore an instrument that admits of being worked either by single or double currents, as magnetized or polarized relays, will give signals twice as strong with double currents as with single with the same battery-power. This applies both to ordinary simplex working and to duplex working. In the latter there is a further advantage in favour of double currents. It is theoretically possible to work duplex with double currents when the interfering currents are little less strong than the received currents; for as the received current is always either $+C$ or $-C$, the superposition of any current of less strength than C will not alter the sign of C, whether $+$ or $-$. On the other hand, in single-current working the received current is always either C or zero. In the first case the current C overpowers the tension of a spring or other opposing force; and in the latter the spring is unopposed. The most rapid signalling is to be obtained when the forces moving the armature or tongue of the relay are equal in each direction; and then the retractile force of the spring must be equivalent to a reverse current of the strength $\frac{1}{2}C$. Therefore the interfering currents in duplex working with single currents must never be so great as $\frac{1}{2}C$—thus giving an immense advantage to the double-current system as regards freedom from interruption by inexact balance or other causes, in addition to the advantage before mentioned of giving signals of twice the strength.

It is found by experiment that duplex working (Morse) will not be actually interrupted until the interfering currents are as much as $\frac{1}{3}$ or $\frac{1}{2}$ the strength of the received currents with double currents, and $\frac{1}{6}$ or $\frac{1}{4}$ with single currents—although no hard and fast line can be drawn, owing to the very numerous causes in operation. On an overland wire worked duplex with differential relays and double currents the resistance which gave exact balance was, at one end, 2560 ohms, which could be increased to 3860 or diminished to 1760 ohms without interrupting the working. At the other end the balancing resistance could be varied from 3000 to 6000 ohms without interfering with rapid signalling. The variation allowable above balance is always much greater than below, because the interfering current is inversely proportional to the resistance external to the battery, which is increased when the balancing resistance is increased. In the above example the line was fairly insulated. When the insulation falls the effect is to strengthen the sent and weaken the received currents; consequently the interfering currents bear a larger ratio to the received currents for a given change of balance, and the balance therefore requires nicer adjustment. The extreme case is reached with the very low insulation which prevails in this country in continuous wet weather, when not much difference can be detected between the resistance of the wire whether it is insulated or to earth at one end. Under such circumstances a very small change of balance is sufficient to upset the working. The ratio of the interfering to the received currents may be diminished *ad lib.* by increasing the resistance of the apparatus, or more simply by inserting a constant resistance in the main circuit. As, however, it is only when the insulation is very bad and the received current very weak that the interfering current due to inexact balance attains such a proportional strength as to mutilate the signals, the increase in the resistance of the apparatus would be an evil rather than a benefit, on account of the reduction in the strength of the received signals, already very weak, that would ensue.

The two other systems described in my former paper (*Phil. Mag* June, 1873) are not likely, in accordance with the principle of the survival of the fittest, to come into practical use; and it is therefore unnecessary to enter into details concerning them. But this I may observe, that in both of them the arrangement which produces the strongest signals at both stations is also the most sensitive balance. That this should be the case in four different systems renders it probable that it is universally true for all duplex systems in which some kind of balance is concerned.

IX.—NOTES ON MR. EDISON'S ELECTRICAL PROBLEM.

[*Telegraphic Journal*, May 1st, 1875, p. 102.]

In the *Telegraphic Journal* for January 15, 1875, Mr. Thomas A. Edison submitted the following problem to its readers for solution :—"Transmit alternately positive and negative currents within a closed circuit from a battery all the poles of which are connected in the ordinary manner, using an ordinary Morse key, to which no extra point or appliances whatever is to be added. No device other than the battery, key, and connecting wires is to be used."

None of the readers of this journal have as yet come forward with any solution. Why is this? It is certainly not because there is nobody in the British Isles who takes an interest in such matters, and I can only suppose that an excess of modesty has prevented many of the readers of this journal from sending a solution for publication. As the problem is of a highly interesting nature, I think it should not be allowed to drop out of mind, and so send a few remarks on the problem and its solution. Perhaps others will then come forward with improved methods.

The practical telegraphist who has been accustomed to the use of the Morse key for sending single currents, and a "double-current" key for sending reversed currents, will probably be inclined, on a first perusal, to consider the problem a sort of electrical conundrum, not admitting of any legitimate solution; but such is certainly not the case.

I must, however, in the first place, point out that it is an impossibility on the face of it to reverse the current in a *closed circuit* containing a single battery all the poles of which are connected in the ordinary manner: the current in the battery itself has necessarily always the same direction, and a second battery of greater strength would be required to reverse the current in the first. All we can do is to reverse the direction of the current, in some or all of the conductors, in the circuit which lie outside the battery. I assume, therefore, that this is what Mr. Edison means is to be done, and on this assumption we can proceed further with the problem.

The restriction contained in the enunciation that all the poles are to be connected in the ordinary manner, I take to mean that the battery is to be joined up "for intensity," to use the convenient old-fashioned phrase; that is to say, the positive pole of one cell is to be connected with the negative pole of the next, and so on all through the battery. This restriction, however, does not forbid us to make a connection by means of a wire between our Morse key and any intermediate pole of the battery, as this will not interfere with all the poles being connected among themselves in the ordinary manner; hence we have the following arrangement, answering every condition of the problem (Fig. 1):—The battery *f* has its two terminal poles connected with the back and front stops of the key respectively, and any intermediate pole is connected through the external resistance, *e*, with the lever of the key. This will obviously produce alternately positive and negative currents in the external resistance when the key is worked, and is too simple to require any further explanation. This system was in use many years ago for signalling on underground or submarine wires, and may possibly be still used.

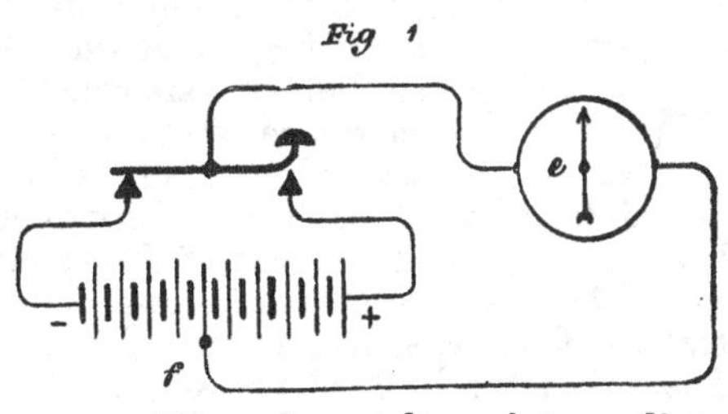

It will be observed that in the above system the whole battery is never in circuit at once; in fact, we are practically employing one battery for the positive currents and another for the negative. If we wish to employ the whole battery both for positive and negative currents we must seek some other plan. Mr. Edison lays no restriction on the *resistance* of the connecting wires, so that, practically speaking, he allows the use of resistance coils. This contradicts the statement in the problem that no device other than the key, battery, and connecting wires is to be used, but we may produce harmony again by uncoiling the wires of the resistance coils. Or, if we have a galvanometer of sufficient sensitiveness, we may use short pieces of wire. Fig. 2, then, shows a second solution of the problem. The external resistance, *e*, is connected between the back and front stops of the Morse key; one pole of the battery is connected with the lever of the key and the other with the junction of two wires, *a*, *b*, the other ends of which go to the back and front stops of the key. When the key is in the position shown in Fig. 2 the current from the + pole of the battery divides so that the greater portion goes through b, and the remainder through a and e, to the back stop of the key, and so to the − pole of the battery. When, however, the key is depressed, so that the lever is in contact with the front instead of the back stop, the current from the + pole divides so that the greater part goes through a, and the remainder through b and e, to the front stop of the key, and so to the − pole of the battery. The current is thus reversed in e. It is obvious that we can give any relative strengths to the + and − currents in e by suitably changing the resistances of a and b, and that when a and b are equal the reversed currents in e are equal. The galvanometer, e, may of course be replaced by a line. The currents sent to line will naturally be less than if the battery were connected direct to line, as in the ordinary double-current key. How much less we must call in the aid of Ohm's laws and algebra to determine.

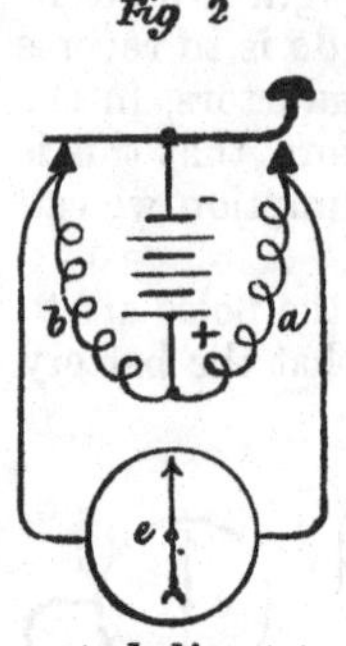

Let E be the current in the line, P the electromotive force of the battery, f its resistance, and a and b the resistances of the two wires. Then when the lever of the key rests on its front and back stops the currents sent to the line are

$$\frac{Pa}{f(a+b+e)+a(b+e)} \text{ and } \frac{Pb}{f(a+b+e)+b(a+e)}$$

respectively; and when $a=b$ each of these becomes

$$E=\frac{P}{\frac{ef}{a}+a+e+2f}. \qquad (1)$$

To have as strong signals as possible with any given line and battery we must make E a maximum subject to the variation of a. Now the denominator of (1) is a minimum when $a=\sqrt{ef}$; therefore E is then a maximum, and its expression is

$$E=\frac{P}{2\sqrt{ef}+e+2f}.$$

Let us take a numerical case. Let the resistance of the line, including the apparatus at the other end, be $e=5000$ ohms, the battery resistance $f=50$ ohms, then $a=\sqrt{50\times 5000}=500$ ohms, and

$$E=\frac{P}{2\times 500+5000+2\times 50}=\frac{P}{6100}.$$

Now, if the battery were joined direct to line, the total resistance in the circuit would be 5050, and the current would be $\frac{P}{5050}$, which is greater than the former result in the proportion of 6100 to 5050; so that the plan of signalling reversals as in Fig. 2 would be attended with a loss of strength of current amounting to about $\frac{1}{6}$ in this particular case. This difference is not very great, but a further disadvantage is that the battery is much harder worked in the system of Fig. 2 than in the ordinary system. These disadvantages would, no doubt, effectually preclude the use of the Morse key for signalling reversals by this particular arrangement; but, on the other hand, it may be adapted to form a system of signalling reversed currents having some advantages over the ordinary method. The principal points of this plan are as follows:—

(1) The reversals are produced by a Morse key.
(2) The sending station works his own instrument, so that he may hear or register his signals at pleasure.
(3) Each station can interrupt the others' sending.

Let the two resistances *a* and *b*, in Fig. 2, be the two coils of a Siemens relay, or any other polarised receiving instrument; replace the galvanometer *e* by the line, and put the front stop of the key to earth. Then we have the arrangement shown in Fig. 3, where the two points *s* and *t* may be joined or separated by means of a switch, or any other contrivance. When they are joined and the key is worked reversals are sent to line, just as in Fig. 2 reversals are sent through *e*. The currents sent split unequally between the two coils *a* and *b*, most going through one coil when the key is depressed, and most through the other coil when it is elevated, and the consequence of this unequal alternate division of the current will be that the armature of the instrument will exactly repeat the movements of the key. This is point (2). When it is desired to receive, separate *s* and *t* by means of the switch and the battery will be cut off, and all received currents will pass through both coils in the *usual* manner. Furthermore, as stated above, the receiving station can interrupt, for the sender's signals will then be no longer correctly repeated by his own instrument. Possibly Mr. Edison is perfectly well acquainted with this extension, or rather application, of his problem, to the discussion of which we may now return.

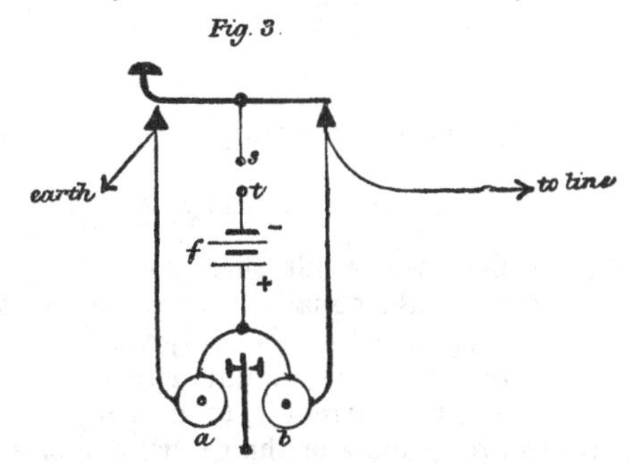

If, in the arrangement in Fig. 2, we make the battery and galvanometer change places, we get another—though somewhat similar—method of sending reversals through *e*. If *e* is greater than *f* the currents will be weaker, but if *f* is greater than *e* they will be stronger. Otherwise this arrangement is so similar to Fig. 2 as to call for no further comment.

In the previous three methods both stops of the key have been used. In the following only one is used. In Fig 4, *e* is the galvanometer or other resistance through which reversals are to be sent; *a*, *b*, and *c* are three resistances; and *f* the battery. The back stop of the key is not connected with any part of the arrangement. When the key is at rest the current from the + pole divides at A, through the two roads ACB and AB, which join at B, and the circuit is completed through c. When the key is depressed the current divides both at A and at B, and the current in e is reversed. We may also change the positions of the battery and galvanometer in Fig. 4 and still have reversals in the galvanometer. No doubt there are other methods, more or less simple, of obtaining reversals in a conductor by means of a Morse key, and, now that a beginning has been made, they ought to pour in from all parts of the United Kingdom. Mr. Edison's own solutions would also be very acceptable.

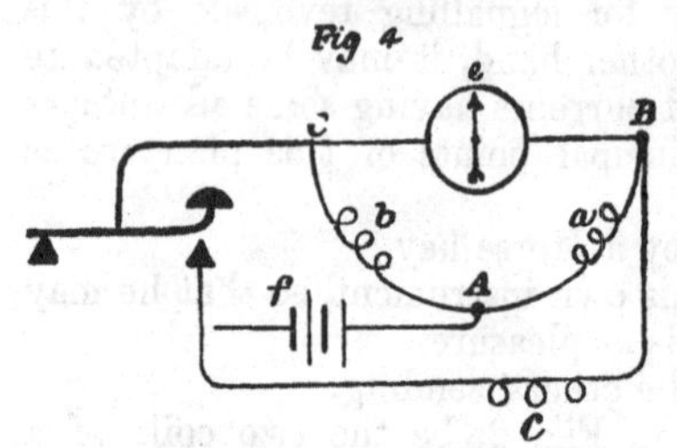

X.—ON THE RESISTANCE OF GALVANOMETERS.

[*Jour. Soc. Tel. Eng.*, April 28, 1880, vol. 9, p. 202.]

THE well-known result that the resistance of a galvanometer coil of given size should equal the external resistance in order to obtain the greatest magnetic force is easily verified. If G is the magnetic force at the centre of the coil for unit current in the coil, and M the magnetic force due to the current γ, then $M=\gamma G$. Also, by Ohm's law, if R and r are the resistances of the galvanometer and of the rest of the circuit, and E the electromotive force, $\gamma = E \div (R+r)$. Whence

$$M=\frac{EG}{R+r}. \qquad (1)$$

Here $G=gl$, where l is the length of wire in the coil, and g the mean value of G per unit of length throughout the space occupied by the coil, and therefore the same for different sizes of wire. Now, if we neglect the thickness of the covering of the wire, it is easily seen that the resistance of the coil varies as the square of the length of the wire. Thus, in (1) $G \propto l$ and $R \propto l^2$, and therefore M is a maximum when $R=r$.

In the next place, suppose the thickness of the covering is constant: let the radius of the wire $=y$ and of the covered wire $=y+b$, then the volume V of the coil is

$$V=4l(y+b)^2.$$

If ρ = the specific resistance of the wire

$$R=\frac{\rho}{\pi}\frac{l}{y^2}.$$

Also $G = gl$, as before. Therefore in (1), $G \propto (y+b)^{-2}$, and $R \propto y^{-2}(y+b)^{-2}$, and M is a maximum when

$$R : r = y : y + b,$$

or the resistance of the galvanometer should be to the external resistance as the radius of the bare wire is to the radius of the covered wire. (Maxwell, II., Art. 716.)

But if we suppose that the radius of the covered wire bears a constant ratio to the radius of the wire itself, the result is again $R = r$. For let the radius of the covered wire $= \beta y$, then

$$R = \frac{\rho}{\pi} \frac{l}{y^2}, \qquad V = 4l\beta^2 y^2.$$

Thus, in (1), $G \propto y^{-2}$ and $R \propto y^{-4}$, and therefore M is a maximum when $R = r$.

In the above, the form of the channel in which the wire is wound is arbitrary, and the thickness of the wire the same throughout the coil. But when the windings are circles there is a certain form of coil which gives the greatest magnetic force at the centre of the coil for a given length of wire. The wire should be wound in layers on surfaces defined by the polar equation

$$r^2 = x^2 \sin \theta,$$

where r is the distance of a circle of wire from the centre of the coil, θ the angle between r and the axis of the coil, and x a constant determining the linear dimension of the layer. With this form of coil, if the ratio of the radius of the covered to the radius of the bare wire is constant, the diameter of the wire in any layer should vary as the linear dimension of the layer to get the greatest electro-magnetic effect. (Maxwell, II., Art. 719.)

Under these circumstances, what should the resistance of the coil be? Professors Ayrton and Perry asked this question, and their answer was, $R = r$ again (*Journal Society Telegraph Engineers*, vol. vii., p. 297). For so simple a result to arise out of such complexity is rather striking, and, being lately occupied with a similar question, I looked for the reason of this result. It appears not to depend on the particular form of coil considered, nor on the particular law governing the diameter of the wire in the different layers, but solely upon the assumption of a constant ratio between the radius of the covered and of the bare wire.

Thus, let y = variable radius of the wire itself, and z = variable radius of the covered wire. Then

$$R = \frac{\rho}{\pi}\int \frac{dl}{y^2}, \qquad V = \int 4z^2 dl, \qquad G = \int \frac{\sin\theta}{r^2} dl = gl.$$

But since the coil is a figure of revolution

$$V = 2\pi \int\int r^2 \sin\theta dr d\theta.$$

Let the limits of integration for r be $x_0 f(\theta)$ and $x_1 f(\theta)$. Then

$$V = 2\pi \int_0^\pi \{f(\theta)\}^3 \sin\theta d\theta \frac{x_1^3 - x_0^3}{3} = \tfrac{1}{3}N(x_1^3 - x_0^3),$$

where N is a numerical constant. Therefore

$$dV = Nx^2dx = 4z^2dl,$$

where dV is the volume of the layer corresponding to dx, and dl the length of wire in the layer. Thus,

$$R = \frac{\rho}{\pi}\int\frac{Nx^2dx}{4y^2z^2}, \qquad G = g\int\frac{Nx^2dx}{4z^2},$$

and

$$M = Eg\frac{l}{R+r} = Eg\frac{\int\frac{Nx^2dx}{4z^2}}{\frac{\rho}{\pi}\int\frac{Nx^2dx}{4y^2z^2}+r}.$$

Let now $z = \beta y$, where β is constant, then

$$M = Eg\frac{l}{R+r} = Eg\frac{\frac{1}{\beta^2}\int\frac{Nx^2dx}{4y^2}}{\frac{1}{\beta^2}\frac{\rho}{\pi}\int\frac{Nx^2dx}{4y^4}+r}.$$

If y is constant with respect to x, $l \propto y^{-2}$ and $R \propto y^{-4}$, so that M is a maximum when $R = r$. If $y = ax$, the same result follows. But y may be any function of x; say $y = \alpha\phi(x)$, where ϕ determines the law of variation of the radius of the wire from layer to layer, and α fixes the actual size of the wire. Then

$$M = Eg\frac{l}{R+r} = Eg\frac{\frac{1}{\alpha^2\beta^2}\int\frac{Nx^2dx}{4\phi^2}}{\frac{1}{\alpha^4\beta^2}\frac{\rho}{\pi}\int\frac{Nx^2dx}{4\phi^4}+r}.$$

Let α vary, then, as before, $R = r$ makes M a maximum; for $l \propto \alpha^{-2}$ and $R \propto \alpha^{-4}$.

Thus $R = r$ makes M a maximum when the diameter of the wires in different layers is arbitrary and the form of the layers arbitrary (except that they are similar surfaces of revolution), provided that $z = \beta y$. Other relations between z and y of course give other results.

The following is more general: Take a long wire of circular section, whose radius varies continuously along its length, and let it be covered so that the thickness of the covering along its length varies in the same manner: *i.e.*, $z : y =$ constant everywhere. Now wind this wire into a coil of any shape and section. It will have a certain resistance, and the unit current in it will produce a certain magnetic force at any point. Now, if the radius of the wire is everywhere reduced to $\frac{1}{n}$th part, and the same space is filled, we have everywhere n^2 wires instead of one; therefore the magnetic force due to the unit current in an element of length of the original wire is increased n^2 times by the unit current now passing in n^2 wires instead of one, and the resistance is increased n^4 times by the change. Since the same is true for each element of length of the original wire, it follows that the magnetic force due to the unit current in the whole coil varies inversely as the square of the radius of the wire, and the resistance of the coil inversely as the fourth power of the same. Therefore,

$$M = \frac{EG}{R+r} = E\frac{A\alpha^{-2}}{B\alpha^{-4}+r},$$

where A and B are constants depending on the form and dimensions of the coil, and α determines the actual radius of the wire at any part. Vary α, then M is a maximum when $R=r$ as before.

XI.—ON A TEST FOR TELEGRAPH LINES.

[*Phil. Mag.*, Dec. 1878, S. 5, vol. 6, p. 436.]

THE true conduction and insulation resistances of a uniform line may be found from the potential and current at the ends, when a constant electromotive force acts at one end. Suppose at one end A of the line there is a battery of electromotive force E, and a galvanometer, the two together of resistance R_1; also at the other end B of the line a galvanometer of resistance R_2, the circuit being completed through the earth. If the potential at distance x from A, where $x=0$, is v, the current at the same point γ, the conduction and insulation resistance k and i respectively per unit of length, then

$$\frac{d^2v}{dx^2} = h^2 v,$$

where

$$h^2 = \frac{k}{i};$$

and

$$\gamma = -\frac{1}{k}\frac{dv}{dx};$$

whence

$$\left.\begin{aligned} v &= a\epsilon^{hx} + b\epsilon^{-hx}, \\ \gamma &= -\frac{1}{\sqrt{ki}}(a\epsilon^{hx} - b\epsilon^{-hx}), \end{aligned}\right\} \quad \text{......................} \quad (1)$$

where a and b are undetermined constants.

If now the potential and current at A are v_1 and γ_1, and the same at B are v_2 and γ_2, then it may easily be shown from equations (1) that

$$ki = \frac{v_1^2 - v_2^2}{\gamma_1^2 - \gamma_2^2}. \quad \text{..............................} \quad (2)$$

Since the length of the line does not appear in (2), the relation therein expressed applies to any two points of the line. The reason is that the product of the conduction and insulation resistances is the same for any length, the one varying directly and the other inversely as the length. Now the insulation of land-lines is in this country very variable, while the real conduction resistance (*i.e.* its resistance if it were perfectly insulated) is nearly constant. It follows that (2) may be used for determining i, considering k as constant. In (2),

$$\left.\begin{aligned} v_1 &= E - R_1\gamma_1, \\ v_2 &= R_2\gamma_2. \end{aligned}\right\} \quad \text{..........................} \quad (3)$$

R_1 and R_2 being interposed resistances are, of course, known; so that three quantities have to be observed, viz., E, γ_1, and γ_2; or equivalent information must be obtained. To make the test in its simplest form, let the resistances R_1 and R_2 be small compared with the line resistance. Also let equally sensitive tangent-galvanometers be used, and let n_1 and n_2 be the deflections corresponding to γ_1 and γ_2, and n_3 the deflection E gives through 1000 ohms. Then (2) becomes

$$ki = \frac{n_3^2}{n_1^2 - n_2^2} \times 10^6, \qquad (4)$$

where k and i are both in ohms; or if k is in ohms and i in megohms, the 10^6 must be cancelled.

If R_1 and R_2 are taken into account, then instead of (4) we have

$$ki = \frac{(10^3 n_3 - R_1 n_1)^2 - (R_2 n_2)^2}{n_1^2 - n_2^2};$$

and if the galvanometers are not equally sensitive, the deflection n_2 must be multiplied by the ratio of the sensitiveness of the galvanometer at B to that at A.

Using formula (4), the test can be easily made, though it is obvious that the line must be long enough to make an appreciable difference between the sent and received currents.

We may also determine k and i separately from the same data. If l is the length of the line, then

$$\left.\begin{aligned} kl &= \sqrt{ki}\,\log\frac{v_1 + \gamma_1\sqrt{ki}}{v_2 + \gamma_2\sqrt{ki}}, \\ \frac{i}{l} &= \sqrt{ki} \div \log\frac{v_1 + \gamma_1\sqrt{ki}}{v_2 + \gamma_2\sqrt{ki}}. \end{aligned}\right\} \qquad (5)$$

It is to be observed that these formulae give the true conduction and insulation resistances. The measured resistances, or those deduced from observations with the bridge, differential galvanometer, etc., at the battery-end alone, are very different from the true, when the line is long and badly insulated. The measured is always less than the true conduction resistance, and the measured always greater than the true insulation resistance; while the measured conduction resistance can never be greater than $\sqrt{ki}$, and the measured insulation resistance never less.

XII.—ON THE ELECTROSTATIC CAPACITY OF SUSPENDED WIRES.

[*Journ. Soc. Tel. Eng.*, 1880, vol. 9, p. 115.]

SUPPOSE, in the first place, we have a single wire suspended in empty space, and charged—no matter how—with a quantity q of electricity per unit of length. The resultant force at a point whose perpendicular distance from the centre of the wire at any point A is r, due to the elementary charge qdx at distance x from A is

$$qdx \div (x^2 + r^2),$$

and this resolved in the direction normal to the wire is

$$qdx \div (x^2 + r^2) \times r \div (x^2 + r^2)^{\frac{1}{2}};$$

therefore the resultant force due to the whole charge is

$$\int_{-\infty}^{\infty} \frac{qrdx}{(x^2 + r^2)^{\frac{3}{2}}} = \frac{2q}{r}.$$

Since the resultant force is the rate of decrease of the potential, the potential is

$$V = 2q \log \frac{z}{r},$$

where z is a constant. If the potential at an infinite distance is zero or constant, the potential of the wire itself is infinite; or, in other words, an infinite amount of work must be done to charge the wire—that is, it would be impossible to charge it. This may be made more intelligible in another way. The capacity of a wire becomes smaller and smaller the further it is removed from other conductors, and in the limit, when the wire is alone in space, it vanishes.

Suppose, now, there is another wire parallel to the first at distance $2h$, and charged with $-q$ per unit of length; the potential due to its charge at distance r' is

$$-2q \log \frac{z'}{r'},$$

where z' is another constant; consequently the potential due to both charges is

$$2q \log \frac{r'}{r},$$

for z and z' both disappear on being made infinite. Therefore, if d_1 and d_2 are the diameters of the wires, their potentials are

$$2q \log \frac{4h}{d_1} \text{ and } 2q \log \frac{d_2}{4h}.$$

Thus the charge divided by the difference of potentials is

$$q \div \left(2q \log \frac{16h^2}{d_1 d_2}\right) = \frac{1}{2 \log \frac{16h^2}{d_1 d_2}},$$

and this is the mutual capacity per unit of length of the two wires in space.

The potential is zero at all points where $r = r'$, that is, in a plane equidistant from the two wires, whose shortest distance from them is h; and the difference of potential between either wire and this plane is half that between the two wires. It follows that the capacity of a wire of diameter d suspended *alone* at height h above the ground is

$$c = \frac{1}{2 \log \frac{4h}{d}}$$

per unit of length, in electrostatic measure. (F. Jenkin, *Electricity and Magnetism*, p. 332.)

If $h=3$ metres and $d=4$ millimetres, $c=\cdot 0624$. To bring into electromagnetic measure this must be divided by $(28\cdot 8\times 10^9)^2$; to bring the result into microfarads, multiply by 10^{15}; and lastly, multiply by the number of centimetres in a mile to find the capacity in microfarads per mile. The result is ·0121 microfarads per mile.

Next, suppose the line consists of two wires, 1 and 2, of radii r_{11} and r_{22}. Let r_{12} = distance between the centres of 1 and 2, s_{11} the distance between the centres of 1 and of its image, s_{12} the distance between the centres of 1 and the image of 2, or of 2 and the image of 1, and s_{22} the distance between the centres of 2 and its image. Also let V_1 and V_2 be the potentials of 1 and 2, and q_1, q_2, their charges per unit of length. Then

$$V_1 = 2q_1 \log\frac{s_{11}}{r_{11}} + 2q_2 \log\frac{s_{12}}{r_{12}},$$

$$V_2 = 2q_1 \log\frac{s_{12}}{r_{12}} + 2q_2 \log\frac{s_{22}}{r_{22}},$$

express the potentials in terms of the charges. For $2q_1 \log\frac{s_{11}}{r_{11}}$ is the potential of 1 due to its own charge and the opposite charge of its image, and $2q_2 \log\frac{s_{12}}{r_{12}}$ the potential of 1 due to the charge of 2 and the opposite charge of its image, and similarly for V_2. From these we deduce

$$q_1 = c_{11}V_1 + c_{12}V_2, \quad q_2 = c_{21}V_1 + c_{22}V_2,$$

where $$c_{11} = \frac{2}{R}\log\frac{s_{22}}{r_{22}}, \quad -c_{12} = \frac{2}{R}\log\frac{s_{12}}{r_{12}}, \quad c_{22} = \frac{2}{R}\log\frac{s_{11}}{r_{11}},$$

and $$R = 2\log\frac{s_{11}}{r_{11}} \cdot 2\log\frac{s_{22}}{r_{22}} - \left(2\log\frac{s_{12}}{r_{12}}\right)^2.$$

Here c_{11} is the capacity per unit length of wire 1, c_{22} the capacity of wire 2, and c_{12} the mutual capacity of 1 and 2.

Suppose the wires have the same radius r, and their distance apart is d, at the same height above the ground. Then

$$r_{11} = r_{22} = r; \quad r_{12} = d; \quad s_{11} = s_{22} = 2h; \quad s_{12} = \sqrt{d^2+4h^2};$$

and $$c_{11} = c_{22} = \left(2\log\frac{2h}{r}\right) \div R; \quad -c_{12} = \left\{2\log\frac{(d^2+4h^2)^{\frac{1}{2}}}{d}\right\} \div R;$$

where $$R = \left(2\log\frac{2h^2}{r}\right) - \left(2\log\frac{\sqrt{d^2+4h^2}}{d}\right)^2.$$

The capacity of each wire is increased by the presence of the other. If the height, as before, is $h=3$ metres, the radius $r=\cdot 002$ metre, and the distance apart $d=\cdot 5$ metre, then

$$c_{11} = c_{22} = \cdot 0691, \; -c_{12} = \cdot 0215,$$

in electrostatic measure. Or

$$c_{11} = c_{22} = \cdot 0134, \; -c_{12} = \cdot 00417$$

microfarads per mile.

As the capacity of the single wire of the same radius and at the same height was 0121 microfarads per mile, the presence of the other wire increases its capacity about 11 per cent. If one of the wires is charged by a battery and the other is to earth, then about $\frac{3}{10}$ths of the opposite charge will be on the second wire and $\frac{7}{10}$ths on the earth.

The formulæ for the capacities of any number of wires may be easily obtained, though the subsequent numerical calculations become complex. Suppose the wires have radii r_{11}, r_{22}, r_{33}, ..., potentials V_1, V_2, V_3, ..., and charges q_1, q_2, q_3, ... per unit of length. Let the distance between the centres of any two wires m and n be denoted by r_{mn}, and the distance between the centre of any wire m and the image of any wire n by s_{mn}. Then the potentials are expressed in terms of the charges by

$$V_1 = 2q_1 \log \frac{s_{11}}{r_{11}} + 2q_2 \log \frac{s_{12}}{r_{12}} + 2q_3 \log \frac{s_{13}}{r_{13}} + \ldots,$$

$$V_2 = 2q_1 \log \frac{s_{21}}{r_{21}} + 2q_2 \log \frac{s_{22}}{r_{22}} + 2q_3 \log \frac{s_{23}}{r_{23}} + \ldots,$$

$$V_3 = 2q_1 \log \frac{s_{31}}{r_{31}} + 2q_2 \log \frac{s_{32}}{r_{32}} + 2q_3 \log \frac{s_{33}}{r_{33}} + \ldots..$$

To find the capacity of any wire, say 1, with respect to itself and the rest, express q_1 in terms of the potentials,

$$q_1 = c_{11}V_1 + c_{12}V_2 + c_{13}V_3 + \ldots$$

Then c_{11} is the capacity of 1, c_{12} the mutual capacity of 1 and 2, and so on.

If there are four wires of the same radius, one pair, 1 and 2, at one height, and the other pair 3 and 4 vertically beneath the first pair at another height, 3 being under 1, and 4 under 2, then we have the following relations amongst the distances:—

$$r_{11} = r_{22} = r_{33} = r_{44}\,; \quad r_{12} = r_{34}\,; \quad r_{13} = r_{24}\,; \quad s_{11} = s_{22}\,; \quad s_{33} = s_{44}\,;$$

$$s_{13} = s_{31} = s_{24} = s_{42}\,; \quad s_{14} = s_{41} = s_{23} = s_{32}.$$

Let

$$\log \frac{s_{11}}{r_{11}} = a\,; \ \log \frac{s_{12}}{r_{12}} = b\,; \ \log \frac{s_{13}}{r_{13}} = c\,; \ \log \frac{s_{14}}{r_{14}} = d\,; \ \log \frac{s_{33}}{r_{11}} = e\,; \ \log \frac{s_{34}}{r_{12}} = f.$$

Then

$$c_{11} = c_{22} = \{a(e^2 - f^2) + d(cf - de) + c(df - ce)\} \div R$$

$$c_{33} = c_{44} = \{e(a^2 - b^2) + d(bc - ad) + c(bd - ac)\} \div R$$

$$-c_{12} = \{d(df - ce) + c(cf - de) + b(e^2 - f^2)\} \div R$$

$$-c_{24} = -c_{13} = \{c(c^2 - d^2) + b(de - cf) + a(df - ce)\} \div R$$

$$-c_{23} = -c_{14} = \{b(df - ce) + a(de - cf) + d(c^2 - d^2)\} \div R$$

$$-c_{34} = \{d(bd - ac) + c(bc - ad) + f(a^2 - b^2)\} \div R$$

where

$$R = (a^2 - b^2)(e^2 - f^2) + (c^2 - d^2)^2 + 2(ac - bd)(df - ce) + 2(ad - bc)(cf - de).$$

These are the whole of the capacity coefficients for the four wires, which may now be numerically calculated.

Let the height of the top pair, 1 and 2, be $3\frac{1}{8}$ metres, of the lower pair, 3 and 4, $2\frac{5}{8}$ metres, and let the horizontal distance from 1 to 2 and from 3 to 4 be ·5 metre. Then

$$a = \log 3166{\cdot}6 \quad = 3{\cdot}5006023, \quad \log a = {\cdot}5441428$$
$$b = \tfrac{1}{2}\log 161{\cdot}4 \quad = 1{\cdot}1040115, \quad \log b = {\cdot}0429695$$
$$c = \log 18 \quad = 1{\cdot}2552725, \quad \log c = {\cdot}0987379$$
$$d = \tfrac{1}{2}\log 100{\cdot}384 = 1{\cdot}0008344, \quad \log d = {\cdot}0003622$$
$$e = \log 2833{\cdot}3 \quad = 3{\cdot}4522977, \quad \log e = {\cdot}5381080$$
$$f = \tfrac{1}{2}\log 129{\cdot}4 \quad = 1{\cdot}0560417, \quad \log f = {\cdot}0236810$$

Here common logarithms are used. The results are

$$c_{11} = 15{\cdot}7863 \div R; \quad c_{33} = 15{\cdot}924 \div R; \quad -c_{12} = 2{\cdot}9876 \div R;$$
$$-c_{13} = 4{\cdot}1992 \div R; \quad -c_{14} = 2{\cdot}4558 \div R; \quad -c_{34} = 2{\cdot}8517 \div R;$$
$$R = 88{\cdot}4668,$$

which must be multiplied by ·4343 for the change of logarithms, making

$$c_{11} = {\cdot}0775; \; c_{33} = {\cdot}0782; \; -c_{12} = {\cdot}0147; \; -c_{13} = {\cdot}0206; \; -c_{14} = {\cdot}0120;$$
$$-c_{34} = {\cdot}0140$$

in electrostatic measure, which are equivalent to

$$c_{11} = {\cdot}01503; \; c_{33} = {\cdot}01517; \; -c_{12} = {\cdot}00285; \; -c_{13} = {\cdot}00399;$$
$$-c_{14} = {\cdot}00232; \; -c_{34} = {\cdot}00271$$

microfarads per statute mile.

Suppose one of the top wires, say 1, is charged by a battery, while the remaining three wires are to earth. Then wire 1 will receive a charge = ·01503 microfarads per volt per mile, and wires 2, 3, and 4 will receive from the battery opposite charges proportional to ·00285, ·00399 and ·00232. The sum of the latter being ·00916, and the capacity of the first wire ·01503, it follows that about $\frac{9}{15}$ths of the opposite charge goes to the three wires, and the other $\frac{6}{15}$ths to the surface of the earth.

As there are only four wires considered, it is evident that with a large number of wires the proportion of the opposite charge on the surface of the earth becomes quite small, nearly the whole going to the neighbouring wires.

F. Jenkin (*Electricity and Magnetism*, p. 332) says there is experimental reason to believe that the actual capacity of a suspended wire is about double the amount, calculated on the supposition of there being no other wires on the same poles, owing to the induction between the wires and the posts and insulating supports; but as the posts only occur at intervals, it seems reasonable to suppose that a great part of the difference is rather due to the neighbourhood of other wires. At any rate, a second wire increases capacity about 11 per cent., and with three more the increase is about 24 per cent. according to the above figures, and a greater number will of course produce still further increase.

XIII.—ON TELEGRAPHIC SIGNALLING WITH CONDENSERS.

[*Phil. Mag.*, June 1874, S. 4, vol. 47, p. 426.]

GIVEN an insulated conductor called the *line* connecting two places, there may be said to be in present use two distinct methods by which signals made at one end of the line are observed at the other. The first, which is that in most general use, is to connect the line with one end of the coils of an instrument affected by electric currents, and the other end of the coils with earth. The battery at the sending-end being also placed between the line and the earth, a *circuit* is established, through which a current will flow so long as the battery and instrument remain undisturbed. This current will in a short time after the first moment of contact with the battery become approximately constant at any one part of the line—and, if there be no leakage, will attain the same strength at every part of the circuit, including the battery and receiving instrument. The second method, first introduced by Mr. Varley,* and now in pretty general use on submarine lines, is somewhat different. The end of the coils of the receiving instrument, which in the first method is connected with the earth, is now joined to one armature or inductive surface of a so-called condenser, properly speaking an electrical accumulator, the other armature of which is to earth; or, which comes to the same thing, the condenser is placed between the line and the receiving instrument. As there is now no longer a complete conductive circuit, no permanent current can flow through the receiving instrument, or indeed through any part of the line, if the insulation be perfect.

Imagine the condenser to be a continuation of the cable, in fact a length of cable having the same capacity as the condenser, insulated at its further extremity, and the receiving instrument connecting the main

FIG. 1.

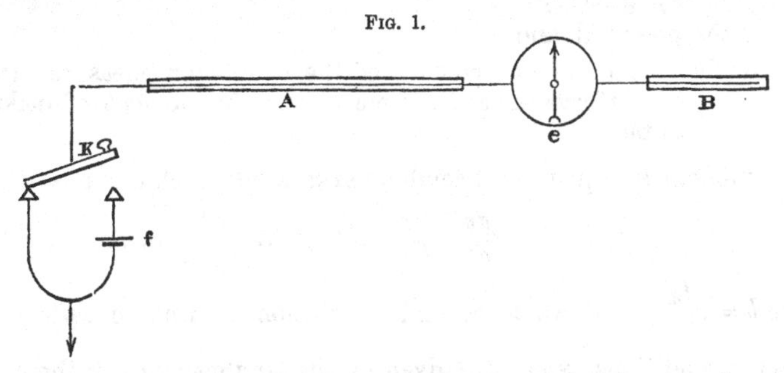

cable with its imaginary continuation, as shown in Fig. 1; where f is the battery at the sending-end of the line, one pole of which is to earth, K a key for making contact between its other pole and the

* [See, however, Mr. Willoughby Smith's claims in his paper, "Working of Long Submarine Cables," *Jour. Soc. Tel. Eng.*, vol. 8, p. 63, and the discussion thereon.]

cable A, and e the receiving instrument placed between A and the continuation B.

Then when contact is made at K it is evident that only so much electricity will pass through e as will charge B up to the potential of the further end of A. The current through e will therefore be transient, rising to a maximum and then dying away. This method of representation would be perfect if we could neglect the resistance of the conductor inside B; as, however, in practice the capacity of the condenser is only a fraction of that of the line, there will be little difference due to this cause. And if the capacity of B be very small, we may consider the flow of current through e to be strictly dependent on the rise or fall of potential of the end of A.

To find an expression for the potential and the current at any point of a cable insulated at one end, at any time after contact is made with a battery at the other end, the only way, as far as I am aware, is to follow the method given by Sir William Thomson in 1855 (*Proc. Roy. Soc.*), making the necessary alterations to suit the changed conditions of the problem. It is to express the actual potential at any time as the difference of two functions, one being the known final distribution of potential, and the other the departure from the final potential, the latter being expressed by an infinite convergent series every term of which is of the form

$$\sin x \, . \, \epsilon^{-t}.$$

Let l be the length of the line,

- k the electrical resistance of the conductor per unit of length,
- c its electrostatic capacity per unit of length,
- k_1 the resistance of the dielectric per unit of length to conduction in a radial direction,
- V the electromotive force of the battery, the resistance of which is neglected,
- v the potential, and
- C the current at any point x of the conductor, measured from the battery-end, at the time t from the moment of making contact.

The differential equation of conduction in a telegraphic line is

$$ck\frac{dv}{dt} = \frac{d^2v}{dx^2} - h^2 v, \qquad (1)$$

where $h = \sqrt{\frac{k}{k_1}}$; and we must find a solution of this to satisfy the following conditions, which are given by the circumstances of the case.

1. $v = V$ when $x = 0$.
2. $\frac{dv}{dx} = 0$ when $x = l$.
3. $v = 0$ when $t = 0$, except when $x = 0$.
4. $v = f(x)$ when $t = \infty$.

To find the function $f(x)$ expressing the permanent distribution of potential after an infinite time, make $\frac{dv}{dt}=0$ in (1), and integrate the resulting equation

$$\frac{d^2v}{dx^2}=h^2v,$$

subject to the first and second conditions. We thus obtain

$$f(x)=V\,.\,\frac{\epsilon^{h(l-x)}+\epsilon^{-h(l-x)}}{\epsilon^{hl}+\epsilon^{-hl}} \quad \text{.......................} \quad (2)$$

for the final distribution.

In expanding (2) in a series of sines we must remember that $\frac{dv}{dx}=0$ when $x=l$, and accordingly use the expansion

$$f(x)=\frac{2}{l}\Sigma_1^\infty \sin\frac{(2i-1)\pi x}{2l}\int_0^l f(x')\sin\frac{(2i-1)\pi x'}{2l}dx';$$

which gives

$$f(x)=V\frac{\epsilon^{h(l-x)}+\epsilon^{-h(l-x)}}{\epsilon^{hl}+\epsilon^{-hl}}$$

$$=4\pi V\Sigma_1^\infty\frac{2i-1}{(2i-1)^2\pi^2+4l^2h^2}\,.\,\sin\frac{(2i-1)\pi x}{2l};$$

consequently the required solution is

$$v=V\frac{\epsilon^{h(l-x)}+\epsilon^{-h(l-x)}}{\epsilon^{hl}+\epsilon^{-hl}}$$

$$-4\pi V\,.\,\epsilon^{-\frac{h^2t}{ck}}\Sigma_1^\infty\frac{2i-1}{(2i-1)^2\pi^2+4l^2h^2}\,.\,\sin\frac{(2i-1)\pi x}{2l}\,.\,\epsilon^{-\frac{(2i-1)^2\pi^2t}{4ckl^2}}\,. \quad \text{....} \quad (3)$$

When the insulation is perfect and $h=0$, this becomes

$$v=V-\frac{4V}{\pi}\Sigma_1^\infty\frac{1}{2i-1}\,.\,\sin\frac{(2i-1)\pi x}{2l}\,.\,\epsilon^{-\frac{(2i-1)^2\pi^2t}{4ckl^2}}\,. \quad \text{..............} \quad (4)$$

As the current equals $-\frac{1}{k}\frac{dv}{dx}$, we have, by differentiating (3) and (4),

$$C=\frac{Vh}{k}\,.\,\frac{\epsilon^{h(l-x)}-\epsilon^{-h(l-x)}}{\epsilon^{hl}+\epsilon^{-hl}}$$

$$+\frac{2V}{kl}\,.\,\epsilon^{-\frac{h^2t}{ck}}\Sigma_1^\infty\frac{(2i-1)^2\pi^2}{(2i-1)^2\pi^2+4l^2h^2}\,.\,\cos\frac{(2i-1)\pi x}{2l}\,.\,\epsilon^{-\frac{(2i-1)^2\pi^2t}{4ckl^2}}; \quad \text{......} \quad (5)$$

and when $h=0$,

$$C=\frac{2V}{kl}\Sigma_1^\infty\cos\frac{(2i-1)\pi x}{2l}\,.\,\epsilon^{-\frac{(2i-1)^2\pi^2t}{4ckl^2}}\,. \quad \text{..................} \quad (6)$$

We can employ (5) and (6) to determine the flow through the receiving instrument, by giving x a value something less than l; but it is preferable to use the series for $\frac{dv}{dt}$ obtained from (3) and (4) by differentiation.

$$\left.\begin{aligned}\frac{dv}{dt}=\frac{4V}{\pi}\cdot\frac{h^2}{ck}\cdot\epsilon^{-\frac{h^2t}{ck}}\Sigma_1^\infty\frac{1}{1+\dfrac{4l^2h^2}{(2i-1)^2\pi^2}}\cdot\frac{1}{2i-1}\cdot\sin\frac{(2i-1)\pi x}{2l}\cdot\epsilon^{-\frac{(2i-1)^2\pi^2t}{4ckl^2}}\\ +\frac{V\pi}{ckl^2}\cdot\epsilon^{-\frac{h^2t}{ck}}\Sigma_1^\infty\frac{1}{1+\dfrac{4l^2h^2}{(2i-1)^2\pi^2}}\cdot(2i-1)\cdot\sin\frac{(2i-1)\pi x}{2l}\cdot\epsilon^{-\frac{(2i-1)^2\pi^2t}{4ckl^2}}.\end{aligned}\right\}\quad(7)$$

This becomes, when $h=0$,

$$\frac{dv}{dt}=\frac{V\pi}{ckl^2}\cdot\Sigma_1^\infty(2i-1)\cdot\sin\frac{(2i-1)\pi x}{2l}\cdot\epsilon^{-\frac{(2i-1)^2\pi^2t}{4ckl^2}}.\;\ldots\ldots\ldots\ldots\;(8)$$

A unit of time of a very convenient magnitude for practical calculations is

$$a=\frac{ckl^2}{10\pi^2}\log_\epsilon 10.$$

Employing this unit, we have the following series for v and $\frac{dv}{dt}$ when $x=l$ and $h=0$:—

$$v=V-\frac{4V}{\pi}(10^{-\frac{t}{40a}}-\frac{1}{3}\cdot10^{-\frac{9t}{40a}}+\frac{1}{5}\cdot10^{-\frac{25t}{40a}}-\text{etc.}).\;\ldots\ldots\ldots\;(9)$$

$$\frac{dv}{dt}=\frac{V\pi}{ckl^2}(10^{-\frac{t}{40a}}-3\cdot10^{-\frac{9t}{40a}}+5\cdot10^{-\frac{25t}{40a}}-\text{etc.}).\;\ldots\;\ldots\ldots\;(10)$$

The "arrival-curve" for v, calculated from equation (9), is shown in Fig. 2.

Fig. 2.

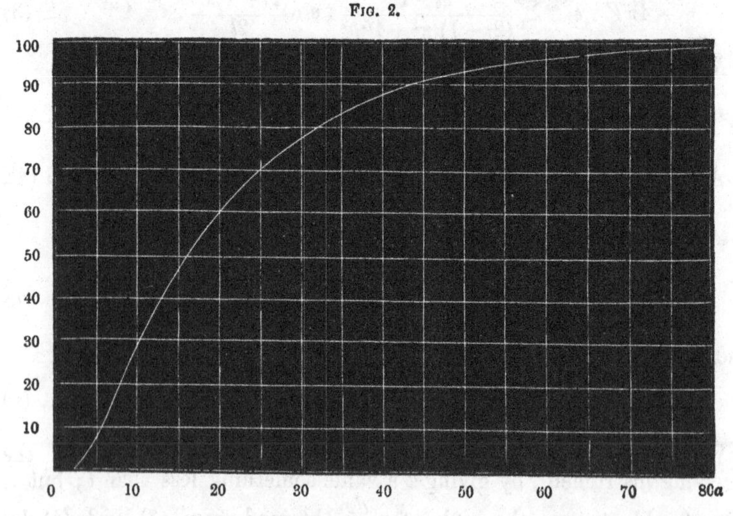

By comparison with the arrival-curve for the *current* at the remote end when to earth, we see that, broadly speaking, it takes about four times as long for the potential to nearly attain its maximum when the end is insulated as it takes for the current to nearly attain its maximum when the end is to earth. Thus, when the end is to earth, the current reaches 98 per cent. of its maximum strength in $20a$; and when the end is insulated, the potential reaches 98 per cent. of its maximum in $80a$. This relation does not hold good throughout the whole extent of the curves; but there is a general similarity. We may conclude that signalling by means of an electrometer connected with the insulated end of a cable would be much slower than the ordinary plan of a galvanometer or recording instrument worked by the current.

Fig. 3 represents $\frac{dv}{dt}$ from $t=0$ to $t=80a$ calculated from equation (10), and is closely the same as the arrival-curve for the current in condenser signalling. It will be seen that $\frac{dv}{dt}$ reaches its maximum in $7a$.

FIG. 3.

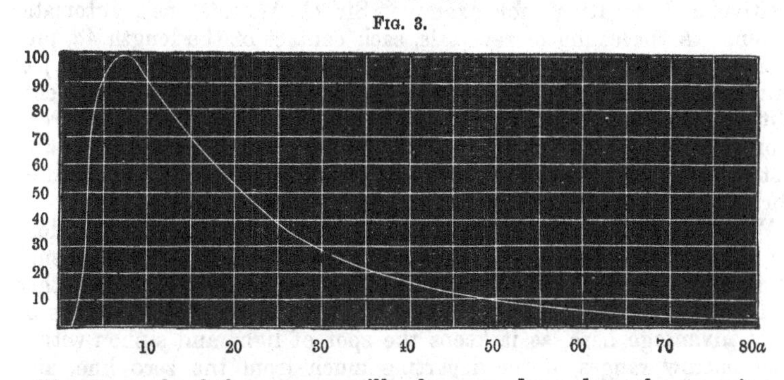

The strength of the current will of course depend on the capacity of the condenser, and will be proportional thereto so long as it is small compared with the capacity of the line. As, however, an increase in the capacity is equivalent to lengthening the line, the maximum strength of the current will not be so soon reached with the larger capacity: although the signals will be stronger, they will be more retarded. Hence the best capacity to be used on any line, which should theoretically be as small as possible, must be determined by the sensitiveness of the instrument and the battery-power employed. When the capacity of the condenser is one-seventeenth part of that of the line, the maximum strength of a signal is about one-tenth of the permanent current which would flow were the end of the line put to earth.

The effect of a condenser on the signals varies according to the description of instruments used for observing the signals, on the system of currents forming the signals, and on the value of the unit of time a for the particular line under consideration. Thus on an overland line worked on the Morse system with single currents, where the signalling consists in alternately connecting the sending-end of the line with one pole of a battery and to earth, it would be impossible to make *dashes*, because on land lines the length of a contact is always so great compared with the unit a, even in rapid signalling, that the current at the receiving-end would have come and gone long before the contact at the sending-end was finished. On cables of any considerable length it would be different. There would then be only a shortening of the dashes, its extent depending on the length of the contact; and it would consequently lead to an increase in the speed of working. There would similarly be very little direct advantage, save immunity from earth currents, in using on land lines the condenser with polarized relays and reversing-keys, although there would be no shortening of the marks, as the armature of a polarized instrument will, if properly adjusted, remain in the position in which it is placed by a transient current. On cables worked with the same instruments, a considerable increase in the speed of working results.

But the condenser is peculiarly applicable to those systems of signalling where the currents sent are of equal duration and alternately positive and negative; for example, Sir C. Wheatstone's automatic system. A succession of reversals, each contact of the length $4a$, produces through the receiving instrument reversals alternately 50 per cent. *plus* and 50 per cent. *minus*, being a whole range of 100 per cent. Without the condenser the whole amplitude of variation of the current is only 24·42 per cent. On a certain circuit worked by the automatic system, on which the value of a was about 0·0175 second, the speed of working with condenser was 75 per cent. greater than without.

With Sir W. Thomson's mirror and recording instruments there does not at first sight appear any reason why the speed of working should be much raised, as they indicate in the one case and record in the other every change in the strength of the current. Yet the condenser is of great advantage here, as it keeps the spot of light and siphon within very narrow ranges, never departing much from the zero line, and naturally the signals are much more distinct.

It will be found on examination that $\frac{dv}{dt}$, when the end of the line is to earth, reaches its maximum in $3a$, as against $7a$ when the end of the line is insulated. Thus it would appear that a considerable increase in

FIG. 4.

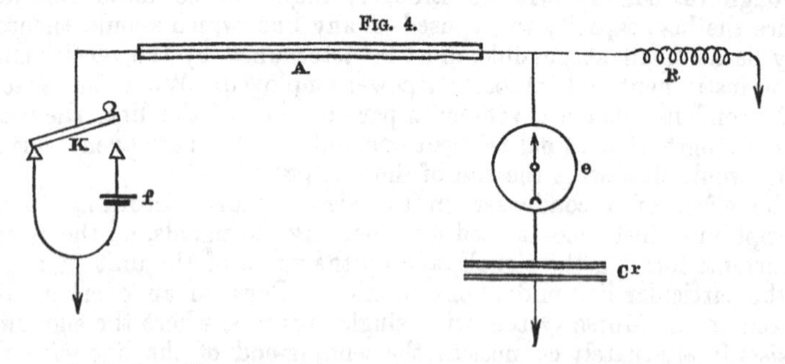

the speed of signalling should result from connecting the line to earth through a resistance, as shown in Fig. 4, where f is the battery, A the cable, e the receiving instrument, C^r the condenser, and R the resistance introduced between the end of the line and earth. This resistance is necessary in order to raise the potential of the end of the cable, and give signals of a workable strength. That this plan does increase the speed, I have verified on the circuit before mentioned, using a polarized instrument and double current key. The increase in speed was about 20 per cent., compared with the speed obtained by making R infinite. The signals were, of course, weaker in the former than in the latter case.

The effects of defective insulation may be traced by giving h different small values in equation (7). In the first place the signals are weakened; and next, the effect of loss is to decrease the time required for the signal to arrive at its maximum strength. Thus when $h=\frac{\pi}{l}$, the maximum is reached in $4a$, as against $7a$ when $h=0$. This is an extreme case, and would nearly correspond to the French Atlantic cable if its insulation-resistance were 3 megohms per knot, instead of more probably 300. Whether loss does or does not increase the speed of working depends on a great many circumstances. It is an undoubted fact that, under some systems of signalling, a cable with a very bad fault in it has worked quicker than when it was perfect; and it is also a demonstrable fact that under other systems the effect of loss is greatly to diminish the speed. In a particular case which I have examined theoretically and practically, the fact that each signal should be more quickly made through a faulty than a perfectly insulated cable is quite consistent with the fact that the speed of working is reduced.

Another method of receiving signals has been tried, though not adopted anywhere. The current is passed through the primary wire of an induction-coil, in the secondary circuit of which is the receiving instrument. The signal here depends on $\frac{dC}{dt}$, the rate of increase of the current; and the arrival-curve has its maximum at about $3{\cdot}5a$, while with condenser pure and simple it is $7a$. It does not, therefore, appear evident why, as Mr. Varley states, he found the condenser more satisfactory. Mr. G. K. Winter reinvented this system, and reports very favourably on its effect (British Association, Brighton, 1872).

XIV.—ON THE EXTRA CURRENT.

[*Phil. Mag.*, Aug. 1876, S. 5, vol. 2.]

Let a wire possessing uniform electrical properties throughout be of length l, resistance kl, electric capacity cl, and let its coefficient of self-induction be sl. Further, let P and Q denote the two ends of the wire, and x the distance of any point from the end P. Let v be the electric potential at the point x at the time t, and Q the quantity of electricity that has passed that point from the time $t=0$, so that $\dot{Q}$ is the current.

The differential equation of the potential of the wire may be found from the following two equations :—

$$-\frac{dQ}{dx}=cv, \quad\text{.......................................}\quad (1)$$

$$-\frac{dv}{dx}=k\dot{Q}+s\ddot{Q}. \quad\text{...................................}\quad (2)$$

The first expresses the fact that the quantity of electricity existing on the surface of the wire between sections at x and $x+dx$ at any moment is the product of the potential and of the capacity of the portion of wire considered. The second expresses that the E.M.F. at the point x at any moment is the sum of the E.M.F. producing the current $\dot{Q}$ and the rate of increase of the momentum of that current. By eliminating Q we obtain

$$\frac{d^2v}{dx^2}=ck\dot{v}+sc\ddot{v}. \quad\text{...................................}\quad (3)$$

If $s=0$, the above equation becomes

$$\frac{d^2v}{dx^2}=ck\dot{v},$$

the differential equation of the linear flow of heat, or of electricity in a submarine cable, the practical solution of which for a wire of finite length can only be accomplished with the assistance of Fourier's theorem. And if $k=0$, we have

$$\frac{d^2v}{dx^2}=cs\ddot{v},$$

which is of the same form as the equation of motion of a vibrating wire, the solution also requiring the use of Fourier's theorem. It is therefore probable that the same method must be adopted to solve the differential equation under consideration, viz.,

$$\frac{d^2v}{dx^2}=ck\dot{v}+cs\ddot{v}. \quad\text{...................................}\quad (3)$$

Let the potential of the wire at any moment be

$$v=V\frac{\sin}{\cos}\frac{i\pi x}{l}f(t), \quad\text{...................................}\quad (4)$$

where $f(t)$ is a function of t only, and V is constant. From (4), by differentiation,

$$\frac{d^2v}{dx^2}=-\frac{i^2\pi^2}{l^2}v,$$

therefore, by (3),

$$cs\ddot{v}+ck\dot{v}+\frac{i^2\pi^2}{l^2}v=0;$$

the solution of which is

$$v=\epsilon^{-t/2\alpha}\left\{A\epsilon^{(t/2\alpha)\left(1-4i^2\pi^2\frac{\alpha}{\beta}\right)^{\frac{1}{2}}}+B\epsilon^{-(t/2\alpha)\left(1-4i^2\pi^2\frac{\alpha}{\beta}\right)^{\frac{1}{2}}}\right\},$$

if $4i^2\pi^2\frac{\alpha}{\beta}<1$; and

$$v=\epsilon^{-t/2a}(A'\cos+B'\sin)\frac{t}{2a}\left(4i^2\pi^2\frac{a}{\beta}-1\right)^{\frac{1}{2}},$$

if $4i^2\pi^2\frac{a}{\beta}>1$. Here A, B, A', B' are constants, and $a=\frac{s}{k}$, and $\beta=ckl^2$, both time-constants. Therefore if the potential when $t=0$ is

$$v=V\frac{\sin}{\cos}\frac{i\pi x}{l},$$

the potential of the wire at time t is

$$v=V\frac{\sin}{\cos}\frac{i\pi x}{l}\epsilon^{-t/2a}\left\{A\epsilon^{(t/2a)\left(1-4i^2\pi^2\frac{a}{\beta}\right)^{\frac{1}{2}}}+(1-A)\epsilon^{-(t/2a)\left(1-4i^2\pi^2\frac{a}{\beta}\right)^{\frac{1}{2}}}\right\}, \quad\ldots (5)$$

or $$v=V\frac{\sin}{\cos}\frac{i\pi x}{l}\epsilon^{-t/2a}(\cos+B'\sin)\frac{t}{2a}\left(4i^2\pi^2\frac{a}{\beta}-1\right)^{\frac{1}{2}}, \quad\ldots\ldots (6)$$

according as $4i^2\pi^2\frac{a}{\beta}<$ or >1. The remaining constants A and B' must be determined from the value of the current at some fixed time. By solving (2), where $-\frac{dv}{dx}$ is to be found from (5) and (6), we shall find

$$\dot{Q}=C\epsilon^{-t/a}\mp\frac{Vi\pi}{kl}\frac{\cos}{\sin}\frac{i\pi x}{l}\epsilon^{-t/2a}\left\{\frac{2A\epsilon^{t/2a\left(1-4i^2\pi^2\frac{a}{\beta}\right)^{\frac{1}{2}}}}{1+\left(1-4i^2\pi^2\frac{a}{\beta}\right)^{\frac{1}{2}}}+\frac{2(1-A)\epsilon^{-\cdots}}{1-(\ldots)^{\frac{1}{2}}}\right\},$$

or $$\dot{Q}=C\epsilon^{-t/a}\mp\frac{Vi\pi}{kl}\frac{\cos}{\sin}\frac{i\pi x}{l}\epsilon^{-t/2a}\frac{\beta}{2ai^2\pi^2}$$

$$\times\left\{\left(1-B'\sqrt{4i^2\pi^2\frac{a}{\beta}-1}\right)\cos+(B'+\sqrt{\ldots})\sin\right\}\cdot\frac{t}{2a}\sqrt{4i^2\pi^2\frac{a}{\beta}-1},$$

where C is a constant current. Let the initial current be C, then

$$A=\frac{1+\sqrt{1-4i^2\pi^2\frac{a}{\beta}}}{2\sqrt{1-4i^2\pi^2\frac{a}{\beta}}}, \qquad B'=\frac{1}{\sqrt{4i^2\pi^2\frac{a}{\beta}-1}},$$

therefore the expressions for the potential and the current become

$$v=V\frac{\sin}{\cos}\frac{i\pi x}{l}\cdot\frac{\epsilon^{-t/2a}}{2m}\{(1+m)\epsilon^{tm/2a}-(1-m)\epsilon^{-tm/2a}\}, \quad\ldots (7)$$

or $$v=V\frac{\sin}{\cos}\frac{i\pi x}{l}\epsilon^{-t/2a}\left(\cos+\frac{1}{m'}\sin\right)\frac{tm'}{2a}, \quad\ldots (8)$$

where $m=\sqrt{1-4i^2\pi^2\frac{a}{\beta}}$, $\quad m'=\sqrt{4i^2\pi^2\frac{a}{\beta}-1}$;

and $$\dot{Q}=C\epsilon^{-t/a}\mp\frac{Vi\pi}{kl}\frac{\cos}{\sin}\frac{i\pi x}{l}\frac{\epsilon^{-t/2a}}{m}(\epsilon^{tm/2a}-\epsilon^{-tm/2a}), \quad\ldots (9)$$

or $$\dot{Q}=C\epsilon^{-t/a}\mp\frac{Vi\pi}{kl}\frac{\cos}{\sin}\frac{i\pi x}{l}\frac{2\epsilon^{-t/2a}}{m'}\left(\sin\frac{tm'}{2a}\right). \quad\ldots (10)$$

In the intermediate case, when $m=m'=0$,

$$v=V\frac{\sin}{\cos}\frac{i\pi x}{l}\epsilon^{-t/2a}\left(1+\frac{t}{2a}\right), \qquad (11)$$

$$\dot{Q}=C\epsilon^{-t/a}\mp\frac{Vi\pi}{kl}\frac{\cos}{\sin}\frac{i\pi x}{l}\epsilon^{-t/2a}\cdot\frac{t}{a}. \qquad (12)$$

The current $C\epsilon^{-t/a}$ does not influence the potential in any way. The above solutions suppose that the initial current is C and the initial potential $v=V\frac{\sin}{\cos}\frac{i\pi x}{l}$, and give the potential and current at any time after. When $\sin\frac{i\pi x}{l}$ is taken the potential at the ends of the wire is always zero, and when $\cos\frac{i\pi x}{l}$ is taken the current is always zero at the ends.

After this preliminary we can pass to more practical cases. In the first place, let a constant current V/kl be flowing through the wire, caused by a battery of negligible resistance and E.M.F. V, and let the potential of the wire be $V(1-x/l)$, so that it is V at the end P and 0 at the end Q. By Fourier's theorem

$$V(1-x/l)=\frac{2V}{\pi}\Sigma_1^\infty\frac{1}{i}\sin\frac{i\pi x}{l};$$

therefore if the end P is put to earth at the time $t=0$ the potential at time t is

$$v=\frac{2V}{\pi}\Sigma\frac{1}{i}\sin\frac{i\pi x}{l}\epsilon^{-t/2a}\left\{\frac{1+m_i}{2m_i}\epsilon^{tm_i/2a}-\frac{1-m_i}{2m_i}\epsilon^{-tm_i/2a}\right\}$$

$$+\frac{2V}{\pi}\Sigma\frac{1}{i}\sin\frac{i\pi x}{l}\epsilon^{-t/2a}\left(\cos+\frac{1}{m_i'}\sin\right)\frac{tm_i'}{2a}, \qquad (13)$$

by (7) and (8); where the first series includes all integral values of i which make $4i^2\pi^2\frac{a}{\beta}-1$ negative, and the second all the rest up to $i=\infty$. And, by (9) and (10),

$$\dot{Q}=\frac{V}{kl}\epsilon^{-t/a}-\frac{2V}{kl}\Sigma\cos\frac{i\pi x}{l}\frac{\epsilon^{-t/2a}}{m_i}(\epsilon^{tm_i/2a}-\epsilon^{-tm_i/2a})$$

$$-\frac{2V}{kl}\Sigma\cos\frac{i\pi x}{l}\frac{2\epsilon^{-t/2a}}{m_i'}\sin\frac{tm_i'}{2a} \qquad (14)$$

expresses the current at time t. If the wire is originally everywhere at potential zero, and without current, the potential v' and current $\dot{Q}'$ at time t after the end P is raised to potential V, the end Q being to earth, are

$$v'=V\left(1-\frac{x}{l}\right)-v,$$

$$\dot{Q}'=\frac{V}{kl}-\dot{Q},$$

where v and Q have the values given in (13) and (14).

Suppose $4\pi^2\frac{\alpha}{\beta} > 1$, then the first series in (13) and (14) disappears, and we have

$$v = \frac{2V}{\pi}\epsilon^{-t/2\alpha}\Sigma_1^\infty \frac{1}{i}\sin\frac{i\pi x}{l}\left(\cos + \frac{1}{m_i'}\sin\right)\frac{tm_i'}{2\alpha}, \quad (15)$$

$$\dot{Q} = \frac{V}{kl}\epsilon^{-t/\alpha} - \frac{4V}{kl}\epsilon^{-t/2\alpha}\Sigma_1^\infty \frac{1}{m_i'}\cos\frac{i\pi x}{l}\sin\frac{tm_i'}{2\alpha}. \quad (16)$$

The extra current is exhibited in (16) as consisting of two parts. One a current $\frac{V}{kl}\epsilon^{-t/\alpha}$, uniform at all parts of the wire, which dies away without oscillations with a rapidity proportional to $\frac{1}{\alpha}$. This current is due entirely to the momentum of the original current V/kl. The other part,

$$-\frac{4V}{kl}\epsilon^{-t/2\alpha}\Sigma\frac{1}{m_i'}\cos\frac{i\pi x}{l}\sin\frac{tm_i'}{2\alpha},$$

is due entirely to the original charge of the wire, and consists at any point x of an infinite series of currents alternately positive and negative, which die away with only half the rapidity. The oscillations are of greatest intensity at the end P and least at the end Q. They are insensible both when α/β is very small and when it is very large. In the former case only the higher terms in (13) and (14) are periodic with respect to the time, and in the latter case they become very rapid and weak in the same proportion. But when the time-constants α and β are not very different, the oscillations are of considerable strength, and may become observable by proper means. Suppose α/β of such magnitude that $\sqrt{4i^2\pi^2\frac{\alpha}{\beta} - 1}$ is appreciably $= 2i\pi\left(\frac{\alpha}{\beta}\right)^{\frac{1}{2}}$, then the time of a complete oscillation, including a positive and a negative current at any point, is nearly $2\sqrt{\alpha\beta}$, so that there are $\sqrt{\frac{\alpha}{\beta}}$ complete oscillations in the time 2α. The strength of these oscillations is proportional to $\sqrt{\frac{\beta}{\alpha}}$, so that the larger α/β the weaker the oscillations, they being at the same time more rapid in the same proportion.

The time-integral of the extra current is

$$\frac{V\alpha}{kl} - \frac{Vcl}{2}\left(\frac{x^2}{l^2} - \frac{2x}{l} + \tfrac{2}{3}\right),$$

where the first part is the same at all points, and is due entirely to the momentum of the initial current. The second part is the excess of the positive over the negative currents due to the initial charge, and is twice as great at the end P as at Q. This is the same when $s = 0$, or there is no self-induction.

The work done in the wire by the extra current is

$$\int_0^\infty kl(\dot{Q})^2 dt,$$

when $\dot{Q}$ is the same at all points, and

$$\int_0^l\int_0^\infty k\dot{Q}^2 dx dt,$$

when $\dot{Q}$ varies with x. Hence the amount of work done by the first part of the current in equation (16) is $\frac{sl}{2}\left(\frac{V}{kl}\right)^2$, and by the second part $\frac{V^2cl}{6}$, which was the energy of the initial charge,

$$= \tfrac{1}{2}\int_0^l V^2c(1-x/l)^2 dx.$$

As another example, suppose that before the time $t=0$ a uniform current V/kl existed in the wire, with potential $v=V(1-x/l)$, and that at the time $t=0$ both ends of the wire are instantaneously and simultaneously insulated without allowing a spark to pass. Then we have $\dot{Q}=0$ at P and Q. Let us first consider v and $\dot{Q}$ resulting from the initial charge, supposing $4\pi^2\frac{\alpha}{\beta}>1$. By Fourier's theorem

$$V(1-x/l) = \tfrac{1}{2}V + \frac{2V}{\pi^2}\Sigma_1^\infty \frac{1-\cos i\pi}{i^2}\cos\frac{i\pi x}{l},$$

where $\frac{1}{2}V$ is the final potential. Therefore, by (8), that part of the potential due to the initial charge is

$$\frac{V}{2} + \frac{2V}{\pi^2}\epsilon^{-t/2\alpha}\Sigma_1^\infty \frac{1-\cos i\pi}{i^2}\cos\frac{i\pi x}{l}\left(\cos + \frac{1}{m_i'}\sin\right)\frac{tm_i'}{2\alpha}; \quad \text{.........} (17)$$

and by (10) that part of the current due to the initial charge is

$$\frac{2V}{\pi kl}\epsilon^{-t/2\alpha}\Sigma_1^\infty \frac{1-\cos i\pi}{i}\sin\frac{i\pi x}{l}\,\frac{2}{m_i'}\sin\frac{tm_i'}{2\alpha}. \quad \text{..................} (18)$$

To find the potential and current due to the initial current, we have

$$\frac{V}{kl} = \frac{2V}{\pi kl}\Sigma_1^\infty \frac{1-\cos i\pi}{i}\sin\frac{i\pi x}{l};$$

therefore $$\dot{Q} = \frac{2V}{kl\pi}\epsilon^{-t/2\alpha}\Sigma_1^\infty \frac{1-\cos i\pi}{i}\sin\frac{i\pi x}{l}(A_i\cos + B_i\sin)\frac{tm_i'}{2\alpha},$$

and $$v = V' + \frac{2V}{\pi^2}\Sigma_1^\infty \frac{1-\cos i\pi}{i^2}\cos\frac{i\pi x}{l}\epsilon^{-t/2\alpha}$$

$$\times\left(\frac{A_i}{2}\cos + \frac{B_i}{2}\sin + \frac{B_i m_i'}{2}\cos - \frac{A_i m_i'}{2}\sin\right)\frac{tm_i'}{2\alpha};$$

where V', A_i, and B_i are constants. The conditions to determine them are that $v=0$ when $t=0$ and when $t=\infty$. Also $\dot{Q}=V/kl$ when $t=0$.

Therefore $\qquad V'=0, \qquad A_i=1, \qquad B_i=-\frac{1}{m_i'}.$

Thus $$\dot{Q}=\frac{2V}{\pi kl}\Sigma_1^\infty \epsilon^{-t/2a}\frac{1-\cos i\pi}{i}\sin\frac{i\pi x}{l}\left(\cos-\frac{1}{m_i'}\sin\right)\frac{tm_i}{2a}, \qquad (19)$$

$$v=-4V\frac{a}{\beta}\epsilon^{-t/2a}\Sigma\frac{1-\cos i\pi}{m_i'}\cos\frac{i\pi x}{l}\sin\frac{tm_i'}{2a}. \qquad (20)$$

The actual potential is the sum of (17) and (20), and the actual current the sum of (18) and (19). When a/β is large the initial charge may be neglected altogether. Considering only the potential and current due to the initial current we find that the current in the wire consists of a series of decreasing waves in opposite directions, causing corresponding changes in the potential of the wire. At the first moment after disconnection the potential at the end Q becomes positive $=V\sqrt{\frac{a}{\beta}}$ nearly, and the end P negative to an equal extent. Provided the E.M.F. suddenly developed is not sufficient to cause a spark, this state of things is rapidly reversed, the end P becoming positive and the end Q negative, which is followed by another reversal, and so on, till the energy of the initial current is all used up against the resistance of the wire.

It is obvious that the somewhat complex form of the above formulæ must be considerably departed from in all practical cases that occur, as, in the above, c and s are assumed to be the same for every unit of length of the wire, which cannot be true, except perhaps in an uncoiled submarine cable. But we may be sure that, in virtue of the property of the electric current which Professor Maxwell terms its "electro-magnetic momentum," whenever any sudden change of current or of charge takes place in a circuit possessing an appreciable amount of self-induction, the new state of equilibrium is arrived at through a series of oscillations in the strength of the current, which may be noticeable under certain circumstances. It is naturally difficult to observe such oscillations with a galvanometer, but some telegraph instruments show them very distinctly. For instance, there is Wheatstone's "alphabetical indicator." The pointer of this instrument is moved one letter forward round a dial by every current passing through it, provided the currents are alternately positive and negative. Now if an insulated straight wire a few miles in length is suddenly raised to a high potential by means of a single current of very short duration from a magneto-electric machine, and then immediately discharged to earth through an indicator, the pointer does not merely move one step forward, as it would if the discharge consisted of a single current, but several steps, indicating a succession of reverse currents. The same thing occurs when a condenser of small capacity is first charged to a high potential and then discharged through the instrument. Expressed in popular language what happens is as follows:—The first discharge is at first retarded by the self-induction of the coils, and then acquiring momentum carries to earth a greater quantity of electricity than the line or condenser originally contained, thus reversing the potential of the line. Hence a reverse current follows to restore the equilibrium, which in its turn carries more than enough electricity to restore the deficiency; hence another current from line to earth, and so on, till the currents are too weak to produce any observable effect.

By supposing that the current at any moment is of the same strength in all parts of the coils, the theory of the alternating currents when a charged condenser is discharged through the coil is much simplified. Let Q_0 be the initial charge and V the initial potential of the condenser, whose capacity is c, and let R be the resistance and L the coefficient of self-induction of the coil. Thus, if Q is the charge and v the potential of the condenser at time t, the current in the coil is

$$Q = -cv,$$

and

$$v = R\dot{Q} + L\ddot{Q},$$

since v is the difference of potential between the ends of the coil. Therefore

$$cL\ddot{v} + cR\dot{v} + v = 0;$$

the solution of which satisfying the conditions $v = V$ when $t = 0$, and $\dot{Q} = 0$ when $t = 0$, is

$$v = \frac{V\epsilon^{-t/2a}}{2\left(1-4\frac{a}{\beta}\right)^{\frac{1}{2}}}\left\{\left(1+\sqrt{1-4\frac{a}{\beta}}\right)\epsilon^{t/2a\left(1-4\frac{a}{\beta}\right)^{\frac{1}{2}}} - \left(1-\sqrt{1-4\frac{a}{\beta}}\right)\epsilon^{-\cdots}\right\},$$

or

$$v = V\epsilon^{-t/2a}\left(\cos + \frac{\sin}{\left(4\frac{a}{\beta}-1\right)^{\frac{1}{2}}}\right)\frac{t}{2a}\left(4\frac{a}{\beta}-1\right)^{\frac{1}{2}},$$

according as $1 - \dfrac{4a}{\beta}$ is + or −. And the current in the coil is

$$\dot{Q} = \frac{V}{R}\,\frac{\epsilon^{-t/2a}}{\left(1-4\frac{a}{\beta}\right)^{\frac{1}{2}}}\left(\epsilon^{t/2a\left(1-4\frac{a}{\beta}\right)^{\frac{1}{2}}} - \epsilon^{-\cdots}\right),$$

or

$$\dot{Q} = \frac{2V}{R}\epsilon^{-t/2a}\frac{\sin\frac{t}{2a}\left(4\frac{a}{\beta}-1\right)^{\frac{1}{2}}}{\sqrt{4\frac{a}{\beta}-1}},$$

where $a = L/R$ and $\beta = cR$. In the first case, when $1 > 4a/\beta$, the potential and current are never reversed, but in the second case, when $4a/\beta > 1$, they are reversed an infinite number of times, the successive charges of the condenser decreasing in geometrical proportion. The current changes sign when t is any multiple of $\dfrac{2a\pi}{(4a/\beta - 1)^{\frac{1}{2}}}$, and has its maximum or minimum values when

$$\cos\frac{t}{2a}\left(4\frac{a}{\beta}-1\right)^{\frac{1}{2}} = \frac{1}{2}\left(\frac{\beta}{a}\right)^{\frac{1}{2}}.$$

The quantity Q' of electricity conveyed in the first current is

$$Q' = Q_0\left(1 + \epsilon^{-\frac{\pi}{(4\alpha/\beta - 1)^{\frac{1}{2}}}}\right),$$

where Q_0 is the initial charge of the condenser. As α/β is increased, Q' approaches $2Q_0$ as its limit, *i.e.*, when the resistance of the coil is reduced, or its magnetic capacity increased, the quantity of electricity conveyed by any current increases until it is nearly double the charge of the condênser at the commencement of that current, and the oscillations are more slowly diminished. The amount of energy expended by the first current is

$$\frac{V^2c}{2}\left(1 - \epsilon^{-\frac{2\pi}{(4\alpha/\beta - 1)^{\frac{1}{2}}}}\right),$$

where $\frac{V^2c}{2}$ is the energy of the original charge Q_0, which becomes indefinitely small as α/β increases. The integral current, irrespective of sign, is

$$\frac{Q_0}{1 - \epsilon^{-\frac{\pi}{(4\alpha/\beta - 1)^{\frac{1}{2}}}}},$$

which increases indefinitely with α/β. From the number of oscillations in a given time L may be determined in terms of R and c. For if the current is reversed n times per second, then

$$L = (2c\pi^2n^2)^{-1}(1 + \sqrt{1 - c^2R^2\pi^2n^2}).$$

Electrical vibrations due to induction occur under various circumstances. For example, the false discharge from a coiled submarine cable; the oscillatory phenomena described by M. Blaserna and others; and Mr. Edison's "etheric force" experiments.*

XV.—ON THE SPEED OF SIGNALLING THROUGH HETEROGENEOUS TELEGRAPH CIRCUITS.

[*Phil. Mag.*, March 1877, S. 5, vol. 3, p. 211.]

When the first trials of speed of working were made on the Anglo-Danish cable, then recently laid (September 1868), it was found that a considerably higher speed could be reached in one direction than in the other. The "line" portion of the circuit consisted of a land-line on the English side of 240 ohms resistance, then a cable of 2500 ohms resist-

* [The oscillatory nature of a condenser discharge in association with self-induction was first discovered by Joseph Henry (1842), whose work has been somewhat overshadowed by Faraday's; and the theory of the reaction between a condenser and coil was given by Sir W. Thomson in his paper "On Transient Electric Currents," *Phil. Mag.*, June 1853. The effect of self-induction in association with the electrostatic capacity of a telegraph line was first considered by Kirchhoff (1857), working on the basis of Weber's electrodynamic theory.]

ance and capacity 120 microfarads, and a land-line on the Danish side of 1250 ohms—all approximate. The circuit was completed through a battery of 150 ohms at one end and a Wheatstone's receiver of 750 ohms at the other, the circuit being worked on the earth-to-earth principle, *i.e.*, without condensers. But although the battery and receiver at each end were the same, or nearly so, the maximum speed obtained with Wheatstone's transmitter, making mechanically exact signals, was 40 per cent. higher from England to Denmark than from Denmark to England.* This unexpected result was abundantly confirmed by the subsequent experience of every-day practice, which proved the existence of a difference in working-speed in opposite directions varying from 20 to 40 per cent. at different times, mainly according to the state of insulation of the land-lines.

Later on the same instruments were introduced between London and Amsterdam, on a circuit consisting of a land-line of 130 miles on the English side, then a cable of 120 miles, and on the Dutch side a land-line of 20 miles (Culley, *Journ. Soc. Tel. Eng.*, vol. i.). In this case the maximum speed obtained was 50 per cent. higher from Amsterdam to London than *vice versa.* Again, on the London-Dublin circuit, consisting of cable 66 miles and land-lines 266 and 10 miles, the longer line being on the English side, the speed from Dublin to London was double that obtained in the reverse direction, viz., 80 and 40 words per minute respectively. Similarly between London and Belfast.

In all these cases it is to be observed that the station nearest the cable receives the most slowly, and that the greater the inequality of resistance of the land-lines the greater is the difference in the working-speeds. This seems to point directly to the conclusion that the uncentrical position of the cable in the circuit actually causes the retardation to be greater in one direction than in the other. The fact that the cable receives a much larger charge of electricity when the battery is connected to the end of the shorter than to the end of the longer land-line might, on a cursory examination, seem to corroborate this conclusion. But when the light of theory is thrown upon this view of the matter it is at once found to be untenable.

It is easily shown that if condensers be distributed in any arbitrary manner along a line which is to earth at each end, dividing it into sections having any resistances, and the condensers be all initially discharged, the introduction of an electromotive force in the first section will cause the current to rise in the last section, in the same manner as the same electromotive force in the last section will cause the current to rise in the first section. Furthermore, it may be shown that if leaks be introduced on the line in any arbitrary manner, the same property will hold good. (The differential equation of the current, which is linear and of the same degree as the number of condensers, is the same for the

* It may be interesting to state the actual speeds obtained on this circuit with different instruments. Morse, 60 to 75 letters per minute; Wheatstone's transmitter and receiver, 90 to 140 letters per minute; Wheatstone's transmitter and Thomson's recorder, 300 to 360 letters per minute—in all cases without condensers.

first and last sections; and the conditions to determine the arbitrary constants are the same.) Now every telegraph-line, however irregular it may be in its resistance, capacity, and insulation in different places, may be considered as such a system of condensers and leaks, infinite in number if necessary; whence it follows that on any line there is absolutely no difference in the retardation in either direction, meaning by *retardation* the time required for an electromotive force at one end to cause the current at the other end to reach any stated fraction of its maximum. Therefore, to account for the facts, which cannot be gainsaid, we must look outside the line and fix our attention on the sending and receiving apparatus. The actual cause or causes must, however, be of such a nature that they only come into operation when the capacity, or the leakage, is unsymmetrically situated in the circuit. No perceptible difference in working speed was observed on the Anglo-Danish circuit when the correspondence was maintained between the two ends of the cable itself. Now, since in all the cases described Wheatstone's transmitter was employed, it is natural to inquire whether the difference is due to any peculiarity in the method of making the signals with that instrument. If so, then we need not expect any difference to exist when simple reversals are made. But, in fact, it exists even then. An instance bearing this out was described by Mr. Varley before the Submarine-Cable Committee (Sub. Report, p. 156). Experimenting with his "wave-bisector" on the underground lines between London and Liverpool, Mr. Varley found that the introduction of resistance at the battery-end of the line lowered the speed to a greater extent than its introduction at the receiving-end, where indeed it made little difference. Here the speed was inversely as the retardation, since the wave-bisector made simple reversals. Mr. Varley attributed the difference to the leakage; but this is in direct contradiction to the theoretical result, that neither leakage nor irregularity in distribution of capacity can, acting alone, cause any difference. Also the difference existed on the Anglo-Danish circuit when simple reversals were made with the transmitter, but apparently to a smaller extent. It was quite perceptible (10 or 20 per cent.) with key-sending, using a common reversing key—though the exact amount of the difference could then not be exactly estimated, since operators differ nearly as much in their hand-signalling as in their hand-writing. Although, therefore, in the case of Wheatstone's transmitter the difference in working-speed may be, and I believe is, mainly due to a peculiarity of that instrument, yet when plain reversals are sent, there must actually be a difference in the retardation in opposite directions; and this I believe is due to the fact, which comes out on closer inspection, that it is not the same circuit which is being worked when the direction of working is reversed.

Let the line consist of a cable of resistance c, having land-lines of resistances a and b attached to its ends, and let the battery and receiver resistances be f and g respectively. Then Fig. 1 shows the arrangement when A sends to B. Further, suppose for simplicity, and to avoid analytical calculations, that the cable's resistance is small compared with the total resistance of the circuit. Then we may obtain tolerably accurate results by considering the cable's capacity as collected at its

Fig. 1.

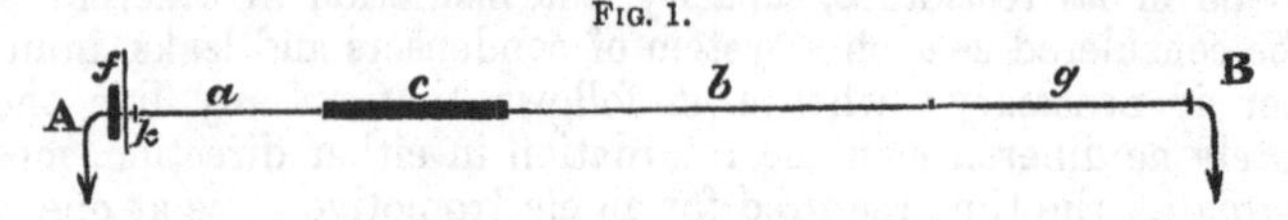

centre. Then, by the theory of the condenser, when A applies his battery to the line, the current rises at B according to the formula

$$C=\frac{E}{R}(1-\epsilon^{-\frac{t}{T}}),$$

where C is the current, E the electromotive force, R the total resistance between A and B, t the time, and

$$T=\frac{S}{R}\left(\frac{c}{2}+a+f\right)\left(\frac{c}{2}+b+g\right),$$

where S is the cable's capacity. Thus the magnitude of T determines the slowness of the rise of the current, and we may therefore call it the retardation. (In the time T, the current reaches about 63 per cent. of its maximum.) Now when B sends to A, f and g change places, producing the arrangement shown in Fig. 2. If C' is the current B produces at A,

Fig. 2.

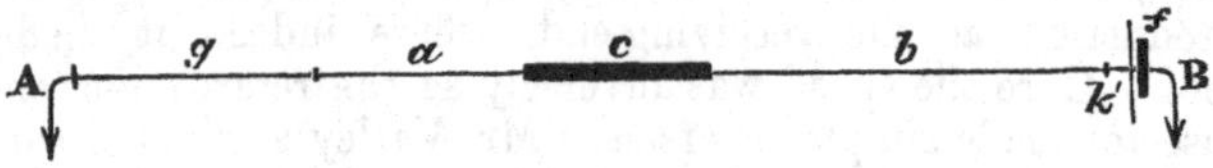

$$C'=\frac{E}{R}(1-\epsilon^{-\frac{t}{T'}}),$$

where
$$T'=\frac{S}{R}\left(\frac{c}{2}+a+g\right)\left(\frac{c}{2}+b+f\right).$$

Comparing the values of T and T', we shall find that if $a=b$, $T=T'$; also if $f=g$, $T=T'$; but if $a<b$, as in the figures, $T<T'$ if $f<g$, and $T>T'$ if $f>g$. Or, in plain English, the retardation is the same in both directions if the land-lines have equal resistances, whatever may be the resistances of the battery and receiver; it is also the same in both directions if the battery and receiver have equal resistances, whatever may be the resistances of the land-lines; but if the resistances of the land-lines are unequal, the retardation is greatest when the station nearest the cable is receiving, if at the same time the battery is less than the receiver resistance, and least in the contrary case. Now if the battery is always in circuit, as in making signals with a reversing key, the effect of any arbitrary signals may be calculated by the same formula, and the maximum working-speed (always provided it be within the reach of the apparatus) will be least when the station nearest the cable receives, if the battery is less than the receiver resistance, and greatest in the contrary case. Generally, the more centrally the capacity is situated the greater the retardation.

The influence of leakage or faults may be readily determined in a similar manner, since the retardation is proportional to the resistance through which the charge in the cable discharges to earth. In all cases the retardation is reduced by a fault, and the more so the nearer the fault is to the centre of capacity. If a fault be introduced on the long land-line b, the difference of the retardation in opposite directions is the same as before as regards direction, while its percentage amount is increased. The influence of the natural leakage of the land-lines is the same, since nearly all the loss will, under ordinary circumstances, take place on the long land-line. But if a fault be introduced on the short land-line, the percentage difference is reduced instead of being increased, and its direction may even be reversed.

We have thus found that on any circuit consisting of a cable with land-lines of unequal resistance at its ends, a difference in the retardation in opposite directions is necessarily introduced when the battery and receiver have not the same resistance. Suppose, in Figs. 1 and 2, $f=1$, $a=10$, $c=10$, $b=100$, $g=10$; then the retardation from A to B is to the retardation from B to A as 184 : 265, *i.e.* 44 per cent. greater from B to A than from A to B; and the natural leakage of the land-line increases this difference. But with Wheatstone's transmitter the observed difference is greater than can be thus accounted for, and exists even when there is no inequality in resistance of the battery and receiver. This is due to a peculiarity in the method of making the signals with that instrument, which is at the same time the cause of two other anomalies, viz. :—reduction of working-speed by leakage, although the retardation is thereby reduced; and increase of working-speed by the addition of resistance, although the retardation is thereby increased. To understand this, it is necessary to examine the way the sending-end of the line is operated upon. The point k in Fig. 1, or k' in Fig. 2, is always connected with the positive or negative pole of the battery, or it is insulated. Currents of equal duration follow each other, alternately + and −, separated either by no interval, or by intervals equal to twice, four, or six times the time of a current.[1] The armature of the receiver is adjusted neutral, so as to remain on the side any current sends it to, until an opposite current reverses its position. Lines of two lengths are thus made :—a "dot" by first a + current immediately followed by a − current to terminate it, thus + −; and a "dash" of three times the length by first a + current, then an interval of insulation for twice as long, and lastly a − current to terminate it, thus +00−. At a speed much below the limiting speed the sent signals are reproduced at the receiving-end without sensible alteration; but as the speed of working is increased and the currents have not time to reach their full strength, irregularities show themselves, which increase rapidly as the length of each contact is reduced, until at length

[1] Mr. Culley's *Handbook* contains a full description of the apparatus.

a limiting speed is reached at which some of the signals miss fire altogether. Consider the succession of signals

$$\overset{a}{+}\overset{}{-}\overset{b}{+}-\overset{c}{+}-\overset{d}{+}\overset{e}{-}0000\overset{f}{+}\overset{g}{-}\overset{h}{+}00\overset{i}{-}\overset{j}{+}-0000$$

(illustrating a typical failure), consisting of a series of dots, followed, after an interval of insulation, by a dot, a dash, and a dot. If the receiver is adjusted so as to record the dots *a*, *b*, *c*, *d* perfectly, the signals *g* and *j* will fail. *g* will fail because the − current *e* has time to die away during the interval of no sent current 0000, thus making the succeeding + current *f* too strong; and *j* will fail because the + current *h* has time to die away during the interval of no sent current 00, thus making the − current *i* too strong. In the first case the dot is continued on to the dash, in the second the dot is lost. Thus, although generally, to get the greatest possible working-speed, the retardation should be as small as possible, yet in this system of contacts of equal duration to make lines of unequal length, it is important that *some* of the currents, viz. those commencing dashes or spaces, should not die away too quickly. They are prevented from doing so, in a great measure, by the insulation of the line at the sending-end during the intervals of no sent current, which, by closing up the path at one end for the charge to escape, prolongs the current at the other. (The compensation currents, sent by an improved form of transmitter, have for their object to still further lengthen out the currents.) Now it will be seen from the figures that when *A* insulates the line at *k*, Fig. 1, the charge of the cable discharges through the resistance $\frac{c}{2}+b+g$, and that when *B* insulates at *k'*, Fig. 2, it discharges through the smaller resistance $\frac{c}{2}+a+g$. Therefore the current dies away more quickly in the latter case, and, by reason of the before-mentioned peculiarities, the station *A* nearest the cable receives more slowly than *B*. The explanation of the reduction of speed by leakage is similar. The leakage lessens the retardation and consequently quickens the signals. If every signal were quickened in the same proportion, as would happen were the circuit always complete, it is evident that the speed of working must be increased; but it is easily seen that the decrease in the retardation caused by the loss is proportionally much less when the circuit is complete than when the line is insulated at the sending-end, thus increasing the irregularity in the received signals due to the unequal intervals between the sent signals, and consequently lowering the working-speed. Again, the addition of resistance at the receiving-end, as at *A* in Fig. 2, when *B* sends to *A*, may increase the working-speed. Now, since the addition of resistance obviously increases the retardation, nothing could result save a decrease of speed if the retardation of every signal were increased in the same ratio. But this is not the case; for the retardation is increased in a greater ratio when the line is insulated at the sending-end than when the circuit is complete—exactly the opposite to what occurs with leakage: then the working-speed was lowered; now it is increased. (This reasoning will not, of course, apply to other-systems of transmission.) On the other hand, the speed is lowered by inserting resistance at the sending-end, B, Fig. 2; for the retardation is unaltered with line insulated, and increased with complete circuit.

To ascertain the exact amount of retardation produced by resistance at either or both ends of a submarine cable, each case must be calculated separately, because the form of the curve of arrival of the current is altered, the law of the squares only holding good when exactly similar systems are compared.

A B C D

Let BC be a cable of length l, resistance k per unit of length, capacity c per unit of length; and let AB and CD be resistances equal to mkl and nkl respectively, connected to the cable at B and C, and to earth at A and D. Let v be the potential of the conductor of the cable at distance x from B at the time t. Then, according to Sir W. Thomson's theory, v must satisfy

$$\frac{d^2v}{dx^2} = ck\frac{dv}{dt}$$

between $x = 0$ and $x = l$. The general solution is

$$v = \Sigma A \sin\left(\frac{ax}{l} + b\right)\epsilon^{-\frac{a^2t}{T}}, \quad \text{.........................} \quad (1)$$

where $T = ckl^2$, if v vanishes for $t = \infty$. Three sets of constants, A, a, and b, have to be determined from the terminal conditions for x and t. In AB and CD the current follows Ohm's law. Therefore

$$-\frac{v}{mkl} = -\frac{1}{k}\frac{dv}{dx} \text{ when } x = 0,$$

and

$$\frac{v}{nkl} = -\frac{1}{k}\frac{dv}{dx} \text{ when } x = l,$$

for all values of t. Therefore, by (1),

$$\sin b = ma\cos b, \text{ or } \tan b - ma = 0,$$

and

$$\sin(a + b) = -na\cos(a + b),$$

or

$$\tan(a + b) + na = 0.$$

Hence, eliminating b,

$$\tan a = \frac{(m + n)a}{mna^2 - 1},$$

from which the a's can be found when m and n are given. The b's are already known in terms of the a's, and the A's can be found by integration if the potential of every part of the conductor of the cable is given

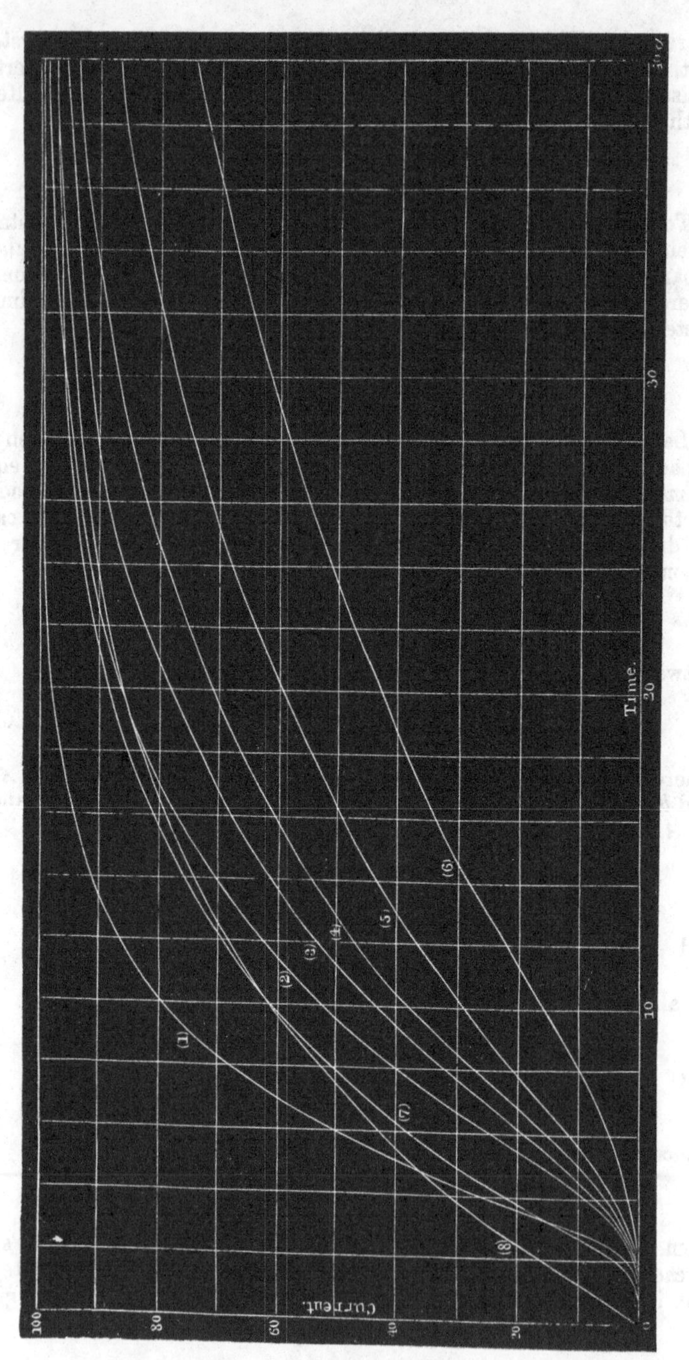

FIG. 3.

for $t=0$. Let it be that produced by an electromotive force E in AB, *i.e.*

$$v=E\,.\,\frac{l(1+n)-x}{l(1+m+n)};$$

then, by integration,

$$A=\frac{\dfrac{2E\cos b}{a}}{1+\dfrac{m}{1+m^2a^2}+\dfrac{n}{1+n^2a^2}};$$

and finally, the potential at time t is

$$v=\sum_{i=1}^{i=\infty}\frac{2E}{a_i}\frac{\dfrac{1}{1+m^2a_i^2}}{1+\dfrac{m}{1+m^2a_i^2}+\dfrac{n}{1+n^2a_i^2}}(\sin+ma_i\cos)\frac{a_ix}{l}\epsilon^{-\frac{a_i^2t}{T}},$$

from which the arrival-curves of the current may be found by making $x=l$. In the diagram six cases are shown. The abscissas represent time, from $t=0$ to $t=40a$, the unit being $a=\dfrac{ckl^2}{10\pi^2}\log_\epsilon 10$. The ordinates represent the arrived current, the maximum strength being in all cases $=100$.

1. $m=0$, $n=0$. Let N be the percentage amount of received current at time t, then

$$\frac{N}{100}=1+\frac{2}{4}\Sigma\cos i\pi\,\epsilon^{-\frac{i^2\pi^2t}{T}}$$

2. $m=0$, $n=\frac{1}{2}$.

$$\tan a+\frac{a}{2}=0,$$

$$\frac{N}{100}=1+\Sigma\frac{12\cos a}{5+\cos 2a}\epsilon^{-\frac{a^2t}{T}}.$$

3. $m=0$, $n=1$.

$$\tan a+a=0,$$

$$\frac{N}{100}=1+4\Sigma\frac{\epsilon^{-\frac{a^2t}{T}}}{\cos a+\sec a}.$$

4. $m=0$, $n=2$.

$$\tan a+2a=0,$$

$$\frac{N}{100}=1+\Sigma\frac{6\cos a}{2+\cos 2a}\epsilon^{-\frac{a^2t}{T}}.$$

5. $m=0$, $n=\infty$.

$$\frac{N}{100}=1+\frac{4}{\pi}\Sigma\frac{\cos i\pi}{2i-1}\epsilon^{-\frac{(2i-1)^2\pi^2t}{4ckl^2}}.$$

6. $m=1$, $n=1$.

$$\tan a = \frac{2a}{a^2-1},$$

$$\frac{N}{100} = 1 - \Sigma \frac{6\epsilon^{-\frac{a^2t}{T}}}{3+a^2}.$$

Curve (1) is the arrival-curve when no resistance is inserted at either end; curve (2) when a resistance equal to one half the cable's resistance is inserted at either end; curve (3) when a resistance equal to the cable's is inserted at either end; and curve (4) when twice the cable's resistance is inserted at either end. (5) shows the curve of arrival of the potential at the insulated end of a cable when the other end is raised to a constant potential; (6) shows the arrival-curve when a resistance equal to the cable's is inserted at each end.

It will be observed from an inspection of the curves that, when resistance is added at one end of a cable only, the effect in increasing the retardation is very great when the added resistance is small, but as more and more resistance is added there is not much further effect. The limit is reached in curve (5). But the insertion of resistance at both ends has a much greater retarding influence, which increases without limit. Compare (4) with (6): in (4) we have twice the cable's resistance at one end and none at the other; in (6) the same resistance is equally divided at each end, and the retardation is very greatly increased.

With respect to the change in the form of the arrival-curves, it will be seen that, when resistance is inserted, the first part of the arrived current is proportionally less retarded than the later parts. Thus, comparing (1) with (6), when there is no resistance inserted the current reaches 5 per cent. of its maximum in $2{\cdot}45a$, whereas (6) takes $6a$, or 2·4 times as long; to reach 10 per cent. (6) takes 3·3 times as long as (1); to reach 40 per cent. it takes 3·7 times as long, and to reach 70 per cent. 4·5 times as long.

Curves (1), (7), and (8) show the effect of different distributions of the same amount of capacity in a line of given resistance. (8) shows the arrival-curve when the capacity is all collected at the centre of the line as a single condenser, (7) when the capacity is uniformly distributed over the middle third of the line, and (1) when it is uniformly distributed over the whole length. The more the capacity is spread the longer is the time taken for the current to reach a sensible strength, whereas the current rises rapidly the moment contact is made when the capacity is collected at one place. Curve (7) is the same as (6) with the abscissas of the latter reduced in the ratio 3 : 1; and curve (8) is the limiting form of the arrival curve when very great equal resistances are inserted at both ends of the cable, the abscissas being reduced in the same proportion as the resistance of the circuit is increased. Its equation is

$$\frac{N}{100} = 1 - \epsilon^{-\frac{4t}{T}}.$$

XVI.—ON THE THEORY OF FAULTS IN CABLES.

[*Phil. Mag.*, July and August, 1879, S. 5, vol. 8.]

1. The only kind of fault to be here considered is either a local defect in the insulation, or an artificial connection between the conductor of a cable and the earth. When a fault occurs in a submarine cable, its most manifest effect on the working is to increase the strength of current leaving the sending-end, because the resistance is reduced; while at the same time the strength of current arriving at the distant end is reduced, the loss of current through the fault being greater than the increase in the current leaving the sending-end. Another effect is to increase the speed at which signals can be made through the cable. A cable may be considered electrically as a long cylindrical condenser, or as a conductor having a great number of condensers of small capacity connected at equidistant points to it on the one side and to earth on the other. When an electromotive force is applied at one end, to establish a permanent current in the circuit these condensers have to be charged, an operation requiring time for its fulfilment; and before the current can cease when the electromotive force is removed the charge must be got rid of: in fact, the current results from the discharge of the cable's electricity. If there is a fault the discharge of the cable is facilitated; for there is not only a smaller quantity of electricity to be discharged, but more paths are open to it. Similarly the charging of the cable is facilitated, as will be seen by supposing the cable when uncharged to contain two exactly equal and opposite charges. Let one of these discharge itself. The cable will then become charged with the other; and since the discharge of the first is facilitated, the charging of the cable by the second is also facilitated. With a fault a smaller quantity of electricity is required in order to produce the permanent state of electrification when an electromotive force is applied at one end of the line than when there is no fault; and also, other things being the same, a given fraction of the final permanent state is more quickly reached in the former than in the latter case. Similarly the effect of a given signal is more rapidly dissipated in the former than in the latter case; and consequently from both these causes signals can be packed more closely together when the cable is faulty; or, in other words, the speed of working can be increased with equal legibility.

2. Before proceeding to the mathematics of the subject, I give some of the calculated arrival-curves in simple cases. Referring to fig. 1, suppose in the first case the cable is perfectly insulated and free from charge, and that both ends are to earth. At a given time $t=0$, introduce a constant electromotive force E at one end P of the cable. Then the well-known curve of arrival of the current at the distant end Q is represented by curve 1. Time is measured to the right, and current strength upwards. The unit of time is

$$a=\frac{ckl^2}{10\pi^2}\log_\epsilon 10=\frac{ckl^2}{42{\cdot}86},$$

where l is the cable's length, and c, k its capacity and resistance per unit of length. For a cable 2000 miles long, with $c=\frac{1}{3}$ microfarad, and $k=6$ ohms, we have $ckl^2=8$ seconds, and $a=\cdot 1866$ sec. The current, though calculable from the first instant, is relatively insensible for some little time. Thus, when $t=1\cdot 5a$, it has only reached ·0047 of its final strength, but, thereafter increasing much more rapidly, reaches half its final strength in $6a$, ·98 in $20a$, and its final strength $\frac{E}{kl}$ after the lapse of a theoretically infinite time.

FIG. 1.

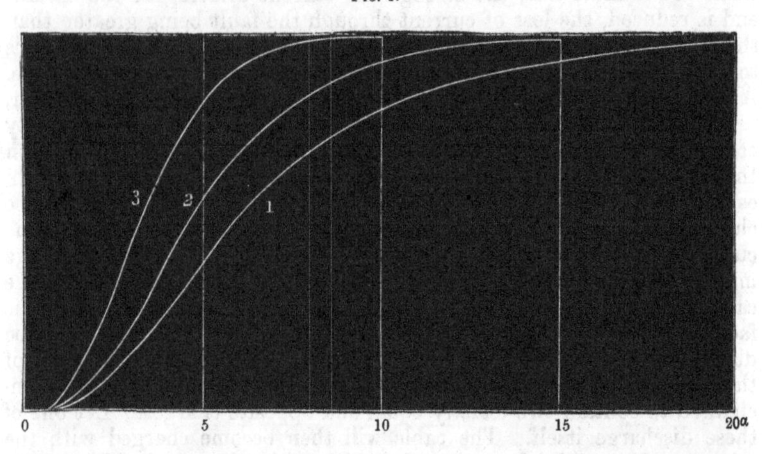

Now compare curve 1 with curve 3. In the latter all the circumstances are the same, with the exception that there is a fault of infinitely small resistance situated at the middle of the cable. Of course such a fault could not be worked through, since no current would arrive at the receiving-end. Nevertheless this is not by any means a case of an unpractical nature; for it is quite possible to work, and very well too, a cable containing a fault of *next* to no resistance. It will be seen that with the fault of no resistance the current becomes sensible sooner, and increases much more rapidly. It reaches ·0017 of its final strength in $1a$, ·044 in $1\frac{1}{2}$, ·1318 in $2a$, ·4274 in $3a$, ·68 in $4a$, ·8357 in $5a$, and ·9826 in $8a$. Half the final strength is reached in $3\cdot 3a$, as against $6a$ with no fault.

When the fault has a finite resistance the arrival curve of the current is intermediate between curve 1 and curve 3. The one shown by curve 2 corresponds to the case of a fault having a resistance equal to one-fourth of the cable's. This makes the final strength of current $=\frac{E}{2kl}$, one half its value when there is no fault. In $2a$ the current reaches ·0429 of its final strength, half its final strength is $4\cdot 7a$, and ·9845 in $14a$.

3. Now, referring to Fig. 2, suppose both ends of the line to be insulated, and the cable free from charge. At any time $t=0$ let a small

charge be instantaneously communicated to one end of the cable. This corresponds to working with condensers at both ends when the capacities of the terminal condensers are very small, and terminal resistances negligible. The charge thus communicated then diffuses itself along the cable, becoming finally equally distributed. Sir W. Thomson's mathematical theory indicates that the potential at any point x rises in exactly the same manner as the current rises at the same point when both ends of the line are to earth and a constant electromotive force operates at one end. Therefore the arrival-curve of the potential at the distant end in working with condensers at both ends is the same as the arrival-curve of the current shown by 1, Fig. 1. It is reproduced in 1, Fig. 2, for comparison with the curves for a fault.

FIG. 2.

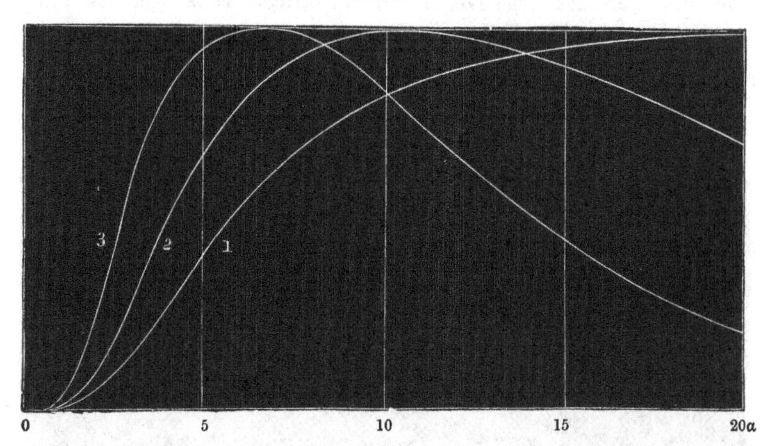

When there is a fault, or merely general loss through the insulator, there is conductive connection between the conductor and earth; consequently the charge initially communicated to the beginning of the line must ultimately all escape, reducing the potential everywhere to zero. Therefore, although the current as shown by curve 1, Fig. 1, never reaches its full strength, yet, since insulation is never absolutely perfect, the potential, as shown by curve 1, Fig. 2, must sooner or later reach a maximum and then fall to zero. As the leakage increases the time taken to reach the maximum decreases. The maximum is reached in $10{\cdot}3a$, as shown by curve 2, Fig. 2, when there is a fault in the middle of the line of one-fourth the resistance of the latter; and with a fault in the middle of infinitely small resistance, the maximum is reached in $6{\cdot}5a$, as shown by curve 3, Fig. 2. It cannot be reached sooner with a single fault.

4. In condenser working, the working current is the current entering the receiving condenser—that is, the rate of increase of its charge, and therefore proportional to the rate of increase of the potential at the insulated receiving end when terminal resistances and capacities are

neglected. The arrival-curve of the current when there is no fault is shown in Fig. 3, curve 1. This curve was given by Sir W. Thomson in 1854, not however in connection with condenser working (for that method was not then invented), but as showing the current at the distant end produced by an infinitely short contact with an infinitely powerful battery at the beginning, both ends being kept to earth. The current reaches a sensible proportion of its maximum much more rapidly than without condensers. It reaches its maximum in 3·93a, and then decreases.

With a fault of infinitely small resistance in the middle of the line, other things being the same, the arrival-curve of the current is shown by 3, Fig. 3. It reaches ·0081 of its maximum in ·8a, ·0523 in 1a, and its maximum in 2·6a nearly. It then falls to zero, which it reaches in 6·5a, and becomes negative, as the electricity runs back to escape through the fault to earth.

Curve 2, Fig. 3, is similar. It corresponds to the fault in the middle having one-fourth the cable's resistance. The maximum is reached in 3·45a, and zero in 10·3a.

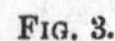

Fig. 3.

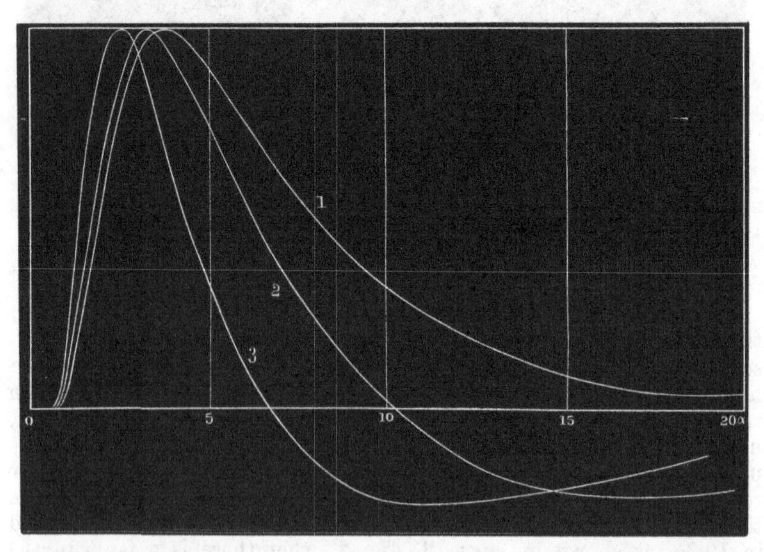

5. The influence of a fault on the amplitude of reversals may be readily calculated. In the first place, without condensers. Let contacts, alternately + and −, be made with a battery at the beginning of the line, while the distant end is to earth. If the reversals are sufficiently rapid, the resulting received current is nearly a simple harmonic function of the time. Let c be the capacity and k the resistance per unit of length of cable of length l, having a fault in the middle of resistance zkl. Also let τ be the period of a wave, or the time

occupied by a pair of contacts. Then the maximum strength Γ of the received-current waves is

$$\Gamma = \frac{E}{kl} \cdot \frac{8n\sqrt{2}}{\pi}(\epsilon^n + \epsilon^{-n} - 2\cos n)^{-\frac{1}{2}}\Big\{\epsilon^n + \epsilon^{-n} + 2\cos n + \frac{1}{2nz}(\epsilon^n - \epsilon^{-n} + 2\sin n) + \frac{1}{8n^2z^2}(\epsilon^n + \epsilon^{-n} - 2\cos n)\Big\}^{-\frac{1}{2}},$$

where $n = \sqrt{\dfrac{ckl^2\pi}{\tau}}$, and E is the electromotive force of the battery. Or, approximately,

$$\Gamma = \frac{E}{kl} \cdot \frac{8n\sqrt{2}}{\pi}\epsilon^{-n} \cdot \left(1 + \frac{1}{2nz} + \frac{1}{8n^2z^2}\right)^{-\frac{1}{2}}.$$

Here $\dfrac{E}{kl}$ is the current that would be produced in the line if perfectly insulated, and permanent contact made with the battery. $\dfrac{8n\sqrt{2}}{\pi}\epsilon^{-n}$ is the reversal-factor, and $\left(1 + \dfrac{1}{2nz} + \dfrac{1}{8n^2z^2}\right)^{-\frac{1}{2}}$ the fault-factor. Now, if Γ_0 is the greatest current possible with the fault,

$$\Gamma_0 = \frac{E}{kl\left(1 + \frac{1}{4z}\right)};$$

therefore

$$\frac{\Gamma}{\Gamma_0} = \rho \cdot \phi(n),$$

where

$$\rho = \frac{1 + \frac{1}{4z}}{\sqrt{1 + \frac{1}{2nz} + \frac{1}{8n^2z^2}}},$$

and $\phi(n)$ is the reversal-factor. $\dfrac{\Gamma}{\Gamma_0}$ represents the proportion of the maximum received current which is arrived at by the reversals, or, for brevity, the proportional amplitude. If $z = \frac{1}{4}$,

$$\rho = \frac{2}{\sqrt{1 + \frac{2}{n} + \frac{2}{n^2}}}.$$

Let $n = 10$, which would make the time τ of a pair of contacts $\tau = \dfrac{ckl^2\pi}{n^2} = 1{\cdot}347a$, where a is the unit previously used; then

$$\rho = \frac{2}{\sqrt{1{\cdot}22}} = 1{\cdot}82 \text{ nearly.}$$

Thus the fault increases the proportional amplitude for this speed 82 per cent. If $z = \frac{1}{100}$ and $n = 10$, then ρ is rather more than 6; and a fault of infinitely small resistance makes

$$\rho = \frac{n}{\sqrt{2}}.$$

6. Now for condenser working. Let everything be the same as in the last paragraph, with the addition of a condenser of capacity $r_1 cl$ at the sending-end and another of capacity $r_2 cl$ at the receiving-end, r_1 and r_2 being extremely small. We shall now have

$$\Gamma = \frac{E}{kl} \cdot \frac{16 n^3 \sqrt{2}}{\pi} r_1 r_2 \epsilon^{-n} \left(1 + \frac{1}{2nz} + \frac{1}{8n^2 z^2}\right)^{-\frac{1}{2}}.$$

The fault-factor is the same as before; and if the maximum received current possible were also the same as before, we should arrive at exactly the same conclusions as regards the influence of the fault on the proportional amplitude of the received current. But the comparison is here faulty, since

$$\frac{E}{kl\left(1 + \frac{1}{4z}\right)}$$

is the maximum current possible with both ends to earth, and the condensers do not allow the received current to reach such a strength, except in the imaginary case of condensers of infinite capacity; for a condenser of infinite capacity is mathematically equivalent to a conductor of no resistance if there is no difference of potential between the coatings to start with, or to a battery of no resistance and electromotive force E if there is a difference of potentials E. But, as is shown later on, the maximum strength of the received current with condensers becomes proportional to

$$\frac{E}{kl\left(1 + \frac{1}{4z}\right)}$$

when is very small; so that for a fault of small resistance the same results follow as before for its effect on the proportional amplitude.

7. Since the proportional amplitude is increased by the fault for the same speed, a higher speed is obtained with the same proportional amplitude. Thus, with the ends of the cable to earth, as in paragraph 5, if n_1 is the value of n when there is no fault, then, to have the same proportional amplitude with a fault of resistance zkl in the centre, we must increase n to n_2, so that

$$n_1 \epsilon^{-n_1} = n_2 \epsilon^{-n_2} \frac{1 + \frac{1}{4z}}{\left(1 + \frac{1}{2n_2 z} + \frac{1}{8n_2^2 z^2}\right)^{\frac{1}{2}}}.$$

Now the speed is inversely proportional to τ, and therefore directly proportional to n^2; therefore the percentage increase in the speed is

$$100\left(\frac{n_2^2}{n_1^2} - 1\right).$$

For $n_1 = 8$, 9, and 10 we shall find $n_2 = 9{\cdot}7$, $10{\cdot}7$, and $11{\cdot}7$, and the increase in the speed 47, 41, and 37 per cent., if $z = \frac{1}{32}$, which would make the greatest possible received current $\frac{E}{9kl}$.

For a fault of no resistance, $z = 0$, and

$$n_1 \epsilon^{-n_1} = \frac{n_2^2}{\sqrt{2}} \epsilon^{-n_2}.$$

With $n_1 = 8$, 9, and 10 this gives $n_2 = 10{\cdot}2$, $11{\cdot}3$, and $12{\cdot}4$; and the increase in the speed is 62, 57, and 53 per cent. These values of n_1, namely 8, 9, and 10, are chosen on account of their nearness to the values in the working of long cables. The corresponding values of τ are $2{\cdot}10$, $1{\cdot}66$, and $1{\cdot}34a$.

8. When a natural fault, or local defect in the insulation is developed in a cable, it tends to get worse—a phenomenon, it may be observed, not confined to cable-faults. Under the action of the current the fault is increased in size and reduced in resistance, and, if it be not removed in time, ends by stopping the communication entirely. Hence the directors and officials of submarine-cable companies do not look upon faults with favour, and a sharp look-out is kept by the fault-finders for their detection and subsequent removal. But an artificial fault, or connection by means of a coil of fine wire between the conductor and sheathing, would not have the objectionable features of a natural fault. If properly constructed it would be of constant resistance, or only varying with the temperature, would contain no electromotive force of polarization, would not deteriorate, and would considerably accelerate the speed of working. The best position for a single fault would be the centre of the line; and perhaps $\frac{1}{32}$ of the line's resistance would not be too low for the fault.

9. In the cable, the potential v at any point x has to satisfy the differential equation

$$\frac{d^2v}{dx^2} = ck\frac{dv}{dt}, \qquad\qquad (1)$$

and the current at x is

$$\gamma = -\frac{1}{k}\frac{dv}{dx}.$$

The particular solutions in paragraphs 5 and 6 regarding the strength of the received current when reversals are made with a battery at the sending-end are derived from the simple harmonic solution

$$v = \epsilon^{\frac{nx}{l}}(A\cos + B\sin)\left(\frac{2\pi t}{\tau} + \frac{nx}{l}\right) + \epsilon^{-\frac{nx}{l}}(A'\cos + B'\sin)\left(\frac{2\pi t}{\tau} - \frac{nx}{l}\right)$$

of the above equation (1). When there are faults, each of the sections into which they divide the cable has a solution of the above form. In the case of a single fault, there are four conditions (namely, two for the fault and one for each end of the line) which suffice to determine the eight constants. But to determine the maximum strength of the received current, it is only necessary to find the sum of the squares of two of the constants. This shortens the labour, which is again greatly shortened by neglecting ϵ^{-n} in comparison with 1.

10. The calculation of arrival-curves demands an entirely different method of proceeding. The general problem may be thus stated. Given a cable with faults in it, also the connections at the ends, resistances, condensers, etc., and given also the electrical state of the whole system at a certain time: to find its state at any time after, the system being left to itself, and the action of the known laws regulating the potential, current, etc. In Fig. 4 let PQ be a cable, of length l, resistance k, and electrostatic capacity c per unit of length. Also, let the terminal connections be as shown, viz. at the beginning P a resistance R_1 and a condenser of capacity C_1 shunted by a resistance S_1, with a similar arrangement at the end Q. This includes the cases of signalling either with or without condensers, shunted or unshunted, at either or both ends. Let the signalling be from P to Q; then R_1 is the battery resistance, and R_2 the receiving-instrument's resistance. Let the electromagnetic capacity of the latter be L. Further, let there be n faults of resistances Z_1, Z_2, ... at distances x_1, x_2, ... from the beginning P, where $x=0$.

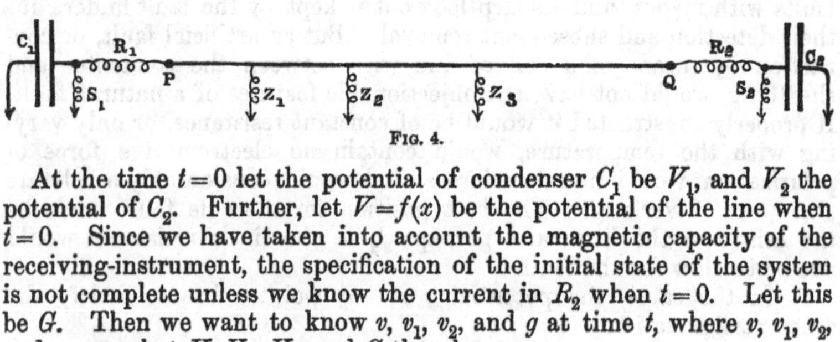

Fig. 4.

At the time $t=0$ let the potential of condenser C_1 be V_1, and V_2 the potential of C_2. Further, let $V=f(x)$ be the potential of the line when $t=0$. Since we have taken into account the magnetic capacity of the receiving-instrument, the specification of the initial state of the system is not complete unless we know the current in R_2 when $t=0$. Let this be G. Then we want to know v, v_1, v_2, and g at time t, where v, v_1, v_2, and g are what V, V_1, V_2, and G then become.

11. Between any two faults let the initial potential be expanded in a convergent series of the form

$$\Sigma A \sin\left(\frac{ax}{l}+b\right).$$

This can be effected in an infinite number of ways. Then

$$\Sigma A \sin\left(\frac{ax}{l}+b\right)\epsilon^{-\frac{a^2t}{T}}, \quad\text{.............................}\quad (2)$$

where $T=ckl^2$, satisfies the partial differential equation (1), and will therefore represent the potential at time t between the same limits, provided the sets of constants A, a, and b are so determined as to make (2) satisfy the conditions imposed by the presence of the faults and the terminal connections. This, of course, can be done only in one way.

At each of the faults two conditions are imposed. First, the potential must be continuous at the fault; secondly, the current in the line going to the fault on the left side exceeds the current coming from the fault on the right side by the current in the fault itself from the conductor to earth; and the latter is, by Ohm's law, equal to the potential of the line at the fault divided by the resistance of the latter.

Let V_{0x_1} be the initial potential between $x=0$ and $x=x_1$, $V_{x_1x_2}$ between $x=x_1$ and $x=x_2$, and so on. Then the first condition is satisfied at all the faults if

$$\left.\begin{aligned} V_{0x_1} &= \Sigma A \sin\left(\frac{ax}{l}+b\right), \\ V_{x_1x_2} &= V_{0x_1} + \Sigma B \sin\frac{a(x-x_1)}{l}, \\ V_{x_2x_3} &= V_{x_1x_2} + \Sigma C \sin\frac{a(x-x_2)}{l}, \end{aligned}\right\} \qquad (3)$$

and so on. The second condition is satisfied at all the faults by making

$$B_i = \frac{A_i}{z_1 a_i}\sin\left(\frac{a_i x_1}{l}+b_i\right),$$

$$C_i = \frac{A_i}{z_2 a_i}\sin\left(\frac{a_i x_2}{l}+b_i\right) + \frac{B_i}{z_2 a_i}\sin\frac{a_i(x_2-x_1)}{l},$$

$$= \frac{A_i}{z_2 a_i}\left\{\sin\left(\frac{a_i x_2}{l}+b_i\right) + \frac{1}{z_1 a_i}\sin\left(\frac{a_i x_1}{l}+b_i\right)\sin\frac{a_i(x_2-x_1)}{l}\right\},$$

$$D_i = \frac{A_i}{z_3 a_i}\sin\left(\frac{a_i x_3}{l}+b_i\right) + \frac{B_i}{z_3 a_i}\sin\frac{a_i(x_3-x_1)}{l} + \frac{C_i}{z_3 a_i}\sin\frac{a_i(x_3-x_2)}{l},$$

and so on, where $z_1 = \dfrac{Z_1}{kl}$, $z_2 = \dfrac{Z_2}{kl}$, ...

12. The terminal arrangements have next to be considered. By the theory of the condenser, at the beginning P we have, at time t,

$$-\frac{v_1}{S_1} - C_1\frac{dv_1}{dt} = \frac{v_1 - v}{R_1} = -\frac{1}{k}\frac{dv}{dx}.$$

Hence, if $m_1 = \dfrac{R_1}{kl}$, $n_1 = \dfrac{S_1}{kl}$, $r_1 = \dfrac{C_1}{cl}$,

we have

$$v_1 = v - m_1 l\frac{dv}{dx} \qquad (4)$$

as the relation between the potentials of C_1 and the beginning of the line, and

$$0 = v - (m_1+n_1)l\frac{dv}{dx} + n_1 r_1 l^2\frac{d^2v}{dx^2} - m_1 n_1 r_1 l^3\frac{d^3v}{dx^3} \qquad (5)$$

as the equation to be satisfied by the potential at $x=0$.

At the end Q we have

$$-\frac{1}{k}\frac{dv}{dx} = g = \frac{v_2}{S_2} + C_2\frac{dv_2}{dt} \qquad (6)$$

$$v - v_2 = gR_3 + L\frac{dg}{dt}. \qquad (7)$$

Therefore if $m_2=\frac{R_2}{kl}$, $n_2=\frac{S_2}{kl}$, $r_2=\frac{C_2}{cl}$, $s=\frac{L}{kl\,.\,ckl^2}$,

we have
$$v_2=v+m_2l\frac{dv}{dx}+sl^3\frac{d^3v}{dx^3}, \quad\text{.........................}\quad (8)$$

giving v_2 in terms of v at $x=l$, and

$$0=v+(m_2+n_2)l\frac{dv}{dx}+n_2r_2l^2\frac{d^2v}{dx^2}+(s+m_2r_2n_2)l^3\frac{d^3v}{dx^3}+n_2r_2sl^5\frac{d^5v}{dx^5} \quad\text{.....}\quad (9)$$

as the condition held by the potential at $x=l$. In (5), (8), and (9), for $\frac{d}{dt}$ has been substituted $\frac{1}{ck}\frac{d^2}{dx^2}$.

13. Now the law of formation of the a's and b's can be found. From $x=0$ to $x=x_1$,

$$V_{0x_1}=\Sigma A\sin\left(\frac{ax}{l}+b\right);$$

and from the last fault at $x=x_n$ to $x=l$,

$$V_{x_nl}=\Sigma A\sin\left(\frac{ax}{l}+b\right)+\Sigma B\sin\frac{a(x-x_1)}{l}+\Sigma C\sin\frac{a(x-x_2)}{l}+\ \dots$$

Inserting the first of these in (5) and the second in (9), and then making $x=0$ in the first case and $x=l$ in the second, we find

$$\tan b_i=\frac{(m_1+n_1)a_i-m_1n_1r_1a_i^3}{1-n_1r_1a_i^2}, \quad\text{.......................}\quad (10)$$

$$\frac{\sin(a_i+b_i)+q_i'\sin a_i\left(1-\frac{x_1}{l}\right)+q_i''\sin a_i\left(1-\frac{x_2}{l}\right)+q_i'''\sin a_i\left(1-\frac{x_3}{l}\right)+\dots}{\cos(a_i+b_i)+q_i'\cos a_i\left(1-\frac{x_1}{l}\right)+q_i''\cos a_i\left(1-\frac{x_2}{l}\right)+q_i'''\cos a_i\left(1-\frac{x_3}{l}\right)+\dots}$$

$$=-\frac{(m_2+n_2)a_i-(s+m_2n_2r_2)a_i^3+n_2r_2sa_i^5}{1-n_2r_2a_i^2}, \quad\text{..............}\quad (11)$$

where $q_i'=\frac{B_i}{A_i}$, $q_i''=\frac{C_i}{A_i}$, $q_i'''=\frac{D_i}{A_i}$, etc.

Equations (10) and (11) serve to determine the a's and b's.

14. Now only the A's remain to be found. This is to be effected by an integration along the line from $x=0$ to $x=l$, with a similar process applied at P and Q to the potentials of C_1 and C_2 and the current in R_2. Collecting the expressions for the separate divisions, we have, from $x=0$ to $x=x_1$,

$$V_{0x_1}=\Sigma A_i\sin\left(\frac{a_ix}{l}+b_i\right)=\Sigma A_iM_i', \text{ say}; \quad\text{..........................}\quad (12)$$

from $x=x_1$ to $x=x_2$,

$$\dots V_{x_1x_2}=\Sigma A_i\left\{\sin\left(\frac{a_ix}{l}+b_i\right)+q_i'\sin\frac{a_i(x-x_1)}{l}\right\}=\Sigma A_iM_i'', \text{ say}; \quad\text{......}\quad (13)$$

from $x=x_2$ to $x=x_3$,

$$V_{x_2x_3}=\Sigma A_i\left\{\sin\left(\frac{a_ix}{l}+b_i\right)+q_i'\sin\frac{a_i(x-x_1)}{l}+q_i''\sin\frac{a_i(x-x_2)}{l}\right\}$$

$$=\Sigma A_iM_i''', \text{ say}; \quad\ldots\ldots\ldots \quad (14)$$

and so on to the end of the line.

At the beginning P, by (4) we have

$$V_1=\Sigma A_i(\sin b_i-m_1a_i\cos b_i)=\Sigma A_iN_i', \text{ say.} \quad\ldots\ldots \quad (15)$$

At the end Q, by (8) we have

$$V_2=\Sigma A_i\left[\left\{\sin(a_i+b_i)+q_i'\sin a_i\left(1-\frac{x_1}{l}\right)+q_i''\sin a_i\left(1-\frac{x_2}{l}\right)+\ldots\right\}\right.$$

$$\left.+(m_2a_i-sa_i^3)\left\{\cos(a_i+b_i)+q_i'\cos a_i\left(1-\frac{x_1}{l}\right)+q_i''\cos a_i\left(1-\frac{x_2}{l}\right)+\ldots\right\}\right]$$

$$=\Sigma A_iN_i'', \text{ say.} \quad\ldots\ldots \quad (16)$$

Also, let $V_3=Gkl$; then by (6),

$$V_3=\Sigma A_i\times -a_i\left\{\cos(a_i+b_i)+q_i'\cos a_i\left(1-\frac{x_1}{l}\right)+q_i''\cos a_i\left(1-\frac{x_2}{l}\right)+\ldots\right\}$$

$$=\Sigma A_iN_i''', \text{ say.} \quad\ldots\ldots \quad (17)$$

To find A_i, the ith value of A. Multiply both sides of each one of the last equations, (12) to (17), by the coefficient of A_i in that particular equation; *e.g.* multiply (12) by M_i', (13) by M_i'', and so on. Next integrate each side belonging to the line between the limits for which it is true. Thus (12) from $x=0$ to $x=x_1$, etc. Apply a similar process to V_1, V_2, and V_3 by multiplying them by r_1l, r_2l, and sl respectively. Finally add together all the results, right and left sides respectively, excepting for V_3, which must be subtracted, and then equate the two sums. The result is

$$\int_0^{x_1}V_{0x_1}M_i'dx+\int_{x_1}^{x_2}V_{x_1x_2}M_i''dx+\int_{x_2}^{x_3}V_{x_2x_3}M_i'''dx+\ldots+V_1r_1lN_i'+V_2r_2lN_i''-V_3slN_i'''$$

$$=\sum_{i'=0}^{i'=\infty}\left\{\int_0^{x_1}A_{i'}M_i'M_{i'}'dx+\int_{x_1}^{x_2}A_{i'}M_i''M_{i'}''dx+\int_{x_2}^{x_3}A_{i'}M_i'''M_{i'}'''dx+\ldots+A_{i'}r_1lN_i'N_{i'}'\right.$$

$$\left.+A_{i'}r_2lN_i''N_{i'}''-A_{i'}slN_i'''N_{i'}'''\right\}. \quad\ldots\ldots \quad (18)$$

It will be found, on making the substitutions in (18) of the expressions for the M's and N's, and effecting the necessary reductions, that in the summation on the right-hand side of (18), the complete coefficient of every one of the A's vanishes identically, by reason of equations (10) and (11), *except* for A_i; whence

$$A_i=\frac{\int_0^{x_1}V_{0x_1}M_i'dx+\int_{x_1}^{x_2}V_{x_1x_2}M_i''dx+\ldots+V_1r_1lN_i'+V_2r_2lN_i''-V_3slN_i'''}{\int_0^{x_1}M_i'^2dx+\int_{x_1}^{x_2}M_i''^2dx+\ldots+r_1lN_i'^2+r_2lN_i''^2-slN_i'''^2} \quad\ldots \quad (19)$$

This completes the solution; and the state of the whole system is determined for any time t.

15. When the initial potentials V_{0x_1}, etc., of the line in the different sections are given explicitly as functions of x, the sum of the integrals in the numerator of (19) may be written

$$\int_0^l V_{0x_1} \sin\left(\frac{a_i x}{l} + b_i\right) dx + \int_{x_1}^l V_{x_1 x_2} q_i' \sin\frac{a_i(x - x_1)}{l} dx + \int_{x_2}^l V_{x_2 x_3} q_i'' \sin\frac{a_i(x - x_2)}{l} dx + \ldots$$

There is a great simplification when the initial state of the system is, not arbitrary, but such as would be finally produced by a constant electromotive force E acting at P (Fig. 4). Then the complete numerator of (19) reduces to

$$\frac{El \cos b_i}{a_i}$$

for any number of faults and for all the terminal arrangements that can be made out of those shown in Fig. 4. The denominator of (19) is a function of a_i and b_i. Thus

$$A_i = \frac{El \cos b_i}{a_i \phi(a_i, b_i)}. \qquad (20)$$

16. There is no difficulty in finding formulæ from the preceding results which will correspond to any particular example considered. Such formulæ, however, have, save to the mathematically curious, little value or interest unless they are interpreted numerically. Even then the labour involved is, save in special cases, out of proportion to the derived benefit. I shall confine myself to the simple cases of direct working without condensers, and with condensers, with a single fault in the centre of the line.

Suppose the signalling is made by means of a battery at P and a receiving instrument at Q, both of negligible resistance, and to earth direct. Then

$$0 = m_1 = m_2 = n_1 = n_2.$$

Let there be a single fault of resistance zkl at the centre of the line. Then

$$\tan b = 0,$$

$$\sin a + \frac{1}{za} \sin^2 \frac{a}{2} = 0,$$

by (10) and (11). The latter splits up into

$$\sin \frac{a}{2} = 0 \text{ and } \tan \frac{a}{2} = -2za. \qquad (21)$$

Therefore, when i is even, $a_i = i\pi$; and when i is odd, a_i lies between $i\pi$ and $(i+1)\pi$. The denominator of (19) is

$$\phi(a, b) = \int_0^{\frac{l}{2}} \sin^2 \frac{ax}{l} dx + \int_{\frac{l}{2}}^l \left\{ \sin\frac{ax}{l} + \frac{\sin\frac{a}{2}}{za} \sin a\left(\frac{x}{l} - \frac{1}{2}\right) \right\}^2 dx = \frac{l}{2}\left(1 - \frac{\sin a}{a}\right).$$

Therefore, by (20), $$A_i = \frac{2E}{a_i - \sin a_i}.$$

Hence the potential v at time t after the electromotive force E which produced the initial state is removed is, from $x=0$ to $x=\frac{l}{2}$,

$$v = 2E\Sigma \frac{\sin\frac{ax}{l}}{a - \sin a} \epsilon^{-\frac{a^2 t}{T}}; \quad (22)$$

and from $x=\frac{l}{2}$ to $x=l$,

$$v = 2E\Sigma \frac{\sin\frac{ax}{l}}{a - \sin a} \epsilon^{-\frac{a^2 t}{T}} + 2E\Sigma \frac{\sin\frac{a}{2}}{za} \frac{\sin a\left(\frac{x}{l} - \frac{1}{2}\right)}{a - \sin a} \epsilon^{-\frac{a^2 t}{T}},$$

which may be transformed into

$$v = 2E\Sigma \frac{-\cos i\pi}{a - \sin a} \sin\frac{ax'}{l} \epsilon^{-\frac{a^2 t}{T}} \quad (23)$$

(where $x' = l - x$) by making use of (21).

Let Γ be the current at Q. Then

$$\Gamma = \frac{2E}{kl} \Sigma \frac{-\cos i\pi}{1 - \frac{\sin a}{a}} \epsilon^{-\frac{a^2 t}{T}}.$$

If Γ_0 is the initial current,

$$\Gamma_0 = \frac{E}{kl\left(1 + \frac{1}{4z}\right)};$$

therefore $$\frac{\Gamma}{\Gamma_0} = \frac{1+4z}{2z} \Sigma \frac{-\cos i\pi}{1 - \frac{\sin a}{a}} \epsilon^{-\frac{a^2 t}{T}}, \quad (24)$$

from which the arrival-curve of the current may be calculated; for $1 - \frac{\Gamma}{\Gamma_0}$ is the proportion of the final current received at Q at time t after contact has been made with the battery at P.

17. The most easily calculated cases are $z=\infty$ and $z=0$. When $z=\infty$ there is no fault, $a_i = i\pi$, (22) and (23) both become

$$v = \frac{2E}{\pi} \Sigma_1^\infty \frac{1}{i} \sin\frac{i\pi x}{l} \epsilon^{-\frac{i^2 \pi^2 t}{T}},$$

and (24) becomes

$$\frac{\Gamma}{\Gamma_0} = 2\Sigma_1^\infty - \cos i\pi\, \epsilon^{-\frac{i^2 \pi^2 t}{T}}. \quad (25)$$

This equation (25) corresponds to curve 1, Fig. 1 (p. 62), and is well known.

To find the limiting form of the arrival-curve when $z=0$. By (23), when z is finite,

$$v=2E\Sigma-\frac{\cos i\pi}{a-\sin a}\sin\frac{ax'}{l}\epsilon^{-\frac{a^2t}{T}},$$

from $x'=0$ to $x'=\frac{l}{2}$. The initial potential v_0 between the same limits is

$$v_0=\frac{Ex'}{l}\frac{4z}{1+4z}.$$

Therefore
$$\frac{v}{v_0}=\frac{l}{x'}\frac{1+4z}{2z}\Sigma\frac{-\cos i\pi}{a-\sin a}\sin\frac{ax'}{l}\epsilon^{-\frac{a^2t}{T}}. \quad (26)$$

The $(2i-1)$th and $2i$th terms are

$$\frac{l}{x'}\frac{1+4z}{2z}\left(\frac{\sin\frac{a_{2i-1}x'}{l}\epsilon^{-\frac{a^2_{2i-1}t}{T}}}{a_{2i-1}-\sin a_{2i-1}}-\frac{1}{2i\pi}\sin\frac{2i\pi x'}{l}\epsilon^{-\frac{(2i\pi)^2t}{T}}\right),$$

where a_{2i-1} lies between $(2i-1)\pi$ and $2i\pi$, and ultimately becomes $2i\pi$ when z is indefinitely reduced, so that the last expression takes the form $\frac{0}{0}$. Evaluating in the usual manner, remembering that

$$z=-\frac{\tan\frac{a_{2i-1}}{2}}{2a_{2i-1}},$$

the $(2i-1)$th and $2i$th terms become

$$-2\left(\cos\frac{2i\pi x'}{l}-\frac{4i\pi lt}{x'T}\sin\frac{2i\pi x'}{l}\right)\epsilon^{-\frac{(2i\pi)^2t}{T}}$$

Consequently (26) becomes, when $z=0$,

$$\frac{v}{v_0}=2\Sigma_1^\infty\left(\frac{4i\pi lt}{x'T}\sin\frac{2i\pi x'}{l}-\cos\frac{2i\pi x'}{l}\right)\epsilon^{-\frac{(2i\pi)^2t}{T}}$$

Now, when x' is indefinitely reduced, $\frac{v}{v_0}$ is the same as $\frac{\Gamma}{\Gamma_0}$; therefore, when $x'=0$,

$$\frac{\Gamma}{\Gamma_0}=\Sigma_1^\infty\left\{\frac{(4i\pi)^2t}{T}-2\right\}\epsilon^{-\frac{(2i\pi)^2t}{T}}. \quad (27)$$

From (27), curve 3, Fig. 1, is calculated. The intermediate curve 2, Fig. 1, for which $z=\frac{1}{4}$, is calculated from equation (24). It is necessary in this instance to first find the odd a's from the second equation (21) and Tables.

18. Now for working with condensers at both ends. Let

$$m_1=0, \qquad m_2=0, \qquad n_1=\infty, \qquad n_2=\infty,$$

and let r_1 and r_2 be both very small. At time t after the introduction of E at P, the potential of the line is

$$v = 2Er_1\Sigma\frac{\cos\frac{ax}{l}}{1+\frac{\sin a}{a}}\epsilon^{-\frac{a^2t}{T}} \quad\text{.........................} \quad (28)$$

from $x=0$ to $x=\frac{l}{2}$, and

$$v = 2Er_1\Sigma\frac{\cos i\pi\cos\frac{ax'}{l}}{1+\frac{\sin a}{a}}\epsilon^{-\frac{a^2t}{T}} \quad\text{.....................} \quad (29)$$

from $x'=0$ to $x'=\frac{l}{2}$, where $x'=l-x$.

The a's are the positive roots of

$$\sin a - \frac{1}{za}\cos^2\frac{a}{2} = 0\ ; \quad\text{..........................} \quad (30)$$

or

$$\cos\frac{a}{2}=0, \qquad \tan\frac{a}{2}=\frac{1}{2za},$$

zkl being the resistance of the fault in the centre.

The current Γ arriving at $x=l$ is $\Gamma = r_2cl\frac{dv}{dt}$; that is,

$$\Gamma = \frac{2E}{kl}r_1r_2\Sigma\frac{-a^2\cos i\pi}{1+\frac{\sin a}{a}}\epsilon^{-\frac{a^2t}{T}}.. \quad\text{...................} \quad (31)$$

When there is no fault, $z=\infty$, $a_i=i\pi$, and equations (28) and (29) both become

$$v = Er_1 + 2Er_1\Sigma_1^\infty\cos\frac{i\pi x}{l}\epsilon^{-\frac{i^2\pi^2t}{T}}. \quad\text{..................} \quad (32)$$

Here Er_1 is placed outside the Σ, because $a_0=0$, and the value of $\frac{1}{1+\frac{\sin a_i}{a_i}}$ is $\frac{1}{2}$ for a_0 and 1 for the rest. The current leaving $x=0$ is $-r_1cl\frac{dv}{dt}$; or

$$\Gamma_{x=0} = \frac{2E}{kl}r_1^2\pi^2\Sigma_1^\infty i^2\epsilon^{-\frac{i^2\pi^2t}{T}}\ ; \quad\text{.............................} \quad (33)$$

and the current arriving at $x=l$ is

$$\Gamma_{x=l} = \frac{2E}{kl}r_1r_2\pi^2\Sigma_1^\infty - i^2\cos i\pi\,\epsilon^{-\frac{i^2\pi^2t}{T}}. \quad\text{..................} \quad (34)$$

19. To find the limiting forms of the solutions when $z=0$. In equation (29), when i is odd, $a_i=i\pi$; and when i is even, including 0, and z

finite, a_i lies between $i\pi$ and $(i+1)\pi$, and ultimately becomes $(i+1)\pi$ when $z=0$. The $2i$th and $(2i+1)$th terms in (29) are

$$E2r_1\left\{\frac{\cos\frac{a_{2i}x'}{l}}{1+\frac{\sin a_{2i}}{a_{2i}}}\epsilon^{-\frac{(a_{2i})^2t}{T}} - \cos\frac{(2i+1)\pi x'}{l}\epsilon^{-\frac{(2i+1)^2\pi^2t}{T}}\right\}. \quad \text{........ (35)}$$

This vanishes when $z=0$, and (29) takes the form

$$v=0+0+0+\ldots,$$

each 0 representing a pair of terms. Now, when z is infinitely small

$$a_{2i}=(1-4z)(2i+1)\pi,$$

by (30). Expanding (35) in powers of z, neglecting squares, etc., it becomes

$$2Er_1z\,.\,4a\left\{\frac{x'}{l}\sin\frac{ax'}{l}-\frac{1}{a}\cos\frac{ax'}{l}+\frac{2at}{T}\cos\frac{ax'}{l}\right\}\epsilon^{-\frac{a^2t}{T}},$$

where a stands for $(2i+1)\pi$. The same result is reached by finding the limiting ratio of the expression (35) to z when $a_{2i}=(2i+1)\pi$, making

$$z=\frac{1}{2a}\cot\frac{a}{2},$$

and multiplying the result by z. Hence (29) finally becomes

$$v=2Er_1z\Sigma_0^\infty\left\{\frac{4ax'}{l}\sin\frac{ax'}{l}+\left(\frac{8a^2t}{T}-4\right)\cos\frac{ax'}{l}\right\}\epsilon^{-\frac{a^2t}{T}}, \quad \text{.......... (36)}$$

where $a_i=(2i+1)\pi$.

The potential v_2 of the receiving condenser is

$$v_2=16Er_1z\Sigma_0^\infty\left\{\frac{(2i+1)^2\pi^2t}{T}-\frac{1}{2}\right\}\epsilon^{-\frac{(2i+1)^2\pi^2t}{T}}; \quad \text{.................... (37)}$$

and the current Γ entering the receiving condenser is

$$\Gamma=\frac{16E}{kl}r_1r_2z\Sigma_0^\infty\left\{-\frac{(2i+1)^4\pi^4t}{T}+\frac{3}{2}(2i+1)^2\pi^2\right\}\epsilon^{-\frac{(2i+1)^2\pi^2t}{T}}. \quad \text{..... (38)}$$

Curve 3, Fig. 2, is calculated from (37), and curve 3, Fig. 3, from (38); curve 2, Fig. 2, from (29), making $x'=0$; and curve 2, Fig. 3, from (31). In the last two $z=\frac{1}{4}$, and the even a's are found by Tables.

20. The two important solutions

$$v=E\left(1-\frac{x}{l}\right)-\frac{2E}{\pi}\Sigma\frac{1}{i}\sin\frac{i\pi x}{l}\epsilon^{-\frac{i^2\pi^2t}{T}}, \quad \text{................ (39)}$$

and

$$v=Er_1+2Er_1\Sigma\cos\frac{i\pi x}{l}\epsilon^{-\frac{i^2\pi^2t}{T}}, \quad \text{........................ (40)}$$

where, in (39), v is the potential at x at time t after the introduction of E at $x=0$, both ends being to earth, and in (40) v is the same when con densers of very small capacities r_2cl and r_1cl are interposed at the ends, there being no fault, may be both deduced from the corresponding formula when the condensers are of finite capacity. Suppose initially the condenser at $x=0$ to be charged to potential E, and the potential of the line and the condenser at $x=l$ to be zero, with no impressed electromotive force in the system. Then at time t the solution is

$$v=\Sigma A \sin\left(\frac{ax}{l}+b\right)\epsilon^{-\frac{a^2t}{T}},$$

where $$\tan b=-\frac{1}{r_1 a}, \qquad \tan(a+b)=\frac{1}{r_2 a},$$

and therefore $$\tan a=\frac{(r_1+r_2)a}{r_1 r_2 a^2-1}. \quad\text{.............} \quad (41)$$

Also

$$A=\frac{-\frac{Er_1 l\cos b}{r_1 a}+\int_0^l 0\times\sin\left(\frac{ax}{l}+b\right)dx+0\times\frac{r_2 l\cos(a+b)}{r_2 a}}{r_1 l\frac{\cos^2 b}{r_1^2 a^2}+\int_0^l \sin^2\left(\frac{ax}{l}+b\right)dx+r_2 l\frac{\cos^2(a+b)}{r_2^2 a^2}}$$

The result is

$$v=\frac{Er_1}{1+r_1+r_2}-2E\Sigma\frac{r_1 a\sin\frac{ax}{l}-\cos\frac{ax}{l}}{1+\frac{1}{r_1}(1+r_1^2a^2)\left(1+\frac{r_2}{1+r_2^2a^2}\right)}\epsilon^{-\frac{a^2t}{T}}, \quad\text{........} \quad (42)$$

where the constant term arises from the zero root of (41). Now, when $r_1=r_2=0$, the other + roots of (41) are π, 2π, 3π, ...; and (42) then becomes the same as (40). But when $r_1=r_2=\infty$, the roots are the same with the addition of a second zero root. In the general term of (42) make

$$r_1=r_2=\frac{1+\cos a}{a\sin a},$$

which follows from (41); and find the limit when $a=0$. The result is

$$E\left(\frac{1}{2}-\frac{x}{l}\right),$$

This, added to $\frac{E}{2}$, what the constant term in (42) becomes when $r_1=r_2=\infty$, makes

$$E\left(1-\frac{x}{l}\right),$$

which is the constant term in (39). The remainder of (39) is immediately deducible from (42) by making $r_1=r_2=\infty$.

21. The solution (40) for the potential in condenser working could be deduced from that for the *current* in working without condensers. For, in the latter case, the final result of the introduction of an electromotive force at $x=0$ is a current in the line of the same strength everywhere, and $v=0$ at $x=0$ and $x=l$; and in the former the final result is that the potential of the line is the same everywhere, and $\frac{dv}{dx}=0$ at $x=0$ and $x=l$. Both the current and the potential must satisfy the same partial differential equation. Hence the current in the latter case at x

at time t must rise in the same manner as the potential in the former. Now

$$\gamma=\frac{E}{kl}\left\{1+2\Sigma\cos\frac{i\pi x}{l}\epsilon^{-\frac{i^2\pi^2 t}{T}}\right\}\text{.......................} (43)$$

is the solution for the current in working without condensers, where $\frac{E}{kl}$ is the final uniform current. In the condenser-problem the final uniform potential is $\frac{Er_1cl}{cl}=Er_1$, substituting which for $\frac{E}{kl}$ in (43), and changing γ into v, equation (40) results without a separate investigation. It is also very remarkable that (40) and (43) are capable of expression in an entirely different form, leading to the identity

$$\epsilon^{-x^2}+\epsilon^{-(x-a)^2}+\epsilon^{-(x+a)^2}+\epsilon^{-(x-2a)^2}+\epsilon^{-(x+2a)^2}+\dots$$

$$=\frac{2\sqrt{\pi}}{a}\left(\frac{1}{2}+\epsilon^{-\frac{\pi^2}{a^2}}\cos\frac{2\pi x}{a}+\epsilon^{-\frac{4\pi^2}{a^2}}\cos\frac{4\pi x}{a}+\epsilon^{-\frac{9\pi^2}{a^2}}\cos\frac{6\pi x}{a}+\dots\right),$$

well known to mathematicians.

When $t=0$, the current as given by (43) is zero everywhere, except at $x=0$, where it is infinite; and in (40) the potential is zero everywhere when $t=0$, except at $x=0$, where it is infinite. These impossible infinite values arise from the neglect of the battery-resistance in the one, and the condenser's capacity in the other instance. All mathematical investigations of physical questions are approximative; and being such, impossible results arise in extreme cases. If R is the battery-resistance, the current at $x=0$ when $t=0$ cannot be greater than $\frac{E}{R}$; but since there is always self-induction, the current, when $t=0$, is mathematically zero, rising in an extremely short time to $\frac{E}{R}$, and then falling to its final strength. The actual rise of the current is more complex, on account of electromagnetic oscillations. Thus, from infinity we have got down to zero for the current at $x=0$ when $t=0$.

22. When we introduce the coefficient $s=\frac{L}{kl\,.\,T}$, calculations become complicated by the presence of imaginary roots. That there must be imaginary terms in the solutions will be evident when it is considered that electromagnetic induction imparts inertia to the electric current, thus causing oscillations, and that

$$v=\Sigma A\sin\left(\frac{ax}{l}+b\right)\epsilon^{-\frac{a^2t}{T}}$$

cannot contain oscillatory terms with real values of a. When there is a pair of terms in which A, a, and b are imaginary, their addition causes the elimination of the imaginary parts, and the result is real, as indeed it must be if the problem has physical reality. It is also evident that if in a physically real problem we have a single imaginary root, it must be of the form $a=0\pm n\sqrt{-1}$, which makes a^2 real.

Taking a simple example, let the line be to earth direct at $x=0$, and to earth through a coil of resistance mkl and electromagnetic capacity L at $x=l$. Also let there be initially a potential distribution

$$E\left(1-\frac{x}{l(1+m)}\right)$$

in the line, and a current

$$\frac{E}{kl(1+m)}$$

through the whole circuit. This state would be produced finally by E at $x=0$. At $x=0$, $v=0$, and at $x=l$,

$$0=v+ml\frac{dv}{dx}+sl^3\frac{d^3v}{dx^3}.$$

At time t,

$$v=\Sigma A\sin\frac{ax}{l}\epsilon^{-\frac{a^2t}{T}},$$

where the a's are the + roots, including imaginary roots with + real parts, of

$$\frac{\tan a}{a}=-m+sa^2,$$

and

$$A=\frac{\frac{El}{a}}{\int_0^l \sin^2\frac{ax}{l}dx-sla^2\cos^2 a}=\frac{2E}{a\left(1-\frac{3\sin 2a}{2a}-2m\cos^2 a\right)}.$$

For simplicity, put $m=0$, then

$$v=\Sigma\frac{2E}{a\left(1-\frac{3\sin 2a}{2a}\right)}\sin\frac{ax}{l}\epsilon^{-\frac{a^2t}{T}}, \qquad (44)$$

where

$$\frac{\tan a}{a}=sa^2. \qquad (45)$$

When s is large, there is no trouble with imaginary roots. There is a root of (45) a little above zero, another a little under $\frac{\pi}{2}$; and the rest are nearly $\frac{3\pi}{2}, \frac{5\pi}{2}, \ldots$ Hence, when s is large, (45) becomes

$$1=sa^2$$

to determine the lowest root, or

$$a^2=\frac{1}{s}=\frac{T.kl}{L}.$$

Therefore (44) is nearly the same as

$$v=-\frac{Ex}{l}\epsilon^{-\frac{klt}{L}}+\Sigma_1^\infty\frac{2E}{(i-\frac{1}{2})\pi}\sin\frac{(i-\frac{1}{2})\pi x}{l}\epsilon^{-\frac{(i-\frac{1}{2})^2\pi^2t}{T}},$$

and the current nearly the same as

$$\gamma = \frac{E}{kl}\epsilon^{-\frac{klt}{L}} + \ldots.$$

This case corresponds to a short land-line, the self-induction of the receiving instrument causing greatly more retardation than the electrostatic capacity of the line. The current at $x=l$ is always $+$. At $x=0$ it is first $-$ for a very short time, and thereafter $+$. Except at first, the current is of the same strength throughout the circuit. Of the line's initial charge of potential $E\left(1-\frac{x}{l}\right)$, a portion of potential E constant everywhere discharges quickly, nearly as if the line were insulated at $x=l$. The other part of potential $-\frac{Ex}{l}$ disappears exactly as the current decays, after the first moment. Or, more simply, the inertia of the current in the electromagnet causes the current at $x=l$ at any time to be stronger than it would have been without self-induction, in which case the current would be simply due to the line's charge. This charge, therefore, cannot supply enough electricity for the current; and the line becomes negatively charged, first at the end $x=l$, and afterwards all along. When this has happened the line-current is constant everywhere, and the $-$ charge and $+$ current die away uniformly.

As s decreases, the two roots of (45) lying between 0 and $\frac{\pi}{2}$ approach each other. When s reaches 1·47, they both become $=1·1396$, and simultaneously

$$1 = \frac{3\sin 2a}{2a};$$

so that in the solution (44) the first term becomes $-\infty$, the second $+\infty$, their sum remaining finite. As s sinks below 1·47, the pair of roots become imaginary, and the first two terms of (44) may be put in a rather complicated mixed real form, indicating oscillations. When s reaches zero, the cable discharges in the ordinary way.

From (44) it follows that the potential at time t after introducing an electromotive force E at $x=0$ is

$$v = E\left(1-\frac{x}{l}\right) - \Sigma\frac{2E}{a\left(1-\dfrac{3\sin 2a}{2a}\right)}\sin\frac{ax}{l}\epsilon^{-\frac{a^2t}{T}}. \quad \ldots\ldots\ldots\ldots\ldots (46)$$

The electromagnet is here at $x=l$. Suppose now it is transferred to $x=0$, other things being the same; then instead of (46) we shall have

$$v = E\left(1-\frac{x}{l}\right) + \Sigma\frac{2E\cos a}{a\left(1-\dfrac{3\sin 2a}{2a}\right)}\sin a\left(1-\frac{x}{l}\right)\epsilon^{-\frac{a^2t}{T}}. \quad \ldots\ldots (47)$$

Except when $s=0$, the permanent state of charge is arrived at in an entirely different manner in the two cases. v in (46) is generally greater than v in (47) at any time. In the extreme, when s is large, the potential of the line according to (46) becomes nearly E everywhere, and afterwards settles down to $E\left(1-\frac{x}{l}\right)$, thus,

$$v=E-\frac{Ex}{l}(1-\epsilon^{-\frac{klt}{L}})+\ldots,$$

whereas according to (47) it rises, thus

$$v=E\left(1-\frac{x}{l}\right)(1-\epsilon^{-\frac{klt}{L}})+\ldots.$$

In spite, however, of this great difference in the phenomena of the charge, the current at $x=l$ rises in precisely the same manner in both instances, as will be seen on differentiating (46) and (47), and making $x=l$.

23. In the following example we have to deal with a single imaginary root. Suppose the line is initially charged to potential $\frac{Ex}{l}$, that the end $x=0$ is to earth, and that the current entering the cable at $x=l$ after $t=0$ is simply proportional to the potential there at any moment. That is, $v=0$ at $x=0$, and $v=ml\frac{dv}{dx}$ at $x=l$, where m is a + constant. At time t the solution is

$$v=\Sigma\frac{2E(m-1)\cos a}{a(1-m\cos^2 a)}\sin\frac{ax}{l}\epsilon^{-\frac{a^2t}{T}}, \qquad (48)$$

where

$$\tan a=ma.$$

There is one particular case where the potential remains unchanged, viz. when $m=1$. All terms in the expression for v in (48) vanish except the first, for which $a=0$. The limiting value of

$$A\sin\frac{ax}{l}=\frac{2E(\sin a-a\cos a)\sin\frac{ax}{l}}{a(a-\frac{1}{2}\sin 2a)}$$

when $a=0$ is $\frac{Ex}{l}$; so that (48) is simply

$$v=\frac{Ex}{l}$$

when $m=1$. If m is greater than 1, v ultimately vanishes; but if m is less than 1, an imaginary root $a=n\sqrt{-1}$, where n is the + root of

$$\frac{\epsilon^n-\epsilon^{-n}}{\epsilon^n+\epsilon^{-n}}=mn,$$

comes into operation. The first term of (48) then increases with t without limit, the rest ultimately vanishing.

24. In general, the conditions imposed at the ends of a cable, when there are no impressed electromotive forces, are of the following form:—

At $x=0$, $$0=v+m_1l\frac{dv}{dx}+m_2l^2\frac{d^2v}{dx^2}+\ldots; \qquad (49)$$

at $x=l$, $$0=v+n_1 l\frac{dv}{dx}+n_2 l^2\frac{d^2v}{dx^2}+\ldots \qquad (50)$$

Here $m_1, \ldots, n_1, \ldots,$ are constants, and v is the potential at any time.

Supposing there are no intermediate conditions, there is a single solution of the form

$$v=\Sigma A\sin\left(\frac{ax}{l}+b\right)\epsilon^{-\frac{a^2t}{T}}, \qquad (51)$$

provided that the right-hand side of (51) can be made to satisfy (49) and (50), and to equal $f(x)$, an arbitrary function of x when $t=0$.

It follows from (49) and (50) that

$$\tan b=-\frac{m_1 a-m_3 a^3+m_5 a^5-\ldots}{1-m_2 a^2+m_4 a^4-\ldots}, \qquad (52)$$

$$\tan(a+b)=-\frac{n_1 a-n_3 a^3+n_5 a^5-\ldots}{1-n_2 a^2+n_4 a^4-\ldots}; \qquad (53)$$

and from these $\tan a$ can be expressed similarly, say

$$\tan a=-\frac{h_1 a-h_3 a^3+h_5 a^5-\ldots}{1-h_2 a^2+h_4 a^4-\ldots}; \qquad (54)$$

and the a's required are the + roots, real and imaginary, of this equation.

Let $$u=\frac{1}{l}\int_0^l \sin\left(\frac{a_1 x}{l}+b_1\right)\sin\left(\frac{a_2 x}{l}+b_2\right)dx,$$

where a_1, b_1, a_2, b_2 are any two pairs of values of a and b. Then, by integration,

$$u=\frac{a_1 a_2}{a_1^2-a_2^2}\cos(a_1+b_1)\cos(a_2+b_2)\left\{\frac{\tan(a_1+b_1)}{a_1}-\frac{\tan(a_2+b_2)}{a_2}\right\}$$
$$-\frac{a_1 a_2}{a_1^2-a_2^2}\cos b_1\cos b_2\left\{\frac{\tan b_1}{a_1}-\frac{\tan b_2}{a_2}\right\} \qquad (55)$$

Substituting in (55) the values of $\tan(a+b)$ and $\tan b$ from (53) and (52), the bracketed quantities are always divisible by $a_1^2-a_2^2$, and u is expressible as

$$u=r_1\phi_1(a_1, b_1)\phi_1(a_2, b_2)+r_2\phi_2(a_1, b_1)\phi_2(a_2, b_2)+\ldots; \qquad (56)$$

i.e. in the form of the sum of a number of products, each being a function of a_1 and b_1 multiplied by the same function of a_2 and b_2 and by a constant r.

Then assuming
$$E_1=\Sigma A\phi_1(a, b), \qquad E_2=\Sigma A\phi_2(a, b)\ldots,$$
it follows that

$$A=\frac{\frac{1}{l}\int_0^l f(x)\sin\left(\frac{ax}{l}+b\right)dx-r_1E_1\phi_1(a, b)-r_2E_2\phi_2(a, b)-\ldots}{\frac{1}{l}\int_0^l \sin^2\left(\frac{ax}{l}+b\right)dx-r_1\{\phi_1(a, b)\}^2-r_2\{\phi_2(a, b)\}^2-\ldots}. \ldots (57)$$

When there are intermediate conditions, producing discontinuity in v or $\frac{dv}{dx}$, etc., at certain points x_1, x_2, etc., each section must have its own series of the form (51). The a's are the same for every section, being determined by the resultant of all the conditions. The A's and b's are different for each section. Thus

$$f(x) = \Sigma A \sin\left(\frac{ax}{l} + b\right) \text{ from } x = 0 \text{ to } x = x_1,$$

$$= \Sigma A' \sin\left(\frac{ax}{l} + b'\right) \quad ,, \quad x_1 \quad ,, \quad x_2,$$

$$= \Sigma A'' \sin\left(\frac{ax}{l} + b''\right) \quad ,, \quad x_2 \quad ,, \quad x_3.$$

The intermediate conditions enable A', b', A'', b'', ... to be expressed in terms of A, a, and b. If

$$u' = \frac{1}{l}\int_0^{x_1} \sin\left(\frac{a_1 x}{l} + b_1\right) \sin\left(\frac{a_2 x}{l} + b_2\right) dx + \int_{x_1}^{x_2} \frac{A'_2}{A_2} \sin\left(\frac{a_1 x}{l} + b'_1\right) \sin\left(\frac{a_2 x}{l} + b'_2\right) dx + \dots,$$

then u' may, as before in the case of u, be put in the form (56), and the value of A follows:—

$$A = \frac{\frac{1}{l}\int_0^{x_1} f(x) \sin\left(\frac{ax}{l} + b\right) dx + \frac{1}{l}\int_{x_1}^{x_2} \frac{A'}{A} f(x) \sin\left(\frac{ax}{l} + b'\right) dx + \dots - \Sigma E r \phi(a, b)}{\frac{1}{l}\int_0^{x_1} \sin^2\left(\frac{ax}{l} + b\right) dx + \frac{1}{l}\int_{x_1}^{x_2} \sin^2\left(\frac{ax}{l} + b'\right)\left(\frac{A'}{A}\right)^2 dx + \dots - \Sigma r\{\phi(a, b)\}^2}. \quad (58)$$

The arbitrary quantities E_1, E_2, ... in (57) and (58), or rather, as many of them as turn out to be independent, are easily found to depend on the initial electromotive forces residing in those parts of the system in connection with the cable, either at the ends or intermediate, which influence v at time t independently of its value $f(x)$ when $t = 0$.

If, for example, we join two points x_1 and x_2 through a coil, its self-induction will introduce one E; and if this coil have a closed circuit near it, a second independent E will be introduced.

25. Considering the line as of infinite length both ways, it will be found that if

$$v = f(x) = \Sigma A \sin\left(\frac{ax}{l} + b\right), \quad \dots\dots\dots\dots (59)$$

where the a's are determined from

$$\tan a = -\frac{h_1 a - h_3 a^3 + h_5 a^5 - \dots}{1 - h_2 a^2 + h_4 a^4 - \dots}, \quad \dots\dots\dots\dots (60)$$

then will v satisfy the differential equation

$$\left(1 + h_1 l \frac{d}{dx} + h_2 l^2 \frac{d^2}{dx^2} + h_3 l^3 \frac{d^3}{dx^3} + \dots\right) f(x + l) = \left(1 - h_1 l \frac{d}{dx} + h_2 l^2 \frac{d^2}{dx^2} - h_3 l^3 \frac{d^3}{dx^3} + \dots\right) f(x - l) \quad \dots (61)$$

everywhere, thus expressing the relation between the values of $f(x)$ at any two points separated by a distance $2l$. Or, which is the same thing,

$$0 = k_1 l \frac{dv}{dx} + k_3 l^3 \frac{d^3 v}{dx^3} + k_5 l^5 \frac{d^5 v}{dx^5} + \dots, \qquad (62)$$

where

$$k_1 = 1 + h_1,$$

$$k_3 = \frac{1}{\lfloor 3} + \frac{h_1}{\lfloor 2} + h_2 + h_3,$$

$$k_5 = \frac{1}{\lfloor 5} + \frac{h_1}{\lfloor 4} + \frac{h_2}{\lfloor 3} + \frac{h_3}{\lfloor 2} + h_4 + h_5,$$

$$\cdot \quad \cdot \quad \cdot \quad \cdot \quad \cdot \quad \cdot \quad \cdot \quad \cdot$$

In the particular case $h_1 = 0$, $h_2 = 0$, ..., equation (61) reduces to

$$f(x+l) = f(x-l), \qquad (63)$$

which simply expresses that $f(x)$ is periodic, repeating itself at intervals $2l$.

Starting from this equation, or an equivalent one, Mr. O'Kinealy (*Phil. Mag.*, August 1874) proves Fourier's theorem for periodic functions; that is, solving the linear equation (63), its solution is found to be

$$f(x) = \Sigma A \sin\left(\frac{i\pi x}{l} + b\right). \qquad (64)$$

Hence it is concluded that an arbitrary function $f(x)$ *may* be expanded in such a series as the right-hand side of (64), though this proof of the possibility does not tell us how to do it. Mr. O'Kinealy, however, completes the solution in the usual way, leading to

$$f(x) = \frac{1}{2l}\int_0^{2l} f(x)dx + \frac{1}{l}\Sigma \cos\frac{i\pi x}{l}\int_0^{2l} f(x)\cos\frac{i\pi x}{l}dx$$
$$+ \frac{1}{l}\Sigma \sin\frac{i\pi x}{l}\int_0^{2l} f(x)\sin\frac{i\pi x}{l}dx. \qquad (65)$$

Similarly, if we start from equation (61), which is linear, with constant coefficients, and includes the above case, we may easily prove that its solution is (59), with the condition that the a's therein are the + roots, real and imaginary, of (60), the A's and b's being undetermined. Or we may get the same result from (62), the a's being now found from

$$0 = k_1 a - k_3 a^3 + k_5 a^5 - \dots. \qquad (66)$$

It will be observed that (60) or (66) have numerically equal + and − roots, each pair of which go to a single term of (59).

Here again the proof, if it may be now called a proof, gives us no information as to how to find the coefficients settling the amplitudes; and even the phases are undefined without further knowledge. But in working out practical problems requiring arbitrary functions to satisfy certain conditions when expanded in a harmonic series, the physical nature of a particular problem will usually suggest, step by step, the necessary procedure to render the solution complete, as in the last paragraph 24; and the completion of a solution is of far greater importance than any proof that the solution is possible.

With respect to the periodic series (65), it is only applicable to a cable when the ends are joined so as to make a closed circuit, changing $2l$ into l; and there must be no external electrical connections with the cable. If there are connections at a point, or at several points, even without interrupting the continuity of the cable, although the potential of the cable will now repeat itself every time x is increased by l or $2l$, etc., yet the periodic form (65) will obviously not be suitable. The proper series are of course more general, and pass into the form (65) in limiting cases.

XVII.—ON ELECTROMAGNETS, etc.

[*Jour. Soc. Tel. Eng.*, 1878, vol. vii., p. 303.]

1. The following investigations have reference to the magnetic induction of electromagnets and suspended iron wires, especially as regards its influence on the speed of working. The resistance of electromagnets to obtain the greatest magnetic force from reversals is also considered, as well as other matters which may be useful to the members of the Society.

2. Suppose we have a circuit containing a battery and an electromagnet, and that a constant current is flowing through the circuit, which is so far removed from other circuits, etc., that there is no appreciable induction between them. If the electromotive force is removed without breaking the circuit, say by shunting the battery, or if a new circuit is made containing the electromagnet, the current, which has now no impressed electromotive force to support it, nevertheless does not cease immediately, but continues to flow in the same direction with continuously decreasing strength, until it is stopped by the resistance of the circuit. We may compare the electric current under these circumstances to a material current, as of water flowing through a pipe. If it be set in motion by external force, and the latter be then removed, the water will continue to flow until it is stopped by frictional resistance. There is an exact analogy if we suppose that the water meets with a resistance exactly proportional to its velocity. Suppose the pipe to be of unit section, M the whole mass of water in the pipe, and v its velocity, at any time t. Its momentum is Mv. Let the whole frictional resistance, which is a force acting against the stream, be Rv, proportional to the velocity. Then the equation of motion, when external force is removed, is

$$M\frac{dv}{dt} = -Rv,$$

whence

$$v = V\epsilon^{-\frac{Rt}{M}};$$

which gives the velocity v, at time t, compared with V, the starting velocity. The total quantity that flows past every section of the pipe is MV/R.

3. In the electric circuit the electromotive force arising from magnetic induction is proportional to the rate of decrease of the current, and to a constant depending on the form and position of the coils, cores, etc. If γ is the current at time t, and L the coefficient of self-induction or electromagnetic capacity of the circuit, and R its resistance, the equation of electromotive force is

$$L\frac{d\gamma}{dt} = -R\gamma.$$

Therefore the current at time t is

$$\gamma = \Gamma\epsilon^{-\frac{Rt}{L}},$$

where Γ is the initial current; and the integral extra-current Q, or the amount of electricity that flows in the circuit after the electromotive force that produced the current in the first place is removed, is

$$Q = \int_0^\infty \gamma dt = \frac{L\Gamma}{R}.$$

These equations are exactly similar to those used in the waterpipe analogy. $L\Gamma$ is the electromagnetic momentum of the circuit containing the current Γ, corresponding to MV, the momentum of the water. Also $\frac{1}{2}MV^2$ is the kinetic energy of the fluid, and $\frac{1}{2}L\Gamma^2$ the electrokinetic energy of the current, which, however, does not reside merely in the wire, as the kinetic energy of the water is confined in the pipe, but in the surrounding space as well. The fluid by friction produces an amount of heat $= \frac{1}{2}MV^2$ before it is brought to rest, and the electric current produces an amount of heat $= \frac{1}{2}L\Gamma^2$ in the wire before it ceases. For, by Joule's law, the rate of generation of heat is $R\gamma^2$, therefore the whole amount is

$$\int_0^\infty R\gamma^2 dt = R\int_0^\infty \Gamma^2\epsilon^{-\frac{2Rt}{L}} dt = \frac{1}{2}L\Gamma^2.$$

The analogy between the electric current and the flow of a material fluid, which is a very useful one, may be carried much further if required. As an example, if a pipe containing water connect two reservoirs of limited capacity, and a difference of pressure be established between them, a state of equilibrium will be arrived at through a series of oscillations of the water through the pipe. The first current from the higher level to the lower will not cease when the levels are equalized, for the water in the pipe must keep moving on till its momentum is destroyed, partly by frictional resistance and partly by the excess of pressure produced in the reservoir to which the water flows. This excess of pressure causes a reverse current to set in, and the process is repeated forwards and backwards until all the potential energy due to the original difference of level is used up, a portion being converted into the kinetic energy of heat during each oscillation, the kinetic energy of the moving water being its intermediate form. An exactly parallel case is produced by charging a condenser, *i.e.*, causing a difference of potential between the two coatings, and then discharging it through a coil. The first current, from the higher potential to the lower, as it acquires momentum, carries more than enough electricity to restore equilibrium, thus causing a reverse current, and so on. Thus there may be a series of currents, each in the opposite direction to and carrying less electricity than the preceding. The electrostatic energy of the original difference of potential is finally wholly converted into heat in the wire (if no external work has been done), a portion during each oscillation, the electrokinetic energy of the current being its intermediate form. The analogy must not, however, be carried too far, for the starting or stopping of a material current in one pipe does not cause any current in a neighbouring pipe, as the starting or stopping of an electric current in one wire does in a neighbouring wire.

4. Maxwell (vol. ii.) gives the necessary information for the calculation of L from the form of the circuit, etc.; also how to measure it experimentally by comparison with the capacity of a condenser, using the Bridge arrangement. Or, it may be roughly determined by observing the integral extra-current Q. Send a known current Γ through the electromagnet whose electromagnetic capacity is required, and calculate Q from the throw of the needle of a galvanometer through which the extra current is then made to flow.

Then
$$L = RQ/\Gamma,$$
where R is the resistance of the circuit through which the extra current passes, starting from the electromagnet. The electromagnetic capacity of the galvanometer does not affect the result, though the motion of the galvanometer magnet introduces an error. Neither will it be affected by shunting the galvanometer, whatever may be the self-induction of the shunt, or the induction, if there be any, between the galvanometer coil and the shunt, for the integral extra-current will divide between the galvanometer and shunt in the inverse proportion of their resistances. Of course Q is increased by the shunt, R being at the same time equally reduced; hence it is necessary to know R. Mr. Preece, in his lecture on "Shunts," has described numerous observations of the extra currents from electromagnets under various circumstances, but we cannot calculate L from them, even proportionately, since R is not given.

5. When an electromotive force is introduced into a circuit, the current rises from zero to its final strength, in the same manner as it falls from it when the electromotive force is removed and the circuit unbroken. If E is the impressed electromotive force, as of a battery inserted in the circuit, then

$$E = R\gamma + L\frac{d\gamma}{dt}. \qquad (1)$$

Thus, when the current is rising, at time t, a portion of E, viz., $R\gamma$, is employed in maintaining, according to Ohm's law, the current γ already established; the other portion of E, viz., $L\frac{d\gamma}{dt}$ is employed in increasing the electromagnetic momentum $L\gamma$. The solution of (1) is

$$\gamma = \frac{E}{R}\left(1 - \epsilon^{-\frac{Rt}{L}}\right). \qquad (2)$$

The current rises in the same manner as it does in a wire connecting the two terminals of a condenser, which allows the determination of L by comparison with the capacity of a condenser, as before mentioned.

If the circuit be broken at any point when a current flows through it, the current does not cease quite instantaneously, but, as there is no conductive circuit, + electricity accumulates at one of the broken ends, and − at the other. The electrostatic capacity of the ends being extremely small, a high difference of potential is produced between the ends, and the dielectric breaks down, with the well-known spark as a result. This is analogous to the bursting of a pipe by the great pressure produced by the sudden stoppage of the flow of a fluid through it.

If a suspended wire, especially an iron wire, as is usual, form a part of the circuit, there may be oscillations in the current during its establishment and decay. They are due to the combined action of electrostatic and electromagnetic induction, for the wire is not only a conductor but a condenser as well, or rather one coating of a condenser. The establishment of the permanent state of the potential of the wire may take place with oscillations, and is quite a different sort of phenomenon to what occurs in a long submarine cable similarly acted upon by an electromotive force at one end. The presence of an electromagnet in the circuit, however, has a material influence.

6. Let there now be a simple harmonic variation of electromotive force

$$E \sin mt$$

in the circuit of resistance R and electromagnetic capacity L; then

$$E \sin mt = R\gamma + L\frac{d\gamma}{dt},$$

where γ is the current at time t. The solution is

$$\gamma = \frac{E}{\sqrt{R^2 + L^2m^2}} \sin\left(mt - \tan^{-1}\frac{mL}{R}\right),$$

neglecting a vanishing term. Thus the amplitude of the current waves is reduced from E/R, what it would be were there no retardation, to

$$\Gamma = \frac{E}{\sqrt{R^2 + L^2m^2}}, \quad \text{..........................} \quad (3)$$

where Γ signifies the maximum strength of current.

If Lm is large compared with R, Γ is a small fraction of E/R. Also, variations in the value of R cause much smaller variations in Γ.

7. This has application to the Bell telephone. This most sensational application of electricity appears to be very indifferent to resistance (sometimes), it being said to be sufficient merely to make earth through the boot and a blade of grass. Let

$$m = 2\pi/T,$$

then T is the period of a complete wave. L/R is also a time-interval. Suppose $T = \frac{1}{1000}$ second, and $L/R = \frac{1}{100}$ second, then a simple calculation applied to equation (3) will show that R must be increased about 110 times to reduce the current to a half, and about 625 times to reduce the current to a tenth part.

Since m is proportional to the pitch, for sufficiently high pitches Γ is inversely proportional to the pitch. Hence it is impossible for a receiving telephone to give forth the same sounds as produced at the sending end, irrespective of mechanical and acoustical difficulties, except in the case of a single pure tone. For in any tone the second partial will be weakened twice as much as the first, or fundamental, the third three times as much, and so on; thus producing a want of brilliancy.

We do not deal in such rapid reversals when using ordinary recording telegraphs. The above value of T, viz., $\frac{1}{1000}$ second, would, with the Morse code, produce at least 2500 words per minute, or $42\frac{1}{2}$ per second, which is considerably faster than the most rapid speaker can talk. As, however, the telephone is sensible to very much more rapid reversals that 1000 per second, the enormous speeds possible on short lines is easily conceivable, could the action be sufficiently magnified and recorded, so as to appeal to the eye instead of the ear.

8. Let now R and L belong to the electromagnet alone, and R_1 and L_1 be the resistance and electromagnetic capacity of the remainder of the circuit. Then

$$\Gamma = \frac{E}{\sqrt{(R+R_1)^2 + m^2(L+L_1)^2}}. \qquad (4)$$

Suppose the diameter of the wire of the electromagnet to be variable. Let n be the number of turns in unit of length, or number of layers in unit of thickness. Then the magnetic force will vary as n^2, while both R and L vary as n^4. This makes the strength of the signals capable of a maximum, dependent on the variation of n; which by (4) will be when

$$R^2 + L^2m^2 = R_1^2 + L_1^2m^2. \qquad (5)$$

The left side refers to the electromagnet, the right to the remainder of the circuit. We may write (5) thus:—

$$\frac{R}{R_1} = \sqrt{\frac{1 + L_1^2m^2/R_1^2}{1 + L_2^2m^2/R_2^2}}.$$

Now L_1/R_1 is constant for the same line wire, whatever its length, since both L_1 and R_1 vary as the length of the line. Also L/R is constant for the same solenoidal coil, if only the diameter of the wire is variable, since L and R both vary as n^4. But the time-interval L/R for the electromagnet is in general much greater than the time-interval L_1/R_1 for the line wire, whence it follows that R must be much less than R_1 to produce the maximum magnetic force when the speed is considerably high; and the higher the speed, which is proportional to m, the smaller should the resistance of the electromagnet be.

The calculation of L_1/R_1 is easy, as the line wire is long, straight, and parallel to the earth; but the calculation of L/R is not so easy, owing to the variety of shapes assumed by electromagnets used for telegraphic purposes, with their cores, pole-pieces, armatures, etc., which all influence the electromagnetic capacity, though they do not influence the resistance. It is therefore impossible to enunciate a general law, that the resistance of an electromagnet should be such or such a fraction of the external resistance to obtain the maximum effect, for the result will be different not only for different speeds, but also for different constructions of the electromagnet.

9. Taking the case of a solenoidal electromagnet, approximate results are easily obtainable. Let its length be l, external radius x, internal radius y, with an iron core of radius z. Its electromagnetic capacity is

$$L=\int_y^x 4\pi n^2 l dr\left\{\int_r^x n^2dr'(\pi r^2+4\pi\kappa\pi z^2)+\int_y^r n^2dr'(\pi r'^2+4\pi\kappa\pi z^2)\right\}$$

or $$L=\tfrac{2}{3}\pi^2 ln^4(x-y)^2(x^2+2xy+3y^2+24\pi\kappa z^2), \qquad (6)$$

(Maxwell, vol. ii., p. 283), where κ is the coefficient of magnetisation. The resistance is

$$R=\pi\rho ln^4(x^2-y^2),$$

where ρ is the resistance of unit of length of wire, of unit diameter. Therefore

$$\frac{L}{R}=16\pi^2\frac{\kappa}{\rho}\frac{x-y}{x+y}z^2, \qquad (7)$$

omitting $x^2+2xy+3y^2$ in the expression for L, as small compared with $24\pi\kappa z^2$, which is a large number, unless the core is very small. Let $\kappa=32$; also, if the specific resistance of copper is taken at 1·7 microhms $=1700$ c.g.s., then $\rho=1700\times 4/\pi$, and

$$\frac{L}{R}=2{\cdot}33\frac{x-y}{x+y}z^2 \text{ seconds.} \qquad (8)$$

10. Maxwell (vol. ii., p. 289) gives the coefficient of self-induction of a straight wire, when the circuit is completed by a parallel straight wire. The same method of calculation is applicable to any number of straight parallel wires by finding the integral

$$T=\tfrac{1}{2}\iiint Hw\,dxdydz,$$

where T is the kinetic energy of the system, and H, w, are the vector-potential and the current at the point x, y, z. Thus, for parallel straight wires of length l, conveying currents C_1, C_2, C_3, ..., of radii a_1, a_2, ..., specific magnetic capacities μ_1, μ_2, ...; then, representing the distance between the centres of two wires m and n by b_{mn}, we shall have

$$\frac{2T}{l}=\tfrac{1}{2}(\mu_1C_1^2+\mu_2C_2^2+\dots)-2\mu_0(C_1^2\log a_1+C_2^2\log a_2+\dots)$$
$$-4\mu_0(C_1C_2\log b_{12}+C_1C_3\log b_{13}+C_2C_3\log b_{23}+\dots), \qquad (9)$$

with the sole condition

$$0=C_1+C_2+C_3+\dots.$$

From the last two equations the coefficients of self and mutual induction may be found. Let there be only four wires, forming two circuits, 1 and 3 for one circuit, 2 and 4 for the other; then $C_3 = -C_1$, and $C_4 = -C_2$. Substituting in (9),

$$\frac{2T}{l} = C_1^2\left(\frac{\mu_1+\mu_3}{2} + 2\mu_0 \log \frac{b^2_{13}}{a_1 a_3}\right) + C_2^2\left(\frac{\mu_2+\mu_4}{2} + 2\mu_0 \log \frac{b^2_{24}}{a_2 a_4}\right)$$
$$+ 2C_1 C_2 \times 2\mu_0 \log \frac{b_{14} b_{23}}{b_{13} b_{24}}. \quad \ldots\ldots\ldots (10)$$

The coefficient of C_1^2 in (10) is the coefficient of self-induction per unit of length of the circuit conveying the current C_1. Similarly for C_2; and the coefficient of $2C_1C_2$ is the coefficient of mutual induction of the two circuits per unit of length.

From (10) we may find the coefficients of self and mutual induction of two suspended wires, the circuits being completed through the earth. Let M be the coefficient of mutual, and L_1, L_2 the coefficients of self-induction of two wires of length l, radii a_1, a_2, heights above ground h_1, h_2, horizontal distance apart d, and specific magnetic capacities μ_1, μ_2; then *

$$\left.\begin{aligned} \frac{L_1}{l} &= \frac{\mu_1}{2} + 2\log\frac{2h_1}{a_1}, \\ \frac{L_2}{l} &= \frac{\mu_2}{2} + 2\log\frac{2h_2}{a_2}, \\ \frac{M}{l} &= \log\frac{d^2+(h_1+h_2)^2}{d^2+(h_1-h_2)^2}, \end{aligned}\right\} \quad \ldots\ldots\ldots\ldots\ldots\ldots (11)$$

where μ_0 has been put $= 1$.

As a special case, let

$$h_1 = h_2 = 3 \text{ metres}, \; d = \cdot 5 \text{ metre},$$
$$a_1 = a_2 = \cdot 002 \text{ metre},$$
$$\mu_1 = \mu_2 = 1 + 4\pi\kappa = 315, \text{ if } \kappa = 25,$$

then $$L_1 = L_2 = 173, \text{ and } M = 5, \text{ approximately}.$$

Also, if the resistance per mile is 13 ohms, the resistance per centimetre is 80778 c.g.s., therefore

$$\frac{L_1}{R_1} = \frac{173}{80778} = \cdot 00214 \text{ seconds}. \quad \ldots\ldots\ldots\ldots\ldots (12)$$

11. This time-interval being in general very small compared with L/R for the electromagnet, we may neglect it, and then

$$\frac{R}{R_1} = \frac{R}{Lm} \text{ approximately}. \quad \ldots\ldots\ldots\ldots\ldots (13)$$

The resistance of the electromagnet is thus inversely as the speed to

* [These are derived by the method of images. The return currents are assumed to spread over a thin conducting sheet on the earth's surface. The calculated inductances in (11) are therefore minimum values, by reason of the ignoration of the magnetic force in the earth. But as regards the first terms, depending upon the inductivity of the wires, they are maximum values, implying full penetration of the current into the wires, a matter considered in later papers.]

obtain the maximum strength of signals, except for low speeds. Inserting in (13) the value of L/R given in (8),

$$\frac{R}{R_1}=\frac{T}{14\cdot 64\,\frac{x-y}{x+y}z^2},$$

where $T=2\pi/m$.

At 100 words per minute, Morse code, $T=$ about $\frac{1}{40}$ second, therefore at this speed

$$\frac{R_1}{R}=585\cdot 6\,\frac{x-y}{x+y}z^2.$$

Suppose $x=2$, $y=z=1$ centimetre, then

$$R_1/R=195\cdot 2,$$

or the resistance of the electromagnet is $\frac{1}{195}$th of the external resistance to obtain the maximum magnetizing force. This is increased to $\frac{1}{174}$th when the self-induction of the suspended wire is taken into account.*

12. Having made the magnetizing force a maximum for a given speed and dimensions of electromagnet, we may next find the ratio between the outer and inner radius of the coil to make the attractive force on a soft iron armature a maximum. We have

$$\Gamma=\frac{E}{\sqrt{(R+R_1)^2+L^2m^2}},$$

neglecting L_1; where

$$L=\tfrac{2}{3}\pi^2 ln^4(x-y)^2(x^2+2xy+3y^2+24\pi\kappa z^2),$$

and

$$R=\pi\rho ln^4(x^2-y^2);$$

also

$$F=\Gamma G,$$

where F is the magnetizing force, and

$$G=4\pi n^2(x-y).$$

To make F a maximum we found

$$R^2+L^2m^2=R_1^2,$$

therefore

$$F=\frac{EG}{R_1\sqrt{2(1+R/R_1)}}.$$

Substituting

$$4\sqrt{\frac{\pi(x-y)R}{\rho l(x+y)}}$$

for G, we have

$$F^2=\frac{\dfrac{8\pi(x-y)E^2}{\rho l(x+y)}}{R_1(1+R_1/R)}.$$

* [These very low estimates arise from the largeness of the time-constant of the type of electromagnet considered. Short telegraphic electromagnets have far smaller time-constants. It may also be mentioned that the assumed constancy of the inductance of an electromagnet with core under reversals implies that its core is non-conducting or is properly divided. The nature and effects of the currents induced in conducting cores is considered later on.]

Now $R_1/R = Lm/R$ approximately,

$$= \frac{2m\pi}{3\rho}\frac{x-y}{x+y}(x^2+2xy+3y^2+24\pi\kappa z^2);$$

therefore $$F^2 = \frac{12E^2/R_1}{lm(x^2+2xy+3y^2+24\pi\kappa z^2)+\dfrac{3\rho l}{2\pi}\dfrac{x+y}{x-y}}.$$

Now the magnetization of the core is proportional to the magnetizing force, and the attractive force between the core and a soft iron armature placed close to it is proportional to the square of the magnetization and to the cross section of the core. Therefore, if A is the attractive force,

$$A \propto \frac{z^2}{lm(x^2+2xy+3y^2+24\pi\kappa z^2)+\dfrac{3\rho l}{2\pi}\dfrac{x+y}{x-y}}.$$

This increases with z, so let $z=y$, the inner radius of the coil. Let x be constant and y variable, then A is a maximum when

$$\frac{x^2}{y^2}+\frac{2x}{y}+\frac{3\rho}{2\pi y^2 m}\frac{x+y}{x-y}$$

is a minimum; *i.e.*, when

$$\frac{2\pi m}{3\rho}x^2 = \frac{y/x-(1-y^2/x^2)}{(1-y/x)(1-y^2/x^2)}.$$

The least value of y/x is

$$\frac{y}{x} = \frac{\sqrt{5}-1}{2} = \cdot 618.$$

Using the former value of ρ, viz., $1700 \times 4/\pi$; also $m=80\pi$, and $x=2$ centimetres,

$$y/x = \cdot 7 \text{ nearly}.$$

y/x increases very slowly as x and m increase.

If y be constant and x variable, smaller values of y/x are obtained.

The attractive force also varies inversely as the length of the coil; that is to say, if the length of the coil is halved, preserving its other dimensions, as well as its resistance constant, the attractive force is doubled. Although this result is only true for long coils, it points in the direction of short coils being the best, especially as the attractive force is increased by increasing the transverse dimensions for any fixed ratio of y/x. We have, however, neglected the increase in the self-induction due to the armature.

13. In the determination of the resistance of an electromagnet in paragraph 11 and before, only one electromagnet is considered to be in the circuit. The results are inapplicable when there is another in circuit, used for sending the signals for example. Let two similar electromagnets, each of resistance R and electromagnetic capacity L, be used telephonically, on Bell's principle. Let $E \sin mt$ be the electro-

motive force induced in the sending electromagnet, then the maximum strength of the currents in the circuit is, by equation (3),

$$\Gamma = \frac{E}{\sqrt{(2R+R_1)^2 + 4L^2m^2}},$$

where R_1 is the line-resistance. Let F be the magnetizing force in the receiving coil, then

$$F = \frac{EG}{\sqrt{(2R+R_1)^2 + 4L^2m^2}},$$

where G is the magnetizing force of the unit current. Now, suppose the thickness of the wire in both coils variable together, then E and G both vary as the number of turns, *i.e.*, as n^2, while R and L both vary as n^4. Let $G=n^2g$, $E=n^2e$, $R=n^4r$, $L=n^4l$, then

$$F = \frac{eg}{\sqrt{\left(2r+\frac{R_1}{n^4}\right)^2 + 4l^2m^2}},$$

whence it may be seen that F increases with n, solely by the reduction of the term R_1/n^4; but the increase is very slow after passing certain limits, as will be seen from the following example. Let $R_1=1000$ ohms, $L/R=\frac{1}{10}$, $m=1200$. Then for the following values of R, viz., $\frac{10}{16}$, 10, 160, ∞, ohms, we have the following proportional value of F, $\frac{1}{16200}$, $\frac{1}{2600}$, $\frac{1}{2401}$, $\frac{1}{2400}$. Thus under 10 ohms the increase is rapid; after that, next to nothing.

14. Two suspended circuits, A and B, have resistances R_1 and R_2, electromagnetic capacities L_1 and L_2, mutual capacity M. If reversals are made by an electromotive force $E \sin mt$ in the primary circuit A, and there is no electromotive force in B except that induced by changes of the current in A, then

$$E \sin mt = \left(R_1 + L_1\frac{d}{dt}\right)\gamma_1 + M\frac{d\gamma_2}{dt},$$

$$0 = \left(R_2 + L_2\frac{d}{dt}\right)\gamma_2 + M\frac{d\gamma_1}{dt};$$

where γ_1 and γ_2 are the currents in A and B. From these we shall find, if Γ_1 and Γ_2 are the amplitudes of the currents in A and B,

$$\Gamma_1 = \frac{E\sqrt{R_2^2 + L_2^2m^2}}{\sqrt{\{R_1R_2 - m^2(L_1L_2 - M^2)\}^2 + m^2(R_1L_2 + R_2L_1)^2}}$$

and
$$\frac{\Gamma_2}{\Gamma_1} = \frac{Mm}{\sqrt{R_2^2 + L_2^2m^2}}.$$

This ratio cannot be greater than M/L_2. If the wires are as in the special case of paragraph 10, $M=5$, and $L_2=173$; therefore $M/L_2=\frac{1}{34}$, expresses the greatest value of the ratio of the induced to the inducing currents.

Let each of the lines A and B be 10 miles in length, of resistance 130 ohms, and let the secondary circuit have an electromagnet at each end of resistance 65 ohms. If $L_2 = L'_2 + L''_2$, where L'_2 belongs to the electromagnets, and L''_2 to the line, then

$$L''_2 = 173 \times 160{,}934 \times 10 = 278{,}415{,}820,$$

since there are 160,934 centimetres in a mile. And

$$M = 5 \times 160{,}934 \times 10 = 8{,}046{,}700.$$

Therefore $$\frac{\Gamma_2}{\Gamma_1} = \frac{M}{L'_2 + L''_2} = \frac{8{,}046{,}700}{L'_2 + 278{,}415{,}820}.$$

If, further, the time-constant for each coil $= \frac{1}{10}$ second

$$L'_2 = \tfrac{1}{10} \times 65 \times 10^9 \times 2 = 130 \times 10^8,$$

and $$\frac{\Gamma_2}{\Gamma_1} = \frac{8}{13278} = \frac{1}{1659}.$$

If the time-constant is as small as $\frac{1}{100}$ second,* then

$$L'_2 = 130 \times 10^7,$$

and $$\frac{\Gamma_2}{\Gamma_1} = \frac{8}{1578} = \frac{1}{197}.$$

Increasing the length of the lines, or of the portions in proximity, increases the ratio of the induced to the inducing currents; and decreasing the electromagnetic capacity of the instruments in the secondary circuits does the same. Sudden changes in the primary current of course cause greater induced currents.

15. If a condenser be discharged through more than one circuit simultaneously, in what manner does magnetic induction affect the division of the charge? Suppose we have a condenser of capacity c, charged to potential E with a charge $Q = Ec$, and that the condenser is discharged through any number of circuits in parallel arc, of resistances R_1, R_2, ..., coefficients of self-induction L_1, L_2, ..., and of mutual induction M_{12}, M_{23}, M_{13}, ... If v is the potential of the condenser at time t after the commencement of the discharge, we have a system of equations equal to the number of the circuits, viz.:—

$$v = R_1\gamma_1 + \frac{d}{dt}(L_1\gamma_1 + M_{12}\gamma_2 + M_{13}\gamma_3 + M_{14}\gamma_4 + \ldots),$$

$$v = R_2\gamma_2 + \frac{d}{dt}(L_2\gamma_2 + M_{12}\gamma_1 + M_{23}\gamma_3 + M_{24}\gamma_4 + \ldots),$$

$$v = R_3\gamma_3 + \frac{d}{dt}(L_3\gamma_3 + M_{13}\gamma_1 + M_{23}\gamma_2 + M_{34}\gamma_4 + \ldots),$$

etc., where γ_1, γ_2, γ_3, ... are the currents at time t in R_1, R_2, R_3, ...

Integrating both sides of all the equations with respect to t between

* [The time-constant of most instruments is a good deal smaller, so that the ratio of induced to inducing current is greater. Also, the inductance of the iron wire in the example will be less than stated (on account of imperfect penetration), still further increasing the ratio.]

the limits $t=0$ and $t=\infty$, we have

$$\int_0^\infty vdt = R_1Q_1 = R_2Q_2 = R_3Q_3 = \ldots,$$

where $$Q_1 = \int_0^\infty \gamma_1 dt, \quad Q_2 = \int_0^\infty \gamma_2 dt, \ldots,$$

since the currents are zero both for $t=0$ and $t=\infty$. It follows that Q_1, Q_2, ..., which are the integral currents through R_1, R_2, ..., are inversely proportional to the resistance, or that the total charge Q divides between the circuits in the inverse proportion of their resistance. This only applies to the whole current, for at any particular moment the currents in the different circuits do not bear the same proportion. In fact, the current in one circuit may be from, and in another to, the condenser at a certain time. The current in any circuit at any time may be calculated by finding the roots (all negative, or imaginary with real parts negative), of an algebraical equation of the $(n+1)$th degree, n being the number of circuits. The equation needed to be added to the above system of n equations is

$$-c\frac{dv}{dt} = \gamma_1 + \gamma_2 + \gamma_3 + \ldots,$$

or the current leaving the condenser = sum of currents in the wires.

In a similar manner it may be shown that if, instead of the charge of a condenser, the extra current of a coil be discharged through any number of parallel circuits, the total quantity passing through any circuit will be inversely proportional to its resistance.

16. As a special case,* suppose the charge Q of a condenser of capacity c is discharged through a single coil of resistance R and electromagnetic capacity L. Then, v being the potential of the condenser and γ the current in R at time t,

$$v = R\gamma + L\frac{d\gamma}{dt},$$

$$-c\frac{dv}{dt} = \gamma;$$

whence $$0 = v + Rc\frac{dv}{dt} + Lc\frac{d^2v}{dt^2},$$

$$0 = \gamma + Rc\frac{d\gamma}{dt} + Lc\frac{d^2\gamma}{dt^2};$$

therefore $$\gamma = A\epsilon^{at} + B\epsilon^{bt},$$

where A and B are constants, and a and b are the roots of

$$Lcx^2 + Rcx + 1 = 0,$$

or $$x = -\frac{R}{2L} + \sqrt{\frac{R}{4L^2} - \frac{1}{Lc}};$$

therefore $$\gamma = \epsilon^{-\frac{Rt}{2L}}(A\sin + B\cos)t\sqrt{\frac{1}{Lc} - \frac{R^2}{4L^2}}.$$

* [See footnote on p. 61.]

Now when $t=0$, $\gamma=0$, therefore $B=0$; and when $t=0$, $E=L\frac{d\gamma}{dt}$, E being the initial potential of the condenser; therefore

$$A=\frac{E}{\sqrt{\frac{L}{c}-\frac{R^2}{4}}}$$

and
$$\gamma=\frac{E}{\sqrt{\frac{L}{c}-\frac{R^2}{4}}}\epsilon^{-\frac{Rt}{2L}}\sin t\sqrt{\frac{1}{Lc}-\frac{R^2}{4L^2}}.$$

Let $L/R=\alpha$, and $Rc=\beta$, both time-intervals, then

$$\gamma=\frac{E}{R}\frac{\epsilon^{-t/2\alpha}}{\sqrt{\frac{\alpha}{\beta}-\frac{1}{4}}}\sin\frac{t}{2\alpha}\sqrt{\frac{4\alpha}{\beta}-1};$$

which may be put in the exponential form if $4\alpha/\beta<1$. In the latter case the discharge is continuously in one direction, but if $4\alpha/\beta>1$, the discharge is oscillatory.

Let the condenser have a capacity of 1 microfarad $=10^{-15}$ c.g.s. and $L/R=10^{-2}$, then

$$\frac{4\alpha}{\beta}=\frac{4L}{R^2c}=\frac{4\times 10^{-2}}{r\times 10^{-6}}=\frac{4\times 10^4}{r},$$

if r is the number of ohms in R, since the ohm $=10^9$ c.g.s. Thus the discharge is oscillatory if R is less than 40,000 ohms, and continuous if it is greater than that amonnt.

Suppose $R=100$ ohms, then $4\alpha/\beta=400$, and

$$\gamma=\frac{E}{R}\cdot\frac{2\epsilon^{-50t}}{\sqrt{399}}\sin 50t\sqrt{399}$$
$$=\frac{E}{10R}\epsilon^{-50t}\sin 1000t,\text{ approximately.}$$

The period of an oscillation is $2\pi/1000$ second $=\cdot006$ sec. The quantity in the first current is $Q\,(1+\epsilon^{-\pi/20})$, a little less than twice the original charge. In the next current (in the reverse direction) it is a little less again. The total current, irrespective of its direction, is

$$\frac{Q}{1-\epsilon^{-\frac{\pi}{\sqrt{4\alpha/\beta-1}}}}=\frac{Q}{1-\epsilon^{-\frac{\pi}{20}}}\text{ nearly.}$$

The discharge is practically over in $\frac{1}{25}$ second.

If the coil is shunted by a coil of resistance S, and no self-induction, other things being the same as before, the current γ at time t in the first coil will be found to be

$$\gamma=\frac{2E\alpha'}{L\sqrt{\frac{4\alpha'}{\beta'}-1}}\epsilon^{-\frac{t}{2\alpha'}}\sin\frac{t}{2\alpha}\sqrt{\frac{4\alpha'}{\beta'}-1},$$

where $$\alpha' = \frac{L}{R + L/Sc}, \qquad \beta' = \frac{cR + L/S}{1 + R/S}.$$

Although the total currents through the coil and its shunt are in the inverse proportion of their resistances, yet the same is not true of the heat produced in the wires. The energy converted into heat in the coil R is

$$\int_0^\infty R\gamma^2 dt = \tfrac{1}{2}cE^2 \cdot \frac{S}{R+S} \cdot \frac{RSc}{RSc + L}.$$

Here $\frac{1}{2}cE^2$ is the electrostatic energy of the original charge, and $S/(R+S)$ the shunt factor. Since the remaining factor is less than unity, it follows that the amount of heat produced in the wire R is always less than in the inverse proportion of the resistances.

17. Let us next examine the influence of a fault on rapid reversals on a land-line. Let R be the resistance from one end of the circuit up to the fault, L its electromagnetic capacity; let R' and L' be similar quantities for the other section of the circuit, and S the resistance of the fault itself. Let γ, γ', and γ'' be the currents in R, R', and S; and let $E \sin mt$ be the electromotive force in R, and v the potential of the wire at the fault. Then

$$\left.\begin{aligned} E \sin mt - v &= \left(R + L\frac{d}{dt}\right)\gamma, \\ v &= \left(R' + L'\frac{d}{dt}\right)\gamma', \\ v &= S\gamma'', \\ \gamma' + \gamma'' &= \gamma, \end{aligned}\right\}$$

by the conditions of the problem; from which, for the current in R', we have

$$E \sin mt = \left(R + R' + \frac{RR'}{S}\right)\gamma' + \left(L + L' + \frac{LR' + L'R}{S}\right)\frac{d\gamma'}{dt} + \frac{LL'}{S}\frac{d^2\gamma'}{dt^2};$$

whence the amplitude Γ' of the waves in R' is

$$\Gamma' = \frac{E}{\sqrt{\left(R + R' + \dfrac{RR' - LL'm^2}{S}\right)^2 + m^2\left(L + L' + \dfrac{LR' + L'R}{S}\right)^2}}.$$

To find the effect of the fault, we may compare this expression with its value when $S = \infty$, or no fault.

1st Case.—Electromagnet at one end only. $R = R'$, or the fault in the middle of the circuit, $L = 0$. Then

$$\Gamma' = \frac{E}{\sqrt{\left(2R + \dfrac{R^2}{S}\right)^2 + m^2L'^2\left(1 + \dfrac{R}{S}\right)^2}}.$$

Let $S = \frac{1}{2}R$. This would reduce the strength of the current received in R' from a constant electromotive force in R to one-half. In the above expression, however, the change of S from ∞ to $\frac{1}{2}R$ doubles $(2R + R^2/S)$ and trebles mL' $(1 + R/S)$; so that the fault weakens the strength of rapid reversals much more than it weakens a permanent current.

2nd Case.—Electromagnet at each end. $R=R'$, $L=L'$, and $Lm/R=n$.

$$\Gamma'=\frac{E}{\sqrt{1+n^2}\sqrt{4\left(1+\frac{R}{S}\right)+\frac{R^2}{S^2}\left(1+n^2\right)}}.$$

When there is no fault, or $S=\infty$,

$$\Gamma'=\frac{E}{2R\sqrt{1+n^2}},$$

and when $S=\frac{1}{2}R$,

$$\Gamma'=\frac{E}{2R\sqrt{1+n^2}\sqrt{4+n^2}}.$$

Thus the fault reduces the current received in R' to less than an $1/n$th part, and for high speeds n may be a large number, whereas a permanent current is only reduced to one-half. This applies to a telephonic circuit with a fault in the middle, and the result shows that leakage has a most prejudicial effect. We may also conclude that circuits worked by magneto-electric transmitters are more affected by leakage than when worked in a similar manner from a battery.*

18. Suppose the receiving instrument has resistance R and electromagnetic capacity L, shunted by a coil of resistance S and capacity L'. Let the line resistance be A. First let there be a constant E.M.F. in A. The shunt reduces the strength of the final current in R

$$\text{from } \frac{E}{A+R} \text{ to } \frac{E}{A+\frac{RS}{R+S}}\times\frac{S}{R+S}=C_1, \text{ say.}$$

At the same time the shunt alters the manner in which the current rises in the electromagnet. If the shunt has no self-induction, or $L'=0$, the current γ in R rises according to the equation

$$\gamma=C_1(1-\epsilon^{-\frac{t}{a}}),$$

where
$$a=\frac{L}{R+\frac{AS}{A+S}}.$$

The time the current takes to reach any stated fraction of its final strength is proportional to a. This time-interval is increased by the shunt of no capacity

$$\text{from } \frac{L}{R+A} \text{ to } \frac{L}{R+\frac{AS}{A+S}}.$$

* [We are not here concerned with a line where electrostatic charge is important. See Art. xiii., p. 53; Art. xv., p. 61; and Art. xvi., p. 71.]

While the current is rising in R, it is falling in S from the strength

$$\frac{E}{A+S},$$

which is almost instantaneously reached, to

$$\frac{E}{A+\dfrac{RS}{R+S}} \times \frac{R}{R+S},$$

its final strength. When the electromotive force is removed, the end of the line being put to earth, the currents in R and S fall to zero in a similar manner, *i.e.*, in R the current is continuously in the same direction as at first; whereas in the shunt it is immediately reversed. The integral extra current in R is

$$C_1 a = \frac{ESL(A+S)}{(AR+RS+AS)^2}.$$

This is greatest, as depending on the resistance of the shunt, when

$$R = \frac{AS}{A+S},$$

or when the resistance of the electromagnet = resistance external to it. Since, when the current is falling in the electromagnet it draws electricity through the line and the shunt, the potential of the line is negative,

$$= -x/l \times \text{current in line},$$

at any point distant x from the beginning of the line of length l. Similarly for the shunt.

Now let the shunt have electromagnetic capacity L'. The differential equation for the current γ in R is

$$ES = \gamma(RA+SA+RS) + \frac{d\gamma}{dt}\Big\{L(A+S) + L'(A+R)\Big\} + \frac{d^2\gamma}{dt^2}LL'.$$

For simplicity let $A=R=S$. Then, when the shunt is not on, the extra current in R is

$$\gamma = \frac{E}{2A}\epsilon^{-\frac{t}{a_1}}, \quad \text{where} \quad a_1 = \frac{L}{2A}.$$

With the shunt on, of no self-induction,

$$\gamma = \frac{E}{3A}\epsilon^{-\frac{t}{a_2}}, \quad \text{where} \quad a_2 = \frac{2L}{3A}.$$

Thus a is increased in the ratio $\frac{1}{2}$ to $\frac{2}{3}$, or $3:4$. Now let $L' = \frac{1}{2}L$, then

$$\gamma = \frac{E}{3A}\epsilon^{-\frac{3At}{L}}\left(\tfrac{1}{2}\epsilon^{\frac{\sqrt{3}\,At}{L}} + \tfrac{1}{2}\epsilon^{-\frac{\sqrt{3}\,At}{L}}\right);$$

and, since the bracketted expression is > 1, $a > L/3A$.

When $L' = L$,

$$\gamma = \frac{E}{3A}\epsilon^{-\frac{t}{a_3}}, \quad \text{where} \quad a_3 = \frac{1}{3}\frac{L}{A},$$

and the current rises in the line in the same manner as it would if for the electromagnet and shunt were substituted a single coil of resistance $\frac{1}{2}A$ and magnetic capacity $\frac{1}{2}L$. At the same time the shunt reduces the retardation from $\alpha_2 = 2L/3A$ to $\alpha_3 = L/3A$, or as $2:1$, as L' increases from 0 to L. But since $\alpha = L/2A$ when there is no shunt at all, the shunt of equal capacity to the receiver's only reduces the retardation in the ratio $3:2$.

$L' = 2L$. Here the extra current in the receiver is

$$\gamma = \frac{E}{3A} \epsilon^{-\frac{3At}{2L}} \left\{ \frac{\sqrt{3}+1}{2} \epsilon^{-\frac{\sqrt{3}\,At}{2L}} - \frac{\sqrt{3}-1}{2} \epsilon^{\frac{\sqrt{3}\,At}{2L}} \right\}.$$

This becomes zero when

$$t = \frac{L}{\sqrt{3}A} \log(2+\sqrt{3}),$$

and a minimum negative when

$$t = \frac{L}{\sqrt{3}A} \log(7+4\sqrt{3}),$$

Thus when the shunt has a greater capacity than the receiver, when the current is put on the current in the receiver first rises above, and then falls to its final strength. When the battery is removed and earth put on at the sending end, the current in the receiver falls through zero, becomes reversed, and then rises to zero again. But we cannot exalt the current from an electromagnetic shunt so as to send back a current *to the line* immediately after each signal, as has been stated. When, as above, the extra current in the receiver becomes reversed in direction, this reverse current does not go to line, but goes round by the shunt.

Joining the two coils of a relay in parallel arc has the effect of quartering the resistance and quartering the capacity of the relay considered as a whole; or rather, it would be so if the coils were at a distance apart instead of being close together with the cores connected by an armature, which lessens the reduction in the retardation. But if, instead of joining the coils in parallel to reduce the retardation, we wind the coils with thicker wire, we get much more advantageous results.

19. With the same notation, let us examine the influence of the shunt on rapid reversals. Suppose the E.M.F. in A is $E \sin mt$, then the amplitude of the currents in the electromagnet R is

$$\Gamma = \frac{E\sqrt{(L'm)^2 + S^2}}{\sqrt{\{(AR+RS+AS) - m^2LL'\}^2 + m^2\{A(L+L') + RL' + SL\}^2}}$$

Let $R = S = A$, then

$$\Gamma = \frac{E\sqrt{A^2 + L'^2m^2}}{\sqrt{(3A^2 - m^2LL')^2 + 4m^2A^2(L+L')^2}}.$$

First, with no shunt at all, or $S = \infty$, and $R = A$,

$$\Gamma = \frac{E}{A\sqrt{n^2+4}}$$

where $n = Lm/A$. Now put on shunt without capacity, $S = A$, $L = 0$, and

$$\Gamma = \frac{E}{2A\sqrt{n^2 + 9/4}}.$$

The current is thus reduced about a half. Next let the shunt be a similar coil to the receiver, then $R = S = A$, $L' = L$, and

$$\Gamma = \frac{E}{A\sqrt{n^2 + 9}},$$

nearly the same as without any shunt. Thus the strength of the currents in the receiver is scarcely affected by putting on a similar electromagnet as a shunt, while the initial retardation is reduced in the ratio 3 : 2, as we found in the last paragraph. Further increase of L' has little influence on the magnitude of the signals.*

XVIII.—MAGNETO-ELECTRIC CURRENT GENERATORS.

[*Jour. Soc. Tel. Eng.*, June 1881, vol. 10, p. 271.]

PERHAPS the simplest specimen of a magneto-electric current generator is a coil rotating with uniform velocity in a uniform field of magnetic force. To get the greatest effect, the axis of rotation must be at right angles to the lines of force. The variation of the amount of induction through the coil induces a simple harmonic E.M.F. in it, and the result, when the initial effect has subsided, is a simple harmonic current. But the phase of the current is behind that of the E.M.F., owing to the self-induction of the circuit, which also diminishes the amplitude of the current waves.

Otherwise, we may consider the current at any moment to be that due to the actual E.M.F. round the circuit at that moment, according to Ohm's law, remembering that the actual E.M.F. is the algebraical sum of the E.M.F. due to the motion, and that due to the variation of the current itself.

Symbolically, let M be the induction through the coil when its plane is at right angles to the lines of force of the external field, ωt the angle turned through from this plane at time t, the angular velocity of rotation being ω; then $M\omega \sin \omega t$ is the impressed E.M.F. in the coil. And if R is its resistance, R_1 the external resistance, L the coefficient of self-induction of the coil, L_1 the external ditto, the equation of the current is

$$M\omega \sin \omega t = (R + R_1)\gamma + (L + L_1)\dot{\gamma};$$

* [The above paper incorporates one "On the Resistance of Electromagnets in Telegraphy," *Phil. Mag.*, Sept. 1878, S. 5, v. 56, which is therefore not reprinted here.]

consequently [if D stand for d/dt],

$$\gamma = \frac{M\omega \sin \omega t}{(R+R_1)+(L+L_1)D} = \frac{R+R_1-(L+L_1)D}{(R+R_1)^2+(L+L_1)^2\omega^2} M\omega \sin \omega t$$

$$= \frac{M\omega \sin(\omega t - \theta)}{\sqrt{(R+R_1)^2+(L+L_1)^2\omega^2}},$$

where $\theta = \tan^{-1}\frac{L+L_1}{R+R_1}\omega.$

The amplitude of the current is therefore

$$\frac{M\omega}{\{(R+R_1)^2+(L+L_1)^2\omega^2\}^{\frac{1}{2}}},$$

and the angular displacement of the zero is θ.

The above is applicable, or nearly so, to any magnetic machine with a single coil: for example, a Siemens' armature revolving between the poles of powerful magnets, the effect of the iron armature being in the main simply to increase M and L.

By reversing the connections of the coil with the external circuit at the moments of zero current—that is, something between 0 and $\frac{1}{4}$ revolution after zero E.M.F., according to the speed, etc.—the external current is put into one direction; but since the current varies greatly in strength, the effective E.M.F. of the machine must be taken as the product of the mean external current into the resistance of the circuit, supposing, of course, that there are no other E.M.F.'s acting in the circuit than already considered.

If Γ is the mean external current,

$$\Gamma = \frac{2}{\pi} \frac{M\omega}{\{(R+R_1)^2+(L+L_1)^2\omega^2\}^{\frac{1}{2}}},$$

and the effective E.M.F. of the machine is

$$\Gamma(R+R_1).$$

The mean current increases directly as the speed at first, but afterwards more slowly, and its limiting strength is

$$\frac{2}{\pi} \cdot \frac{M}{L+L_1};$$

that is, the ratio of the mean amount of external induction through the coil to the self-induction of the circuit per unit current.

The theory of multiple coils is quite similar, and resembles that of galvanic cells joined up in series or in multiple arc. If any number n of similar coils rotate simultaneously in the same magnetic field, and are equally acted upon by varying induced E.M.F.'s, all in the same phase, the coils may obviously be joined up all for "quantity" or "intensity" without any interference, a simple harmonic current resulting equivalent to that from a single coil with constants nM, nR, and nL in the intensity case, and M, R/n, and L/n in the other. And reversing the coils all at the same moment, the external current is put in one direction, as before, though of very varying strength.

With only one coil, we cannot get rid of this great variation, but using many coils we may reduce the variations as much as we please, multiplying their frequency at the same time, by making the phases of the induced E.M.F.'s differ by equal amounts and reversing every coil at the moment of zero current for that coil, supposing the others not to act. And, curiously enough, the resultant mean external current is not affected by the changes.

Thus, to fix ideas, suppose we have any number of coils arranged at equiangular intervals round a circle, and revolving together in a uniform field of force. It is unnecessary to specify any particular form of machine. The two coils at opposite ends of any diameter have exactly equal E.M.F.'s acting on them at any moment, so they may be joined together permanently, and treated as a unit in the arrangement. Thus we have, say, n pairs of coils, which have all equal simple harmonic E.M.F. acting on them, but at different times.*

Joining them up in series by means of an n-fold commutator, which reverses the coils one after another in proper order, let M, R, and L be the constants used before, but now referring to a pair of coils, and suppose No. 1 pair to act alone. We have

$$M\omega \sin \omega t = (nR + R_1)\gamma + (nL + L_1)\dot{\gamma};$$

therefore
$$\gamma = \frac{M\omega \sin(\omega t - \theta)}{\{(nR + R_1)^2 + (nL + L_1)^2\omega^2\}^{\frac{1}{2}}},$$

where
$$\theta = \tan^{-1}\frac{nL + L_1}{nR + R_1}\omega.$$

Here γ is the current in No. 1 pair, due to its own motion, but as it is reversed at the moments of zero current for itself the mean external current due to pair No. 1 is

$$\frac{2}{\pi}\frac{M\omega}{\{(nR + R_1)^2 + (nL + L_1)^2\omega^2\}^{\frac{1}{2}}}.$$

Now, letting all the coils work, and superimposing the currents, we have a mean external current Γ, where

$$\Gamma = \frac{2}{\pi}\frac{nM\omega}{\{(nR + R_1)^2 + (nL + L_1)^2\omega^2\}^{\frac{1}{2}}};$$

the same as from a single coil with constants nM, nR, and nL, but with the difference of having many small variations from its mean strength in place of few large ones.

It is to be remarked here that although every pair is successively reversed at the moments when its own current is zero, yet, since it has to carry the currents from the remaining $n - 1$ pairs, this current is necessarily reversed suddenly: so that whilst externally we have a nearly steady current, and also in the coils in series, considered as a whole, yet a regular succession of abrupt reversals of current is going on all through the series.

* [The coils must be well separated, as their mutual induction is assumed to be nil.]

So to minimise the unavoidable sparking we should theoretically subdivide the coils as much as possible, thus making the inertia of the currents to be reversed as small as possible; or make the reversal a continuous operation, instead of intermittent.

Joining the circle of coils in one continuous series, and putting the external circuit on between opposite ends of the neutral diameter, gives an external current equivalent to that from a single coil, with constants $\frac{1}{2}nM$, $\frac{1}{4}nR$, $\frac{1}{4}nL$.

Now, arranging the commutator to connect the n pairs of coils for quantity, we have, considering the first pair alone to act,

$$M\omega \sin \omega t = R\gamma + L\dot{\gamma} + R_1\gamma_1 + L_1\dot{\gamma}_1;$$

and also $$M\omega \sin \omega t = R\gamma + L\dot{\gamma} - \frac{1}{n-1}(R\gamma_2 + L\dot{\gamma}_2);$$

where M, L, R, L_1, R_1 are as before, but now γ is the current in No. 1 pair, γ_1 the external current due to it, and γ_2 the total current in the remaining $n-1$ pairs due to No. 1.

Also, by continuity, $\gamma + \gamma_2 = \gamma_1$. Therefore

$$\gamma_1 = \frac{M\omega \sin \omega t}{(R + nR_1) + (L + nL_1)D};$$

$$\gamma_2 = -(n-1)\frac{R_1 + L_1 D}{R + LD}\gamma_1; \qquad \gamma = \gamma_1 - \gamma_2.$$

For the external current we have

$$\gamma_1 = \frac{M\omega \sin(\omega t - \theta)}{\{(R + nR_1)^2 + (L + nL_1)^2\omega^2\}^{\frac{1}{2}}},$$

where $$\theta = \tan^{-1}\frac{L + nL_1}{R + nR_1}\omega.$$

This is due to pair No. 1, supposing it not reversed. But reversing it at the proper times, we have a mean external current $= 2/\pi \times$ amplitude of γ_1; and consequently n times as much when all the coils act, or

$$\Gamma = \frac{2}{\pi}\frac{M\omega}{\left\{\left(\frac{R}{n} + R_1\right)^2 + \left(\frac{L}{n} + L_1\right)^2\omega^2\right\}^{\frac{1}{2}}},$$

the same as from a single coil with constants M, R/n, and L/n, with of course the difference of being nearly steady.

For simplicity of expression we may replace a multiple coil machine by its equivalent single coil machine, say, with constants M, L, R, and with external current:—

$$\Gamma = \frac{2}{\pi}\frac{M\omega}{\{(R + R_1)^2 + (L + L_1)^2\omega^2\}^{\frac{1}{2}}}.$$

Varying the size of the wire, since M varies as the number of turns, whilst R and L vary as its square, the following expression follows as the condition for maximum current with a given speed of rotation:—

$$R^2 + L^2\omega^2 = R_1^2 + L_1^2\omega^2.$$

Neglecting the self-induction of the external wire, the resistance of the machine should be less than the external resistance, and more so with higher speeds.

Professors Ayrton and Perry advocate extremely high speeds. Only make the speed high enough, and R becomes relatively small. At the same time a counterbalancing factor will come into play, namely L_1. Also at excessively high speeds the electrostatic capacity of the line and other things would need consideration, so that on the whole it is perhaps premature to say what the resistance of excessively high speed machines should be. But as the experiments mentioned in Mr. A. Siemens' paper show that the current is nearly proportional to the speed, it is possible that there is still left a wide margin for further increase.

NOTE.—Since the above was in print there has appeared in the *Electrician* (June 18, 1881, p. 70) an article on "The Theory of Alternating Current Machines," relating to M. Joubert's experiments and theoretical conclusions. Finding experimentally that the current from an alternating current machine could be represented exactly by the formula

$$I = \frac{e}{(R^2 + a^2)^{\frac{1}{2}}},$$

where I is the mean current, R the total resistance, a a constant proportional to the speed, and e a constant = quotient by $\sqrt{2}$ of the maximum E.M.F. of the machine with open circuit, measured by a Thomson portable electrometer, M. Joubert sought to justify it by theory, and gives the theory of a revolving coil, similar to that in the early part of the above paper. There is, however, this peculiarity, that although M. Joubert assumes the existence of a simple harmonic induced E.M.F., which implies a uniform external field of force, he only brings in the speed as a factor afterwards as an experimental result, whereas it is a necessary consequence of uniform speed in a uniform field; thus,

$$\text{E.M.F.} = -\frac{d}{dt} M \cos \omega t = M\omega \sin \omega t.$$

Also M. Joubert employs the numerical factor $1/\sqrt{2}$ instead of the $2/\pi$, which I have used.

XIX.—ON INDUCTION BETWEEN PARALLEL WIRES.

[*Jour. Soc. Tel. Eng.*, 1881, vol. 9, p. 427.]

1. ELECTRICAL induction is of two kinds, electrostatic and electromagnetic. The electrification of one conductor is always accompanied with electrification of others that may be in its neighbourhood, and one way of expressing this is to say that the charge on the first conductor

induces an opposite charge on a neighbouring conductor. Also, any change in the amount of magnetic induction passing through a circuit is accompanied by an E.M.F. in the circuit proportional to the rapidity of the change, which E.M.F. produces a current in the circuit. The change in the amount of induction may be due to a change in the current flowing through another circuit in the neighbourhood, or to relative motion of the circuits, or to changes in the magnetic field due to other causes, as the motion of a magnet for instance. In any case, the transient current accompanying the change is said to be induced. Both inductions are in action together, which considerably complicates the matter, but either one or the other may be frequently ignored for the time, without serious loss of accuracy.

To illustrate the general nature of induction between parallel wires it is sufficient to consider two wires. Suppose that both wires were originally free from charge and at potential zero, and that we then put a battery on at the beginning of the first wire, whose remote end is to earth, as are both ends of the second wire. After a little time a steady current is found to be flowing through the first wire and no current through the other. The value of the steady current is E/R, where E is the potential at the beginning of the first wire and R its resistance. But before this steady current is reached a somewhat complex state of things exists, due to the action of electrostatic and electromagnetic induction. Considering the electrostatic alone in the first place, the surface of the first wire forms one coating of a condenser, of which the other coating is the surface of the earth and of the second wire, which is in connection with the earth. Now one pole of the battery is connected to the first line and the other pole to earth, and therefore with the second line. The first line receives a charge, say, positive, while the earth and second wire receive an equal negative charge, the amount of these charges depending on the size and position of the two wires. As the first wire is being charged, a positive current flows in from the battery to do it. The negative charge on the earth and second wire may be considered as resulting from a negative current from the battery to earth and the second wire. Or we may say, using old-fashioned language, that the + electricity on the first wire attracts − from the earth to the second wire. Or that the + charge on the first wire induces a − charge on the earth and second wire. Or that the potential of the second wire due to the + charge on the first is +, therefore a + current must flow from the second wire to earth until its potential is brought to zero, leaving it negatively charged. Or, more accurately, because more comprehensively, we may consider all the elementary circuits, partly conductive and partly inductive, from one pole of the battery to the first wire, and from the latter to earth direct, and also *viâ* the second wire to the other pole of the battery, in every one of which circuits a + current flows, producing electrical polarization of the dielectric, whose residual polarization appears as a + charge on the first wire, and a − charge on the second wire and the earth. But whatever mode of expression be used the result is the same. The final distribution of potential and charge when equilibrium is reached may be easily stated. By Ohm's law, the final potential v_1 of the first wire is $E(1-x/l)$, x being the distance from the battery to any point, and l the whole length. And the density ρ_1 per unit of length of the first wire is

$$c_1v_1 = c_1E(1-x/l),$$

where c_1 is its electrostatic capacity per unit of length, The final potential of the second wire is of course zero. The linear density ρ_2 of its charge is

$$c_{12}v_1 = c_{12}E(1-x/l);$$

c_{12} being the coefficient of mutual electrostatic capacity of the two wires per unit of length. c_{12}, it should be remembered, is $-$, so that ρ_2 is $-$. (The ratio $-c_{12}:c_1$ may be taken roughly at about $\frac{1}{4}$ when the wires are suspended at the shortest usual distance apart: the exact value may be readily calculated.)* Thus the potential and density of the charge on the first wire fall uniformly from their greatest at the beginning to zero at the distant end, while the density of the charge on the second wire rises from its greatest $-$ value at the beginning to zero at the far end, the densities at corresponding points being roughly as 4 : 1 at the most. Thus the second wire has $\frac{1}{4}$ the opposite charge of the first, and the earth the remaining $\frac{3}{4}$.

2. Now removing the battery and putting earth on instead, the wires are discharged. The charge of the first wire flows out at both ends, reducing its potential to zero, twice as much going out at the battery end as at the other. As its potential falls, that of the second wire also falls: from zero it becomes $-$. For its potential due to its own $-$ charge is $-$, and this was only neutralized by the equal $+$ potential due to the original charge of the first wire; so as the latter is reduced the former comes into play. During the whole time the first wire is discharging the potential of the second wire is therefore $-$, causing a $+$ current from earth at both ends, which lasts until enough electricity has entered to cancel its original $-$ charge and bring it to potential zero again. Or we may say equivalently that its $-$ charge flows out at both ends simultaneously with the discharge of the first wire, and in the same proportion, $\frac{2}{3}$ at the battery end and $\frac{1}{3}$ at the other.

3. Now insulate the second wire at both ends, and again apply the battery to the first. It becomes charged, and after a little time a steady current E/R flows through it, and its potential is $E(1-x/l)$ as before. But as the second wire is now cut off from earth it cannot receive any charge, that is, its total charge must be zero. Likewise its potential must be $+$, and of the same value all along.

If v_1, v_2 are the potentials, ρ_1, ρ_2 the linear densities at distance x; c_1, c_2 the linear electrostatic capacities, and c_{12} the linear mutual electrostatic capacity, the potentials and densities are connected by the equations

$$\rho_1 = c_1v_1 + c_{12}v_2, \qquad \rho_2 = c_2v_2 + c_{12}v_1.$$

* [See Art. xii., p. 42 *ante*.]

Now, since v_2 is constant, and the second line has no charge on the whole,

$$c_2 v_2 l + c_{12}\int v_1 dx = 0; \quad \text{where} \quad v_1 = E\left(1 - \frac{x}{l}\right).$$

From these data it follows that

$$v_2 = -\frac{c_{12}}{2c_2}E; \quad \rho_1 = c_1 E\left(1 - \frac{x}{l}\right) - \frac{c_{12}^2}{2c_2}E; \quad \rho_2 = Ec_{12}\left(1 - \frac{x}{l}\right) - \frac{c_{12}}{2}E.$$

Thus the potential of the second wire becomes uniformly about one-eighth of the potential at the battery end of the first, on the former assumption that $-4c_{12} = c_2$. The density of the charge of the first wire falls uniformly from $E(c_1 - c_{12}^2/2c_2)$ at the beginning to $-(c_{12}^2/2c_2)E$ at the end, thus dividing the line into a positively and a negatively charged portion, the length of the latter being (on the same assumption) $\frac{1}{32}$ of the whole length. And the density of the charge on the second wire rises uniformly from $\frac{1}{2}c_{12}E$ at the beginning to $-\frac{1}{2}c_{12}E$ at the far end; thus the second half is positively, the first negatively charged to the same amount. The total charge of the first is $\frac{1}{2}El(c_1 - c_{12}^2/c_2)$; of the second, zero.

Thus during the establishment of the steady current in the first wire there is a current in the second in the same direction, a transfer of electricity from the first half of the line to the second, leaving the former negatively charged and the latter positively. Removing the battery and earthing the first wire, the discharge of the first wire will make the potential of the second wire less in the first half than in the second, so that the disappearance of the first wire's charge is accompanied by a $-$ current in the second, restoring it to zero potential again.

4. These examples are perhaps sufficiently elucidative of the part that electrostatic induction plays during the establishment of a current in a wire, and of its influence on the final potentials and densities. All disturbing influences have been of course ignored; perfect earth connections have been supposed, also perfect conductivity of the earth, perfect insulation, and absence of earth currents and atmospheric electricity.

When the number of parallel wires is not limited to two, the phenomena, though more complex, are essentially of the same nature. The final states of potential and charge assumed by any number of wires with batteries applied to one or more of them may be found from the general equations—

$$\left.\begin{aligned} \rho_1 &= c_1 v_1 + c_{12}v_2 + c_{13}v_3 + c_{14}v_4 + \dots, \\ \rho_2 &= c_{12}v_1 + c_2 v_2 + c_{23}v_3 + c_{24}v_4 + \dots, \\ \rho_3 &= c_{13}v_1 + c_{23}v_2 + c_3 v_3 + c_{34}v_4 + \dots, \\ &\cdot \quad \cdot \quad \cdot \quad \cdot \quad \cdot \quad \cdot \end{aligned}\right\} \quad \dots\dots\dots\dots (1)$$

Here c_1, c_2, c_3, ... are the electrostatic capacities of wires 1, 2, 3, ..., c_{12} the mutual capacity of 1 and 2, c_{13} that of 1 and 3, and so on, all per unit of length; $v_1, v_2, v_3, \ldots$ the potentials, and $\rho_1, \rho_2, \ldots$ the densities per unit of length. (See Maxwell's *Electricity*, vol. 1, ch. 3.) These equations will be referred to later on.

5. Now there is electromagnetic induction to be considered. For simplicity suppose it to act alone; and as before, take the case of two parallel wires, both earthed, and apply a battery to the first. During the establishment of the current in the first wire (supposed to take place uniformly all along its length), a current in the *opposite* direction is *induced* in the second (also uniform all along), which ceases when the current in the first reaches its steady strength. And on the cessation of the current in the first wire a current is *induced* in the second in the *same* direction. But this, though sufficient for many, is but a very rudimentary statement of the case.

According to Thomson and Maxwell's theory, the electric current is a kinetic phenomenon, involving matter in motion, and the motion is not confined to the wire alone, but is to be found wherever the magnetic force of the current extends. As matter has to be set in motion when a current is in course of establishment, inertia has to be overcome, the real inertia of moving matter, having the negative property of remaining in the state of motion it may have. So that the current cannot be established instantaneously, but rises gradually. And if the current be left to itself without any impressed E.M.F. to support it, it does not cease instantaneously, but gradually decays in the same manner as it was set up, in virtue of the real momentum of the moving matter. That it decays at all is due to the production of heat by the current, which is inseparable from its existence, *i.e.*, the kinetic energy of the current is degraded into the kinetic energy of heat. Thus when the source of energy is cut off by removing the battery the momentum of the current begins immediately to fall off, drawing all the while upon its reserve store of energy to maintain it. Now respecting the currents induced in neighbouring conductors. The momentum exists in all parts of the field, and on the removal of the E.M.F. becomes visible in all of them, the energy becoming degraded into heat in all. Granting this, the currents induced must be all in the same direction, viz., as that in the primary wire; and it follows immediately that on setting up a current the opposite occurs, currents in the opposite direction to that set up being caused in all the wires. In the secondary wires it is evident as such; in the primary it is evident as retarding the rise of the current.

Not knowing the actual mechanism of the current and of the magnetic force, we cannot know what the actual amount of real momentum is, although the amount of energy, the connecting link between all forces, may be calculated. But, in a dynamical system, it is not at all necessary that the mechanism should be known completely. If the state of the system is completely defined by the values of a certain number of variables the relations between forces, momenta, etc., corresponding to these variables may be calculated on strictly dynamical principles. Thus Maxwell's electromagnetic momentum of a circuit bears the same relation to the impressed E.M.F. in the circuit that momentum does to force in ordinary dynamics. Ohm's law, however, remains an experimental fact, and is taken as such alone.

In the case of two circuits, the equations of motion are

$$E = R_1\gamma_1 + \frac{d}{dt}(L_1\gamma_1 + M\gamma_2),$$

$$0 = R_2\gamma_2 + \frac{d}{dt}(L_2\gamma_2 + M\gamma_1).$$

Here γ_1 and γ_2 are the currents at any moment in the circuits 1 and 2 of resistances R_1 and R_2, and E the impressed E.M.F. in circuit 1. $L_1\gamma_1 + M\gamma_2$ is the electromagnetic momentum of the first circuit, and $L_2\gamma_2 + M\gamma_1$ that of the second; L_1, L_2, and M being constants depending on the form and position of the circuits. In the first circuit the E.M.F. E is employed partly in maintaining the current γ_1 against the resistance R_1, and partly in increasing the momentum of the first circuit. In the second circuit, where there is no impressed force, the induced E.M.F. is $-\frac{d}{dt}(L_2\gamma_2 + M\gamma_1)$, that is, the rate of decrease of its electromagnetic momentum. Further than this it is only necessary to mention here that the setting up of the current E/R_1 in circuit 1 is accompanied by an integral current ME/R_1R_2 in the opposite direction in circuit 2, and the decay of the current in 1 by an equal integral flow in 2 in the + direction.

6. We may now compare together the integral currents of electrostatic and electromagnetic induction in circuit 2, circuits 1 and 2 being two parallel suspended wires for definiteness, earthed at their ends.

If Q_1 is the integral current due to changing magnetic induction,

$$Q_1 = \frac{ME}{R_1R_2},$$

where M is the mutual electromagnetic capacity of the two wires.

If Q_2 is the electrostatic charge received by the second wire,

$$Q_2 = \frac{c_{12}lE}{2};$$

therefore

$$\frac{Q_2}{Q_1} = \frac{R^2}{2}\frac{c_{12}l}{M},$$

supposing the wires have the same resistance. Thus Q_2/Q_1 increases as the square of the length. One-third of this must be taken for the ratio at the distant end, where Q_2 and Q_1 are in opposite directions.

If $c_{12} = \cdot 003$ microf. per mile, $M = ml$, and $m = 3$ per centimetre,* or 482,802 per mile; $R = kl$, and $k = 15$ ohms per mile,

$$\frac{1}{3}\frac{Q_2}{Q_1} = \frac{l^2}{4291}.$$

Thus when $l =$ about 65 miles the integral currents at the receiving end are equal. For a shorter length the electrostatic is overpowered by the

* [This is rather a low estimate for wires at usual distance. See (10) and (11), p. 101.]

electromagnetic and the reverse for a greater length. This calculation is quite a rough one, but will do approximately for two suspended wires at the shortest usual distance apart.

7. Having considered the general nature of electrostatic and electromagnetic induction in the simple case of two wires, and the final distribution of electricity on the wires in equilibrium, the next step is to more accurately determine the manner in which the potential and the current behave in the variable state that intervenes between one state of equilibrium or steady flow and another. In the first place, electromagnetic induction will be ignored, as it greatly complicates the theory, while it is of quite secondary importance on lines not less than a certain short length.

The fundamental equations, whether we consider one line or many, with or without electromagnetic induction, are

$$\gamma'_1+\dot{\rho}_1=0\,;\quad \gamma'_2+\dot{\rho}_2=0\,;\quad \gamma'_3+\dot{\rho}_3=0\,;\ \text{etc.} \qquad (2)$$

where γ stands for current, ρ for linear density, and the suffix denotes the wire referred to. In the following $'$ will be generally used to indicate differentiation with respect to x and $\dot{}$ differentiation with respect to t the time. Thus any one of (2) written fully is

$$\frac{d\gamma}{dx}+\frac{d\rho}{dt}=0.$$

This is nothing more than the equation of continuity, and it may be proved thus:—If γ is the current at x at time t then $\gamma+\gamma' dx$ is the current at $x+dx$ at time t, and the excess of the current at x over that at $x+dx$ is $-\gamma' dx$. Therefore in the time dt the excess of the quantity of electricity that has passed x over what has passed $x+dx$ is $-\gamma' dxdt$, and this quantity must have been added to the charge of dx. At time t the latter is ρdx and at time $t+dt$ it becomes $\rho dx+\dot{\rho} dxdt$; whence another expression for the increase in time dt of the elementary charge of dx is $\dot{\rho} dxdt$, and the above equation follows.

If the wire exists alone we have $\rho=cv$. Also by Ohm's law, $\gamma=-k^{-1}v'$, where k is the resistance per unit of length. Whence

$$v''=ck\dot{v}, \qquad (3)$$

with the same equation for ρ or γ. (It is practically best to work with v.) This is Sir Wm. Thomson's well-known equation of the potential in a submarine cable, or of course in any uniform wire unaffected by the induction of others. The same equation was given by Ohm for the "tension" of a uniform wire, but singularly enough it was arrived at by an entirely erroneous assumption, viz., that a wire had a capacity or power of absorbing electricity into its substance, just as a conductor of heat has a "capacity" for heat. In fact, he applied to electricity Fourier's equations for the diffusion of heat by conduction. Mathematically considered, it amounts to exactly the same thing for the purpose of calculating the propagation of signals, whether the electricity is detained in the substance of a wire, or forms a superficial charge, connected by tubes of induction with an equal opposite charge on other conductors separated from it by a dielectric, the so-called surface charge being merely the residual polarization of the dielectric, within which the potential energy of electrification is stored; yet the difference between Ohm's capacity and the real thing is very striking.

The general solution of (3) adapted for suiting limiting conditions is

$$v = \Sigma A \sin\left(\frac{ax}{l} + b\right) e^{-a^2t/ckl^2}; \quad \text{..................} \quad (4)$$

that is, the sum of any number of terms of this general form, where A, a, and b are arbitrary.

Suppose that the ends of the wire are connected to apparatus, resistances, condensers, etc., in a given manner, and that the conditions thus imposed are expressed in an analytical form. Substitution in the general term of (4) will give rise to two equations,

$$\tan(a+b) = \phi_1(a), \qquad \tan b = \phi_2(a);$$

and between them we can eliminate b, obtaining

$$\tan a = \phi(a).$$

These determine the admissible values of a and b consistent with the nature of the terminal connections; and if the electrical state of the system is known at any moment the coefficients A can be determined so that the right hand side of (4) expresses the potential of the line at that moment and subsequently.

If $v = U$ when $t = 0$, where U is an arbitrary function of x, then will the general value of A be

$$A = \frac{\dfrac{2}{l}\displaystyle\int_0^l U \sin\left(\frac{ax}{l} + b\right) dx}{1 - \cos^2 a \dfrac{d}{da}\phi(a)}. \quad \text{......................} \quad (5)$$

This is on the supposition that at $t = 0$ none of the energy of the system resided in the terminal apparatus. For example, if there is a condenser, it must be uncharged; otherwise, additional terms, which are easily found, must be added to the numerator of (5), the denominator remaining the same.*

In fact

$$U = \frac{2}{l}\Sigma \sin\left(\frac{ax}{l} + b\right) \frac{\displaystyle\int_0^l U \sin\left(\frac{ax}{l} + b\right) dx}{1 - \dfrac{d}{da}\tan^{-1}\phi(a)} \quad \text{...............} \quad (5a)$$

is an identity. If the line be infinitely long, it becomes *

$$f(x) = \frac{2}{\pi}\int_0^\infty du \int_0^\infty dy \sin(ux + b) \sin(uy + b) f(y), \quad \text{............} \quad (5b)$$

* [For detailed information regarding the construction of these formulae see previous articles, especially Art. xvi. for special examples, and §§ 24, 25 (p. 91 *ante*) of the same for the general theory, and regarding the terminal arbitrary functions omitted in the above. Equation (57) there (p. 93 *ante*) is the same as (5) above, by omission of the terminal arbitraries and an important simplification of the denominator. See also the next, Art. xx., for another establishment of (5) and (5a) above, and (§§ 8 to 13) for the supplementary integral required in (5b) above when the terminal arrangement at the beginning of the line is such as to cause the determinantal equation to have imaginary roots.]

where $f(x)$ is any function of x and $\sin b = \cos b \times \phi_2(u)$ where $\phi_2(u)$ is of the form

$$\phi_2(u) = h_1 u + h_3 u^3 + h_5 u^5 + \ldots.$$

The above is an extended form of Fourier's Theorem as given in Treatises on the Integral Calculus.

8. With any number of wires we have

$$\gamma_1 = -\frac{1}{k_1} v'_1, \quad \gamma_2 = -\frac{1}{k_2} v'_2, \text{ etc.}, \quad \ldots\ldots\ldots\ldots\ldots\ldots (6)$$

where $k_1, k_2, \ldots$ are their resistances per unit of length. Therefore by eliminating $\gamma_1, \gamma_2, \ldots$ between (2) and (6),

$$v''_1 = k_1 \dot{\rho}_1, \quad v''_2 = k_2 \dot{\rho}_2, \text{ etc.}, \quad \ldots\ldots\ldots\ldots\ldots\ldots (7)$$

from which $\rho_1, \rho_2, \ldots$ may be eliminated by means of (1), leading to

$$\left.\begin{aligned} v''_1 &= k_1(c_1\dot{v}_1 + c_{12}\dot{v}_2 + c_{13}\dot{v}_3 + \ldots), \\ v''_2 &= k_2(c_{12}\dot{v}_1 + c_2\dot{v}_2 + c_{23}\dot{v}_3 + \ldots), \\ &\cdot \quad \cdot \quad \cdot \quad \cdot \quad \cdot \end{aligned}\right\} \ldots\ldots\ldots\ldots\ldots\ldots (8)$$

the exact solutions of which for any given terminal conditions and given initial distributions may be readily obtained.*

Beginning with the simple case of two wires of the same length, exactly similar, and at the same height from the ground, so that

$$c_1 = c_2 = c, \qquad k_1 = k_2 = k.$$

Here $\quad v''_1 = ck\dot{v}_1 + c_{12}k\dot{v}_2, \quad v''_2 = ck\dot{v}_2 + c_{12}k\dot{v}_1.$

Now choose two other dependent variables w_1 and w_2, so that

$$v_1 = w_1 + w_2, \qquad v_2 = w_1 - w_2,$$

and substitute; then

$$w''_1 = (c + c_{12})k\dot{w}_1, \qquad w''_2 = (c - c_{12})k\dot{w}_2.$$

By this substitution we have got two equations of the same form as (3), that for a wire uninfluenced by the induction of others. Thus, imagine two fresh lines of capacities $c + c_{12}$ and $c - c_{12}$ per unit of length, which have *no* induction on one another, and let their potentials at any moment be half the sum and half the difference of those of the real lines at the same moment at corresponding points; let these imaginary lines discharge independently, then the potentials of the real lines at any subsequent moment will be the sum and the difference of what the potentials of the imaginary lines have then become.

For instance, if the first line had at $t = 0$ a steady current E/kl in it and potential $v = E(1 - x/l)$, while in the second line $v_2 = 0$; and the battery is then removed and both lines earthed at both ends, the subsequent potentials are

$$v_1 = \frac{E}{\pi} \Sigma \frac{1}{n} \sin \frac{n\pi x}{l} (e^{D_1 t} + e^{D_2 t}),$$

* [See Sir W. Thomson's article "On Peristaltic Induction of Electric Currents," *Proc. R. S.*, May 1856, or Reprint, vol. 2, Art. lxxv., p. 79, for the earliest treatment of this subject; that is, with magnetic induction ignored.]

$$v_2 = \frac{E}{\pi}\Sigma\frac{1}{n}\sin\frac{n\pi x}{l}(e^{D_1t} - e^{D_2t});$$

where $$D_1 = -\frac{n^2\pi^2}{(c+c_{12})kl^2}, \quad D_2 = -\frac{n^2\pi^2}{(c-c_{12})kl^2},$$

and the summations include all integral values of n from 1 to ∞.

At the end remote from the battery the currents are

$$\gamma_1 = \frac{E}{kl}\Sigma(-\cos n\pi)(e^{D_1t} + e^{D_2t}),$$

$$\gamma_2 = \frac{E}{kl}\Sigma(-\cos n\pi)(e^{D_1t} - e^{D_2t}).$$

The arrival curve of the current at the remote end of a line earthed at both ends and not exposed to induction from other wires is given in a readily accessible form in Professor Jenkin's *Electricity and Magnetism*, with a table of the value of the current at different times after contact was made.* If the curve there given be drawn on two different scales as regards time, viz., proportional to $c + c_{12}$ and $c - c_{12}$, and from these two curves are constructed two new ones by making the ordinates of the latter equal to half the sum and half the difference of the former, the latter will be the arrival curves for the case just considered; the big curve being the current on line 1, and the smaller one, above the zero line, the induced arrival curve on line 2.

9. Again two wires, dissimilar, but with the same terminal conditions. Here, by (8)

$$\left.\begin{aligned} v''_1 &= c_1k_1\dot{v}_1 + c_{12}k_1\dot{v}_2, \\ v''_2 &= c_2k_2\dot{v}_2 + c_{12}k_2\dot{v}_1. \end{aligned}\right\} \qquad (9)$$

To find elementary simultaneous solutions, eliminate either v_1 or v_2; thus

$$(\nabla^2 - c_1k_1D)(\nabla^2 - c_2k_2D)v = c_{12}^2k_1k_2D^2v,$$

where ∇ stands for d/dx and D for d/dt. Let ∇^2 be any numerical quantity; the above is then an algebraical quadratic equation in D, whose roots are

$$D_{\frac{1}{2}} = \frac{\nabla^2}{2k_1k_2(c_1c_2 - c_{12}^2)}\{(c_1k_1 + c_2k_2) \pm \sqrt{(c_1k_1 + c_2k_2)^2 - 4k_1k_2(c_1c_2 - c_{12}^2)}\};$$

therefore

$$\left.\begin{aligned} v_1 &= \sin(i\nabla x + b)(Ae^{D_1t} + Be^{D_2t}), \\ v_2 &= \sin(i\nabla x + b)(r_1Ae^{D_1t} + r_2Be^{D_2t}); \end{aligned}\right\} \qquad (10)$$

where i stands for $\sqrt{-1}$. A, B, ∇, and b are arbitrary, while r_1, r_2 are constants which may be found by inserting these solutions (10) in (9), giving

$$r_1 = \frac{\nabla^2 - c_1k_1D_1}{c_{12}k_1D_1}, \quad r_2 = \frac{\nabla^2 - c_1k_1D_2}{c_{12}k_1D_2}.$$

The sum of an infinite number of elementary solutions such as those in (10) may be made to represent initially any given distribution of

* [See curve 1, Fig. 1, p. 72 *ante.*]

potential in the wires, and also to satisfy the terminal conditions. The latter, when given, settle the admissible values of ∇ and b. D_1 and D_2 are known multiples of ∇; and, finally, the values of $A+B$ and r_1A+r_2B, and therefore of A and B, can be found by integration, so as to make the complete solutions represent any initial arbitrary potentials.

In the simple case of direct earth connection at both ends of both lines, and $v_1=E(1-x/l)$, $v_2=0$, initially, we have $b=0$ and $i\nabla_n=n\pi/l$,

$$A_n+B_n=\frac{2E}{n\pi}, \qquad r_1A_n+r_2B_n=0;$$

therefore
$$A_n=\frac{r_2}{r_2-r_1}\,\frac{2E}{n\pi}, \qquad B_n=-\frac{r_1}{r_2-r_1}\,\frac{2E}{n\pi};$$

therefore
$$v_1=\frac{2E}{\pi}\Sigma\frac{1}{n}\sin\frac{n\pi x}{l}(r_2e^{D_1t}-r_1e^{D_2t})\div(r_2-r_1),$$
$$v_2=\frac{2E}{\pi}\Sigma\frac{1}{n}\sin\frac{n\pi x}{l}(e^{D_1t}-e^{D_2t})\times\frac{r_1r_2}{r_2-r_1}$$

represent the potentials at time t.

10. It will be observed that the differential equations (8) each contain the differential coefficients with respect to t of the potentials of all the lines. But by a linear transformation, thus:—

$$\begin{aligned} v_1&=w_1+w_2+w_3+\dots,\\ v_2&=a_1w_1+a_2w_2+a_3w_3+\dots,\\ v_3&=b_1w_1+b_2w_2+b_3w_3+\dots,\\ &\cdots \end{aligned}$$

and by properly determining the constants $a_1, \dots, b_1, \dots$, equations (8) may be brought to the form

$$w''_1=f_1\dot{w}_1; \qquad w''_2=f_2\dot{w}_2; \text{ etc.},$$

each of which contains only one dependent variable, so that we may represent the actual potentials of the lines as linear functions of the potentials of an equal number of imaginary lines having no mutual influence, with the same terminal conditions, and charged initially in a known manner. But there is no advantage in commencing thus, the following process being simpler:—

Equations (8) may be written

$$\left.\begin{aligned} 0&=p_{11}v_1+p_{12}v_2+p_{13}v_3+\dots,\\ 0&=p_{21}v_1+p_{22}v_2+p_{23}v_3+\dots,\\ 0&=p_{31}v_1+p_{32}v_2+p_{33}v_3+\dots,\\ &\cdots \end{aligned}\right\}$$

where
$$p_{11}=-\nabla^2+c_1k_1D; \qquad p_{12}=c_{12}k_1D, \text{ etc.}$$

By eliminating all the variables except one, we get for the equation of the potential of any line $Pv=0$, where P is the determinant

$$P=\begin{vmatrix} p_{11} & p_{12} & p_{13} & \cdots \\ p_{21} & p_{22} & p_{23} & \cdots \\ p_{31} & p_{32} & p_{33} & \cdots \end{vmatrix}.$$

The terminal conditions being the same for all the lines, the elementary solutions will all contain the same function of x; then, assuming ∇^2 to be a numerical quantity, the equation $P=0$ is an algebraical equation in D/∇^2 of the same degree as the number of lines. Let the roots for D be $D_1, D_2, \ldots$. These are all known multiples of ∇^2, and the elementary solutions are

$$v_1 = \sin(i\nabla x + b)\{Ae^{D_1t} + Be^{D_2t} + Ce^{D_3t} + \ldots\},$$
$$v_2 = \sin(i\nabla x + b)\{r_1Ae^{D_1t} + r_2Be^{D_2t} + \ldots\},$$
$$v_3 = \sin(i\nabla x + b)\{s_1Ae^{D_1t} + s_2Be^{D_2t} + \ldots\},$$

etc. The quantities $r_1, \ldots, s_1, \ldots$ may be found by inserting these solutions in equations (8). As before, the terminal conditions settle the different values of ∇ and b; and integrations serve to determine A, B, C, ..., so that the sum of infinite series of the above elementary solutions may represent the initial potentials.

11. But if the terminal conditions are not limited to be the same for all the lines at $x=0$, and also all the same at $x=l$, the elementary solutions must be more general. First, for two lines.

Here $$(\nabla^2 - c_1k_1D)(\nabla^2 - c_2k_2D)v = c_{12}^2k_1k_2D^2v,$$

as before, for either line. Treating D as a constant, this is a quadratic equation in ∇^2, whose roots are

$$\nabla^2 = \tfrac{1}{2}D\{(c_1k_1 + c_2k_2) \pm \sqrt{(c_1k_1 + c_2k_2)^2 - 4k_1k_2(c_1c_2 - c_{12}^2)}\},$$

therefore an elementary solution is

$$\left.\begin{aligned} v_1 &= \{(A_1\sin + B_1\cos)i\nabla_1x + (A_2\sin + B_2\cos)i\nabla_2x\}e^{Dt}, \\ v_2 &= \{r_1(A_1\sin + B_1\cos)i\nabla_1x + r_2(A_2\sin + B_2\cos)i\nabla_2x\}e^{Dt}, \end{aligned}\right\} \ldots\ldots \quad (11)$$

where $$r_1 = \frac{\nabla_1^2 - c_1k_1D}{c_{12}k_1D}, \qquad r_2 = \frac{\nabla_2^2 - c_1k_1D}{c_{12}k_1D},$$

by inserting (11) in (9).

The four terminal conditions, viz., two for each line, give the ratios $A_1 : B_1 : A_2 : B_2$, and also furnish an equation for the admissible values of D.

We have thus the complete solutions

$$v_1 = \Sigma Au_1e^{Dt}, \qquad v_2 = \Sigma Au_2e^{Dt},$$

where the functions u_1 and u_2 are completely known; also the values of D. If when $t=0$, $v_1 = U_1$, and $v_2 = U_2$, these being arbitrary functions of x, we must find the values of the constants A so that

$$\left.\begin{aligned} U_1 &= \Sigma Au_1, \\ U_2 &= \Sigma Au_2. \end{aligned}\right\} \ldots\ldots\ldots\ldots\ldots\ldots \quad (12)$$

This may be done by availing ourselves of the conjugate property possessed by two different possible distributions of potential and density. Let u_{1i}, u_{2i} represent one, and u_{1j}, u_{2j} represent another elementary distribution of potential, corresponding to D_i and D_j respectively. The first decays at one rate, the second at another; they may co-exist without interference, and their mutual potential energy is zero. If E_{ij} is their mutual potential energy,

$$E_{ij}=\int(u_{1i}\rho_{1j}+u_{2i}\rho_{2j})dx=\int(u_{1j}\rho_{1i}+u_{2j}\rho_{2i})dx,$$

where ρ_{1i}, ρ_{2i} are the densities corresponding to u_{1i}, u_{2i}, and ρ_{1j}, ρ_{2j} those corresponding to u_{1j}, u_{2j}. This reciprocal property is an identity and does not depend on the distributions being independent. That $E_{ij}=0$ for independent distributions may be thus proved. We have

$$u''_{1i}=D_ik_1\rho_{1i},\ldots\ldots(13)_1 \qquad u''_{2i}=D_ik_2\rho_{2i},\ldots\ldots(13)_3$$
$$u''_{1j}=D_jk_1\rho_{1j},\ldots\ldots(13)_2 \qquad u''_{2j}=D_jk_2\rho_{2j},\ldots\ldots(13)_4$$

by (7). Therefore

$$E_{ij}=\int(u_{1i}\rho_{1j}+u_{2i}\rho_{2j})dx=\frac{1}{D_jk_1}\int u_{1i}u''_{1j}dx+\frac{1}{D_jk_2}\int u_{2i}u''_{2j}dx,$$

by $(13)_2$ and $(13)_4$;

$$=\frac{1}{D_jk_1}[u_{1i}u'_{1j}-u'_{1i}u_{1j}]+\frac{1}{D_jk_1}\int u''_{1i}u_{1j}dx+\frac{1}{D_jk_2}[u_{2i}u'_{2j}-u'_{2i}u_{2j}]+\frac{1}{D_jk_2}\int u''_{2i}u_{2j}dx$$

by double integration by parts;

$$=\frac{1}{D_jk_1}[u_{1i}u'_{1j}-u'_{1i}u_{1j}]+\frac{1}{D_jk_2}[u_{2i}u'_{2j}-u'_{2i}u_{2j}]+\frac{D_i}{D_j}\int u_{1j}\rho_{1i}dx+\frac{D_i}{D_j}\int u_{2j}\rho_{2i}dx,$$

by $(13)_1$ and $(13)_3$. Therefore, by the reciprocal property,

$$(D_j-D_i)E_{ij}=\frac{1}{k_1}[u_{1i}u'_{1j}-u'_{1i}u_{1j}]+\frac{1}{k_2}[u_{2i}u'_{2j}-u'_{2i}u_{2j}]. \quad\ldots\ldots\ldots (14)$$

The right-hand side is to be taken between limits, and vanishes when i and j are different if $u/u'=$ constant; that is, if the lines are put to earth at the ends, or insulated, or put to earth through mere resistances. When $i=j$, we find, by taking the limit and discarding i and j,

$$\int(u_1\rho_1+u_2\rho_2)dx=\frac{1}{2nk_1}\left[u'_1\frac{du_1}{dn}-u_1\frac{du'_1}{dn}\right]+\frac{1}{2nk_2}\left[u'_2\frac{du_2}{dn}-u_2\frac{du'_2}{dn}\right],(15)$$

where $n^2=-D$. By applying (14) and (15) to (12) the value of any coefficient A is found to be

$$A=\frac{\int U_1u_1\rho_1dx+\int U_2u_2\rho_2dx}{\frac{1}{2nk_1}\left[u'_1\frac{du_1}{dn}-u_1\frac{du'_1}{dn}\right]+\frac{1}{2nk_2}\left[u'_2\frac{du_2}{dn}-u_2\frac{du'_2}{dn}\right]},$$

the limits for the integrations being 0 and l, and the quantities in square brackets being taken between the same. The form of A requires modification when u/u' not = constant.

12. The extension to any number of wires m is obvious. In § 10, in the equation $Pv=0$, let D be any numerical quantity; then $P=0$ is an algebraical equation in ∇^2/D of the mth degree. Let its roots for ∇^2 be ∇_1^2, ∇_2^2, etc., then

$$v_1=u_1e^{Dt}, \quad v_2=u_2e^{Dt}, \quad v_3=u_3e^{Dt}, \text{ etc.},$$

where $u_1 = (A_1 \sin + B_1 \cos) i\nabla_1 x + (A_2 \sin + B_2 \cos) i\nabla_2 x$
$+ (A_3 \sin + B_3 \cos) i\nabla_3 x + \ldots,$
$u_2 = r_1(A_1 \sin + B_1 \cos) i\nabla_1 x + r_2(A_2 \sin + B_2 \cos) i\nabla_2 x$
$+ r_3(A_3 \sin + B_3 \cos) i\nabla_3 x + \ldots,$
etc., etc.,

is a system of elementary solutions; the ratios $r_1, \ldots$ being found by insertion in (8). The constants $A_1, B_1, \ldots$ are $2m$ in number. The terminal conditions are also $2m$ in number, and serve to determine the $2m-1$ ratios $A_1 : B_1 : A_2 : B_2 : \ldots$, and give besides an equation for D. Thus the complete solutions are

$$v_1 = \Sigma A u_1 e^{Dt}, \qquad v_2 = \Sigma A u_2 e^{Dt}, \text{ etc.} \qquad (16)$$

where $u_1, u_2, \ldots$ and the values of D are completely known. The constants A may be found by the conjugate property, which is now more complex, though essentially the same. $u_{1i}, u_{2i}, u_{3i}, \ldots$ and $u_{1j}, u_{2j}, u_{3j}, \ldots$ being two different independent potential distributions, and $\rho_{1i}, \ldots, \rho_{1j}, \ldots$, the corresponding densities, the reciprocal property is

$$E_{ij} = \int (u_{1i}\rho_{1j} + u_{2i}\rho_{2j} + u_{3i}\rho_{3j} + \ldots)dx = \int (u_{1j}\rho_{1i} + u_{2j}\rho_{2i} + u_{3j}\rho_{3i} + \ldots)dx;$$

and the conjugate property may be shown in a similar manner as before to be

$$(D_j - D_i)E_{ij} = \frac{1}{k_1}[u_{1i}u'_{1j} - u'_{1i}u_{1j}] + \frac{1}{k_2}[u_{2i}u'_{2j} - u'_{2i}u_{2j}] + \ldots \qquad (17)$$

Thus $E_{ij} = 0$ when $u/u' = \text{constant}$. And

$$\int (u_{1i}\rho_{1i} + u_{2i}\rho_{2i} + u_{3i}\rho_{3i} + \ldots)dx$$
$$= \int (c_1 u_1^2 + c_2 u_2^2 + c_3 u_3^2 + \ldots + 2c_{12}u_1u_2 + 2c_{13}u_2u_3 + \ldots)dx$$
$$= \frac{1}{2nk_1}\left[u'_1 \frac{du_1}{dn} - u_1 \frac{du'_1}{dn}\right] + \text{similar terms with other suffixes,} \ldots (18)$$

where $n^2 = -D$.

The left-hand member is twice the potential energy of $u_1, u_2, u_3, \ldots$. By these properties (17) and (18), applied to (15), the constant A may be found so as to make

$$v_1 = U_1, \qquad v_2 = U_2, \qquad v_3 = U_3, \text{ etc.,}$$

when $t = 0$, where $U_1, \ldots$ are arbitrary functions of x.

13. In paragraphs 7 to 12 inclusive only electrostatic induction has been taken into account. A considerably greater complication arises when we attempt to fully exhibit the joint action of both inductions. We have still the fundamental equations of continuity,

$$\gamma'_1 + \rho_1 = 0, \qquad \gamma'_2 + \rho_2 = 0, \text{ etc.}$$

and also the same relations between potential and density,

$$\rho_1 = c_1 v_1 + c_{12} v_2 + c_{13} v_3 + \ldots,$$

etc., but the equations connecting the potential and current are now no longer

$$\gamma_1 = -\frac{1}{k_1}v'_1, \quad \gamma_2 = -\frac{1}{k_2}v'_2, \text{ etc.,}$$

but are of the much more complex forms

$$\left.\begin{aligned} -v'_1 &= k_1\gamma_1 + s_1\dot{\gamma}_1 + s_{12}\dot{\gamma}_2 + s_{13}\dot{\gamma}_3 + \ldots, \\ -v'_2 &= k_2\gamma_2 + s_{12}\dot{\gamma}_1 + s_2\dot{\gamma}_2 + s_{23}\dot{\gamma}_3 + \ldots, \\ -v'_3 &= k_3\gamma_3 + s_{13}\dot{\gamma}_1 + s_{23}\dot{\gamma}_2 + s_3\dot{\gamma}_3 + \ldots, \\ &\cdot \quad \cdot \quad \cdot \quad \cdot \quad \cdot \quad \cdot \end{aligned}\right\} \quad \ldots\ldots\ldots\ldots\ldots (19)$$

Here* s_1, s_2, s_3, ... are the coefficients of self-induction (or electromagnetic capacity) of the wires per unit of length, while s_{12} is the coefficient of mutual capacity of 1 and 2 per unit of length, s_{23} the same for 2 and 3, etc., and γ_1, γ_2, ... are the currents in the wires. (*See* vol. ii., Maxwell's *Electricity*.) The impressed E.M.F. $-v'$ in wire 1 is not only employed in maintaining the current in that wire, but also in varying the currents in all the rest.

Eliminating the currents, we have

$$\left.\begin{aligned} v''_1 &= k_1\dot{\rho}_1 + s_1\ddot{\rho}_1 + s_{12}\ddot{\rho}_2 + \ldots, \\ v''_2 &= k_2\dot{\rho}_2 + s_{12}\ddot{\rho}_1 + s_2\ddot{\rho}_2 + \ldots, \\ v''_3 &= k_3\dot{\rho}_3 + s_{13}\ddot{\rho}_1 + s_{23}\ddot{\rho}_2 + \ldots, \\ &\cdot \quad \cdot \quad \cdot \quad \cdot \quad \cdot \quad \cdot \end{aligned}\right\} \quad \ldots\ldots\ldots\ldots\ldots\ldots\ldots (20)$$

and since the densities are linear functions of the potentials, the former may be eliminated, giving rise to a set of simultaneous partial differential equations, each containing all the dependent variables v_1, ... and d^2/dx^2, d/dt, and d^2/dt^2. By eliminating the potentials, the resulting equations in the densities are somewhat simpler as regards the coefficients.

Beginning, for more completeness, with the case of a single wire, we have

$$v'' = ck\dot{v} + cs\ddot{v}, \quad \ldots\ldots\ldots\ldots\ldots\ldots\ldots\ldots\ldots\ldots (21)$$

or

$$\nabla^2 v = (ckD + csD^2)v,$$

whose general solution is

$$v = \Sigma \sin(i_\nabla x + b)(Ae^{D_1 t} + Be^{D_2 t}),$$

where D_1 and D_2 are the roots of

$$csD^2 + ckD - \nabla^2 = 0,$$

and ∇^2 is any numerical quantity. That is,

$$D_{\substack{1\\2}} = \frac{1}{2cs}\left(-ck \pm \sqrt{c^2k^2 + 4cs\nabla^2}\right).$$

For simplicity, let the wire be earthed at both ends, then $i_\nabla = n\pi/l$, where n is any integer, and $b = 0$, so that the elementary solution is

$$v = \sin\frac{n\pi x}{l}(Ae^{D_1 t} + Be^{D_2 t});$$

* [For developed formulae, see p. 101.]

or, if D_1 and D_2 are imaginary,

$$v = \sin\frac{n\pi x}{l}e^{-pt}(A\sin + B\cos)qt,$$

where
$$p = \frac{k}{2s}, \quad q = \frac{k}{s}\sqrt{\frac{n^2\pi^2}{l^2}\cdot\frac{s}{ck^2} - \frac{1}{4}}.$$

The latter form will be used preferably. If q be imaginary, D_1 and D_2 are real and negative, and the simple harmonic distribution of potential $\sin n\pi x/l$ decreases asymptotically towards zero. But if q be real it oscillates at a rate proportional to q, the amplitude of the oscillations decreasing asymptotically towards zero at a rate proportional to p. Whether v is oscillatory or not depends on the values of l, n, s, c, and k. Let $c = {\cdot}02$ microf. per mile, $k = 15$ ohms per mile. The value of s for iron wire depends mainly on the quality of the iron, and is therefore rather indefinite in the absence of actual measurement. Let $s = 15 \times 10^6$ per mile (electromagnetic units).* Then

$$\frac{s}{k} = \frac{15 \times 10^6}{15 \times 10^9} = 10^{-3}\text{ sec.},$$

and
$$ck = {\cdot}02 \times 10^{-15} \times 15 \times 10^9 = 3 \times 10^{-7}\text{ sec.};$$

therefore
$$q = 10^3\sqrt{\frac{n^2}{l^2}\frac{10^5}{3} - \frac{1}{4}}\text{ approximately.}$$

Now the first value of n is 1, so for the first term to be oscillatory $l^2 < \frac{4}{3}10^5$, or $l < 365$ miles. Therefore if the length of the line is less than 365 miles, all the terms are oscillatory.

For 100 miles

$$q = 10^3\sqrt{\frac{10}{3}n^2 - \frac{1}{4}}.$$

The period of an oscillation is $2\pi/q$; for the first term this amounts to $2\pi/1755$ second, or about $\frac{1}{280}$ second. But since $s/k = \frac{1}{1000}$ second, the amplitude falls very quickly, the second oscillation being small compared with the first. It is only on short lines that there can be many oscillations without much diminution of amplitude.

The physical reason of the oscillation will be readily understood from § 5. Suppose that at $t = 0$, $v = \sin\pi x/l$, so that the density is $\rho = c\sin\pi x/l$, that the charge is at rest at that time, and is then left to itself. The E.M.F. of the charge is at first $-\pi/l\cos\pi x/l$; this sets the charge in motion symmetrically from the centre. Self-induction retards the outward flow at first, but once in motion the current requires force to stop it. Therefore the current does not cease when the line is discharged, but continues, thus producing a negative electrification similar to the first positive, though smaller, and the E.M.F. of this − charge brings the current to rest. Then the current sets in the other way, producing

* [Iron wires were exclusively used for overland telegraphs when this was written. This value of s is, of course, far too high for copper wires. Even for iron it is too high when rapid reversals are in question, since no allowance has been made for the reduction due to imperfect penetration.]

similarly a + charge, and this goes on until the resistance of the line uses up all the original energy.

Suppose that the line is originally free from charge and connected to earth at both ends, and that at the time $t=0$ an electromotive force E is put on at $x=0$. The course of the potential is then

$$v=E\left(1-\frac{x}{l}\right)-\frac{2E}{\pi}\Sigma\frac{1}{n}\sin\frac{n\pi x}{l}e^{-pt}\left(\frac{p}{q}\sin+\cos\right)qt, \quad \text{.........} \quad (22)$$

and of the current

$$\gamma=\frac{E}{kl}(1-e^{-2pt})+\frac{2E}{kl}\Sigma\cos\frac{n\pi x}{l}\cdot\frac{2p}{q}e^{-pt}\sin qt, \quad \text{.............} \quad (23)$$

These formulæ are not very intelligible without some study. A rough idea of their meaning may, however, be obtained without calculating them exactly. When the line is quite short the $\frac{1}{4}$ under the radical sign in the value of q is small compared with the preceding term, and p/q is also small compared with 1; therefore, by neglecting $p/q \sin qt$ in (22), and taking $q=\dfrac{n\pi}{l\sqrt{cs}}$ by neglecting the $\frac{1}{4}$, that equation becomes

$$v=E\left(1-\frac{x}{l}\right)-\frac{2E}{\pi}\Sigma\frac{1}{n}\sin\frac{n\pi x}{l}e^{-pt}\cos\frac{n\pi t}{l\sqrt{cs}}. \quad \text{...........} \quad (24)$$

Now it may be shown that if we cancel the factor e^{-pt} common to all the terms in the summation in (24), then

$$v=0, \quad \text{from } t=0 \text{ to } t=x\sqrt{sc},$$

and

$$v=E, \quad \text{from } t=x\sqrt{sc} \text{ to } t=l\sqrt{sc}.$$

Or,

$$v=E \quad \text{when} \quad 0<x<\frac{t}{\sqrt{sc}},$$

$$v=0 \quad \text{when} \quad \frac{t}{\sqrt{sc}}<x<l.$$

This is best interpreted graphically. Divide the period $2l\sqrt{sc}$ into say eight parts equal to τ; let abscissæ represent distance along the line, and ordinates the potential, then the following show the progress of the potential during one period.

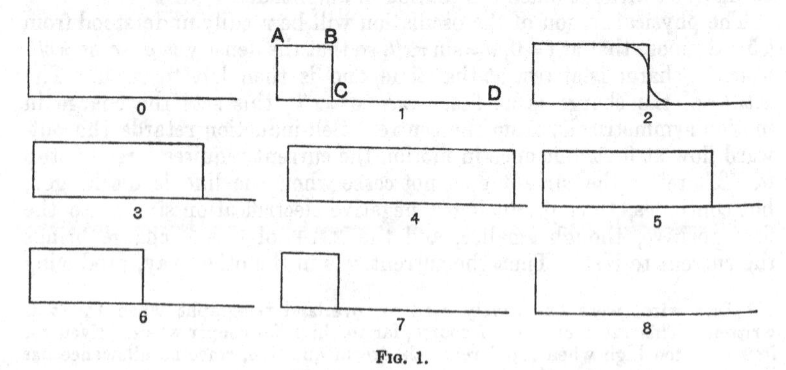

FIG. 1.

The potential E here travels to and fro along the line with uniform velocity $1/\sqrt{sc}$. At $x=\frac{1}{2}l$, $v=0$ from $t=0$ to $t=2\tau$; is then E from $t=2\tau$ to $t=6\tau$; then 0 from $t=6\tau$ to $t=10\tau$; and so on, thus being alternately E and 0 for equal periods, its mean value being $\frac{1}{2}E$. At $x=\frac{1}{4}l$, $v=0$ from $t=0$ to $t=\tau$; is then E from $t=\tau$ to $t=7\tau$; then 0 from $t=7\tau$ to $t=9\tau$; and so on, being three times as long at E as at 0, so that its mean value is $\frac{3}{4}E$. At $x=\frac{3}{4}l$, v is also alternately 0 and E, but is three times as long at 0 as at E, its mean value being $\frac{1}{4}E$. Similarly for any other point, so that the mean value of v is $E(1-x/l)$, what it ultimately becomes in the real case.

Now the abrupt discontinuities are due to the assumptions made, so that to pass to the real case we must first of all round off the sharp corners B and C, and slightly curve the straight lines, as shown in the figure marked 2. And next, we must introduce the factor e^{-pt}. The effect of this is to make the summation in (24), *i.e.*, the amount by which v depart from $E(1-x/l)$, the final potential, decrease rapidly with the time. As the wave progresses the upper portion continuously falls, while the lower portion continuously rises, and the double turn in the curve becomes less prominent, and is finally wiped out altogether, $ABCD$ thus becoming ultimately a straight line joining A and D, representing $v=E(1-x/l)$, the final state.*

* [The above description of the oscillatory mode of establishment of the steady state due to an impressed force at the beginning of the line (which is a development of p. 57 *ante*) is obtained by first ignoring the resistance of the line, and afterwards introducing it as an attenuating factor. The results are therefore specially applicable when the resistance really enters as a small factor, and are readily understandable in terms of the distortionless circuit whose theory is worked out later in "Electromagnetic Induction and its Propagation," sections 40 to 47. I have also, in my paper "On Electromagnetic Waves," given the complete solution of the above problem in terms of the successive waves, so that their true shapes, and their attenuation to nothing, for any value of the resistance can be obtained, thus allowing one to examine the transition to the diffusion formulae which hold when resistance is paramount. It is interesting to observe how far the above tentative process is correct. The course of events shown in Fig. 1 would repeat itself over and over again if there were no resistance. It is the nature of the modification due to resistance which does not appear, except very roughly, in the above. When the first wave reaches the end of the line (as in 4, Fig. 1), it is somewhat attenuated all along, mostly of course at its front, say to ·9 for example. Now to obtain the state of things in the next semi-period we must imagine the first wave to keep going on, and superimpose a wave of negative potential travelling the other way of initial strength $-{\cdot}9$, so that it is attenuated to $-{\cdot}81$ when it gets to the beginning of the line. In the third semi-period we must superimpose a wave in the original direction, starting from the beginning of the line at strength $+{\cdot}81$; and so on, *ad inf.* In the first wave the attenuation makes the curve of potential concave upward. This is, of course, not shown in the figure. Now the full theory indicates that when the resistance of the wire is constant the wave front is quite abrupt, without the rounding off shown. But, as a matter of fact, surface conduction causes the resistance to be far greater at and near the wave front than where penetration has taken place, so there must be greater attenuation at and near the wave front, with a rounding off of the abruptness somewhat as shown, though the wave front must be reckoned at the extreme limit of the disturbance.]

The rapidity with which the wave decreases being proportional to p is the same for any length; the velocity of the wave is also constant. But as the time the wave takes to reach the distant end is $l\sqrt{sc}$, the longer the line is, the less does the wave become before it has made a single journey from 0 to l. For 100 miles, $p=2s/k=1/500$ second (see above) and $l\sqrt{sc}=\sqrt{3}/1000$, so that even on its first journey the wave is greatly diminished. For 10 miles, $l\sqrt{sc}=\sqrt{3}/10000$, so that the wave may make 10 journeys before it decays as much as in one journey on a line of 100 miles.

On removing E, there is a similar wave travelling to and fro whilst v falls from $E(1-x/l)$ to 0. Corresponding to the former figures, we have the following:—

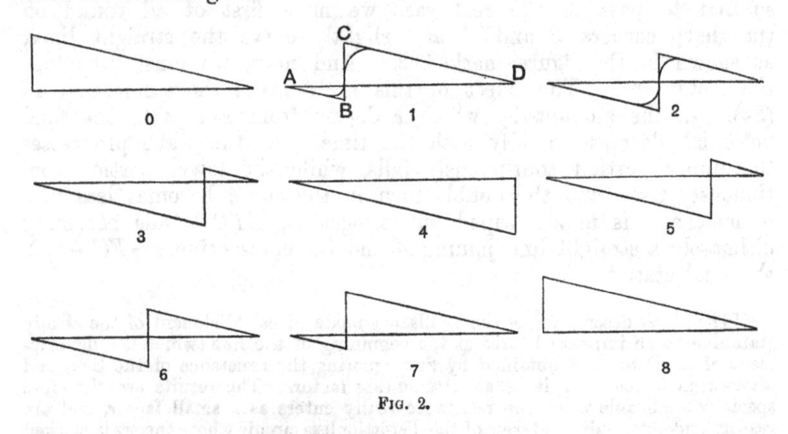

FIG. 2.

showing the changes during the period $2l\sqrt{sc}$, divided into eight equal intervals. Here v at any point x is alternately $E(1-x/l)$ and $-Ex/l$, the times during which it has these values being such that the mean value of v is 0. Thus at $x=\frac{1}{4}l$, v is three times as long at $-\frac{1}{4}E$ as at $+\frac{3}{4}E$, and at $x=\frac{3}{4}l$, v is three times as long at $\frac{1}{4}E$ as at $-\frac{3}{4}E$. As before, the lines $ABCD$ must be rounded, and the wave supposed to diminish rapidly as it progresses.

The current-wave may be similarly treated. Neglecting the $\frac{1}{4}$ in q the summational part of (23) becomes

$$\frac{2E}{\pi}\sqrt{\frac{c}{s}}\Sigma\frac{1}{n}\cos\frac{n\pi x}{l}e^{-pt}\sin\frac{n\pi t}{l\sqrt{sc}};$$

and, cancelling the factor e^{-pt}, this represents

$$E\sqrt{\frac{c}{s}}\left(1-\frac{t}{l\sqrt{sc}}\right), \text{ from } x=0 \text{ to } x=\frac{t}{\sqrt{sc}},$$

and $$-E\sqrt{\frac{c}{s}}\frac{t}{l\sqrt{sc}}, \qquad \text{from } x=\frac{t}{\sqrt{sc}} \text{ to } x=l.$$

The changes during the semi-period $l\sqrt{sc}$ are shown in the following, where the ordinates now represent current.

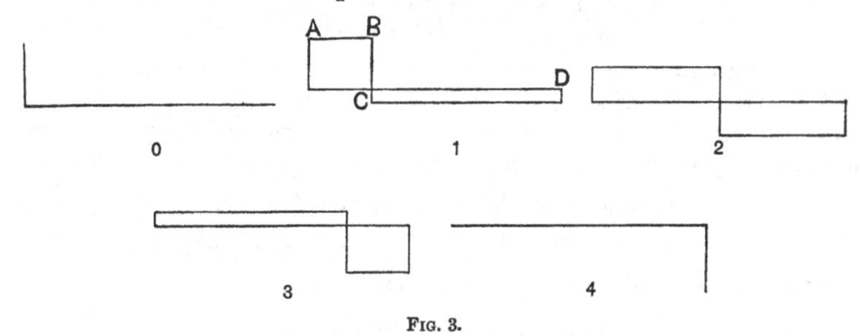

FIG. 3.

In the second semi-period, 5 is the same as 3, 6 the same as 2, 7 the same as 1, and 8 the same as 0. As before, the lines $ABCD$ must be rounded at B and C, and the wave as it progresses supposed to diminish rapidly. This wave is superimposed on the steadily rising current $\frac{E}{kl}(1-e^{-2pt})$. As the latter is increasing in strength the oscillations are decreasing, and the final current is E/kl.*

The ratio of the maximum strength of the current-wave to the final current is $kl\sqrt{\frac{c}{s}}=\frac{l\sqrt{3}}{100}$. For 100 miles this is 1·73, for 10 miles ·173, varying as the length of the line. The wave is strongest at the moment of starting from $x=0$, where it may be taken to represent the well-known *charge* at the moment contact is made, which varies as the length of the line.

It should be remembered that this mode of representation applies to short lines only. [On very long lines the effect of magnetic momentum tends to be confined to the battery end.]

14. The simplest case for two wires is when they are alike, that is

$$c_1=c_2=c, \qquad k_1=k_2=k, \qquad s_1=s_2=s,$$

and their terminal conditions are the same. By (20) our equations are

$$v_1''=ck\dot{v}_1+(cs+c_{12}s_{12})\ddot{v}_1+c_{12}k\dot{v}_2+(cs_{12}+c_{12}s)\ddot{v}_2,$$
$$v_2''=ck\dot{v}_2+(cs+c_{12}s_{12})\ddot{v}_2+c_{12}k\dot{v}_1+(cs_{12}+c_{12}s)\ddot{v}_1.$$

Here we may separate the dependent variables by the transformation

$$v_1=w_1+w_2, \qquad v_2=w_1-w_2,$$

* [For "same as," here occurring four times, read "negative of." The accuracy of this correction may be checked by superimposing the current $(E/kl)(1-e^{-2pt})$, the value of which is Et/sl when k is negligible. The result is to turn the curves in Fig. 3 to the corresponding first five in Fig. 1, so that the resultant current and potential are in the same phase and constant ratio. But after that, in the second semi-period, although the potential is annulled by the reflected wave the current is doubled; hence in 5, 6, 7, 8 (Fig. 1), the current and potential (not allowing for attenuation) are in the same phase only where the potential is not cancelled; elsewhere there is current without potential.]

so that the resulting equations

$$w''_1 = (c + c_{12})\{k\dot{w}_1 + (s + s_{12})\ddot{w}_1\},$$
$$w''_2 = (c - c_{12})\{k\dot{w}_2 + (s - s_{12})\ddot{w}_2\}$$

each contain only one dependent variable. These are what we should find for two wires of electrostatic capacity $c + c_{12}$ and $c - c_{12}$ and electromagnetic capacity $s + s_{12}$ and $s - s_{12}$, with the same resistance as the real wires, having no mutual inductive action; and elementary solutions are

$$w_1 = \sin(nx + b)e^{-p_1 t}(A_1 \sin q_1 t + B_1 \cos q_1 t),$$
$$w_2 = \sin(nx + b)e^{-p_2 t}(A_2 \sin q_2 t + B_2 \cos q_2 t);$$

where $-p_1 \pm iq_1$ and $-p_2 \pm iq_2$ are the roots of

$$(c + c_{12})(s + s_{12})D_1^2 + (c + c_{12})kD_1 + n^2 = 0,$$
$$(c - c_{12})(s - s_{12})D_2^2 + (c - c_{12})kD_2 + n^2 = 0,$$

and n is any numerical quantity. Or,

$$p_1 = \frac{k}{2(s + s_{12})}; \quad q_1 = \frac{\sqrt{4n^2 \frac{s + s_{12}}{c + c_{12}} - k^2}}{2(s + s_{12})};$$

and p_2, q_2 are found by changing the signs of s_{12} and c_{12} in p_1 and q_1. The solutions in the last paragraph may be immediately applied. Thus if both wires are to the earth at both ends and E is applied to the first at $x = 0$, we shall have

$$w_1 = \frac{E}{2}\left(1 - \frac{x}{l}\right) - \frac{E}{\pi}e^{-p_1 t}\Sigma\frac{1}{n}\sin\frac{n\pi x}{l}\left(\frac{p_1}{q_1}\sin + \cos\right)q_1 t,$$
$$w_2 = \frac{E}{2}\left(1 - \frac{x}{l}\right) - \frac{E}{\pi}e^{-p_2 t}\Sigma\frac{1}{n}\sin\frac{n\pi x}{l}\left(\frac{p_2}{q_2}\sin + \cos\right)q_2 t$$

for the imaginary wires, and v_1, v_2, the potentials of the real wires, will be the sum and difference of w_1 and w_2. And if η_1, η_2 are the currents corresponding to w_1 and w_2,

$$\eta_1 = \frac{E}{2kl}(1 - e^{-2p_1 t}) + \frac{E}{kl}e^{-p_1 t}\Sigma\cos\frac{n\pi x}{l}\,\frac{2p_1}{q_1}\sin q_1 t,$$
$$\eta_2 = \frac{E}{2kl}(1 - e^{-2p_2 t}) + \frac{E}{kl}e^{-p_2 t}\Sigma\cos\frac{n\pi x}{l}\,\frac{2p_2}{q_2}\sin q_2 t,$$

and γ_1, γ_2 will be the sum and difference of η_1 and η_2.

15. More generally, if the lines are not exactly similar,

$$v_1'' = a_1\dot{v}_1 + b_1\ddot{v}_1 + d_1\dot{v}_2 + f_1\ddot{v}_2,$$
$$v_2'' = a_2\dot{v}_2 + b_2\ddot{v}_2 + d_2\dot{v}_1 + f_2\ddot{v}_1,$$

where a_1, b_1, d_1, f_1, ... may be found by inspecting (2) and (8). If the terminal conditions are the same, assume $d^2/dx^2 = -n^2$, then

$$v_1 = \sin(nx + b)(Ae^{D_1 t} + Be^{D_2 t} + Ce^{D_3 t} + De^{D_4 t}),$$
$$v_2 = \sin(nx + b)(r_1 Ae^{D_1 t} + r_2 Be^{D_2 t} + r_3 Ce^{D_3 t} + r_4 De^{D_4 t});$$

where D_1, ... are the roots of

$$(n^2+a_1D+b_1D^2)(n^2+a_2D+b_2D^2)=(d_1D+f_1D^2)(d_2D+f_2D^2),$$

r_1 is given by

$$0=n^2+a_1D_1+b_1D_1^2+r_1(d_1D_1+f_1D_1^2),$$

and r_2, r_3, r_4 by similar equations containing D_2, D_3, D_4 instead of D_1.

After finding the admissible values of n and b from the terminal conditions, D_1, ... are known in terms of n, and by means of four integrations, A, B, C, D may be determined so as to make the sum of an infinite series of elementary solutions represent initially arbitrary distributions of potential and current.

16. Still more generally, any number m of wires with the same terminal conditions. The result of eliminating from equations (20) all the variables but one is $Pv=0$ (see § 10); the elements of P are now

$$\begin{aligned}
p_{11} &= -\nabla^2+c_1k_1D+(c_1s_1+c_{12}s_{12}+c_{13}s_{13}+c_{14}s_{14}+\dots)D^2,\\
p_{12} &= c_{12}k_1D+(c_{12}s_1+c_2s_{12}+c_{23}s_{13}+c_{24}s_{14}+\dots)D^2,\\
p_{13} &= c_{13}k_1D+(c_{13}s_1+c_{23}s_{12}+c_3s_{13}+c_{34}s_{14}+\dots)D^2,\\
&\cdot\quad\cdot\quad\cdot\quad\cdot\quad\cdot\quad\cdot\quad\cdot\\
p_{21} &= c_{12}k_2D+(c_1s_{12}+c_{12}s_2+c_{13}s_{23}+c_{14}s_{24}+\dots)D^2,\\
p_{22} &= -\nabla^2+c_2k_2D+(c_{12}s_{12}+c_2s_2+c_{23}s_{23}+c_{24}s_{24}+\dots)D^2,\\
p_{23} &= c_{23}k_2D+(c_{13}s_{12}+c_{23}s_2+c_3s_{23}+c_{34}s_{24}+\dots)D^2,\\
&\cdot\quad\cdot\quad\cdot\quad\cdot\quad\cdot\quad\cdot\quad\cdot
\end{aligned}$$

etc.,

where ∇ stands for d/dx and D for d/dt. Putting $-\nabla^2=n^2$, any numerical quantity, $P=0$ is an algebraical equation in D of the $2m$th degree. Let its roots be D_1, D_2, ..., then a system of simultaneous elementary solutions is

$$\begin{aligned}
v_1 &= \sin(nx+b)(A_1e^{D_1t}+A_2e^{D_2t}+A_3e^{D_3t}+\dots),\\
v_2 &= \sin(nx+b)(B_1e^{D_1t}+B_2e^{D_2t}+B_3e^{D_3t}+\dots),\\
v_3 &= \sin(nx+b)(C_1e^{D_1t}+C_2e^{D_2t}+C_3e^{D_3t}+\dots),\\
&\cdot\quad\cdot\quad\cdot\quad\cdot\quad\cdot\quad\cdot
\end{aligned}$$

Here the ratios $A_1:B_1:C_1:\dots$ will be settled by the equations

$$\begin{aligned}
0 &= p_{11}A_1+p_{12}B_1+p_{13}C_1+\dots,\\
0 &= p_{21}A_1+p_{22}B_1+p_{23}C_1+\dots,\\
0 &= p_{31}A_1+p_{32}B_1+p_{33}C_1+\dots,\\
&\cdot\quad\cdot\quad\cdot\quad\cdot\quad\cdot
\end{aligned}$$

where in the values of p_{11}, ... are substituted $-n^2$ for ∇^2 and D_1 for D. Similarly the ratios $A_2:B_2:C_2:\dots$, by similar equations with D_2 for D, instead of D_1. So after finding the admissible values of n and b from the terminal conditions, the elementary solutions above only contain $2m$ arbitrary constants. Therefore, as there are m lines, $2m$ integrations in the ordinary manner, viz., two for every line, will determine the values of these constants for any term in the complete solution. The given data may be the initial values of v and $\dot{v}$, or of v and γ, or of something equivalent, for every line.

17. The influence of defective insulation will now be briefly considered. As every English telegraphist knows, on long lines in very

bad weather the leakage is so great that hardly any current is received from the sending station, sometimes putting a stop to communication altogether. But although leakage when carried too far is a great evil, it nevertheless has its good points and important practical uses. It quickens the signalling, or would do so were the speeds attained anything like what the amount of inductive retardation due to the lines would admit of (such speeds as can be got with Bain's recorder), which is far from being the case with the instruments in present use. But leakage has still its value, viz., in reducing the magnitude of the foreign currents in receiving instruments due to the induction of neighbouring wires. Charging one line, and so raising the potential of a parallel one, causes a flow of electricity from the latter to earth. Now defective insulation not only makes the inducing charge less, and therefore the induced charge less, but also allows a portion of the latter to pass between earth and line through the poles direct, instead of through the receiving instruments.

When a line is permanently charged, the equation of its potential is

$$v'' = h^2 v,$$

where $h^2 = k/i$, and k and i are the resistances per unit of length of the wire and its insulation. The general solution being

$$v = Ae^{hx} + Be^{-hx},$$

if the potential is that produced by an E.M.F., E at $x = 0$, and the wire is earthed at $x = l$, its potential is then

$$v = E \cdot \frac{e^{h(l-x)} - e^{-h(l-x)}}{e^{hl} - e^{-hl}},$$

and the total charge is therefore

$$\int_0^l vcdx = \frac{Ecl}{2} \cdot \frac{2}{hl} \cdot \frac{e^{hl} + e^{-hl} - 2}{e^{hl} - e^{-hl}},$$

instead of $\frac{1}{2}Ecl$ when $h = 0$, and there is perfect insulation.

The influence of leakage is of course greatest on long lines. Let $l = 400$ miles, $k = 15$ ohms, and $i = 1$ megohm per mile, then it will be found that the charge is ·84 of the normal charge $\frac{1}{2}Ecl$ with perfect insulation. And if the insulation is as low as $\frac{1}{4}$ megohm per mile, the charge is only ·59 of the normal charge. The induced charges on neighbouring wires are reduced in the same proportion.

The current is

$$-\frac{v'}{k} = \frac{E}{kl} \cdot hl \cdot \frac{e^{h(l-x)} + e^{-h(l-x)}}{e^{hl} - e^{-hl}}.$$

It is increased from its normal amount E/kl in the first part of the line and reduced in the other part, but it will be found that its mean strength is still E/kl.

When the line is discharged by removing E and earthing at $x = 0$, the integral current of discharge at any point x is

$$\frac{Ecl}{2}(e^{hl} - e^{-hl})^{-1}\left\{ -\frac{1}{hl}(e^{h(l-x)} + e^{-h(l-x)}) + \frac{x}{l}(e^{h(l-x)} - e^{-h(l-x)}) + 2\frac{e^{hx} + e^{-hx}}{e^{hl} - e^{-hl}}\right\},$$

which, for perfect insulation, becomes

$$\frac{Ecl}{2}\left(-\frac{2}{3}+2\frac{x}{l}-\frac{x^2}{l^2}\right).$$

In the line just considered, when $i=1$ megohm, and the charge is ·84 of the normal charge $\frac{1}{2}Ecl$, the discharge at $x=0$ is ·4 of the normal charge, and at $x=l$ it is only ·2 of the normal charge, the remaining ·24 passing to earth direct. And if the insulation is $i=\frac{1}{4}$ megohm, the discharge at $x=0$ is $\cdot 32\times\frac{1}{2}Ecl$, and at $x=l$ it is $\cdot 06\times\frac{1}{2}Ecl$; and as the total charge was $\cdot 59\times\frac{1}{2}Ecl$, the amount of leakage is $\cdot 21\times\frac{1}{2}Ecl$. Thus, with $i=1$ megohm, the leakage reduces the proportion of the discharge at the receiving end from $\frac{1}{3}$ of the original charge to less than $\frac{1}{4}$, and with $i=\frac{1}{4}$ megohm, from $\frac{1}{3}$ with perfect insulation to about $\frac{1}{10}$.

18. Of the three sets of equations, those of continuity, those between the potentials and densities, and those between the currents and electromotive forces, only the first require to be modified to bring in leakage. As in § 7, one expression for the excess of current at x over that at $x+dx$ is $-\gamma' dx$. The other is now $(\dot\rho+v/i)dx$, where the additional term $(v/i)dx$ is the leakage current of dx, viz., the potential of dx divided by its insulation resistance, which is i/dx. Thus the equation of continuity becomes

$$\gamma'+\dot\rho+v/i=0,$$

and the changes in the equations for the potential are easily found.

Equation (21), § 13, for the potential of an isolated line, now becomes

$$v''=\left(k+s\frac{d}{dt}\right)\left(\frac{v}{i}+c\dot v\right) \qquad (25)$$

First putting $s=0$, or ignoring electromagnetic induction compared with electrostatic, which may be done for long lines, (25) becomes

$$v''=h^2v+ck\dot v,$$

whose elementary solution is

$$v=Ae^{-t/ci}\sin(nx+b)e^{-n^2t/ck};$$

thus the general solution for any arbitrary initial potential is the same as (4), § 7, if the summation there be multiplied by $e^{-t/ci}$.

To estimate the relative rapidity of disappearance of an assumed initial potential $A\sin\pi x/l$ with and without leakage, let $l=400$, $k=15$ ohms, $i=1$ megohm, as previously, and $c=\cdot 02$ microfarad. Then $ck=3\times 10^{-7}$ and $ci=\cdot 02$ second. Therefore

$$\frac{ckl^2}{\pi^2}=\frac{48\times 10^{-3}}{\pi^2}=\text{about }\frac{1}{200},$$

and the rates of decay are as 250 : 200, or 5 : 4 in favour of the leakage. If $l=800$ miles, the ratio is 2 : 1. The higher terms in the expansion for v are little affected.

It is worthy of notice that if we neglect electrostatic in comparison with electromagnetic induction, an equation of the same form is obtained. Putting $c=0$ in (25),

$$v''=h^2v+(s/i)\dot v.$$

Here ck has become s/i, and the solution is therefore

$$v = \Sigma A e^{-kt/s} \sin(nx+b) e^{-in^2t/s}.$$

But s/i is of course very small compared with s/k, so that the potential on putting on a battery is established almost instantaneously. The rise of the current, however, depends almost entirely on the time-constant s/k, and is consequently much slower.

Now, reckoning both the inductions, the elementary solution of (25) is

$$v = \sin(nx+b) e^{-pt}(A \sin + B \cos) qt,$$

where

$$p = \frac{k+s/ci}{2s}; \qquad q = \frac{k}{s}\sqrt{\frac{s}{ck^2}\left(n^2+\frac{k}{i}\right) - \frac{1}{4}\left(1+\frac{s}{cki}\right)^2}.$$

Comparing with the corresponding formulæ in § 13 without leakage, it is seen that n^2 becomes $n^2 + k/i$, and k becomes $k + s/ci$. The first change is insignificant, the second of some magnitude. The rate of decay being proportional to p is increased in the ratio $1 : 1 + s/cki$. Taking $s/k = 1/1000$, and $ci = {\cdot}02$, the increase amounts to 5 per cent.

For two parallel wires, the potential equations are

$$v''_1 = \left(k_1 + s_1\frac{d}{dt}\right)\left(\frac{v_1}{i_1} + c_1\dot{v}_1 + c_{12}\dot{v}_2\right) + s_{12}\frac{d}{dt}\left(\frac{v_2}{i_2} + c_2\dot{v}_2 + c_{12}\dot{v}_1\right),$$

$$v''_2 = \left(k_2 + s_2\frac{d}{dt}\right)\left(\frac{v_2}{i_2} + c_2\dot{v}_2 + c_{12}\dot{v}_1\right) + s_{12}\frac{d}{dt}\left(\frac{v_1}{i_1} + c_1\dot{v}_1 + c_{12}\dot{v}_2\right),$$

i_1 and i_2 being the insulation resistances.

If the lines are quite similar,* so that their coefficients are the same, we may use the transformation

$$v_1 = w_1 + w_2, \qquad v_2 = w_1 - w_2,$$

* [Since this paper was written looped wires have come largely into use, that is, a pair of parallel wires are looped to make a single circuit. The equations of potential and current for one or any number of such circuits are particular cases of the equations in this article. When the two wires of each circuit are similar in all respects considerable simplification results. Thus, for a single pair of wires looped, let

$V = v_1 - v_2$ = transverse potential difference of the wires,
C = current in each,
$R = 2k$ = resistance per unit length of circuit,
$S = \frac{1}{2}(c - c_{12})$ = permittance per unit length of circuit,
$L = 2(s - s_{12})$ = inductance ,, ,,
$K = 1/2i$ = leakage-conductance ,, ,,

We shall then have, by the equations in the text,

$$-\frac{dV}{dx} = \left(R + L\frac{d}{dt}\right)C, \qquad -\frac{dC}{dx} = \left(K + S\frac{d}{dt}\right)V,$$

which are the equations connecting the space and time variations of current and transverse voltage; and the characteristic equation is

$$\frac{d^2V}{dx^2} = \left(R + L\frac{d}{dt}\right)\left(K + S\frac{d}{dt}\right)V,$$

which is of the same simple form as for an isolated line (see equation (25) in text). That is, propagation along looped wires may be treated as a case of propagation along a single wire, when the members of the loop are equal. This is the system

obtaining
$$w''_1 = \left(k + (s + s_{12})\frac{d}{dt}\right)\left(\frac{w_1}{i} + (c + c_{12})\dot{w}_1\right),$$
$$w''_2 = \left(k + (s - s_{12})\frac{d}{dt}\right)\left(\frac{w_2}{i} + (c - c_{12})\dot{w}_2\right),$$
which are of the same form as (25) for an isolated wire. Thus, supposing there to be two imaginary lines of capacities $s + s_{12}$, $c + c_{12}$, and $s - s_{12}$, $c - c_{12}$, and having the *same* insulation and conductivity as the real, and without any inductive action on one another, the potential of the real lines may be expressed as the sum and the difference of the potentials of the imaginary lines.

XX.—CONTRIBUTIONS TO THE THEORY OF THE PROPAGATION OF CURRENT IN WIRES.

[Written in 1882, but now first published.]

1. THE four papers in the *Philosophical Magazine*, viz., "On Condenser Working of Cables" (June, 1874); "On the Extra Current" (August, 1876); "On Signalling through Heterogeneous Telegraph Circuits" (1877); "On Faults in Submarine Cables" (1879); and the paper in the *Journal of the Soc. of Tel. Eng. and Electricians*, "On Induction between Parallel Wires" (1881), may, to a great extent, be considered as forming one series, since the subjects are allied and a nearly uniform notation is employed throughout. The present paper may be regarded as a continuation of the last of those mentioned above, and, so far as is convenient, I shall employ the same notation as therein. For brevity it will be referred to as the last paper.*

Let there be a single perfectly insulated wire of length l, uninfluenced by foreign induction, whose electrical constants are c, k, and s; c being the electrostatic capacity, k the resistance, and s the coefficient of self-induction, all per unit length. Let v, γ, and ρ be the potential, current, and linear density of free electricity (charge of the wire per unit length), at a point distant x from the beginning of the line, $(x=0)$. As in § 7 of the last paper, the equation of continuity of the current is
$$\frac{d\gamma}{dx} + \frac{d\rho}{dt} = 0.$$
We have also
$$\rho = cv.$$

of working introduced by me in my paper "On the Self-Induction of Wires," and in "Electromagnetic Induction and its Propagation," to follow later on. It does away with a great deal of complexity, but is subject to certain limitations of application.]

* [It is Art. xix., p. 116. The others referred to are Arts. xiii., p. 47; xiv., p. 53; xv., p. 61; and xvi., p. 71. The present article will be found to be a sort of missing link between the earlier articles on propagation and the later ones, in which the subject is discussed on the basis of Maxwell's theory of the ether as a dielectric.]

The electric force is the sum of that arising from difference of potential, which is $-dv/dx$, and from electromagnetic inertia, which is $-sd\gamma/dt$. We have therefore, by Ohm's law,

$$\left.\begin{aligned} \left(k+s\frac{d}{dt}\right)\gamma &= -\frac{dv}{dx}, \\ \text{and, by continuity,} \qquad \frac{d\gamma}{dx} &= -c\frac{dv}{dt}. \end{aligned}\right\} \qquad (1)$$

To obtain the equation of the current, eliminate v. We obtain

$$\frac{d}{dx}\frac{1}{c}\frac{d}{dx}\gamma = \frac{d}{dt}\left(k+s\frac{d}{dt}\right)\gamma. \qquad (2)$$

To obtain the equation of the potential, eliminate γ. This gives

$$\left(k+s\frac{d}{dt}\right)\frac{d^2v}{dx^2} - \left(k+s\frac{d}{dt}\right)^2 c\frac{dv}{dt} = \left(\frac{dk}{dx}+\frac{ds}{dx}\frac{d}{dt}\right)\frac{dv}{dx}. \qquad (3)$$

As the latter is far more complex than the former it may be ignored. Consider instead the two equations (1); or else (2) for the current, and derive v from γ by means of the first of (1).

It will be understood that k, s, and c, so far, have been treated as variable along the line, functions of x, that is to say. In a submarine cable k and c are constants, whilst s also may be taken as a constant, to a first approximation.

2. The above refers to the line only. The terminal conditions will come in later, as their necessity naturally presents itself. To obtain normal solutions, so that we may determine the variable states arising from any initial states, or the effect of impressed forces, let

$$v = u\epsilon^{Dt}, \qquad \gamma = w\epsilon^{Dt},$$

D being constant (the negative reciprocal of the time-constant of subsidence of the normal system when it happens to be real), and u and w functions of x, but not of t the time. The equation for finding w is (2), turning γ into w and treating d/dt as a constant, viz. $= D$. Its solution is a function of x and D, with two arbitrary constants of integration; that is, there are two independent solutions, whose sum, taken in any proportions, gives w. From it u may be found by the second of (1), writing u for v and treating d/dt as the constant D. This does not bring in any more constants.

Thus, if ∇ stand for d/dx,

$$\left.\begin{aligned} \nabla c^{-1}\nabla w &= D(k+sD)w \quad \text{finds } w, \\ \text{and} \qquad cDu &= \nabla w \quad \text{finds } u \text{ from } w. \end{aligned}\right\} \qquad (4)$$

The admissible values of D are those which are consistent with the terminal conditions, as yet unstated. Leaving them for the present in the background, let D_1 and D_2 be any two admissible values belonging to the pair of normal systems u_1, w_1, and u_2, w_2, which only differ in the value ascribed to D in them, being D_1 in the first and D_2 in the second. We have then, by (4),

$$(a) \quad (k+sD_1)w_1 = D_1^{-1}\nabla c^{-1}\nabla w_1; \qquad (c) \quad cD_1u_1 = -\nabla w_1;$$
$$(b) \quad (k+sD_2)w_2 = D_2^{-1}\nabla c^{-1}\nabla w_2; \qquad (d) \quad cD_2u_2 = -\nabla w_2.$$

Here multiply equation (c) by u_2, and (d) by u_1; subtract the second result from the first, and integrate to x between the limits 0 and l. We get

$$(D_1-D_2)\int cu_1u_2dx=\int(u_1\nabla w_2-u_2\nabla w_1)dx$$

$$=\int\left(\frac{\nabla w_1\cdot\nabla w_2}{cD_2}-\frac{\nabla w_1\cdot\nabla w_2}{cD_1}\right)dx, \qquad (5)$$

the second form on the right being got by again using (c) and (d).

In a similar manner, multiply the equation (a) by w_2 and (b) by w_1; subtract, and integrate; and we obtain

$$(D_1-D_2)\int sw_1w_2dx=\int\left(\frac{w_2}{D_1}\nabla c^{-1}\nabla w_1-\frac{w_1}{D_2}\nabla c^{-1}\nabla w_2\right)dx$$

$$=\left[\frac{w_2}{D_1}c^{-1}\nabla w_1-\frac{w_1}{D_2}c^{-1}\nabla w_2\right]-\int\left(\frac{\nabla w_1\cdot\nabla w_2}{cD_1}-\frac{\nabla w_1\cdot\nabla w_2}{cD_2}\right)dx, \qquad (6)$$

the second line being got by performing one integration by parts. The last integral being the same as that in (5), we have, by subtracting equation (6) from (5),

$$(D_1-D_2)\int_0^l(cu_1u_2-sw_1w_2)dx=[v_2w_1-v_1w_2]_0^l. \qquad (7)$$

Here the square brackets indicate that the quantity within them is to be taken at the limits of integration, 0 and l. In another form

$$\int_0^l(cu_1u_2-sw_1w_2)dx=\frac{[w_1w_2(u_1/w_1-u_2/w_2)]_0^l}{D_1-D_2}. \qquad (8)$$

If the normal systems be one and the same, $D_1=D_2=D$, etc., and (8) gives, on going to the limit,

$$\int_0^l(cu^2-sw^2)dx=\left[w^2\frac{d}{dD}\frac{u}{w}\right]_0^l. \qquad (9)$$

Equation (8) expresses the excess of the mutual potential over the mutual kinetic energy of the two normal systems, so far as the line is concerned, in terms of the potential and current at the ends of the line. But the full interpretation is this:—The mutual potential energy of two normal systems equals their mutual kinetic energy, when the normal systems are completed by taking into account the terminal "apparatus." So that the right member of (8) is the excess of the mutual kinetic over the mutual potential energy of the corresponding two normal arrangements of potential and current in the terminal apparatus.

Since the mutual potential and kinetic energies of two complete normal systems are equal at every moment, the mutual dissipativity is derived from their rates of decrease *equally*. This gives us two other forms of the conjugate property (8); one, containing c and k instead of c and s, expressing that the mutual dissipativity equals half the time-rate of decrease of the mutual potential energy; the other, containing k and s, expressing that the mutual dissipativity equals half the time-rate of decrease of the mutual kinetic energy. We shall not require to write out the mathematical expressions of these statements, as, though entirely different in the operations concerned, they are merely more complex expressions of the same fundamental fact. (It will be found to be of the greatest utility to bear in mind the physical interpretations of results, such enabling us often to avoid useless and tedious work.) If we do not take electromagnetic induction into account, then the mutual potential energy is zero, and consequently also the mutual dissipativity. If it be electrostatic induction that is ignored, the mutual kinetic energy is zero, and also the mutual dissipativity.

3. Equations (7) or (8), of which (8) is the most useful, show us in what form we should get the terminal conditions. Let the potential and the current be subjected to the differential equation

$$Q_0 v = P_0 \gamma \quad \text{.............................} \quad (10)$$

at the $x=0$ end, and to a similar equation at the $x=l$ end. Here P_0 and Q_0 stand for functions of d/dt and constants. This is the general form. (10) is in fact the linear differential equation of the potential and current at the junction of the end of the line with the terminal arrangements, derived solely from the latter themselves. Applying to a normal system, putting D for d/dt, we have

$$Q_0 u = P_0 w \quad \text{at} \quad x=0, \quad \text{........................} \quad (11)$$

$$Q_1 u = P_1 w \quad \text{at} \quad x=l, \quad \text{........................} \quad (12)$$

P_0, P_1, Q_0, and Q_1 being functions of D, which may be either finite or infinite series.

Now, as before remarked, the u and w functions contain two arbitrary constants. One is indeterminate, fixing the size; the other is got rid of thus :—Let

$$u = aG + bH, \qquad w = aI + bJ,$$

where I and J are the two solutions of the equations for w, G and H the corresponding u solutions—equations (4)—and a, b are constants. Taking these expressions for u and w in (11) and (12) we get

$$Q_0(aG_0 + bH_0) = P_0(aI_0 + bJ_0),$$

$$Q_1(aG_1 + bH_1) = P_1(aI_1 + bJ_1);$$

to $_0$ and $_1$ affixed to G, H, I, J, indicating their values at the limits. Or,

$$\left.\begin{aligned} a(Q_0G_0 - P_0I_0) &= b(P_0J_0 - Q_0H_0), \\ a(Q_1G_1 - P_1I_1) &= b(P_1J_1 - Q_1H_1), \end{aligned}\right\} \quad \text{..................} \quad (13)$$

Cross multiplication gives us an equation free from a and b altogether, viz.,

$$(Q_0G_0 - P_0I_0)(P_1J_1 - Q_1H_1) = (P_0J_0 - Q_0H_0)(Q_1G_1 - P_1I_1), \quad \text{......} \quad (14)$$

containing D only. It is the determinantal equation of D, whose roots give us the admissible rates of subsidence. Further, the ratio a/b is known by (13). Therefore, when we say that the normal functions are u and w, ignoring G, H, etc., there is nothing arbitrary about them except a multiplier to fix their absolute magnitude. Thus Au, Aw, the constant A being the same in both, only needs the value of A to be fixed to become definite.

4. If the terminal arrangements be given in detail, and not merely the resultant equations (10) and its companion, to which they lead, we may easily write down the expressions for the parts of the mutual energy of the two normal systems contained in the apparatus, and so immediately obtain the conjugate property (8) in a fully developed form, and employ it to decompose an arbitrarily given initial state of potential and current into normal systems of absolutely determined magnitude. Thus, if

$$U = A_1 u_1 + A_2 u_2 + A_3 u_3 + \dots,$$
$$W = A_1 w_1 + A_2 w_2 + A_3 w_3 + \dots,$$

U and W being the functions of x which express the potential and the current in the line at the time $t=0$; u_1 and w_1, u_2 and w_2, etc., the different normal solutions; and A_1, A_2, ..., constants fixing their magnitudes; the conjugate property will enable us to find the A's, provided that besides U and W being given the initial state of the terminal apparatus, if there be energy therein that can be communicated to the line, is given. The process will be the carrying out of this statement:—The excess of the mutual potential over the mutual kinetic energy of the given initial state and a normal system, equals twice the excess of the potential over the kinetic energy of the normal system itself. Having thus determined the A's, we need merely to introduce the time factor ϵ^{Dt} to obtain the whole subsequent history.

But without any knowledge of the terminal arrangements in detail, but only of the resultant equation (10), we may obtain the complete solution, so far as regards the line, if it be given in addition that there is no energy in the terminal arrangements initially that can be communicated to the line. Besides that, we can, from the form of the terminal equations, determine in what manner it is possible for the terminal apparatus to influence the potential and current in the line, without being able to find whether there is any such influence (except when it is explicitly given, as in the last sentence). That is to say, from the terminal equations we can find the number of degrees of freedom of the terminal apparatus so far as it can affect the line, and the nature of these freedoms as regards energy, and specify them by a number of variables of definite form which may have any initial values. Calling them E_1, E_2, etc., these may be got out of (11) and (12); if they be zero initially, the potential and current in the line is determinate solely from its own initial state, which also determines the subsequent values of the E's themselves. Otherwise, if no information be given regarding the first values of the E's, the future state of the line is indeterminate to just this extent.

Go back to equation (8). In the right member use the equations (11), (12). It becomes

$$\int_0^l (cu_1u_2 - sw_1w_2)dx = \left[\frac{w_1w_2}{D_1 - D_2}\left(\frac{P_{01}}{Q_{01}} - \frac{P_{02}}{Q_{02}}\right)\right]_0^l, \quad \dots\dots\dots\dots (15)$$

P_{01} and P_{02} being the values of P_0 when in it D is made first D_1 and then D_2; and similarly for the Q's. On examining the forms of P and Q, equations (11), (12), we shall find that the division by $D_1 - D_2$ can be effected. When this has been done, let the quotient be arranged in the form of a series of products, every product having one factor a function of D_1 and the other the *same* function of D_2. Thus, let the right member of (15) be arranged in the form

$$r_1 f_1(D_1)f_1(D_2) + r_2 f_2(D_1)f_2(D_2) + r_3 f_3(D_1)f_3(D_2) + \ldots = \Sigma r f(D_1)f(D_2), \quad (16)$$

wherein the r's are constants and the f's functions of D. Some of these will belong to one end of the line, the rest to the other end, according to the nature of the terminal equations. We thus get (15) to the form

$$\int(cu_1u_2 - sw_1w_2)dx = \Sigma r f(D_1)f(D_2). \quad \ldots\ldots\ldots\ldots (17)$$

As regards the above decomposition, which is easily enough effected in simple cases, since it may be carried out in various ways, the functions f obtained may not be all independent, but a comparison of their forms will enable us to reduce them to a certain number of independent functions. This is best studied in the concrete application.

A practical way of decomposing

$$w_1w_2\frac{P_{01}/Q_{01} - P_{02}/Q_{02}}{D_1 - D_2}$$

into products is not to do it, but to decompose

$$w^2\frac{d}{dD}\frac{P_0}{Q_0}$$

into sums of squares, say

$$r_1 f_1^2(D) + r_2 f_2^2(D) + r_3 f_3^2(D) + \ldots,$$

which is more easily effected (perhaps at sight); then we only require for $f^2(D)$ to take $f(D_1)f(D_2)$ to obtain the required form (16).

5. Having now cleared the ground, let

$$\left.\begin{aligned} U &= \Sigma A_n u_n, \\ W &= \Sigma A_n w_n, \\ E_1 &= \Sigma A_n f_1(D_n), \\ E_2 &= \Sigma A_n f_2(D_n), \\ E_3 &= \Sigma A_n f_3(D_n), \\ &\cdot\quad\cdot\quad\cdot\quad\cdot \end{aligned}\right\} \quad \ldots\ldots\ldots\ldots\ldots\ldots (18)$$

Here the summations include all values of D. The nth A is A_n, the nth u is u_n, and so on. U and W are the initial potential and current in the line, and E_1, E_2, etc., are quantities defined by the above expressions. Apply the conjugate property to the quantities U, W, and the E's in this way:—Multiply the first equation by cu and integrate to x from 0 to l; the second by sw and integrate similarly; the third by $r_1 f_1(D)$; the fourth by $r_2 f_2(D)$, etc.; add all the results, taking however all except the first negatively. Thus

$$\int (cUu - sWw)dx - E_1 r_1 f_1(D) - E_2 r_2 f_2(D) - \dots$$
$$= \Sigma A_n \int (cuu_n - sww_n)dx - \Sigma A_n r_1 f_1(D) f_1(D_n) - \Sigma A_n r_2 f_2(D) f_2(D_n) - \dots .$$

Remembering here that u, w, D refer to a definite, whilst u_n, etc., refer to *any* normal system, and that the summations are with respect to n; and, lastly, that equation (17) makes all terms cancel one another except those containing squares, belonging to the definite normal system, the equation reduces to

$$\int (cUu - sWw)dx - E_1 r_1 f_1(D) - E_2 r_2 f_2(D) - \dots$$
$$= A\left\{\int (cu^2 - sw^2)dx - r_1 f_1^2(D) - r_2 f_2^2(D) - \dots\right\}, \quad \dots\dots (19)$$

which gives us the value of A explicitly in terms of the initial state U, W, and of the arbitrary quantities E.

We may at once put the quantity in the { } in (19) into a much simpler form. For the value of

$$\int (cu^2 - sw^2)dx \quad \text{is} \quad \left[w^2 \frac{d}{dD}\frac{u}{w}\right]_0^l,$$

and that of

$$\Sigma r f^2(D) \quad \text{is} \quad \left[w^2 \frac{d}{dD}\frac{P}{Q}\right]_0^l,$$

so that (19) may be written

$$A = \frac{\int_0^l (cUu - sWw)dx - E_1 r_1 f_1(D) - E_2 r_2 f_2(D) - \dots}{\left[w^2 \frac{d}{dD}\left(\frac{u}{w} - \frac{P}{Q}\right)\right]_0^l}. \quad \dots\dots\dots (20)$$

If therefore the expansions (16) are possible, this value of A given by (20) used in (16) makes them identities. They represent the initial state. At the time t later we shall have

$$v = \Sigma A u \epsilon^{Dt}, \qquad \gamma = \Sigma A w \epsilon^{Dt}, \qquad e_1 = \Sigma A f_1(D) \epsilon^{Dt}, \text{ etc.}, \dots\dots\dots (21)$$

v and γ being the potential and current in the line at x, and the small e's what the big E's then become.

At the first moment the E's have no effect on the values of U and W. That is, if we leave out all the auxiliary terms in the numerator of (20) we shall still have

$$U = \Sigma A u, \qquad W = \Sigma A w,$$

identically. So far as the decomposing of U and W into normal functions is concerned, we need not trouble ourselves at all about the terminal conditions except just so far as to find the determinantal equation of the D's and the ratio a/b (§ 3), the value of A then being (20) with the E's put $= 0$.

It follows that if, for brevity, M denote the denominator in (20),

$$\Sigma f_1(D)u/M=0, \qquad \Sigma f_2(D)u/M=0, \text{ etc.,}$$

and
$$\Sigma f_1(D)w/M=0, \qquad \Sigma f_2(D)w/M=0, \text{ etc.,}$$

identically. For instance, the part contributed by E_1 to U is

$$-E_1r_1\Sigma f_1(D)u/M,$$

and is, of course, zero. These identities are true for values of x between 0 and l, but not at the limits themselves. When the t factor is introduced these summations no longer vanish, except when $t=0$. The part contributed to v by E_1 is

$$-E_1r_1\Sigma f_1(D)u\epsilon^{Dt}/M.$$

On the other hand, although the E's may be zero at first, the e's, which equal them then, immediately become finite, by the influence of the charge and current in the line.

6. It now becomes necessary to say a few words on the subject of the assumption made in the preceding section, that the expansions (18) are possible. It is assumed to be possible to expand any initial U in an infinite series of normal u's, and any initial W in a series of w's; and at the same time it is assumed that the summations E, not containing x, but only the terminal values of u and w, can be made to have any values we please. For instance, one E might be ΣAu_1, if u_1 for the moment denote the value of u at $x=l$; another might be $\Sigma Au_1/D$, and so on. Not only then has ΣAu to have a prescribed value at every point between the limits, but at the limits themselves it, and the sums of various functions of u_1, are to have prescribed values.

Supposing it to be stated thus, simply. Given a function u, in which D has to have a series of values in succession, can ΣAu be made to represent any single valued function? Considered merely from the mathematical point of view, and without other information, it is difficult to return either a positive or a negative reply, unless an explicit demonstration be framed showing the complete transition of U into ΣAu, which would be in general very difficult to arrange.

If, on the other hand, we regard the question from the physical point of view, the possibility of the expansion does not seem to require any demonstration. We are investigating a real physical problem, in which we know the detailed relations to be dynamically consistent, and which has necessarily a solution. When therefore we arrive at the solution in the form of a series, satisfying all the conditions completely and requiring that U should be identically made $=\Sigma Au$, and knowing how to do it if it be possible, why should it not be possible? Familiarity with the working out of physical problems breeds contempt for the idea of requiring a special demonstration of the possibility of what seems to be necessary.

But the argument that when the complete solution of a physical problem requires a certain expansion to be effected, such expansion must be possible, does not cover the whole ground. For, as a matter of fact, the expansions are possible when the constants involved have values given to them which are physically impossible, making negative resistances for example. It is also necessary to state what terminal conditions are, and what are not admissible. For example, why should it not be possible to expand U in the series $\Sigma A \sin ax/l$, subject to $\cos a = n$, a constant? This is not possible; yet, *a priori*, it would seem just as likely to be possible as when a is subjected to some other equation. Now when the enormous number of different kinds of normal functions is considered, as well as the enormous number of different ways of expanding a function in a series of one kind of normal function, it is plainly taking too narrow a view of the matter to devise special proofs of possibility, or to think they are needed.

The shortest way to arrive at thorough conviction, logical and legitimate, is to go to the theory of ordinary linear differential equations, and observe the manner in which the genesis of partial differential equations and their solutions takes place. Instead of a line along which capacity, conductivity, and self-inductive power are continuously distributed, consider a number of coils, joined in sequence, whose terminals are joined to the earth through condensers. That is, we replace an infinite number of degrees of freedom by a finite number. We have one degree of freedom for every condenser, and one for every coil; say n altogether. Given the state of this system at a given moment, to find its state at any time after, when left to itself without impressed force, requires the solution of a linear ordinary equation of the nth degree, in which t, the time, is the independent variable. There are n rates of subsidence, and n normal arrangements of charge and current, any one of which, subsiding at its proper rate, is a possible solution. We require, then, to decompose the initial state of charge and current into these n arrangements. That it is possible requires no proof; it merely means finding the values of n constants out of n equations. Or, simply use the conjugate property of the normal solutions. A normal solution here means the solution belonging to a single D, whether real or imaginary.

Now by subdividing the coils and condensers, having any resistances and self-inductive powers and capacities, we can, still keeping the number of degrees of freedom finite, approximate as nearly as we please to any continuous distribution of k, c, and s, along a line. It is therefore true that we can decompose the initial state into normal solutions however nearly we approach the conditions belonging to the continuous distributions. Therefore, since a breach of continuity is out of the question, it must be true in the limit, when the ordinary linear differential equation becomes of infinite degree, and is equivalent to a partial differential equation adapted to certain conditions. We may now well ask, not whether the decomposition is possible, but what is there to prevent it?

It is possible if the normal functions are quite comprehensive, and represent every mode of subsidence that the electrical conditions permit. Just as $\cdot 333\ldots$ must be continued to infinity to reach $\frac{1}{3}$, so every normal solution is, in general, needed to make up the complete solution. Be sure then that we have got all the normal solutions, and possibility becomes certainty.

As regards the data initially required, they are exactly those quantities which, in the case of a finite number of degrees of freedom, are then required; viz., the current, wherever there is electromagnetic induction, and the charge, wherever there is electrostatic induction; or equivalent information. The terminal arbitraries E merely represent these quantities, or functions of them, in the terminal apparatus.

The terminal conditions themselves must be expressible in the form (10), or

$$\phi_1\left(\frac{d}{dt}\right)v = \phi_2\left(\frac{d}{dt}\right)\gamma,$$

ϕ_1 and ϕ_2 containing first powers only of the differentiators d/dt, d^2/dt^2, etc. For any combination of coils and condensers gives rise to this form ultimately. The terminal arrangements may include continuous distributions of capacity, etc., involving partial differential equations, but the above form can be got out of them. Why the above-mentioned expansion $U = \Sigma A \sin(x/l \cos^{-1} n)$ is not possible, is because the condition $\cos a = n$ cannot arise from the above form of terminal condition. The constants in (10) may, however, be positive or negative, irrespective of physical reality. This will be understandable by the consideration of a finite number of degrees of freedom.

Special cases, involving discontinuities, equal roots, etc., require special treatment. Strictly, U and W must be single-valued at every point along the line. U, for instance, cannot have two values at a point, however rapidly it may change value. If then we say that $U=1$ from $x=0$ to $x=\frac{1}{2}l$, and $U=2$ from $x=\frac{1}{2}l$ to $x=l$, we must suppose that the double value 1 and 2 at $x=\frac{1}{2}l$ really means an infinitely steep rise from the value 1 just before to 2 just after $x=\frac{1}{2}l$, so that at that point the mean value is $1\frac{1}{2}$. Discontinuities of this kind may exist in any number at the first moment. It is the function of the infinitely numerous normal solutions with large values of D to enable us to represent discontinuities.

If the determinantal equation of the D's have equal roots, the denominator in the expression (20) for A always vanishes, making A apparently infinite for the equal D's. But when by changing the value of a constant we make two roots approach to equality, the corresponding A's do not also approach, but separate infinitely widely, viz., to $+\infty$ and $-\infty$, if the numerator in (20) remains finite. Just before vanishing the denominator of one A is positive and of the other negative. An example of this occurs in the theory of a cable with its end earthed through a coil. (That equality of roots makes the denominator vanish may be seen by comparing the form of the denominator with that of the terminal equations, out of which the determinantal equation arises; remembering that for $\phi(D)=0$ to have equal roots, $\frac{d}{dD}\phi(D)$ must also vanish.) In such a case we must unite the terms corresponding to the equal roots, before equalising them, to obtain the proper result. But there are cases requiring different treatment in which equal roots make the normal functions, and therefore the numerator, vanish as well as the denominator of A.

7. *Special Case. c and k constant, s = 0. Fourier Series.*

The above being general, with c, k, s functions of x, must be taken as an example of method. In the complexity and irregularity of special cases we are apt to lose sight of general principles if already known, or not to recognise them readily if not known. Hence the value of general investigations. Let us next deduce from the preceding the solution in the case of a long uniform line in which electrostatic induction predominates, by taking c and k constant, and $s = 0$.

Equations (1) become

$$\nabla\gamma = -c\dot{v}, \qquad k\gamma = -\nabla v;$$

and the equation of v is

$$\nabla^2 v = ck\dot{v},$$

that of γ being the same. Hence the equation of the normal functions is

$$\nabla^2 w = ckDw.$$

Thus, if $D = -a^2/ckl^2$, we shall have

$$\left.\begin{aligned} u &= \sin(ax/l + b), \\ w &= -\frac{\nabla u}{k} = -\frac{a}{kl}\cos(ax/l + b); \end{aligned}\right\}$$

a and b being any constants. Hence

$$\left.\begin{aligned} v &= \Sigma A \sin(ax/l + b)\epsilon^{Dt}, \\ \gamma &= -\frac{1}{k}\nabla v \end{aligned}\right\} \qquad (22)$$

give the solution at time t, wherein a, b, and D have to be settled from the terminal conditions, and A from the initial state. Only the potential requires to be given, for, on account of $s = 0$, γ is settled by v.

Here
$$\frac{u}{w} = -\frac{kl}{a}\tan\left(\frac{ax}{l} + b\right),$$

so that we have, by (11), (12),

$$\left.\begin{aligned} \tan b &= -\frac{a}{kl}\frac{P_0}{Q_0} = \phi_0(a), \text{ say,} \\ \tan(a + b) &= -\frac{a}{kl}\frac{P_1}{Q_1} = \phi_1(a), \text{ say,} \end{aligned}\right\} \qquad (23)$$

P_0, Q_0, P_1, Q_1 being functions of D, and therefore of a^2. On account of the multiplication by a we see that ϕ_0 and ϕ_1 are always odd functions of a.

Here

$$\tan a = \tan\{\tan^{-1}\phi_1(a) - \tan^{-1}\phi_0(a)\} = \frac{\phi_1(a) - \phi_0(a)}{1 + \phi_1(a)\phi_0(a)} = \phi(a) \text{ say.} \qquad (24)$$

This is also an odd function of a.

Draw the curves $y = \tan a$, and $y = \phi(a)$; their intersections show where the real roots are. The imaginary roots are certainly not in the plane of the diagram, and it is often difficult to find them, or even to suspect their existence. A pair of imaginaries near the origin ($a = 0$) may have a most important influence. The introduction of a is convenient, instead of D, on account of the circular functions. But since D varies as a^2, and the roots of $\tan a = \phi(a)$ are symmetrically placed with respect to the origin, we need only take one half of them, say the + half.

The numerator of A, (20), becomes

$$\int Udx - \Sigma Erf(a),$$

if we divide by c; and the denominator, similarly divided by c, using

$$\frac{d}{dD} = -\frac{ckl^2}{2a}\frac{d}{da},$$

and effecting some reductions, becomes

$$\frac{l}{2}\left(1 - \cos^2 a \frac{d}{da}\phi(a)\right);$$

hence the value of A is

$$A = \frac{\int U \sin(ax/l + b) - \Sigma Erf(a)}{\frac{l}{2}\left(1 - \cos^2 a \frac{d}{da}\phi(a)\right)}. \quad \ldots\ldots\ldots\ldots\ldots\ldots (25)$$

This, used in (22), makes the solution complete, the a's and b's being subject to (24) and the first of (23).

The formula (25) was given in the last paper,* equation (5), omitting the undetermined E's. But the later formula (5b), referring to an infinitely long line, requires additional terms not mentioned in that paper, should there happen to be imaginary roots. To show how the additions are obtained it will be convenient to go over a portion of the general investigation, in terms of a instead of D.

8. On account of $\nabla^2 = ckD$, we may write the terminal equations (11), (12) thus,

$$0 = h_0 v + h_1 l v' + h_2 l^2 v'' + h_3 l^3 v''' + \ldots \quad \text{at} \quad x = l,$$
$$0 = k_0 v + k_1 l v' + k_2 l^2 v'' + k_3 l^3 v''' + \ldots \quad \text{at} \quad x = 0;$$

i.e., we put double differentiations to x in place of singles to t. Take

$$v = u = \sin(ax/l + b),$$

and we obtain

$$\left.\begin{aligned} \tan(a+b) &= -\frac{h_1 a - h_3 a^3 + h_5 a^5 - \ldots}{h_0 - h_2 a^2 + h_4 a^4 - \ldots} = \phi_1(a), \\ \tan b &= \text{ same function of } k\text{'s } = \phi_0(a). \end{aligned}\right\} \ldots\ldots\ldots\ldots (25a)$$

Now the special form of (17) is

$$\frac{1}{l}\int_0^l \sin(a_1 x/l + b_1)\sin(a_2 x/l + b_2)dx$$
$$= \frac{a_1 a_2 \cos(a_1 + b_1)\cos(a_2 + b_2)}{a_1^2 - a_2^2}\left(\frac{\tan(a_1 + b_1)}{a_1} - \frac{\tan(a_2 + b_2)}{a_2}\right)$$
$$- \frac{a_1 a_2 \cos b_1 \cos b_2}{a_1^2 - a_2^2}\left(\frac{\tan b_1}{a_1} - \frac{\tan b_2}{a_2}\right).$$

* [Art. xix., p. 123, *ante.*]

Here, in the right member, we may put for the quantities in the large brackets their values in terms of ϕ_0 and ϕ_1, and then get it into the form

$$r_1 f_1(a_1,b_1)f_1(a_2,b_2) + r_2 f_2(a_1,b_1)f_2(a_2,b_2) + \dots .$$

This being done, let

$$U = \Sigma A \sin(ax/l + b), \qquad E_1 = \Sigma A f_1(a,b), \qquad E_2 = \Sigma A f_2(a,b), \text{ etc.,}$$

the summations to include all the a's; then apply the already obtained conjugate property,

$$\frac{1}{l}\int u_1 u_2 dx = \Sigma r f(a_1,b_1) f(a_2,b_2).$$

We arrive at

$$v = \Sigma \sin(ax/l + b) \frac{\int U \sin(ax/l + b)dx - \Sigma r E f(a,b)}{\frac{l}{2}\left(1 - \cos^2 a \frac{d}{da}\phi(a)\right)} \epsilon^{Dt}. \quad \dots\dots\dots (26)$$

Examine now what happens when the length of the line is indefinitely increased. As we shall ultimately have only one end of it ($x = 0$) at disposal to operate on by terminal appliances, we may take the $x = l$ condition to be $v = 0$. This makes $\tan(a + b) = 0$, and

$$\tan a = -\tan b = -\phi_0(a) = \phi(a).$$

Next put $a/l = z$. We shall then have the second of (25a) become

$$\tan lz = c_1 z + c_3 z^3 + c_5 z^5 + \dots = Z, \text{ say,} \quad \dots\dots\dots\dots\dots\dots (27)$$

wherein the c's do not contain l; or else Z may be any odd function of z.

Drawing the curves $y = \tan lz$, and $y = Z$, y being the ordinate and z the abscissa, the Z curve will extend to infinity both ways, though only the positive half need be considered. On increasing l indefinitely the tangent curves get packed infinitely closely together, making the Z curve cut them at (ultimately) equidistant infinitely small intervals $dz = \pi/l$. That is to say, any value of z is then a root of the determinantal equation (27), so that z varies continuously from zero to infinity, and the summation (26) becomes a definite integral.

This takes no count of the imaginary roots, but only of the real roots with real intersection of the Z and tangent curves. Put $\tan lz = Z$ in the exponential form

$$(\epsilon^{lzi} - \epsilon^{-lzi}) = Zi(\epsilon^{lzi} + \epsilon^{-lzi}),$$

and make l infinite. It becomes

$$\epsilon^{lzi}(1 - Zi) = 0,$$

i standing for $(-1)^{\frac{1}{2}}$. Here $\epsilon^{lzi} = 0$ must correspond to the real roots, and

$$Zi = 1, \quad \text{or} \quad Z = -i, \quad \dots\dots\dots\dots\dots\dots\dots\dots (28)$$

to the imaginaries. We have therefore a summation, not a definite integral, for the imaginary roots, should there have been any when l was finite.

We may remark here that it is quite indifferent whether the line be assumed to be earthed or insulated at its distant end. In the former case we shall have

$$\left.\begin{aligned}\sin(lz+b)&=0,\\ \tan b&=-Z,\end{aligned}\right\}\quad\therefore\ \tan lz=Z,$$

Z depending on the $x=0$ connections; and in the latter case,

$$\left.\begin{aligned}\cos(lz+b)&=0,\\ \tan b&=-Z,\end{aligned}\right\}\quad\therefore\ -\cot lz=Z.$$

There is a great difference when l is finite; but on making l infinite the exponential forms become

$$\epsilon^{lzi}(1-Zi)=0,\quad\text{in first case,}$$
$$\epsilon^{lzi}(i+Z)=0,\quad\text{in second,}$$

which are equivalent. It is also indifferent, in the final results, whether we take $Z=i$ or $-i$.

The general term of the expansion of U may be written, by taking $t=0$ in (26), expanding the sines, and using (27),

$$(\sin zx-Z\cos zx)\frac{\int U(\sin zx-Z\cos zx)dx}{\frac{1}{2}\left(\frac{l}{\cos^2 b}-\frac{dZ}{dz}\right)},\ \ldots\ldots\ldots\ldots\ (29)$$

when l is finite, if the E's be zero. Here

$$1/\cos^2 b=1+\tan^2 b=1+Z^2,$$

thus making the denominator of (29) be

$$\tfrac{1}{2}\left\{l(1+Z^2)-\frac{dZ}{dz}\right\}.$$

In the case of the real roots $l^{-1}=dz/\pi$, and in case of the imaginaries $l(1+Z^2)=0$ (because $Zi=1$); so that the denominator of (29) is

$$\text{(real)}\quad\frac{\pi}{2}\cdot\frac{1+Z^2}{dz}-\frac{1}{2}\frac{dZ}{dz},\quad\text{or else}\quad-\frac{1}{2}\frac{dZ}{dz}\quad\text{(imaginary)};$$

giving

$$U_r=\frac{2}{\pi}\int_0^\infty dz\int_0^\infty dx_1U_1(\sin zx-Z\cos zx)\frac{\sin zx_1-Z\cos zx_1}{1+Z^2}\ \ldots\ldots\ldots\ (30)$$

as the part of U depending on the reals (the E's being zero), and

$$U_i=-2\Sigma\left(\frac{dZ}{dz}\right)^{-1}(\sin zx-Z\cos zx)\int_0^\infty U_1(\sin zx_1-Z\cos zx_1)dx_1\quad(31)$$

as the part depending upon the imaginaries. Or, more concisely, since in the latter $Z=-i$,

$$U_i=2\Sigma\left(\frac{dZ}{dz}\right)^{-1}\epsilon^{-zxi}\int_0^\infty U_1\epsilon^{-zx_1i}dx_1.\ \ldots\ldots\ldots\ldots\ldots\ldots\ (32)$$

The sum of (30) and (32) gives $U = U_r + U_i$, the value of the initial potential at x; whilst under the integral sign, x_1 being the variable, U_1 is the value at x_1. In (32), the summation has only to include the imaginary roots of $Z = -i$, not however counting the single, if it be a cubic for example, but only the pair.

The corresponding extra terms on account of the auxiliaries E can naturally only be fully written down when their actual expressions are known. We are only concerned with those belonging to $x=0$, therefore functions of $\sin b$, or $\cos b$, or of z. Taking $E_1 = \Sigma A f_1(z)$, when l is finite, let V_1 be the part of U depending upon it (zero in value); then, when l is infinite,

$$V_1 = -2E_1 r_1 \int f_1(z)(\sin zx - Z\cos zx)dz - 2E_1 r_1 \Sigma f_1(z)\left(\frac{dZ}{dz}\right)^{-1} \epsilon^{-zxi}, \quad (33)$$

the definite integral coming from the real, and the summation from the imaginary roots. At the time t, v at x will be given by the sum of (30), (32), (33) and its companions, after the requisite introduction of the time factor $\epsilon^{-z^2 t/ck}$ under the integral sign. Some examples follow.

9. *Resistance R at $x=0$.* First suppose the line is earthed direct at $x=l$, and through a resistance R at $x=0$. Here we shall have

$$v=0 \text{ at } x=l; \qquad v=\frac{R}{k}v' \text{ at } x=0;$$

so, if $m = R/kl$, we get

$$\sin b = ma \cos b, \quad \text{and} \quad \sin(a+b)=0,$$

from which

$$\tan a = -ma$$

determines the admissible a's. And

$$U = \frac{2}{l}\Sigma \frac{\sin zx + ma\cos zx}{1+m+m^2a^2}\int U_1(\sin zx_1 + ma\cos zx_1)dx_1,$$

the expansion when l is finite, becomes, if $n = R/k$,

$$U = \frac{2}{\pi}\iint dz dx_1 \frac{U_1}{1+n^2z^2}(\sin zx + nz\cos zx) \text{ (ditto with } x_1),$$

when l is infinite; which, on introducing the time factor, gives v at x at time t. The values $n=0$ and $n=\infty$ give the common

$$U = \frac{2}{\pi}\iint dz dx_1 U_1 \sin zx \sin zx_1,$$

and the same with cos instead of sin. The roots are all real in this example. There are also no terminal arbitraries, physically because there is neither potential nor kinetic energy concerned in the mere resistance; mathematically because

$$\frac{d}{da}\frac{\tan a}{a} = 0.$$

10. *Condenser discharged into line.* Put a condenser of capacity C between line and earth at $x=0$; then

$$\left.\begin{array}{ll} C\dot{v}=k^{-1}v' & \text{at} \quad x=0, \\ v=0 & \text{at} \quad x=l, \end{array}\right\} \therefore \quad lv'=rl^2v'', \quad \text{if} \quad C/cl=r,$$

from which $\tan a = 1/ra$ finds the a's. Here

$$A=\frac{\frac{1}{l}\int U_1 \sin(zx_1+b)dx_1+rE\sin b}{\frac{1}{2}(1+r\sin^2 a)},$$

if E be the initial potential of the condenser, or

$$E=\Sigma A\sin b.$$

As the roots are in this case also all real, the denominator of A becomes $\frac{1}{2}$ when $l=\infty$; and then, if $C/c=n$, we have

$$U=\frac{2}{\pi}\iint dzdx_1\, U_1 \sin(zx+b)\sin(zx_1+b)+\frac{2}{\pi}\int dz\sin(zx+b)\,.\,nE\sin b,$$

subject to $\tan b=-1/nz$. If the line be originally uncharged there is only the second term to consider, and we have

$$e=nE\frac{2}{\pi}\int_0^\infty dz\sin^2 b\,\epsilon^{-z^2t/ck}=nE\frac{2}{\pi}\int_0^\infty\frac{\epsilon^{-z^2t/ck}}{1+n^2z^2},$$

showing what E becomes at time t (viz., value of v at $x=0$), after commencement of the discharge. The current entering the line is

$$\gamma_0=\frac{2}{\pi}\int dz\frac{nE}{k}\frac{nz^2}{1+n^2}\epsilon^{-z^2t/ck}.$$

The integral current should equal the original charge Enc, which is easily verified. The integral current at any point should also equal Enc. At x we have

$$\gamma=\frac{2}{\pi}\frac{nE}{k}\int\frac{dz}{1+n^2z^2}(z\sin zx+nz^2\cos zx)\epsilon^{-z^2t/ck},$$

giving
$$\int_0^\infty\gamma dt=\frac{2}{\pi}ncE\int\frac{dz}{1+n^2z^2}\left(n\cos zx+\frac{\sin zx}{z}\right);$$

which, by the known integrals,

$$\int_0^\infty dz\frac{n\cos zx}{1+n^2z^2}=\frac{\pi}{2}\epsilon^{-x/n},\qquad\int_0^\infty\frac{dz}{z}\frac{\sin zx}{1+n^2z^2}=\frac{\pi}{2}\left(1-\epsilon^{-x/n}\right),$$

gives the required result, ncE.

The sum of the condenser's charge and the whole charge in the line at any moment should also come to ncE. The former is

$$Q_1=n^2cE\frac{2}{\pi}\int dz\sin^2 b\,\epsilon^{-\cdots}=n^2cE\frac{2}{\pi}\int dz\frac{\epsilon^{-z^2t/ck}}{1+n^2z^2},$$

and the latter is

$$Q_2=ncE\frac{2}{\pi}\iint dzdx_1\sin(zx_1+b)\sin b\,\epsilon^{-z^2t/ck}.$$

Also $Q_2=0$ when $t=0$, and $=ncE$ when $t=\infty$. The first is easily tested. The other needs interpretation; for, when $t=\infty$, the time factor vanishes, and $v=0$ everywhere. So we must, to have consistency, keep t finite and integrate to t before making it infinite. The value of Q_1, if found, would give that of Q_2 at any moment. But Q_1 is only integrable in a series. In terms of another simpler integral,

$$Q_1 = ncE\frac{2}{\pi}\left(\sqrt{\pi}\,\epsilon^{t/ckn^2}\int_{t_1}^{\infty}\epsilon^{-z^2}dz\right),$$

the lower limit t_1 being $=(t/ckn^2)^{\frac{1}{2}}$.

11. *Condenser and Coil at* $x=0$.

Let $C=rcl$ be the capacity of the condenser, $R=mkl$ the resistance of the coil, and $L=skl\,.\,ckl^2$ the coefficient of self-induction of the coil, one end of which is joined to the line and the other end to one terminal of the condenser, which is earthed at the other terminal. At $x=0$ we shall have, if v_1 be the potential of the condenser on the side next the coil, at time t,

$$\left.\begin{aligned} k^{-1}v' &= rcl\dot{v}_1, \\ v - v_1 &= (R+LD)k^{-1}v', \end{aligned}\right\} \qquad (34)$$

which lead to

$$0 = lv' - rl^2v'' + rml^3v''' + rsl^5v''''',$$

and give us the determinantal equation

$$\tan b = -1/ra + ma - sa^3,$$

supposing the line earthed at $x=l$. ($\tan b = -\tan a$, when this is the case, always.)

The solution being

$$v = \Sigma A \sin(zx+b)\epsilon^{-z^2t/ck} \qquad (35)$$

for the potential at x at time t, the value of A is

$$A = \frac{\frac{1}{l}\int_0^l U\sin(zx+b)dx + rE_1/ra\cos b - sE_2a\cos b}{\frac{1}{2}\{1+\cos^2 b(1/ra^2+m-3sa^2)\}},$$

by the general method. Here

$$E_1 = \Sigma A/ra\cos b, \qquad E_2 = \Sigma Aa\cos b,$$

which may be got by considering the energy of coil and condenser; or, without energy considerations, thus:—

$$\frac{1}{l}\int_0^l \sin(z_1x+b_1)\sin(z_2x+b_2)dx = -\frac{a_1a_2\cos b_1\cos b_2}{a_1^2-a_2^2}\left(-\frac{1}{ra_1^2}+\frac{1}{ra_2^2}-sa_1^2+sa_2^2\right)$$

$$= -r\,.\,\frac{\cos b_1}{ra_1}\,.\,\frac{\cos b_2}{ra_2} + s\,.\,a_1\cos b_1\,.\,a_2\cos b_2,$$

by following method (25*a*); which we see is put into the right form, as there described. To find what E_1 means, we have

$$v_1 = v - \left(R+\frac{L}{ck}\frac{d^2}{dx^2}\right)k^{-1}\frac{dv}{dx} = -\Sigma\frac{A\cos b}{ra},$$

by (34) and (35). Hence E_1 is the negative of the initial potential of the condenser. Similarly we may show that

$$-E_2 = Gkl,$$

if G be the initial current in the coil (reckoned positively from condenser to line).

Thus, in the expression for A, E_1 and E_2 show the effect of the condenser and the coil, if originally charged or with current respectively, on the subsequent state. Putting $-E_1 = V$, and making $l = \infty$, we obtain, U_r being the part of U arising from real roots,

$$U_r = \frac{2}{\pi}\iint dz dx_1 \sin(zx+b)\sin(zx_1+b)U_1$$

$$-\frac{2}{\pi}\int dz \sin(zx+b)\frac{V\cos b}{z} + \frac{2}{\pi}\int dz \sin(zx+b)\frac{LG}{ck}z\cos b, \quad \text{......} \quad (36)$$

the limits of all the integrals being 0 and ∞ as usual; the double integral arising from the original charge in the line, the next from the condenser, and the third from the coil. They are subject to

$$\tan b = -1/n_1 z + n_2 z - n_3 z^3,$$

where $$n_1 = C/c, \qquad n_2 = R/k, \qquad n_3 = L/ck^2.$$

The above is the complete U only if $L=0$. There are no imaginaries then; but with L finite there will be additional terms, to be later brought in. At present let $L=0$, and the line be initially uncharged, so that only the second integral in (36) is wanted. Then

$$v = -\frac{2V}{\pi}\int_0^\infty dz \sin(zx+b)\cos b\frac{\epsilon^{-z^2t/ck}}{z},$$

$$\gamma = \frac{2V}{k\pi}\int_0^\infty dz \,.\, \cos(zx+b)\cos b \,.\, \epsilon^{-z^2t/ck},$$

represent the potential and current at time t, resulting from the discharge of the condenser to the line. The integral current should be Vn_1c. Examining this we find that it requires

$$n_1 = \frac{2}{\pi}\int_0^\infty \frac{dz}{z^2} \,.\, \frac{\cos zx - \sin zx(-1/n_1 z + n_2 z)}{1+(-1/n_1 z + n_2 z)^2}.$$

n_1 and n_2 are as before, *i.e.*, any positive constants. This must be true for all positive values of x, n_1, and n_2. When $n_2 = 0$, it may be verified by the last example. Taking $x=0$ for a special case, the last integral is equivalent to

$$1 = \frac{2}{\pi}\int_0^\infty \frac{dz}{z^2+(jz^2-1)^2},$$

if j be any positive quantity. The geometrical interpretation is curious.

12. *Special Case of* § 11. *L finite.*

When L is finite the self-induction of the coil brings in at least one, and sometimes two, pair of roots, which may or may not be imaginary. Supposing these roots got, the completion of the solution (36) by the addition of U_i is easily carried out, as in § 8. But the practical examination of imaginary roots being difficult, simplify the last example by having direct earth on to coil instead of through a condenser; or, what comes to the same thing, take $C=\infty$ and $V=0$. Next, take $R=0$, which, though unreal, will not alter the theory greatly, as the coil is not on short circuit, but in circuit with the whole cable.

We shall then have

$$\tan a = sa^3, \qquad (37)$$

and the general solution gives us, when l is finite,

$$v = -\Sigma \frac{2sE_2 a \cos b \sin(zx+b)}{1-3sa^2\cos^2 b}\epsilon^{-z^2t/ck}$$

as the potential at time t due to the terminal arrangement. (That is, when $U=0$.) Or,

$$v = \frac{2LG}{ck}\Sigma \frac{z(\sin zx - n^3z^3\cos zx)}{l(1+n^6z^6-3n^3z^2/l)}\epsilon^{-z^2t/ck},$$

if G be the initial coil current (see § 11), $z=a/l$, and $n^3=sl^3$. (n is constant, independent of l.)

Now the equation (37) has a pair of roots which are real when s is greater than 1·47, become equal for that value of s, and then imaginary when s is less. (Case in paper "On Faults,* etc.," before referred to.) But $s=n^3/l^3$, and is therefore necessarily less than the critical value when the line is long enough, and the pair of imaginaries must be still in operation when the line is infinitely long. They are then given by

$$\tan lz = (nz)^3, \quad \text{or} \quad (nz)^3 = -i,$$

whose roots are

$$nz = i, \quad \text{and} \quad nz = -\frac{i}{2} \pm \sqrt{\frac{3}{4}}.$$

The first we have nothing to do with, as it is a pair we want. It gives a term increasing with the time, so can have no business here. The others must be the required pair.

Thus, if z_1 and z_2 be the roots, our additional terms are

$$\frac{2LG}{ck}\left(z_1\frac{\sin z_1x + i\cos z_1x}{-3n^3z_1^2}\epsilon^{-z_1^2t/ck} + \text{same with } z_2\right),$$

or,

$$v_i = \frac{4}{3}\frac{LG}{ckn^2}\epsilon^{-\frac{1}{2}(x/n+t/ckn^2)}\cos\left\{\left(\frac{3}{4}\right)^{\frac{1}{2}}(x/n - t/ckn^2)+\frac{\pi}{3}\right\}; \qquad (38)$$

v_i meaning the part of v depending on imaginaries. Corresponding to this, we have

$$\gamma_i = \frac{4}{3}\frac{LG}{ck^2n^3}\cos\left(\frac{3}{4}\right)^{\frac{1}{2}}(x/n - t/ckn^2)\,.\,\epsilon^{-\frac{1}{2}(x/n+t/ckn^2)} \qquad (38a)$$

for the part of the current depending on imaginaries.

* [Art. xvi., § 22, p. 88 *ante*.]

The corresponding v and γ depending on reals are, by (36), third term on right,

$$v_r = +\frac{LG}{ck}\frac{2}{\pi}\int_0^\infty dz\sin(zx+b)\,.\,z\cos b\,\epsilon^{-z^2t/ck}$$

$$= +\frac{LG}{ck}\frac{2}{\pi}\int_0^\infty zdz\frac{\sin zx - n^3z^3\cos zx}{1+n^6z^6}\epsilon^{-z^2t/ck}, \qquad (38b)$$

($\tan b = -(nz)^3$ here); and, by differentiation, ($\gamma k = -\nabla v$),

$$\gamma_r = -\frac{LG}{ck^2}\frac{2}{\pi}\int_0^\infty z^2dz\frac{\cos zx + \sin zx\,.\,n^3z^3}{1+n^6z^6}\epsilon^{-z^2t/ck}. \qquad (38c)$$

These, with v_i and γ_i, make the complete solution.

In order to obtain verification of accuracy, calculate the integral current at x. This we may do by means of

$$\int e^{cx}\frac{\cos}{\sin}ax\,dx = \frac{e^{cx}}{a^2+c^2}\left(c\frac{\cos}{\sin}ax \pm \frac{\sin}{\cos}ax\right).$$

Applying this formula to (38a), we shall arrive at

$$\int_0^\infty \gamma_i dt = \frac{2}{3}\frac{LG}{nk}\epsilon^{-x/2n}(\cos + \sqrt{3}\sin)\frac{x}{r}\sqrt{\frac{3}{4}}. \qquad (39)$$

This is the part of the current crossing the point x from $t=0$ to $t=\infty$, so far as depends on the imaginaries. But evidently, since the line is in conductive connection with earth, the total current at any point must be zero. Consequently the just obtained result must be exactly cancelled by the integral current arising from the reals. Now the time-integration in (38c) can be immediately effected, and merely requires us to write ck/z^2 for the t term. Doing this, we must have

$$\frac{2LG}{k\pi}\int_0^\infty \frac{dz}{1+n^6z^6}(\cos zx + n^3z^3\sin zx) = \text{result in (39)},$$

and this is exactly verifiable by the partial fraction decomposition method, if no easier way present itself.

It may, perhaps, be questioned whether the integral current ought to be zero. Although it is clear that the potential at every point continuously decreases after the maximum of the wave has reached it; yet, as the wave is always travelling onward, the total charge in the line may not tend to zero. To this the answer is, that in the first place, if the line had no capacity there would be no current at all in it, because the resistance is infinite; and next, that if the line be charged in any manner with a finite charge, and left in conductive connection with earth, not merely the potential but the total charge will fall to zero. As shown * in a paper in the journal of the S. T. E. and E. ("On Electromagnets," etc.), if a condenser be discharged through several conductive paths open to it, although the current at any moment will not in general divide inversely as their resistances, yet it will be true of the time integral of the current. This applies to the elementary

* [Page 105, § 15.]

charges at any moment in the line; they have a finite resistance on one side through which to discharge, and infinite on the other side; hence all ultimately goes out; the travelling wave tending to have an infinitely small total charge, although extending over an infinite distance.

Returning to the special question, there is a case in which the integral current from the coil is finite, viz., when we increase the capacity of the line infinitely. For this is equivalent to short-circuiting the coil. The integral current must then be LG/R. The above solution, however, is useless for showing this, on account of our having taken $R=0$ to get at the imaginary roots easily. It is true that the above shows that when c is finite, however great, the integral current is zero; the interpretation then of its being finite when $c=\infty$ is that in the latter case an infinitely long time would have to elapse before the coil ceased sending a current in its original direction to the line (to its commencement only), and that we ignore its return to earth through the coil because we can never reach its beginning. We have, therefore, a quite different problem.

13. *Induction Coil and Condenser.*—Let the line be connected to earth through a coil and a condenser, in sequence, and the coil be under the inductive influence of another coil. Let R and C be the resistance and capacity of the condenser, *i.e.*, R is the resistance of a shunt to it if it have no conductivity itself. Let L_1, L_2, and M be the coefficients of self and mutual induction of the two coils. Let v and γ be the potential and current at $x=0$, v_1 the potential of the condenser, and γ_2 the current in the coil R_2. This is the secondary, closed upon itself, and in proximity to the coil R_1 between the line and the condenser.

The current entering the line is the same as that through the primary coil R_1, which again is the same as that through the condenser and its shunt. This gives

$$\gamma=\frac{v_1-v}{R_1}=-\left(\frac{v_1}{R}+CDv_1\right). \quad \text{.....} \quad \text{...............} \quad (40)$$

The equations of E.M.F. in the first and second coils are

$$v_1-v=(R_1+L_1D)\gamma+MD\gamma_2,$$
$$0=(R_2+L_2D)\gamma_2+MD\gamma.$$

If we eliminate v_1 and γ_2 we obtain

$$\frac{v}{\gamma}=-(R_1+L_1D)+\frac{M^2D^2}{R_2+L_2D}-\frac{R}{1+RCD}. \quad \text{............} \quad (41)$$

Put $\sin b$ for v, and $-(z/k)\cos b$ for γ (these being the $x=0$ representatives of $\sin(zx+b)$ and the corresponding current); and put for D its equivalent $-z^2/ck$; we then obtain

$$\tan b=\frac{R/k}{1-RCz^2/ck}z+\frac{R_1}{k}z-\frac{L_1}{ck^2}z^3-\frac{M^2/c^2k^3}{R_2-L_2z^2/ck}z^5. \quad \text{.......} \quad (41a)$$

If there be a similar arrangement at the other end of the line we shall obtain the same equation (41), with changed coefficients, of course, but with $-\gamma$ instead of γ; because the current in the line is reckoned positive always from 0 to l, so that leaving the line at one end corresponds to entering it at the other. And the following equation will become, with changed values of the coefficients, the same as (41*a*), with $-\tan(zl+b)$ instead of $\tan b$. Thus

$$\tan b = \phi_0(z), \qquad \tan(zl+b) = \phi_1(z),$$

$\phi_0(z)$ being given in (41*a*), and $\phi_1(z)$ being the negative of $\phi_0(z)$ with changed coefficients. Between them we get equation $\tan zl = Z$, finding the values of z.

To obtain the complete solution we may either form the expressions for the mutual energy, potential and kinetic, in all parts of the system, of two normal solutions, and then make use of their necessary equality to decompose the given initial state into normal solutions (the method followed in the first part * of paper "On Faults, etc.") ; or else, without going into the details of the terminal arrangements, get the solution out of the resultant equation (41) or (41*a*), (the method used at the conclusion † of that paper).

By the first method, find the expressions for V, G, and Γ, the initial potential of the condenser, and the currents in coils R_1 and R_2. Also for U the initial potential of the line. They are

$$\left.\begin{aligned} U &= \Sigma A \sin(zx+b), & V &= \Sigma A \frac{z \cos b}{1 - RCz^2/ck} \cdot \frac{R}{k}, \\ G &= \Sigma - A \frac{z \cos b}{k}, & \Gamma &= \Sigma - \frac{Az \cos b}{k} \frac{Mz^2/ck}{R_2 - L_2 z^2/ck} \end{aligned}\right\} \quad \ldots\ldots (41b)$$

by the detailed equations. Here V, G, and Γ belong to the apparatus at the $x=0$ end. There are similar equations for them at the $x=l$ end of the line, with the $-$ sign made $+$ and the b changed to $zl+b$.

The mutual potential energy of the systems corresponding to the roots z_1 and z_2, in the line, is

$$c\int_0^l dx \sin(z_1 x + b_1) \sin(z_2 x + b_2), \quad \ldots \ldots\ldots\ldots\ldots (41c)$$

(with unit A's). The mutual potential energy of the corresponding condenser charges at $x=0$ is

$$C \cdot \frac{z_1 \cos b_1}{1 - RCz_1^2/ck} \frac{R^2}{k^2} \frac{z_2 \cos b_2}{1 - RCz_2^2/ck}. \quad \ldots\ldots\ldots\ldots\ldots (41d)$$

The mutual kinetic energy of the corresponding currents in the coils R_1 and R_2 is

$$L_1 G_1 G_2 + L_2 \Gamma_1 \Gamma_2 + M G_1 \Gamma_2 + M G_2 \Gamma_1,$$

if G_1, Γ_1 be one set, G_2, Γ_2 the other. Or,

$$\frac{L_1}{k^2} z_1 \cos b_1 \,.\, z_2 \cos b_2 + \frac{L_2}{k^2} z_1 \cos b_1 \,.\, z_2 \cos b_2 \frac{Mz_1^2/ck}{R_2 - L_2 z_1^2/ck} \frac{Mz_2^2/ck}{R_2 - L_2 z_2^2/ck}$$
$$+ \frac{M}{k^2} z_1 \cos b_1 \,.\, z_2 \cos b_2 \frac{Mz_2^2/ck}{R_2 - L_2 z_2^2/ck} + \frac{M}{k^2} z_1 \cos b_1 \,.\, z_2 \cos b_2 \frac{Mz_1^2/ck}{R_2 - L_2 z_1^2/ck},$$

which admits of reduction to the simpler form

* [§ 14, p. 80.] † [§ 24, p. 91.]

$$\frac{L_1 - M^2/L_2}{k^2} z_1 \cos b_1 \,.\, z_2 \cos b_2 + \frac{M^2 R_2^2}{L_2 k^2} \frac{z_1 \cos b_1}{R_2 - L_2 z_1^2/ck} \frac{z_2 \cos b_2}{R_2 - L_2 z_2^2/ck}. \quad (41e)$$

The conjugate property is therefore obtained by equating the sum of (41c), (41d), and the corresponding $x=l$ term, to the sum of (41e) and the corresponding $x=l$ term; this gives the complete value of A by the equation

$$\frac{cl}{2}\left(1 - l^{-1}\cos^2 zl \frac{dZ}{dz}\right) A = c\int_0^l U \sin(zx+b)dx + V\frac{z\cos b}{1 - RCz^2/ck}\frac{R}{k}$$
$$+ G\left(L_1\frac{z\cos b}{k} + M\frac{z\cos b}{k}\frac{Mz^2}{ckR_2 - L_2 z^2}\right)$$
$$+ \Gamma\left(M\frac{z\cos b}{k} + L_2\frac{z\cos b}{k}\frac{Mz^2}{ckR_2 - L_2 z^2}\right)$$
$$+ \text{corresponding } V, G, \Gamma \text{ terms at } x=l,$$

which value of A used in (41b) completes the solution, when we affix the t factor.

Although this method is very thorough, yet the other is simpler. Ignoring the details of connections, and given only the v/γ equations (39) at $x=0$ and $x=l$, we can easily arrive at the above solution complete; except that V, G, and Γ will not be identified; or, we may arrive at an exactly equivalent form, in which appear other three arbitraries, independent, and therefore fully supplying their place; of course functions of V, G, and Γ.

Differentiate (41) with respect to D; we have

$$\frac{d}{dD}\frac{v}{\gamma} = -L_1 + \frac{R^2 C}{(1+RCD)^2} + \frac{2M^2 D}{R_2 + L_2 D} - \frac{M^2 L_2 D^2}{(R_2 + L_2 D)^2}.$$

Of these four terms the first two are of the proper form, but the third and fourth need rearrangement. This is easily done on uniting them, and gives

$$\gamma^2 \frac{d}{dD}\frac{v}{\gamma} = -\left(L_1 - \frac{M^2}{L_2}\right)\gamma^2 + \frac{R^2 C}{(1+RCD)^2}\gamma^2 - \frac{M^2 R_2^2/L_2}{(R_2 + L_2 D)^2}\gamma^2, \quad \dots (41f)$$

which is in the correct form of sum of squares, showing that there are three arbitraries (easily seen to be independent) at $x=0$ end (with an equal number at the $x=l$ end), which are, since the elementary normal γ is proportional to $z\cos b$ at $x=0$,

$$E_1 = \Sigma Az\cos b, \quad E_2 = \Sigma\frac{Az\cos b}{1 - RCz^2/ck}, \quad \text{and} \quad E_3 = \Sigma\frac{Az\cos b}{R_2 - L_2 z^2/ck}, \quad (42)$$

in their simplest forms. They are, by (41b), proportional to V, G, and (the third) a simple function of G and Γ.

To turn (41f) into the products form, write, on the right side, $w_1 w_2$ for γ^2,

$$\frac{w_1}{1+RCD_1}\cdot\frac{w_2}{1+RCD_2} \quad \text{for} \quad \frac{\gamma^2}{(1+RCD)^2},$$

and

$$\frac{w_1}{R_2 + L_2 D_1}\cdot\frac{w_2}{R_2 + L_2 D_2} \quad \text{for} \quad \frac{\gamma^2}{(R_2 + L_2 D)^2},$$

which makes the right member of (41f) take the form $\Sigma rf(D_1)f(D_2)$. Let E_1, E_2, and E_3 be the values of the three summations (42); we shall have

$$\left.\begin{aligned} f_1 = z\cos b,\quad f_2 = \frac{z\cos b}{1 - RCz^2/ck},\quad f_3 = \frac{z\cos b}{R_2 - L_2 z^2/ck}, \\ r_1 = -\left(L_1 - \frac{M^2}{L_2}\right)\frac{1}{k^2},\quad r_2 = \frac{R^2C}{k^2},\quad r_3 = -\frac{M^2R_2^2}{k^2}, \end{aligned}\right\} \quad \text{........} (43)$$

at the $x=0$ end. The conjugate property is, with these expressions for the r's and f's,

$$\int cu_1u_2dx = r_1f_1(z_1)f_1(z_2) + r_2f_2(z_1)f_2(z_2) + r_3f_3(z_1)f_3(z_2)$$
$$+ \text{the } (x=l) \text{ terms of similar kind,}$$

and the value of A is consequently given by

$$A\frac{cl}{2}\left(1 - \cos^2 zl\frac{dZ}{dz}\Big/l\right) = \int_0^l cUudx - r_1E_1f_1(z) - r_2E_2f_2(z) - \dots,$$

only differing from the formerly obtained value in E_1, E_2, E_3 being no longer exactly V, G, and Γ; but, on account of the similar changes in the r's and f's, the two values are identically the same.

By examination of dimensions we can of course settle what physical interpretations must be given to the E's as defined by (42), and of the constant factors, the r's in (43), and so, from (41) only, on which the present method depends, arrive approximately at the nature of the terminal connections in detail. But this can only be done so far as the terminal apparatus is able to influence the line. Thus, from (41) we may conclude that there is a coil concerned, because we shall find that the current entering the line at the first moment is arbitrary; and also a condenser, because the time integral of the current is also arbitrary. But there might be any number of charged condensers and coils in the terminal arrangements which cannot affect the state of the line. For it is absolutely necessary for the end of the line to be connected to earth, either conductively or through a condenser, in order that the terminal apparatus may have any influence whatever on the state of the line. The terminal condition will be simply $\gamma=0$ equivalent to insulation, if this be not complied with; and this is equivalent to cutting off all connection between the line and apparatus, which will discharge itself according to its own internal constitution. For example, let the terminal arrangement be a closed circuit containing a charged condenser, not connected to earth anywhere. Connection with the line will not influence it in any way that can be taken into account by our equations. There would, however, really be an effect, not by reason of imperfect insulation, but by reason of dielectric displacement generally, through the air, which is entirely ignored in our solutions.

14. *Impressed Forces.*—The whole of the preceding excludes impressed force in the line, and likewise in the terminal arrangements if it can affect the line. We can determine the effect of impressed force thus. Let a steady impressed force E be introduced anywhere in the line, say at x_1. By elementary methods we can find the steady state it will

finally produce—elementary so far as not involving time differentiations. Then remove the force E. We can, by the preceding, find the transient state that results. If X_0 denote the final state, and X what it becomes at time t after E is removed, then $X_0 - X$ represents the state at time t after E is put on. Thus if $U = \Sigma Au$ be the potential set up finally by a unit impressed force,

$$v = X_0 - E\Sigma Au\epsilon^{Dt}$$

gives the potential at time t after putting on E, being zero when $t = 0$, and X_0 when $t = \infty$. No zero root is to be allowed in the summation.

Suppose now we let E last only for the time dt. We can obtain the effect it produces by supposing E to be kept on, and that at the time $t + dt$ we put on an impressed force $-E$, cancelling the former. This latter force will produce potential

$$= -X_0 + E\Sigma Au\epsilon^{D(t-dt)}$$

at time t later. Adding this to the former, we find the effect at time t of E lasting from $t = 0$ to $t = dt$, to be

$$E \,.\, \Sigma Au\epsilon^{Dt}(\epsilon^{-Ddt} - 1) = -\Sigma AuEdtD\epsilon^{Dt}.$$

Consequently the effect of E lasting from $t = t_1$ to $t = t_1 + dt_1$, at the later time t, is

$$-\Sigma AuDEdt_1\epsilon^{D(t-t_1)}.$$

Therefore, by time integration, the potential due to a variable impressed force E at one spot, starting at time t_1, is

$$v = -\Sigma AuD\epsilon^{Dt}\int_{t_1}^{t} E\epsilon^{-Dt_1}dt_1,$$

in which E must be considered a function of t_1.

Similarly the effect of any impressed force at other places may be represented, and, by integrating along the line, the effect of any distribution of impressed force varying with the time; thus

$$v = -\Sigma uD\epsilon^{Dt}\int_0^l\int_{t_1}^{t} AE\epsilon^{-Dt_1}dx_1dt_1, \quad\ldots\ldots\ldots\ldots\ldots\ldots (44)$$

wherein E is a function of both x_1 and t_1, whilst A is a function of x_1, the position of the elementary impressed force.

What we require to know is, therefore, A as a function of x_1. Now ΣAu represents the potential set up by unit E at x_1. We need not, however, go to the great labour of expanding the final state due to E in the series ΣAu by the preceding methods, by line integrations and corresponding terminal operations. For if we imagine the impressed force at x_1 replaced by an inserted condenser whose difference of potential was unity initially, and imagine the capacity of the condenser to be infinitely increased, we shall ultimately get the same result as from the unit impressed force without the condenser.

Let $\Sigma Aw\epsilon^{Dt}$ be the current at time t after the introduction of a condenser of capacity C and charge CV. Then we have, at the place $x = x_1$, if w_1 be the value of w there,

$$-C\dot{V} = \Sigma Aw_1\epsilon^{Dt}.$$

The expansion of V is therefore

$$V = -\Sigma A \frac{w_1}{CD},$$

and the mutual potential energy of VC and a normal system (u, w) is

$$+C.V\left(-\frac{w_1}{CD}\right) = -\frac{w_1}{D}V.$$

But in the numerator of A, equation (20), the only term will be that due to the condenser, or this $-(w_1/D)V$; hence

$$A = -Vw_1/DM$$

is the value of A in $v = \Sigma A u \epsilon^{Dt}$ giving the potential at time t after introduction of the condenser, and due *to* the condenser, if M be twice the excess of the potential over the kinetic energy of the normal system u, w. So far the condenser has been of finite capacity, and consequently u, w, M, and D depend on its capacity as well as upon the terminal conditions. But, on infinitely increasing its capacity, we get mathematical equivalence to direct connection; then the determinantal equation of D is simply that resulting from the terminal conditions, and M has the value in equation (20). Therefore

$$v = -\Sigma \frac{w_1 V/D}{\left[w^2 \frac{d}{dD}\left(\frac{u}{w} - \frac{P}{Q}\right)\right]} u \epsilon^{Dt} \quad \ldots\ldots \quad \ldots\ldots \quad \ldots\ldots (45)$$

is the potential due to V at x_1. This will have a term corresponding to a zero root (arising from the infinite increase of C in the before substituted problem), expressing the final state. Hence, leaving out this term, the summation (45), with sign changed, and with $t = 0$, expresses the final state itself. Thus, taking $V = 1$,

$$+\Sigma \frac{w_1 u}{DM}$$

is the expansion $\Sigma A u$ required to be applied to (44). Put then $A = +w_1/DM$ in that equation, and it becomes

$$v = -\Sigma \frac{u \epsilon^{Dt}}{M} \int_0^l \int_{t_1}^t w_1 E \epsilon^{-Dt_1} dx_1 dt_1, \quad \ldots\ldots\ldots\ldots\ldots\ldots (46)$$

expressing the potential at x at time t due to distributed impressed force E; w_1 being the value of w at x_1, and E that of the impressed force at x_1 at time t.

To obtain the current, write w instead of u.

In the special case of uniform capacity and conductivity, with no magnetic induction, we have, as in § 7,

$$u = \sin(ax/l + b), \qquad w = -\frac{a}{kl}\cos(ax/l + b), \qquad D = -a^2/ckl^2,$$

$$M = \frac{cl}{2}\left(1 - \cos^2 a \frac{d}{da}\phi(a)\right), \qquad \tan a = \phi(a);$$

and (46) becomes

$$v=\frac{2}{ckl^2}\Sigma\frac{a\sin(ax/l+b)}{1-\cos^2 a\frac{d}{da}\phi(a)}\epsilon^{Dt}\int_0^l\int_{t_1}^t E\cos(ax_1/l+b)\epsilon^{-Dt_1}dx_1dt_1. \quad \ldots (47)$$

An important practical case is that of steady impressed force at one end of the line. The solution is, of course, obtainable by inserting a condenser there and afterwards making its capacity infinite, without reference to distributed variable impressed forces; but it will be useful to derive it from (47). Take $x_1=0$, $t_1=0$, $E=$constant, and effect the time-integration. The double integral becomes

$$E\cos b(1-\epsilon^{-Dt})/D.$$

Use this in (47), putting D in terms of a, and we obtain

$$v=2E\Sigma\frac{(\cos b)/a}{1-\cos^2 a\,.\,\phi'(a)}\sin(ax/l+b)(1-\epsilon^{Dt}),$$

or

$$v=X-2E\Sigma\frac{(\cos b)/a}{1-\cos^2 a\,.\,\phi'(a)}\sin(ax/l+b)\epsilon^{Dt}, \quad \ldots\ldots\ldots\ldots (48)$$

if X be the final state of potential.

If the impressed force be at x_1, write $\cos(ax_1/l+b)$ for $\cos b$.

If the line be earthed at both ends, $b=0$, $a=n\pi$, n being a + integer, and the denominator is unity; hence

$$v=X-\frac{2E}{\pi}\Sigma\frac{1}{n}\cos\frac{n\pi x_1}{l}\sin\frac{n\pi x}{l}\epsilon^{Dt},$$

and the current is

$$\gamma=\frac{E}{kl}\left(1+2\Sigma\cos\frac{n\pi x_1}{l}\cos\frac{n\pi x}{l}\epsilon^{Dt}\right).$$

Returning to (46), let the impressed force be simple harmonic

$$E=E_0\sin mt \quad \text{at} \quad x_1;$$

then we may effect the time integration, and obtain, if $t_1=0$,

$$v=-E_0\Sigma\frac{uw_1m}{M(D^2+m^2)}\epsilon^{Dt}+E_0\Sigma\frac{uw_1}{M(D^2+m^2)}(D\sin+m\cos)mt.$$

The first summation ultimately vanishes; the second therefore represents the final simple harmonic variation of potential. If the impressed force be $E_0\cos mt$, change sin to cos and cos to $-$sin in the second, and m to D in the first summation. The second, the final state, admits of being put in a finite form, by finding the general simple harmonic solution of the equation of the normal functions, and subjecting it to the terminal conditions and to that at x_1.

Or, more simply, by developing the resultant differential equation itself connecting the impressed force with the potential (or the current) it produces, and bringing it to the form $v=(a+bD)E$ by means of the property $d^2/dt^2=-m^2$.

15. *R negative in* § 9. This case, which is of some interest, both mathematically and in the physical interpretations, was discussed in

brief in the "On Faults, etc." paper,* so far as a line of finite length is concerned. In the following the solution when the line is infinitely long will be given.

The meaning of R negative is simply this, that whereas if we connect the end of the line to earth through a resistance R, the current *leaving* the line will be v/R at any moment, we now make the current *entering* the line be v/R, which of course requires artificial appliances. The solution is the same as in § 9, with m made negative.

Supposing the line of finite length, it is easy to see that whatever the initial distribution of charge may be, it will all escape, provided $m < -1$; or, more conveniently, let $m = -m_1$, so that

$$\tan a = m_1 a$$

is the determinantal equation, m_1 being positive; then if $m_1 > 1$, the line's charge will all escape. But this equation has a root between 0 and $\frac{1}{2}\pi$, which vanishes when $m_1 = 1$. There is then a term independent of the time. For instance, if the initial state were a steady current E/kl from 0 to l, it would remain unchanged. This is readily seen without examining the solution in detail. And if it have any other initial distribution, it will finally settle down to be a steady current of some other strength. For instance, if v were initially constant, $= E$, the final state will be a steady current of strength $3E/2kl$, from 0 to l.

When m_1 is made less than unity, the zero root becomes imaginary, so that the charge in the line, after initial irregularities have disappeared, increases continuously. Passing to the case of m_1 very small indeed, if we ignored the term depending upon the imaginary root, we should find that our expansion of the initial state $v = E$ constant represented E (very nearly), from $x = l$ up to nearly $x = 0$ (say to $x = x_1$, x_1 being very small), and then suddenly changed from representing E to representing $-E$ from $x = x_1$ to $x = 0$. This departure from E everywhere is of course due to the imaginary, which gives, when m_1 is very small, $v = 0$, practically everywhere except close to $x = 0$, where it gives $2E$. So long as the small charge this represents has any existence at all, *i.e.* so long as m_1 is not actually made quite zero, the charge in the line will increase continuously from this small beginning, all the rest of the initial charge subsiding. When m_1 is made actually zero, the influence of the imaginary is completely gone; and it is gone for good, for on making m_1 negative, or m positive, we have the case of a line discharging through a resistance.

If U be the initial potential, the complete solution is

$$v = \Sigma \frac{\int_0^l U \sin a\left(1 - \frac{x}{l}\right)dx}{\frac{1}{2}l(1 - m_1 \cos^2 a)} \sin a\left(1 - \frac{x}{l}\right)\epsilon^{-a^2t/ckl^2},$$

subject to $\tan a = m_1 a$. Let $m_1 = R_1/kl = n_1/l$, and keep n_1 constant. When we make l infinite, m_1 being less than 1, we must take the imaginary into account. It will be given by

$$\tan lz_0 = n_1 z_0 \quad \text{with} \quad l = \infty, \quad \text{or} \quad n_1 z_0 = -i, \quad \therefore \quad z_0 = -i/n_1.$$

* [See page 91, § 23.]

We shall then find that the corresponding term in the preceding series becomes

$$v_i = \frac{2}{n_1}\epsilon^{-x/n_1}\int_0^\infty U_1 \epsilon^{-x_1/n_1} dx_1 \, . \, \epsilon^{-t/ckn_1^2}, \quad \text{..... ....} \quad (49)$$

v_i being the part of the potential of the line at x at the time t depending on the imaginary; whilst the reals give

$$v_r = \frac{2}{\pi}\iint dz dx_1 \frac{\sin zx - n_1 z \cos zx}{1 + n_1^2 z^2} U_1 (\sin zx_1 - n_1 z \cos zx_1)\epsilon^{-z^2 t/ck}, \text{...} \quad (49a)$$

in fact, the same as in § 9, except in the change from n to n_1.

Observe that when n_1 is made zero, $v_i = 0$, or the influence of the imaginary disappears. We must then keep it zero, as n_1 is made negative, although according to (49) v_i comes on again then. In fact, if n_1 be negative the solution is given by v_r only, (49a). If n_1 be positive it is given by $v_i + v_r$. The reason is that we constructed the solution v_i on assumption of n_1 positive. Start with n_1 negative and it has no existence. $\tan a = n_1 a$ has no imaginary root if n_1 be negative; if positive there is one when $n_1 < 1$. (In speaking of one root, instead of the necessary pair, remember that $D = -a^2/ckl^2$, so that only one need be considered.) Thus we account for the at first sight puzzling fact that in the § 9, where no imaginary can possibly be concerned, the equation $\tan lz = +n_1 z$ occurring there, with n_1 negative, *has* an imaginary root when the line is infinitely long. It really belongs to the problem of the present section, n_1 positive. Similarly, in § 12, where we have an extra imaginary besides the pair we want, we may conclude that it is required to give correct expansions in case of s negative, equivalent to a reversal of the law of induction, interpretable by the aid of an impressed force of properly adjusted strength. In the case treated in § 12, of a condenser discharged into a cable, we have a similar result. There is no imaginary involved when C, the capacity of the condenser, is positive, but there is one when it is negative.

16. *Cable*. $\phi(a) = m \sin a$.—By a cable is to be here understood a line in which c and k are constant and $s = 0$. Let the $x = 0$ end be earthed, so that $v = 0$ is the condition there; this requires that

$$U = \Sigma A \sin ax/l, \quad \text{..........................} \quad (50)$$

the constant b being absent. Let also the $x = l$ condition give rise to

$$\tan a = m \sin a, \quad \text{..........................} \quad (51)$$

m being any constant. This requires that

$$\frac{v}{\gamma} = klm\left(1 + \frac{1}{3!}TD + \frac{1}{5!}T^2D^2 + \dots\right) \quad \text{at } x = l,$$

if $T = ckl^2$. The determinantal equation (51) splits up into

$$\sin a = 0, \quad \text{and} \quad \cos a = m^{-1},$$

giving two series of roots,

$$\left.\begin{array}{lllllll} & 0, & \pi, & 2\pi, & 3\pi, & 4\pi, & \text{etc.,} \\ \text{and} & \theta, & 2\pi - \theta, & 2\pi + \theta, & 4\pi - \theta, & 4\pi + \theta, & \text{etc.,} \end{array}\right\} \text{...} \quad (52)$$

if θ be the least value of $\theta = \cos^{-1} 1/m$. Owing to this simplicity we can obtain easily verifiable results.

Let the terminal arbitraries be

$$E_1 = \Sigma Aa \cos a, \qquad E_3 = \Sigma Aa^3 \cos a, \qquad E_5 = \Sigma Aa^5 \cos a, \text{ etc.};$$

then we shall find, by the general method, that, U representing the initial potential, the value of A in (50) is given by

$$A\frac{1 - m\cos^3 a}{2m} = \frac{1}{ml}\int_0^l U \sin\frac{ax}{l}dx + E_1\frac{\cos a}{a^2}\left(a - \sin a\right)$$
$$+ \frac{E_3}{a^4}\cos a\left(a - \frac{1}{3!}a^3 - \sin a\right) + \dots \qquad (53)$$

Thus, the E series give

$$\Sigma\frac{\cos a}{N}\sin\frac{ax}{l} \cdot \frac{a - \sin a}{a^2} = 0, \text{ etc.};$$

and generally,

$$\Sigma\frac{\cos a}{N} \cdot \sin\frac{ax}{l}\frac{1}{a^{2n}}\left(\sin a - \text{first } n \text{ terms of } \sin a\right) = 0,$$

n being any + integer, and N the coefficient of A on the left side of (53).

Let the E's = 0. Then, if we take $U = V$, constant, we get, as the full form of (50),

$$V = V\frac{2}{\pi}\Sigma\frac{1}{n}\sin\frac{n\pi x}{l}\frac{1 - \cos n\pi}{1 - m\cos^3 n\pi} + \frac{2Vm}{m+1}\Sigma\frac{1}{a}\sin\frac{ax}{l}, \qquad (54)$$

the first series referring to the first set of roots (52), going from $n = 1$ to $n = \infty$, the second to the other set. The first lot comes to $V/(1+m)$. Consequently the second gives

$$\tfrac{1}{2} = \Sigma\frac{1}{a}\sin\frac{ax}{l}, \qquad (55)$$

subject to $\cos a = 1/m$, whatever m may be.

When $m = 1$ the zero root of $m \cos a = 1$ needs attention. Its term (first term of the second (54) series) equals Vx/l, and the rest of the series $\frac{1}{2}V(1 - 2x/l)$; whilst the first series comes to $\frac{1}{2}V$. This is at the first moment. All terms ultimately vanish except the Vx/l, depending upon the zero root. That is to say, the initial state of constant potential V along the line is ultimately replaced by a steady current of strength V/kl from l to 0. The transient state and final result are identically the same as if, starting with the line insulated at $x = 0$ and charged to potential V by a battery of no resistance at $x = l$, we suddenly put the insulated end of the line to earth, without removing the battery at the other end. Thus the terminal condition, in this case, simulates a steady impressed force. But should the initial state of the line be $U = 0$, then $v = 0$ later also, unless one of the E's be not zero.

When m lies between 1 and -1, the above zero root becomes imaginary, and exponential forms are required to present results in a real form.

The E's are proportional to γ and its time differential coefficients $\dot{\gamma}$, $\ddot{\gamma}$, etc., at $x=l$, at the first moment. Thus, to expand $U=0$ in the series (50), subject to (51), is nugatory. But if, for instance, we say in addition that γ shall equal Γ at $x=l$, and $\dot{\gamma}$, $\ddot{\gamma}$, etc. $=0$, the required expansion of $U=0$ is

$$U=\Gamma\Sigma \cos a \sin\frac{ax}{l} \cdot \frac{a-\sin a}{a^2} \Big/ \frac{1-m\cos^3 a}{2m},$$

by (53), taking $U=0$, $E_1=\Gamma$, and the remaining E's$=0$. We can thus expand a function in a series of normal functions when its value is zero everywhere, provided not all its odd differential coefficients (dU/dx, d^3U/dx^3, etc.) are zero at $x=l$. This is, of course, a special result; in general, in order that we may expand $U=0$ under any given possible terminal conditions, it is necessary that at least one of the terminal E's shall be finite; and it is further to be understood that v and γ are to be subjected only to their natural relations *between* the limits.

Suppose, however, that, instead of (51), we take

$$\sec a = m$$

as the determinantal equation, so that we have only the second series of roots (52). We know beforehand that we cannot expand U in the series $\Sigma A \sin ax/l$, subject to this condition. For (53) results from a legitimate v/γ relation, to comply with which both series of roots are required. But $\sec a=m$ cannot be derived from any legitimate v/γ relation of the form (10), § 3. On expanding U we shall not obtain U, but something else. Let

$$f(x)=U=f_1(x)+f_2(x),$$

$f_1(x)$ being the function represented by the expansion in which the first set of roots (52) is used, and $f_2(x)$ corresponding to the second set.

We may easily show that

$$f_1(x)=\frac{f(x)-mf(l-x)}{1-m^2};$$

consequently
$$f_2(x)=\frac{-m^2f(x)+mf(l-x)}{1-m^2}.$$

This, then, is what we should arrive at, following $\sec a=m$; $f_2(x)$ instead of U.

The condition $\sec a=m$ by itself, or $-a/kl\cos a=-a/klm$, or $\gamma=-\sqrt{-TD}/klm$, is clearly a meaningless terminal condition.

17. *Condition* $\tan a=m\tan na$.—This is clearly legitimate, the right member being an odd function of a.

Here
$$\cos^2 a\frac{d}{da}\phi(a)=mn\frac{\cos^2 a}{\cos^2 na}=mn\cos^2 a+\frac{n}{m}\sin^2 a.$$

Hence, if U be the initial potential in the line, earthed at $x=0$, we shall have

$$U=\Sigma\frac{\int U\sin ax/l \,.\, dx}{\frac{1}{2}l(1-mn\cos^2 a/\cos^2 na)}\sin ax/l; \quad\ldots\ldots\ldots\ldots\ldots (56)$$

and, if in the arrangements at $x=l$ there be nothing to influence the state of the line (the yet undetermined E's $=0$), the introduction of the time factor in (56) will give us the potential at time t.

As regards the interpretation of $\tan a = m \tan na$, we find that it results from

$$v/\gamma = -klm/\sqrt{-TD} \tan n\sqrt{-TD},$$

which, expanded, is of the correct form. The terminal "apparatus" at $x=l$ is another cable of uniform capacity and conductivity, earthed at its distant end. For if c_1, k_1, l_1, refer to a second line thus earthed, we can show that

$$v/\gamma = k_1 l_1/\sqrt{-T_1 D} \tan\sqrt{-T_1 D}$$

is the differential equation connecting v and γ in it at a distance l_1 from the earthed end. It is therefore the terminal condition at $x=l$ of the first line, where they join. So we have

$$m = -(ck_1/c_1k)^{\frac{1}{2}}, \qquad n = (T_1/T)^{\frac{1}{2}}.$$

But the theory of combinations of lines of the same or different types is best treated separately. The present section is merely to illustrate the treatment of terminal conditions. It will be observed that (56) is subject not only to the equation $\tan a = m \tan na$, but also to $v=0$ at $x=0$. Should we impose other $x=0$ conditions we shall obtain quite different results; every change at $x=0$ involving corresponding changes in the $x=l$ condition, in order to keep the determinantal equation the same.

Since

$$mn = -k_1 l_1/kl, \qquad n/m = -c_1 l_1/cl,$$

we see that if the constants be so adjusted that $\sin^2 a = 0$ the expansion (56) becomes

$$U = \Sigma \frac{k\int_0^l U \sin ax/l \,.\, dx}{\frac{1}{2}(kl + k_1 l_1)} \sin ax/l. \qquad \text{(56}a\text{)}$$

But caution is here necessary. Thus, for example, let the two lines have the same total resistance and the same total capacity. Then $m=-1$, $n=1$, and the determinantal equation is

$$\tan a = -\tan a.$$

Superficially this means simply $\tan a = 0$, or $a = i\pi$. But this is wrong. The two curves $y=\tan a$ and $y=-\tan a$ (the latter $-\tan a$ being a special form of $\phi(a)$), coincide not only when $a=i\pi$, but midway between these places, at the infinite positive and negative values of y. So the set of roots in the expansion (56a) is given by $a=\frac{1}{2}i\pi$. This we may at once corroborate by observing that the second line, although its type is different from that of the first (not being necessarily of the same length), is yet precisely equivalent to an exact copy of the first line, to which our solution refers, so far as its influence on the first line is concerned.

18. $s=0$. *c and k^{-1} proportional to x. Bessel functions.*

Let the capacity and the conductivity vary as the distances from one end of the line. Thus, let

$$c=c_0x, \qquad k=k_0/x,$$

c_0 and k_0 being constant. The differential equation of v is then

$$c_0k_0\dot{v}=\frac{1}{x}\frac{d}{dx}x\frac{dv}{dx}.$$

Put $c_0k_0D=-n^2$, then the equation of the normal potential functions u is

$$\frac{1}{x}\frac{d}{dx}x\frac{du}{dx}+n^2u=0\,; \qquad (57)$$

therefore

$$u=J_0(nx)=1-\frac{n^2x^2}{2^2}+\frac{n^4x^4}{2^24^2}-\dots, \qquad (58)$$

and the corresponding current function is

$$w=-\frac{x}{k_0}\frac{du}{dx}=\frac{nx}{k_0}J_1(nx), \quad \text{if} \quad J_1(nx)=-\frac{dJ_0(nx)}{d(nx)}. \qquad (58a)$$

The u of (58) is only one of the two solutions of (57). The other is

$$K_0(nx)=J_0(nx)\log nx+\frac{n^2x^2}{2^2}-(1+\tfrac{1}{2})\frac{n^4x^4}{2^24^2}+(1+\tfrac{1}{2}+\tfrac{1}{3})\frac{n^6x^6}{2^24^26^2}-\dots. \qquad (59)$$

If our boundary limits are at any two places of positive conductivity (say x_1 and x_2, both positive), we shall require to use both solutions (58) and (59), and we may impose any legitimate v/γ conditions at these limits. But should one of the limits be $x=0$, where the conductivity is zero, not only are we restricted to use the (58) function only (because the other is infinite at $x=0$), but it is no longer any use imposing a v/γ condition there. No current can pass. Hence, taking 0 and l as our boundaries, the solution is

$$v=\Sigma AJ_0(nx)\epsilon^{Dt}, \qquad (60)$$

subject to boundary v/γ condition at $x=l$ only. (If the boundaries are both on the negative side, both solutions are required again; and if one boundary be on the negative, the other on the positive side, the problem splits into two perfectly independent problems, owing to the nonconducting barrier at $x=0$. Of course the negative c and k cases are not physically real, at least in terms of electrical capacity and conductivity.) If

$$Qv=P\gamma, \quad \text{or} \quad Qu=Pw,$$

be the boundary condition, and U the initial potential, the value of A is

$$A=\frac{\int cUJ_0(nx)dx-\dots}{w^2\dfrac{d}{dD}\left(\dfrac{u}{w}-\dfrac{P}{Q}\right)},$$

where the quantities in the denominator have the $x=l$ values; the other limit $x=0$ not being concerned. The unrepresented terms $-\dots$ in the numerator refer to the terminal arbitraries. The value of

$$w^2\frac{d}{dD}\frac{u}{w}, \quad \text{or} \quad \int_0^l cu^2dx, \quad \text{is} \quad \frac{c_0l^2}{2}\{J_0^2(nl)+J_1^2(nl)\}.$$

We will take just one special case. Let the cable be joined to earth through a coil and condenser in sequence at $x=l$; R and L the coil constants, C the condenser's capacity. Then the terminal condition is

$$\frac{v}{\gamma}=\frac{1}{CD}+R+LD; \qquad (61)$$

and, if v_1 be the potential of the condenser at its junction with the coil, $CDv_1=\gamma$. So if, initially,

$$\left.\begin{array}{l} U=\Sigma AJ_0(nx) \text{ represents the potential of the line,} \\ G=\Sigma Anl/k_0 \,.\, J_1(nl) \text{ represents the coil current,} \\ V=\Sigma Anl/k_0 \,.\, J_1(nl)/CD \text{ represents the condenser's potential,} \end{array}\right\}(t=0);$$

from these we find that

$$A=\frac{\int cUJ_0(nx)dx+CV \,.\, nl/k_0 \,.\, J_1(nl)/CD-LG \,.\, nl/k_0 \,.\, J_1(nl)}{\int cu^2dx+\{nl/k_0 \,.\, J_1(nl)\}^2(1/CD^2-L)}, \quad \dots (62)$$

except in the case of the $n=0$ root, which gives

$$A_0=\frac{\int cUdx+CV}{\frac{1}{2}c_0l^2+C},$$

i.e., the total initial charge divided by the total capacity, or the mean potential.

The equation of n is, by (61) and (58),

$$CD\,J_0(nl)=\{1+CD(R+LD)\}nl/k_0 \,.\, J_1(nl). \qquad (63)$$

Increase C to ∞. Then A_0 becomes V, and

$$A=\frac{\int cUJ_0(nx)dx+nl/k_0 \,.\, J_1(nl)(V/D-LG)}{\int cu^2dx-L\{nl/k_0 \,.\, J_1(nl)\}^2}.$$

Hence the potential due to V only, or

$$v=V-Vc_0l\Sigma J_0(nx)\frac{J_1(nl)/n}{\int cu^2dx-\dots}\epsilon^{Dt}, \qquad (64)$$

represents the effect at time t after putting on a steady impressed force V at $x=l$, it taking the place of the condenser. The n's are now the roots of

$$J_0(nl)=nl/k_0 \,.\, J_1(nl)(R+LD).$$

The accuracy of (64) may be tested by its having to give $v=0$ when $t=0$.

19. *Murphy's and Legendre's Functions.* A peculiarity of the last example was that the conductivity was zero at a certain place. As results, one of the normal solutions assumes an infinite value there, and

it is excluded from the solution of any problem in which one of the limits is at that place, where terminal conditions are nugatory. In the present example there are two places of zero conductivity, producing a further development of the above specialities. Let the capacity be constant, whilst the conductivity is given by

$$\frac{1}{k}=\frac{1}{k_0}\frac{x}{l}\left(1-\frac{x}{l}\right),$$

being therefore positive between 0 and l, a maximum midway between them, and vanishing at both $x=0$ and $x=l$. Beyond these limits k is negative, so that for electrical reality we require to keep to 0 and l as the extreme limits. The equation of the normal potential functions, if $ck_0l^2D=-a^2$, is

$$\frac{a^2}{l^2}u+\frac{d}{dx}\left\{\frac{x}{l}\left(1-\frac{x}{l}\right)\frac{du}{dx}\right\}=0\text{ ; } \qquad (65)$$

and that of the normal current functions is

$$\frac{x}{l}\left(1-\frac{x}{l}\right)\frac{d^2w}{dx^2}+\frac{a^2}{l^2}w=0. \qquad (66)$$

Here u and w are connected by

$$w=-\frac{x}{k_0l}\left(1-\frac{x}{l}\right)\frac{du}{dx}, \quad \text{and} \quad u=\frac{k_0l^2}{a^2}\frac{dw}{dx}.$$

One solution of (65), in rising powers of x, is

$$u=1-\frac{x}{l}a^2+\frac{1}{2^2}\frac{x^2}{l^2}a^2(a^2-1.2)-\frac{1}{2^23^2}a^2(a^2-1.2)(a^2-2.3)+\dots. \qquad (67)$$

This being the potential function, the corresponding current function is

$$w=\frac{a^2}{k_0l}\left\{\frac{x}{l}-\frac{1}{2}\frac{x^2}{l^2}a^2+\frac{1}{2^23}\frac{x^3}{l^3}a^2(a^2-1.2)\right.$$
$$\left.-\frac{1}{2^23^24}\frac{x^4}{l^4}a^2(a^2-1.2)(a^2-2.3)+\dots\right\}.$$

It may be shown that u is finite (for any value of a^2) between the limits 0 and l, but infinite at $x=l$ itself, except for special values of a^2 Also that the second solution corresponding to u in (67) is infinite both at $x=0$ and $x=l$.

If then one of our limits be $x=0$, and the other at $x=x_1$, less than l, the case resembles the last, zero conductivity at one end. No terminal condition can be imposed at $x=0$, but any legitimate v/γ condition may hold at $x=x_1$, so that from it the admissible a's may be obtained.

There are cases of particular simplicity, viz., when $x_1=\frac{1}{2}l$, or the limits are the places of zero and of maximum conductivity, and at the latter place the line be either earthed or insulated. Also if $x_1=l$, or the conductivity vanish at both ends. Then all terminal conditions are nugatory, and only those values of a^2 can be admitted which keep u finite. They are $a^2=0.1$, or 1.2, or 2.3, or 3.4, etc., making the u series stop abruptly.

Then the first four u functions are

$$P_0 = u_{01} = 1,$$
$$P_1 = u_{12} = 1 - 1.2\, x/l,$$
$$P_2 = u_{23} = 1 - 2.3\, x/l + 2.3\, x^2/l^2,$$
$$P_3 = u_{34} = 1 - 3.4\, x/l + 30\, x^2/l^2 - 20\, x^3/l^3.$$

If we make $l = 1$, these are Murphy's P's. Not praties, but the functions invented by Murphy, given at the commencement of his "Electricity and Magnetism," a work that is rather out of date electrically, but containing a good introduction to spherical harmonics, the modern harmonies of the spheres. Murphy said any function could be expanded in P's, but seemed to think the possession of the conjugate property a sufficient proof. Of course this is quite inadequate. That property is

$$\int P_m P_n dx = 0,$$

the limits being 0 and l. That of the u function (or of it and its companion solution) is

$$\frac{a_1^2 - a_2^2}{l^2}\int_0^l u_1 u_2 dx = \left[\left(\frac{x}{l} - \frac{x^2}{l^2}\right)\left(u_1\frac{du_2}{dx} - u_2\frac{du_1}{dx}\right)\right]_0^l,$$

a_1^2 and a_2^2 being any two a^2's. At first sight the right member appears to vanish for any values of a_1^2 and a_2^2, on account of the vanishing of $(x/l - x^2/l^2)$; but this ignores the infiniteness of u and du/dx, and, in fact, of $(x/l - x^2/l^2)u\dfrac{du}{dx}$, at the limit $x = l$, for all values of a^2 except those belonging to Murphy's P's.

Physically there is no difficulty in understanding why, when the limits are at places of zero conductivity, and terminal conditions are nugatory, we should be absolutely restricted always to one set of values of a^2.

All the odd P's vanish at the centre of the line, and the slope of all the even P's vanishes there. Consequently, if our limits be 0 and $\frac{1}{2}l$, and the line be insulated at the latter, making $\gamma = 0$, we are restricted to the even P's. And if the line be earthed there, we are restricted to the odd P's only.

Suppose now our limits are both between 0 and l, and therefore at places of finite positive conductivity. We require two normal solutions to be able to satisfy the v/γ conditions we may impose at the ends. It is best now to take the place of maximum conductivity as origin. Let

$$y = x - \tfrac{1}{2}l,$$

then y is the distance from the maximum conductivity, and $y = \pm\frac{1}{2}l$ are the places of zero conductivity. In terms of y, we have

$$k^{-1} = k_0^{-1}(1 - 4y^2/l^2) = k_0^{-1}(1 - y^2/l_1^2),$$

if $l_1 = \frac{1}{2}l$. The equation of the current functions is

$$(1 - y^2 l_1^2)\frac{d^2 w}{dy^2} + \frac{a^2}{l_1^2}w = 0,$$

and the u functions are got by

$$u=\frac{kl_1}{a^2}\cdot\frac{dw}{dy}.$$

We can now have two solutions in ascending powers of y. Those for u are

$$p=1-\frac{1}{2!}\frac{y^2}{l_1^2}a^2+\frac{1}{4!}\frac{y^4}{l_1^4}a^2(a^2-2.3)-\ldots,$$

$$q=\frac{y}{l_1}-\frac{1}{3!}\frac{y^3}{l_1^3}(a^2-1.2)+\frac{1}{5!}\frac{y^5}{l_1^5}(a^2-1.2)(a^2-3.4)-\ldots,$$

and we see at once that $a^2=0.1$, 2.3, 4.5, etc., in p give us the even P's, whilst $a^2=1.2$, 3.4, 5.6, etc., in q give us the odd P's, except as regards constant multipliers fixing the absolute magnitude.

As p and q are quite independent, we may now take our limits at any places of finite conductivity, and then introduce any legitimate v/γ conditions from which to determine what values of a^2 to use to be able to expand the initial potential in a series

$$\Sigma A(p+bq),$$

and thus obtain the subsequent history.

If $y=0$ be one limit (the place of maximum conductivity) we are restricted to the q series if the line be there earthed, and to the p series if it be insulated, the condition at the other limit settling the a's. Except if the second limit be at $y=l_1$ or $-l_1$, when we are further restricted to those particular values of a^2 which make the series stop abruptly, p and q being infinite at $y=\pm l_1$ for all other values of a^2. This, however, is merely a corroboration of the former conclusions regarding the P's.

The current functions corresponding to the potential functions p and q are

$$+\frac{a^2}{k_0l_1}\left\{\frac{y}{l_1}-\frac{1}{3!}\frac{y^3}{l_1^3}a^2+\frac{1}{5!}\frac{y^5}{l_1^5}a^2(a^2-2.3)-\ldots\right\},$$

and

$$-\frac{1}{k_0l_1}\left\{1-\frac{1}{2!}\frac{y^2}{l_1^2}a^2+\frac{1}{4!}\frac{y^4}{l_1^4}a^2(a^2-1.2)-\frac{1}{6!}\frac{y^6}{l_1^6}a^2(a^2-1.2)(a^2-3.4)+\ldots\right\}$$

If one limit be at x_1 between $x=0$ and $x=l$, and the other be at x_2 beyond l, the problem breaks up into two; one, electrically real, from $x=x_1$ to l, to be treated as above; the other, electrically impossible, from $x=l$ to $x=x_2$, in which a single normal function is concerned. If both limits be beyond l, two normal functions are wanted. Terminal v/γ conditions may be imposed at both limits, except in the former case of one being at a place of zero conductivity. As regards the functions to be used we need say nothing, but refer to works on spherical harmonics. Our object is only to point out the physical peculiarities connected with the different solutions. I may remark, however, in passing, that I believe the difficulties attendant upon the purely analytical treatment of spherical harmonics and similar subjects would be greatly lightened to the student by having some physical basis to rest upon.

20. Another case, physically very different as regards the distribution of c and k, in which we come upon the same functions, is this:—Let the capacity and the conductivity be both variable, given by

$$c = c_0 \sin\frac{mx}{l}, \qquad k^{-1} = k_0^{-1} \sin\frac{mx}{l},$$

so that, for example, if $m = \pi$, the capacity and the conductivity will both vanish when $x = 0$ and l, and be positive between them, having maximum values midway.

Here, if $D = -a^2/c_0k_0l^2$, the normal potential equation is

$$\frac{a^2}{l^2}\sin\frac{mx}{l}u + \frac{d}{dx}\left(\sin\frac{mx}{l}\frac{du}{dx}\right) = 0.$$

Introduce the new variable $\mu = \cos(mx/l)$, and we shall have

$$\frac{a^2}{m^2}u + \frac{d}{d\mu}\left\{(1-\mu^2)\frac{du}{d\mu}\right\} = 0,$$

the equation of Legendre's coefficients when $a^2/m^2 = 0.1$, 1.2, 2.3, etc., and requiring the same functions as before.

To show the relation of this case to the sphere, cut out a thin strip of uniform thickness from the surface of a spherical shell extending from pole to pole, bounded by two close meridional lines. Let the strip be a conductor for heat, and be perfectly insulated all round. Its heat capacity and conductivity both vary as the cosine of the latitude. Hence the above form of equation as regards the diffusion of heat. We may of course also interpret the case electrically, since if the strip be very thin its conductivity (electric) will vary as the cosine of the latitude, and if it be properly coated with a dielectric, so will its electric capacity.

All cases in which c is constant, and $s = 0$, come under, by equation (4),

$$f(x)\frac{d^2w}{dx^2} + n^2w = 0,$$

$f(x)$ being proportional to the variable conductivity. The u function is got at once by differentiating the w function.

All cases in which k = constant, and $s = 0$, come under

$$f(x)\frac{d^2u}{dx^2} + n^2u = 0,$$

and w is got by differentiation.

21. If in § 19 we make k_0 negative, without other change, the regions of positive conductivity lie outside the limits 0 and l, or $-l_1$ and $+l_1$, between which it was then positive. As regards the solutions, since $ck_0l^2D = -a^2$ we have only to treat a^2 as a negative quantity. Thus, let $a^2 = -n^2$ in (67). This makes

$$u = 1 + \frac{x}{l}n^2 + \frac{1}{2^2}\frac{x^2}{l^2}n^2(n^2+1.2) + \frac{1}{2^23^2}n^2(n^2+1.2)(n^2+2.3) + \dots$$

be the solution suitable for the case in which one of our boundaries is at $x=0$. If the other be on the negative side, where the conductivity is positive, the problem is electrically possible. If it be on the positive side, between 0 and l, since the only difference is that a^2 is reversed in sign, we must effect the decomposition of the initial potential into normal distributions in the same manner as in § 19. But now the t term, on account of the positivity of D in ϵ^{Dt}, will increase indefinitely with the time. Of course there is no such absurdity when we keep to the regions of positive conductivity, between $x=0$ and l in § 19, or outside those limits in the present case.

22. Since we have by equations (4), when $s=0$,

$$\nabla c^{-1}\nabla w = Dkw, \qquad -cDu = \nabla w,$$
$$\nabla k^{-1}\nabla w = Dcu, \qquad -kw = \nabla u;$$

if we make c constant, say $=c_0$, the equation of w, the current function, is

$$\nabla^2 w = c_0 Dkw;$$

and if, on the other hand, it be k that is constant, $=k_0$ say, we have

$$\nabla^2 u = k_0 Dcu.$$

If then, c_0 being constant in the one case and k_0 in the other, the distributions of c and of k be similar, the same functions which serve for the current in the one case will serve for the potential in the other. Therefore the peculiarities attendant upon vanishing conductivity in the one case (with their mathematical difficulties, requiring different forms of solutions in different regions) will be repeated in the second case on account of the vanishing of the reciprocal of the capacity, or the inductive resistance.

It is easy to see, by physical considerations, the reason of this. If the conductivity vanish anywhere it completely cuts off current connection between the contiguous parts of the line, dividing it into perfectly independent sections of finite conductivity. Infinite capacity at a point would act similarly. In one case the current is made zero, in the other it is the potential that is made zero. But in both cases no current can pass the place. It is equivalent either to disconnecting or to putting on earth at the point. In one case no current can pass the place because there is no current to pass—in the other case because the current is all cut off by the condenser of inexhaustible capacity whose potential can never be raised above zero.

XXI.—DIMENSIONS OF A MAGNETIC POLE.

[*The Electrician*, June 3, 1882, p. 63.]

I OBSERVE (in *The Electrician*, May 27, 1882, p. 27) that Professor Clausius objects to Maxwell's dimensions of Magnetism in the electrostatic system of units, and that he appears to be supported by Professor Everett, the author of a valuable work on units.

The argument, as stated by the latter, is that since

$$\text{Pole} \times \text{Length} = \text{Current} \times \text{Area},$$

we should have the dimensions of Pole $= L \times$ dimensions of Current. This comes out right in the electromagnetic system, but quite wrong in the electrostatic system of units.

The error appears to lie in the neglect of the distinction between magnetic force and magnetic induction. The latter is μ times the former, μ being the magnetic permeability of the medium. Now, a small plane electric current and a small magnet produce similar fields of force in a particular medium, so we may choose a certain strength of current that shall make the fields of the same strength externally. But a remarkable difference comes in when we change the medium. In the case of the closed current the magnetic force remains always the same, the induction, therefore, varying directly as μ. But with the magnet it is the induction that remains the same, whilst the force varies inversely as μ.

So we have

$$\text{Magnetic force} = \frac{\text{Current} \times L}{L^2} = \frac{\text{Current}}{L},$$

and also

$$\text{Magnetic force} = \frac{\text{Magnetism}}{L^2 . \mu}.$$

Consequently,

$$\text{Magnetism} = \text{Current} \times L\mu,$$

or

$$\text{Pole} \times L = \text{Current} \times L^2\mu,$$

or

$$\text{Pole} \times \text{Length} = \text{Current} \times \text{Area} \times \mu,$$

so that the magnetic moment of the current is really μ times that of Professors Clausius and Everett.

Maxwell determines the dimensions of the quantities from relations in which μ does not appear, and from the ratio of two of them, viz., the magnetic induction and the magnetic force, the dimensions of μ are found to be $L^{-2}T^2$ in the electrostatic system, and *nil* in the electromagnetic. This naturally agrees with

$$\text{Pole} = \text{Current} \times L\mu.$$

A very similar thing occurs in the electro*magnetic* system if we do not attend to K, the coefficient of dielectric capacity. Thus, from the dimensions of electricity being $L^{\frac{1}{2}}M^{\frac{1}{2}}$, and from

$$\text{Electric Potential} = \frac{\text{Electricity}}{\text{Distance}},$$

we might hastily conclude that the dimensions of potential were $L^{-\frac{1}{2}}M^{\frac{1}{2}}$, whereas Maxwell gives $L^{\frac{3}{2}}M^{\frac{1}{2}}T^{-2}$. The explanation is that a given charge in a medium of capacity K gives a force $= \frac{1}{K} \cdot \frac{\text{Electricity}}{L^2}$, and, consequently, Potential $= \frac{\text{Electricity}}{KL}$. K being a numeric in the electrostatic system, its neglect does not cause error there, but in the electromagnetic system relations not containing K should be used. Like μ in the electrostatic system K in the other system has dimensions $L^{-2}T^2$, the product $K\mu$ being, therefore, $L^{-2}T^2$ in both systems alike, the reciprocal of the square of Maxwell's velocity of propagation of disturbances.

In Ampère's theory of magnetism, magnetised molecules are simply molecules with currents flowing round them, *i.e.*, small closed currents. But these currents are not of constant strength, but vary according to the magnetic force passing through them from external causes, so as to preserve the amount of induction through them constant. This follows from the currents meeting with no resistance. If a magnetised molecule were replaced by an unchangeable closed current, they would only produce equal external force in some particular medium, and could not be generally exchangeable.

XXII.—THEORY OF MICROPHONE AND RESISTANCE OF CARBON CONTACTS.

[*The Electrician*, Feb. 10, 1883, p. 293.]

THAT a joint between two wires required to be firmly made in order to make good continuity for the current was probably learnt by the very earliest experimenters in galvanic electricity. It certainly became matter of common knowledge when telegraphs were practically introduced, and it may have been often noticed that the current passing a contact was decreased when the pressure relaxed before becoming actually interrupted. Now that we know how easily this fact lends itself to the electrical transmission of speech, it seems (after the event) surprising that it was not done generations ago. The nature of acoustical vibrations was known, and with the knowledge of the variation of current with pressure little more was required. We can now make a Morse key transmit speech—very badly, it is true. But the best microphone transmitters are bad enough, so what can one expect from a Morse key?

However, it seems to be established that Reis utilised the phenomenon in his transmitter, long erroneously supposed to work only by makes and breaks.* Indeed, presuming that he was at first unacquainted with it, and to be adjusting his transmitter very finely, it would have been difficult for him to have *not* noticed that the current passed continuously with very light contact; and the wonderful change produced thereby, a harsh, disagreeable tone being replaced by a soft and smooth one, would be unmistakable. And, in fact, he did transmit speech in this way with unbroken current. If Reis had but employed carbon contacts instead of metallic, there can be little doubt that the practical introduction of telephony would have been much accelerated.

* [See Professor Silvanus Thompson's memoir on Reis and his work.]

But, not to enter upon matters historical or controversial, or the labours of subsequent experimenters and inventors, why does the current passing a light contact vary with the pressure? In a sense the theory of the microphone as a transmitter is not affected by the answer, for, only granting the fact, this theory follows immediately:—That we make a light contact in a circuit, vary the pressure between its members by setting it in vibration, vary the current in a somewhat similar manner, and vary the attraction on the telephone armature, and set it vibrating. With so many transformations from the speaker to the telephone disc, it would not be fair to expect an exact reproduction of his voice. And we do not get it. It is recognisable, and even intelligible, if one has good ears. It is said that some of the Middle Age artists used to write such remarks as "This is a horse" by the side of their representations, to prevent any misconception. Something of the same kind is really wanted with some microphone transmitters in extensive use. A stranger requires to be told what the funny noise is meant for, and then he may understand it. Some never do.

In a certain transmitter, which is, relatively speaking, a good one, one contact piece is fixed to the centre of a thin board, the sound board in fact, and the other is mounted upon a spring and presses gently against the first. This is only the Morse key arranged for continuous light contact, arranged to be worked by the air vibrations instead of by hand. The sound board is spoken to, it moves to and fro and varies the pressure, and there is, or should be, no jolting, or scraping, or interruption of the current.

Professor Blyth's arc microphone * seems to be a thing *sui generis.* There is no visible contact; a strong current is employed to maintain an arc, and it can be made to transmit speech. Granting that, according to Professor Blyth, the action is really direct action of the air vibrations on the arc (though how the electrodes are to be stopped vibrating by making the fixtures heavy is not very evident), it yet seems an extraordinary step to take to conclude that the same thing happens in contact microphones; that the tremor breaks contact, and sets up arcs, which are then acted upon directly by the air vibrations. The facts that a strong current is required in the arc arrangement, and a circuit of low resistance, alone seem sufficient to invalidate the conclusion. But when we consider that in a contact microphone the slightest discontinuity, the least amount of sparking, at once spoils the microphonic action, the whole ground of the explanation seems undermined.

Granting, now, that there is no arc in the ordinary sense, as there could not be with such relatively considerable pressures as can be employed between the contact pieces without so much enfeebling the action that no sound is given out by the telephone, there is yet much to be learnt concerning the nature of the apparent resistance which is presented by a contact. I made some time ago numerous experiments on this subject, which led me to the conclusion that it was principally

* [*The Electrician*, Nov. 25, 1882, p. 2.]

the air that was concerned, and I am confirmed in this belief by Mr. Berliner's interesting communication (*The Electrician*, Dec. 23, 1882, p. 135), in which he states a good case against the arc theory, and that a Blake transmitter in vacuo had only $\frac{2}{10}$ths of an ohm resistance, which was that of the air left.

There is, or seems to be, a cushion of badly-conducting air between two pieces of carbon in contact, which air is partially squeezed out by increasing the pressure, and the resistance is simultaneously reduced, more points or a larger surface of carbon being brought into real contact, if there be such a thing. At any rate, this is in harmony with the difficulty of ensuring a really good connection between wires, considerable pressure being required. In soldering, also, the solder displaces the air, and, besides enlarging the area of contact, of course stops the oxidation that the displaced air would cause.

But there is uncertainty here. We do not know whether the conduction takes place between those parts of the carbon pieces which we may assume to be in real contact—conduction of the metallic nature; or whether there is conduction through the air elsewhere; or whether there is any real contact at all (except under considerable pressure), so that the conduction is invariably through a layer of air, apparently continuously and without disruption, though if the discharges were only sufficiently rapid, we should know nothing of them.

If we have two carbon pieces with flat surfaces, place them in contact exactly parallel, and have them mounted so as to be capable of to-and-fro motion; and if we assume that with increased pressure more points are brought into contact, which is reasonable enough, and, further, that the conduction is of the metallic nature, we can deduce these results. The contact resistance should vary inversely as the area with the same pressure, and be independent of the current strength. Two similar contacts, separately adjusted to exactly the same resistance, should present double the resistance when put in sequence, and half the resistance when abreast; and similarly for other combinations. We should be able to arrange contacts like battery cells, in tandem, and abreast, so that when all were similarly acted upon with the same vibrations we should obtain the greatest variation of current in the circuit. The best arrangement for a definite variation of pressure would be that in which the resistance of the external circuit, supposed constant, and containing a constant E.M.F., equalled the geometrical mean of the highest and lowest resistance of the combination of contacts under the given change of pressure, which we may consider as approximately equal to the resistance when in quiescence.

But all this is wrong. The resistance is not independent of the current for the same pressure, but varies considerably when the pressure is light, so that all conclusions based upon Ohm's law, $E=RC$, with R independent of C and E, are erroneous. The contact conduction can therefore be only partly of the metallic nature, mostly so when the pressure is not very light, for there is then the least departure from constancy of resistance; and the air therefore probably plays an important part in conducting the current, besides serving to prevent metallic contact, and a part of the apparent resistance is of the nature of a back E.M.F.

To make exact observations, contacts, though light, should not be loose. They must be definite, capable of easy regulation, and of exact reproduction as regards a given state. For this there seems nothing better than to fix one piece rigidly to the sound board, and the other piece on the end of a light flat spring, whose further end is securely fixed to some place where there is comparatively little amplitude of vibration. If the spring be horizontal, the most feeble pressure may be obtained by first adjusting to nearly give contact, and then placing a small weight on the spring; and, by altering its position, the pressure may be varied by very small amounts. The connections should be so

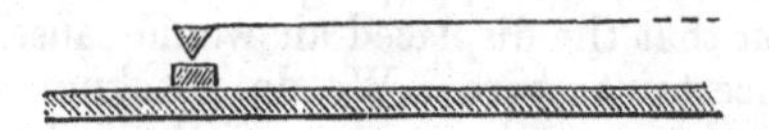

arranged that there is no straining of the spring by wires, or of the board either, as great differences in the current passing may arise from shifting wires.

The elementary contact is that between a point and a *plane*, not between two points. The point, to localise the discharge; the plane, because it does not require any adjustment to meet the point, which should only have motion perpendicularly to the plane. So one piece of dense carbon may be ground to a cube, or, at any rate, with two flat parallel surfaces; the other to a point, with a flat side opposed. All dust should be removed, as it introduces irregularities.

Join up with a battery, a telephone, and a galvanometer (say a tangent, to measure current-strength). The galvanometer is indispensable, for the telephone gives us little information as to current strength; and the telephone should be used constantly, as it tells us a great deal about the state of the contact that the galvanometer cannot. No bridge, and no induction coil. The bridge is of no use for such variable resistances as are presented by very light contacts, and especially so when the resistance varies with the current strength. To measure roughly the resistance r of a contact, observe the current before it is introduced into the circuit and after, say, C and c; then

$$r=\frac{E}{C}-\frac{E}{c}=\frac{E}{G}(\cot\theta_2-\cot\theta_1),$$

where G is the galvanometer constant, E the E.M.F. of the battery, θ_1 and θ_2 the deflections. Thus, r is proportional to the difference of the cotangents. To obtain suitable falls of deflection, vary the resistance of the circuit, or vary E.

If we have, say, a dozen contacts, as above, and observe their resistance and microphonic action under different pressures, no two of them will behave exactly alike. There is, however, in the majority of cases, a general resemblance, and the following I found to be typical. If the point be heavily weighted, there is no appreciable resistance; with very light pressure there is a considerable fall in the current. If the contact be then let alone, the current will remain nearly steady for long periods, with only slow variations of not great amount. Using one cell Leclanché in a circuit of 72 ohms resistance without the contact, with deflection 54°, when the deflection fell to 50° by reducing the pressure, indicating a contact resistance of 11·2 ohms, the sound of a watch placed upon the sound-board as a constant source of sound was just audible in the telephone. The sound increased, though not very rapidly, until 35° was reached, indicating a contact resistance of 70 ohms. From 35° down to 20°, when the resistance was 202 ohms, the sound increased much faster, and was now about three times as loud as that of the watch itself when placed close to the ear instead of the telephone. Below 20° the microphonic action became imperfect, due to slight breaks mixed up with the proper continuous action; and at 15° (resistance 300 ohms) the contact broke suddenly, and could not be permanently maintained at that pressure. If the deflection were above 20° it could be maintained nearly constant for a long time, with perfect microphonic action without sparking. (Watch with face upwards; for when placed the other way, glass upon board, there was a violent kick at every second tick, which broke contact.)

The point at which the watch became audible was naturally very variable, as it depended so much on quietude. Again, the point at which the break occurred was variable with different contacts, as any slight noise was sufficient to cause the break. But below a certain point, the current once broken would not remake itself. It required the assistance of a touch, or another noise. This was independent of any permanent change of pressure.

Some contacts, however, though seemingly just the same as the rest, would not behave regularly. The sound was as usual in the earlier stages, though not so good as it should have been; but when the later stages were reached, say at 30°, there was a stoppage of the regular sequence. Poor action. Left to itself, the contact changed so that the current through it, instead of remaining steady, decreased rapidly, the deflection sometimes going down to zero (or less than $\frac{1}{2}$° at any rate); and whereas the normal contacts were at their best, microphonically, nearly up to their breaking point, these abnormal contacts got worse and worse, the sound becoming feebler and feebler, though even with no visible current passing the watch was still faintly audible. This "bad" behaviour also varied capriciously, the current undergoing changes somewhat like earth-currents (except as regards reversal of direction), *i.e.*, gradual, but great. I was never able to settle decisively the cause of this; sometimes carbon dust produced a similar effect; sometimes repointing the point or replaning the plane removed the evil, but frequently it did not. So, although bad contact in one sense is essential to microphonic action, yet there must be good contact in another sense. All such bad contacts were rejected in making comparisons later, only the normal ones being used. Sometimes a contact which had been good for many days got wrong, in some unknown manner, and behaved in the bad fashion, and stuck to it too.

Of course a contact when adjusted so finely as described would not do for speaking purposes, as it would break at every syllable; but it answers for distant and faint sounds, whether speaking or not; a German band down the street, for example, or distant church bells. There must be no sparking whatever if the continuous microphonic action is to be preserved. Sparking is generally easily distinguishable by the peculiar sound it causes, and when the telephone indicates its existence it may be detected by ocular inspection, unless it be very minute. The microphonic action is loudest and best (for feeble sounds) when the current is least (excepting in the case of the "bad" contacts). The deflection is the same whether the sound-board be vibrating or not, except that there is a slight fall for loud sounds, even when they do not break the contact, thus showing that the average current is somewhat reduced. At the same time it is noticeable that the watch-sound gets louder when there is another sound going on, without any particular fall in the current. What the extent of variation of current is, the galvanometer will not say. It may be roughly guessed from the intensity of sound produced by breaking the circuit, and this only tells us that the variations are very small when speaking to a transmitter with coarse adjustment, and very great, perhaps as much as 30 to 50 per cent., when the adjustment is very fine, and the sound-board motion is nearly sufficient to break the contact. The disc of the telephone (a flexible disc) could also be felt to be strongly vibrating when a finely adjusted transmitter was spoken to, care being taken not to break contact.

Now regarding other forms of contact. If we round the point slightly like a worn lead-pencil point, we get substantially the same results, both as regards resistance and the corresponding intensity of sound; but it is not so perfect in the finer stages; and it is the same with any kind of contact substituted for the point, if the current passes always at one place. On the whole, though, they are more irregular than points, and the more so the flatter they are.

Flat contacts are peculiar. If the two surfaces be not exactly parallel, they will only touch at a corner, or along a portion of an edge of one of them, and the behaviour is not much different from that of a point and plane; but if care be taken to grind them quite flat, and to mount them so that, on slightly separating them, sparking can be seen to go on quite irregularly at various places, so that we may be sure that the surfaces are practically parallel, then, when brought into contact by slightly weighting the upper one, we know that there is contact at many places, and we find that it behaves perfectly abominably. The regularity of action is quite gone as regards the state of fine adjustment, and it is vastly inferior to a point microphonically. The resistance, instead of being less, is, with light pressure, usually much higher; and the current varies, as with a "bad" point. Therefore flat contacts were rejected.

With respect to the resistance of a contact, it varies according to the current passing. But there are two principal ways of considering the resistance. We may calculate it as resistance by Ohm's formula, and we then find that it decreases greatly as the current rises, with the same pressure; the current being made to vary by inserting or removing resistance from the circuit, or by varying the number of cells in the battery. Or we may reckon it as a back E.M.F. without resistance, say e, so that if E is the battery E.M.F., $(E-e)$ is the actual E.M.F. in the circuit, on the assumption that the real resistance is that of the rest of the circuit. We now find that e rises with the current, though not nearly so fast as the resistance falls on the former assumption. There is probably both a back E.M.F. and resistance at a light contact, but it is not easy to separate them, owing to the variations going on. As far as mere calculations go it is simpler to employ e, especially with combinations of contacts.

The resistance r on the assumption of no back E.M.F., and the back E.M.F. e on the assumption of no resistance, are thus connected. Let E and R be the E.M.F. of battery and resistance of circuit without contact, and let the current fall from C to c on inserting the contact. Then

$$c=\frac{E-e}{R}=\frac{E}{R+r}=\frac{e}{r}.$$

In the example before mentioned of a typical point and plane contact we have

$c=$	20·6	17·8	10·5	5·5	4	milliams.
$e=$	0	·2	·75	1·09	1·2	volt.
$r=$	0	11·2	70	202	300	ohms.

This is with E and R constant, and pressure varying.

Now, keeping the contact at constant pressure, and varying the current; if the pressure be not too light the deflection keeps steady (*i.e.*, with a particular E and R), but when very light care is required that the contact does not vary much during a series of observations, so that it should be repeated backwards and forwards two or three times to see if consistent results are obtained, all anomalous series rejected, and only those chosen which give nearly the same result on returning to the same E and R. A good series was the following:—

Contact out,	$20\frac{1}{2}°$	35°	45°	52°	56°
,, in,	13°	25°	36°	44°	50°

These give

$c=$	3·4	7	10·9	14·4	17·8	milliams.
$e=$	·57	·93	1·09	1·25	1·16	volt.
$r=$	166	133	100	87	65	ohms.

c is, of course, the current corresponding to the lower deflection. Very roughly, the apparent resistance varies inversely as the square root of the current. No particular importance can be attached to the figures as regards exactness, for it was a very light contact (and therefore somewhat variable), in order to get a good fall of deflection, which is necessary, because 1° makes a considerable difference when the fall is small.

The work done by the current at the contact is ec, or rc^2, or $e_1c+r_1c^2$, if e_1 and r_1 are the real back E.M.F. and resistance at the contact. In the above case we find

$ec=$	·0019	·0065	·0118	·0180	·0206

The loudness of the sound given out by the telephone is not proportional to the work done at the contact, for if so it would be a maximum when $e=\frac{1}{2}E$, which is not true. In the case previously mentioned (varying pressure) $e=\cdot 8E$ gave the loudest sound, just before breaking, and there is never any falling off except with "bad" contacts. With "good" contacts the sound is always greatest with the least current, provided it be continuous (the battery being of course kept the same), and its intensity is roughly proportional to e, except when e is above $\frac{1}{2}E$, for in the later stages the sound increases much faster than e.

Thus, if we adjust two similar contacts carefully to the same resistance (both planes on middle of board and points on separate springs), and then put them in series, the sound is nearly doubled if e is, say, below $\cdot 2E$ for each contact, and we find that the back E.M.F. of the two is a little less than the sum of the separate E.M.F.'s, and the resistance of the two not much greater than the sum of their separate resistances. But if e for each contact is greater, say $\cdot 5E$, the sum of the E.M.F.'s shows a large falling off, and the sum of the resistances a large increase, and the sound is much less than double that of either.

If we have six similar contacts, each by itself giving the same resistance and sound, putting them in series increases the sound up to, say, three contacts, after which there is little perceptible increase. The current falls to a certain extent on adding a fresh contact, but comparatively much more for the later additions than for the earlier. This may be understood by remembering that every contact added reduces the current, and that the resistance of all of the contacts increases simultaneously, and the more so as the current gets smaller. The more sensitive the contacts are the less advantage there is in putting them in series.

When two contacts are unequally strong, and e is small, they add their effects, both as regards e and the sound. Thus $e_1=\cdot 17E$, $e_2=\cdot 22E$; $e_1+e_2=\cdot 36E$. Even if e is great for one and small for the other there is usually a slight increase, or, at any rate, no decrease. If both are made as sensitive as possible, and as equal as possible, putting them in sequence usually increases the sound very little, but this varies according to the resistance of the circuit, on which depends the amount of fall of current on inserting the second contact (the first being already in).

Thus
$$e_1=e_2=\cdot 6E, \qquad e_1+e_2=\cdot 81E;$$
$$e_1=\cdot 61E, \qquad e_2=\cdot 16E, \qquad e_1+e_2=\cdot 67E;$$
$$e_1=\cdot 33E, \qquad e_2=\cdot 38E, \qquad e_1+e_2=\cdot 59E.$$

On the whole, I found that the calculated value of e was a sort of guide, though not by any means a perfect one, to the intensity of the microphonic action. Sensitive contacts in series require separate examination in turns (by short-circuiting all except one) occasionally, to see that they keep steady. To go by the apparent resistance is very misleading.

But there is also the battery E.M.F. to be considered. The loudness appeared to be about proportional to the number of cells used, from

1 up to 5, with circuit resistance 150 ohms; so that the product Ee would be roughly proportional to the intensity of the sound, or, otherwise, proportional to the product of the current without the contacts into the fall of current on inserting them. This is true in a large number of cases, but must not be carried to extremes. Thus if one contact brought deflection from 42° down to 10° with one cell, putting it in sequence with a similar one brought deflection down to 5° or less, but with no increase of loudness—in fact, a slight decrease; whereas in making the same experiment with 5 cells there was a slight increase. It was also verified many times that if two contacts adjusted equal in the first place, and then put in sequence, bring deflection down to 10° (with 1 cell), there is less sound than with either of them alone when adjusted so as to cause the same fall of current. Also, with many contacts in sequence, the sound ultimately decreases, the later additions having apparently much higher resistances than the earlier; the sound increases by addition of two or three, is then about constant with one or two more (all separately giving good sound), and finally decreases on further additions being made. Now here e, calculated for the whole as for a single contact, continually increases, although it becomes practically constant, because the reduction of current for, say, the 6th contact, although it may be considerable compared with the current already reduced by the first five, is quite small compared with the original current.

Contacts in parallel arc behave, in the main, as might be expected from the relation between e and the current passing observed in the case of a single contact. The total current is always increased by putting two contacts abreast; *i.e.*, it is greater than through either of them when the other was disconnected, and this is true whether the contacts are separately equal or not. As for the sound, it is never increased, and is generally reduced. If the contacts be unequal, the joint sound is intermediate between their separate sounds. When one is very weak it shuts up the other, however good it may be, which is sufficiently plain, because nearly all the current goes through the weak one. Should there be a difference in character in the sound of the contacts taken separately, such difference will be also recognisable to a certain extent when they are put abreast. Otherwise there is simply a weakening of intensity, unless both are in a very sensitive condition, in which case there is little perceptible difference acoustically between A and B singly and A and B in parallel arc. Now in this last case the current through each is halved when they are put abreast, which involves a reduction in e (or larger increase in r), and at the same time the total current is increased.

Owing to the unknown real resistance, it is desirable to make contacts exactly equal first, before putting them abreast, so that we know that the current goes half through each, whatever their resistance may be. If they could be regarded as battery cells of constant resistance and E.M.F., observations of the fall of current in sequence and abreast and cut out would allow us to reckon the real resistance. But there is no harmony in the results. Four deflections, taken in all possible pairs, gave for the resistance of a single contact from 32, through 42, 46, 55, 61, up to 76 arbitrary units.

On the other hand, calculating e from the results and comparing with the corresponding current, and making allowance for the halving of current when abreast, a fall of e with reduced current was obtained, similar to that observed with a single contact on varying the external resistance.

In concluding this abstract [of experimental notes], I will merely add that a great deal of patience is necessary when working with very light contacts, and that when combinations are made something more than patience is required, and many precautions must be taken, to be found by experience, else very contradictory results may be arrived at.

XXIII.—THE EARTH AS A RETURN CONDUCTOR.

[*The Electrician*, Nov. 11, 1882; p. 605.]

THE daily newspapers, as is well known, usually contain in the autumn time paragraphs and leaders upon marvellous subjects which at other times make way for more pressing matter. The sea-serpent is one of these subjects. This year, however, that interesting animal has not been so observable, which is, perhaps, the reason why an equally wonderful and not less time-honoured phenomenon has come to the fore again. There appeared lately an account of the performance of "an innocent boy" with a stick of wood, which, being held in the hands of the operator as he walked about in a field, twisted and turned itself so as to prove the existence and point out the situation of water beneath the surface. This may or may not be. Never having studied the action of divining rods makes me an incompetent judge; but on further reading the explanation of a philosopher of the cause of the phenomenon, viz., that the water was a conductor of electricity, and, therefore, the electric currents deflected the wand, I was at once reminded (perhaps strangely, for there is hardly any connection between the theories) of a theory of the action of the earth as a return conductor that I first read some fifteen years ago in a Handbook which has since passed through many editions.

It was to this effect: that if two insulated conductors were connected to the poles of a battery, a certain quantity of electricity would pass to charge them; that the larger the conductors the longer time the charging would take; and, finally, that if they were infinitely large, it would take so long that the current would pass as if the poles of the battery were directly connected. The application of this theory to the earth's circuit was that the earth was practically infinitely large, and so the current passed continuously.

Both the theory and its application seem erroneous. For, by the laws of electrostatics, the charge the two insulated conductors would receive would depend upon the size of their opposed surfaces, and upon their proximity, but not upon their bulk *per se* at all. If we had two worlds side by side, and joined them by a wire with a battery inserted in it, quite a small quantity of electricity would suffice to charge the system; and if the bulk of the auxiliary world were increased indefinitely on the side furthest from the real, it would make scarcely any difference. The way to increase the charge would be to make the second world surround the first, and bring it as close as possible, when the capacity would be greatly increased, though not to any such extent as to give rise to a permanent current.

And the application of the theory to the case of the earth is faulty, because although it may be regarded for certain purposes as an infinitely large conductor, yet it is only one conductor, not two conductors insulated from one another.

There seems, in fact, no hypothesis at all wanted to explain why, when the ends of the wire containing a battery are put to earth, the current continues to flow. The earth is a conductor, and completes the circuit, and what more is wanted? It would be very extraordinary if the current did not continue to flow indefinitely, or until the battery got used up, or a disconnection occurred somewhere. The very existence of such a theory, however, shows that there must have been considerable doubt as to the real action of the earth.

Another theory, a very popular one, is much more satisfactory. The earth is so large, and contains so much electricity, that it may be regarded as an infinite store, to which all charges we may add or take away are utterly insignificant. Allied with this is the theory that compares the earth to a immense reservoir of water, or the sea. We may pump out or pour in as much water as we please without making any appreciable difference. This last form, by proper limitations, has the advantage of being easily converted into an exact analogy.

Let there be a reservoir of water, large or small, and let the water be completely enclosed on all sides in a tight-fitting envelope, so as to completely fill it. Let, further, the water be absolutely incompressible, instead of nearly so, and let a pipe, also completely filled with water, make connection between two parts of the reservoir, where of course are corresponding openings. We have then a quantity of fluid occupying a certain space, and which must always continue to occupy the same space. Set the water in the pipe in motion, then it follows that the current crossing every section of the pipe is the same, that an equal current leaves the pipe at one end, and at the same time an equal current necessarily enters the pipe at its other end. Furthermore, the lines of flow of the water in the reservoir itself are perfectly definite, depending only on the shape of the reservoir and the position of the pipe terminations, or source and sink.

For the reservoir of water substitute a conductor of electricity of the same size and shape, and for the pipe a conducting wire similarly terminated conveying a current of electricity, and the analogy is quite complete, so far as *steady* currents are concerned, and the lines of flow of the electric current are the same as those of the real fluid in the former case.

In fact, for purposes of analogy, we may regard the earth and a wire joining two points of it as being always filled with electricity, whose quantity may be as small or as great as we please, and that it is incompressible, and must follow the same law of continuity as a real, incompressible fluid, if such existed. Such hypothetical electricity would not, however, be the electricity of the definition $e_1e_2/r^2 =$ force between two charges e_1 and e_2 at distance r, because it has no statical action. Statical actions do occur, owing to surface charges, but we have no concern with them here, having only the current under consideration.

Or, without any hypothesis as to the universal existence of the electricity, simply regard the conductors (wire and earth) as being capable of bearing at every point an electric *current*, simply subject in its distribution to the law of continuity of an incompressible fluid. There are other quantities in physics having the same property, and there is no occasion whatever to consider electricity as a fluid at all, except for purposes of illustration.

Although the mere size of the earth has nothing to do with the permanent flow of the current, it has an important influence upon the resistance and lines of flow. To study the matter more in detail we may start with the simplest case imaginable, a long thin wire buried in an infinitely extended conducting mass of uniform conductivity, the wire being, of course, insulated except at the ends, which we may suppose terminated in spherical electrodes. Let there be a steady current, C, in the wire. A current, C, leaves the + electrode and enters the − at the same time. Considering one electrode only, say the + (let, for the time, the − be at an infinite distance), then, owing to symmetry, the current C spreads out equably in all directions, so that its strength is $C/4\pi r^2$ at distance r from the electrode, and is everywhere radial. (This is the current density, or current across unit area perpendicular to lines of flow.) The lines of flow are straight lines, starting from the electrode and uniformly distributed.

The same is true for the − electrode, except that the direction of the current is to it and not from it. And to get the lines of flow in the real case of finite distance between electrodes, we have only to find the resultant of the two systems. At a point distant r_1 from the + and r_2 from the − electrode, the current density is the resultant of $C/4\pi r_1^2$ in direction r_1, and of $-C/4\pi r_2^2$ in direction r_2.

In the neighbourhood of the electrodes the lines of flow are, as before, straight, radial, and uniformly distributed, but all the lines from the + ultimately curve round and join those belonging to the − electrode.

The distribution of lines of flow is the same as that of the lines of force between two small spherical conductors with equal opposite charges, or as the lines of magnetic force between the two poles of a long, thin solenoidal magnet.

It is easily seen that the form of the electrodes is of no importance so far as the lines of flow at a considerable distance from them are concerned, and it is only near them that changes in the form of the electrodes must alter the lines of flow. We choose spheres for simplicity.

Observe that there is no current *across* any plane section through the electrodes, and that such a plane separates the current system into two symmetrical halves. We may therefore take one of these halves, completely disregarding the other, and we have the correct distribution of lines of flow in the case of two hemispherical electrodes with their flat sides flush with the plane surface of a conductor infinitely extended on one side only. This conductor we may consider as the earth, and the electrodes the earth plates, to which a wire conveying a current C is connected at the ends. Then, presuming uniform conductivity of the earth, the current density at distance r_1 from the + and r_2 from the − electrode is the resultant of $C/2\pi r_1^2$ along r_1 and $-C/2\pi r_2^2$ along r_2. (We now have 2π instead of 4π, on account of the one-sided radiation.)

We may also easily find the form of the lines of flow when the electrodes are buried at some depth from the surface. We have only to double the system by introducing another earth above the real, with electrodes in it to correspond, as much vertically above the plane of separation as the real ones are below it—images of them, in fact. We have then two + and two − electrodes in an infinite conductor, and the system of lines of flow is the resultant of four radial systems; and since the plane of separation has no current across it, the lines of flow are unaltered in the lower system when the upper is removed. Thus in the real case of buried electrodes the lines of flow are deflected by the surface of the earth in the same manner as if the images had a real existence.

Similarly, we may find the lines of flow for any system of electrodes by superimposing the different elementary systems, employing images when necessary.

Resistance of Earth.

A wire being put to earth at its two ends, the resistance of the circuit formed may clearly be separated into three portions, that of the wire right up to the electrode (with which we have nothing here to do), that of the electrode itself, and that of the earth between the electrodes. The last may be readily found.

Imagine a single spherical electrode, say +, of radius a, in an infinite conductor of specific resistance K, supposed uniform, *i.e.*, $K=$ resistance of a cubic centim. In consequence of the uniform diffusion of the current, the equipotential surfaces are concentric spherical surfaces, and the resistance between any two of them of radii r and $r+dr$ is

$$K\times\frac{\text{thickness}}{\text{area}}=\frac{Kdr}{4\pi r^2}.$$

Therefore, if R is the whole resistance outwards from the electrode,

$$R=\int_a^\infty\frac{Kdr}{4\pi r^2}=\frac{K}{4\pi a}.$$

In the case of the earth, with hemispherical electrodes of radii a_1 and a_2, the resistance between them is

$$K/2\pi a_1 + K/2\pi a_2 \qquad \text{(A)}$$

But if the electrodes are spherical, and deeply buried, the resistance is

$$K/4\pi a_1 + K/4\pi a_2 \qquad \text{(B)}$$

nearly.

In (A) the resistance is double that of (B), on account of one-sided diffusion, and in both we add the resistances calculated for each electrode separately, because it is practically confined to their neighbourhood. And this is why in (B) we have 4π, the same as for an infinitely extended conductor.

To exemplify this point—the distribution of the resistance—compare the resistance between a and $2a$, $2a$ and $4a$, etc., for one hemispherical electrode.

Between $r = a$ and $r = 2a$,

$$R = \int_a^{2a} \frac{K dr}{2\pi r^2} = \frac{K}{2\pi}\left(\frac{1}{a} - \frac{1}{2a}\right) = \frac{1}{2} \cdot \frac{K}{2\pi a}.$$

Similarly, between $r = 2a$ and $r = 4a$, the resistance is $\frac{1}{4}(K/2\pi a)$, and from $r = 4a$ to $r = 8a$, it is $\frac{1}{8}(K/2\pi a)$. Now, $K/2\pi a$ is the whole resistance; so one half of it lies between a and $2a$. It may be easily shown that about 99 per cent. lies between a and $100a$, and the remaining 1 per cent. beyond.

The difference of potential between two electrodes is, of course, $R \times C$, where R is the total resistance between them. It is customary, and generally convenient, to consider the earth as being at the same potential; but of course the return current could not flow if such were really the case.

As regards the amount of resistance, it depends essentially on the size of the electrode in the first place, varying inversely as its radius; so that it may be as great or as small as we please from this cause alone. And it is directly proportional to the specific resistance K. Thus (hemispherical), with a radius of 1 metre, the resistance is $K/628$. With $a = 1/2\pi$ centim., and $K = 1$ ohm, R is also 1 ohm. Even with a specific resistance of 1 megohm, we can bring the earth resistance down to 1 ohm by taking the radius of electrode $= 10^6/2\pi$ centim., or nearly 1 mile.

Naturally, we may include the electrodes in the earth resistance by letting the end of the earth wire be the electrode; then we see that from its smallness the conductivity immediately around it is a matter of the greatest importance, and since common earth is badly conducting, the utility and necessity of "earth plates," whether plates or not. Also, that around the earth plates for a considerable distance the conductivity is of importance, and we cannot get a good earth in a rock without very large earth plates. But as we proceed further away the conductivity becomes of less and less importance, and we may say that it hardly matters; with this proviso, that the very badly conducting material does not completely shut off one electrode from its fellow at the other extremity of the line wire. The earth may be perfectly insulating for hundreds of miles, and the current will go round the impermeable mass with no sensible increase of resistance. In the extreme case of a non-conducting screen, only leave a practicable opening in it, and the total resistance of earth will be merely increased by the resistance of the matter in the opening, and a little way on each side of it, where the current converges and diverges. If the non-conducting screen were quite complete no permanent current could flow from one electrode to the other, and signalling would have to be carried on by transient currents, the two sides of the screen forming the two poles of a large condenser. Such a case is hardly likely to arise in practice.

XXIV.—THE RELATIONS BETWEEN MAGNETIC FORCE AND ELECTRIC CURRENT.*

SECTION I. THE UNIVERSAL RELATION BETWEEN A VECTOR AND ITS CURL.

EVERY one knows that electric currents give rise to magnetic force, and has a general notion of the nature of distribution of the force in certain practical cases, as within a galvanometer coil, for example. Further than this few go. The subject is eminently a mathematical one, and few are mathematicians. There are, however, certain higher conceptions, created mainly by the labours of eminent mathematical scientists, from Ampère down to Maxwell, which are usually supposed to be within the reach of none but mathematicians, but which I have thought could be to a great extent stripped of their usual symbolical dress, and in their naked simplicity made to appeal to the sympathies of the many. Let not, however, the reader (if he belong to the many) imagine that thinking can be dispensed with; there is no royal road to knowledge, and hard thinking and rigid fixation of ideas are required. Even the machinery of the mathematician, so great an assistance when made to work, requires severe training on the part of the operator to make it work. But earnest students, if they will not or cannot learn the mathematical methods, need not therefore be discouraged, for the name of Faraday will shine forth to the end of time as a beacon of hope and encouragement to them. He was no mathematician, yet achieved results apparently only attainable by such methods. It need not be supposed that he had the peculiar brains of a calculating boy, able to do long sums "in his head" by special methods of his own. The work

* [*The Electrician*, section I., Nov. 18, p. 6; section II., Nov. 25, p. 32; section III., Dec. 2, p. 55; section IV., Dec. 30, 1882, p. 151; section V., Jan. 6, 1883, p. 175; section VI., Dec. 16, 1882, p. 102.]

was of quite a different kind, and probably Faraday could never have made an ordinary mathematician, with the best of training. In fact, mathematical reasoning does not necessarily involve any calculating in the usual sense, though it is, of course, greatly assisted thereby sometimes; and as for the use of symbols, they are merely a sort of shorthand to assist the memory, which even those who openly contemn mathematical methods are glad to use so far as they can make them out—in the expression of Ohm's law for instance, to avoid spinning a long yarn.

To introduce the subject, we start with the case of a long, straight cylindrical wire, conveying a steady current C. The magnetic force is known to be of intensity $2C/r$ in electromagnetic measure at distance r from the axis outside the wire, and its direction to be perpendicular to r and to the axis. This, however, does not settle in which direction along the perpendicular the force acts, and so a rule becomes necessary. Look along the axis in the direction the current is going; the magnetic force is then in the direction of right-handed rotation about the axis. This rule, or any equivalent one, is the key to all the directional relations in electromagnetism. The hands of a watch, viewed from the front (they cannot be seen from the back), revolve right-handedly, so that if we imagine the watch-face to be a section of the wire, and the current to go from face to back, the magnetic force is in the direction of the rotation of the hands.

The force being of the same intensity at the same distance from the axis, a line of force is a circle embracing the axis, and the axis is perpendicular to its plane. Confining ourselves now to a single plane normal to the axis, all circles centered upon the point where the axis cuts the plane are therefore lines of force. But if we wish to show graphically the intensity of force as well as its direction at any place, the best way is to draw the lines so that their density or closeness together shall be proportional to the intensity of force at the place. In our present case this gives us the rule that the radii of successive circles should increase geometrically.

Within the wire the magnetic force is known to be of intensity $2Cr/a^2$ at distance r from the axis, a being the radius of the wire, and, like the external force, perpendicular to r and to the axis, with right-handed rotation. At the surface the two expressions become the same, viz., $2C/a$. The force being now directly as r, a different rule is required to make the density of lines of force proportional to the force. The squares of the radii of successive circles must form an arithmetical progression. Further, the scale outside and within should be the same, so we require a relation between the common ratio of consecutive radii outside and the common difference of squares of consecutive radii within. If ρ is the external ratio and d the internal common difference, $d = 2a^2 \log \rho$. For let a_1 be the radius of first circle outside, and a_0 that of first inside, a_0, a, and a_1 being thus consecutive, we must have, k being any constant,

$$\frac{k}{a_1 - a} = \frac{1}{a_1 - a}\int_a^{a_1} \frac{2C\,dr}{r}, \qquad \therefore k = 2C \log \frac{a_1}{a} = 2C \log \rho.$$

Also $$\frac{k}{a-a_0}=\frac{1}{a-a_0}\int_{a_0}^{a}\frac{2Cr}{a^2}dr, \quad \therefore\ k=C\frac{a^2-a_0^2}{a^2}=C\frac{d}{a^2}.$$

Therefore $$d=2a^2\log\rho.$$

In the figure the common difference d is chosen $=1$ and the radius $a=2$, so that the three lines within the wire (which is bounded by the thick circle) have radii 1, $\sqrt{2}$, $\sqrt{3}$. The circle $a=\sqrt{4}$ is also a line of force, and the rest have radii 2ρ, $2\rho^2$, $2\rho^3$, etc., where $1=8\log_\epsilon\rho$.

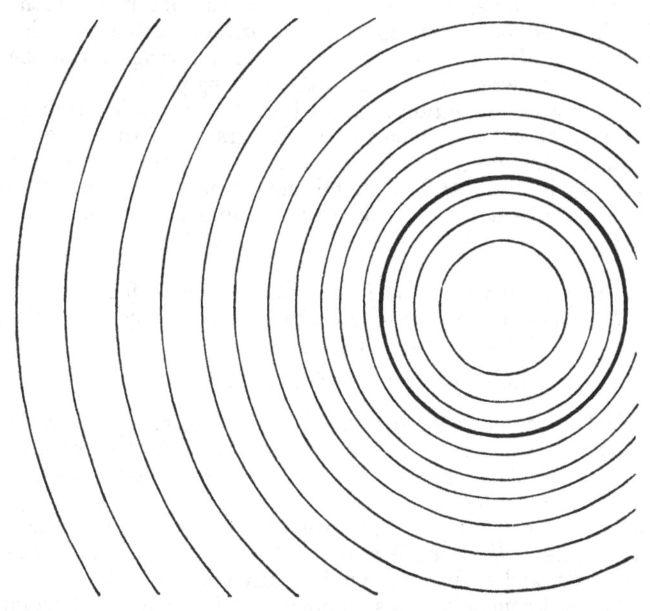

Let us now examine a general property of this system of magnetic force. If we place a unit magnetic pole at any point in the external field, and suppose that it can move freely under the influence of the magnetic force of the current, and that no other forces act upon it, the pole will evidently describe a circle about the axis, and the work done by the force on the pole during a complete revolution will be

$$\text{force} \times \text{distance}=(2C/r)\times 2\pi r=4\pi C.$$

And evidently, in moving through any stated fraction of the complete circle, the work will be the same fraction of $4\pi C$. Observe that this is the same at any distance from the wire, and the work depends only upon the angle turned through. Also, if the pole be moved either radially from the axis or parallel to the axis there is no work done, because there is no force in those directions. Let now the pole be carried from any one place to any other by any path, its motion at any point of the path may be compounded of a motion *from* the axis, a second *parallel* to it, and a third *round* it. Since the last alone involves the performance of work, it follows that the work done on the pole between any two points by any path is $2C \times$ angle turned through in the positive or right-handed direction, between the first and last positions.

Now, limiting our consideration to motion in completely closed paths, if the curve embraces the wire once, the line integral of the force once round the closed curve is $4\pi C$, and if it goes round n times (in + direction) it is $n \times 4\pi C$. But if the closed path does not embrace the current, or if embracing it a certain number of times in the + it embraces it the same number of times in the − direction, so that if the path were a string it could be drawn off the wire without cutting it, the line-integral is *nil*. In this case the angle turned through about the axis comes to nothing when we return to the starting point.

Transferring our attention next to the current passing through the closed curve, we see that when the line-integral is *nil* the current is *nil*, and when the line-integral is $4\pi C$ the total current through the curve is C, and when it is $n \times 4\pi C$ the total current is nC. This may be all summed up in one statement. The line-integral of the magnetic force once round any closed curve equals $4\pi \times$ total current through the curve.

Let us now see whether the force within the wire follows the same rule. First let the unit pole follow a line of force once round the axis at distance r. The work done by the force

$$= \text{force} \times \text{distance} = 2Cr/a^2 \times 2\pi r = 4\pi Cr^2/a^2.$$

Now, $C/\pi a^2$ is the current-density, and πr^2 is the area enclosed by the line of force, so that Cr^2/a^2 is the current through the closed curve; and, as before, line-integral of force $= 4\pi \times$ current enclosed. And, since there is no force parallel to, or radially from the axis, the same statement is true for any closed path whatever within the wire; and, being true outside the wire, is also true for any path partly within and partly without, and so, in fact, is universally true.

Now we shall generalise the statement. It is not merely true for the magnetic force of a straight current, but also for any possible system of magnetic force. It is the fundamental relation between magnetic force and electric current. Given, then, a system of magnetic force, we may find the corresponding current system by the following process:— Required the current at any point P. Through P draw any straight line, and describe a small plane closed curve above it as an axis. Find the line-integral of the magnetic force along the curve once round the axis, or the work done upon a unit pole during one rotation. The result is 4π times the current through the curve. Divide by the area enclosed, and we obviously get 4π times the component of the current-density in the direction of the axis.

Let the axis turn about the point P as fulcrum into any other position, the closed curve moving with it as if rigidly attached; the line-integral in the new position will be $4\pi \times$ current through curve in the new position. There is a certain direction of the axis for which the line-integral is a maximum; this direction is that of the actual current at P, and the maximum line-integral divided by 4π times the area is the actual current-density. We may conveniently take the area for unit area, so as to get the current-density at once. Or, we may find the current-density by the above process in any three rectangular directions, and their resultant will be the actual current-density.

This process, by which we derive current from the magnetic force, is of great importance in physics. Especially so in electromagnetism, where there are several quantities bearing to one another the same relation. When one vector or directed quantity, **B**, is related to another vector, **C**, so that the line-integral of **B** *round* any closed curve equals the integral of **C** *through* the curve, the vector **C** is called the curl of the vector **B**. The term curl was proposed by Maxwell, though he does not appear to have used it much. Its appropriateness is evident on considering the method by which we derive **C**, the curl of **B**, from **B** itself; and, as a name for the operation is wanted for descriptive purposes, it will be used in the following. Thus, current $= 4\pi \times$ curl of magnetic force.*

The presence of the factor 4π is due to the definition of a unit magnetic pole. If we defined the unit pole so that the unit amount of force emanated from it, the force at distance r would be $1/4\pi r^2$, since $4\pi r^2$ is the area of the spherical surface over which the force is spread at distance r. Then we should have current = curl of magnetic force. And if we had a similar definition of the unit of free electricity, we should have the electric force at a surface numerically equal to the surface density, instead of, as at present, 4π times as much. But the actual definitions chosen make the force at distance r from a unit pole be $1/r^2$. This looks simpler, but it leads to the awkward result that mathematical investigations, both in electrostatics and in electromagnetism, are filled with 4π's and $1/4\pi$'s. Sometimes we multiply, at other times divide. They would mostly be got rid of by defining electric and magnetic forces as fluxes in the same manner as the electric current; for a current, C, spreading from a centre produces current-density $C/4\pi r^2$ at distance r, and not C/r^2. At the same time 4π would make its appearance in certain cases where it is now absent, such as in spherical problems; and its presence there would be perfectly natural 4π being the area of the unit sphere.

*[As this is the first use of vectors in this Reprint, it may be appropriately mentioned here that the algebra and analysis of vectors is introduced very gradually. At first the same type was used both for vectors and scalars, but I found later that it was a matter of some practical importance to facilitate the reading and ease the stress on the memory by employing a special type for vectors. So, the German type used by Maxwell being utterly unpractical, I introduced Clarendon type for the purpose in the *Phil. Mag.*, August, 1886, and later papers, and now do the same in these earlier papers to harmonize. It will be found to be a particularly suitable type, being very neat, easily read, and well adapted for use in formulæ along with ordinary type, roman and italic. When only the tensor (or size) of a vector is concerned, the ordinary type is used. Thus C is the tensor of **C**.

In MS. work special letters for vectors need not be used, but ordinary letters only. The tensor may then be C_0. Or the letters may be marked in some conventional way to indicate that they stand for vectors. This, of course, becomes necessary when the MS. is to be "copy" for the printer.]

To show that the idea of the curl is not without practical utility, even to non-mathematicians, we may employ the process to find the distribution of current for fresh systems of magnetic force, starting from the one already treated. Call the magnetic force of a straight current **B**, and the current **C**, this being now the current-density. We know that $4\pi\mathbf{C}$ = curl of **B**. What is, next, the curl of **C**? This is easily found. Outside the wire **C** is *nil*, therefore so is its curl. Within the wire **C** is uniform, so that its curl is also *nil* there. But in passing through the surface, **C** suddenly changes from **C** in the wire to 0 outside, so there *is* surface-curl. Let

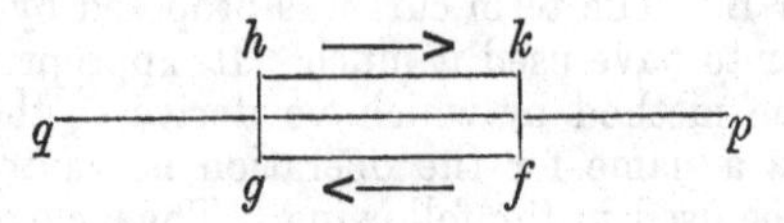

pq be a straight line upon the surface in the direction of **C**. Let the closed curve of integration be the rectangle $fghk$, consisting of two straight lines fg and hk parallel to the surface, and very near it, with two connecting pieces gh and kf; and let the rectangle be perpendicular to the surface. We have to find the line-integral of **C** once round $fghk$. Evidently the portion fg contributes $fg \times C$, and the rest nothing; the connecting pieces because they are perpendicular to **C**, the other because it is outside the current. Therefore $fg \times C$ is the total amount of the curl of **C** passing through the rectangle, *i.e.*, it is of strength C per unit of length along pq. Shorten the connecting pieces indefinitely; we have still the same result, so that the curl of **C** is finally a vector quantity of amount C per unit of length of pq, drawn upon the surface at right angles to the current. Let the current go from right to left, then its curl is directed downwards through the paper, that is, in the same direction as the magnetic force. The same is true for every point of the surface, so that the lines of the curl of **C** are circles upon the surface, centered upon the axis.

Let, now, **C** represent the magnetic force in a new system; *i.e.*, the magnetic force is confined to the space within a long cylinder, and is everywhere parallel to its axis, and of uniform intensity C. We have found the curl of **C**, so we know the corresponding current distribution. It consists of a cylindrical current-sheet, the current circulating round the axis, its amount per unit length of cylinder being $C/4\pi$. This is the case of a long solenoidal coil of a single layer of wire; if the current per unit length be C, the magnetic force is *nil* outside, and of uniform intensity $4\pi C$ within, parallel to the axis. Put on more layers of wire, and we have simply to add on the additional magnetic force, and the result is that the magnetic force is $4\pi C$ (where C is the *total* current round unit of length) everywhere inside the innermost layer, and falls from $4\pi C$ to 0 in passing through the layers of current to the external space.

For another example, let us find the curl of the current in the last case. Start with a cylindrical current of strength Ct per unit length, and let the small thickness of the current layer be t, so that the current-density is C. We found before that to deduce the curl of a vector **C**, when it suddenly changed from **C** to 0 in passing through a surface, we had merely to turn it through a right angle upon the surface, and we obtained the new vector. The direction of rotation must, however, be carefully attended to. If we look down upon the surface beyond which **C** exists, and its direction is as shown, then its curl, **D**, goes from left to right. Now our current-sheet has two surfaces, inside and outside. In passing from the current to outer space, we therefore rotate one way through a right angle, thus bringing us parallel to the axis; and in passing from the current to inner space we rotate the other way through a right angle, bringing us also parallel to the axis, but pointing in the reverse direction. Since we get the same result everywhere, the curl of **C** consists of two tubular vector systems separated by the thickness t, of equal strength, but oppositely directed.

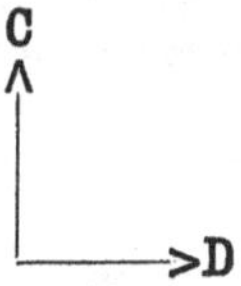

If, then, **C** represents the magnetic force in a new system, we have the following:—Two thin concentric tubes, distance between them $= t$, with a current of total strength

$$C \times 2\pi a / 4\pi = \tfrac{1}{2} Ca,$$

where a is the mean radius of the tubes, straight along the outer tube, returning by the inner one. There is no magnetic force within the inner or without the outer tube. Between them the magnetic force is in circles about the common axis, and is of mean strength C.

Further application of the same process only gives rise to repetitions in a more complex form of the last two examples, multiplications of cylindrical and straight tubular currents alternately, and we need only notice the first of the series. If we find the curl of the current in the last case, and then transform current into magnetic force, we obtain four cylindrical current-sheets. In the outer and innermost the current circulates round the axis one way, in the two intermediate the other way. The magnetic force is parallel to the axis between the first and second, and also between the third and fourth, but oppositely directed, and is *nil* everywhere else.

We see that the process of deriving fresh distributions from a known one by curling may be continued. We have a series of vector systems, **A**, **B**, **C**, etc., of which any one is the curl of the preceding; and, taking any one to represent magnetic force, the following one is the corresponding current, excepting a constant factor. It will readily suggest itself that the series may be continued the other way. This is true, but is not so easily managed. The reverse operation to finding the curl of a vector, viz., to find the vector whose curl is a given vector, is more difficult than the direct, though of not less interest or importance. This matter, and some other relations between current and magnetic force, will form our subject later.

SECTION II. THE POTENTIALS OF SCALARS AND VECTORS.

In the preceding section, the meaning of the "curl" of a vector was explained and illustrated in the case of a steady current in a long straight wire. Also, from that solution were derived, by the operation of curling, the distributions of current corresponding to other possible arrangements of magnetic force. We now come to some other properties of magnetic force and current, and more generally of any similarly related vector quantities. To fix ideas, let **B** be a possible distribution of magnetic force, and **C** the corresponding electric current. We know that **C** = curl **B**/4π, and we may similarly find the curl of **C**, which we may call **D**. Thus,

$$\begin{aligned} &\mathbf{B} = \text{magnetic force, given.} \\ &\left.\begin{aligned} &\mathbf{C} = (4\pi)^{-1} \text{ curl } \mathbf{B} = \text{current}, \\ &\mathbf{D} = \text{curl } \mathbf{C}, \end{aligned}\right\} \text{deduced.} \end{aligned}$$

Supposing, however, that it is the current distribution that is given, how shall we find the magnetic force? Considering the fundamental relation only, that by finding the line-integral of **B** once *round* any axis we get $4\pi \times$ current-component *along* that axis, if we reverse the operation we discover at once that it fails to work in a suitable manner. We do indeed know from the given value and direction of the current at a given place what the line-integral of magnetic force round it is, but that does not tell us the magnetic force at different points along the line of integration. Some other method is, therefore, wanted.

There are different ways of obtaining the magnetic force from the current. We shall commence with that one of them which has the advantage of telling us immediately in a great many cases the general nature of the magnetic force. This method is expressed in the following statement:—The magnetic force is the vector-potential of the curl of the current. Here we introduce another concept, that of the potential of a vector quantity, and in order to render it intelligible, some explanation becomes necessary.

The meaning of potential in electrostatics is well known, therefore we need here merely remind the reader that the potential of a charge e at distance r therefrom is e/r, where we suppose e to be at a point, or, at least, within a very small space surrounding the point from which r is measured. It is the work that must be done to bring a unit charge from an infinite distance to the place considered. For the intensity of electric force is by definition e/r^2, and the work done in bringing a unit charge from distance r_2 to distance r_1 from e is

$$\text{Mean force} \times \text{distance} = \frac{e}{r_1 r_2} \times (r_2 - r_1) = \frac{e}{r_1} - \frac{e}{r_2};$$

and when the distance r_2 is infinite, the work becomes simply e/r_1. The potential of any system of free electricity is the sum of the potentials of the elementary charges into which it may be divided, and we may write it

$$\Sigma \frac{e}{r} = \frac{e_1}{r_1} + \frac{e_2}{r_2} + \frac{e_3}{r_3} + \dots,$$

e_1, e_2, ... being the charges, and r_1, r_2, ... their distances from the point where the potential is to be found. The electric force in any direction is the rate of decrease of the potential in that direction. But we may also consider it as the sum of the forces due to the elementary charges.

Here, however, we must, since force is a vector, not simply add the numerical values of the forces together, but make the proper allowance for their being in different directions; that is to say, we must find their resultant. This is most conveniently expressed by saying that we find the vector sum of the separate forces. This we may denote by $\Sigma(e/r^2)\mathbf{r}_1$, where e/r^2, the intensity of force due to the charge e, is multiplied by $\mathbf{r}_1$, which signifies a *unit vector* drawn along the line from the charge to the point under consideration, thus making $(e/r^2)\mathbf{r}_1$ be the *vector* force due to e. The sign Σ signifies summation.

Now, in finding the potential of a vector quantity, such as current, we add together the potentials of the elements into which the current-system may be divided, *i.e.*, we find $\Sigma\mathbf{C}/r$; but we must do it exactly as in the last-mentioned case of electric force, that is, find the vector sum. Free electricity is scalar or directionless, and so is its potential, therefore simple addition of the numerical values of the potentials of the elements gives the value of the whole potential, and there is nothing else to consider. But if, for example, one current-element, $\mathbf{C}$, is directed from right to left, and another equal one from left to right, the sum of their potentials at a given point is the difference of their separate potentials, and is directed parallel to the greater.

In certain cases, however, the process is simplified. In a straight current, for example, the current-elements all point the same way; the potential of any element, $\mathbf{C}$, at a point distant r, is $\mathbf{C}/r$, and is parallel to $\mathbf{C}$; every element of the total potential is parallel to $\mathbf{C}$, and so is the total potential, and its value is the scalar sum of the potentials of the elements. We should therefore find the potential exactly as for free electricity occupying the same space as the current, and then make it a vector by giving it direction parallel to the current. Thus the potential of a system of parallel straight currents at any point is a vector drawn from that point parallel to the current; and, in general, the potential of one vector-system is another vector-system. To distinguish it from ordinary scalar potential it is sometimes called the vector-potential. In the same way as we represent systems of magnetic force or current by means of lines, we may represent their vector-potentials by systems of lines, the direction of a line showing that of the vector-potential, and the density of the lines its magnitude at any place.

Thus the lines of vector-potential of a straight current are straight lines parallel to it, closely packed in the wire and near it, and falling off in density as we recede from the wire, according to the logarithm of the distance.

Consider next a circular current. It is easily seen, in the first place, that the vector-potential at any point, P, upon the axis of the circle is *nil*, for all the current-elements are equidistant from P, and for any element pointing one way there is just one other (at the opposite end of the diameter) pointing the other way, so that the vector-potential of one half of the circular current is annulled by that of the other half. Now, let P be no longer on the axis. It is, consequently, nearer one side of the circle than the other. The vector-potential of the nearer side preponderates, and the result is a vector at P drawn parallel to the nearest part of the circle. Let P describe a circle about the axis, parallel to the current; at every point of this circle, from symmetry, the same statement holds good, consequently the circle described by P is a line of vector-potential. Thus the lines of vector-potential of a circular current are parallel circles, centred upon the axis, closely packed in the current and close around it, and falling off in density both towards the axis in the neighbourhood of the circle, and away from it outside it.

If the current flows in a plane sheet, the current lines are closed curves in the sheet. The lines of vector-potential must be in planes parallel to the sheet, since there is no current perpendicular to it, and they are also closed curves.

In all cases we may find the vector-potential by means of three scalar sums instead of the vector sum; this is most conveniently done by forming the scalar sums of the components in three rectangular directions, and then compounding them. But, though convenient for calculations, this method often very much obscures the matter under consideration.

Now, we stated that the magnetic force was the vector-potential of the curl of the current. Take a straight current, for instance. As explained in the last section, the curl of the current is confined entirely to the surface of the wire, its strength is numerically equal to the current-density, and its direction is perpendicular to the current, so that the lines of the current-curl are equidistant circles on the surface, enclosing the current. From symmetry, together with the late remarks on the vector-potential of a circular current, we see that the vector-potential of the curl of the current is also in circles, in planes perpendicular to the axis, and centred thereon—that this must be the case both within and without the wire, and that they must be closely packed near the surface, and fall off in density both ways, *i.e.*, towards the axis within the wire, and from it outside. These characteristics will be readily recognised to be those of the magnetic force of a straight current.

Again, consider a current-system in a plane sheet, and derive the general nature of the magnetic force as far as we can. The current-density is supposed to be constant through the small thickness of the sheet, so that the current has no curl within the sheet and parallel thereto. But the current will in general vary from place to place in the plane of the sheet; it has, therefore, curl perpendicular to it, to be found by the line-integral of the current round a small area in its plane. This curl is a vector drawn perpendicular to the sheet, at some places it may be from the + side, at other places from the − side. Its vector-potential is the component of magnetic force perpendicular to the sheet, and we see immediately that it has the same strength and direction at corresponding points on opposite sides, and that it is continuous at the sheet itself. If *to* the sheet on one side, it is *from* the sheet at the nearest point on the other side. But, besides this, there is the surface-curl of the current to be considered. As explained in the last section, it is got by turning the vector representing the current-density through a right angle upon the surface. Since the sheet has two sides, we have two systems of surface-curl, both exactly alike, but oppositely directed. The lines of surface-curl intersect the current-lines at right angles. The vector-potential of this double system is the component of magnetic force parallel to the sheet. Now, on either side of the sheet, the nearer side preponderates over the other in contributing to the vector-potential, and since the surface-vectors only differ in being oppositely directed, it follows that the vector-potential is numerically the same at corresponding points on the + and − sides, but oppositely directed. Thus there is a remarkable difference between the components of magnetic force perpendicular and parallel to the sheet. The former is continuous, the latter discontinuous, as may be thus diagrammatically represented:—

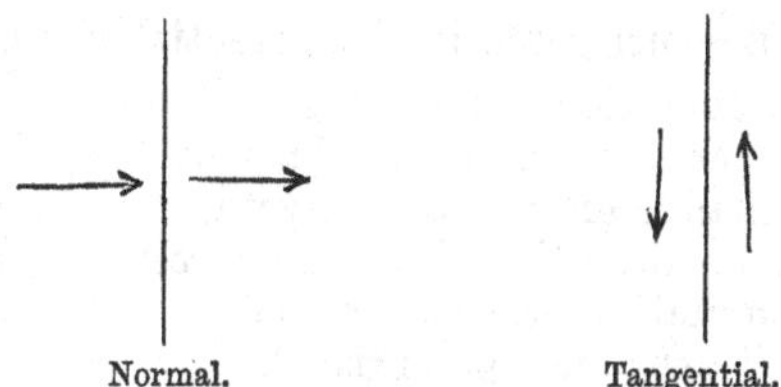

Returning now to the general relations, it will be observed that whilst **D**, the curl of the current, is derived from **B**, the magnetic force, by double curling (with division by 4π), on the other hand **B** is derived from **D** by finding its vector-potential. That is to say, the operation of finding the vector-potential is exactly annulled by double curling and division by 4π (the last operation being a question of units). Thus,

$$4\pi\mathbf{D} = \text{curl curl potential } \mathbf{D}.$$

This suggests that if we form a new vector, viz., the vector-potential of **C**, *its* curl will be the magnetic force. Thus, calling the new vector **A**, and at the same time introducing another vector **E** at the other end of the series, we have

$$\mathbf{A} = \text{potential of } \mathbf{C}.$$
$$\mathbf{B} = \text{curl } \mathbf{A} = \text{potential of } \mathbf{D} = \text{magnetic force}.$$
$$\mathbf{C} = (4\pi)^{-1} \text{ curl } \mathbf{B} = \text{potential of } \mathbf{E} = \text{current}.$$
$$\mathbf{D} = \text{curl } \mathbf{C}.$$
$$\mathbf{E} = \text{curl } \mathbf{D}.$$

Of the series, only **B** and **C** represent quantities having undoubted existence as physical realities; the rest are purely concepts. But the last introduced quantity **A**, the vector-potential of the current, is of as much importance in electromagnetism as the potential of free electricity is in electrostatics, being in fact its exact counterpart. **D** is also very useful sometimes; the rest of the possible series up or down cannot be said to be more than curiosities. **A** is found from the current-distribution in an analogous manner to scalar potential from free electricity, only compounding the component parts like velocities, or taking the vector sum instead of the scalar sum. And as we derive electric force from the static potential by finding the direction in which the potential decreases fastest, and the rate of its decrease, the former giving the direction and the latter the magnitude of the electric force, so we derive magnetic force from the vector-potential of the current by curling, that is, by finding the direction of the axis round which the line-integral of the vector-potential is greatest, and its amount; the first being the direction of the magnetic force, the second proportional to its magnitude. This is the most usual way in electromagnetic investigations; it does not, however, usually give so much information without calculation as the method previously described, as may be illustrated in the case of a straight current.

If we have found the vector-potential of a current-system, we know the magnetic force, viz., its curl. But it is also the vector-potential of the curl of current, so we have

$$\mathbf{B} = \text{curl potential } \mathbf{C} = \text{potential curl } \mathbf{C}.$$

Thus the two operations of curling and finding the potential are reversible, or rather commutative. This putting the cart before the horse is a slight change only in words, yet makes a vast difference when we come to carry it out, although the final result is the same.

From this commutative property we have

$$\begin{aligned} 4\pi\mathbf{C} &= \text{curl curl potential } \mathbf{C}, \\ &= \text{curl potential curl } \mathbf{C}, \\ &= \text{potential curl curl } \mathbf{C}, = \text{potential } \mathbf{E}. \end{aligned}$$

As may be seen from the last list, the whole series from **A** to **E** is involved in the above, which is the expression of a characteristic property of the class of functions to which magnetic force and current belong. They have all the property of continuity which distinguishes the electric current, viz., that of flow in closed paths like an incompressible fluid. It is of the greatest assistance to conceive them as fluxes like the current, not merely as quantities having certain values at certain places. Given one of them, and then forming the others, any one of them may be taken to represent a possible system of either current or magnetic force.

The vector-potential **A**, for instance, of a straight current **C** is a flux parallel to the current, its strength proportional to $C \log (a/r)$, where a is a constant, and r the distance from the axis. Let this flux **A** be a system of magnetic force; its curl is the corresponding current, which must therefore be **B**; or the lines of current flow are exactly those of the magnetic force of a straight current.

Similarly, if we find the vector-potential of **B**, say $\mathbf{A}_0$, we know that its curl is proportional to **A**, so that if $\mathbf{A}_0$ represent magnetic force, the corresponding current is proportional to $\mathbf{C} \log (a/r)$. And so on.

SECTION III. CONNECTED GENERAL THEOREMS IN ELECTRICITY AND MAGNETISM.

In mathematical investigations relating to electromagnetism, it often happens that the equations assume such a very complex form that the real meaning of the relations expressed by them becomes hidden away, as it were, beneath a tangled mass of x, y, z's, and can only be recognised by groping about from one equation to another, comparing them, selecting certain equations as important, rejecting others as needless, and, finally, from the few selected main equations, serving as successive stepping-stones, determining the essential nature of the relations under investigation. That there are so many quantities involved in electromagnetism is one reason for this complexity and obscurity, but it is immensely increased by the circumstance that they are usually vectors, or directed quantities, requiring three specifications, instead of the one which is sufficient for a scalar quantity; and from the ordinary x, y, z system only recognising magnitudes; so that a set of three equations is required to fully exhibit a single relation between a pair of vectors. Thus we may say, in words, that the current is the curl of the magnetic force, but if we have to express this symbolically we require three equations, which express the same thing for the components in three rectangular directions.

A very remarkable system of mathematics was invented by Sir W. Hamilton, called Quaternions, which may be described as the calculus of vectors. Owing to the universal presence of vectors in physical science, it is exactly fitted to express physical relations. Instead of breaking up vectors into three components, working with them as scalars, and then, when required, compounding them again to get back to vectors (a most roundabout method), in the calculus of vectors we may fix our attention upon the vectors themselves, and work with them direct. One equation takes the place of three. Investigations are greatly shortened. The real relations between the quantities are not lost sight of, and this again serves to annihilate a lot of useless work that might be done in the scalar system owing to obscurity.

The calculus of Quaternions ought then, one would say, to speedily supplant the ordinary methods in physical applications; in fact, it should have done so already. But it has not. Does this arise from mere Conservatism—the hatred of having to leave the old ways even for better? Although this may be partly true, it cannot be the whole truth. Against the above stated great advantages of Quaternions has to be set the fact that the operations met with are much more difficult than the corresponding ones in the ordinary system, so that the saving of labour is, in a great measure, imaginary. There is much more thinking to be done, for the mind has to do what in scalar algebra is done almost mechanically. At the same time, when working with vectors by the scalar system, there is great advantage to be found in continually bearing in mind the fundamental ideas of the vector system. Make a compromise; look behind the easily-managed but complex scalar equations, and see the single vector one behind them, expressing the real thing.

An easily-grasped example of the importance of considering the vector itself may be here given. In a field of electric force, what relation does the total amount of force passing outward through a completely closed surface bear to the amount of free electricity enclosed by it? Consider a single charge e within the surface. By definition, the electric force is radial, of strength $e \div r^2$ at distance r; and since $4\pi r^2$ is the area of the spherical surface of radius r, the total force through it is $(e \div r^2) \times 4\pi r^2 = 4\pi e$. This is perfectly clear when the closed surface is that of a sphere, and the charge is at its centre. But put the charge at some other part of the enclosed space. The force is now of different strength at different parts of the surface; and, moreover, it does not go straight through, perpendicular to it, and it is quite a complex matter to sum up the total force by algebraical methods. But if we fix our attention upon the system of force, which we may do by imagining a set of equably distributed radial lines drawn from the charge, inseparably connected with it, we see at once that the same amount of force as before, viz., $4\pi e$, goes through the spherical surface at whatever part of the enclosed space the charge may be placed, just as the same amount of light would come from a candle substituted for the charge; and, moreover, that the same is exactly true for any other closed surface, however complex its form may be, and however many times a radius drawn from the charge cuts the surface before finally emerging never to enter again. The only essential thing is that the surface must *enclose* the charge. Now this being true for one charge is true for any number, so that the total amount of force passing out through a closed surface is always $4\pi \times$ total charge within, however it may be distributed.

If the surface be unclosed, we may, in the case of a single charge, substitute for it a portion of a spherical surface bounded by the same lines of force, whence it follows that the amount of force passing through the unclosed surface is $4\pi e \times$ ratio of area of the portion of spherical surface to the whole, that is, $e \times$ solid angle subtended at e by the line bounding the unclosed surface, by the definition of a solid angle.

An immediate consequence is that the electric force at the surface of a charged conductor is $4\pi\sigma$, where σ is the surface-density. For, by the foregoing, $4\pi\sigma$ is the amount of force coming from unit of area of the surface. Now there is no force within the surface, otherwise there could not be equilibrium, therefore $4\pi\sigma$ is the external force; it is perpendicular to the surface, because if it were not, there would be force along the surface, which is against equilibrium again.

Besides the complexity above referred to, there is in general working a frequent repetition of the same succession of operations in different places. These operations may themselves be of a complex nature, yet, if they are carried out in one investigation, the results may be transferred to another, perhaps relating to a quite different matter. Hence the utility of general theorems. The three allied theorems which we will now discuss are of great assistance in electricity and magnetism. Analytical proofs of them may be, and are given, which are only to be followed with some difficulty, especially as regards Theorem (B) below; one I have seen in a German work of Theorem (A), which is comparatively simple, was about six pages long. This is by no means necessary, for without losing the character of exactness, the demonstrations may be given in words; although in the final expression of the theorems symbols are desirable.

(A.) *Theorem of Divergence.*—The first theorem relates to the surface-integral of a vector over a closed surface, and we require to substitute for it a volume-integral taken throughout the enclosed space. We have already had one example of this, viz., for electric force; the quantity integrated throughout the space being $4\pi\rho$, where ρ is the volume-density of electricity. We now require it in a more general form, applicable to any vector. But, to fix ideas, we may conveniently think of electric force. What we really require is the excess of the amount of force leaving the surface at some places over that entering it at other places. We may go all over the surface, dividing it into little bits, find the flow through each little bit, and add the results together. If **R** is the force, dS an element of surface, ϵ the angle between **R** and the normal outwards, the force through dS is $R\cos\epsilon\, dS$, and the surface-integral is $\iint R\cos\epsilon\, dS$ over the whole surface. We now require to express this in the form of a volume-integral taken throughout the enclosed space.

In the first place, we may easily prove that this is capable of being done. For divide the volume V into two parts, V_1 and V_2. The act of division creates a fresh surface, or, rather, a pair of twin surfaces, born at identically the same moment, like + and − electricity. Now consider the integrals of the vector over the two complete surfaces of V_1 and V_2 The surface-integral over V_1 consists of the integral over that portion of the surface of V which the piece V_1 possesses, and of the integral over the new surface, one of the twins. Similarly, the integral over surface of V_2 consists of the integral over the other portion of the surface of V, and of the integral for the other twin. Now regard only the twin surfaces. Put the surfaces together, so as to get the original volume; the two integrals for the twins are exactly equal, for the surfaces coincide. But they are reckoned positively in opposite directions; therefore, by addition, they cancel. Therefore the integral over the closed surface of V equals the sum of the two surface-integrals over the complete surfaces of the two pieces V_1 and V_2 into which we divided V. The same thing is obviously true if we divide each of our two volumes into two more; the original integral now equals the sum of the four complete surface-integrals, the newly created twin integrals always cancelling when taken together. And since this may be carried on to any extent, we may divide the volume V into an indefinitely great number of elements, each $= dV$, and our surface-integral is exactly equal to the sum of all the surface-integrals outwards over the complete surfaces of all the elementary volumes. Thus the volume-integral is possible; and further, we know what the element of the volume-integral is. It is the integral of force outward over the surface of the element of volume.

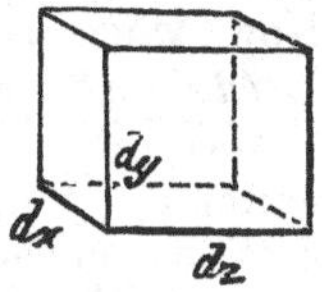

It now only remains to put this element of the volume-integral into a symbolical form in terms of the force. Let the element of volume be a cube, edges dx, dy, and dz. Let X, Y, Z

be the components of **R** at the corner where dx, dy, and dz meet. The amount of force leaving the cube is the sum of the six amounts leaving it through the six plane faces; each of these is force × surface, therefore they are

$$-Xdydz \quad \text{and} \quad \left(X+\frac{dX}{dx}dx\right)dydz,$$

$$-Ydzdx \quad \text{and} \quad \left(Y+\frac{dY}{dy}dy\right)dzdx,$$

$$-Zdxdy \quad \text{and} \quad \left(Z+\frac{dZ}{dz}dz\right)dxdy.$$

Adding these together, we get

$$\left(\frac{dX}{dx}+\frac{dY}{dy}+\frac{dZ}{dz}\right)dV,$$

where $dV=dxdydz$, the element of volume. This is the element of the volume-integral; consequently we have

$$\iint R\cos\epsilon\, dS=\iiint\left(\frac{dX}{dx}+\frac{dY}{dy}+\frac{dZ}{dz}\right)dV \quad\ldots\ldots\ldots\ldots\ldots \text{(A)}$$

for the complete expression of the theorem. The left side is the integral of **R** over the surface of V; the right side is the equivalent volume-integral, to be extended throughout V.

Although we have spoken of force, it is true for any vector. If **R** is the velocity of an incompressible fluid at any point, and X, Y, Z its components, then, since the same quantity of fluid must enter as leaves any closed surface, the surface-integral is *nil*, and

$$\frac{dX}{dx}+\frac{dY}{dy}+\frac{dZ}{dz}=0$$

is true everywhere. This is the equation of continuity. If **R** is current-density, the same equation is true. If **R** is electric force, the same equation applies wherever there is no electrification.

Where there is electrification, let ρ be its volume-density, then $\rho\, dV$ is the amount within the element of volume dV. We know that 4π times this is the surface-integral of force over the element. Consequently,

$$4\pi\rho\, dV=\left(\frac{dX}{dx}+\frac{dY}{dy}+\frac{dZ}{dz}\right)dV,$$

or

$$4\pi\rho=\frac{dX}{dx}+\frac{dY}{dy}+\frac{dZ}{dz};$$

which is the general relation between electric force and electrification. Now, when ρ is $-$, lines of force converge to the element, whence $-\left(\frac{dX}{dx}+\frac{dY}{dy}+\frac{dZ}{dz}\right)$ was called by Maxwell the "convergence" of the vector **R**. Electric force has no convergence save where there is electrification; on the other hand, electric currents have no convergence *anywhere*, neither has their magnetic force, nor their vector-potential, nor any of the quantities considered in Section II.

(B.) *Theorem of Version.*—We now come to unclosed surfaces, or surfaces bounded by a closed curve. If we take any closed curve, then, by Theorem (A), the surface-integral of electric force over any two surfaces bounded by the curve is the same, provided there be no free electricity between them, the integrals being reckoned in the same direction through the curve. And this is true for any vector-quantity which has no convergence. The surface-integral thus depends on the form of the bounding curve only, and we may naturally expect that it can be put in the form of a line-integral taken round the curve. But it is convenient for the demonstration to commence with the line-integral. Let **R** be any vector, which we may now consider as magnetic force; the line-integral of **R** once round the closed curve is then the amount of work done by the magnetic force on a unit magnetic pole when it is caused to move once round the curve. If ds is an element of the curve, the work for the path ds is $ds \times$ component of force along ds, or $R\cos\epsilon\, ds$, where ϵ is the angle between **R** and ds; consequently, the line-integral is expressed by $\int R\cos\epsilon\, ds$. We require now to express this in the form of an equivalent surface-integral of another vector over any surface bounded by the curve, and to find what relation the new vector bears to **R**.

The proof that the surface-integral is always possible is easy. Imagine any surface S bounded by the closed curve. As in the first theorem we divided the volume we started with into two, so now we divide the surface S into two, S_1 and S_2, by means of a line joining any two points of the boundary of S. We thus form two closed curves, each having the new line in common. Now reckon the line-integral of force once round both of the closed curves, rotating the same way. In going round the first we move one way along the common path, and in going round the second we move the other way; therefore, the sum of the two integrals is equal to that round the original closed curve, for the integrals for the common portion cancel when added. This device, which is due to Ampère, may be naturally extended. Divide the whole area S into an indefinitely great number of elements, dS; find the line-integral round every element, add together all the results, and the sum is equal to the line-integral once round the bounding curve of S, owing to all the inner line-integrals cancelling. The direction of rotation must, of course, be the same for every element of surface. The elementary part of the equivalent surface-integral is, therefore, nothing more than the line-integral of **R** round an element of surface. To find the exact relation between this and **R**, it is best to suppose in the first place that the surface S is made up entirely of plane portions, all of which are either parallel to the planes of y,z, or of z,x, or of x,y. Take an element parallel to the plane of y,z; axis along x. The + direction of rotation is right-handed, and the + direction of x downwards through the paper. Now, X, Y, Z being the components of **R** at the corner P, we have nothing to do with the component X, because

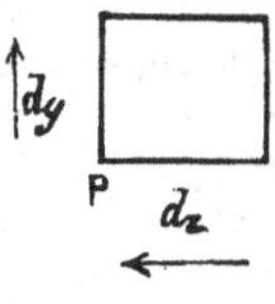

dy P dz

it is at right angles to the closed curve. The integrals for the four sides of the square are, starting from P,

$$Ydy, \quad \left(Z+\frac{dZ}{dy}dy\right)dz, \quad -\left(Y+\frac{dY}{dz}dz\right)dy, \quad -Zdz.$$

Adding these all together we get simply

$$\left(\frac{dZ}{dy}-\frac{dY}{dz}\right)dydz = X_1 dydz, \text{ say.}$$

Similarly, the line-integrals round elements of surface $dzdx$ and $dxdy$ are

$$\left(\frac{dX}{dz}-\frac{dZ}{dx}\right)dzdx = Y_1 dzdx, \text{ say, and } \left(\frac{dY}{dx}-\frac{dX}{dy}\right)dxdy = Z_1 dxdy, \text{ say.}$$

If, then, we assume X_1, Y_1, Z_1 to be the components of a new vector $\mathbf{R}_1$, we have line-integral of $\mathbf{R}$ = surface-integral of $\mathbf{R}_1$, provided that the surface is made up as above stated.

But if we differentiate the three components of $\mathbf{R}_1$ to x, y and z, and then add the results, we find

$$\frac{dX_1}{dx}+\frac{dY_1}{dy}+\frac{dZ_1}{dz}=0,$$

which shows that the new vector has no convergence anywhere, and consequently its surface-integral over *any* surface bounded by the closed curve is the same. Consequently, the theorem (attributed to Stokes) is proved for any surface. We may write it simply

$$\int R\cos\epsilon\, ds = \iint R_1 \cos\epsilon_1 dS, \quad \text{.... ....} \quad \text{(B)}$$

the left side being the line-integral of $\mathbf{R}$, the right the surface-integral of $\mathbf{R}_1$; ϵ the angle between $\mathbf{R}$ and the element of curve ds; ϵ_1 the angle between $\mathbf{R}_1$ and the normal to the surface-element dS; and the relation between the components of $\mathbf{R}_1$ and those of $\mathbf{R}$ being those above given.

If $\mathbf{R}$ is magnetic force, $\mathbf{R}_1$ is $4\pi \times$ current-density.

If $\mathbf{R}$ is the vector-potential of current, $\mathbf{R}_1$ is magnetic force. And, in general, $\mathbf{R}_1$ is the curl of $\mathbf{R}$.

(C.) *Theorem of Slope.*—There is yet one more connected theorem, which relates to *un*closed curves. If a vector, $\mathbf{R}$, has no curl, the line-integral once round any closed curve is zero, by the last theorem. Such a vector is electrostatic force. Taking, then, two points A and Z on the closed curve, the line-integral of $\mathbf{R}$ from A to Z by one path is cancelled by that from Z to A by another; or the line-integral of $\mathbf{R}$ from A to Z is the same along any path; it therefore depends only upon the *position* of A and Z, the common terminations of the paths. Let, then, at every point of space the value of a scalar quantity P be given, its values at A and Z being P_A and P_Z, and similarly for any other point. Divide any line joining A and Z into pieces AB, BC, CD, etc., YZ. Then, evidently,

$$P_A - P_Z = (P_A - P_B) + (P_B - P_C) + \text{etc.} + (P_X - P_Y) + (P_Y - P_Z),$$

or the difference in the values of P at A and Z = sum of the differences for all the pieces into which the whole path AZ is broken. $P_A - P_Z$ is, therefore, expressible as a line-integral along the path; if ds is an element thereof, the element of the integral is

$$Pds - \left(P + \frac{dP}{ds}\right)ds = -\frac{dP}{ds}ds.$$

Consequently,

$$P_A - P_Z = \int_A^Z -\frac{dP}{ds}ds = \int_A^Z R\cos\epsilon ds. \quad \text{.................} \quad (C)$$

This theorem is sufficiently obvious. It is added for the sake of completeness. If P be the electrostatic potential, **R** is the electric force, and its component in any direction **s** is $-dP/ds$. If P be the magnetic potential of electric currents **R** is the magnetic force, provided that the line of integration does not pass through electric current. For the magnetic force has no curl where there is no current. Magnetic force is, therefore, derivable from a scalar potential, like electrostatic force, everywhere except where there is current. But it must be added in this case that $P_A - P_Z$ is only the same for the two paths, provided they do not embrace a current, for if they do, the line-integral of magnetic force once round the closed curve formed by the two paths will equal the surface-integral of the curl of the magnetic force over any surface bounded by the curve, and this surface will have current through it at some places, and there the curl is not zero, but $4\pi \times$ current-density.

A comparison of the three theorems is instructive. Starting with the third theorem (C), we find that the difference in the values of a *scalar* function at the two points = line-integral of a derived *vector* along any connecting line.

Join the points by a second line, thus forming a closed curve, and the second theorem (B) tells us that the line-integral of a *vector* once round it = surface-integral of a related *vector* over any surface bounded by it.

Lastly, let two surfaces have the same bounding curve, so that, taken together, they form a closed surface; the first theorem (A) tells us that the surface-integral of a *vector* over it = volume-integral of a related *scalar* throughout the enclosed space.

Here we stop. Space has but three dimensions, and we cannot make a fourth to please anybody. Starting with a scalar for points, we have vectors for lines and surfaces, and a scalar again for a volume. I am not able to fathom the nature of the corresponding magnitude to be integrated in four dimensions, but probably it does not matter.

SECTION IV. THE CHARACTERISTIC EQUATION OF A POTENTIAL, AND ITS SOLUTION.

In sections I. and II. we have explained the nature of two operations of frequent occurrence in electromagnetism, viz., from a given vector-function to derive its "curl," and also to find its vector-potential; and some of the relations of these functions were mentioned. But that was for descriptive purposes merely; the only proof that curling the magnetic force gave rise to the current was that derived from the known magnetic force of a straight current, together with the assumption of a certain distribution of force within the current itself.

We have now, to make the matter a little more complete, to prove that the properties ascribed to the vector-potential of current and its curl are really true, and to further develop them. It need not be thought that a knowledge of mathematical analysis of a high order is required, for it so happens that nearly the whole of investigations in electromagnetism (although they may be of a very involved nature, the causes whereof were briefly alluded to in Section III.) are, when resolved into their simplest forms, merely the application in various shapes of the three theorems given in Section III.; and these theorems are of such a nature that they may be reasoned out mentally, without symbolical aid save for final expression; and, in fact, in thinking them out in this way we may obtain a far clearer conception of what the theorems really mean than by working them out analytically, blindfolding the mind, so to speak, and working mechanically. No disparagement is intended of strictly symbolical demonstrations, or of the labours of the great mathematical electricians. To use a simile which may be readily applied, when once a bold explorer has reached the top of a hitherto inaccessible mountain, by circuitous paths and with the expenditure of great labour, others can do it after him, and easier routes may be discovered; and in course of time people may travel up comfortably seated in railway carriages. And, of course, it is still easier to start from the top and travel down. (A fall down a precipice, in seeking a short cut, may be compared to an abrupt breach of continuity in the reasoning.) A sound knowledge of the fundamental *principles* of the Calculus is desirable, but, so far as general results go, apart from special, very little of the *practice* thereof, for differentiations and integrations are usually merely indicated, not performed.

We have, in the first place, to clearly conceive what is meant by a flux of no convergence, and this is best done by means of a material analogy. Think of an incompressible fluid filling all space. Within a given volume there must be always the same amount of fluid. If, then, we suppose the volume fixed in space and the fluid to be in motion, the same amount of fluid as enters it through its surface at some places must leave it at others, in any given time; or, algebraically, the total flux through the closed surface from within outwards must be *nil*. This is self-evidently true for any closed surface anywhere situated, and however the fluid may be moving. The flux through the surface of the element of volume is

$$=\left(\frac{dX}{dx}+\frac{dY}{dy}+\frac{dZ}{dz}\right)\times \text{volume, per second,}$$

if X, Y, Z are the component velocities in the directions of the axes of x, y, z (Theorem A); consequently

$$\frac{dX}{dx}+\frac{dY}{dy}+\frac{dZ}{dz}=0$$

in the case of an incompressible fluid.

If we have any vector or directed quantity, **R**, components X, Y, Z, and we say that

$$\frac{dX}{dx}+\frac{dY}{dy}+\frac{dZ}{dz}=0,$$

or that it has no convergence, we simply imply that the distribution of **R** is similar to that of velocity in a possible case of motion of an incompressible fluid, so that we may, without going wrong, guide ourselves by the material analogy.

We may represent the fluid motion by means of lines, their direction showing that of the motion, their density the velocity. Or, rather, this is a convenient method of imagining the motion, because, if we draw the lines on a sheet of paper, we can only strictly represent in this way the nature of the motion in a section of the fluid bounded by two close parallel planes, and we may require a different diagram for another part of the fluid. We may, indeed, represent on paper the distribution in three dimensions according to Maxwell's system, but this has the disadvantage of requiring interpretation, for in many cases it makes the lines densest where they (in a plane section) would be furthest apart. Similarly, we may employ lines to indicate the distribution through space of any vector-magnitude, **R**. When there is no convergence anywhere the lines form closed curves. We may, indeed, imagine lines of force to go out to infinity, and so to be lost sight of. Thus we may imagine an indefinitely long, straight current coming from $-\infty$, and going to $+\infty$, and the corresponding lines of vector-potential will behave similarly. But this is not realisable. However long the straight portion of the current may be, it must ultimately return to the starting point, when the lines of flow will form closed curves; and, at the same time, the lines of vector-potential will also do so.

We have next to consider vectors that have convergence. In an incompressible fluid this would imply that at a place of convergence more fluid arrived than left in a given time, or *vice versâ*, and that would mean that fluid was perpetually going out of or coming into existence at the place—an annihilation or creation of fluid. Places of convergence are, therefore, sources or sinks. In a field of electric force, *e.g.*, that between two charges $+e$ and $-e$ on insulated conductors, the lines of force start from the surface of one and terminate on that of the other. They do not form closed curves. In the dielectric between the conductors, the force has no convergence. Imagine any closed surface entirely in the dielectric; every line of force which enters the enclosed space leaves it again. Such a line is the representative of a tube of force, across every section of which the same amount of force passes. But if the closed surface be partly within and partly without one of the conductors, say the $+$, a certain number of lines of force leave it on one side and *none* enter it upon the other side; there is, therefore, negative convergence, and similarly at the $-$ conductor there is $+$ convergence. The amount of convergence per unit of surface is a measure of the surface-density of the negative electrification. In any vector-system **R**, the quantity $-\left(\frac{dX}{dx}+\frac{dY}{dy}+\frac{dZ}{dz}\right)$ is a measure of the convergence at any place, expressed per unit volume.

The corresponding expression for unit of surface is

$$(R_1 \cos \epsilon_1 + R_2 \cos \epsilon_2),$$

where R_1 and R_2 are the values of **R** on opposite sides of the surface, and ϵ_1, ϵ_2 the angles $\mathbf{R}_1$ and $\mathbf{R}_2$ make with the normals drawn *to* the surface from opposite sides. $R_1 \cos \epsilon_1$ is the amount of **R** reaching the unit of surface from one side, and $R_2 \cos \epsilon_2$ from the other side. In the case of no convergence these quantities are numerically equal, but of opposite signs, so that

$$R_1 \cos \epsilon_1 + R_2 \cos \epsilon_2 = 0$$

$$\text{corresponds to} \quad \frac{dX}{dx} + \frac{dY}{dy} + \frac{dZ}{dz} = 0.$$

To further illustrate the distinction between converging and non-converging vectors, consider a linear current with its ends immersed in a conducting medium. The lines of flow in the medium from the + to the − end are exactly similar to the lines of force between the two charges $+e$ and $-e$ before considered, so that if we were to ignore the linear current there would be + convergence at the place where the current enters the wire and − at the other end. Similarly, if we ignored the current in the medium, we should have convergence at the ends of the wire of opposite sign to the last. But, considering the complete system, there is no convergence anywhere, the above convergences cancelling. In the static electrical case there is no channel joining the conductors to make a complete circuit for the force, so there is really convergence.

It will be convenient, preliminarily, to alter the definition of potential. Instead of the potential of any agent e at distance r being e/r, let it be $e/4\pi r$. If e is a scalar, so is its potential; if a vector, its potential is also a vector, to be drawn parallel to direction of **e**. In the case of electrostatic force, this definition of potential implies that the force at distance r is of strength $e/4\pi r^2$, so that the whole amount of force leaving e is numerically equal to e. Now, the whole amount of force leaving any space is independent of the distribution of the free electricity within it, and is numerically expressed by the amount of the latter. Expressed for a unit-volume, we now have, by Theorem (A),

$$\frac{dX}{dx} + \frac{dY}{dy} + \frac{dZ}{dz} = \rho, \quad \dots\dots\dots\dots\dots\dots \quad (1)$$

ρ being the volume-density, so that $-\rho$ measures the convergence.

We shall now prove an important property of converging vectors having no curl. Let a vector be denoted by **R**, and suppose it is given in direction and magnitude throughout all space. Measure its convergence everywhere; *i.e.*, divide all space into elements of volume, find the excess of the amount of **R** leaving any one of them over that entering it, and denote it by ρ per unit volume. Then we have equation (1) again, where, however, ρ does not now mean the density of free electricity, but merely the divergence of the vector **R**. ρ is then known everywhere, being 0 at those places where there is no convergence.

Next, let **R** have no *curl* anywhere. The line-integral of **R** once round any closed curve is then zero, by Theorem (B), therefore, the integral of **R** from any one point A to any other Z is the same by any path, being determined solely by the positions of A and Z; therefore, by Theorem (C), the integral of **R** from A to Z is $P_A - P_Z$ where P is a scalar function of position, and the component of **R** in any direction **s** is $-dP/ds$. The direction of the resultant vector **R** is that in which P decreases fastest, and its magnitude is the rate of decrease. We may denote the resultant vector by $-\nabla P = \mathbf{R}$, with components $-dP/dx$, $-dP/dy$, $-dP/dz$ in the three rectangular directions of x, y, z.

The question now arises, how is the quantity P to be determined, starting with ρ given, and with the condition that **R** has no curl; *i.e.* what is the explicit relation of P to ρ? If we consider a single element of volume, a unit cube for example, taking the unit indefinitely small, ρ is the excess of the amount of **R** leaving it over that which enters. Now, disregarding altogether the manner in which **R** really enters and leaves the space, let the quantity ρ give rise to a vector **R**$'$ in the same manner as free electricity gives rise to electrostatic force; that is, let $\mathbf{R}' = \rho/4\pi r^2$ at distance r from ρ, and be directed along **r** from ρ outwards, so that the whole amount of **R**$'$ leaving ρ is numerically $= \rho$. Let the same thing take place wherever there is convergence, and then find the resultant, *i.e.*, the vector-sum of $\rho/4\pi r^2$, or,

$$\mathbf{R}' = \Sigma \frac{\rho}{4\pi r^2} \mathbf{r}_1,$$

where $\mathbf{r}_1$ is a unit vector drawn from P along the line **r** whose length is r. This new vector **R**$'$ obviously satisfies the condition expressed by equation (1) above; by the manner of its construction the convergence anywhere is $-\rho$, for those of the lines of **R**$'$ which proceed from centres lying *outside* any one of the elements of volume which enter the latter leave it again.

Now,
$$\frac{\rho}{4\pi r^2} = -\frac{d}{dr}\frac{\rho}{4\pi r};$$

i.e., the flux from the element ρ is derived from the potential $\rho/4\pi r$, for the latter plainly decreases fastest along **r** outwards, and its rate of decrease is the strength of the assumed flux. Form next the sum of the potentials as above for all the elements, or $\Sigma\rho/4\pi r$, and call it P; then since

$$-\nabla\Sigma\frac{\rho}{4\pi r} = -\Sigma\frac{d}{dr}\frac{\rho\mathbf{r}_1}{4\pi r} = \Sigma\frac{\rho}{4\pi r^2}\mathbf{r}_1,$$

therefore
$$\mathbf{R}' = -\nabla P.$$

We have thus found a scalar potential whose variation in space is a vector **R**$'$ which has the same convergence as **R**, and we may now easily prove that **R**$'$, like **R**, has no curl. For, by the general equations, Theorem (B), connecting a vector and its curl, the x component of the curl of **R**$'$ is

$$\frac{dZ'}{dy} - \frac{dY'}{dz} = \frac{d}{dy}\left(-\frac{dP}{dz}\right) - \frac{d}{dz}\left(-\frac{dP}{dy}\right) = 0,$$

identically; and similarly for the y and z components.

Now, is it possible for **R**$'$ to differ from **R**? If they do differ, let **R**$''$

be the difference. It must have *no* convergence, because **R** and **R**′ have the *same* convergence. But if **R**″ had convergence, some of the lines of **R**″ would terminate at certain places; therefore, having none, the lines of **R**″ must be closed curves [if **R**″ vanishes at infinity]. But if we integrate **R**″ once round one of these closed curves, we obtain a finite quantity, because the direction of **R**″ at any place coincides with that of the curve. But **R** and **R**′ have no curl, hence their difference **R**″ has none, consequently the line-integral of **R**″ round *any* closed curve is *nil*. Thus the lines of **R**″ cannot be closed, and since, also, they cannot terminate anywhere, it follows that they do not exist. Therefore, **R**″ is non-existent, and $\mathbf{R}'=\mathbf{R}$ identically. That is,

$$\mathbf{R}=\Sigma\frac{\rho}{4\pi r^2}\mathbf{r}_1 \quad\ldots\ldots\ldots\ldots\ldots\ldots\ldots\ldots\ldots\ldots \quad (2)$$

is the complete solution of (1) with the condition that **R** has no curl [and that it vanishes at an infinite distance from the sources, that is, from the places where ρ exists. For example, $\mathbf{R}=$ constant over all space gives no curl and no divergence anywhere; but it does not vanish at infinity].

Now, if in (1) we put for X, Y, Z their values in terms of P, viz., $-dP/dx$, $-dP/dy$, $-dP/dz$, we obtain

$$\frac{d^2P}{dx^2}+\frac{d^2P}{dy^2}+\frac{d^2P}{dz^2}=-\rho. \quad\ldots\ldots\ldots\ldots\ldots\ldots\ldots \quad (3)$$

Therefore, we have also found *a* solution of this last equation, viz. :—

$$P=\Sigma\frac{\rho}{4\pi r}. \quad\ldots\ldots\ldots\ldots\ldots\ldots\ldots\ldots\ldots\ldots \quad (4)$$

It may be shown that any other solution of (3) can only differ from this one by a constant quantity. For if there be another solution, say $P+P'$, then $-\nabla P'$ must have no convergence anywhere, because the convergence is fully given by that of $-\nabla P$ or **R**; and it must have no curl anywhere, because like **R** it is derived from a scalar potential; therefore by the same reasoning as that proving that **R**″ was non-existent, we see that $-\nabla P'$ is non-existent, and hence P' must be constant everywhere. This constant allows us to locate the zero of P where we like; it is most convenient to put $P'=0$, so that $P=0$ at an infinite distance from places where ρ exists, and then (4) is the complete solution of (3).

SECTION V. RELATIONS OF CURL AND POTENTIAL, DIRECT AND INVERSE. SCALAR POTENTIAL OF A VECTOR.

Coming now to the electromagnetic problem, we have two distinct things to do. First to prove the properties stated in Sections I. and II., as applied to the system of vectors therein defined, quite apart from any physical application—(this will be comparatively easy with the aid of (4) and (3) above, which are the characteristic equations of the potential function, (3) giving ρ in terms of the second derivatives of P, and (4) giving P in terms of ρ and distances)—and next, to show that magnetic force and electric current belong to the system.

Let **A** be any vector of no convergence, to be imagined by the fluid analogy. Its components will be represented by A_1, A_2, A_3, and similarly for other vectors. Derive from **A** its curl **B**. The fundamental relation between **A** and **B** is that the line-integral of **A** round any closed curve equals the amount of **B** passing through it. The components of **B** are expressed in terms of those of **A** by means of the equations

$$B_1 = \frac{dA_3}{dy} - \frac{dA_2}{dz}, \quad B_2 = \frac{dA_1}{dz} - \frac{dA_3}{dx}, \quad B_3 = \frac{dA_2}{dx} - \frac{dA_1}{dy}; \quad \ldots\ldots \quad (5)$$

by theorem (B). If we differentiate B_1, B_2, B_3 with respect to x, y, z and add, we find that **B** has no convergence, for the sum is zero. Similarly, derive from **B** its curl, **C**, and from **C** its curl, **D**, and so on. They are all vectors of no convergence. Now the equations for the components of **C**, similar to (5), are

$$C_1 = \frac{dB_3}{dy} - \frac{dB_2}{dz}, \quad C_2 = \frac{dB_1}{dz} - \frac{dB_3}{dx}, \quad C_3 = \frac{dB_2}{dx} - \frac{dB_1}{dy}.$$

In these, substitute the values of B_1, B_2, B_3 from (5), and we obtain

$$C_1 = \frac{d}{dy}\left(\frac{dA_2}{dx} - \frac{dA_1}{dy}\right) - \frac{d}{dz}\left(\frac{dA_1}{dz} - \frac{dA_3}{dx}\right).$$

Re-arranging the terms, and adding and subtracting $\frac{d}{dx}\frac{dA_1}{dx}$, this becomes

$$C_1 = -\left(\frac{d^2A_1}{dx^2} + \frac{d^2A_1}{dy^2} + \frac{d^2A_1}{dz^2}\right) + \frac{d}{dx}\left(\frac{dA_1}{dx} + \frac{dA_2}{dy} + \frac{dA_3}{dz}\right).$$

Here we at once recognise the quantity in the second ( ) to be the divergence of **A**. But **A** has no divergence, so we are reduced to the quantity in the first ( ). Compare with equation (3) above. For A_1 write P, and for C_1 write ρ. Since (4) is the solution of (3), or P the potential of ρ, therefore in the present case A_1 is the potential of C_1. Similarly we shall find A_2 to be the potential of C_2, and A_3 of C_3. Now compound A_1, A_2, and A_3, and we obtain the vector **A**. Similarly compound C_1, C_2, and C_3, and we get the vector **C**. Therefore, **A** is the *vector*-potential of **C**.

Similarly **B** is the vector-potential of **D**, and so on.

Thus, whilst the operation of curling carries us one step down the scale, that of finding the vector-potential carries us two steps up. This we may express by

$$\text{curl (curl } \mathbf{A}) = \mathbf{C}, \quad \text{or} \quad (\text{curl})^2\mathbf{A} = \mathbf{C};$$

also **A** = potential **C**, therefore

$$\text{curl curl potential } \mathbf{C} = \mathbf{C}.$$

We may now show that the operations curl and potential are commutative, or curl potential **C** = potential curl **C**. For

$$\begin{aligned} \mathbf{B} &= \text{potential } \mathbf{D}\text{; just proved.} \\ &= \text{potential curl } \mathbf{C}\text{, by definition of } \mathbf{D}. \\ &= \text{curl } \mathbf{A}\text{, by definition of } \mathbf{B}. \\ &= \text{curl potential } \mathbf{C}\text{; just proved.} \end{aligned}$$

Therefore, $\mathbf{B} = \text{potential curl } \mathbf{C} = \text{curl potential } \mathbf{C}.$

We may thus find any one of the series from the next following by either first going up two steps by the potential and then down one by curling, or else by first going down one step and then up two. *E.g.*, given **C**, to find **B**, we may go up to **A** and down to **B**, or down to **D** and up two steps to **B**.

These roundabout methods may be replaced by a single operation—very complex, however, though for single elements easily enough understood; that is to say, we may find **B** direct from **C** without reference to the other vectors. The operation has no name; since its effect is the inverse of curling, we might denote it by $(\text{curl})^{-1}$, just as we might denote vector-potential by $(\text{curl})^{-2}$, thus indicating its property.

The vector-potential of an element of **C**, say **c**, at a point Q distant r therefrom, is $\mathbf{a} = \mathbf{c}/4\pi r$ directed parallel to **c**. (Small letters when referring to elements.) Let the direction of **c** be parallel to the z-axis. Then

$$a_1 = 0, \qquad a_2 = 0, \qquad a_3 = c/4\pi r$$

express the components of **a** at Q. Find the curl of **a** by equations (5) above; thus,

$$b_1 = \frac{c}{4\pi}\frac{dr^{-1}}{dy} = -\frac{cy}{4\pi r^3}, \qquad b_2 = -\frac{c}{4\pi}\frac{dr^{-1}}{dx} = \frac{cx}{4\pi r^3}, \qquad b_3 = 0.$$

Now let **h** be the perpendicular from Q on to the axis of the current (the axis of z), and let the x-axis be parallel to **h**, so that $y = 0$ and $x = h$, the origin being at the element. Since $b_3 = 0$, **b** is in the plane of x, y; and since $y = 0$, therefore $b_1 = 0$, and

$$b = \frac{ch}{4\pi r^3}, \quad \text{perpendicular to } \mathbf{r} \text{ and to } \mathbf{h}, \quad \ldots\ldots\ldots\ldots (6)$$

with right-handed rotation about the axis of the current when we look along it the way the current is going.

The lines of **b** are, therefore, circles about the axis of **c** in planes perpendicular to it, and the strength of **b** varies inversely as the square of the distance from **c**, and is at the same time proportional to h/r, which is the sine of the angle between **r** and the axis of **c**. Every element of **c** must be understood to produce such a system **b**, and the vector-sum of the whole will give the complete value of **B**. Thus,

$$\mathbf{B} = \Sigma \frac{ch}{4\pi r^3}\mathbf{b}_1 = \Sigma - \frac{c}{4\pi}\frac{dr^{-1}}{dh}\mathbf{b}_1, \quad \ldots\ldots\ldots\ldots\ldots\ldots (7)$$

where $\mathbf{b}_1$ is the *unit* vector of the elementary **b**, inserted in order to vectorize its magnitude $ch/4\pi r^3$. In an exactly similar manner **A** may be expressed in terms of **B**, or **C** in terms of **D**. Thus, to express **A** in terms of **B**, change **B** to **A** and c to b in (7), and also change $\mathbf{b}_1$ to $\mathbf{a}_1$.

(The above method of expressing $(\text{curl})^{-1}\mathbf{C}$ was obtained by finding the curl of the vector-potential of an element of **C**. We shall get exactly the same result by finding the vector-potential of the curl of an element of **C**. Thus, take a cylindrical element, of length ds, circular section πa^2, radius a. Current-density **c**, directed parallel to axis ds. The surface-curl of **c** is (Section I.) of strength $c/\pi a^2$, and is directed in circles about the axis on the curved portion of cylinder.

Now, the vector-potential of a unit circle of radius a, when a is infinitely small, may be obtained from equation (9), Section VI., by taking the first term only of the series, viz.,

$$h/r^3 \times 2\pi a^2 \times \tfrac{1}{2} = \pi a^2 h/r^3.$$

Or, without making use of the equation (9) for a finite, let a/r in the previous equation (6), Section VI., be infinitely small, and expand $1/\rho$ by the binomial theorem; thus

$$\frac{1}{\rho} = \frac{1}{r} + \frac{ha \sin \phi}{r^3};$$

then we get
$$\int_0^{2\pi} \frac{a \sin \phi}{\rho} d\phi = \int_0^{2\pi} \frac{ha^2}{r^3} \sin^2\phi \, d\phi = \pi a^2 \frac{h}{r^3},$$

as before. Multiply by $c/\pi a^2$ and we get the intensity of the required vector-potential of curl of **c**, viz., ch/r^3, which, when divided by 4π to suit our present definition of potential, agrees with (6) above.)

At a distance r from a small plane closed curve great compared with its linear dimensions, its form is immaterial. The vector-potential is of intensity $h/r^3 \times$ area, and is directed perpendicular to **h** and **r**.

We are now able to make another step. We proved before that when a vector has no curl anywhere, the vector itself is the space-variation of P, and P is the potential of the divergence of the vector. Now, our vectors **A**, **B**, **C**, ... have curl, but not necessarily everywhere, and by Theorems (B) and (C) a vector may be derived from a scalar potential function wherever it has no curl. But **A**, **B**, **C** have no convergence anywhere, so the scalar potential cannot be found in the same manner for a vector which has curl in some parts of space and not in others, as for a vector which has no curl anywhere (in which case (4) is the solution). To find the nature of the scalar potential for a vector of no convergence, let the vector **c** be confined to the boundary of a small plane area dS, *i.e.*, let **c** circulate round the closed curve bounding dS, and be nothing everywhere else. The vector **b**, of which **c** is the curl, must be derivable from a scalar potential in all space except in **c** itself. Now, we know the vector-potential of **c**, viz.,

$$\mathbf{a} = \frac{ch\,dS}{4\pi r^3}\mathbf{a}_1, \quad \text{perpendicular to } \mathbf{r} \text{ and } \mathbf{h},$$

where **r** is the straight line from **c** to the place where **a** is measured, and **h** the perpendicular from the latter on to the axis of **c**; and that its three components are

$$a_1 = \frac{c}{4\pi}\frac{dr^{-1}}{dy}dS, \quad a_2 = -\frac{c}{4\pi}\frac{dr^{-1}}{dx}dS, \quad a_3 = 0.$$

From these derive **b** by means of equations (5). Thus

$$b_1=\frac{c}{4\pi}\frac{d}{dx}\frac{dr^{-1}}{dz}dS,\quad b_2=\frac{c}{4\pi}\frac{d}{dy}\frac{dr^{-1}}{dz}dS,\quad b_3=-\frac{c}{4\pi}\left(\frac{d^2r^{-1}}{dx^2}+\frac{d^2r^{-1}}{dy^2}\right)dS.$$

Here consider b_3. Since $c/4\pi r$ is the scalar potential of a quantity of "matter" c collected at the origin, we have, by equation (3),

$$\frac{c}{4\pi}\left(\frac{d^2r^{-1}}{dx^2}+\frac{d^2r^{-1}}{dy^2}+\frac{d^2r^{-1}}{dz^2}\right)=0,$$

everywhere, except at the origin. Therefore,

$$b_3=\frac{c}{4\pi}\frac{d}{dz}\frac{dr^{-1}}{dz}dS.$$

Thus, b_1, b_2, b_3 are the components of the space variation of a scalar Ω given by

$$\Omega=-\frac{c\,dS}{4\pi}\frac{dr^{-1}}{dz}=\frac{c\,dS}{4\pi r^2}\cos\epsilon,$$

if ϵ is the angle **r** makes with the axis of **c**. That is,

$$b_1=-\frac{d\Omega}{dx},\quad b_2=-\frac{d\Omega}{dy},\quad b_3=-\frac{d\Omega}{dz}.$$

This scalar potential Ω is of a singular nature. It varies inversely as the square of the distance, and as the cosine of the angle ϵ. It is evidently not the potential of any collection of matter at the origin, either wholly + or wholly −, but we can show that it is the potential of equal amounts of + and − matter infinitely near each other.

Coat the small plane area dS with matter of surface density m, so that its quantity is mdS. Its potential at Q, distant r, is evidently $mdS/4\pi r$. Now shift this matter through a small space dz in the + direction along the axis of **c**. Its potential at the same point Q will now be

$$\frac{mdS}{4\pi}\left(\frac{1}{r}-\frac{dr^{-1}}{dz}dz\right).$$

It is therefore increased by the amount $-\frac{m}{4\pi}dSdz\frac{dr^{-1}}{dz}$ or $\frac{mdSdz}{4\pi r^2}\cos\epsilon$.

Hence if we coat the surface dS with a quantity mdS on its + side and with $-mdS$ on its − side, at the small distance dz apart, the potential of the combination will be $\frac{mdSdz}{4\pi r^2}\cos\epsilon$. This becomes identically the same as Ω if we make $mdz=c$.

Up to the present the vectors **A**, **B**, **C** have been treated quite apart from whether any of them represented physical quantities (although they may all be possible cases of velocity distribution in an incompressible fluid). The properties are true whether electric current or magnetic force exist or not. But the last step made, the nature of the scalar potential whose space-variation gives **b**, brings us into contact with magnetism, and will enable us to fit current and magnetic force into the system. One experimental truth suffices. A small plane electric current and a small magnet produce exactly similar fields of magnetic force, not too near them, when we reckon the + direction of the axis of the magnet to be from its S. to its N. pole, and that of the current to be such that when we look along the axis in the + direction the current circulates right-handedly about it. If the magnet is very small, and also the current area, the similarity exists everywhere except very close to them. We may go *through* a current, but not *into* a magnet. Let the magnet and the current area be infinitely small, then we may say there is identity of magnetic force at any finite distance.

Now, the magnetic force of a magnet may be conceived to arise from an imaginary substance, free magnetism, having the same self-repulsive power as the imaginary free electricity, with the same law of force, viz., that of the inverse square, so that the magnetic force at distance r from a charge m of magnetism is $m \div 4\pi r^2$, and that if this acts upon a charge m', the repulsion is $mm'/4\pi r^2$.

If, then, our vector **c** stands for electric current, the "matter" producing the potential Ω must be magnetism, and since the unit of magnetism is expressible in terms of the fundamental units of length, mass, and time, we may define the unit current so that

$$(m\,dS)dz = c\,dS,$$

or, $$\text{strength of pole} \times \text{length} = \text{current} \times \text{area},$$

either of which may be called the magnetic moment (of the magnet, or of the current). (Having nothing to do here with the electrostatic system of units, we do not introduce the factor μ for induced magnetization.)

After having thus made use of magnetism, we might dismiss it for good. In fact, we might replace every magnetised particle by its equivalent electric current, according to Ampère's theory of magnetization. This theory generally strikes one on first acquaintance as very fanciful, the idea of electric currents flowing round molecules being so difficult to swallow. But it explains magnetism—one mystery takes the place of two; and since we neither know what an electric current is nor what magnetism is, it is well to abolish one of them, and there can be no question as to which it should be.

Having now identified **C** with electric current and **B** with its corresponding magnetic force, it will be as well to point out the modifications needed to suit the adopted unit magnetic pole and the common definition of potential. They were temporarily discarded in the consideration of the properties of **A**, **B**, **C**, in order to have a uniformity of relations, for otherwise the 4π's would not come in regular order. We must now remember that the force of a magnetic pole is of strength m/r^2 at distance r, but that a current c spreading from a centre has density $c/4\pi r^2$. On the other hand, potential follows the rule m/r or c/r in both cases.

We therefore, have

$$\text{Potential of } \mathbf{C} = \mathbf{A} = \Sigma\frac{\mathbf{c}}{r} = \Sigma\frac{bh}{4\pi r^3}\mathbf{a}_1.$$

$$\text{Magnetic force} = \mathbf{B} = \Sigma\frac{\mathbf{d}}{r} = \Sigma\frac{ch}{r^3}\mathbf{b}_1 = \text{curl } \mathbf{A}.$$

$$\text{Current} = \mathbf{C} = \frac{1}{4\pi}\text{curl } \mathbf{B}.$$

$$\mathbf{D} = \text{curl } \mathbf{C}.$$

From the uniformity of properties of **A**, **B**, **C**, as originally defined, any one of them may be current, the one preceding it being its magnetic force, of which examples were given in Sections I. and II. The same is, of course, true in the last system, except that the 4π's require shifting occasionally.

There is one more remark to be made. We started with **A**, a vector of no convergence, and of course **C** has also none. But we identified **C** with current. This is admitted to have no convergence in conducting circuits, and in Maxwell's theory it has none anywhere. Much confusion would arise were convergence allowed. All vectors below **C** are therefore also non-converging, but we have not shown that those above it are. For proof it is sufficient to observe that the vector-potential of a small plane closed current (components given above) is non-converging, and that any current-system may be built up of such closed currents. This makes the matter square.

SECTION VI. MAGNETIC FORCE OF RETURN CURRENT THROUGH THE EARTH, AND ALLIED MATTER.

The magnetic force of a current-system can only be completely known when the current is everywhere known (or equivalent information must be given), and no part of the system can in strictness be neglected. If, then, the circuit of a linear current is completed through the earth, the earth-current itself contributes something to the magnetic force. This, as will be seen, is quite a small matter; yet its investigation is both interesting in itself as affording an excellent example of the relations between the current, its vector-potential and magnetic force, and on account of a current-sheet analogue that presents itself.

The magnetic force of a straight current C at distance h from the axis is of intensity $2C/h$, and is perpendicular to the plane containing **h** and

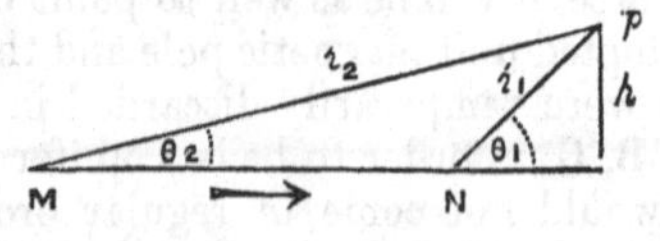

the axis, provided the point chosen be not near the ends of the wire, nor at a very great distance from the wire. But in our present case we shall require the correction for the ends.

Let MN be the wire, the current $\mathbf{C}$ going from M to N; and let P be any point distant r_1 from N, and r_2 from M, and h from the axis or its continuation. Let θ_1 and θ_2 be the angles $\mathbf{r}_1$ and $\mathbf{r}_2$ make with the axis. Then the magnetic force at P of the wire current alone (vector-potential of curl of current) is

$$\frac{C}{h}(\cos\theta_2 - \cos\theta_1) = \frac{C}{h}(1 - \cos\theta_1) - \frac{C}{h}(1 - \cos\theta_2), \quad \ldots\ldots\ldots\ldots(1)$$

and its direction is perpendicular to the plane of $\mathbf{r}_1$, $\mathbf{r}_2$, and $\mathbf{h}$, being, in the diagram, upwards through the paper. When the point P is near the wire, but not near the ends, we may put $\theta_1 = \pi$, $\theta_2 = 0$, and we then obtain the former expression $2C/h$. Confining ourselves to the neighbourhood of one end, say N, we may put $\theta_2 = 0$; consequently the magnetic force is of intensity

$$\frac{C}{h}(1 - \cos\theta_1). \quad \ldots\ldots\ldots\ldots\ldots\ldots\ldots\ldots\ldots\ldots(2)$$

Just over the end, $\theta_1 = \frac{1}{2}\pi$, and the force is C/h only; and along the continuation of the wire it is zero.

Now, although the expression (1) is the vector-potential of the curl of the wire current only, it is also the complete magnetic force of the closed circuit which would be formed by immersing the ends of the wire in an infinitely extended conducting medium, so that the current C leaving N spread out uniformly, the tubes of flow afterwards bending round and finally converging uniformly to M. *I.e.*, (1) expresses the magnetic force of the straight current, together with a radial current of density $C/4\pi r_1^2$ parallel to $\mathbf{r}_1$, and of another of density $-C/4\pi r_2^2$ parallel to $\mathbf{r}_2$.

For consider the first mentioned radial current. Its three components are

$$u = \frac{Cx}{4\pi r_1^3}, \qquad v = \frac{Cy}{4\pi r_1^3}, \qquad w = \frac{Cz}{4\pi r_1^3},$$

if x, y, z are the components of $\mathbf{r}_1$. The x-component of the curl is, by Theorem (B),

$$\frac{dw}{dy} - \frac{dv}{dz} = \frac{C}{4\pi}\left\{z\frac{d}{dy}\left(\frac{1}{r^3}\right) - y\frac{d}{dz}\left(\frac{1}{r^3}\right)\right\} = 0;$$

and similarly the y and z components are *nil*, therefore the curl of the radial current is *nil*, and likewise the vector-potential of the curl; hence the radial currents contribute nothing to the magnetic force.

Here we may notice that any radial vector-function has no curl. For, let $f(r)$ parallel to $\mathbf{r}$ be such a function, then $f(r)x/r$, $f(r)y/r$, $f(r)z/r$ are its components, and the x-component of its curl is

$$f(r)\left(z\frac{dr^{-1}}{dy} - y\frac{dr^{-1}}{dz}\right) + \frac{1}{r}f'(r)\left(z\frac{dr}{dy} - y\frac{dr}{dz}\right) = 0.$$

But in the actual case of return through earth the diffusion is one-sided only. The curl of current within the earth is *nil*, as before, but since there is now a bounding surface, and consequently surface-curl, the earth current does give rise to magnetic force. If $\mathbf{D}$ is the surface-curl, its strength is $C/2\pi a^2$ at distance a from the electrode (meaning the point N on the earth's surface), and its direction is perpendicular to a (Section I.). *I.e.*, the lines of surface-curl are concentric circles drawn upon the earth's surface, centred at the electrode. The vector-potential of **D** is the magnetic force of the earth current. Calling it **B**, since **D** has no vertical component, neither has **B**; and evidently **B** has the same value and direction at two corresponding points above and below the surface (one being the image of the other). From the symmetry about the electrode, we see that the lines of magnetic force are circles in planes parallel to the earth's surface, and centred upon the vertical axis through the electrode. As we recede from the latter, the strength of force rapidly diminishes.

The strength of force may be determined indirectly thus:—Let us denote by $\mathbf{C}_1$ the uniformly-diffused radial current of density $C/4\pi r_1^2$. If to this we add another radial current of the same density everywhere as $\mathbf{C}_1$, but directed *to* the electrode above the earth, and from it below —which system we may call $\mathbf{C}_2$—their sum will be 0 above and $C/2\pi r_1^2$ below the earth's surface, and this is the density of the real earth-current.

Now, if we take the curl of the vector (2), we shall obtain the current $\mathbf{C}_1$ (disregarding the wire current). We may do this by using the components of (2); or, more simply, thus:—The line integral of (2) round the circle of radius h perpendicular to the axis is

$$\frac{C}{h}(1-\cos\theta_1)\times 2\pi h=\frac{C}{r_1^2}\,.\,2\pi r_1^2(1-\cos\theta_1).$$

This is $4\pi\times$ current through the circle (Section I.). Now,

$$2\pi r_1^2(1-\cos\theta_1)$$

is the area of the portion of spherical surface with centre at N and radius r_1 bounded by the circle; consequently $C/4\pi r_1^2$ is the current-density.

Now, the curl of the magnetic force **B** of earth-current must be $4\pi\times\mathbf{C}_2$. But $\mathbf{C}_2$ and $\mathbf{C}_1$ are identical within the earth, therefore **B** must be of the same form as (2), *i.e.*,

$$B=\frac{C}{h_1}(1-\cos\phi), \qquad\qquad (3)$$

where h_1 is the distance of the point considered from the axis of the earth-current (not of the wire), *i.e.*, the vertical through the electrode, and ϕ the angle between $\mathbf{r}_1$ and this axis (positive direction downwards). But, above the earth, $\mathbf{C}_2$ and $\mathbf{C}_1$ are oppositely directed; **B** must, therefore, be so changed that its curl becomes reversed. This may be done by making

$$B=\frac{C}{h_1}(1+\cos\phi) \qquad\qquad (4)$$

We may express (3) and (4) in one formula, thus:—

$$B=\frac{C}{h_1}(1-\sqrt{\cos^2\phi}); \qquad\qquad (5)$$

i.e., we take the numerical value of $\cos\phi$, whether it is + or −.

The actual magnetic force near the electrode is the vector-sum of (2) and (5). One case may be noticed, where it simplifies. Let $\theta_1 = \phi$ and $h = h_1$, so that the wire enters the earth perpendicularly. Above the earth, we shall have

$$B = \frac{C}{h}(1 - \cos\theta_1) + \frac{C}{h}(1 + \cos\theta_1) = \frac{2C}{h},$$

perpendicular to $\mathbf{r}_1$ and $\mathbf{h}$. And below the surface,

$$B = \frac{C}{h}(1 - \cos\theta_1) + \frac{C}{h}(1 - \cos\theta_1) = \frac{2C}{h}(1 - \cos\theta_1)$$

We may now examine an interesting analogy which has a realisable existence, and may serve to clarify the above. Suppose that we have an infinite thin plane sheet of conducting material; choosing any point N in it for origin, let currents circulate in the sheet in circles about N, the strength being $C/2\pi a^2$ at distance a. If when we view the sheet we see the current circulate right-handedly, we see the negative side, and the + direction of the axis of the currents is from the negative to the positive side of the sheet. Now we have found the vector-potential of this current-system; it is expressed by equation (5), using the same notation as regards h_1 and ϕ. The magnetic force is the curl of (5), and is therefore radial, of intensity C/r^2 on the + side and $-C/r^2$ on the − side. That is, the magnetic force of the current-sheet is that which would be produced by a magnetic pole of strength C placed at the origin, with this difference, that it acts as a + pole on the + side and

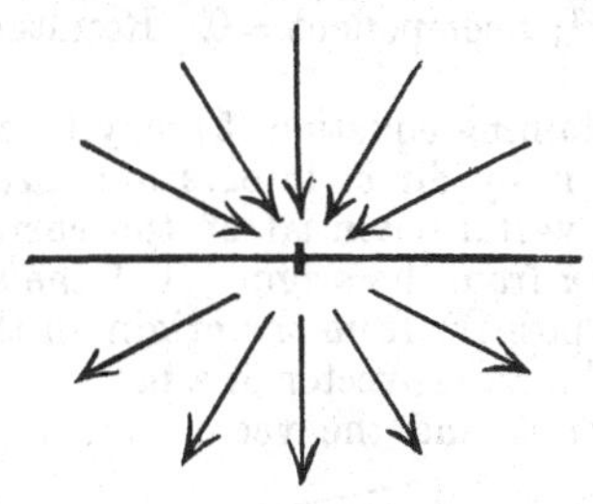

as a − pole on the − side of the sheet, as shown in the diagram annexed. (We have here an illustration of the continuity of the normal and discontinuity of the tangential component of magnetic force referred to in Section II.)

How produce such a current system? Easily enough; permanently, if we had a sheet of infinite conductivity, transiently and imperfectly with a resisting sheet.

It is a cardinal truth in electrostatics that, when an uncharged insulated conductor is placed in an electric field, it becomes charged in such a manner that the force due to its charge exactly annuls that of the undisturbed field for all internal points, so that there is no electric force within the conductor.

Now, there is a corresponding theorem in electromagnetics. If a closed sheet of infinite conductivity, having no current in it initially, be placed in a magnetic field, a system of currents will be set up in it of such a nature that the magnetic force due to it exactly annuls that of the undisturbed field within the space bounded by the sheet (or, in the case of an infinite sheet, on the further side from the source of magnetic force). Practically, with a resisting sheet, bringing it rapidly into the magnetic field, or, bringing the field to it, would set up the currents approximately, to be quickly killed by the resistance.

Let us then bring a magnetic pole of strength C suddenly to the origin on the + side. Its force is C/r^2 radial, on both sides. But there must be no force on the − side, by the above-mentioned theorem. So the magnetic force of the currents set up will be $-C/r^2$ radial on the − side, and on account of the reversal of the tangential component in passing through the sheet, it must be $+C/r^2$ radial on the + side. But since this field is exactly that produced by the distribution of current assumed before, it must be the current set up by the magnetic pole. The actual magnetic force of pole and current together is 0 on the − and $2C/r_1^2$ on the + side, *i.e.*, similar to the distribution of the current entering earth.

A magnetic shell which would produce the same field of force would have strength

$$\int_a^\infty \frac{C\,da}{2\pi a^2} = \frac{C}{2\pi a}.$$

This would be the magnetic moment per unit area.

(*Test.* Curl of intensity of magnetisation = vol. density of equivalent current. Taking the curl of $C/2\pi r \parallel \mathbf{z}$, we find x-component $= -Cy/2\pi r^3$; y-component $= Cx/2\pi r^3$; z-component $= 0$. Resultant, $C/2\pi r^2 \perp \mathbf{r}$ in plane of sheet.)

As the mode of obtaining equation (5) may be considered obscure, it is desirable to check it by direct integration. It comes to the same thing if we find the vector-potential of the current-sheet, of density $C/2\pi a^2 \perp \mathbf{a}$ at distance a from the origin. Let the sheet be in the plane of x, y, and measure z positive from the origin on the + side. Let θ be the angle between $\mathbf{r}$ the radius-vector of a point P and $\mathbf{z}$, and x, y, z the co-ordinates of P. First find the vector-potential at P of a circular

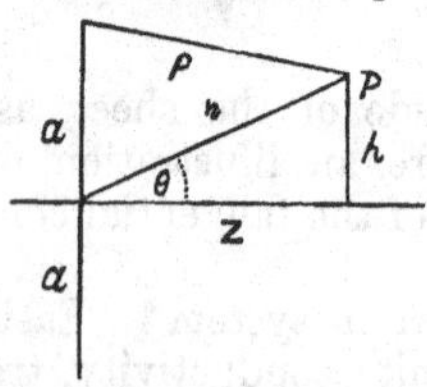

unit-current of radius a. If $\mathbf{h}$ is parallel to the x-axis, the y-components of the current contribute nothing to the vector-potential; the x-component of the element $a\,d\phi$ (measuring ϕ from the x-axis) is $a \sin\phi\, d\phi$; therefore the vector-potential of the circular unit-current is

$$\int_0^{2\pi} \frac{1}{\rho} a \sin\phi\, d\phi, \qquad\qquad (6)$$

where ρ is the distance of P from the element $a\,d\phi$, and is given by

$$\rho^2 = r^2 + a^2 - 2ha \sin\phi.$$

The complete vector-potential of the current-sheet is therefore

$$B = \frac{C}{2\pi}\int_0^\infty\int_0^{2\pi}\frac{1}{a\rho}\sin\phi\, da\, d\phi. \qquad\qquad (7)$$

Let the limits for a be first a and ∞; a to be afterwards made 0. Then, by integration to a, we find

$$B = -\frac{C}{2\pi}\int_0^{2\pi}\frac{1}{r}\sin\phi\log\left(\frac{1}{r}-\frac{h\sin\phi}{r^2}\right)d\phi + \frac{C}{2\pi}\int_0^{2\pi}\frac{1}{r}\sin\phi\log\left(\frac{1}{a}-\frac{h\sin\phi}{r^2}+\frac{\rho}{ar}\right)d\phi$$

$= B_1 + B_2$ say.

Now the definite integral

$$\frac{1}{2\pi}\int_0^{2\pi}\sin\phi\log(m+n\sin\phi)\,d\phi = \frac{m-(m^2-n^2)^{\frac{1}{2}}}{n};$$

consequently, $$B_1 = C\frac{r-(r^2-h^2)^{\frac{1}{2}}}{rh}.$$

Now, in the second integral B_2, a has to be ultimately 0. The quantity under the radical, *i.e.*, in the expression for ρ, therefore varies infinitely little from r^2, whilst ϕ goes from 0 to 2π. The quantity to be integrated then becomes of the same form as in B_1; but also, we shall have $m=\infty$, $\therefore$ $B_2 = 0$. Thus

$$B = B_1 = C\frac{r-(r^2-h^2)^{\frac{1}{2}}}{rh} = C\frac{1-\sqrt{\cos^2\theta}}{h} \qquad\qquad (8)$$

is the complete solution, showing that (5) is correct.

The integral (6) does not admit of finite expression, being elliptic (see Maxwell's "Electricity," Vol. II., Art. 701). Its expansion in spherical harmonics will furnish us with another proof of (8). Substitute for the circular unit-current of radius a an equivalent plane magnetic shell of unit strength, so that there is magnetism of unit density on its + and of density -1 in its $-$ face. Let V be the potential of the + magnetism, then $+dV/dy$ and $-dV/dx$ will be the x and y components of the vector-potential. First find V for a point upon the axis, then

$$V = \int_0^a\int_0^{2\pi}\frac{a\,da\,d\phi}{\sqrt{a^2+z^2}} = 2\pi\{(a^2+z^2)^{\frac{1}{2}} - z\}.$$

Expand this by the Binomial Theorem in powers of z, and call it V_1 when $z<a$ and V_2 when $z>a$. Then

$$V_1 = -2\pi z + 2\pi a\left(1+\frac{1}{2}\frac{z^2}{a^2}-\frac{1}{2.4}\frac{z^4}{a^4}+\frac{1.3}{2.4.6}\frac{z^6}{a^6}-\dots\right),$$

$$V_2 = \qquad 2\pi z\left(\quad\frac{1}{2}\frac{a^2}{z^2}-\frac{1}{2.4}\frac{a^4}{z^4}+\frac{1.3}{2.4.6}\frac{a^6}{z^6}-\dots\right).$$

Next let the point be not on the axis; we may generalise the last by writing rQ_1 for z, r^2Q_2 for z^2, etc., in V_1, and Q_n/r^{n+1} for $1/z^{n+1}$ in V_2, where the axis of Q's is along z. Thus,

$$V_1 = 2\pi a\left(1-\frac{r}{a}Q_1+\frac{1}{2}\frac{r^2}{a^2}Q_2-\frac{1}{2.4}\frac{r^4}{a^4}Q_4+\dots\right),$$

$$V_2 = 2\pi a\left(\frac{1}{2}\frac{a}{r}Q_0-\frac{1}{2.4}\frac{a^3}{r^3}Q_2+\frac{1.3}{2.4.6}\frac{a^5}{r^5}Q_4-\dots\right).$$

Consequently the x-component of the vector-potential is

$$\frac{dV_1}{dy}=2\pi\frac{y}{a}\left(-\frac{1}{2}+\frac{1}{2.4}\frac{r^2}{a^2}\cdot\frac{dQ_3}{d\mu}-\frac{1.3}{2.4.6}\frac{r^4}{a^4}\frac{dQ_5}{d\mu}+\dots\right),$$

or

$$\frac{dV_2}{dy}=2\pi a^2\frac{y}{r^3}\left(-\frac{1}{2}+\frac{1}{2.4}\frac{a^2}{r^2}\frac{dQ_3}{d\mu}-\frac{1.3}{2.4.6}\frac{a^4}{r^4}\frac{dQ_5}{d\mu}+\dots\right),$$

according as r is $<$ or $>$ a. In getting the last expressions we have made use of the relations

$$\frac{d}{dy}(r^nQ_n)=-yr^{n-2}\frac{dQ_{n-1}}{d\mu},\qquad \frac{d}{dy}\left(\frac{Q_n}{r^{n+1}}\right)=-\frac{y}{r^{n+3}}\frac{dQ_{n+1}}{d\mu},$$

where $\mu=\cos\theta$. The y-component is got by writing $-x$ for y, so that the actual magnitude of the vector-potential of the circular unit-current is

$$\left.\begin{aligned}\frac{2\pi h}{a}\left(+\frac{1}{2}-\frac{1}{2.4}\frac{r^2}{a^2}\frac{dQ_3}{d\mu}+\frac{1.3}{2.4.6}\cdot\frac{r^4}{a^4}\frac{dQ_5}{d\mu}-\dots\right)&=A_1 \text{ say,}\\ \text{or}\quad 2\pi a^2\cdot\frac{h}{r^3}\left(+\frac{1}{2}-\frac{1}{2.4}\frac{a^2}{r^2}\frac{dQ_3}{d\mu}+\frac{1.3}{2.4.6}\cdot\frac{a^4}{r^4}\frac{dQ_5}{d\mu}-\dots\right)&=A_2 \text{ say.}\end{aligned}\right\}\dots(9)$$

Now, since the current strength in the sheet is $C/2\pi a^2$, the complete vector-potential is

$$B=C\int_r^\infty\frac{A_1da}{2\pi a^2}+C\int_0^r\frac{A_2da}{2\pi a^2}.$$

Substituting the above values of A_1 and A_2, and performing the integrations,

$$\begin{aligned}B&=\frac{Ch}{r^2}\left(+\frac{1}{2}\cdot\frac{1}{2}-\frac{1}{2.4}\cdot\frac{1}{4}Q_3'+\frac{1.3}{2.4.6}\cdot\frac{1}{6}Q_5'-\dots\right)\\&+\frac{Ch}{r^2}\left(+\frac{1}{2}-\frac{1}{2.4}\cdot\frac{1}{3}Q_3'+\frac{1.3}{2.4.6}\cdot\frac{1}{5}Q_5'-\dots\right)\\&=\frac{Ch}{r^2}\left(\frac{3}{4}-\frac{7}{3.4}\cdot\frac{1}{2.4}Q_3'+\frac{11}{5.6}\cdot\frac{1.3}{2.4.6}Q_5'-\dots\right),\end{aligned}\qquad\dots\dots\dots\dots(10)$$

where Q'_n is written for $dQ_n/d\mu$.

We must now prove that (10) is the expansion of (8) Any function F of μ may be expanded in $dQ/d\mu$'s or Q's; thus

$$F=F_1Q'_1+F_2Q'_2+\dots,$$

and the value of any coefficient F_n is given by

$$F_n=\frac{2n+1}{2n(n+1)}\int_{-1}^1F(1-\mu^2)Q'_n\,d\mu.$$

Now in the present case B in (8) is

$$B=\frac{C}{h}(1-\sqrt{\mu^2})=\frac{C}{r}\cdot\frac{1-\sqrt{\mu^2}}{\sin\theta},$$

and $1-\sqrt{\mu^2}$ is such a function of μ that

$$F(\tfrac{1}{2}\pi+a)=F(\tfrac{1}{2}\pi-a).$$

Also Q'_n is an even function of μ when n is odd, and an odd function when n is even. Therefore, since from $\theta=0$ to $\theta=\frac{1}{2}\pi$ we have

$$B=\frac{C}{r}\,\frac{1-\mu}{\sin\theta}=\frac{C}{r}\,\frac{\sin\theta}{1+\mu},$$

if we expand $(1+\mu)^{-1}$ in odd Q''s between $\mu=0$ and $\mu=1$, the same expression resulting will give the proper value between $\mu=-1$ and $\mu=0$. So we shall have

$$B=\frac{C\sin\theta}{r}\cdot\frac{1}{1+\mu}=\frac{C\sin\theta}{r}(F_1Q'_1+F_3Q'_3+\ldots),$$

where
$$F_n=\frac{2n+1}{n(n+1)}\int_0^1(1-\mu)\,Q'_n d\mu,$$

and, the limits being halved, the outer factor has been doubled, and only odd values of n are taken. Carrying this out, we shall obtain the expression for B in (10).

XXV.—THE ENERGY OF THE ELECTRIC CURRENT.*

SECTION I.—THE MUTUAL POTENTIAL ENERGY OF MAGNETIC SHELLS AND LINEAR CURRENTS.

AS was remarked towards the close of Section 5, Art. xxiv., p. 223, if we adopt Ampère's theory of magnetism, and substitute for every magnetised particle a closed current flowing round it, we may do away with free magnetism, or imaginary magnetic matter, and have nothing but electric currents to deal with. It is, indeed, not necessary to introduce magnetism at all. We did that in order to reach the definition of a unit current in the electromagnetic system, which is based upon that of a unit magnetic pole. We might, for example, obtain an equivalent result from the mechanical force between the terminations of two long solenoids, which varies inversely as the square of the distance; or in various other ways.

We found that the magnetic force of a small plane closed current could be derived from a scalar potential, that the force was what would be produced by equal quantities of positive and negative matter very close together—in fact supposed to be infinitely near each other—the quantity of matter being at the same time infinitely great, but so that the product of the quantity into the distance was finite. If magnets were unknown, this matter would be certainly purely imaginary; quite as much so as the matter which, quite similarly, would give rise to a scalar potential, whose derivatives would express the vector-potential of electric currents in space where there was no magnetic force; or, to come nearer home, the matter whose potential derivatives express the electric force wherever it has no curl, which last matter is the imaginary matter of free electricity. The fact that magnets exist does not make the magnetism any the more real, considered merely as a self-repulsive

* [*The Electrician*, 1883; Section i., Jan. 20, p. 233: ii., Feb. 23, p. 342; iii. March 9, p. 390; iv., March 23, p. 437.]

substance existing wholly on the surface of a solenoidal magnet, and also within its substance when not solenoidal; and the same may be said of the electricity once supposed to reside upon the surface of conductors.

But it happens to be a considerable assistance to make use of the idea of magnetism (or of magnetic polarisation, which comes to the same thing) in reasoning regarding electric currents, and it is particularly useful when we are considering the energy of currents. In fact, it would certainly be employed for this purpose were magnetisation unknown, (presuming that this would not involve the non-existence of current,) in the form of a magnetic shell, for the idea presents itself in a perfectly natural manner.

Suppose we have a current circulating in a closed path of any form, which we may, when we do not want to go into the space occupied by the current, regard simply as a closed curve. Such a current is a linear closed current. Let C be the strength of current. If we join any two points of the circuit by means of another line, and let the same current C go both up it and down it at the same time, we form two closed circuits both with equal currents of strength C flowing round them, with the same direction of rotation as regards lines drawn through the original closed curve, which lines we may reckon to be positively directed when the current flows round them righthandedly. Now, since the two circuits have a portion of their length in common, in which equal currents are oppositely directed, there can be no change in the magnetic action at all; hence the two closed currents, when taken together, are exactly equivalent to the original one.

Therefore, by the same reasoning as was employed in the demonstration of Theorem (B), p. 211, if we suppose the original circuit to be the boundary of a surface S, and we divide this surface into an indefinitely great number of small elements of area dS, and let the same current C flow round the boundary of every one of these elements with the same direction of rotation, we know that the whole collection of small closed currents produces identically the same field of magnetic force as the single current C flowing round the boundary of the whole. The form of the surface is quite indifferent; the essential circumstance is that it must be bounded by the real current.

Now, we found [p. 223] that the magnetic force of a current C circulating round the boundary of a small area dS was the same as that due to charges of magnetism of amounts mdS and $-mdS$ placed on opposite sides of dS at the small distance dz apart, the axis of z being that of the current; that is to say, **z** is the normal to dS. The condition connecting m and C is that $mdz = C$, or magnetic moment of magnet = that of the current (per unit of area, here).

Let the surface of every one of the small closed circuits we have imagined to replace the real circuit be similarly covered with magnetism on its two faces, according to the same rule. This is equivalent to covering the complete surface S to surface-density m on one side and $-m$ on the other, with the condition $mdz = C$ everywhere. Here mdz, the magnetic moment per unit area, is called the Strength of the shell. The thickness of the shell must be infinitely small, but it need not be constant. If it vary, m must vary inversely, so as to keep the strength constantly $= C$. The magnetic force of this shell and that of a current C round its edge are identical everywhere, except in the shell itself.

Now, since the mechanical action between a magnet and a small closed current is the same as that between the magnet and a small magnet of the same moment as the current, similarly placed, and since we may build up any closed linear current by putting together an immense number of such little closed currents, it follows that the mechanical action between a magnet and any linear current is the same as that between the magnet and an equivalent magnetic shell. But the magnet may itself be a second magnetic shell; and, since we may substitute for it a current round its edge, it follows that the mechanical action between two closed currents is the same as that between two magnetic shells whose edges are bounded by the currents, and of numerically the same strength as the latter. And the mutual potential energy of the two currents must be the same as that of the shells, which may be found from the distribution of the magnetic matter.

Now, we found [p. 222] the magnetic potential of an element dS of a shell of strength C to be

$$\Omega = -CdS\frac{dr^{-1}}{dz},$$

(now employing the common definition of unit pole), where r is the tensor of $\mathbf{r}$, the vector drawn from dS to the point Q where Ω is measured, and $\mathbf{z}$ the normal to dS. The differentiation in the above is performed at Q. If we perform it at dS, we must write

$$\Omega = +CdS\frac{dr^{-1}}{dz},$$

because if $1/r$ decreases when dS is fixed and Q is shifted in direction $\mathbf{z}$, it increases when Q is fixed and dS shifted in the same direction.

Place a magnetic pole of strength C at the point Q. Its potential at dS is C/r, therefore its force at dS in the direction $\mathbf{z}$ is $-C(dr^{-1}/dz)$; and since $\mathbf{z}$ is normal to dS the whole amount of force passing through dS from its $-$ to its $+$ side, due to the pole C at Q, is $-C(dr^{-1}/dz)dS$, the differentiation being performed at dS. But this, with sign changed, is our expression for Ω, the magnetic potential at Q of the element of the shell at dS. The same applies to every element of the shell; consequently, by summation, it follows that the magnetic potential Ω at Q of the complete shell = amount of force passing through the shell in the positive direction, due to a pole of strength $-C$ placed at Q, or in the negative direction if of strength $+C$. Thus, we may express Ω in terms of a solid angle, as for electrostatic force [on p. 208]; thus, (Ω at Q) $= C \times$ solid angle subtended at Q by the edge of the shell. The magnetic potential of a current C round the edge of the shell is, of course, the same.

The magnetic potential Ω must not be confounded with the vector-potential $\mathbf{A}$. The curl of the latter gives the magnetic force everywhere, both within and without a current; whereas Ω is scalar, and the magnetic force $= -\nabla\Omega$, where ∇ has the same meaning as before [p. 217], but this is only in space where there is no current. Therefore, in currentless space only,

$$\mathbf{B} = \text{curl } \mathbf{A} = -\nabla\Omega;$$

otherwise $\mathbf{B} = \text{curl } \mathbf{A}$, and Ω is non-existent.

Next, consider two shells of strength C_1 and C_2, and let Ω_1 and Ω_2 be their magnetic potentials. The mechanical force between them may be ascribed to the mutual actions of the magnetic matter, and calculated therefrom. To fix ideas, let the shells be circular planes. If we place them parallel to one another, upon the same axis, and so that the N. side of one is next the S. side of the other, there is attraction between them. We may call the N. face of a shell the + face, and the S. the − (N. and S. meaning north-seeking and south-seeking, as applied to a needle), so that the N. face of shell is the same as the + face of the corresponding current round its edge. The + direction through the shell or the current is from the − to the + face. The magnetic force of a closed current goes through it in the + direction; it also goes righthandedly about the current, and the current goes righthandedly about a line of force. These relations, which are sometimes puzzling to others besides beginners, are important to be remembered, for confusion will inevitably arise if we do not keep to one consistent system. In our present case the parallel shells with contrary faces next each other attract. The corresponding currents round their edges are similarly directed, so there is consistency with the fact that parallel currents attract or repel according as their directions are similar or opposed.

Since we (temporarily) ascribe the force between the shells to the attraction of matter, their mutual energy is, on this view, potential energy, and tends to decrease. If one shell be fixed, and the other move through a distance dz parallel to itself, and F be the mechanical force in the direction of motion, the work done by the force is Fdz. This is also the reduction in the amount of potential energy; or $-dM$, if M be the mutual energy. Therefore,

$$F = -\frac{dM}{dz}$$

is the expression for the force. This is, in the case considered, a real force in the dynamical sense, viz., acceleration of momentum. But we may give an extended meaning to the term force (and this is important in the theory of energy). Let dz indicate *any* "displacement," and $-dM$ the corresponding fall in the potential energy, then $F = -dM/dz$ is the "force" in the generalised sense, so that "force" × "displacement" always equals the decrease of potential energy during the displacement.

The two shells with contrary faces nearest, would attract each other from an infinite distance, and their potential energy would decrease all the time; hence, if the energy be taken to be nil at an infinite distance, where the force is evanescent, the energy is negative at any finite distance, and becomes more and more negative as the shells approach. Turn one of them round, so that similar faces are nearest; the potential energy is now positive, the amount of work done in turning the shell round being the whole increase of energy. (Rigid magnetisation is supposed.) Similar remarks apply to the currents round the edges of the shells; but we have not at present under consideration what becomes of energy added to the system by externally performed work.

The mutual energy of two shells or currents admits of tolerably simple expression by means of the theorem connecting line and surface integrals. If a charge m of magnetism be brought from an infinite distance to a place in a magnetic field where the potential is Ω, the work done against the repulsion is $m\Omega$, which is consequently the potential energy of m with respect to the field in which it is placed.

Apply this to a shell placed in the field of a second shell. Let dS be an element of area of the first shell, of strength C, and dz its thickness, z being, therefore, the normal at dS. On the + face there is magnetism to the amount $m\,dS = (C/dz)dS$. Let Ω' be the magnetic potential of the second shell there. The energy of $m\,dS$ with respect to the second shell is then $(C/dz)\Omega'\,dS$. Now, Ω' becomes $\Omega' - (d\Omega'/dz)dz$ when we pass through dS to the other side, where there is a quantity $-(C/dz)dS$ of magnetism, whose energy with respect to the second shell is, therefore, $-(C/dz)dS\,\{\Omega' - (d\Omega'/dz)dz\}$. The energy of the combination is the sum of the separate energies of the two charges of magnetism, that is, $C\,dS(d\Omega'/dz)$.

But $d\Omega'/dz$ is the magnetic force at dS of the second shell in the direction of the normal from the positive to the negative side; therefore, $dS(d\Omega'/dz)$ is the amount of magnetic force arising from the second shell which passes through dS in the negative direction. The same is true for all areas in the first shell; hence, by summing up, we find that the energy of the first shell with respect to the second = strength of first shell × total amount of magnetic force passing through it in the negative direction arising from the second shell. Or, if M denote the energy, and we always reckon the force in the positive direction through a shell, then

$$M = -C \times \text{amount of mag. force through } C \text{ due to } C', \quad\text{.........} \quad (1)$$

if C' is the strength of the second shell. Similarly, by starting with the second shell, we find

$$M = -C' \times \text{amount of force through } C' \text{ due to } C. \quad\text{..............} \quad (2)$$

It is customary to speak of the "number of lines" of force, each line being the representative of a definite "amount" of force, or "surface-integral' of force, to use the mathematical expression. But no one speaks of the "number of lines" of current passing through a surface; the expression "amount" or "quantity" would rather be used. Fractional parts of a line of force are awkward, but there is no such difficulty with amount or quantity. On the other hand, "surface-integral" is rather too scientific.

The right-hand members of equations (1) and (2) are identically equal, and a remarkable reciprocal property of closed currents or shells is disclosed. Make the currents each unity; then it follows that the unit current in the first circuit sends the same amount of force through the second as the unit current in the second circuit sends through the first, whatever may be the form or relative position of the circuits.

Equation (1) does not apply merely to the field of a shell or single circuit. By the manner of its derivation it applies to any magnetic field whatsoever produced by magnets or currents, substituting for "due to C'" the expression "due to the field." This is to exclude the magnetic force of the current C itself.

Equations (1) and (2) may be immediately put in another form. The magnetic force of a current is the curl of its vector-potential, therefore, by Theorem (B), [p. 211], the amount of magnetic force through any circuit = line-integral once round it of the vector-potential, whose curl is the force considered. So let **A** and **A**$'$ be the vector-potentials of the complete currents C and C', then (1) and (2) become

$$M = -C' \times \text{line-integral of } \mathbf{A} \text{ once round } C', \quad\ldots\ldots\ldots\ldots\ldots\ldots\ldots (3)$$
$$= -C \times \text{line-integral of } \mathbf{A}' \text{ once round } C. \quad\ldots\ldots\ldots\ldots\ldots\ldots\ldots (4)$$

These, again, may be put in another form, which will be useful. Consider (3) only. Let ds' be an element of length of the current C'. The portion of M due to it alone is

$$-AC'\,ds'\cos(\mathbf{AC}').$$

Here it will be convenient,—its importance will be seen later,—to introduce the notion of the scalar product of a pair of vectors. If **a** and **b** be any vectors, a and b their tensors or magnitudes, and θ the angle between them, then the scalar $ab\cos\theta$ is called the scalar product of **a** and **b**, and is denoted by **ab** simply. Suppose, for instance, **F** is a force and **v** the velocity of its point of application, then **Fv** is its activity. Should **F** and **v** be parallel, **Fv** reduces to Fv simply, the product of their tensors or magnitudes.

Thus, in our present case, the portion of M due to the element of current **C**$'$ at ds' is $-\mathbf{AC}'\,ds'$.

But **A** is itself a line-integral, viz., once round the current C. Let ds be an element of length of C, then the portion of **A** contributed by it is (by the definition of vector-potential) $(\mathbf{C}/r)ds$ at a point distant r from the element **C**, and its direction is parallel to **C**. Therefore the portion of M due to the pair of elements $\mathbf{C}\,ds$ and $\mathbf{C}'\,ds'$ is

$$-\frac{CC'}{r}\cos(\mathbf{CC}')\,ds\,ds' \quad \text{or} \quad -\frac{\mathbf{CC}'}{r}ds\,ds', \quad\ldots\ldots\ldots\ldots (5)$$

and the complete energy is the sum of all such terms, so as to include the whole of both currents; that is,

$$M = -CC'\iint\frac{\cos\epsilon}{r}ds\,ds', \quad\ldots\ldots\ldots\ldots\ldots\ldots\ldots\ldots (6)$$

where r is the distance between $d\mathbf{s}$ and $d\mathbf{s}'$, and ϵ the angle between their directions, and the integration extends once round each circuit, the direction chosen being that of the current in each case. (6) is one of the simplest forms of Neumann's celebrated formula.

With respect to (5) and (6), the latter, referring as it does to closed circuits, is exact; but as regards the former all we can say at present is that the potential energy of two closed currents is the same *as if* a pair of linear elements, one taken from the first circuit, the other from the second, possessed the mutual energy expressed in (5); for such elements could not exist alone.

SECTION II.—VARIATION OF THE ENERGY WITH THE SIZE OF THE SYSTEM. THE MUTUAL ENERGY OF ANY TWO DISTRIBUTIONS OF CURRENT.

Given two closed circuits of any form with steady currents in them, placed at any distance apart in any relative positions, we have found [Section I, p. 234] that the mechanical force between them may be completely defined by the variation of a certain quantity, M, which has a definite value for every configuration. Any linear displacement dx of one conductor, moving as a whole without rotation, is assisted by a real force arising from the mutual action of strength $-dM/dx$, or the rate of decrease of M per unit displacement; and, extending the meaning of a displacement to signify any small change of position, and at the same time generalising the meaning of force so that its definition shall be contained in the statement "force $\times$ displacement $=$ work done by the force during the displacement," we have the same relation, force $= -dM/dx$.

This expression of the force as the variation of the quantity M is simply a reversal of the statement that M is the integral of the force. For the method of obtaining M is entirely founded upon the fact that the mechanical force between two conductors bearing currents is the same as that between two magnetic shells of the same strengths as the currents, and with coincident edges, which again is reducible to forces acting according to the law of the inverse square of the distance between quantities of magnetic matter, so that M is really the mutual potential energy of the distributions of matter. We may also say that M is the mutual potential energy of the currents, remembering, however, that the currents must not vary; for our equivalent magnetic shells are rigidly magnetised, and we so far ignore the existence of changes in the currents produced by the relative motion of the circuits.

The quantity M for the complete circuits is the same as if any pair of current-elements of strengths C and C' and lengths ds and ds', at distance r apart, one being in the first and the other in the second circuit, contributed the amount $-(\mathrm{CC}'\,ds\,ds'/r)$ [equation (5), p. 236]. From this we may see how the energy and the forces are affected by changes of dimensions in similar systems. For let there be another pair of circuits similar to the first, of n-fold linear dimensions, and similarly situated, so that every line joining corresponding points is magnified n times; divide the circuits into the same number of elements of length as the first pair, thus making each element n times as long; then $(ds\,ds')/r$ is multiplied $(n \times n) \div n = n$ times. Hence if the currents are the same in both cases, the energy is n times as great in the magnified system.

As for the forces, they are quite unaltered. For if dx be a displacement in the first case, it becomes $n\,dx$ in the second; and the changes in M corresponding are dM and $n\,dM$. But the force is the change in M per *unit* displacement, and is therefore simply dM as before. Thus, in similar linear systems with the same currents, the forces are independent of, and the energy varies as the linear dimensions.

This law, that the force in similar systems is unchanged, was made one of the bases of Ampère's theory, as an experimental fact; from it follows, that if the actually observed actions were due to mutual forces between the different parts of the two conductors or currents, such forces must vary as the inverse square of the distance, when the inclinations of elements are the same. This, of course, also follows from $(ds\,ds')/r$; its rate of decrease along r being $ds\,ds'/r^2$.

In varying the linear dimensions we kept the currents constant, so that it did not matter whether we varied the *sections* of the conductors or not. But, varying all the linear dimensions, if we keep the current *densities* constant, so that with a doubled sectional area we have a doubled current, then, since an n-fold increase of linear dimensions multiplies each of the sections n^2 times, the currents are made n^2 times as great; hence in the magnified system the energy is multiplied $n \times n^2 \times n^2$, or n^5 times instead of only n times, whilst the forces are multiplied n^4 times.

This law of the fifth power, which in its application to dynamos has been pointed out by M. Marcel Deprez, is a particular case of a more general law. The potential energy of any similar distributions of matter with the gravitational law of inverse squares varies as the fifth power of the linear dimensions and as the squares of the densities. (Thomson, "Electrostatics and Magnetism," p. 435.) The matter may be of the electric or magnetic kind, repulsion between like, and attraction between unlike kinds. If ρ is the volume-density and P the potential, $\frac{1}{2}\Sigma P\rho$ is the potential energy, the summation including all the matter. Here $P = \Sigma\,\rho'/r$ as usual; hence the energy is $\frac{1}{2}\Sigma\Sigma\,\rho\rho'/r$, where ρ and ρ' are the quantities of matter in any two elements of volume and r their distance apart. Now, in a system of n-fold dimensions, divided into the same number of elements of volume, each of them is n^3 times as large, and their distances apart are all multiplied n times; hence the energy is multiplied $n^3\,n^3/n = n^5$ times. And, of course, when the density varies, the energy varies as its square, since ρ and ρ' are both similarly increased.

But applying this law to similar magnets of different dimensions, but of the same *intensity of magnetization* at corresponding places, the law becomes that of the cube, not of the fifth power. This may be easily seen by comparing two similar magnets, solenoidally magnetised. If their intensities of magnetisation are the same, the surface-densities of free magnetism are the same. But the law of the fifth power applies to *volume*-densities, hence there is a loss of two dimensions. Or, replace the surface-density, say σ, by a volume-density σ/t, t being the small thickness of the layer. An n-fold increase of linear dimensions with volume-density constant would increase the energy n^5 times, as before; but then, since t would be multiplied n times, σ would experience a like increase to keep σ/t the volume-density constant; consequently, the intensity of magnetisation, which is numerically the same as the surface density, would also vary as the linear dimensions. Thus, with the intensity of magnetisation kept constant, an n-fold increase of linear dimensions of a magnet makes the volume-density (replacing the surface-density) vary inversely as n. Hence, remembering that the energy varies as the square of the volume-density, we have the energy multiplied only n^5/n^2 or n^3 times.

Yet, although the n^3 law holds for magnets with intensity of magnetisation constant, it does not hold for current systems, with the density of current kept constant, even when we replace the currents by magnetic shells producing the same fields of force. For the strength of the equivalent shell in any case is its intensity of magnetisation multiplied by its thickness, and the strength must be numerically equal to the strength of the current it replaces. Now, the latter varies as n^2 with constant current-density; hence, if the thickness of the shell becomes n times as great in the magnified system, its intensity of magnetisation or the surface-density must be also multiplied n times to keep pace with the current it replaces. Thus we have the case of similar magnets whose intensities of magnetisation vary as the linear dimensions; or of similar distributions of matter of constant volume-densities, and therefore the n^5 law holds. I observe, however, that Sir W. Thomson, in the paper above referred to (*loc. cit.*, p. 435), makes the cube law hold both for magnets ("polar magnets") with constant intensities of magnetisation, *and* for current systems ("electromagnets," or systems of magnetic force due entirely to electric currents) with constant densities of current (or "intensities"). This may be a slip, or I may not catch the meaning correctly, or the n^5 law may not apply to currents as it appears to do.

As regards the *magnetic* force in similar systems of current, since at a point P it is the sum of $(Ch/r^3)\mathrm{b}_1 ds$ for each element of a linear circuit [p. 224, *ante*], r being the distance of P from the element, and h its distance from the axis of the current, h, r, and ds are all multiplied n times in a system of n-fold linear dimensions with the same integral current C in corresponding places; the magnetic force is therefore multiplied $n.n/n^3$ or $1/n$ times, and thus varies *inversely* as the linear dimensions. But under the same circumstances, with density of current constant, so that C varies as n^2, the magnetic force varies as n^2/n or n; *i.e.*, *directly* as the linear dimensions.

Now, the energy is expressible, as will be seen later, also in the form $\Sigma(8\pi)^{-1}(\text{force})^2$, the summation being extended throughout all space. Here $(\text{force})^2$ varies as n^2 under the last-mentioned circumstances, and the volume as n^3; hence we obtain the n^5 law again with density of current constant.

Leaving now the question of dimensions, let us pass on to the more general case of the mutual energy of two arbitrary current systems.

If we have a closed linear current C in presence of any number of others, C', C'', etc., we know exactly the value of M for our selected circuit C and any one of the others, say C', if the rest of the system were non-existent. It is numerically $C \times$ amount of magnetic force through it due to C', or conversely. Also, $-M$ is the amount of mechanical work that would have to be done by external force against the electromagnetic forces to separate C and C' from their actual positions to an infinite distance apart, keeping the currents constant. Now, this being the case for C and C' alone, and for C and C'' alone, when C' and C'' exist together, the mechanical work required to remove C is simply the sum of the amounts of work done separately in the former cases. It is, indeed, obvious that the amount of magnetic force that C' and C'' when co-existent send through C is the sum of the amounts they would separately send through C; and that the corresponding proposition is true for the work done follows from the mutual nature of the mechanical forces between the conductors, and is most easily seen to be true by replacing the currents by magnetic shells, by which we are reduced to mutual forces between sets of attracting or repelling points. In the case of the current, we may either remove C, fixing C' and C'', or fix C and remove C' and C'', or remove both C and the pair; but the pair must move as a rigid body without relative motion; or, if there be any, it must be cancelled later. And, in general, the energy of C with respect to any current-system is the sum of its energies with respect to the individual members of the system. And here remark that we have no concern with the mutual energies of the members of the system amongst themselves.

This being true for one closed current in presence of a system, is true for a second, or for any number; and, therefore, we have the result that the mutual energy of two arbitrary current systems (arbitrary in all except that they must consist entirely of closed currents) is the sum of the mutual energies of all the pairs that can be made, taking one member of a pair from one system, and the other from the second; and, since in neither system are we concerned with the internal energy, their mutual energy is the work to be done to bring them together to their actual positions moving as rigid bodies, without internal relative motion; or, if there be such, we must not count work so performed.

For the expression of this and other results in a simple manner, it will be convenient to introduce volume-elements of current, instead of the linear elements previously used. If we have a current in a wire, and consider the latter as a mere line, we have, of course, infinite concentration of current. There is no appreciable error caused by replacing wires by lines so long as the wires are not very close together, and there will be no error if we locate the lines properly; but when we are considering currents in general, it is best to take things as they are, viz., finite currents across finite areas.

A current system may be split up into tubes of flow, which are all closed. Any tube is independent of the rest, in so far that there is no current from it to its neighbours, for it is bounded by lines of flow.

Now, by splitting up the tubes into an indefinitely great number, we at the same time reduce their sections indefinitely, and also the currents they carry; hence in the limit we may treat every current-tube as a closed linear circuit carrying an infinitely small current. Let ds be an element of length of such a tube, dS its section anywhere, then $ds\,dS$ is an element of volume, say dV. Let, now, C be the current-density, *i.e.*, $C\,dS$ = current crossing dS, so that C would be the current across unit of area normal to the current were its density all over the unit area the same as at dS. Now, our linear element [p. 236] was integral current × element of length; this now becomes $(C\,dS)ds$ or $C\,dV$. Further, we may conveniently take the element of volume to be the unit volume, since the latter may be as small as we please. We now have, instead of equation (5) [p. 236], the following:—The mutual energy of two closed currents is the same as if every pair of unit volumes, one belonging to the first, and the other to the second circuit, contributed the amount

$$-\mathbf{CC}'/r = -\mathbf{AC}' = -\mathbf{A}'\mathbf{C}. \qquad (1)$$

Here, $\mathbf{A} = \mathbf{C} \div r$ is the vector-potential of $\mathbf{C}$ at $\mathbf{C}'$, and $\mathbf{A}' = \mathbf{C}' \div r$ is the vector-potential of $\mathbf{C}'$ at $\mathbf{C}$, r being their distance apart.

Now, apply (1) to two arbitrary current-systems. We know that (1) is true when summed up for two closed tubes of indefinitely small section, and that our systems may be split up into such tubes. It follows that when we divide space into elements of volume (ignoring the tubes completely), and apply (1) to every pair of volume-elements that can be made between the two systems, and sum up, we shall obtain the mutual energy, for as soon as we have finished the summation we have closed all the tubes.

Therefore, the mutual potential energy M of two systems of densities $\mathbf{C}$ and $\mathbf{C}'$, is

$$M = -\Sigma\Sigma\,\mathbf{CC}'/r = -\Sigma\,\mathbf{AC}' = -\Sigma\,\mathbf{A}'\mathbf{C}, \qquad (2)$$

where $\mathbf{A}$ is the complete vector-potential of the first, and $\mathbf{A}$ of the second system. In one form we have a double summation, viz., to all pairs of current-elements; in the other forms there is but one summation, because the vector-potential itself is obtained by a summation. Thus, in the form $-\Sigma\,\mathbf{AC}'$, we sum up $\mathbf{AC}'$ for every element of the second system; here $\mathbf{A}$ expresses a summation over the first system, viz., $\Sigma\,\mathbf{C}/r$ (vector sum).

Corresponding to (2) these are exactly analogous expressions for the mutual energy of two systems of electrification, and the comparison is instructive. ρ and ρ' being the volume-densities of free electricity, P and P' the potentials, and M the energy, we have

$$M = \Sigma\Sigma\,\rho\rho'/r = \Sigma\,P\rho' = \Sigma\,P'\rho. \qquad (3)$$

Here ρ, the density of electrification, corresponds to $\mathbf{C}$, the current-density in (2), and P the scalar potential to $\mathbf{A}$, the vector-potential. Both ρ and P are scalars, and their product is found as in algebra. But in (2) both the currents and their potentials are vectors, which is how it comes about that we have to multiply by the cosine of the angle between their directions. As for the negative sign in (2) and positive in (3), it arises from similarly directed currents attracting, whilst similar electrifications repel.

The following rule is useful in adding the scalar products of vectors. If $\mathbf{A}_1$, $\mathbf{A}_2$, $\mathbf{A}_3$, ... are any vectors whose vector-sum is $\mathbf{A}$, and $\mathbf{C}$ is any other vector, then

$$\mathbf{A}_1\mathbf{C} + \mathbf{A}_2\mathbf{C} + \mathbf{A}_3\mathbf{C} + \ldots = \mathbf{AC}.$$

[Similarly $\mathbf{C}$ may be split up into any number of vectors $\mathbf{C}_1$, $\mathbf{C}_2$, ... ; from which we see that in the formation of scalar products of pairs of vectors, each of which is the sum of any number of other vectors, we proceed precisely as in common algebra. For example,

$$\begin{aligned}(\mathbf{A}_1 + \mathbf{A}_2)(\mathbf{C}_1 + \mathbf{C}_2) &= \mathbf{A}_1\mathbf{C}_1 + \mathbf{A}_1\mathbf{C}_2 + \mathbf{A}_2\mathbf{C}_1 + \mathbf{A}_2\mathbf{C}_2\\ &= \mathbf{C}_1\mathbf{A}_1 + \mathbf{C}_2\mathbf{A}_1 + \mathbf{C}_1\mathbf{A}_2 + \mathbf{C}_2\mathbf{A}_2.\end{aligned}$$

The practical utility of the algebra of vectors is easily seen.]

SECTION III. THE SELF-ENERGY OF A CURRENT-SYSTEM.

We have next, in natural order, to consider the potential energy of a current-system on itself, as distinguished from its energy with respect to another system. The latter being called the mutual energy, we may, for brevity, term the former the self-energy, just as the induction of a current on itself is called self-induction. The meaning of self-energy follows plainly from that of mutual energy; and hence, after the method previously given [p. 240,] we may define the self-energy (potential) of a current-system to be the amount of work spent by externally applied forces against the electromagnetic forces in building up the system by bringing together to their required positions from an infinite distance all the infinitely fine tubes of current into which the system may be split up, each infinitely small current being of constant strength. Now this amount is always negative (as is plain in simple cases, and is proved for an arbitrary system below), though the mutual energy of two systems may be positive or negative; hence the self-energy (potential), with sign changed, is the work needed to pull the system to pieces against the electromagnetic forces, finally bringing it to a state of division into infinitely fine tubes, infinitely widely separated.

The calculation may be made by the same method of pairing tubes as for the mutual energy, with one important difference, however, which, when attended to, allows the method to give us the formula for the self-energy. Divide the system into, say, n tubes, and remove them, one at a time, to an infinite distance. Or, simply turn any tube in the act of removal into such a position that the total amount of magnetic force through it arising from that part of the system which remains is *nil*, and then cancel it. The required total amount of work performed is obviously the sum of the mutual energies, first of No. 1 tube with respect to the remaining $(n-1)$; then of No. 2 with respect to the remaining $(n-2)$, and so on. Thus, in pairing the tubes, we count each once only with respect to all the rest, or we form all the different pairs that can be made in the system.

Now, imagine there to be another system, exactly equal to the first, and coincident therewith. In calculating their *mutual* energy, we should form all the pairs that can be made, taking one member from the first, and the other from the second system. Now, we have, for example, the mth tube in the first paired with the nth in the second system, and also the nth in the first paired with the mth in the second. But the energies corresponding to these two pairs are equal, by the assumption that the systems are exactly equal. So the pairing is the same as if, in the previous case (for the self-energy), we had counted every pair twice over, with only this difference, that we have additional pairs 1,1′, 2,2′, 3,3′, etc., the accents referring to the imagined system. But the sum of the energies corresponding to these pairs, when the subdivision is carried to extremes, becomes infinitely small compared with the sum of all the rest. Hence we have the result that the self-energy of a system is exactly one-half its energy with respect to an exactly equal coincident system. The latter we know to be $\Sigma\,\mathbf{AC}$, where $\mathbf{C}$ is the current and $\mathbf{A}$ the vector-potential of either system. Consequently, if M_1 be the self-energy, we have

$$-M_1 = \Sigma\,\tfrac{1}{2}\mathbf{AC}. \qquad (1)$$

Any other method, properly carried out, will lead to the same result. Thus, first divide the system into two coincident systems, each of half the strength, and, therefore, having the vector-potential halved as well as the density of current, and separate them. The work done is $\frac{1}{2}\,.\,\frac{1}{2}\Sigma\,\mathbf{AC}$, being their mutual energy, with sign changed. Next, similarly divide the two halves into systems of a quarter the current-density and vector-potential, and separate as before. The work done is $2\,.\,\frac{1}{4}\,.\,\frac{1}{4}\Sigma\,\mathbf{AC}$. In the third similar operation the current and vector-potential are $(\mathbf{C}\div 8)$ and $(\mathbf{A}\div 8)$ and there are four pairs; hence the work done amounts to $4\,.\,\frac{1}{8}\,.\,\frac{1}{8}\Sigma\,\mathbf{AC}$. Carry this on *ad infinitum*, and sum up the amounts of work spent. The result is

$$\Sigma\,\mathbf{AC}(\tfrac{1}{4}+\tfrac{1}{8}+\tfrac{1}{16}+\dots) = \Sigma\,\tfrac{1}{2}\mathbf{AC}, \text{ as before,}$$

since the sum of the series $=\frac{1}{2}$.

Collecting results, let M_1 and M_2 be the self-energies, M_{12} the mutual energy of two arbitrary systems, in which $\mathbf{C}_1$ and $\mathbf{C}_2$ are the currents, and $\mathbf{A}_1$ and $\mathbf{A}_2$ the vector-potentials, and let M be the complete energy. Then we have

$$-M = -(M_1+M_2+M_{12}) = \Sigma\,\tfrac{1}{2}\mathbf{A}_1\mathbf{C}_1 + \Sigma\,\tfrac{1}{2}\mathbf{A}_2\mathbf{C}_2 + \Sigma\,\mathbf{A}_1\mathbf{C}_2, \;\dots\dots \quad (2)$$

where, instead of the last term, we may write $\Sigma\,\mathbf{A}_2\mathbf{C}_1$.

We may easily show that (2) is consistent with (1) by regarding the two systems as a single one. For the actual vector-potential at any point is the vector-sum of $\mathbf{A}_1$ and $\mathbf{A}_2$, say $\mathbf{A}$. Now if we substitute for the last summation in (2) one half of $(\Sigma\,\mathbf{A}_1\mathbf{C}_2 + \Sigma\,\mathbf{A}_2\mathbf{C}_1)$, which we may do because the two sums are equal, and rearrange terms, we get

$$-M = \Sigma\,\tfrac{1}{2}(\mathbf{A}_1\mathbf{C}_1+\mathbf{A}_2\mathbf{C}_1) + \Sigma\,\tfrac{1}{2}(\mathbf{A}_1\mathbf{C}_2+\mathbf{A}_2\mathbf{C}_2) = \Sigma\,\tfrac{1}{2}\mathbf{AC}_1 + \Sigma\,\tfrac{1}{2}\mathbf{AC}_2.$$

[See rule on p. 242]. Therefore, if the systems nowhere overlap, this is reduced to $\Sigma \frac{1}{2}\mathbf{AC}$, extending the summation over both systems. And when they do overlap it is the same. For in the overlapping parts the actual current is the vector-sum of $\mathbf{C}_1$ and $\mathbf{C}_2$, hence we are reduced to $\Sigma \frac{1}{2}\mathbf{AC}$ always, agreeing with equation (1) for the self-energy of a system.

There is another very important and significant expression for the energy, which involves a transformation of an extraordinary character. Expressed generally, it amounts to this:—Let **A** be any vector function of no convergence, **B** its curl, and **C** that of **B**, then

$$\Sigma\,\mathbf{AC} = \Sigma\,\mathbf{B}^2, \qquad (3)$$

the summations being taken throughout all space; or, which comes to the same thing, the first summation extended to all places where **C** exists, and the second over all space where **B** exists. **A**, **B**, **C** are here defined exactly as for the volume-element of a current, viz., amount crossing unit of area × volume, and taking the volume-element as the unit volume. By making use of the fundamental theorem connecting a vector and its curl, and by following the tubes of flow in our summations, the proposition admits of proof in a manner which makes evident how the transformation takes place.

Since **B** and **C** are non-converging, these systems may be completely divided into tubes of flow (*i.e.*, of flow as applied to a liquid), which are all closed. We shall first prove (3) in the case of a single tube of **C**, which may have any form, but is of indefinitely small section; *i.e.*, $C=0$ everywhere except along a certain closed channel, across every section of which we have a constant amount of C, which we may call C_1. Let B_1 and A_1 be the corresponding values of A and B. ($\mathbf{A}_1$ is the vector-potential of $\mathbf{C}_1$, and $\mathbf{B}_1$ the curl of $\mathbf{A}_1$, as before shown.) We know by the fundamental relation of Theorem (B), [p. 211], that the line-integral of $\mathbf{B}_1$ once round any closed curve = whole amount of $\mathbf{C}_1$ through the curve; hence the line integral is zero unless the tube of $\mathbf{C}_1$ goes through it (the closed curve). Now, the system $\mathbf{B}_1$ is wholly made up of closed tubes, and since, if we integrate $\mathbf{B}_1$ once round one of them, we always get a finite result (because the direction of $\mathbf{B}_1$ is along the tube), it follows that *all* the tubes of $\mathbf{B}_1$ pass through the circuit of $\mathbf{C}_1$ and complete their circuits outside.

Let any surface be described whose edge is bounded by the circuit of $\mathbf{C}_1$; then, by what was said last, it cuts all the tubes of $\mathbf{B}_1$. Next, let it be so situated as to cut them at right angles, and describe a series of other surfaces indefinitely close together, all similar to the first in that they are bounded by the circuit of $\mathbf{C}_1$ and cut the tubes $\mathbf{B}_1$ perpendicularly. This is possible, because in space where $\mathbf{B}_1$ has no curl, that is, wherever $C=0$, or everywhere except within the tube of $\mathbf{C}_1$, $\mathbf{B}_1$ may be derived from a scalar potential; and these surfaces are the equipotential surfaces. (We need not consider any correction in the following due to $\mathbf{B}_1$ within $\mathbf{C}_1$, because we take that space infinitely small.)

We have to prove that $\Sigma\,\mathbf{A}_1\mathbf{C}_1 = \Sigma\,\mathbf{B}_1^2$. Now the summation of $\Sigma\,\mathbf{A}_1\mathbf{C}_1$ extends only to the tube of $\mathbf{C}_1$. Hence

$$\Sigma\,\mathbf{A}_1\mathbf{C}_1 = C_1 \times \text{line-integral of } \mathbf{A}_1 \text{ once round } \mathbf{C}_1; \qquad (4)$$

(not meaning round $\mathbf{C}_1$ in the sense that a tube of $\mathbf{B}_1$ goes round $\mathbf{C}_1$, but *along* the tube of $\mathbf{C}_1$). But $\mathbf{B}_1 = \text{curl } \mathbf{A}_1$, therefore (4) becomes, by Theorem (B), [p. 211],

$$\Sigma \mathbf{A}_1\mathbf{C}_1 = C_1 \times \text{integral of } \mathbf{B}_1 \text{ over any surface bounded by } \mathbf{C}_1. \quad \ldots (5)$$

But again, $\mathbf{C}_1 = \text{curl } \mathbf{B}_1$, therefore, by Theorem (B) again,

$$\mathbf{C}_1 = \text{integral of } \mathbf{B}_1 \text{ once round any tube of } \mathbf{B}_1. \quad \ldots\ldots\ldots\ldots (6)$$

Hence, by substituting (6) in (5) we find

$$\Sigma \mathbf{A}_1\mathbf{C}_1 = \text{line-integral of } \mathbf{B}_1 \times \text{surface-integral of } \mathbf{B}_1. \quad \ldots\ldots\ldots (7)$$

It is important to notice that *any* tube of $\mathbf{B}_1$ will do for the line-integral, and *any* surface bounded by $\mathbf{C}_1$ for the surface-integral. But all the tubes of $\mathbf{B}_1$ are cut by the surface, and when it is an equipotential one they are cut perpendicularly. The same being true for the whole series of equipotential surfaces, it follows that the product of the line and surface-integrals in (7) includes every place where $\mathbf{B}_1$ exists; hence we may ignore the tubes completely, and write

$$\Sigma \mathbf{A}_1\mathbf{C}_1 = \Sigma \mathbf{B}_1^2,$$

extending summations through all space; whence the proposition is proved in the case of an infinitely fine tube of $\mathbf{C}$.

Now, in the general case of an arbitrary system, divide $\mathbf{C}$ into its tubes, and consider one of them, say, $\mathbf{C}_1$, as before. We have $\mathbf{A}$ and $\mathbf{B}$ instead of $\mathbf{A}_1$ and $\mathbf{B}_1$, and $\Sigma \mathbf{A}\mathbf{C}_1$ instead of $\Sigma \mathbf{A}_1\mathbf{C}_1$; hence, by the previous reasoning,

$$\begin{aligned} \Sigma \mathbf{A}\mathbf{C}_1 &= C_1 \times \text{line-integral of } \mathbf{A} \text{ round } \mathbf{C}_1, \\ &= C_1 \times \text{surface-integral of } \mathbf{B}, \quad \ldots\ldots\ldots\ldots\ldots\ldots (8) \end{aligned}$$

the latter being over any surface bounded by $\mathbf{C}_1$. Divide space as previously by means of the equipotential surfaces of $\mathbf{B}_1$ and the tubes of $\mathbf{B}_1$ (not of $\mathbf{B}$). The tubes of $\mathbf{B}$ will not, save exceptionally, cut the surfaces at right angles, hence we have to take the normal component of $\mathbf{B}$ in the surface integral in (8); that is, $B \cos (\mathbf{B}\mathbf{B}_1)$. Also $\mathbf{C}_1$ has the same relation to $\mathbf{B}_1$ as before, therefore, by (6) and (8),

$$\left.\begin{aligned} \Sigma \mathbf{A}\mathbf{C}_1 = \text{line-integral of } \mathbf{B}_1 \\ \times \text{ surface-integral of } B \cos (\mathbf{B}\mathbf{B}_1). \end{aligned}\right\} \quad \ldots\ldots\ldots\ldots (9)$$

But $BB_1 \cos (\mathbf{B}\mathbf{B}_1) = \mathbf{B}\mathbf{B}_1$, and the space integrated over is as before; therefore,

$$\Sigma \mathbf{A}\mathbf{C}_1 = \Sigma \mathbf{B}\mathbf{B}_1 \ldots\ldots\ldots\ldots\ldots\ldots\ldots\ldots (10)$$

Similarly for a second tube of $\mathbf{C}$, say $\mathbf{C}_2$, we shall have

$$\Sigma \mathbf{A}\mathbf{C}_2 = \Sigma \mathbf{B}\mathbf{B}_2;$$

and when we do the same for all the tubes of $\mathbf{C}$, and add, we get

$$\Sigma \mathbf{A}\mathbf{C} = \Sigma (\mathbf{B}\mathbf{B}_1 + \mathbf{B}\mathbf{B}_2 + \mathbf{B}\mathbf{B}_3 + \ldots).$$

But $\mathbf{B}$ is the vector sum of $\mathbf{B}_1$, $\mathbf{B}_2$, &c.; whence the bracketed quantity in the last equation $= \mathbf{B}\mathbf{B} = \mathbf{B}^2$; therefore, finally,

$$\Sigma \mathbf{A}\mathbf{C} = \Sigma \mathbf{B}^2 \ldots\ldots\ldots\ldots\ldots\ldots\ldots\ldots (11)$$

for any arbitrary system.

I think it was a philosopher who propounded the theory that men always thought in some language; an Englishman in English, for example; always, if he knew no other language; otherwise he might think in any one he knew. Not to raise the obvious objection that persons dumb from birth should, according to this, have no thoughts at all, the theory is certainly proved to be false by an examination of such a transformation as the above. As regards the case of a single tube of **C**, if only the geometrical conditions are pictured in the mind, the division of space into small cubes by the tubes of $\mathbf{B}_1$ cut across by the equipotential surfaces of $\mathbf{B}_1$, the transformation becomes as self-evident as an axiom, and no form of words or sentences is necessary. The less one is cumbered with them the better. And although, the extension to an arbitrary system is less easy, it is still easier to be pictured than logically demonstrated. The transformation might have been seen quite intuitively; it is only when one has to prove it to some one else that clothing the thoughts in words becomes necessary; and, even then, the clothes do not correspond to the original thoughts, but to those arising in the act of description, and both words and thoughts require to be readjusted, perhaps two or three times, before they will mutually fit with any decency. The Cartesian transformation, breaking up each of the vectors **A**, **B**, **C** into three rectangular components, is short enough, but is gifted with a total absence of visible reason and significance.

In applying (11) to electromagnetism, we have to remember that, **B** standing for the magnetic force, **C** for current, and **A** for the vector-potential, we have $\mathbf{B} = \text{curl } \mathbf{A}$, but $4\pi\mathbf{C} = \text{curl } \mathbf{B}$. Consequently the potential self-energy of a current-system is, by (1) and (11)

$$-\Sigma \tfrac{1}{2}\mathbf{AC} = -\Sigma \mathbf{B}^2/8\pi. \qquad (12)$$

Now, go back to (10). The left-hand member is the mutual potential energy with sign changed of an arbitrary current-system whose vector-potential is **A**, and of a single tube of current $\mathbf{C}_1$. In the same manner as (11) followed from (10) we may easily show that the mutual potential energy of two arbitrary systems denoted by the suffixes 1 and 2 is

$$-M_{12} = -\Sigma \mathbf{A}_1\mathbf{C}_2 = -\Sigma \mathbf{A}_2\mathbf{C}_1 = -\Sigma \mathbf{B}_1\mathbf{B}_2/4\pi. \qquad (13)$$

Hence, in terms of the magnetic force, equation (2), for the complete energy of two systems, becomes

$$-M = \Sigma \mathbf{B}_1^2/8\pi + \Sigma \mathbf{B}_2^2/8\pi + \Sigma \mathbf{B}_1\mathbf{B}_2/4\pi,$$

which is obviously reducible to $\Sigma(\mathbf{B}_1 + \mathbf{B}_2)^2/8\pi$ or $\Sigma \mathbf{B}^2/8\pi$, if **B** be the actual resultant magnetic force.

The expression $\Sigma \mathbf{B}^2/8\pi$, that is, the potential energy with sign changed, may be proved, though not without reference to induction phenomena, to be the Energy of a current-system, or its capacity for performing work in various ways. The form in terms of the square of the magnetic force, due originally to Sir W. Thomson, is of great significance in the theory of action through an intervening medium, as opposed to action at a distance.

Section IV.—Probable Localization of the Energy. Division of any Vector into a Circuital and a Divergent Vector.

We have arrived at three distinct expressions for the potential energy of currents, involving line, surface, and volume integrations. Thus, confining ourselves first to a pair of linear circuits, we started with a surface-integral, derived from the properties of magnetic shells, viz., the integral amount of magnetic force passing through one circuit due to the current in the other. This amount, multiplied by the strength of current in the circuit for which the integral is reckoned, is the quantity upon whose variations the mutual forces depend. Now, the idea of magnetic force as a vector or directed quantity has become so widely spread and utilised that it might appear that the expression of the energy in terms of the number of lines of force through a circuit was a very natural one. In all modern explanations relating to magneto and dynamo machines, the lines of force are much employed. Yet this form for the energy is very artificial, and, in that respect, is like some other forms. For, although we may frequently merely consider the lines of force of the field in immediate proximity to a conductor, and the manner in which they cross the wire when it is moved, or when the source of the field is moved, thus producing variations in the integral amount of force (or of magnetic induction, to use the more general term, when induced magnetisation, notably that of iron, not here considered, is operative) through a circuit, yet to reckon up the integral amount a knowledge of the strength and direction of the magnetic force just about the wire is not sufficient. We require strictly to know the magnetic force all over an imagined surface, with its edge coinciding with the linear circuit. Choose one side for positive, one direction through it for the positive direction, and then by integration over the surface find the excess of the amount of force going through it in one over that in the reverse direction. This process may become singularly complex when the circuit has not a simple form, owing to the extraordinarily involved character of the surface, through which a selected line of force may pass again and again. In practical applications, however, we may simplify the process. Thus, in the case of a coil of closely-packed windings, say n in number, since each winding forms nearly a closed curve in itself, we may imagine a separate surface for every turn of wire, thus reducing the problem without sensible error to that of n distinct circuits of simple forms placed in a magnetic field. The sum of the amounts of force through the individual surfaces will plainly be almost precisely the same as through a single surface whose edge is bounded by the real circuit.

By a transformation we next obtain the mutual energy in terms of line-integrations following the course of the current, with a result most succinctly expressed in Neumann's formula, equation (6), [p. 236], and we are able to recognize the very simple manner in which the different current elements may be considered to contribute to the result; and, following this up, we arrive at simple expressions for the mutual and self-energy of arbitrary distributions of current in terms of the currents and their vector-potentials, which are exactly analogous to the formulæ for electrostatic energy. And, by another transformation, equation (12), [p. 246], we obtain the energy in terms of the magnetic force alone.

Now, if we change the sign of the quantity we have called the potential energy of a current, which is always negative (being $= -\Sigma \mathbf{B}^2/8\pi$, equation (12), [p. 246] we obtain a quantity which is necessarily always positive. Calling it T, so that we have

$$T = \Sigma \tfrac{1}{2}\mathbf{AC},$$

where $\mathbf{C}$ is the current and $\mathbf{A}$ the vector-potential, it will be shown later that T expresses the capacity the system has for performing work, which may be of various kinds, in virtue of the existence of the current, so that T may be strictly called the Energy of the system. Taking this for granted for the present, in order not to enter upon inductive phenomena, we may, however, remark in passing that we calculated the potential energy $(-T)$ in exactly the same manner as electrostatic energy may be calculated, from the mutual forces, to wit. P being the electrostatic potential, and ρ the density of electrification, $\Sigma \frac{1}{2}P\rho$ is the potential energy of the electrification, corresponding to $-\Sigma \frac{1}{2}\mathbf{AC}$. Now, $\Sigma \frac{1}{2}P\rho$ really represents the capacity the electrified system has for performing work, that is, its energy. It is always and necessarily positive, and, when referred to the dielectric medium, may be considered to express the potential energy arising from elastic forces due to displacements of some kind. The fact that the potential energy M, in the electromagnetic case, is negative (resulting from attraction of similarly directed currents) is sufficient to show that it cannot represent the capacity for work of the current-system. Yet the variations of M do represent the mechanical forces. So do those of T, any increase of one being equivalent to an equal decrease of the other. Therefore T might be the energy. But it could not be potential energy, for that tends to decrease: for instance, two movable circuits tend to move so that their potential energy decreases, with currents constant, and, therefore, so as to increase T. If, then, T be the energy, and it cannot be potential, it must be kinetic. And it is true that, considered as kinetic energy, it follows from strictly dynamical principles that T would tend to increase (with unchanged currents) by the motions resulting from the actions upon movable parts. Without, then, any knowledge of inductive phenomena, we might hazard the conjecture that T represented the energy, and that it was kinetic; and from the properties of kinetic energy all the laws of induction would follow.

At present, however, let us merely examine how the different expressions for T tend to locate the energy. The original form (surface-integrals) is quite out of the question, even for a single linear circuit, to say nothing of an arbitrary system of current, whether in wires or diffused in large conductors.

The form $T = \Sigma \frac{1}{2}\mathbf{AC}$, however, is more definite. Here any unit volume where the density of the current is $\mathbf{C}$ contributes $\frac{1}{2}\mathbf{AC}$ to the total, and of course space where there is no current contributes nothing. The summation extends throughout the current. Now, we know that there is something going on in a wire conveying a current; we know, for instance, that there is a transfer of energy from a battery or other source of electricity; therefore, so far, it would not be unreasonable to suppose the energy resides where the current resides. But observe that, upon this view, the energy at any spot would depend not merely upon the current there, but also upon the vector-potential there, and the latter depends upon the state of the whole system. Now, the energy at any place, whether potential arising from elastic displacements of matter tending to return to neutral positions, or kinetic, due to a motion going on, would obviously depend on the displacements or motions at the place considered, and not upon those in all parts of the system. Evidently, then, the expression $\Sigma \frac{1}{2}\mathbf{AC}$ does not locate the energy properly. Notice, also, that although the total is positive, yet the portion $\frac{1}{2}\mathbf{AC}$ in any particular unit volume would be positive or negative, according as the current and vector-potential were similarly or oppositely directed, and zero should they be perpendicular. We may therefore dismiss the idea of $\frac{1}{2}\mathbf{AC}$ representing the energy per unit volume, or the density of the energy.

But we have still the form $\Sigma \mathbf{B}^2/8\pi$ at disposal. This is identically the same in amount as $\Sigma \frac{1}{2}\mathbf{AC}$, but indicates a very different distribution of energy. In a unit volume we have the amount $\mathbf{B}^2/8\pi$, where $\mathbf{B}$ is the strength of magnetic force there. Being a square, it is always positive. Thus, not only have we the total positive, but every element of the sum is positive as well. And the energy at any place depends only upon the square of the magnetic force at the place, and not upon the state of all the rest of the system.

The conclusion is irresistible that we have got an expression for the energy which may correctly locate it in amount at different places. As to its distribution, although particular arrangements of current may be such as to leave certain spaces without magnetic force, yet, in general, the latter extends throughout all space. The portion of the energy residing in the space occupied by the current may be only a small fraction of the whole amount. And, even then, it depends on the magnetic force, and not upon the current-density; although, of course, there is connection between the force and current in other ways.

Examples.—In the case of a closed solenoid there is no external magnetic force. And if the current-layer be thin compared with the diameter nearly all the energy resides in the space enclosed by the current, and next to none in the current itself. In the case of a long, straight solenoid the internal energy is very great compared with the external; the force is uniform, of strength $4\pi C$, where C = current across unit length of the solenoid, except near the ends where the force falls off; hence the energy in any part of the internal space not near the ends is proportional to the volume of the part.

When, then, we set up a current in a conductor, a transmission of energy outwards at once begins, and not until it is completed does the current get steady. Theoretically it never gets quite steady; space is boundless, and the transmission of energy outwards never quite ceases; but, in general, the permanent state is, practically, reached very quickly, and then we have a definite amount of energy in every place where the magnetic force extends, falling off in density of distribution rapidly as we recede from the current. A medium of some kind to receive the energy is, of course, necessary.

If the energy be kinetic some kind of motion must be going on where there is magnetic force. One suggested form is a rotation of matter about the lines of force as axes. This should be of a frictionless character, for there is no loss of energy in a steady current except what can be accounted for in heating the conductor or by other work done. It is as if we had a frictionless fly-wheel, or immense number of fly-wheels somehow set moving by the establishment of a current in a circuit, and possessing in virtue of their motion a store of energy which can be afterwards utilised. The supply of energy from the battery may be stopped, and then the reserve store comes into action, is returned to the same wire, or to other conductors in its neighbourhood, creating the phenomena of induced currents.

But in the absence of reason to think the energy kinetic, we could equally well consider it potential, with the same distribution in space. We might, for instance, replace the steady motions of rotation last mentioned by displacements in the direction of the magnetic force, or by the same rotations as before, but stopped by counteracting elastic forces brought into play. The energy would be similar to that of a bent spring. On the removal of the E.M.F. that kept up the current, the forced state would be relaxed, the displacement cease, and the energy be set free again.

We may now notice some remarkable properties of the energy considered merely as expressing the quantity $\Sigma\,\mathbf{B}^2/8\pi$ or $\Sigma\,\frac{1}{2}\mathbf{AC}$.

Let there be a given distribution of current $\mathbf{C}_1$; corresponding thereto we have a definite distribution of magnetic force $\mathbf{B}_1$, and also of vector-potential $\mathbf{A}_1$. Briefly, they are related thus,

$$\mathbf{A}_1 = \Sigma\,\mathbf{C}_1/r,$$
$$\mathbf{B}_1 = \text{curl}\,\mathbf{A}_1 = \Sigma\,\mathbf{D}_1/r,$$

(if $\mathbf{D}_1 = \text{curl}\ \mathbf{C}_1$); and

$$4\pi\mathbf{C}_1 = \text{curl}\,\mathbf{B}_1.$$

All these relations have been fully explained. In addition we have $T_1 = \Sigma\,\mathbf{B}_1{}^2/8\pi$, and $\mathbf{B}_1$ being definite, so is T_1.

Consider the effect upon T_1 of altering the magnetic force in any arbitrary manner, that is, $\mathbf{B}_1$ at any place is to be changed to some other value, say $\mathbf{B}$, differing in general from $\mathbf{B}_1$ both in direction and magnitude. We may make the alteration by bringing in another magnetic field, say $\mathbf{B}_2$, of an arbitrary nature. At any point the actual magnetic force $\mathbf{B}$ will be the vector-sum of $\mathbf{B}_1$ and $\mathbf{B}_2$. Now T_1 becomes T, where

$$T = \Sigma(\mathbf{B}_1 + \mathbf{B}_2)^2/8\pi = \Sigma\,\mathbf{B}_1^2/8\pi + \Sigma\,\mathbf{B}_2^2/8\pi + \Sigma\,\mathbf{B}_1\mathbf{B}_2/4\pi, \quad \text{........} \quad (1)$$
$$= \quad T_1 \quad + \quad T_2 \quad + \quad T_{12} \text{ say.}$$

The energy is increased by T_2+T_{12}. Here T_2 is the energy of the system $\mathbf{B}_2$ by itself, and is essentially a positive quantity. But T_{12} containing products may be either positive or negative, so that whether the energy is increased or decreased is quite indefinite so far.

But now, instead of letting $\mathbf{B}_2$ be quite arbitrary, let it be subjected to the condition that it shall not alter the current. We can then evaluate T_{12}. The original current $\mathbf{C}_1$ must be still $\mathbf{C}_1$, after the superposition of the second field, and no fresh current must be introduced in other places. $\mathbf{C}_1$ being completely defined by the curl of $\mathbf{B}_1$, it is evident that the condition imposed upon $\mathbf{B}_2$ is that it shall have no curl anywhere.

If we have two *current* systems, we have seen that [equation (13), p. 246],

$$\Sigma\, \mathbf{B}_1\mathbf{B}_2 = 4\pi\, \Sigma\, \mathbf{A}_1\mathbf{C}_2,$$

where $\mathbf{B}_1$ and $\mathbf{B}_2$ are the magnetic forces, $\mathbf{A}_1$ the vector-potential of one current, and $\mathbf{C}_2$ the other current. Therefore, if $\mathbf{C}_2=0$, which involves curl $\mathbf{B}_2=0$, we shall have $\Sigma\, \mathbf{B}_1\mathbf{B}_2=0$. This is suggestive, but does not exactly correspond to the circumstances considered in equations (1) above. For $\mathbf{C}_2=0$ involves $\mathbf{B}_2=0$; whereas in (1) $\mathbf{B}_2$ is not to vanish, but merely to have no curl. Otherwise it may be arbitrary. Nevertheless, it may be readily shown that $\Sigma\, \mathbf{B}_1\mathbf{B}_2$ really vanishes in (1) by considering the imposed property of $\mathbf{B}_2$. When we say that its curl is nothing everywhere, we imply that nowhere can any closed curve be described so that the integral of $\mathbf{B}_2$ once round it differs from zero. Now, $\mathbf{B}_1$ consists entirely of closed tubes. Select one of them, and let it be of infinitely small section. The portion of $\Sigma\, \mathbf{B}_1\mathbf{B}_2$ belonging to this tube is

$$B_1 \times \text{section of tube} \times \text{integral of } \mathbf{B}_2 \text{ once round its length.}$$

But $B_1 \times$ section is a constant for the same tube, and the last factor is zero, hence the portion of $\Sigma\, \mathbf{B}_1\mathbf{B}_2$ is zero for that tube. Similarly it is zero for any other, and for all, and it follows that the whole summation

$$\Sigma\, \mathbf{B}_1\mathbf{B}_2=0, \qquad \text{or} \qquad T_{12}=0.$$

Consequently (1) becomes

$$T=T_1+T_2.$$

Hence, if we alter the magnetic force of a current in any manner consistent with keeping the current the same, the energy is invariably *increased;* for T_2 is the sum of squares, and the products have gone out; *i.e.*, out of the infinite number of distributions of magnetic force which have the same curl $4\pi\mathbf{C}_1$, the real distribution is that one which makes the energy an absolute minimum.

There are remarkable differences between the two fields of force signified by $\mathbf{B}_1$ and $\mathbf{B}_2$. $\mathbf{B}_1$ has no convergence anywhere; that is its fundamental property. Its lines of force are all closed curves. It necessarily has curl somewhere. Quite apart from our having specified $\mathbf{B}_1$ to have curl, we may show that if it have no convergence it must have curl. For, if the integral of $\mathbf{B}_1$ be taken once round a closed line of force, thus always going with the force, or always against it, and its amount, necessarily differing from zero, be noted, and then we do the same for another closed line of force close to but within the first, the amount must be the same or different. If different, there is evidently curl in the space passed over in transferring one line of force to the other, and we need proceed no further. If the same, there is no curl on the whole. If so, we may go on to a third line of force within the second, and so on till we either find the integral change its value, or in the extreme case find ourselves reduced to a curve bounding an infinitely small area, with the same finite value of the integral, so that we have infinite density of curl. In electromagnetism, the curl corresponds to $4\pi \times$ current-density, so that the existence of closed lines of force involves the existence of current somewhere, in fact passing through the closed lines.

Now consider $\mathbf{B}_2$. It must have no curl anywhere. Describe any closed curve in the field. It cannot be made to coincide with a single line of force, for that would give curl at once. The integral of $\mathbf{B}_2$ must be *nil*. Hence, since there is not a single closed line of force in the field, every line of force has a beginning and an end, and the system may be completely divided into tubes which are all terminated, or unclosed, or open. (In order to exclude getting infinite values for the energy, it should be understood that $\mathbf{B}_1$ and $\mathbf{B}_2$ vanish at infinity. And in the case of unclosed lines of force going out to infinity, we may terminate them upon an imaginary large surface enclosing the practical field of force.)

The field $\mathbf{B}_2$ corresponds to that of a permanent magnet on the theory of magnetic matter, or north and south magnetism. This theory gives but an imperfect view of the force of a magnet, but just answers our purpose here. In a bar magnet uniformly magnetised there is positive magnetism at one end and negative at the other; and, in general, there may with irregular magnetisation be both surface and internal distributions of magnetism. The lines of force go from the positive to the negative magnetism always, and are thus all terminated or unclosed. The line-integral of force round any close curve is always *nil*. The amount of magnetism anywhere is measured by the amount of convergence or divergence of the force, which convergence and divergence really constitute all the evidence there is of the existence of the matter.

Now, we may arrange our magnetism anyhow, the only restriction being that there is just as much positive as negative, which, interpreted, means that a line which is not closed on itself must have two ends (which nobody can deny); hence we can produce a field of force $\mathbf{B}_2$ which is quite arbitrary, save the restriction that $\mathbf{B}_2$ has no curl anywhere. And since we can arrange electric currents anyhow with the sole restriction that any line of current must have no ends, or be closed, we can produce another field of force $\mathbf{B}_1$, which is arbitrary, subject to $\mathbf{B}_1$ having no convergence. Now superimpose the two fields, the currents and the magnets adding their forces. The result is an unconditionally arbitrary field of force, which has both curl and convergence.

Given, then, an arbitrary field **B** of unknown source, we may immediately divide it into two fields—one due to current, the other to magnetism. There is only one way of effecting the division, and no other course. For, measure the curl of **B** everywhere and construct the field, say $\mathbf{B}_1$, which has just that curl and no convergence. Deduct the field $\mathbf{B}_1$ from the field **B**; the residual field, say $\mathbf{B}_2$, has obviously no curl, and has, therefore, convergence. Or we may start by finding the convergence of **B**, and construct the field, which has the same convergence but no curl; this will be the same $\mathbf{B}_2$ as before, and, deducting it from **B**, will leave a field $\mathbf{B}_1$, which has no convergence, and which, therefore, has curl. $\mathbf{B}_1$ is due to current, $\mathbf{B}_2$ to magnetism.

Now, the arbitrary field **B** might be divided into two fields in any number of ways, and in general equation (1) above would hold, it being, however, now permissible for both $\mathbf{B}_1$ and $\mathbf{B}_2$ to have curl and convergence as well as **B**, and T_{12} would not vanish. But of all these ways there is just one that makes T_{12} vanish, and when that is got there is a perfect separation of the closed from the unclosed tubes of force, the curl is confined to $\mathbf{B}_1$ and the convergence to $\mathbf{B}_2$. These properties do not belong to magnetic force only, but apply to any continuous vector functions, displacements, velocities, etc., and have extensive applications in physics. The work spent against magnetic force in carrying a unit pole from A to Z in the field $\mathbf{B}_2$ is independent of the path followed, therefore depending only on the positions of A and Z, and hence we have a scalar potential. On the other hand, a vector-potential is appropriate to the field $\mathbf{B}_1$, in which the work spent is not independent of the path, although in space not occupied by current a scalar potential may be used under restrictions. Compare Theorems (B) and (C) [pp. 211, 212].

In proving the minimum property above we applied it to a field of force due to currents. But it is easily seen to apply equally well to a field due to magnetism. Thus, let $\mathbf{B}_2$ be given without curl, representing the field of a magnet, and consider how the quantity $\Sigma\,\mathbf{B}_2^2$ is affected by altering the field in any way that does not introduce fresh magnetism. The auxiliary field $\mathbf{B}_1$ must have no convergence, hence $\Sigma\,\mathbf{B}_1\mathbf{B}_2 = 0$, and $\Sigma\,\mathbf{B}_2^2$ becomes $\Sigma\,\mathbf{B}_2^2 + \Sigma\,\mathbf{B}_1^2$, and is, therefore, always increased. Hence $\Sigma\,\mathbf{B}_2^2$ for the magnet is the least possible.

An application of the minimum property is to prove that one, and only one, solution exists for the magnetic force or the vector-potential when the current is given (or similarly for the force or scalar potential when magnetism is given). Merely to illustrate the general course of argument, take the case of a given system of current. Prove that there is one, and only one, distribution of magnetic force. Here we define magnetic force so that its curl shall be $4\pi \times$ current, and that it shall be without convergence. Assume that $\mathbf{B}_1$ is a solution. We can show that, if we alter the magnetic force without altering the current, we always increase the quantity $\Sigma\,\mathbf{B}_1^2$, and that $\Sigma\,\mathbf{B}_1^2$ is, therefore, the least possible. That is to say, if $\mathbf{B}_1$ is a solution, then $\Sigma\,\mathbf{B}_1^2$ is a minimum. But $\Sigma\,\mathbf{B}_1^2$, being always positive, is capable of being made a minimum, hence there must be a solution. That there is only one follows obviously by assuming there to be another, and showing that it is the same as the first.

It may be asked, and very naturally, what is the use of this when we know there is a solution, and have been working with it all along? Not much, certainly, in the present case. But when we are considering induced magnetisation and various other questions, the equations and conditions to be satisfied may become so complex that it may not be at all evident *à priori* that they are consistent with the existence of a single distribution of force, etc., free from ambiguity and impossibility. In such cases valuable evidence is obtainable by forming the expressions for the energy or analogous quantities, and investigating their minimum or maximum properties.

Consider next the mutual energy of two systems of current. Denoting their densities by $\mathbf{C}_1$ and $\mathbf{C}_2$ respectively we have [equation (13), p. 246],

$$T_{12} = \Sigma \mathbf{B}_1\mathbf{B}_2/4\pi = \Sigma \mathbf{A}_1\mathbf{C}_2 = \Sigma \mathbf{A}_2\mathbf{C}_1.$$

Now here $\mathbf{C}_1$ is the current corresponding to the magnetic force $\mathbf{B}_1$, and $\mathbf{C}_2$ to $\mathbf{B}_2$, and similarly for the vector-potentials. But since all alike have the property of absence of convergence we may equally well let $\mathbf{C}_1$, $\mathbf{C}_2$, $\mathbf{A}_1$, $\mathbf{A}_2$ stand for magnetic force. We have then six fields of force, all different, though related three and three, and we may arrange them in three pairs, so that their mutual energies are equal. As an example, if two coils of any form containing currents be so placed that either current sends no force on the whole through the other, the same will be true of another pair of currents so arranged that the magnetic force of one is represented by the current in one of the original circuits, and the force of the other by the vector-potential of the second of the original currents; and similarly the other pair of currents may be found.

Nor need we stop here, for we may do the same with the newly-obtained systems of magnetic force, and hence construct an unlimited number of pairs of fields which shall have the same mutual energy as a given pair. This may be symbolised thus:—Let $\mathbf{B}_1$, $\mathbf{B}_2$, $\mathbf{B}_3$, $\mathbf{B}_4$, $\mathbf{B}_5$ be fields of force so related that $\mathbf{B}_2 = \text{curl } \mathbf{B}_1$, $\mathbf{B}_3 = \text{curl } \mathbf{B}_2$, and so on; and, similarly, let there be another set, with accents, similarly related. Then, for example, starting with $\mathbf{B}_3$ and $\mathbf{B}'_3$, we have, out of other combinations,

$$\Sigma \mathbf{B}_3\mathbf{B}'_3 = \Sigma \mathbf{B}_2\mathbf{B}'_4 = \Sigma \mathbf{B}_1\mathbf{B}'_5, \quad \text{and also} \quad = \Sigma \mathbf{B}_4\mathbf{B}'_2 = \Sigma \mathbf{B}_5\mathbf{B}'_1,$$

the summations extending in each case over the proper fields. We may have, however, merely superficial distributions to deal with.

To conclude the present article, there is a curious form in which the energy may be expressed, which I have not seen noticed, viz., in terms of the vector-potential of current and the scalar potential of free electricity. Let a steady current be set up, say by a battery, so that we may locate the impressed E.M.F.'s distinctly at a certain section or sections of the circuit; let the conductivity k be uniform, P the electrostatic, and $\mathbf{A}$ the vector-potential; then

$$T = \tfrac{1}{2}k \Sigma PA_n,$$

where A_n denotes the normal component of **A** reckoned inwards, and the summation extends over the whole bounding surface of the circuit. This may be verified by Theorem (A) relating to convergence [p. 209]; P being scalar, $P\mathbf{A}$ is a vector, and its surface-integral may be expressed in terms of its convergence within the enclosed space.

Since the potential P is discontinuous at the section of E.M.F. (one, for simplicity), we must cut the circuit there, thus producing an enclosed space which does not enter into itself as the closed conductor does, and reckon P differently for the two new surfaces formed, the values, of course, differing by an amount equal to the E.M.F.

In general, with conductivity not uniform,

$$T=\frac{1}{2}\iint k\,PA_n dS+\frac{1}{2}\iiint P\,.\,\mathbf{A}\nabla k\,dV,$$

where dS and dV are elements of surface and of volume, and ∇k is the vector rate of increase of k.

XXVI.—SOME ELECTROSTATIC AND MAGNETIC RELATIONS.

[*The Electrician*, 1883; §§ 1 to 5, April 14, p. 510; §§ 6 to 8, April 28, p. 558; §§ 9 to 15, May 5, p. 582; §§ 16, 17, June 2, p. 54; §§ 18 to 22, June 9, p. 79.]

Comparison of Divergent and Circuital Vectors.

1. When we confine ourselves to a single dielectric, as air, for example, and do not take into consideration the modifications produced in the distribution of force by the presence in the electric field of dielectrics of specific capacity differing from that of the main body of the field (which amounts practically to having no variations of specific capacity unless it may be in very weak parts of the field, where no sensible effect would be produced), the relations of the principal electrostatic quantities are capable of simple expression, and of more or less easy comprehension, according to circumstances. We have in the first place the electrostatic force, which is by far the most important; and next, two auxiliary functions, the electrification and its potential.

In an electric field in equilibrium a small charged conductor in general experiences force at any place, and the electrostatic force is the force that would act upon a unit positive charge placed at the point considered and supposed not to modify the field [due to the other electrification] by its presence; the ordinary electrostatic unit of electricity being so defined that two charges of amounts e_1 and e_2 at distance r apart experience a repulsion of magnitude e_1e_2/r^2.

Electrostatic force is thus a vector, having direction as well as magnitude; and the electric field becomes completely known by following the direction of the force from one point to another, thus travelling along a line of force; then doing the same for another line of force, and so on; and ultimately dividing the whole field into tubes of force, a tube being everywhere bounded by lines of force, there being no force normal to its surface (save when cut across), and such that the integral amount of force crossing any section of a tube is the same, a constant for the tube. Now these tubes are all unclosed, or start somewhere and terminate somewhere else, usually upon conductors, and at the ends of a tube is supposed to reside free electricity. According to Maxwell's remarkable theory there is a real displacement of electricity all along a tube of force, proportional in amount to the strength of force at any place, and in the same direction in general (*i.e.*, in an isotropic medium, whose specific capacity does not vary in different directions); the whole displacement across one section being the same as that across any other. If this amount be e for a certain tube, then at the commencement of the tube there is an amount e of positive, and at its end an equal amount of negative electrification, or of *free* electricity, to distinguish it from the electricity displaced in other parts of the tube, which gives no indication of its presence, for a similar reason that a uniformly magnetised magnet shows no signs of "magnetism" save at the ends of the lines of magnetisation.

2. But, quite apart from this hypothesis—however probable it may be—which so neatly harmonises the equations of electromagnetism, and may almost be considered as a truth, whose recognition was, perhaps, hindered by the absurd 4π multiplier connecting electric force and surface charge, viz.:—

$$\text{Surface density (or displacement)} = (4\pi)^{-1} \times \text{force} \times \text{spec. capacity},$$

which is just as reasonable as it would be to say that, in a conductor,

$$\text{Current} = (4\pi)^{-1} \times \text{E.M.F.} \times \text{conductivity},$$

we may always determine the distribution of electrification from the convergence or divergence of the force; and from a mathematical point of view, when we are only concerned with the quantitative and directional relations, we may consider the electrification to have no other meaning than to express the amount of such divergence or convergence at any place.

Describe any closed surface, and measure the integral amount of force leaving it. This (divided by 4π) is the measure of the amount of electricity contained in the enclosed volume. Applying this to the unit volume, we see that the volume density ($\times 4\pi$) = excess of amount of force leaving over that entering the volume. If this be positive, there is evidently a divergence of force on the whole at the place considered; if negative, a convergence. Let $\mathbf{R}$ denote the electrostatic force, and ρ the volume-density of electrification, then we may say

$$4\pi\rho = -\operatorname{conv} \mathbf{R}, \quad \text{or} \quad 4\pi\rho = \operatorname{div} \mathbf{R},$$

where we use conv and div as abbreviations to be understood as follows:—In terms of the components X, Y, Z of the force, we have

$$4\pi\rho = \left(\frac{dX}{dx} + \frac{dY}{dy} + \frac{dZ}{dz}\right). \quad \text{(Theorem (A), p. 209, \textit{ante}.)}$$

The expression on the right hand side of this equation (with the − sign prefixed) Maxwell called the "convergence" of the force; it is really the integral amount of force, taken algebraically, *entering* the unit volume; but since + convergence indicates negative electrification, we may as well use the term "divergence" for the same quantity with + sign prefixed, as it appears in the above equation, in fact; thus, if the amount of divergence be positive, it indicates positive electricity. Electrification is thus a scalar (directionless) quantity of one degree lower dimensions than electrostatic force as regards length.

The other auxiliary function to **R**, viz., the electrostatic potential, say P, is one degree higher than **R** as regards length, and is such that

$$P = \Sigma\, \rho/r.$$

Divide the space where there is electricity into small parts; divide the charges in these parts by their distances from a given point, and take the sum of all the quotients; the result is the value of the potential P at the given point. Since ρ is scalar, so is P.

We have also the relation

$$\mathbf{R} = \Sigma(\rho/r^2)\mathbf{r}_1$$

between force and density, the summation being now of vectors drawn radially from the charges, the unit vector $\mathbf{r}_1$ being introduced to vectorise the quantities summed. Likewise there is the very important relation

$$\mathbf{R} = -\nabla P,$$

which, translated into words, says that the force is the vector rate of fastest decrease of the potential, which we may call simply the space-variation of the potential. At any point, find which way the potential falls fastest; it is the direction of the force, and its magnitude is the rate of decrease.

Thus we have, in descending order,

Scalar.	Vector.	Scalar.	
$P = \Sigma\, \rho/r,$	$\mathbf{R} = -\nabla P,$	$4\pi\rho = \mathrm{div}\, \mathbf{R}.$	.... (1)

Although we have only spoken of volume-density, we may easily pass to surface-density by the same method of considering the amount of force leaving a closed surface, which in this case must be intersected by the electrified surface where the density is required.

3. If we compare the relations in equation (1) with the corresponding relations between **A**, **B**, and **C** in a current-system, **C** being the density of current, **B** the magnetic force, and **A** the vector-potential, we observe similarities and differences. For we have

Vector.	Vector.	Vector.	
$\mathbf{A} = \Sigma\, \mathbf{C}/r,$	$\mathbf{B} = \mathrm{curl}\, \mathbf{A},$	$4\pi\mathbf{C} = \mathrm{curl}\, \mathbf{B}.$	.... (2)

In (2) **A**, **B**, and **C** are in descending order as regards length dimensions, as are P, **R**, and ρ in (1). Again, as P is the potential of ρ, so is **A** the potential of **C**. So far there is similarity. But whereas in (2) all the quantities are vectors, in (1) we have two scalars and a vector. Also, whilst in (2) **C** is derived from **B** in the same manner as **B** from

A, in (1) as we pass from the vector **R** to the scalar ρ we have the operation divergence, and from the scalar P to the vector **R** the operation of space-variation. The relations in (2) are thus much more uniform than in (1).

Now, as has been discussed in former articles, if we form additional auxiliaries in the series belonging to the current system, below **C** and above **A**, we have the same properties repeated. Thus, if we form the quantity **D** = curl **C**, we shall have **B** = vector-potential of **D**; just as (4π) **C** = curl **B** and **A** = vector-potential of **C**. And if we form the quantity $\mathbf{A}_0$ = vector potential of **B**, $\mathbf{A}_0$ being thus one degree above **A**, we shall have **A** = curl $\mathbf{A}_0$ (with a 4π factor). And we may get rid of the remaining irregularity of appearance of 4π, and make the relations uniform all along the series, by making the potential of a quantity x be $\Sigma x/4\pi r$ instead of $\Sigma x/r$, which in our present case amounts to making the integral amount of force emanating from a magnetic pole of strength m numerically equal to m.

Extension of Electrostatic Properties.

4. What are now the corresponding properties in the series P, **R**, and ρ, when we form auxiliaries, one below ρ, another above P? Their existence is somewhat masked by the want of uniformity in (1), but they exist nevertheless.

In the first place, form the vector function $\mathbf{R}_0$ one degree above P as regards length dimensions, such that

$$\mathbf{R}_0 = \Sigma\,\mathbf{R}/r.$$

Then we shall have the following relation between $\mathbf{R}_0$ and P, viz.,

$$4\pi P = \text{div}\,\mathbf{R}_0\,;$$

or, in words—The electrostatic potential ($\times\,4\pi$) equals the divergence of the vector-potential of the electrostatic force.

Analytically, if the components of $\mathbf{R}_0$ are X_0, Y_0, Z_0, we have, by the definition of $\mathbf{R}_0$ above,

$$X_0 = \iiint \frac{X}{r}dV, \quad Y_0 = \iiint \frac{Y}{r}dV, \quad Z_0 = \iiint \frac{Z}{r}dV\,; \quad \text{.........} \quad (3)$$

dV representing the element of volume. Therefore, by the definition of divergence, and by (3),

$$\text{div}\,\mathbf{R}_0 = \frac{dX_0}{dx} + \frac{dY_0}{dy} + \frac{dZ_0}{dz} = -\iiint \left(X\frac{dr^{-1}}{dx} + Y\frac{dr^{-1}}{dy} + Z\frac{dr^{-1}}{dz} \right) dV,$$

where we introduce the − sign on transferring the place of differentiations from the point where $\mathbf{R}_0$ is measured to dV itself, at the other end of **r**. Hence, by integrating by parts,

$$\text{div}\,\mathbf{R}_0 = -\iint \frac{R_n}{r}dS + \iiint \frac{1}{r}\left(\frac{dX}{dx} + \frac{dY}{dy} + \frac{dZ}{dz}\right) dV\,;$$

where R_n is the outward normal component of **R** at the bounding surface; or

$$\text{div}\,\mathbf{R}_0 = 4\pi \iint \frac{\sigma}{r}dS + 4\pi \iiint \frac{\rho}{r}dV = 4\pi P,$$

as was to be shown, σ and ρ representing the surface- and volume-densities of electrification.

5. In the next place, form $\mathbf{R}_1$, a vector representing the space-variation of ρ, or

$$\mathbf{R}_1 = -\nabla\rho;$$

$\mathbf{R}_1$ thus bearing the same relation to ρ as $\mathbf{R}$ to P, and inquire what relation $\mathbf{R}_1$ bears to $\mathbf{R}$. The answer is

$$\mathbf{R} = \Sigma\,\mathbf{R}_1/r,$$

or in words—The electrostatic force is the vector-potential of the space-variation of the electric density.

Take one component of $\mathbf{R}$ at a time. For the x-component X we have

$$X = -\frac{dP}{dx} = -\frac{d}{dx}\iiint\frac{\rho}{r}dV = \iiint\rho\frac{dr^{-1}}{dx}dV,$$

where we originally perform the differentiation at the point to which X belongs, and then, in the last integral, transfer it to dV. Hence, "by parts" integration,

$$X = \iint\frac{l\rho}{r}dS - \iiint\frac{1}{r}\frac{d\rho}{dx}dV, \qquad (4)$$

where dS is an element of the bounding surface, and l the cosine of the angle between the normal outward and the x-axis. Disregarding the surface-integral, by extending the volume-integration over all space, and writing X_1, Y_1, Z_1, for the components of $\mathbf{R}_1$, we have, by (4), and remembering that $X_1 = -d\rho/dx$, etc.,

$$X = \iiint\frac{X_1}{r}dV, \qquad Y = \iiint\frac{Y_1}{r}dV, \qquad Z = \iiint\frac{Z_1}{r}dV.$$

Hence, by compounding the three left-hand members to form $\mathbf{R}$, and X_1, Y_1, Z_1, to form $\mathbf{R}_1$, we obtain simply

$$\mathbf{R} = \Sigma\,\mathbf{R}_1/r,$$

the required result.

6. We may see from the surface integral in (4) what to do in the case of a purely surface distribution of electrification, *i.e.*, a finite quantity of electrification upon a surface with no thickness in the layer. For, substitute for the surface charge a volume charge by endowing the layer with a small uniform thickness, t, so that if σ be the surface-density, we shall have $\sigma = \rho t$. The surface-integral in (4), with the corresponding ones for Y and Z, tell us that we must draw a vector of length ρ normally to the surface of the layer outward on both sides, one such vector for every unit of surface, ρ having the proper value for the portion considered. This constitutes a normal system of vectors. The volume-integral tells us that we must draw a vector tangential to the surface of length $-\nabla\rho$, one for each unit of volume of the layer, and hence we may substitute one of length $-t\nabla\rho$ for each unit of surface. The vector-potential of these two systems of vectors will be the electrostatic force at any point.

Now let Q be a point where $\mathbf{R}$ is required, and let its distance from a point S on the real electrified surface be r, and its distances from the corresponding points S_1 and S_2 of the outer and inner surfaces of the electric layer be r_1 and r_2, the outer surface being that next Q. Here S_2, S, and S_1 lie along the normal to the surface through S. Then, for the unit of surface surrounding S, we have the normal vectors contributing $(\rho/r_1 - \rho/r_2)$ to the value of $\mathbf{R}$ at Q. Now let the thickness of the layer decrease indefinitely, so that S_2 and S_1 approach S; then, since $r_2 = r_1 + t\cos\epsilon$, where ϵ is the angle between $\mathbf{r}$ and the normal $\mathbf{n}$ outward, this expression becomes

$$\rho\frac{r_2 - r_1}{r_1 r_2} = \rho\frac{t\cos\epsilon}{r^2} = \frac{\sigma}{r^2}\cos\epsilon = \sigma\frac{dr^{-1}}{dn}.$$

Also, for the tangential vector, we have for the part of $\mathbf{R}$ contributed by the unit of surface surrounding S,

$$\frac{1}{r}(-\nabla\rho)t = -\frac{t}{r}\nabla\rho = -\frac{1}{r}\nabla_0\sigma.$$

Therefore, divide the electrified surface into unit areas. On each unit area erect a single vector of length $(\sigma/r^2)\cos\epsilon$ in the direction of the normal, and draw the vector $(1/r)\nabla_0\sigma$ tangential to the surface in the direction in which the surface-density σ decreases fastest. The resultant of the two systems of vectors will be the electrostatic force at Q.*

If the surface-density be constant, we have only the normal vectors to deal with, and it might appear that the force due to an electrified *plane* surface with σ constant were always parallel to the normal, which we know to be not true, unless the surface be infinitely extended. In fact, the preceding construction applies only to a closed surface, without a bounding edge, that is to say; or to cases of unclosed surfaces in which the density decreases gradually to zero, either at the edge or before reaching the edge. We may imagine the electrification to terminate suddenly at a certain boundary. If so, we see, by first endowing the layer with thickness, that we have, in the above, neglected the edge, where we should, from every unit of area of the strip forming the edge, draw a vector of length ρ normally outward. Now, when we decrease the thickness of the layer indefinitely, the edge vectors are ultimately to be thus defined. Divide the edge (now a closed curve) into unit lengths, and from each unit length draw a vector of length σ/r, whose direction is perpendicular to the edge and also to the normal to the surface at its edge. It is now the resultant of the three systems, the normal, tangential, and the edge vectors which gives the force at Q. Thus

$$\mathbf{R} = \Sigma\left(\sigma\frac{dr^{-1}}{dn}\right)\mathbf{n} + \Sigma\left(\frac{-\nabla_0\sigma}{r}\right) + \Sigma\frac{\sigma}{r}\mathbf{m}, \quad \ldots\ldots\ldots\ldots\ldots (5)$$

* [The difference between ∇ and ∇_0 lies in this, that ∇_0 ultimately refers to a surface only, instead of three dimensions. Thus $\nabla_0\sigma$, when σ is surface-density, means the vector rate of increase of σ on the surface. In the original, both ∇_0 and ∇ were denoted by ∇.]

where the first two summations are extended over the surface, and the third round its bounding edge, $\sigma(dr^{-1}/dn)\mathbf{n}$ being normal, $(-\nabla_0\sigma)/r$ tangential to the surface, and $(\sigma/r)\mathbf{m}$ perpendicular to the normal to the surface, and to the tangent to the edge. Here $\mathbf{n}$ is the unit normal, and $\mathbf{m}$ the unit edge vector, perpendicular to the normal and to the edge.

Now, we also know that $\mathbf{R} = \Sigma(\sigma/r^2)\mathbf{r}_1$, extending summation over the surface, the vectors being now drawn in the direction of $\mathbf{r}$ from S to Q. This we may decompose into normal and tangential summations, viz.:—

$$\Sigma\frac{\sigma}{r^2}\mathbf{r}_1 = \Sigma\frac{\sigma}{r^2}\cos\epsilon\,\mathbf{n} + \Sigma\frac{\sigma}{r^2}\sin\epsilon\,\mathbf{t},$$

where $\mathbf{t}$ is a unit vector in the plane of the surface. Comparing which with (5), their right hand members being equal, and the first terms of the same identical, we see that

$$\Sigma\frac{\sigma}{r}\,\mathbf{m} = \Sigma\frac{\nabla_0\sigma}{r} + \Sigma\frac{\sigma}{r^2}\sin\epsilon\,\mathbf{t}, \qquad (6)$$

the edge summation thus being expressed in terms of surface summations of two tangential vectors, the first in the direction of greatest increase of σ, the second in the plane of $\mathbf{r}$ and $\mathbf{n}$.

The proof of this theorem in Cartesian co-ordinates, x, y, z, is rather complex, but we may see its truth by means of Ampère's "dodge" of substituting a network of linear currents over a surface for a current round the edge. See Theorem (B), [p. 211], and observe that, although there the line integral is of the resolved part of a vector in the direction of the curve, and that when we substitute for it the sum of a number of other line-integrals (it being allowable to do so because the direction of rotation round the closed curves is the same for all, so that all the interior line-integrals cancel, and nothing is left but the integral round the bounding edge), yet the same method would apply exactly if the line-integral were of a vector drawn perpendicular to the edge in the plane of the surface; for if we join two points of the original closed curve by a line, thus making two circuits with a portion in common, and draw the vectors perpendicular to the curves in the plane of the surface for both circuits, outwards in each case, the vectors for the common portion are oppositely directed and of equal magnitude, and therefore annul. We see, therefore, that the edge summation $\Sigma(\sigma/r)\mathbf{m}$ can be expressed by a surface summation, and that the portion of this summation for the unit of surface is nothing more than the value of $\Sigma(\sigma/r)\mathbf{m}$ taken round its bounding line. We require then to know the resultant all round the line bounding the unit area enclosing any point S, of the vector of length σ/r drawn perpendicular to the bounding line. Its value will depend on the tangential variation of σ in different directions from S radially, and upon the variation of $1/r$ about S. The first is $(\nabla_0\sigma)/r$, the latter $(\sigma/r^2)\sin\epsilon\,\mathbf{t}$, and these are the vectors that appear in equation (6), whose truth is, therefore, verified. Vector integrals are sometimes very troublesome to manage. The above example shows the great aid to be derived from looking at the vector itself, rather than working with its components, introducing long and complex formulæ.

If we imagine a plane surface electrified with constant density, it might be the surface of a dielectric for example, the normal vectors give a force parallel to the normal, and the edge vectors a force parallel to the plane of the surface; their resultant is the actual force. The component parallel to the plane is, of course, of great importance near the edge, making the lines of force curve outwards.

The method by which we passed from the two vectors for volume-density to one normal vector for surface-density, and the calculation of their potential, may be compared with the process of finding the magnetic potential of a normally magnetised shell. The only difference is that we are here concerned with vectors instead of scalars. In the case of the shell we have + magnetism in one side, − in the other, of finite amounts when the thickness is finite, and we find their combined potential at any point. Then decrease the thickness of the shell infinitely, and increase the surface-density of magnetism correspondingly, so that the magnetic moment remains constant, and find the limit to which the potential approaches. It is $(\sigma/r^2)\cos\epsilon$ per unit area. In the above the process is the same, but the quantity operated on is a vector drawn parallel to the normal instead of a scalar. The method is also analogous to that which may be employed [p. 204 *ante*] for finding the magnetic force due to a current sheet by means of the vector-potential of the curl of the current, except that the oppositely directed vectors are tangential in that case, and the single vector is normal.

Complete Scheme of Potentials.

7. We may leave, however, the interpretation of the special forms which results assume for surface distributions, as it is so much easier to work with volume-densities. The relations stated in paragraphs (4) and (5) will become clearer, and a more comprehensive view of the matter will be obtained if we make the intensity of force be $e/4\pi r^2$ at distance r from a charge e, which is equivalent to defining the electrostatic potential $P = \Sigma\,\rho/4\pi r$.

Let P_1, $\mathbf{R}_1$, P_2, $\mathbf{R}_2$, P_3, $\mathbf{R}_3$, P_4, ... be quantities thus arranged. Starting with, say, P_4, let it be an arbitrary scalar; it may be volume-density of electrification. Let P_3 be its potential, P_2 that of P_3, and P_1 that of P_2. The P's are therefore all scalars, and differ two degrees consecutively in length dimensions. Now between them insert $\mathbf{R}_1$, $\mathbf{R}_2$, $\mathbf{R}_3$, ... such that $\mathbf{R}_1$ is the space-variation of P_1, $\mathbf{R}_2$ of P_2, and so on. Thus, the $\mathbf{R}$'s are all vectors, and of intermediate degrees in length dimensions.

It follows from this specification, first, that any $\mathbf{R}$ is the *vector-potential* of the next following $\mathbf{R}$; thus

$$\mathbf{R}_1 = \Sigma\,\mathbf{R}_2/4\pi r,$$

etc.; and, secondly, that any P is the divergence of the preceding $\mathbf{R}$; and there is uniformity throughout. There is no need of additional proofs, as we have already gone through the process for two auxiliary

R's, viz., $\mathbf{R}_1$ below ρ and $\mathbf{R}_0$ above P, in paragraphs 5 and 4 respectively. Two consecutive P's may stand for electrostatic potential and for volume-density, the intermediate $\mathbf{R}$ then representing the corresponding electrostatic force.

8. But we may, and with some advantage, transfer ideas from electrostatic force to the magnetic force of permanent magnets, taking the limited view of the magnetic force that it is due to magnetic "matter" [p. 223 *ante*]. That is, we replace electrification by magnetism, electrostatic force by magnetic force, and electrostatic potential by magnetic potential; and we shall employ the definition of potential as in the last paragraph. The relations are exactly the same mathematically.

Now, besides the force due to magnetism, let there be also ordinary electric currents, and consider the resultant field, which is due to the superposition of a field consisting of closed tubes (due to current) and of another of unclosed tubes (due to magnetism). Let $\mathbf{B}$ be the actual magnetic force, and separate it [as on p. 253 *ante*] into $\mathbf{B}_1$ and $\mathbf{B}_2$, of which $\mathbf{B}_1$ has no convergence, and $\mathbf{B}_2$ no curl; the latter corresponding to the R's of the last paragraph. Let $\mathbf{C}_1$ be the curl of $\mathbf{B}_1$; it is the current-density; let C_2 be the divergence of $\mathbf{B}_2$; it is the density of magnetism; let $\mathbf{A}_1$ be the vector-potential of $\mathbf{C}_1$, and A_2 the scalar potential of C_2. And to exhibit the relations more fully, introduce $\mathbf{D}$ below $\mathbf{C}$, and $\mathbf{Z}$ above $\mathbf{A}$, both vectors; $\mathbf{Z}_1$ the vector-potential of $\mathbf{B}_1$, $\mathbf{Z}_2$ that of $\mathbf{B}_2$; $\mathbf{D}_1$ the curl of $\mathbf{C}_1$, and $\mathbf{D}_2$ the space-variation of C_2.

$$\overbrace{\underset{\text{v.}\quad\text{v.}}{\mathbf{Z}_1+\mathbf{Z}_2}}^{\mathbf{Z}}\qquad \overbrace{\underset{\text{v.}\quad\text{s.}}{\mathbf{A}_1+A_2}}^{(\mathbf{A})}\qquad \overbrace{\underset{\text{v.}\quad\text{v.}}{\mathbf{B}_1+\mathbf{B}_2}}^{\mathbf{B}}\qquad \overbrace{\underset{\text{v.}\quad\text{s.}}{\mathbf{C}_1+C_2}}^{(\mathbf{C})}\qquad \overbrace{\underset{\text{v.}\quad\text{v.}}{\mathbf{D}_1+\mathbf{D}_2}}^{\mathbf{D}}$$

The quantities with the suffix 1 refers to the current-system, with 2 to the magnetism. The letters v. and s. are placed to show whether they are scalar or vector. We know that $\mathbf{Z}_1$, $\mathbf{A}_1$, $\mathbf{B}_1$, $\mathbf{C}_1$, $\mathbf{D}_1$ are related uniformly thus:—

$$\mathbf{A}_1 = \text{curl } \mathbf{Z}_1, \quad \mathbf{B}_1 = \text{curl } \mathbf{A}_1, \quad \text{etc.};$$

and also thus:—

$$\mathbf{Z}_1 = \text{potential } \mathbf{B}_1, \quad \mathbf{A}_1 = \text{potential } \mathbf{C}_1, \quad \text{etc.}$$

Also, in the other series,

$$\mathbf{Z}_2 = \text{potential } \mathbf{B}_2, \quad A_2 = \text{potential } C_2, \quad \mathbf{B}_2 = \text{potential } \mathbf{D}_2;$$

alternately scalar and vector; whilst a scalar in the series is derived from the preceding vector by the operation divergence, and a vector from the preceding scalar by the operation of space-variation.

Observe that we may compound $\mathbf{B}_1$ and $\mathbf{B}_2$ and obtain $\mathbf{B}$, the real magnetic force. Also we may compound $\mathbf{D}_1$ and $\mathbf{D}_2$, since they are both vectors, forming a new vector, $\mathbf{D}$; and likewise $\mathbf{Z}_1$ and $\mathbf{Z}_2$ (both vectors), forming a new vector, $\mathbf{Z}$. And here $\mathbf{Z}$, $\mathbf{B}$, and $\mathbf{D}$ are vectors which have both curl and convergence. Also, $\mathbf{Z}$ is clearly the vector-potential of $\mathbf{B}$, and $\mathbf{B}$ the vector-potential of $\mathbf{D}$; because this is true of their constituents, which are in each case homogeneous—of the same nature, that is to say. But the intermediate A's and C's do not apparently admit of being combined. $\mathbf{A}_1$ is truly the potential (vector) of $\mathbf{C}_1$, and A_2 the potential (scalar) of C_2; but $\mathbf{A}_1$ and A_2 are not homogeneous, nor are $\mathbf{C}_1$ and C_2, so that the magnitudes (A) and (C) produced by their union are of a peculiar nature, demanding consideration.

9. The division of physical magnitudes by Hamilton into scalars and vectors is not merely one of the most useful ideas ever conceived, but is also one that is perfectly intelligible to every one as a natural division. Scalars being such as pressure, temperature, density, and so forth, directionless, and requiring but one specification, viz., magnitude; and vectors being such as displacement, velocity, force, etc., involving direction as well as magnitude, and, therefore, requiring three specifications (as the magnitudes of the components in three rectangular directions; or the magnitude of the vector itself, and two data to specify its direction); it is impossible to confound scalars with vectors. They are distinct and separate entities. The word entity is obviously applicable to a scalar, whilst it is equally applicable to a vector in spite of its three data. The data may be of different kinds, yet the final result is the same, viz., a definite directed magnitude.

But although we cannot combine a scalar with a vector to form a fresh scalar or fresh vector, or a new quantity having an individuality of its own, it is sometimes convenient to pair them, and the result is called a quaternion, the name implying the four data. But a quaternion is always merely a definite scalar and a definite vector paired, and is consequently a purely artificial idea, not having the same naturalness as a scalar or a vector. (A) and (C) are thus quaternions, whilst **Z**, **B**, and **D** are pure vectors. Nevertheless, so far as the potential property goes, there is no occasion to draw any distinction between them, for we have $\mathbf{Z} = \text{potential } \mathbf{B}$, $(\mathrm{A}) = \text{potential } (\mathrm{C})$, and $\mathbf{B} = \text{potential } \mathbf{D}$.

In the quaternion analysis, however, a quaternion assumes more definiteness than in the above, being in fact the ratio of two vectors, *i.e.*, the operation that must be performed upon one vector α to turn it into another β. Imagining them drawn from the same point, we may turn α into β by first rotating α through a definite angle in their common plane until it coincides with β in direction, and then by stretching or shortening it till it is identically the same as β. The angle of rotation and the stretching require each one specification, and the plane of rotation two more, thus making four in all.

Energy Properties.

10. In the next place we may notice that the energy properties of **Z**, (A), **B**, (C), and **D** with the suffix 1 have their parallel in the other set with the suffix 2. We have already discussed the former, relating to the current system, and shown that

$$\Sigma \mathbf{B}_1^2 = \Sigma \mathbf{A}_1\mathbf{C}_1 = \Sigma \mathbf{Z}_1\mathbf{D}_1 = \dots, \qquad (a)$$

the summations extending over all space. Now in the other set we have exactly similar relations amongst the mixed scalars and vectors, viz. :—

$$\Sigma \mathbf{B}_2^2 = \Sigma A_2 C_2 = \Sigma \mathbf{Z}_2 \mathbf{D}_2 = \dots, \qquad \dots\dots\dots\dots \quad (\beta)$$

In equations (α) the quantities are all vectors, and the products are scalar products, the scalar product of two vectors being the product of their tensors and the cosine of the angle between their directions. In (β) we have a similar product in the third summation, because $\mathbf{Z}_2$ and $\mathbf{D}_2$ are vectors, but not in the second, because A_2 and C_2 are scalars, and their product is found by ordinary arithmetic.

We may prove the identity $\Sigma \mathbf{B}_2^2 = \Sigma A_2 C_2$ in a manner somewhat similar to that employed before in proving $\Sigma \mathbf{B}_1^2 = \Sigma \mathbf{A}_1 \mathbf{C}_1$; but the process is now somewhat simpler, owing to the simpler nature of the system of force due to a charge at a point, viz., straight lines radiating equably from the charge, as compared with the closed lines of force of a current.

We have

$$A_2 = \Sigma \frac{C_2}{4\pi r}, \qquad \mathbf{B}_2 = -\nabla A_2, \qquad \text{and} \qquad C_2 = \operatorname{div} \mathbf{B}_2.$$

Select any small volume containing the charge C'_2, say, and consider the portion of $\Sigma A_2 C_2$ belonging thereto, viz., $A_2 C'_2$. Here we have

$$\left.\begin{aligned} A_2 &= \text{line-integral of } \mathbf{B}_2 \text{ from } \infty \text{ to } C'_2, \\ C'_2 &= \text{surface-integral of } \mathbf{B}'_2 \text{ over a surface enclosing } C'_2; \end{aligned}\right\}$$

where $\mathbf{B}'_2$ is the force corresponding to C'_2. The first of these equations follows from the relation $\mathbf{B}_2 = -\nabla A_2$, and the second from $C'_2 = \operatorname{div} \mathbf{B}'_2$. Noticing here that any path from infinity terminating at C'_2 will do for the line-integration, and any surface surrounding C'_2 for the other, and that by expanding the surface, starting at C'_2, we may make it sweep over all space, at the same time that its point of intersection by the line chosen for the line-integral goes from C'_2 to ∞, we see that $A_2 C'_2 = \Sigma \mathbf{B}_2 \mathbf{B}'_2$, over all space. Similarly, $A_2 C''_2 = \Sigma \mathbf{B}_2 \mathbf{B}''_2$, if C''_2 be the charge in another small volume. Hence, by including the whole of C_2 we obtain

$$\Sigma A_2 C_2 = \Sigma (\mathbf{B}_2 \mathbf{B}'_2 + \mathbf{B}_2 \mathbf{B}''_2 + \dots) = \Sigma \mathbf{B}_2 \mathbf{B}_2 = \Sigma \mathbf{B}_2^2,$$

since $\mathbf{B}_2$ is the vector sum of $\mathbf{B}'_2$, $\mathbf{B}''_2$, etc. Interpreted for electrostatics, C_2 being density of electricity, A_2 its potential, $\mathbf{B}_2$ the force, the quantity $\Sigma \mathbf{B}_2^2$ is double the electrostatic energy expressed in terms of the square of the force throughout the whole dielectric with unit capacity, whilst in the equivalent form $\Sigma A_2 C_2$ it is expressed in terms of the charges and their potentials. It ($\Sigma \frac{1}{2}\mathbf{B}_2^2$) is the amount of work expended in setting up the state of electrification, and given out again when the charges are allowed to combine and neutralise by conducting paths, in the form of heat ultimately. We must not, however, hastily conclude that $\Sigma \frac{1}{2}\mathbf{B}_2^2$, when $\mathbf{B}_2$ stands for the magnetic force of a permanent magnet, really represents the energy of the magnetisation, a matter we shall not here touch upon.

11. The other identity $\Sigma A_2 C_2 = \Sigma \mathbf{Z}_2 \mathbf{D}_2$, where $\mathbf{Z}_2$ is the potential of $\mathbf{B}_2$, and $\mathbf{B}_2$ that of $\mathbf{D}_2$, may be thus established. Let F, G, H be the components of $\mathbf{Z}_2$ and F_1, G_1, H_1 those of $\mathbf{D}_2$. Then, because $A_2 = \text{div } \mathbf{Z}_2$, we have

$$\Sigma A_2 C_2 = \iiint C_2 \left(\frac{dF}{dx} + \frac{dG}{dy} + \frac{dH}{dz} \right) dV,$$

and, by integration "by parts" through all space,

$$= - \iiint \left(F \frac{dC_2}{dx} + G \frac{dC_2}{dy} + H \frac{dC_2}{dz} \right) dV.$$

But $-dC_2/dx = F_1$, etc., since $-\nabla C_2 = \mathbf{D}_2$; hence

$$\Sigma A_2 C_2 = \iiint (FF_1 + GG_1 + HH_1) dV = \Sigma \mathbf{Z}_2 \mathbf{D}_2$$

in our notation.

12. We saw [Art. XXV., p. 253,] that the division of the arbitrary magnetic force $\mathbf{B}$ into two fields, $\mathbf{B}_1$ and $\mathbf{B}_2$, the first due to current, the second to magnetism, is effected in such a manner that

$$\Sigma \mathbf{B}^2 = \Sigma \mathbf{B}_1^2 + \Sigma \mathbf{B}_2^2,$$

the products of $\mathbf{B}_1$ and $\mathbf{B}_2$ completely annulling. We now see that the fields thus obtained have precisely the same energy properties, employing the corresponding connected functions in each case, without the necessity of distinguishing between the different natures of the functions.

We shall, of course, find the same relations in the quantities $\mathbf{Z}$ and $\mathbf{D}$, since they are similar to $\mathbf{B}$, being pure vectors, of which $\mathbf{Z}_1$ and $\mathbf{D}_1$ are the closed tube portions, and $\mathbf{Z}_2$ and $\mathbf{D}_2$ the open. Thus

$$\Sigma \mathbf{Z}^2 = \Sigma \mathbf{Z}_1^2 + \Sigma \mathbf{Z}_2^2, \quad \text{and} \quad \Sigma \mathbf{D}^2 = \Sigma \mathbf{D}_1^2 + \Sigma \mathbf{D}_2^2.$$

But we cannot expect to find this property when we take the intermediates (A) and (C), which are not pure vectors. For instance, we must not expect to have $\Sigma (\mathrm{C})^2 = \Sigma \mathbf{C}_1^2 + \Sigma C_2^2$, for this would involve $\Sigma \mathbf{C}_1 C_2 = 0$. This is a vector summation, for $\mathbf{C}_1$ being vector and C_2 scalar, $\mathbf{C}_1 C_2$ in the summation represents a vector of length $C_1 C_2$ in the direction of $\mathbf{C}_1$.

13. But although $\Sigma \mathbf{C}_1 C_2$ does not in general vanish, yet we have $\Sigma C_2 = 0$ and $\Sigma \mathbf{C}_1 = 0$ separately. The former we have assumed from the beginning, for it merely expresses that there is just as much positive as negative magnetism, which is always the case in any arbitrarily assumed state of magnetisation. The same is true for electricity, but we must exclude lines of force going out to infinity without returning, or we may terminate such lines upon a surface enclosing the whole space considered. In fact, $\Sigma C_2 = 0$ separately for every tube of force.

The other equation, $\Sigma \mathbf{C}_1 = 0$, interpreted for current, says that, in any system of closed currents, the current has, on the whole, no preponderance in any direction. As an illustration of the meaning of this, divide the whole space occupied by current into a very large number, say n, of equal small volumes, and draw n vectors from a fixed point, representing the current in each volume. If these were forces acting upon the point, they would exactly balance. This may be seen by first noting that a closed *line* has no preponderating direction, since it returns into itself, and next that we may split up any system of current into closed tubes of infinitely small section, each conveying a definite current, and that, although the section of a tube may vary, yet the density of current then varies exactly inversely; consequently $\Sigma \mathbf{C}_1 = 0$ for any tube, as for a closed line, and hence $\Sigma \mathbf{C}_1 = 0$ universally. Similarly $\Sigma \mathbf{B}_1 = 0$, and the same for all the quantities with the suffix 1, since they have no convergence.

14. But we do not have $\Sigma \mathbf{B}_2 = 0$, since the lines of $\mathbf{B}_2$ are not closed. Let $\mathbf{B}_2$ be electrostatic force, and consider a tube of force of section dS starting from one conducting surface and terminating upon another. $B_2 dS$ is constant, although B_2 and dS may vary along the tube, and this product is numerically equal to the amount of positive and negative electricities at the terminal sections of the tube upon the conducting surfaces. If we join the ends by a tube whose axis follows the shortest distance between them, and let the same amount of force as went from + to − electricity along the real tube return by the other, we have a closed tube, and $\Sigma \mathbf{B}_2 = 0$. Consequently $\Sigma \mathbf{B}_2$ for the real tube is numerically equal to the product of the amount of terminal electricities into the shortest distance, l, between them, and is directed from + to − along $\mathbf{l}$.

As a magnetic illustration, let there be a cylindrical magnet of length l, and of any section, with plane faces perpendicular to $\mathbf{l}$, uniformly magnetised parallel to $\mathbf{l}$, and let $\mathbf{B}_2$ be its magnetic force. Then $\Sigma_2 \mathbf{B}$ throughout all space is numerically equal to the whole magnetic moment, and is directed from the N. to the S. end, parallel to $\mathbf{l}$. We may verify this as follows:—Given that the magnetic force in a cylindrical space similar to that of the magnet is of uniform strength B parallel to $\mathbf{l}$, and directed from left to right say, and that outside the space there is no magnetic force at all, what is the distribution of current and of magnetism that would produce such a field of force? We must find the curl of $\mathbf{B}$ for the one, and its divergence for the other. Plainly the latter is B per unit area of the terminal faces, positive at the left end, negative at the right. Hence the left terminal face has surface density of magnetism $+B$, and the right $-B$. The curl of $\mathbf{B}$ is plainly confined to the curved surface of the cylinder, where $\mathbf{B}$ is tangential, and suddenly changes from $\mathbf{B}$ to zero in passing through the surface from within outward. Turn $\mathbf{B}$ at the surface through a right angle, and we obtain the surface curl. The current, therefore, circulates round the curved surface of the cylinder in planes perpendicular to its length, and is of density B per unit of length of the cylinder. The magnetic fields of this current, say $\mathbf{B}_1$, and of the terminal magnetism, say $\mathbf{B}_2$, together produce the uniform field of strength $\mathbf{B}$ within and zero outside. But for the former we have $\Sigma \mathbf{B}_1 = 0$, and since $\Sigma \mathbf{B}$ is plainly $= \mathbf{B} \times$ volume, and therefore = magnetic moment, and directed from left to right, the same must be true for $\Sigma \mathbf{B}_2$, *i.e.*, for the magnetic force of the magnet with which we started. In fact, what we have done is to make a current flow round the magnet of such a strength and direction that the external field of force is exactly annulled, and of course we make the supposition that the magnetisation is rigid, or unaffected by its exposure to the magnetising force of the current. Now, on the understanding that the force of the magnet is derived from the magnetic potential as well within as without the magnet, we see from the above that the internal force of the magnet is annulled at the same time as the external by the superimposition of the magnetic force of the current, leaving only the uniform field **B**. On the other hand, if we were to employ the "electromagnetic definition" of the internal force of a magnet, we should find that the current field exactly neutralised that of the magnet, both within and without.

The Operator ∇ and its Application.

15. Going back to **Z**, (A), **B**, (C), **D**, we have, in spite of the identity of the potential property of their constituents with suffixes 1 and 2, and of the energy properties as above mentioned, a striking apparent dissimilarity in the mode of derivation of any term from the preceding in the first set, as compared with the second. Thus, in the first set we have $\mathbf{A}_1 = \text{curl } \mathbf{Z}_1$, $\mathbf{B}_1 = \text{curl } \mathbf{A}_1$, and so on throughout; whilst in the second set we have the two operations of divergence when a scalar is derived from the preceding vector, and of space-variation when a vector is derived from the preceding scalar. Now it is very remarkable that (as was discovered by Professor Tait) these three operations of curl, divergence, and space-variation are really only three different forms of the same operation, the effect varying according to the nature of the function under examination.

We have hitherto used the symbol ∇P to express the resultant space-variation of P per unit length, but have applied ∇ only to scalar quantities. Let a scalar function be given, as for example the temperature at every part of space, single-valued at any point. Nothing is needed to specify it but its magnitude, it having no direction. Owing to this, its variation from point to point is one merely of magnitude. Measure its rates of increase, dP/dx, dP/dy, dP/dz, in three rectangular directions, **x**, **y**, **z**, and call them X, Y, Z. In directions **x**, **y**, **z** draw vectors of lengths X, Y, Z, and compound them (as forces, velocities, displacements, etc.). The resultant vector, say **R**, shows the direction and rate of greatest increase of P, and, with Hamilton's symbol, $\mathbf{R} = \nabla P$. The above method of forming **R** is what we are literally told to do when we use the full expression for ∇, and its effect upon a scalar is to give a vector expressing the most rapid space-variation. There being, as before mentioned, merely a variation of magnitude concerned, there is little difficulty in conceiving the nature of the space-variation of a scalar.

Now it appears that when ∇ is applied to a vector, it gives its curl and its convergence respectively. This extraordinary effect of ∇ is not easily to be understood—although symbolically it works out very simply—for there is undeniably a certain amount of mystery about the rules for vector multiplication. But we may gain some insight into the matter by examining in what manner a vector may vary, and by analysing simple cases.

16. In the first place we see that, given a definite vector, as electrostatic force, for every point of space, it may vary as we pass from one point to another as well in its direction as in its magnitude, and it is evidently not an easy matter to form an idea of what its resultant space-variation may be if we endeavour to follow the rule for a scalar in the last section. In fact the construction fails, and obscurity prevails. But let us completely separate change of magnitude from change of direction, by starting with a vector which is everywhere directed the same way, and which can, therefore, suffer only change of magnitude. Let its direction be parallel to **x**, and its magnitude at any point be X. Regarding this as a scalar, it is clear that there is a certain direction in which X increases most rapidly, and that we can find it, and the rate of increase, by the construction for a scalar, viz., the resultant of vectors of lengths dX/dx, dX/dy, dX/dz drawn in the directions of **x**, **y**, and **z**.

But this will give us neither the convergence of **X** nor its curl. To obtain them we must separate the space-variation of **X** into two portions, first variation in its *own* direction, and next perpendicular thereto. We may easily recognise a manifest distinction between these two kinds of variation.

First, let **X** vary in amount only in its own direction, then, in passing from any point through the distance dx parallel to **x**, X becomes $X+(dX/dx)dx$, where dX/dx is the rate of increase. Let there be a cubical element of volume $dV=dx\,dy\,dz$, whose edges are parallel to **x**, **y**, **z**, and consider the amount of **X** entering and leaving the space. Two opposite faces are perpendicular to **X**, and the four others are parallel thereto. The latter may be disregarded, whilst the amount of **X** entering one of the first pair is $X\,dy\,dz$, and leaving the other $\{X+(dX/dx)dx\}dy\,dz$, both faces being of area $dy\,dz$. The excess of the latter over the former amonnt is $(dX/dx)dV$, that is, dX/dx per unit volume. This is the divergence of the vector in the special case taken, since **Y** and **Z** are zero.

Next consider the variation of **X** perpendicular to **x**, *i.e.*, in the plane **y**, **z**, or parallel thereto. If we start from any point and go in different directions in this plane, X may or may not vary, but if it should do so, there will be in general a certain direction in which it increases fastest, and another direction, crossing the first at right angles, in which there is no variation, just as when one is upon the side of a hill there is a direction of greatest slope at any point, and a level direction perpendicular to the first, disregarding singular points requiring special treatment. dX/dy and dX/dz being the rates of increase of X along **y** and **z**, the resultant of vectors of these lengths drawn along **y** and **z** is a vector in the direction of most rapid increase of X in the plane **y**, **z**, of length equal to the rate of increase. Now, rotate this vector through a right angle in the plane **y**, **z**, *i.e.*, about the axis of **x**, so that in its final position it points along the axis of no variation, it will then represent the curl of **X**. It will come to the same thing if we rotate the original vectors through a right angle, and compound them afterwards; and by this we see that the components of the curl of **X** are $+dX/dz$ along **y**, and $-dX/dy$ along **z**, when the rotation is left-handed about the axis of **x**. Let the plane of the paper be the plane of **y**, **z**, and the positive direction of the axis of **x** be downward through the paper at the point

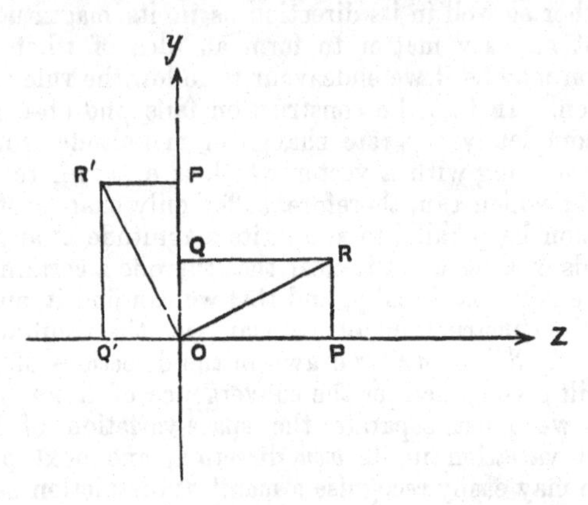

O where the variation of X is estimated. Let OP and OQ be of lengths dX/dz and dX/dy respectively. Their resultant is OR, showing the direction and rate of most rapid increase. Rotate the rectangle $OPRQO$ about the axis of **x** through a right angle into the position $OP'R'Q'O$, then OR' will represent the curl of **X**, and OP' and $-OQ'$ its components along **y** and **z**.

Now let the vector whose space-variation is required be **Y**, everywhere parallel to **y**, and treat it similarly. We shall find its divergence $= dY/dy$ from the variation in its own direction, and its curl to have components $+dY/dx$ along **z** and $-dY/dz$ along **x**, by rotating the resultant space-variation in the plane **z**, **x**, through a right angle about the axis of **y**.

And with a vector function **Z** everywhere parallel to **z** treated similarly, we shall find its divergence $= dZ/dz$, and the components of its curl to be $+dZ/dy$ along **x** and $-dZ/dx$ along **y**.

Finally, if we compound **X**, **Y**, and **Z**, we obtain a vector **R**, which is arbitrary, and, consequently, may vary both in direction and magnitude from point to point. Its divergence will be the sum of the separate former divergences, or $dX/dx + dY/dy + dZ/dz$, which expresses the whole amount of **R** leaving the unit volume, reckoned algebraically. And its curl will be represented by the resultant of the three vectors representing the curls of **X**, **Y**, and **Z** (the first being OR' in the figure, and the second and third two other vectors in planes perpendicular to the plane of the paper and to each other), and its components will be the sum of the former components, and are consequently $dZ/dy - dY/dz$ along **x**, $dX/dz - dZ/dx$ along **y**, and $dY/dx - dX/dy$ along **z**. [Compare with Theorem (B), p. 211.]

17. We thus find the vector curl of **R** by compounding three vectors which represent respectively the rates of greatest increase in three rectangular planes of the components of **R** perpendicular to the planes, each of the three vectors being then turned through a right angle in its proper plane. Now this selection and subsequent rotation is effected mechanically when we use the operator ∇ according to quaternion rules. For, if we denote by **i**, **j**, **k** three rectangular vectors of *unit* lengths parallel to **x**, **y**, and **z**, then $X\mathbf{i}$ will denote a vector of length X parallel to **x**, and similarly for $Y\mathbf{j}$ and $Z\mathbf{k}$, consequently we may write

$$\mathbf{R} = X\mathbf{i} + Y\mathbf{j} + Z\mathbf{k},$$

with the convention that the sign of addition signifies compounding as velocities. Now the full expression for ∇ is

$$\nabla = \mathbf{i}\frac{d}{dx} + \mathbf{j}\frac{d}{dy} + \mathbf{k}\frac{d}{dz};$$

hence

$$\nabla\mathbf{R} = \left(\mathbf{i}\frac{d}{dx} + \mathbf{j}\frac{d}{dy} + \mathbf{k}\frac{d}{dz}\right)\left(X\mathbf{i} + Y\mathbf{j} + Z\mathbf{k}\right).$$

Expand this expression, with the further conventions

$$\mathbf{i}^2 = \mathbf{j}^2 = \mathbf{k}^2 = -1, \quad \text{and} \quad \mathbf{ij} = \mathbf{k}, \quad \mathbf{jk} = \mathbf{i}, \quad \mathbf{ki} = \mathbf{j},$$

and we obtain,

$$\nabla\mathbf{R} = -\left(\frac{dX}{dx} + \frac{dY}{dy} + \frac{dZ}{dz}\right) + \mathbf{i}\left(\frac{dZ}{dy} - \frac{dY}{dz}\right) + \mathbf{j}\left(\frac{dX}{dz} - \frac{dZ}{dx}\right) + \mathbf{k}\left(\frac{dY}{dx} - \frac{dX}{dy}\right),$$

i.e.,

$$\nabla\mathbf{R} = \text{conv}\ \mathbf{R} + \text{curl}\ \mathbf{R}.$$

The meaning of the rules $\mathbf{ij} = \mathbf{k}$, etc., may be interpreted thus:—**j** signifying a unit vector parallel to **y**, and **k** another parallel to **z**, let **i** be taken to mean, not a unit vector parallel to **x**, but a rotation through an angle of 90° about **x** as an axis. Then since **j** rotated 90° about the **x**-axis is turned into coincidence with **k**, we have $\mathbf{ij} = \mathbf{k}$. Similarly for the other products. As for the squares, we may verify $\mathbf{i}^2 = -1$ thus:—Rotate **j** a second time through 90° about the axis of **x**. The first rotation brought **j** into coincidence with **k**, the second brings it into the same line as at first, but pointing the other way. Thus $\mathbf{i}^2\mathbf{j} = -\mathbf{j}$, or $\mathbf{i}^2 = -1$. This use of vectors to denote either lines drawn in definite directions, or else rotations about such lines, is the foundation of the great simplicity and conciseness of quaternion operations and expressions; it is justified by the already mentioned identity of the rules for compounding velocities and rotations, and by its always leading to results which are found to be correct when expanded; but it must be confessed that this double use of the same symbols makes it difficult to establish the elementary parts of quaternions in an intelligible manner.* However, this is merely parenthetical, and we shall

* [See Tait's *Quaternions*, chapter II., for the quaternionic establishment of vector analysis. As the above is the only paper in which I have used the

have no more concern with quaternion expressions beyond noticing that the properties of the quantities defined in § 8, relating to an arbitrary system of magnetic force **B**, divided into two systems $\mathbf{B}_1$ due to current $\mathbf{C}_1$, and $\mathbf{B}_2$ due to magnetism C_2, may be thus formally expressed. Since ∇ operating on a scalar function gives the vector rate of greatest increase, and on a vector gives its convergence and its curl, or its curl only if it has no convergence, and conversely, we have simply

$$\nabla\mathbf{Z}_1=\mathbf{A}_1,\qquad \nabla\mathbf{A}_1=\mathbf{B}_1,\qquad \nabla\mathbf{B}_1=\mathbf{C}_1,\qquad \nabla\mathbf{C}_1=\mathbf{D}_1;$$
$$-\nabla\mathbf{Z}_2=A_2,\qquad -\nabla A_2=\mathbf{B}_2,\qquad -\nabla\mathbf{B}_2=C_2,\qquad -\nabla C_2=\mathbf{D}_2.$$

In the first line, the ∇ signifies "curl" only, because the quantities are vectors of no divergence. In the second line (relating to the magnetism) it is convergence when operating on a vector, and the vector rate of increase when operating on a scalar. The presence of the $-$ sign all along the second line makes it awkward to combine the quantities consistently as they stand. But by simply changing the sign of C_2 and A_2, so that C_2 when positive represents S. magnetism instead of N. we get things straight. $\mathbf{Z}_2$, $\mathbf{B}_2$, and $\mathbf{D}_2$ are unchanged, and denoting $-C_2$ by C'_2 and $-A_2$ by A'_2 we have

$$\nabla\mathbf{Z}_2=A'_2,\qquad \nabla A'_2=\mathbf{B}_2,\qquad \nabla\mathbf{B}_2=C'_2,\qquad \nabla C'_2=\mathbf{D}_2.$$

We may now combine the quantities with suffixes $_1$ and $_2$; thus $\mathbf{B}=\mathbf{B}_1+\mathbf{B}_2$, etc., and we have

$$\nabla\mathbf{Z}=(\mathrm{A}),\qquad \nabla(\mathrm{A})=\mathbf{B},\qquad \nabla\mathbf{B}=(\mathrm{C}),\qquad \nabla(\mathrm{C})=\mathbf{D},$$

where **Z**, **B**, and **D** are pure vectors, **Z** being the potential of **B**, which is the actual resultant magnetic force due to currents and magnetism, and **B** the potential of **D**. On the other hand, (A) and (C) are both scalar and vector, *i.e.*, quaternions, though still (A) = potential (C).

The operator ∇ contains the whole theory of potentials, whether of scalars or vectors. But owing to the remarkably different nature of the effects of ∇ on different functions, it conduces to clearness to distinctly separate the space-variation of a scalar, which is easily grasped, from that of a vector, and to instead speak of the curl or the divergence of the latter, as the case may be, and as we have always done hitherto.

quaternionic ideas and notation, it is perhaps desirable to emphasize the fact that the use was parenthetical. There is great advantage in most practical work in ignoring the quaternion altogether, and also the double signification of a vector above referred to, and in abolishing the quaternionic *minus* sign. The establishment of the algebra of vectors, too, is independent of the difficult theory of quaternions. See especially the articles to follow on "Electromagnetic Induction and its Propagation," and on "The Electromagnetic Wave-Surface" (1885). Professor Willard Gibbs, the author of a valuable work on Vector Analysis, also ignores the quaternion, abolishes the *minus* sign and the double signification of a vector, following Grassmann rather than Hamilton. He has been denounced by Professor Tait in consequence as a retarder of quaternionic progress. Perhaps so; but there is no question as to the difficulty and the practical inconvenience of the quaternionic system.]

DISPLACEMENT AND FLUID MOTION ANALOGIES.

18. What is, however, of greater importance than the mere symbolical identity of the operations is the physical interpretation that may be assigned to the different kinds of space-variation. Operating upon an arbitrary vector function we obtain the scalar convergence and the vector curl; thus from magnetic force we obtain the density of magnetism and of current. That is to say, there are certain invariable relations between the space-variation of the magnetic force about any point, and the density of magnetism and of current at that point, which are most comprehensively stated in saying that the amount of magnetism within any space, which may be large or small, equals the whole amount of magnetic force leaving the space through its surface,- and that the total current passing through any closed curve equals the line-integral, or the circulation of the magnetic force along the curve in making one complete circuit. But instead of such an abstraction as magnetic force, let us suppose that our vector function is of the simplest conceivable type, that it represents the continuous displacements of the particles of a continuous mass. Selecting any particle, the straight line drawn from its original to its final position is its displacement. The displacements of all the points constitute the vector system, or, expressed mathematically, the vector function. In estimating the space-variation of the displacement, we have to examine the manner in which it varies in the neighbourhood of any particle O. In the first place the particle O with the surrounding particles may be displaced as a whole, a bodily translation, in fact. This we disregard in considering the relative displacements, that is, we regard O as fixed. In the next place, the group of particles surrounding O may occupy a greater or less volume in the strained than in the unstrained state, *i.e.*, there may be expansion or compression. The expansion is estimated by finding the additional volume occupied in the strained state by the particles which occupied the unit volume in the unstrained state, and this is plainly to be done by finding the whole displacement outward through the surface of the unit volume; hence "divergence" in general, when applied to the special case of displacements, has the same meaning as the cubical expansion.

Again, the group of particles surrounding O may, during the act of displacement, suffer not merely the translation and the expansion, if any, but a rotation through a definite angle about a definite axis. Going back to the figure, let the displacements **X** be all perpendicular to the paper, say downward, and consider the variation of X about the point O. If O and the particles surrounding it are equally displaced, the matter in the plane of the paper about O is merely transferred bodily, remaining in a plane parallel to the paper; but should there be a greater displacement on one side than on the other, there will be rotation as well. Thus, if OR as before be the direction in which the displacement increases fastest, OR' perpendicular thereto is the axis of rotation of the matter about O. Now OR' represented our vector curl; consequently, interpreted for parallel displacements, the curl is the vector axis of rotation, and by elementary considerations its amount is twice the (small) angle of rotation. Hence, remembering the manner of composition of rotations, we see that the curl of the displacement function is a vector showing by its direction the resultant axis of rotation of the matter surrounding any point, and by its length twice the angle of rotation.

Also, it results from the analysis of the most general continuous displacements of a collection of particles, that the particles which originally occupied a small sphere with centre at O in the unstrained state occupy an ellipsoid after the displacements. The principal axes of the ellipsoid always correspond to a set of three rectangular lines in the original sphere, but there are two distinct cases. The three lines in the sphere may, as the sphere is turned into the ellipsoid, keep their directions unchanged, in which case the strain is pure, and may be produced by three rectangular compressions or elongations acting in directions parallel to the three lines. In the second case the three lines in the sphere do not keep their directions unchanged during the displacement, but are rotated as a whole about some axis. This is a rotational strain, and there is one definite manner of decomposing an arbitrary strain into the simultaneous effects of a pure strain and a rotation.

Thus, corresponding to divergence and curl, we have expansion and rotation. The condition of no divergence means, in the case of displacements, that the strain ellipsoid has the same volume as the corresponding sphere, and the condition of no curl implies that the strain is pure, or unaccompanied by rotation.

We have similar results when we consider not the displacements, but their rates of increase, *i.e.*, the velocities of the particles. Thus, the motion of a fluid may be rotational or irrotational at any place; in the latter case the curl of the velocity is *nil*, in the former its value is twice the angular velocity.

19. Electric current and its related quantities are all characterised by the absence of divergence. Their distributions are therefore similar to possible states of displacement of the particles of an incompressible solid, or of the instantaneous velocity of an incompressible liquid. There is also a common characteristic, that having no divergence anywhere they have necessarily curl somewhere, a property essentially connected with the existence of *closed* lines (of force, current, etc.). Let us suppose that we have a field of magnetic force, $\mathbf{B}_1$, due to currents only, and that the force actually produces a small displacement in the medium where it acts, of amount proportional to the strength of force. A state of strain would be set up. In space unoccupied by current, curl $\mathbf{B}_1 = \mathbf{C}_1 = 0$, and the strain is pure. But where there is current, curl $\mathbf{B}_1 = \mathbf{C}_1$ is finite, and the strain is rotational, its axis is the direction of $\mathbf{C}_1$, and the angle of rotation $\frac{1}{2}C_1$ in amount. In both cases, either within or without the current, div $\mathbf{B}_1 = 0$, or there is no expansion accompanying the strain.

The existence of the rotation (with the above assumption of displacement in the direction of $\mathbf{B}_1$) is easily recognised in the case of a straight current. Let the wire be of circular section. The lines of force are circles in planes perpendicular to the axis, centred thereon. The strength of force increases from *nil* at the axis regularly to a maximum at the surface of the wire, being, anywhere between, proportional to the distance from the axis. If we then take a thin slice of the wire bounded by two parallel planes perpendicular to the axis, and displace its parts through very short distances proportional to and in the direction of the magnetic force at any place, the result will be a rotation of the slice *en masse* about the axis, as a wheel for example. It is easily seen also that the supposed displacements produce relative rotation of the same angular amount of any two points in the slice, and that besides the motion of translation as a whole of the particles surrounding a given point, there is the definite rotation about a line parallel to the axis of the wire. The rotations are equal, because we have assumed the current-density to be uniform throughout the section of the wire. Outside the wire, however, the strength of force falls off, varying inversely as the distance from the axis. This distribution of force makes curl $\mathbf{B}_1 = 0$, and the corresponding strain at any place when small displacements are produced by the force is unaccompanied by rotation, or the displacement is differentially irrotational. (We have used the word "strain" to include the case of mere rotation without distortion.)

20. But by far the most interesting analogue is that presented by the motion of a perfect incompressible liquid. Let it fill all space (since the magnetic force of our rectilinear current extends without limit), and replace lines of force by lines of motion of the liquid, its velocity to everywhere correspond to the magnetic force. We have then a case of vortex motion. The liquid within the cylindrical space corresponding to the wire is rotating, every particle of it, with the same angular velocity. The motion of the liquid outside the vortex, however, although the lines of motion are circles about the vortex, is differentially irrotational—*i.e.*, particles which at a given moment occupy a very small sphere about a given point, at a very short time after occupy an ellipsoid of the same volume without rotation from one to the other. We might, from our knowledge of the relations between magnetic force and current, deduce some of the remarkable properties of vortex motion, so far as they do not involve the dynamics of the subject—*i.e.*, merely characteristics essentially connected with the motion. Current direction corresponds to the axis of a vortex, current-density to twice the angular velocity; vortex lines and tubes, analogous to current lines and tubes, are always re-entrant; the line-integral of magnetic force about a current measuring the amount of current corresponds to the circulation of liquid about a vortex being proportional to the angular velocity of the latter, etc. But, not to go into details, we will merely note, further, the analogue of the identity

$$\Sigma \tfrac{1}{2}\mathbf{B}_1^2 = \Sigma \tfrac{1}{2}\mathbf{A}_1\mathbf{C}_1 ;$$

[p. 244 *ante*]. Here, $\mathbf{B}_1$ being the magnetic force of current $\mathbf{C}_1$, whose potential is $\mathbf{A}_1$, we have two expressions for the quantity that there is reason to believe is the kinetic energy of the system of magnetic force.

Now, let $\mathbf{B}_1$ be the velocity of a perfect liquid of unit density, then $\Sigma\frac{1}{2}\mathbf{B}_1^2$ is really the kinetic energy of the whole motion (sum of mass of each elementary portion $\times\frac{1}{2}$ square of its velocity). We have then the remarkable result that the kinetic energy of the moving liquid can be expressed in terms of its angular velocity in those places where the motion is rotational, and of the corresponding vector-potential. Further, since if $\mathbf{C}_1=0$ everywhere, making $\Sigma\mathbf{A}_1\mathbf{C}_1=0$, it follows that $\Sigma\mathbf{B}_1^2=0$, which (being the sum of squares) involves $\mathbf{B}_1=0$ everywhere. That is to say, if there is no rotation there can be no motion of the liquid [except a uniform motion not vanishing at infinity]. Hence, if the liquid fills all space, its motion, whatever it be, cannot be everywhere irrotational. When, however, the liquid is bounded, as by the surfaces of immersed solids, the motion of the liquid *may* be everywhere irrotational in itself. If so, the rotation is to be sought at the bounding surfaces, and there it will be found, being represented by the difference of the tangential motion of the liquid passing the surface of a solid and of the solid itself at the same place. Or we may reduce this to the former case by imagining the solid to (momentarily) become liquid like the rest, whilst the new liquid has the motion of the solid at the moment considered. There is now continuity of the normal components, and in general discontinuity of the tangential components of the motion of the liquid just within and just without the surface, which was that of the solid before it was melted; and, consequently, differential rotation at the surface. The latter, therefore, constitutes a vortex sheet, analogous to a current sheet, and the vortices are distributed just as the current would be in the current sheet which would produce a field of magnetic force exactly corresponding to the actual state of motion of the liquid at the time.

21. A simple example is that of a spherical solid set in motion in a perfect liquid originally at rest. Let its velocity at any time be $\mathbf{V}$ parallel to $\mathbf{s}$. The motion of the liquid exactly corresponds to the external magnetic force of the system of surface current over the sphere which would produce an internal field corresponding to the given internal motion, *i.e.*, a uniform field of strength V parallel to $\mathbf{s}$. To find this system, we may notice that the curl of the magnetic force given, the internal uniform field, is purely superficial, of amount $V\sin\theta$ per unit area at any part of the surface whose angular distance is θ from the pole or most forward point of the sphere in motion. A system of circular currents over the surface, of surface-density $V\sin\theta$, with positive motion about the axis $\mathbf{s}$, will be easily found to give a uniform internal field parallel to $\mathbf{s}$, but only of strength $\frac{2}{3}V$. Consequently, if we make the strength of current $\frac{3}{2}V\sin\theta$ we get the required internal field of strength V, and the corresponding external field will represent the liquid motion when we translate $\mathbf{V}$ as velocity. The added current corresponds to the external surface curl, of magnetic force or of velocity as the case may be. The external vector-potential of the surface current is $\frac{1}{2}Va^3/r^2\sin\theta$ at distance r, directed perpendicular to $\mathbf{r}$ and $\mathbf{s}$, the lines of vector-potential thus being also circles about $\mathbf{s}$ in planes normal to $\mathbf{s}$. At the surface it becomes $\frac{1}{2}Va\sin\theta$, if a be the radius of

the sphere, and coinciding in direction with the current. Hence we may easily find the energy, viz. :—

$$\Sigma\,\tfrac{1}{2}\mathbf{AC} = \tfrac{1}{2}\iint \tfrac{1}{2}Va\sin\theta \,.\, \tfrac{3}{2}V\sin\theta \,.\, a^2 d\mu d\phi = V^2\pi a^3,$$

[by integration over the spherical surface, between the limits ± 1 for μ, and 0 and 2π for ϕ.] Of this $\frac{1}{3}$ is external, $\frac{2}{3}$ internal, as we may conclude from our having obtained twice as great internal as external surface curl, agreeing with the well known result that the kinetic energy of the liquid is one half that of the solid sphere which moves it, if it has the same density. $V^2\pi a^3$ is in fact the whole energy in the electromagnetic case, and also the whole energy in the case of the moving liquid, including that of the sphere of same density. The external magnetic force, or velocity, is the curl of the above vector-potential $\frac{1}{2}Va^3/r^2 \sin\theta$. The scalar potential, on the other hand, is $\frac{1}{2}Va^3/r^2 \cos\theta$, and its space-variation gives the magnetic force or the velocity at any point. This scalar potential is that of the magnet which would give the same external force. It may be got by substituting a plane magnetic shell for each circular current. The whole collection of shells constitutes a sphere magnetised to intensity $\frac{3}{2}V$ parallel to **s**.

22. The other kind of magnetic force, $\mathbf{B}_2$ above, arising from magnetism, possesses just the same characteristics as $\mathbf{B}_1$ arising from currents, as regards the space external to the magnetised matter. In fact, given a space in which there is a given distribution of force, but quite devoid of current or of magnetisation, and given no knowledge of anything outside the space, it is impossible to say what the origin of the force is, whether magnetisation or current. No necessary distinction can be drawn under the mentioned circumstances. The sources external to the space considered might be either currents or magnetism. The lines of force of $\mathbf{B}_2$, however, in a complete field, as well as in the incomplete field above, are unclosed, and hence arise important differences in the mathematical treatment. But this idea of magnetic force being due to magnetic matter, the residual or unneutralised polarity of the magnetisation, whilst perfectly accounting for the external force, is in all probability entirely erroneous as regards the force within a magnet, and so may be here dismissed, and the more complete theory of internal force will be considered later.

XXVII.—THE ENERGY OF THE ELECTRIC CURRENT.*

SECTION V*a*. THE INDUCTION OF ELECTRIC CURRENTS.

THE mechanical forces between conductors carrying currents, the induction of currents, Joule's law of the generation of heat, and the

*[*The Electrician*, 1883; Section V*a*., June 16, p. 104; V*b*., June 30, p. 149; V*c*., July 14, p. 198; VI*a*., July 28, p. 246; VI*b*., Aug. 11, p. 294; VII., Aug. 25, p. 342; VIII., Sept. 15, p. 414; IX*a*., Oct. 12, p. 510; IX*b*., Oct. 27, p. 558; X., Dec. 1, p. 55; XI., Dec. 22, 1883, p. 127; XII., Jan. 12, 1884, p. 199; XIII., Feb. 2, p. 270; XIV., March 1, p. 367; XV., March 29, 1884, p. 463.]

principle of the conservation of energy are all intimately connected. We cannot, in fact, isolate the mechanical forces from the induced electromotive forces without setting up artificial barriers, to be afterwards taken down. Yet it is necessary, since we cannot treat every part of a subject at the same time, to make a beginning somewhere, by selection of some facts and temporary exclusion of others, and the question presents itself, where to make a starting point for the induction of currents from the theoretical point of view. It is not so much a question of what are the laws of induction, though they are of course involved in the matter, as how to exhibit them and their connections with other electromagnetic phenomena in a complete theory based upon the simplest and least number of experimental laws, and in what order to take the latter.

Now, in building up a consistent theory to embrace a number of facts of a certain class (complete relatively, that is to say, by a process of abstraction) there may be a large choice of methods. As all roads were said to lead to Rome, so all truth is consistent, and by however roundabout a method we go to work we shall arrive at the same results on the way or in the end, if we are working correctly and not trusting to unsound hypotheses. From a group of experimental facts we may divine a certain relation, and by generalising it from the particular cases observed, make it a law, empirical so far, *i.e.*, not deducible from previously known laws at the time, though it may become so later. Such a law is that of gravity; though so long since its discovery, it has not been satisfactorily explained by more easily understood laws. Ohm's law is also without explanation, whilst, on the other hand, the gaseous laws have been explained dynamically, as Newton's law of gravity and Ohm's law may be some day.

In general, several such experimental laws are required to form the framework of a complete theory, though two or three may suffice to make considerable progress. Thus in electrostatics, after learning that electricity is a physical magnitude capable of measurement, that charges may be added and subtracted, the addition to this of the law of force between two concentrated charges, as ascertained by means of Coulomb's torsion balance, generalised to apply to all cases of electricity distributed in a single dielectric, enables us to at once apply the theory of the potential in its general aspect, and to find the distribution of force and the potential energy in the case of any arbitrary distribution of electricity. Add to this, further, the division of bodies into conductors and non-conductors, thus necessitating that in a state of electrical equilibrium there can be no electric force in any part of a conductor, and that therefore the potential of each insulated conductor or of each group of conductively-connected conductors must be a constant, and we have the means of determining the distribution of electricity in practical cases. To this add the further law relating to the difference of capacity of different dielectrics, and we are enabled to calculate the modifying influence on the field of force produced by varying dielectric capacity, and we have got nearly as far as the mathematical theory of electrostatics goes at present. There are plenty of facts outside the theory, but so far as it goes it is complete and consistent, and the facts excluded relate in the main to phenomena observed during the change from one state of equilibrium to another, on the borderland between purely static and purely dynamic phenomena.

Now, we might arrive at just the same results by other methods. Thus, in the theory of the influence of varying dielectric capacity above mentioned, instead of putting Faraday's law at once into a mathematical form, and starting with it, as Sir W. Thomson long ago did (introducing the dielectric constant K into the potential and force equations), we may try to do without it by assuming that the law of force between two charges is exactly the same for any dielectric, but that there are an immense number of small conducting particles embedded in a dielectric. After long travail, we shall find that this hypothesis leads eventually to Faraday's law. The hypothesis may be really true, but the mathematical theory is much simpler without it altogether.

There is an analogous case in the theory of induced magnetisation. The simplest method of laying down the theory is to start with Faraday's idea that different bodies "conduct" magnetic force differently, and to at once introduce a coefficient to express this "conductivity" (the magnetic permeability, etc.). Calling the magnetic force as thus modified, that is $\mu \times$ force, the magnetic induction, we find that the induction is a circuital flux,* or is distributed in closed tubes, and we have magnetic force producing magnetic induction in heterogeneous media, just as electromotive force produces current in similarly heterogeneous conducting media. But we may arrive eventually at the same result, which Maxwell thought best represented Faraday's ideas, by following Poisson, making the induced magnetisation the object of attention in the first place, and not the magnetic induction.

In these and similar cases the best course to pursue is to adopt that initial hypothesis which leads to the general results in the most direct manner, provided, of course, the hypothesis be free from objection.

In the theory of electromagnetism, there is considerable choice of methods. As regards the electromagnetic forces alone, we may find a law of force between a pair of current-elements which will enable us to deduce the force acting upon any complete circuits. The investigations concerning the imaginary force between a pair of elements are usually very complex, and there is besides a grave objection. For we have no reason to believe that current-elements can be isolated in the same way as charges of electricity may be (viz., on insulated conductors separated by a dielectric from all other charges). The current-elements of course exist, but never independently of all other elements, if currents are always closed. As a consequence, there is an indefinite choice of formulæ for the force between a pair of elements, which is essentially

* [To save circumlocution, I here substitute the valuable word "circuital," introduced by Sir W. Thomson in 1890. A circuital flux has no divergence. It has curl, or rotation. The other kind of flux, which has divergence, but no curl, may be similarily termed a "divergent" flux. The words apply to vector distributions in general.]

indeterminate, a definite formula being obtainable only by making some pure assumption.

But a knowledge of the force in question is not required, even if it could be definitely found, and it is preferable to start with closed circuits and keep to closed circuits. Thus, all the mechanical forces between currents and currents, and currents and magnets, may be based upon the law of the equivalence of a small closed current and a little magnet as regards the forces they exert; or we might do without the magnet altogether. Then, with a knowledge that currents are quantitatively expressible, have direction, and follow closed paths, we may, by geometrical or analytical reasoning regarding the properties of lines and surfaces, deduce all the rest, aided in particular by Ampère's device, whereby a finite linear circuit is replaced by any number of other circuits forming a network externally bounded by the original circuit, a process which has many important applications, and is of great assistance in various complex questions. Without further laws to help us, for we have no concern with *how* the currents are obtained, we can develop the whole theory of the quantitative and directional relations of current and magnetic force, and of the vector-potential of current, and calculate the electromagnetic forces, and the potential energy of any distributions of current with respect to those forces. As regards a pair of linear currents, we find that the forces between the conductors may be found from a scalar function whose value depends upon the size, form, and relative position of the circuits, multiplied by the strengths of the currents, say MC_1C_2, where M is the function mentioned. The mutual potential energy of the two currents is $-MC_1C_2$, and its rate of decrease as the circuits are shifted measures the rate of working of the electromagnetic forces during the displacement, and consequently the forces concerned. The most readily intelligible form for the potential energy is in terms of the amount of induction or number of lines of induction through the circuits. The unit current in the first circuit sends the same amount of induction through the second as the unit current in the second does through the first, and this quantity is the M in the potential energy, reckoning the positive direction through a current to be that in which a free N. pole would travel through it under the action of its magnetic force.

From a pair of linear circuits we easily pass to the case of any distribution of currents, and we find the potential energy of two systems of densities $\mathbf{C}_1$ and $\mathbf{C}_2$ to be $-T$, where

$$T = \Sigma \tfrac{1}{2}\mathbf{A}_1\mathbf{C}_1 + \Sigma \tfrac{1}{2}\mathbf{A}_2\mathbf{C}_2 + \Sigma \mathbf{A}_1\mathbf{C}_2,$$
$$= \Sigma \tfrac{1}{2}\mathbf{B}_1^2 + \Sigma \tfrac{1}{2}\mathbf{B}_2^2 + \Sigma \mathbf{B}_1\mathbf{B}_2,$$

and $\mathbf{A}_1$, $\mathbf{A}_2$ are the vector-potentials, $\mathbf{B}_1$, $\mathbf{B}_2$ the magnetic forces, and space is divided into volume-elements. But the most convenient form for two linear circuits is

$$T = \tfrac{1}{2}L_1C_1^2 + \tfrac{1}{2}L_2C_2^2 + MC_1C_2,$$

where M is the same quantity as before, and L_1, L_2 are the corresponding quantities for the circuits taken separately. C_1 and C_2 are now the integral currents, not the current-densities.

We have now to take down the artificial barrier set up in dealing with the electromagnetic or mechanical forces. The currents were presumed to be given constant, and to remain constant during relative motion, and although work done by the electromagnetic forces has been estimated, no account has been taken of other work in the system, as in keeping up the currents, or of the principle of conservation of energy. But currents do not remain in general of constant strength when in motion in a magnetic field, the changes in the currents being the induced currents, and it becomes absolutely necessary to enlarge the field of view.

There are several methods of laying down the laws of induced currents. If we were to choose that method which leads most easily and directly to the required laws, we should employ Maxwell's dynamical method, which exhibits the whole subject in a concise and comprehensive manner, whilst both the electromagnetic actions and the electromotive forces of induction are deduced in the simplest possible mode. But it can scarcely be said that the cardinal assumptions of the dynamical method, that the energy of a current system is kinetic energy, that the system is a dynamically connected system in which currents correspond to velocities, and that the expression for the energy does not contain products of the velocities of the geometrical and the electric variables, are sufficiently simple to justify us in selecting the method to start with. The dynamical method should rather follow other methods, which, if less direct, are more easily understood in the earlier stages.

The laws of induction were first completely quantitatively established by J. Neumann, who took for his basis Lenz's law, which was a generalisation from a comparison of the results of obtaining induced currents by the relative motion of a circuit (the secondary), without a permanent current, and of a primary circuit containing a current kept up by a battery. Now, if we grant that a current is induced in the secondary by relative motion of the circuits, it must be in either one direction or the other, and the electromagnetic action set up between the induced and the primary current must, therefore, be such as either to resist or assist the motion. If the latter were the case, it would follow that the secondary circuit, when at rest, and with, therefore, no current in it (the primary being also supposed at rest), and with no force acting on it, would be in unstable equilibrium. For the least displacement of the secondary circuit, either to or from the primary, would call up forces increasing the displacement, and the secondary would continue to approach or retreat from the primary circuit, according to its initial motion. Such a state of unstable equilibrium being plainly inadmissible, it follows that the direction of the induced current, if there is any, must be such that the mutual forces between it and the primary resist the motion. This principle, applied to the motion of either primary or secondary, constitutes Lenz's law.

In reference to Neumann's investigations, Maxwell remarked that a step of still greater scientific importance was soon after made by Helmholtz and Sir W. Thomson, who showed that the induction of currents could be mathematically deduced from the electromagnetic actions of Oersted and Ampère by the application of the principle of the conservation of energy.

But it must not be concluded from this that, given the electromagnetic forces, and the truth of the general principle named, nothing more is needed to build up the whole science of the induction of currents. The statement sometimes made, that the laws of induction follow of necessity from Ampère's forces and conservation, is of too broad a nature. If we modify the statement, and say that the laws of induction are consistent with Ampère's actions, and with conservation, there will be nothing to be objected to. That this is not a mere difference of tweedledum and tweedledee may be easily seen from the history of the subject, if it be not sufficiently evident by itself. We may, indeed, from the existence of Ampère's forces, and a conviction of the truth of the conservation principle, conclude certainly that some other actions occur, but the principle merely asserts that energy is never lost, that energy put into a system from outside must necessarily be either stored up or make its appearance somewhere in some form or other; but what the form may be depends upon the mechanism—on the dynamical connections—and conservation does not tell us what they are, nor what will happen. There must be other information given.

SECTION V*b*. TRANSFERENCE OF ENERGY. OHM'S LAW.

Given a magnet placed near a closed circuit with no current in it. There is *no* mutual force when they are at rest, and it is not immediately evident that there should be any when the magnet or the conductor is moved. But, following Helmholtz, if we start with a current in the conductor, kept up by a constant E.M.F., we can go further. For now there is force between them, exactly determinate when the data of shape, etc., and strength of current and of magnetisation, are given. Let the positions be such that there is a repulsion of whole amount F, so that a pressure of this amount must be applied to prevent the magnet and conductor separating further. Now push the magnet towards the current through the space dx. The amount of work externally done against the force F during the operation is Fdx, and by the principle of the conservation of energy an equal amount of energy is put into the system, and must be accounted for somewhere. But where to look for it conservation will not tell us, and more facts are needed to assist us to a definite conclusion. As a matter of fact, the work spent is accounted for finally as heat in the conductor. How it gets there is a mystery. The current in the conductor is increased by the motion of the magnet, the rate of transfer of energy from the battery is also increased, the generation of heat which is inseparable from the existence of conduction currents is likewise increased, and this additional generation of heat is the equivalent of the additional energy leaving the battery *and* of the external work spent against the electromagnetic forces. This is sufficiently complex, but the reality is even more so, for the above supposes that the motion is such that the induced current is not changing, that is, that the actual increased current remains steady. We see, however, preliminarily, that the energy put into the system externally by moving the magnet must exist in at least one intermediate form before it becomes heat in the wire, for it has to be transmitted across the intermediate space, and the heat is not generated at the place of external work, as when a button is rubbed upon the coat sleeve. Now, this intermediate energy must also be taken into account when the induced current is not steady, for then there will be no longer equivalence between the rate of generation of heat and the rate of supplying energy from the battery and from the external source, the difference going to the intermediate form.

But to develop this matter it becomes necessary to take a general view of the connection of Ohm's law with Joule's law, and the principle of conservation as concerned in the transfer of energy from a battery in *steady* action. As regards Ohm's law, if there is little to be said that has not been said over and over again, it is certain that a good deal of nonsense has been written about it. Perhaps no scientific law has had so much unscientific discussion, a result to be attributed in the main to its remarkable practical importance bringing it down from the professors to the multitude, which must always contain amongst the great mass who are willing to learn, and are too modest to imagine, if they cannot understand a thing, that the professors are all wrong, a certain number of self-confident paradoxers, whose peculiar conceit is that their views are necessarily right. Self-confidence is, no doubt, an excellent thing in its way, but when coupled with ignorance of the fundamental truths of dynamics (which they should know is an *exact* science), leads to extraordinary jumbles sometimes. Did they only deceive themselves in their delusions little harm would be done, but when they take to writing books for students, then a whole body of blind followers is precipitated into the ditch of mental confusion, from which extrication is so difficult, and whose mud sticks for so long.

In any conductive metallic circuit in which a steady current is flowing, if there be any connection at all between the current C and the whole electromotive force E in the circuit, we may write $E=f(C)$, some function of the current, and, it may be, of other quantities. Or we may say $E=RC$, where R is the ratio of the E.M.F. to the current, and R may be a function of the current and of other quantities. Now, according to Ohm's law, this ratio is constant for a particular circuit, that is, is independent of the current. Increase the E.M.F. in any proportion, and the current will be increased in the same proportion. There is a reservation to be made, viz., that the temperature must be kept constant, otherwise the ratio will slightly change. Also another, of lesser immediate importance, that the long-continued passage of currents may slightly alter the ratio, a result probably due to some structural alteration produced in the conductor. And minute changes may be due to other causes, but practically the ratio is a constant.

What is true for a complete circuit is also true for any portion thereof, the ratio of the E.M.F. in that portion to the current is constant, and this naturally leads us to the most general way of stating Ohm's law, which is to take the unit volume (cube) of a conductor, with two opposite faces perpendicular to the current, and apply the law to it. Let e be the E.M.F. between the faces mentioned, and c the current crossing them and every intermediate section of the tube, then $e = rc$, where r is constant for the *material*. Here e is the E.M.F. per unit of length, or the electric force, c the current per unit area (the current-density), and r the specific resistance.

Another useful form is $c = ke$, where k, the constant ratio of the current to the electric force, is the conductivity, the reciprocal of r.

Corresponding to $c = ke$ in a conductor we have $D = Ke$ in a dielectric, where D is the displacement produced by the electric force e, D referring to unit surface and e to unit length as before, and K being a constant for the material.

Again, we have in induced magnetisation the analogous law $B = \mu H$, where μ is the ratio of the magnetic induction B to the magnetic force H. And similarly $I = \kappa H$, where I is the magnetisation produced by H, and κ their ratio.

But the constancy of K in any dielectric is by no means well assured, whilst in solid dielectrics there is the complication introduced by the phenomenon of "absorption," making the value of K appear to vary according to the time the E.M.F. has been acting on the dielectric, and the displacement already produced, a phenomenon analogous to the effects of imperfect elasticity in bodies under stress.

And κ and μ in magnetisation are only constants (approximately) for weak magnetic forces, their values [after initial augmentation] fall very rapidly when the force is greatly increased, a limiting maximum magnetisation being reached.

On the other hand, Ohm's law is remarkable in that the ratio r remains constant from the lowest to the highest electric forces, a constancy which could not be predicted or expected *à priori* with any certainty in our ignorance of the nature of what is really going on in a body carrying a current.

The constancy of r implies that the existence of one current in a wire does not in any way alter the wire so as to interfere with its capacity for bearing another current if the required additional electric force act, which is very different from the superposition of magnetisations, where the effect of additional magnetic force depends upon the already existing state of magnetisation. An immediate consequence is that any number of electric forces may be superimposed at any point of a conductor, the resulting current being the algebraical sum of the currents the electric forces would separately produce if they act all in the same line; and when their directions are not all the same, the resultant current and resultant electric force have the same direction, and have the same ratio of magnitude r as any of the component electric forces and currents, of which they are the vector sums. [An isotropic conductor is referred to here.]

With the application of Ohm's law to practical circuits and its developments we have no concern here, but a few remarks as to some misunderstandings may not be out of place.

The constant R in $E=RC$ is called the resistance of the circuit (or portion thereof, as the case may be). Notice that it is "called" so. Now, students are sometimes led, or rather misled, by the name and by certain ideas they may possess as to frictional resistance to imbibe the idea that R is really analogous to the frictional resistance to the flow of water through a pipe, when the current of water is compared to the electric current. To credit them with supposing it to be not merely analogous to, but actually frictional resistance, would be to place them in the category of "men of science who are not natural philosophers," described by Maxwell, who seized on the word Fluid as something intelligible, and forthwith endowed electricity, along with fluidity, with mass, inertia, etc.

The real analogue to the frictional resistance of water flowing through a pipe is not R, but RC in the electrical case. Thus, if we set water in a pipe in motion, driving it by means of a constant difference of pressure between its ends, the velocity will increase until the motive force is just balanced by the frictional resistance. Now, let the frictional resistance be exactly proportional to the velocity, as it is said to be approximately for low velocities in thin pipes. Then, if C' be the current, $R'C'$, where R' is some constant multiplier, will be the frictional resistance, and C' will increase until $R'C'=E'$, which gives the steady current corresponding to the motive force E', after which there will be no further acceleration of velocity.

Now, representing the difference of pressure by the E.M.F. in the wire, and the current of water by the electric current, the analogue of frictional resistance in the pipe is RC, and the constancy of R (which is analogous to the coefficient of friction) implies that the resistance is proportional to the current.

Again, if ideas are correct, it is of secondary importance what language is used to express them, but it certainly would appear that, when people say that a certain E.M.F. is required to enable the current to *overcome* the resistance of a wire, it is the idea that is wrong. For there is really no question of overcoming, though there is of coming over. To every E.M.F. corresponds its definite current, neither more nor less. To give a parallel case in mechanics. A body resting on a level surface is set sliding under the application of a constant force; what velocity it will assume depends upon the friction when in motion, since there is no lifting work done. Assuming, for the sake of argument, that it is proportional to the velocity v, say Rv, where R is the coefficient of friction, and that F is the applied force, $Rv=F$ determines the steady velocity, similarly to the case above. The final velocity and the applied force are in a constant ratio. Here there is no question of overcoming, for it is the motion that brings the resistance.

But if the body be at rest, there will be a definite force required to set it moving at all, depending upon the statical friction. There is really a resistance to be overcome before the body will move. But this has no analogue in conduction currents in metals; the least E.M.F. will set up its corresponding current, down to the smallest measurable, and there is no initial resistance to be overcome. In gases, however, there appears to be something of the kind, and naturally we find it expressible as E and not as R. Thus Mr. Varley found that 323 cells were required to start the current in a certain tube, although when set up the current was that due to the excess of the E.M.F. over 304 cells, following Ohm's law thereafter. The initial $E = 323$ Daniells is analogous to the statical friction. For the rest, the analogy is not a good one for our purpose.

Ohm's classical memoir of 1827, of which a translation is to be found in Taylor's "Scientific Memoirs," contains a good deal more than $E = RC$. Therein will be found the laws of distribution of potential in different parts of a circuit, with the well-known zig-zag lines showing its changes in passing through a battery of many cells. Also, if I remember rightly, the method of joining up cells to get the maximum current in a given external resistance, and the corresponding law for the size of wire of a galvanometer. There is, besides, an analytical investigation relating to the propagation of electricity in a wire, wherein, proceeding upon an entirely erroneous assumption regarding the power of a wire for storing up electricity in its substance, like heat, following in fact Fourier's investigation of the conduction of heat, he was led to the true equations for the propagation of potential in a long wire, with electromagnetic induction neglected, afterwards legitimately established by Sir W. Thomson, and he gave the solution in a certain case of constant E.M.F. acting in one part of a circuit. Why he should have come to the right result by a wrong method was simply that, whether electricity is stored up in the substance of a wire, or goes to the surface and stays there, the equations are of exactly the same form.

SECTION Vc. OHM'S LAW AND EOLOTROPY. THE ROTATIONAL PROPERTY.

When we apply Ohm's principle of the constancy of the resistance coefficient in the most general manner possible to bodies which conduct differently in different directions, a remarkable consequence is the establishment of the possibility of existence of a *rotatory* resistance coefficient, or rather of three coefficients which specify a definite axis of rotation. Thus, in order to express without hypothesis the relations between electric force and current at any point, we require no less than nine coefficients of resistance, or an equal number of conductivity, viz., three direct and six transverse; a direct coefficient referring to the current produced in a certain direction by an electric force acting in that direction, and a pair of transverse coefficients to express the component current in the plane perpendicular to the electric force. Taking

x, y, z for any three rectangular axes of reference through the point considered, let an electric force of strength X along x produce currents $k_{11}X$, $k_{21}X$, $k_{31}X$ along x, y, and z respectively, these being the components of the actual current, and employ a similar notation for electric forces acting along y and z. Then, if any electric force **E**, whose components are X, Y, Z, produce a current **C**, whose components are u, v, w, we have the equations of conductivity,

$$u = k_{11}X + k_{12}Y + k_{13}Z,$$
$$v = k_{21}X + k_{22}Y + k_{23}Z,$$
$$w = k_{31}X + k_{32}Y + k_{33}Z;$$

where the k's are the nine coefficients of conductivity, k_{11}, k_{22}, and k_{33} being direct, and the rest transverse. We have, of course, an exactly similar set of equations of resistance, expressing the electric force in terms of the current, thus,

$$X = r_{11}u + r_{12}v + r_{13}w,$$
$$Y = r_{21}u + r_{22}v + r_{23}w,$$
$$Z = r_{31}u + r_{32}v + r_{33}w;$$

where the values of the r's, the coefficients of resistance, may be put in terms of the k's by solving the former equations for X, Y, Z.

Now, these equations are of the same form as those concerned in the transformation of a sphere into an ellipsoid, exemplified in the theory of strains, and we may thus interpret them. Let the electric force at the point considered be of constant intensity, but variable in direction. Representing it by a straight line drawn from the point, the extremity of this line of electric force will obviously travel over the surface of a sphere when we vary its direction. At the same time, the corresponding current will vary both in direction and strength, and if we represent it also by a straight line drawn from the centre of the sphere, then as the extremity of the line of electric force travels over the spherical surface, that of the line of current will travel over the surface of an ellipsoid.

There is, therefore, in general, corresponding to an electric force of constant intensity, but variable direction, a direction of maximum current, one of minimum current, and a third minimax, corresponding to the three principal axes of the ellipsoid. Two axes may be equal, or all three, in which last case the ellipsoid becomes a sphere.

Similarly, let it be the current that is kept of constant strength whilst its direction varies, and represented by a straight line of constant length, whose extremity, therefore, travels over a spherical surface. Now it is the electric force required to produce the current which varies in intensity in different directions; and as the line of current travels over the spherical surface, that of electric force travels over the surface of an ellipsoid. But this second ellipsoid has not in the general case its principal axes in the same direction as those of the first. For this to be the case there must be a certain symmetrical relation, which may be thus expressed. $\mathbf{E}_1$ and $\mathbf{E}_2$ being any two electric forces, and

$\mathbf{C}_1$ and $\mathbf{C}_2$ the currents they produce, we must have $\mathbf{E}_1\mathbf{C}_2 = \mathbf{E}_2\mathbf{C}_1$; or, in terms of components,

$$X_1u_2 + Y_1v_2 + Z_1w_2 = X_2u_1 + Y_2v_1 + Z_2w_1.$$

The interpretation of $\mathbf{E}_1\mathbf{C}_2 = \mathbf{E}_2\mathbf{C}_1$ is that an electric force acting in any direction produces the same component current in any second direction, as the same electric force acting in the second direction does in the first.

In a system of connected linear conductors the corresponding property is that the current produced in a branch R_2 by an electric force in another branch R_1, equals the current in R_1 due to the same electric force in R_2, with similarity of direction. That is, choosing the directions of E in R_1 and of C in R_2 as positive, when E is transferred to R_2, and acts in the assumed positive direction, C will be also in the assumed positive direction in R_1. This reciprocal property is necessary in a system of linear conductors; but there is nothing to prove that the corresponding law $\mathbf{E}_1\mathbf{C}_2 = \mathbf{E}_2\mathbf{C}_1$ is necessary in a non-isotropic conductor. But, supposing it to exist, it follows that the transverse coefficients are equal in three pairs; thus $k_{12} = k_{21}$, etc., and $r_{12} = r_{21}$, etc. The principal axes of the two ellipsoids above mentioned are now coincident in direction, and if we choose them for our axes of reference (x, y, z) we reduce the general equations to the simpler forms

$$u = k_1X, \quad v = k_2Y, \quad w = k_3Z, \quad \text{and} \quad X = r_1u, \quad Y = r_2v, \quad Z = r_3w,$$

where the k's are the principal conductivities, and the r's the principal resistances, viz., along the common axes—and now, which was not the case before, the r's are the reciprocals of the k's.

From these equations we see at once that there are three lines of directional identity of the current and the electric force, viz., the three mutually perpendicular principal axes for which the current is maximum, minimum, or minimax, with an electric force of variable direction but constant intensity. In any other direction there is not coincidence; thus l, m, n being the direction cosines of the electric force, those of the current are lk_1/k, mk_1/k, nk_1/k, where

$$k = \sqrt{l^2k_1^2 + m^2k_2^2 + n^2k_3^2};$$

and the strength of current is $C = kE$.

If we describe an ellipsoid whose axes are along the principal axes, but of lengths proportional to the square roots of the principal resistances, and let the radius vector from the centre to any point of the surface represent the electric force in magnitude and direction, the corresponding current will be parallel to the normal to the surface at the point, therefore along the perpendicular from the centre upon the tangent plane, and its strength will vary inversely as the length of the perpendicular. Similarly, with an ellipsoid whose axes are of lengths proportional to the square roots of the conductivities, the radius vector representing the current, the electric force will be inversely as the perpendicular on the tangent plane and in its direction.

When this symmetry does not exist (k_{12} not $= k_{21}$, etc.), we have rotation. Representing the electric forces by radii of a sphere, the lines of current trace out an ellipsoid, as before, but we no longer have the lines of directional identity of current and electric force for the principal axes of the ellipsoid, for the ellipsoid is rotated as a whole about a definite axis, besides altering somewhat in shape to become another ellipsoid. Similarly, with the sphere for current, the ellipsoid for electric force is rotated, though not about the same axis, save in a special case. There may be still three lines of parallelism of current and electric force, though no longer mutually perpendicular, nor coincident with the principal axes; but also there may be only *one* such direction. And this last case is the special case of identity of rotation axis of resistance and conductivity, and exhibits the rotatory phenomenon in the simplest form. Let a conductor be isotropic in the first place, then $\mathbf{E}=\mathbf{RC}$; or, in terms of the components,

$$X=Ru, \qquad Y=Rv, \qquad Z=Rw,$$

where R is the one resistance coefficient. Now introduce rotation upon the top of this. We may put

$$r_{11}=r_{22}=r_{33}=R$$

in the general equations, and

$$r_{23}=-r_{32}=T_1, \qquad r_{31}=-r_{13}=T_2, \qquad r_{12}=-r_{21}=T_3.$$

We now have

$$X=Ru+X_1, \qquad Y=Rv+Y_1, \qquad Z=Rw+Z_1;$$

where

$$X_1=T_3v-T_2w, \qquad Y_1=T_1w-T_3u, \qquad Z_1=T_2u-T_1v.$$

Multiplying X_1, Y_1, Z_1 first by u_1, v_1, w_1 respectively and adding, and then the same with T_1, T_2, T_3, we find

$$X_1u+Y_1v+Z_1w=0, \qquad \text{and} \qquad X_1T_1+Y_1T_2+Z_1T_3=0.$$

Hence the electric force, say $\mathbf{E}_1$, whose components are X_1, Y_1, Z_1, is perpendicular to the current $\mathbf{C}$, and to a vector, say $\mathbf{T}$, whose components are T_1, T_2, T_3. Also, if θ be the angle between $\mathbf{T}$ and $\mathbf{C}$, we have

$$E_1=TC\sin\theta.$$

$\mathbf{E}$ is therefore the resultant of two electric forces, one of intensity RC in the direction of the current, the other of intensity $CT\sin\theta$, perpendicular to $\mathbf{C}$ and to $\mathbf{T}$. These three electric forces form the three sides of a right-angled triangle, of which $\mathbf{E}$ is the long side; whence, by the famous 47th,

$$E^2=(RC)^2+(TC\sin\theta)^2,$$

which gives

$$E=(R^2+T^2\sin^2\theta)^{\frac{1}{2}}C.$$

Also, if ϕ be the angle between $\mathbf{E}$ and $\mathbf{C}$,

$$\tan\phi=T/R.$$

The one line of parallelism of current and electric force is the axis of $\mathbf{T}$, which indicates a definite direction in the body, independent of the directions of $\mathbf{E}$ or $\mathbf{C}$. Put $\theta=0$, and we have simply $E=RC$ when $\mathbf{E}$

acts along the axis of **T**. Next, let it act straight across the axis. Put $\theta = 90°$, then

$$E = (R^2 + T^2)^{\frac{1}{2}} C,$$

or the specific resistance is apparently increased from R to $(R^2 + T^2)^{\frac{1}{2}}$; but **C** and **E** are not parallel. **C** is still perpendicular to the axis, the same as **E** is, but is rotated through an angle ϕ, such that $\tan \phi = T/R$.

In the general case of **E** inclined at any angle to the rotation axis, the component of **E** along the axis produces current parallel to itself as ordinarily, and the component perpendicular to the axis produces current with resistance $(R^2 + T^2)^{\frac{1}{2}}$ and rotation ϕ. The sphere of electric force becomes a prolate ellipsoid of revolution for the current, the long axis being parallel to **T**, and showing the one line (instead of three) of parallelism of **C** and **E**.

The necessity of nine coefficients in the general case, and hence the possibility of existence of the rotatory effect, was maintained by Sir W. Thomson in 1854; Maxwell said there was reason to believe it did not exist in any known substance, but should be found, if anywhere, in magnets. Its actual existence [if this be the true explanation of the Hall effect] was only demonstrated two or three years ago by Hall's discovery that it is developed in metals when placed in a powerful field of magnetic force.

Let there be a steady current in a straight isotropic wire, and for distinctness let it go from left to right in the plane of the paper, and be kept up by a battery. Now let the lines of force of a magnetic field pass straight through the paper, and therefore perpendicular to the current. Ignoring altogether the ordinary current of induction, examine what the effect of the rotatory resistance vector will be, assuming it to be parallel to the lines of force, say downwards through the paper. The current must be deflected in the plane of the paper, say from $\rightarrow \rightarrow$ it tends to $\nearrow \nearrow$. But this transverse current will alter the distribution of the surface charge, the upper half of the wire will receive a positive, and the lower a negative charge, independent of the original distribution. This will introduce a downward electric force across the wire, tending to decrease the deflection. So long as any transverse current exists this opposing electric force will increase; hence the final result is that it reaches such a value as to keep the current going straight from left to right as before the magnetic force was put on. Thus the deflection of the current can be but momentary; on its cessation the current goes on as before, but now under the influence of an impressed electric force not in its direction, but from left to right with a downward slant, being the resultant of the original electric force and of the transverse downward electric force, whose strength must be TC, if C is the current-density, as before.

Two points of the wire, one above and the other below, which were originally at the same potential, are under the influence of the rotation made of different potentials, the upper being positive to the lower, and a current may be therefore taken off in a shunt wire between the two points. The difference of potential is, of course, greatly magnified by using a very thin sheet instead of a wire, and thus was rendered perceptible.

SECTION VI*a*. THE CONSERVATION OF ENERGY.

After Ohm's law, the Conservation of Energy demands consideration. For Joule's law, which naturally follows Ohm's, is an example thereof, and as it happens that in Current electricity we meet with some of the most important and practical, as well as scientifically interesting applications of this great modern generalisation (made the more interesting by our ignorance of the connecting mechanism) it will save future repetition to here briefly discuss it from the standpoint of theoretical dynamics. One may indeed gain, as most educated people have gained, by often reading about it, or from popular lectures, without previous study of dynamics, a general notion of the conservation of energy, and accept the principle, as an article of faith, and at the same time have very vague ideas as to what is meant by energy, or why it should be conserved. Nothing will supply the deficiency save a careful study of dynamics, a repulsively dry subject to most people, but, owing to the far reaching of its principles, one whose preliminary study is indispensable to those who wish to form correct ideas in electricity and magnetism. Not that they will thereby learn what electricity and its connected functions are, but rather that they will know certainly what they are *not*, and hence be able to avoid the absurdities arrived at by those who ignore dynamical relations.

Now, in theoretical dynamics the conservation of energy is a necessary consequence of Newton's laws of motion, with our definition of what we mean by a force doing work. A force being what causes, or tends to cause motion, is naturally measured by the amount of motion caused. Measuring the force acting upon a free particle of mass m by the rate of acceleration of its momentum, mv, where v is the velocity (relative to a body assumed to be at rest, or to bodies not in relative motion), and calling F the force, we have $F=m\dot{v}$, the dot indicating rate of time-increase. (Newton's notation.)

Also, work is done by a force when its point of application moves with the force, and work is done against the force in the reverse case, which is exemplified when a stone falls to or rises from the ground with or against the force of gravity. A force F acting through a distance x does work of amount Fx. In the general case, F must be taken to be the force acting in the direction of motion, the component in that direction of the actual force. Thus the downward force of gravity does no work on a body moving horizontally.

Since v the velocity is the distance moved per second, with the proper qualification for varying velocity as the distance that would be moved through per second if the velocity kept constant, we have Fv as the work done per second—the rate of working, or the activity. But since $F=m\dot{v}$, we have $Fv=mv\dot{v}$, and the latter is the same as $d/dt(\frac{1}{2}mv^2)$. Calling the quantity in the brackets the kinetic energy, and denoting it by T, we have $Fv=\dot{T}$ always. That is, the rate of working of the force equals the rate of increase of the kinetic energy of the mass moved. Hence the work done by the force in increasing the velocity from v_1 to v_2, however the force may vary in the interval, is exactly equivalent to the whole increase of kinetic energy, $T_2 - T_1$, or $\frac{1}{2}m(v_2^2 - v_1^2)$.

The quantity T, the equivalent of the work done in producing the motion from rest, is conversely the amount of work the body can do in coming to rest by moving against a force, hence the propriety of the term kinetic (*i.e.*, motional) energy. For we have now retardation instead of acceleration of momentum, work done against, instead of by the force, and by a reversal of the previous reasoning the total work done when the body is brought to rest is exactly T, the initial kinetic energy.

Thus, project a body upward with initial velocity v, it will rise to such a height h as to do $\frac{1}{2}mv^2$ work against the downward force of gravity g; that is, $\frac{1}{2}mv^2 = gmh$ gives us the greatest height to which it will ascend. Gravity still acting, the body will return, and on reaching the ground have the same velocity and kinetic energy as at first. That is to say, the original kinetic energy, although wholly lost at a certain height, is completely recoverable on allowing the mass to return. The energy, when it has thus disappeared from the kinetic form, but is recoverable, is called potential energy, or energy of position. To whatever height the body may have ascended at any moment, with a certain loss of kinetic energy, exactly the amount lost is recoverable, and hence is to be considered potential energy. At the greatest height, where the velocity is *nil*, the potential energy equals T_0, the initial kinetic energy; whilst on starting from the ground and on reaching it again the potential energy is *nil*; and in any intermediate position the sum of the kinetic and potential energies is T_0, and remains constant throughout the motion both ways.

Potential energy, or work obtainable in virtue of position, is a more abstract idea, seemingly, than kinetic energy, but the two are quite correlative, and one is as easy or as hard to understand as the other. Force of some kind is equally involved in kinetic and potential energy, when there is change from one to the other. Force produces relative motion and its kinetic energy, and the latter cannot be utilised without force. In the above simple case the conservation of energy only means that when kinetic energy is lost by the body moving against the force, such loss is perfectly recoverable in the return motion, since the force remains the same in the same places. This last remark, indeed, contains the reason why the energy is conserved, or returnable to the kinetic form.

Consider a system of free particles in motion on which no external forces act, the forces being wholly mutual stresses, say attractions or repulsions, which vary only with distance, so that the force between two particles is the same at the same distance apart, whatever be their actual positions. Let the system move from any one configuration through any series of intermediate configurations, back again to the original. During the cycle, any two particles which approached or receded from one another during one part, recede from or approach one another the same distance in the remainder of the cycle. The work done by the mutual force in the first part is therefore exactly equal to the work done against the force in the second part of the cycle, and this applies to every pair of particles. Hence the whole work done by the forces during the cycle is *nil*, and the kinetic energy at the end of the cycle is the same as at the beginning. (To make the particles return to the original configuration without alteration of energy, the artificial plan of frictionless constraints may be adopted, *e.g.*, guide a particle through a perfectly smooth tube of any form desired under the action of the given forces; the constraining force of the tube will be always perpendicular to the direction of motion of the particle, and no work will be done by it.) Further, the kinetic energy in *any* configuration will be a function of the configuration only, *i.e.*, in whatever way the particles move from one configuration to another, the gain of kinetic energy, being the total work done by the forces, will be the same, for the forces depend only on the configuration.

Defining, then, the potential energy V of the system in any configuration A to be the work spent by the forces when the system moves from A to a standard configuration chosen arbitrarily, and T to be the kinetic energy in the state A, it follows that $V+T$ remains constant in every configuration. In the standard state, $V=0$ (or any constant value we like); any departure from that state which is attended by an increase or decrease of kinetic is attended by an equal decrease or increase of potential energy, meaning that the decrease of kinetic energy is regained or the increase lost by letting the system go back to the standard state.

If we have two such systems of particles, each with internal forces alone, their energies are naturally independent and remain constant as above. But should there be force between one system and the other, energy may pass between them, and now it is the two systems as a whole, considered as a single system, that is conservative.

We may pass from systems of particles to the ideal rigid bodies of mechanics, which cannot change shape, by introducing constraints. Let any collection of particles be constrained to always preserve the same relative positions, to become, as it were, a rigid body. No work can be now done by the mutual internal forces if they be pulls or pushes along the joining lines between the particles, since the relative motion of any pair must be always perpendicular to the line joining them. Hence the internal forces wholly disappear from the equations of energy, and only external forces need be considered. Any collection of such rigid bodies with mutual forces preserve the sum of their kinetic and potential energies constant, unless work be done on the system from without, when the amount of such work is the gain in its total energy. Practically, real bodies which do not change their form appreciably under not too great external forces, though their parts may be in irregular motion, come under the same law, from the experimental evidence that the unknown internal actions do not tend to change their state of bodily motion, whether of translation or rotation.

Again, in the ideal perfectly elastic body, the work done by external forces in changing it from one form to another against the internal stresses is independent of the series of intermediate forms, and the energy thus put into the body is perfectly recoverable by letting it return to its original shape through the same or any other series of shapes, the stresses depending on the shape only, with consequent conservation of energy.

But the conservation of energy, in the limited sense in which we have employed the idea above, does not exist in Nature. Thus, when one body is set sliding over the surface of another, its motion is retarded, and it finally comes to rest. The force that brought the body to rest will not act back and restore the kinetic energy, which is apparently lost for good. Make the body move back over the same path, and there is a further loss of energy, for the frictional force is not the same in coming back, but is always against the motion, and, besides, it depends upon the velocity more or less. Again, the stone which, projected upwards with a given amount of energy, loses it all at a certain height, and recovers it all (or nearly all, some being lost in friction against the air) on reaching the ground, suddenly loses it again, for there is only a small fraction in its first rebound. A portion of the energy may be traced in the vibrations set up in the masses in collision, but this is only a portion; whilst the vibrations themselves subside, and leave no trace.

Energy thus disappearing from view was formerly supposed to be lost, or, at any rate, it was disregarded. But the modern principle of conservation of energy teaches that energy is never lost, though it may not be recoverable directly. Being a broad generalisation from innumerable experiments, it must be regarded as an experimental law, whose observed fulfilment in so many cases leads us to believe that its truth is universal, and that in every case of disappearance of energy of one kind there is an equal gain in some other kind or kinds, no doubt ultimately resolvable into the simple kinetic and potential energies of dynamics, but usually of unknown exact nature, as the energy of a distribution of static electrification, probably the potential energy of a strained state of the medium, the energy of electric currents, or of magnetisation, or of chemical affinities. To enumerate all would be to range over all natural phenomena, but as we are confined to Electricity, we need only mention preliminarily that it has been proved, and abundantly so, that heat is energy itself, requiring no multiplication by some other physical quantity to make energy, as is the case with electrification, electric current, magnetisation, etc. A definite amount of heat represents a definite amount of energy, and the very important relation between heat in caloric units and mechanical, after being theoretically, though by what is considered unsound reasoning, calculated by Mayer, was experimentally determined by Joule, who, from a numerous series of experiments, found that 772 foot-pounds of work per pound of water frictionally spent in stirring it raises its temperature 1° F., a result confirmed in many other ways less direct, through other forms of energy. The quantity of heat required to raise the temperature of a pound of water 1° F. is therefore 772 foot-pounds, or of a gramme 1° C. is 42,000,000 ergs.

Now, whatever may be the ultimate nature of electrostatic actions, the quantities termed electrification and electromotive force are such that their product is energy, which is something definite, being work obtainable. That this is so follows from the definitions of electrification (quantity of electricity) and E.M.F., and elementary considerations. Thus, let a static charge Q be carried from a place where the potential is P_1 to a place where it is P_2, the whole E.M.F. along the path being $P_1 - P_2 = E$, say. Then the work done by E upon Q during the transfer is EQ. For the electrostatic force f, or force per unit charge, is the space-variation or rate of decrease per unit distance of the potential; the force upon Q is therefore Qf, and this is a real mechanical force (as in the common expression repulsion $= qq'/d^2$, q is electrification, and q'/d^2 electric force.) Its space-integral from the beginning to the end of the path is $Q(P_2 - P_1)$ or EQ, which is therefore energy.

Energy being thus the product of a quantity of electricity and an E.M.F., of course neither of the factors can be energy. That is, electricity cannot be energy, as heat is. Such a very obvious conclusion it might seem to be impossible to misunderstand, yet there are men of standing who have failed to see the force of the argument, simple as it is. Can their failure have arisen from a want of acquaintance with the fundamentals of dynamics? If not, I can think of no other explanation than that the rapid whirl of their ideas, for they are men of imagination, may have produced some degree of oblateness of the spheroid.

Of course, if either E or Q were known the other would be known, since their product is a known quantity. But concerning the often-asked question, What is electricity? I can attach but little importance to the answer by itself. But the question, What is the mechanism of electrical phenomena? is quite another thing. For, if its answer were known, the functions E and Q would be known, and found to be worth —their full value. It *might* be then found desirable to completely alter the nomenclature of electrical theory, and instead of the present functions to employ others to which a plain dynamical meaning can be assigned. Of course the present established relations would remain true, but for them might be substituted equivalent relations in terms of better-understood quantities. But, naturally, until this desirable consummation is reached we had better keep to the present E and Q.

In $W = EQ$, where W is the work done by the E.M.F. E during the transfer of the quantity Q, we have supposed Q to be a static charge, and to be transferred by convection. That the same relation should hold when the charge is transferred by conduction, or when, as in a galvanic circuit, Q does not appear as a static charge at all, cannot be considered as immediately self-evident. For although experiment proves that a conduction current is virtually equivalent to the transfer of electricity, expressed symbolically by $Q = Ct$, where Q is the quantity transferred by the current C in time t, yet this relation is only a quantitative one. It suggests that there is actually motion of electricity round the circuit, and the current is popularly spoken of as such. But this must not be taken literally. There is an enormous difference between a state of electrification and the electric current which may produce it or be derived from it, and we cannot from electrostatic properties say what properties will be developed when a static charge disappears, and energy is transferred from a dielectric to a conductor. In the resulting phenomenon all trace of the electrostatic properties is lost, there is no external sign of any electricity in the conductor, whilst entirely new properties come into existence. As the static charge disappears, and the action upon static electricity likewise, or rather as the latter occurs and we infer the former, action upon a magnet at once appears in its place. The relation symbolised by $Q = Ct$ does not, without special hypothesis, imply the motion of electricity from place to place, meaning by electricity the quantity we make acquaintance with in electrostatics, in spite of the law that the current is the same in all parts of the circuit, and that it is virtually equivalent to convection of electricity.

The German philosophers seem determined, however, to make an electric current be static electricity in motion round the circuit, and have made elaborate attempts, with much success, to find the law of force between two charges of electricity in motion, to include electrodynamics as well as electrostatics. Weber gets over the difficulty of absolute apparent disappearance of the electricity as such by supposing an electric current to consist of two equal currents, one of positive, the other of negative electricity, in opposite directions, the currents being static electricity in motion. In any finite part of the wire there are always equal amounts of positive and negative electricity, and hence no external electrostatic force can be shown. This is, perhaps, the simplest way of evading the difficulty, and apparently far simpler than Clausius's hypothesis, wherein the very artificial plan is adopted of making a current in, say, the positive direction consist of positive electricity moving one way, with an equal amount of negative held fixed, or of negative moving the other way with an equal amount of positive held fixed, or of any combination of these opposite currents with their corresponding fixed charges, which will make the total current come right and balance the electrostatic force of the moving by that of equal amounts of fixed of the opposite kinds. It will be observed, however, that Clausius's hypothesis gives us much greater latitude than Weber's, which it includes as a particular case (although Clausius considers Weber's hypothesis unthinkable), viz., equal opposite currents, when of course the fixed electricities cancel, not being wanted.

Now, it may be that a conduction current really consists of convection of electricity on charged molecules, with inter-molecular discharges, the continual cancelling of positive and negative charges being quantitatively equivalent to the transfer of electricity round the circuit, with no integral free electricity in any space containing a large number of molecules, but I have not as yet been able to conceive Clausius's hypothesis of the moving and fixed electricities. It will be observed that the German speculations contrast very strongly with the methods of the British school of electricians.

SECTION VI*b*. APPLICATION OF CONSERVATION OF ENERGY TO A STEADY CURRENT.

But although, because $W=EQ$ is true for convection of electricity, it is not therefore immediately seen to be self-evidently true in the case of a conduction current; yet it is found to be necessary when we apply the principle of conservation of energy, according to which the disappearance of an amount of electrostatic energy must always be accompanied by the appearance of an equal amount in other forms of energy. Let there be a distribution of static electricity, the separate charges being kept insulated and unalterable. We know, in the first place, that if we alter the distribution in any manner by convection of the charges the mechanical work required to effect the change of configuration is equal to the increase produced in the quantity $V=\Sigma\frac{1}{2}P\rho$, where ρ is an elementary charge and P its potential, depending on its position with respect to the other charges. The increase produced in V equals the sum of force × distance moved for all the charges; or, in terms of the charges, the sum of electric force × electricity × distance, where the product of the first two factors is mechanical force, and of the first and third electromotive force, or difference of potential; the E.M.F. per unit distance being the electric force, otherwise called the E.M.F. at a point. Of course ordinary force acts on matter, and electric force on electricity. The quantity V, in any state of the system, is the whole work required to set up that state by bringing the elementary charges to their places from an infinitely widely separated state. But (and this view of the matter is instructive in regard to discharge by cancelling of opposite charges), it is also the whole amount of work done in separating the positive half of the electricity from the negative half, supposing the electricity given initially in a state where we have every elementary positive charge paired with an equal negative charge infinitely near it, a state equivalent to no electrification. And, conversely, V is the work done by electrostatic force on the electricity when we reverse the above processes, and either separate the charges infinitely, or let them come together and co-exist in pairs of opposite kinds. To illustrate the latter case, let there be two conductors, A with a positive charge Q, B with an equal negative charge, every elementary portion of A's charge being connected with a corresponding negative charge on B by a tube of displacement. Let A be a hollow shell with a trap door; bring B up to A. By this V is reduced, the force being an attraction. (The value of V is simply $\frac{1}{2}EQ$, where E is the difference of potential of the two conductors.) Open the door, put B inside, and shut the door. During the passage of B through the door most of the electricity on A moved from the external to the internal surface of the shell, and when the door is shut again it is all on the internal surface. Now let B expand and take the same form as the inner surface of A, and be separated therefrom by a thin layer of dielectric. Their difference of potential now, and in any further expansion of B, will be simply proportional to the thickness of the layer (as for two insulated parallel plates with equal opposite charges), and hence becomes infinitely small as B approaches A, and the value of V becomes infinitesimal. Here V has disappeared by mere convection of the charges; a quantity, V, of mechanical work has been done, and V is strictly the potential energy of the state of electrification.

But if, as in Maxwell's development of Faraday's views, we regard a state of electrification to be accompanied by an elastically strained state of the dielectric medium, the potential energy of the strain being V, it becomes more easily comprehended how, since when a strained elastic body is allowed to return to a state of no strain it gives back the work done in straining it, which amount of energy may be convertible into various forms, if we in *any* way cause the electrical phenomenon to disappear, there will be a transfer of an amount V of energy from electrostatic into other forms, whose kind will depend on the circumstances of the disappearance. The appropriate form for V on this view is $\Sigma K\mathbf{R}^2/8\pi$ (identically equal to the former expression in terms of the charges and potentials), $\mathbf{R}$ being the electric force and K the specific inductive capacity of the medium, throughout which the summation extends. This is with the ordinary electrostatic units, but if we choose them so that $e/4\pi r^2$ is the electric force at distance r from a charge e in air, we shall find

$$V=\Sigma\tfrac{1}{2}K\mathbf{R}^2.$$

Here $\mathbf{R}$ is the force, and $K\mathbf{R}$ the displacement it produces, so that the energy per unit volume $=\frac{1}{2}$ force $\times$ displacement, the force and displacement being both electrical; the case is exactly analogous to that of a real force producing a displacement of matter, the work done during the displacement being the displacement $\times$ mean value of the force; *i.e.* when the displacement is proportional to the force, $\frac{1}{2}$ force $\times$ displacement.

The potential energy of a state of electrification may, according to the circumstances of the discharge, be used up as mechanical work, setting bodies in motion with consequent kinetic energy of visible motion or of heat of friction, etc., or as heat through the medium of the kinetic energy of conduction currents, or as the energies of sound, light, magnetisation, etc., the number of possible transformations through the agency of conduction currents being very considerable. In passing to conduction currents we may notice the anomalous character of a fact connected with one kind of discharge. A dielectric, as air, is unable to bear tension above a certain amount. The tension along lines of force is measured by the same quantity, $K\mathbf{R}^2/8\pi$, as before, the tension per unit area being numerically equal to the energy per unit volume. Above the limiting tension we have disruptive discharge, varying from the tiny spark to the magnificent lightning flash that so terrifies the vulgar and charms a Faraday—a break-down of the dielectric, accompanied or followed by light, heat, sound, violent commotion of particles, and other effects. Now, Sir W. Thomson found that a greater tension was required to produce a spark between two close parallel plates (one having very slight curvature to localise the discharge) at small than at greater distances—an extraordinary result, if true generally and not dependent on some unobserved special peculiarity in the experiments.

We have this for a consequence, that two electrified bodies, the tension between which is insufficient to effect disruption, may be discharged by increasing the distance between them. This will obviously be so if the tension be nearly sufficient to effect disruption at the smaller distance, and if in moving to the greater distance the tension remains the same as before, hence becoming sufficient for discharge at the greater distance. And the constancy of tension is attained in the case of two parallel plates equally and oppositely charged by a battery to any desired difference of potential. On removing the battery, and so keeping the charges constant, and increasing the distance between the plates, the work done in separating them to, say, n times the original distance just multiplies the energy n times, and the force and tension remain of constant amount, save near the edges, so long as the distance is a small fraction of the diameter of the plates.

Now, let our system of electrification be merely a charged condenser, and connect its terminals by a conducting wire. The charge rapidly disappears, the rate of disappearance being the measure of the electric current, expressed symbolically by $Q = Ct$ for any interval of time t so small that the current C may be considered constant, Q being the quantity of electricity that has disappeared in that interval. The current at any moment is proportional to the E.M.F. at the moment, which (disregarding correction for electromagnetic induction) equals the difference of potential of the condenser, and hence is proportional at any moment to the charge left. Thus, as time increases arithmetically, the charge left and the current fall geometrically. We have

$$C = -c\dot{E},$$

where C is the current, c the capacity of the condenser, and E the difference of potential at time t. Also, by Ohm's law, $E = rC$, if r be the resistance of the wire. Hence,

$$C = -rc\dot{C},$$

integrating which we find

$$rC = E_0 \epsilon^{-t/rc},$$

which gives us the current at any time t after the commencement of the discharge, E_0 being the initial difference of potential. The time taken in falling from the initial full strength of current to any stated fraction thereof is proportional to rc, which is an interval of time, called by Lord Rayleigh the time of subsidence, really the time required to fall from 1 to ϵ^{-1}, or from 1 to $1/2{\cdot}7$.

Let the condenser be made larger and larger, thus increasing the time of subsidence, and imagine it to become, for the purpose of argument, enormously large, thus containing, with the same difference of potential, an immense store of electricity and energy. The time of subsidence becomes so great that we shall have, on joining the terminals through a wire, a practically steady current with constant E.M.F. In any interval of time t (small compared with the time of subsidence) we have $Q = Ct$, where Q is the charge that has left in that time, and C the current; also $E = RC$; and, finally,

$$W = EQ$$

is the work done by E in driving Q, or, in other words, it is the amount by which the potential energy stored in the condenser falls in the time. By conservation, W is the amount of energy to be accounted for.

Now, in the above, we have been really dealing with the disappearance of static electricity, and so know that $W=EQ$ must be true. But if, instead of discharging our big condenser through a conductor and setting free electrostatic energy, we substitute a galvanic cell for the condenser, we find exactly the same phenomena produced as before. There is mechanical action between the wire and a magnet, between different parts of the wire, power of magnetising iron, of effecting electrolysis, etc., in both cases, and in both cases are the actions steady. Now here, although with the galvanic cell we have no disappearance of static electricity, or, indeed, any sign of a large store of it to be drawn upon, yet the phenomena, being the same as regards the conductor, must be virtually equivalent to the discharge of static electricity, and we may apply the same formulæ exactly as in the case of the condenser,

$$Q=Ct \quad \text{and} \quad W=EQ,$$

which are now truths, though not truisms. The Q is now not electricity *in esse*, but *in posse* (*i.e.*, without hypothesis to account for the absence of electrostatic force from the electricity we may assume to be moving). To verify which, we may insert our big condenser in the circuit, when the phenomena in the wire will be as before, whilst in the time t the condenser will acquire a charge of amount Ct. It will, however, have very little energy compared with EQ, the whole work done whilst it was being charged. And now we may, of course, discharge this Ct of electricity (removing the cell first) through a wire, when it will be entirely done for.

Returning to the cell with steady current and no condenser in circuit, we may measure the E.M.F. maintaining the current in the wire electrostatically, by means of an electrometer connected to its ends, and its strength is RC, where R is the resistance of the external wire. RC is not the whole E.M.F. in the circuit, for the battery has itself the same power of limiting the current as the wire, which is proved by the observed difference of potential rising when the length of the external wire is increased. The limit to which it tends is the complete E.M.F.

But the cell will show the same difference of potential when the circuit is not closed conductively at all. According to Maxwell's theory there has been still a current in a closed circuit, viz., as before in the same direction through the cell, and then from one terminal to the other through the air; the current of "displacement" in the air, however, having the remarkable difference (amongst others) from that in a conductor that the displacement is elastically resisted, and hence will return when it is allowed to; whereas there is no such reversibility in a conduction current. The case is, of course, substantially the same as in charging a condenser, the difference being in the amount of capacity and the quantity of electricity concerned, which are excessively small when the terminals have no large opposed surfaces connected to them. And the same thing occurs when the circuit is closed conductively, there being also other circuits through the air on starting the current, which results in charging the wire.

The observed difference of potential of the disconnected cell should not be confounded with its E.M.F. They are numerically equal, but opposed in direction, as regards their power of producing current in the circuit, viz., cell and air. That this is so is evident on considering that if there were no E.M.F. to keep up the static charges they would unite through the cell itself. It is the same when we have not an ordinary galvanic cell, but merely two metals A and B in contact in air, forming a circuit A, air, B, A. If we find B at a higher potential than A, the E.M.F. that caused it must have been numerically equal to their difference of potential, and have acted in the direction B, air, A, B, with a transfer of electricity in this direction round the circuit, lasting until the difference of potential stops further current. It is no easy matter, however, to settle exactly where the small consumption of energy needed for this state of electrostatic energy comes from, though probably it arises merely from a very minute amount of chemical action. [See Section XIII. later.]

Since EQ of work is done by E during the passage of Q, an amount EC is done per second. The supply of energy obviously comes from the galvanic cell, of which more later, whilst we now consider its destination. In $W=EQ$ insert RC for E and Ct for Q, their equivalents when the circuit is conductively closed, and we obtain

$$W=RC^2t$$

in terms of resistance and current, as the amount of energy to be accounted for. The solution is supplied by Joule's discovery that with a steady current heat is continuously developed in a conductor, its amount being proportional to the square of the current and to the time it has been on. Hence, if H be the heat, expressed as energy to avoid the useless introduction of Joule's coefficient, we have

$$H=R_1C^2t,$$

where R_1 is a necessary quantity required to make R_1C^2t be energy. Now RC^2t is energy, being the same as EQ, consequently R_1 is resistance, and can therefore be only R multiplied by a mere numeric. But if no other work is done than in heating the wire, we must have $H=W$, the numeric $=1$, and $R_1=R$. Hence

$$H=RC^2t$$

expresses Joule's law.

Thus, in $EC=RC^2$, which is the equation of activity, or rate of working in a galvanic circuit when the sole result is heat in the conductor, we may say that EC is the work done by E in driving C (true, whatever other effects than heat may be produced), and RC^2 the equivalent rate of generation of heat. We may here advantageously reintroduce the mechanical analogy before employed, viz., a body set in motion by a constant force F, and opposed by a resisting force rv, simply proportional to its velocity, v. So long as F is greater than rv there is acceleration of velocity, which must cease when $F=rv$, which equation consequently gives us the steady velocity. At the same time Fv is the activity of the applied force F, or its rate of working. Now, if the resisting force rv be frictional, the equivalent of Fvt is found to be heat. Thus

$$Fv = rv^2$$

corresponds exactly to

$$EC = RC^2$$

in the galvanic circuit, and, with the proper limitations, the expression sometimes used, the "frictional generation of heat" by the current is perfectly appropriate. If we like, the body moved may be a fluid in a pipe, moving round and round, as some people think electricity moves.

The heat RC^2 per second is produced under all circumstances in conduction currents, being a necessary part of the phenomenon. Other work may be being done, but, with a given current, it will not affect the generation of heat in the least, which can only be altered by altering the current, unless we count the change in R produced by the heat itself raising the temperature, making R indirectly a function of C or of Q, but really a function of the temperature. But in a dielectric current there is no such development of heat, the energy being potential, and returnable. The only quite universal characteristic of current is probably its relation to magnetic force. The heat resulting from current in a conductor has done with electricity; it is diffused by heat conduction, a comparatively slow intermolecular process, by currents of air, or goes off by the wonderful process of radiation.

It will be interesting to observe the form taken by $EC = RC^2$ in a non-isotropic medium, in which the relations between E.M.F. and current are contained in three linear equations with nine coefficients of resistance, and to note whether the rotational phenomenon has any peculiar influence upon its form. Using the previous notation [p. 287], and referring to the equations of resistance $X = r_{11}u + \ldots$, etc., we have, $\mathbf{E}$ being the electric force with components X, Y, Z, and $\mathbf{C}$ the current-density with components u, v, w,

$$\mathbf{EC} = Xu + Yv + Zw$$
$$= r_{11}u^2 + r_{22}v^2 + r_{33}w^2 + (r_{12} + r_{21})uv + \ldots .$$

Here we see that we are only concerned with the sums of the transverse coefficients, not with their differences, on which the rotation depends. If no rotation, $r_{12} = r_{21}$, etc., and now, choosing the axes of reference to be the principal axes, we have

$$\mathbf{EC} = r_1u^2 + r_2v^2 + r_3w^2,$$

where $r_1, \ldots$, are the principal resistances; which reduces to $EC = RC^2$ when the coefficients are equal, and the medium is isotropic, with $\mathbf{E}$ parallel to $\mathbf{C}$.

In the case considered later, isotropy plus rotation, we have

$$\mathbf{EC} = RC^2,$$

nearly as simple as with perfect isotropy. Dividing by C, we see that the component in the direction of the current of the actual electric force is RC, as may be verified by the formulæ given [p. 289]. When, as later, the transient effect has subsided, and the current goes straight under the influence of the ordinary electric force, which we may now call E_1, and the transverse electric force of strength TC, their resultant being $\mathbf{E}$, the actual electric force,

$$\mathbf{EC} = RC^2$$

becomes, by division by C,

$$E_1 = RC, \quad \text{hence} \quad E_1 C = RC^2,$$

just as without the rotation.

Section VII. The Minimum Heat Property in Conductors, Linear or Continuous.

In electrostatics and electromagnetism there are remarkable minimum properties connected with the energy of distributions of magnetic and electric force. We considered this matter in Section IV. [p. 250] with respect to magnetic force, either arising from closed currents, and therefore consisting of closed tubes, in which case the value of the summation $\Sigma \mathbf{B}_1{}^2$, where $\mathbf{B}_1$ is the force, is the least possible, the force being supposed to vary in any manner consistent with the same currents; whilst if arising from magnetism, say now $\mathbf{B}_2$, the value of $\Sigma \mathbf{B}_2{}^2$ is the least possible, the force varying in any manner consistent with the same distribution of magnetism, the ultimate reason being that $\Sigma \mathbf{B}_1 \mathbf{B}_2$ through all space is zero, when $\mathbf{B}_1$ has no convergence, and $\mathbf{B}_2$ no curl.

In connection with Ohm's and Joule's laws there is a similar property, which, in its present application, is particularly useful and instructive, for, owing to the general better acquaintance with current than with magnetic force, and the certainty that we are really dealing with energy and its distribution, or its rate of transformation from one form to another, practical interpretation is much facilitated, whilst at the same time light is cast upon the more abstract similar properties of magnetic force, etc.

The matter to be discussed is perhaps most easily approached by starting from a very simple case, which, though not comprehensive, will lead up to the more general cases in a natural manner, and with an already attained idea of what to expect or look for, which is as valuable in theory as Faraday found it to be in experimentation. Let there be two wires in parallel arc of resistances R_1 and R_2, and let the difference of potential of their common terminals be E. By Ohm's law the currents are given by

$$E = R_1 C_1 = R_2 C_2.$$

This is the natural division of the whole current $(C_1 + C_2)$ supplied at one terminal and leaving at the other, when there are no intrinsic E.M.F.'s in the two wires, *i.e.*, no E.M.F. in either wire except that arising from the given difference of potential. And the energy supplied per second is $EC_1 + EC_2$, which is accounted for by RC_1^2 of heat per second in one wire, and RC_2^2 in the other.

Now, whilst the potentials of the terminals remain constant, and the supply of current keeps the same, let the current divide in some other than the natural manner. If then C_1 becomes $C_1 + c$, C_2 must become $C_2 - c$ to keep the supply constant. We have, therefore, in addition to the natural current, a current of strength c flowing in the circuit formed by R_1 and R_2. Thus, as a special case, let $c = C_2$; then the current in R_2 is *nil*, and in R_1 it is $C_1 + C_2$; that is, the whole supplied current goes through one wire only.

Consider the change in the amount of heat developed in accordance with Joule's law. It is now $R_1(C_1 + c)^2$ in one wire, and $R_2(C_2 - c)^2$ in the other. The additional heat is therefore

$$R_1c^2 + R_2c^2 + 2R_1C_1c - 2R_2C_2c.$$

But the sum of the third and fourth terms is *nil*, because $R_1C_1 = R_2C_2$. Hence the additional heat is $R_1c^2 + R_2c^2$, which is exactly what would be produced per second by the auxiliary current c if it existed alone. Thus, whether c be positive or negative, small or great, the heat is always increased by any departure from the natural division of the current, which is therefore that which makes the heat a minimum with a given supply and a given difference of potential.

From this case we may pass to that of any system of connected linear conductors. Thus, let there be any number of terminals given, whose potentials are P_1, P_2, etc., and let them be connected together by wires in any manner. Every terminal may be connected to every other, but we need not suppose any two terminals to have more than one wire connecting them, as it would only introduce useless complication. Thus, if there are three terminals, we need not have more than three wires; with four terminals, six wires; and, in general, with n terminals, $\frac{1}{2}n(n-1)$ wires at most. Any junctions between the terminals are inadmissible, such really introducing fresh terminals. Let, in the first place, the system contain no intrinsic E.M.F.'s. Then, under the action of the differences of potential of the terminals (some of which are of course connected otherwise with sources of electricity), current enters the system at some and leaves at others. Let Q_1 be the current supplied at the first terminal, Q_2 at the second, and so on. Since in steady flow as much must leave as enters the system, we have the condition $\Sigma Q = 0$.

Again, current Q_1 is supplied at potential P_1, Q_2 at potential P_2, and so on; hence the whole rate of supply of energy to the system is ΣPQ. But we cannot say that P_1Q_1 of energy is supplied at the first terminal, or P_2Q_2 at the second; this is indeterminate without further knowledge. To be true, Q_1 must leave at potential zero, Q_2 at potential zero, and so on; thus the result is only true in the sum. If but two terminals, we have

$$P_1Q_1 + P_2Q_2 = (P_1 - P_2)Q_1, \quad \text{(because } Q_2 = -Q_1\text{)},$$

and

$$= E_1Q_1,$$

if E_1 is the *difference* of potentials or E.M.F. We may add a constant potential all round without making any difference in the energy supplied. For if p be the additional potential, the increase in ΣPQ is

$$\Sigma pQ = p \Sigma Q = 0, \quad \text{because} \quad \Sigma Q = 0.$$

If the potentials are given at a certain number of terminals, and the supplies at the remainder, the distribution of currents in the system is exactly determinate, however complex it may be. For, if it were possible for there to be two different distributions of current, which should both be consistent with the P's having certain values at some terminals, and the Q's certain values at the remainder, it would be possible for a third distribution, viz., their difference, to exist, with the P's *nil* at some terminals, and the Q's *nil* at the remainder. But in this third distribution we should have $\Sigma PQ = 0$, owing to the vanishing of the P's at one set of terminals, and of the Q's at the remainder, *i.e.*, no energy would be given to the system. Hence, as we have supposed there to be no intrinsic E.M.F.'s, and as electric currents produce heat, there can be no current at all in the third distribution; hence our supposed second distribution is the same as the first, which is therefore unique. Special cases included in the above are when all the potentials and none of the supplies, or all the supplies and none of the potentials, are given. It should also be noticed that we have made use of the principle of Ohm's law in the demonstration of uniqueness, viz., when we formed the third distribution, afterwards proved to be a state of no current.

Now, the current in every conductor being determinate, the heat according to Joule's law is known. If R be the resistance of and C the current in any conductor, the heat is ΣRC^2 per second. Hence, by conservation of energy, we must have

$$\Sigma PQ = \Sigma RC^2,$$

if all the energy supplied goes to generate heat. And since $RC = E$, if E is the difference of potential between the ends of a conductor, we have also

$$\Sigma PQ = \Sigma EC,$$

the first summation referring to the terminals, the second to the conductors.

Although we thus obtain

$$\Sigma PQ = \Sigma EC$$

through Joule's and Ohm's laws and the law of conservation, yet this equation is independent of Ohm's law altogether, and is true for any kind of distribution of the currents in the conductors consistent with the same supplies at the terminals. It is true when Q_1 divides in *any* manner between the conductors connected to the first terminal; Q_2 in any manner into its conductors, with the exception, of course, that in the wire connecting the second and first terminals we do not alter the current already fixed upon; and so on to the rest. Every wire has two terminals; hence for a particular wire joining, say P_1 with P_2, we have

$$P_1 C_{12} + P_2 C_{21} = (P_1 - P_2) C_{12} = E_{12} C_{12},$$

and by extension to all the conductors we obtain

$$\Sigma PQ = \Sigma EC.$$

With the proper distribution according to Ohm's law we have also

$$\Sigma PQ = \Sigma RC^2,$$

and *then* conservation of energy, knowing that the heat is RC^2 for any conductor.

Now, let the currents be altered from their natural distribution, without changing the potentials and supplies, the P's and Q's. Let c_1, c_2, etc., be the currents required to be added to the old to make the new distribution, and let q_1, q_2, etc., have the same relation to c_1, c_2, etc., as Q_1, Q_2, etc., to C_1, C_2, etc. That is, q_1 is the sum of the additional currents entering the system at the first terminal. But the supplies are to be unchanged, therefore $q_1 = 0$, $q_2 = 0$, etc. Hence, $\Sigma Pq = 0$. But $\Sigma PQ = \Sigma EC$, independently of the manner of division of the Q's. Hence also $\Sigma Pq = \Sigma Ec$. Therefore $\Sigma Ec = 0$ also, or the differences of potential do no work upon the new currents.

The heat per second is now

$$\Sigma R(C + c)^2 = \Sigma RC^2 + \Sigma Rc^2 + 2\Sigma RCc.$$

But the last summation $= 2\Sigma Ec$, already proved to vanish; hence the additional heat is precisely that due to the new system of currents (c) alone. The two systems (C) and (c) are quite independent, and when they co-exist the heat equals the sum of their separate heats.

The heat of the system (c) being essentially positive, we see that the natural division of current is the one which makes the heat the least possible of all the distributions consistent with the supply conditions. With this natural distribution $\Sigma PQ = \Sigma RC^2$; in any other case ΣRC^2 is the greater. The excess proves the existence of intrinsic E.M.F.'s, producing a quite independent system of current. The intrinsic E.M.F.'s may be very variously arranged, and they may be so distributed as to give rise to no difference of potential in any part of the system. Thus, select a number of conductors forming a closed chain, and in this chain put E.M.F. of uniform amount per unit of resistance. If r be the whole resistance of the chain, and e the whole intrinsic E.M.F., we shall have $e = rc$, where c is the current to be added to the current due to difference of potential. No change will be produced in the currents in the other branches. Then we may take another chain (part of which may belong to the first), and do the same for it, and similarly for all the chains that may be made up. The final resultant will not alter the potentials or the terminal supplies.

We may pass with less difficulty than might be imagined to the corresponding question when we are concerned, not with a linear system, but with a conducting mass of any shape and size, and of any conductivity, uniform or not, with or without isotropy. Let P be the potential at any point of the bounding surface of the body, in general a function of its position, varying from point to point over the surface. We may conveniently divide the surface into unit areas, and assume the potential of such an area to be its mean potential. A unit area thus corresponds to a terminal in the former sense. Since we may take our unit area as small as we please, we shall clearly arrive at results which are correct in the limit. Let also Q be the current supplied to the body per unit area, corresponding to the former Q. Then, exactly as before, ΣPQ, with the summation extended over the whole surface, expresses the energy communicated to the body per second, the rate of working of the E.M.F.'s on the current. (Although we here have our terminals, for simplicity of diction, solely upon the exterior bounding surface of the conductor, yet what follows will not be affected by causing our conducting body to be bounded by any number of internal surfaces as well, to be counted with the exterior surface; thus we might have an infinitely extended conductor with internal electrodes to be taken for the bounding surface.)

Now, P being given over a portion, and Q over the remainder of the surface, and the resistance of every part of the body being given, and that there are no internal intrinsic E.M.F.'s, the distribution of current is determinate and is unique, as well as the distribution of potential, when Ohm's law, or its linear extensions, are followed, their characteristic being that if we cause an electric force to act when there is an already existing electric force with its corresponding current, the new electric force produces its current just as if the other current did not exist; *i.e.*, the resultant electric force corresponds to the resultant current in the same way as the separate electric forces do to their currents. In the first place, that there must be current is obvious if Q have any value not *nil* at any part of the surface, or if P is not the same all over the surface and the body has any conductivity at all. And that there can be but one distribution of current and of potential throughout the body under the given circumstances may be simply proved as formerly for a linear system. If two systems of current could exist separately with the same P at certain parts of the surface and the same Q at others, their difference would constitute a third system in which there would be no potential at certain parts of the surface and with no current entering the body through any portion of the remainder. Hence $\Sigma PQ = 0$ in this third distribution, and it is a state of no current anywhere, for if there were any, heat would result with no supply of energy to produce it, since there are no intrinsic E.M.F.'s. Thus there can be but one distribution of current possible, and since Ohm's law or its extensions give the corresponding distribution of electric force, if the potential is given at any part of the surface it becomes known everywhere.

Next we have to see what form the equation $\Sigma PQ = \Sigma EC$ takes. Divide the body into unit volumes, and let $\mathbf{E}$ and $\mathbf{C}$ be now the electric force and the current-density, these being not in general in the same direction, and therefore to be treated as vectors. Now Q is the normal component inwards of the surface current; hence, since P is scalar, PQ is the normal component of the vector $P\mathbf{C}$, and we may at once apply the general theorem expressing the surface-integral of a vector as a volume-integral. The energy supplied to the body per second through its surface equals the sum of the energies supplied through the surface of each unit of volume. Hence

$$\Sigma PQ = \Sigma \text{ conv } P\mathbf{C},$$

where the second summation extends throughout the body. Performing the operation of convergence, we have

$$\text{conv } P\mathbf{C} = \mathbf{EC} + P \text{ conv } \mathbf{C}.$$

But the current has no convergence, or does not accumulate anywhere; hence, finally,

$$\Sigma PQ = \Sigma \mathbf{EC},$$

the first summation referring to the surface and the second to the volume it encloses.

This equation is quite independent of the nature of the conductor, or of any relation whatever between electric force and current, following strictly from the property of the current that it has no convergence and of the electric force that it is the rate of decrease of the potential. But we must introduce connection between $\mathbf{E}$ and $\mathbf{C}$. First, if the conductor be isotropic, we have $\mathbf{E} = R\mathbf{C}$, where R is the resistance per unit volume, and therefore

$$\Sigma PQ = \Sigma R\mathbf{C}^2$$

simply, where of course R may be variable from point to point. Should isotropy not prevail, we have, instead, the three linear equations of resistance [p. 287],

$$X = r_{11}u + \ldots, \text{ etc.};$$

consequently now

$$\Sigma PQ = \Sigma (r_{11}u^2 + \ldots),$$

the most general form. [See also p. 302.] But there is no occasion whatever to use the lengthy expression for the heat, it being quite sufficient to write $\Sigma \mathbf{EC}$, with the understanding that $\mathbf{E}$ and $\mathbf{C}$ are related through the linear equations.

Whilst P and Q are unchanged, let the natural current be altered from $\mathbf{C}$ to $\mathbf{C}+\mathbf{C}_1$. This can only be by the addition of systems of internal current that either do not reach the surface at all, or, if so, do it tangentially, so as not to alter the surface supply; and intrinsic E.M.F. must be supplied just sufficient to keep up the new currents. $\mathbf{E}$ becomes $\mathbf{E}+\mathbf{E}_1$, where $\mathbf{E}_1$ bears to $\mathbf{C}_1$ the same relation as $\mathbf{E}$ to $\mathbf{C}$. The total heat becomes

$$\Sigma(\mathbf{E}+\mathbf{E}_1)(\mathbf{C}+\mathbf{C}_1), \quad \text{or} \quad \Sigma \mathbf{EC} + \Sigma \mathbf{E}_1\mathbf{C}_1 + \Sigma \mathbf{EC}_1 + \Sigma \mathbf{E}_1\mathbf{C},$$

by expanding.

We can easily show that the third sum vanishes. For $\mathbf{C}_1$ consists of closed tubes of current entirely within the body. Select one of these, of very small section, and sum up $\Sigma \mathbf{EC}_1$ for it alone. Since C_1 is constant for the tube we obtain $C_1 \times$ line-integral of $\mathbf{E}$ once round the tube, and since the line-integral is the sum of differences of potential round a closed curve it vanishes. Hence $\Sigma \mathbf{EC}_1 = 0$ for any tube, and for all.

As for the fourth sum, if the coefficients of resistance form a symmetrical system, $r_{12}=r_{21}$, etc.; *i.e.*, if there is no rotatory property, we have $\mathbf{EC}_1=\mathbf{E}_1\mathbf{C}$ at every point of the mass [see p. 288]. Hence $\Sigma\,\mathbf{E}_1\mathbf{C}$ also vanishes, and the heat reduces to $\Sigma\,\mathbf{EC}+\Sigma\,\mathbf{E}_1\mathbf{C}_1$. That is, when the systems $\mathbf{C}$ and $\mathbf{C}_1$ coexist, the heat is the sum of their separate heats. The heat added to the original heat by altering the current being positive, the actual distribution of current $\mathbf{C}$ must be that arrangement out of all consistent with the supply conditions that makes the heat a minimum. Then, and only then, it equals $\Sigma\,PQ$.

When the rotatory property exists $\mathbf{E}_1\mathbf{C}$ is not equal to $\mathbf{EC}_1$; hence it is not proved that $\Sigma\,\mathbf{E}_1\mathbf{C}=0$, and it requires separate treatment. Let $\mathbf{C}_2$ be the current that must be added to $\mathbf{C}_1$ to make the current that would correspond to the same electric force $\mathbf{E}_1$ if the transverse coefficients of conductivity changed places, k_{12} becoming k_{21}, etc. Then

$$\Sigma\,\mathbf{E}_1\mathbf{C}=\Sigma\,\mathbf{EC}_2$$

throughout the body,

$$=\Sigma\,PQ_2\text{ (surf.)}+\Sigma\,\dot{P}\text{ div }\mathbf{C}_2\text{ (vol.)},$$

where Q_2 is the normal component of $\mathbf{C}_2$. Here $\mathbf{C}_1$ being real, $\mathbf{E}_1$ is real; but $\mathbf{C}_2$ is not a system of closed currents, or div $\mathbf{C}_2$ does not vanish; neither does Q_2 necessarily vanish. The disappearance of $\Sigma\,\mathbf{E}_1\mathbf{C}$ does not therefore follow.

Section VIII. Thermo-Electric Force. Peltier and Thomson Effects.

We had occasion in Section VII. to consider the distribution of steady current in a conducting body of any conductivity, uniform or variable, and with or without identity of properties as regards the electric current in different directions, such system of current being supposed to be kept up by E.M.F. arising purely from difference of potential, there being supposed to be no internal intrinsic E.M.F.'s. This is, however, an ideal state of things. For, in a perfectly homogeneous and non-crystalline conductor there are intrinsic electric forces unless every part of it be at one temperature. And, whenever there is change of material or of structure there is usually intrinsic electric force, even without change of temperature at the place. Also, in a naturally crystalline, or in an originally isotropic material when strained, the intrinsic electric forces arising from difference of temperature are altered so that the thermo-electric qualities vary in different directions. In general, when an electric current passes from one material to another, or in one material from a hot to a cold place, it produces, besides the ordinary "frictional" generation of heat according to Joule's law, which, varying in amount as the square of the current, is irreversible, or always positive, thermal effects which are reversible with the current, being a heating or a cooling, according to its direction. To the consideration of the theory of these reversible effects we now proceed.

If a conductor forming part of a galvanic circuit, or of a circuit containing any constant source of electrical energy outside the conductor under observation, is wholly of one homogeneous metal, and is all at the same temperature, the heat which is the result of the current is generated uniformly per unit of resistance, the amount being RC^2 per second in a portion of resistance R, with the steady current C. But if the conductor is made up of a number of different metals, still all at one temperature, there will be, besides the above heating, a heating effect at some junctions and a cooling at others, their sum balancing on the whole, so that the whole heat is still in accordance with Joule's law. And when the current is reversed, at any junction where there was previously a heating there will now be a cooling, and conversely. This is the Peltier reversible thermal effect of the current.

To particularise, when an iron wire is inserted between two copper wires, and a current is passed through them, there is a cooling at the junction where the current goes from copper to iron, and a heating where it goes from iron to copper, whilst away from the junctions the heating is as the square of the current. The amounts of these heating and cooling effects are found to vary in simple proportion to the strength of the current (their reversibility shows that they must vary as some odd function of the current—first, third, etc., powers, or combinations). Hence, in the resistance R containing a junction, the heat per second is not RC^2, but $PC + RC^2$, where P is a quantity independent of C. What it does depend on will be seen later. P is positive at the iron-copper junction, and negative at the copper-iron junction (the order of the linked metals showing that of the current), or the extra heat is PC at the iron-copper and the deficiency PC at the copper-iron junction, if we reckon P always positive. We have thus a transference of an amount of heat PC from the copper-iron to the iron-copper junction, when both are at the same temperature, taking place, not by heat conduction, but through the medium of the energy of the electric current we pass through the wire.

Since RC^2t represents an amount of energy, and also PCt, both being heats generated in time t, the quantity P is clearly an E.M.F. In fact, if E is the externally impressed E.M.F. in a portion of the circuit containing one junction, we have

$$EC = PC + RC^2.$$

Hence $$E - P = RC,$$

or the actual E.M.F. is $E - P$. Thus, from the Peltier effect we recognise the existence of at least two E.M.F.'s in the circuit besides that of the battery, which do not, at least immediately, appreciably alter the current strength, from their being equally strong and oppositely directed in the circuit, and whose localities are the two junctions. In short, there is an intrinsic E.M.F. P from copper to iron at both junctions. This is the real contact force of copper and iron. All others are counterfeits. To distinguish it from the apparent contact force of copper and iron in air or other medium, often erroneously referred to the metallic contact, we shall call it the Peltier E.M.F.

Any two metals exhibit the Peltier E.M.F., and to a different extent at different temperatures. Its magnitude is obtainable without ambiguity by measurements, when possible, of the reversible heat effect at a junction. If it is a cooling, the E.M.F. acts with the current; if a heating, against it; and the magnitude of P (being, of course, PCt/Ct), is numerically equal to the extra heat produced during the passage of the unit of electricity through the junction, and is found by dividing the extra heat in time t by the integral current in that time. The extra heat may be found by two measurements of the whole heat generated, one with the current from copper to iron, and the second from iron to copper, their difference being twice the Peltier effect.

Now, we could not expect on cutting out the battery to observe any current in a closed circuit of copper and iron if one E.M.F. continued to exactly balance the other, as happens when the junction temperatures are equal. And, in fact, in a circuit of any number of metals whose junctions are all at one temperature no current is to be observed; hence, just as for two metals we have

$$P_{ab} + P_{ba} = 0,$$

(the order of the small subscript letters indicating the direction of the E.M.F., thus in P_{ab}, from metal a to metal b), so for three metals we must have

$$P_{ab} + P_{bc} + P_{ca} = 0$$

at one temperature, with similar extensions to any number of metals. This summation law of Peltier E.M.F.'s shows that we may refer all metals to one metal as a standard; hence we may drop one letter when the standard is one of the metals. P_a and P_b being the Peltier E.M.F.'s from a and from b to the standard metal, that from a to b, or P_{ab}, is $P_a - P_b$.

But our battery current heated the iron-copper and cooled the copper-iron junction, destroying the equality of temperature, and on cutting out the battery there is a weak current found in the circuit whose direction is opposed to the original current, viz., from copper to iron at the warmer junction, which ceases quickly on restoration of equality of temperature. It may, however, be maintained indefinitely by mere application of heat to one of the junctions to keep up a difference of temperature, and the current will be always from copper to iron at the warmer junction when the temperature does not depart greatly from the ordinary atmospheric temperature. We now have the thermo-electric current of Seebeck and Cumming. As the current in this case is in the reverse direction to that of the battery current which would cause the same difference of temperature, there must now be a transference of heat in the reverse direction, viz., from the warm to the cold junction; heat must be supplied at the warm junction to prevent it cooling, and heat must be taken away at the cold junction to keep it from getting warmer, quite apart from the frictional heat, and the alteration of temperature by thermal conduction, radiation, etc. P_1 and P_2 being the Peltier E.M.F.'s, now no longer equal, we should have

$$P_1 - P_2 = RC,$$

by Ohm's law. Also P_1C of heat is absorbed at the hot junction per second, P_2C generated at the cold, and their difference, which equals RC^2, is frictionally generated. This is in accordance with Ohm's law and the law of conservation of energy, with the assumption that there are no other intrinsic E.M.F.'s than the Peltier in the circuit.

So far, we have no knowledge of how P varies with the temperature, except in this respect: that since P is from copper to iron when both the metals are at the same (ordinary) temperature, and the current is from copper to iron at the warmer junction when its temperature is raised, it follows that if P_2 be the E.M.F. at the cold and P_1 at the warm, both from copper to iron, P_1 must have been increased by the heating, for P_2 remains the same, the temperature of the cold junction being unaltered. But this legitimate conclusion is soon found to be utterly erroneous. For if we keep the cold junction constantly at, say, 0° C., and continuously raise the temperature of the other junction, the current increases up to a maximum (when the hot junction is at about 275° C.), and then decreases to nothing at a higher temperature (about 550° C.), and immediately sets in the reverse way on further heating, viz., from iron to copper at the hot junction.

Now, it has been proved that when a junction is at the temperature 275° C., there is neither absorption nor generation of heat there, *i.e.*, $P_1 = 0$, or the metals, iron and copper, are thermo-electrically neutral at this temperature. Hence, when the hot junction is at this neutral temperature, the current being from copper to iron *there*, and the only known E.M.F. being from copper to iron at the *cold* junction, the current is against the E.M.F. Therefore there must be other E.M.F.'s in the circuit not at the junctions, whose sum is greater than P_2, the Peltier E.M.F. at the cold junction, and opposed to it in direction. Three courses are open: (1) There must be an E.M.F. in copper from cold to hot, with possibly a weaker in the iron, also from cold to hot. (2) Or, E.M.F.'s in the copper and iron from hot to cold, the latter being the greater. (3) Or, an E.M.F. in iron from hot to cold, and in copper from cold to hot. In all three cases their sum to be greater than P_2, and against it in the circuit.

The existence of E.M.F.'s (with reversible thermal effects) in unequally heated wires was theoretically predicted by Sir W. Thomson, reasoning from the behaviour of iron and copper, and he afterwards verified his prediction experimentally, and determined the directions of the E.M.F.'s. The reasoning itself is of the simplest character, but the results thereof may be put into a form likely to cause some bewilderment. Following the analogy of a material fluid moving in a pipe unequally heated, and giving out or receiving heat in its motion, Sir W. Thomson expressed his results in terms of the convection of heat by vitreous or by resinous electricity. If the E.M.F. in a metal is from cold to hot, the specific heat of electricity in that metal is positive, and it is the vitreous electricity that carries heat with it; whilst, if the E.M.F. is from hot to cold, the specific heat is negative, and the resinous electricity conveys the heat. For, imagine the fluid at rest in the pipe in the first place, the fluid having everywhere the temperature of the pipe at the place where it may be. Set the fluid moving; as the fluid moves from colder to warmer places it receives heat from, and, therefore, cools the pipe; this heat it carries with it until it reaches places where the temperature falls, when it gives out the heat and warms the pipe. We have similar heat effects in a wire when the E.M.F. acts from cold to warm, a cooling when the current passes from cold to warm, and a heating when from warm to cold, hence the electric current acts as a real fluid would do. But should the E.M.F. be from hot to cold, the heat convection would be reversed, therefore now it is the resinous electricity that carries heat with it. I had much difficulty in following thermo-electric descriptions in terms of the specific heat of electricity and the convection of heat by resinous or vitreous electricity. However, all that it is necessary to remember is whether the E.M.F. acts from cold to hot or from hot to cold. The corresponding absorptions and generations of heat when a current passes may be of course easily deduced.

Let σ be the E.M.F. from cold to hot per unit rise of temperature in any metal. Then σ is Sir W. Thomson's specific heat of electricity ("without hypothesis, but by an obvious analogy"). It would be very much better if this important quantity had a less misleading name, but I cannot at present think of a suitable word to convey a connection between an E.M.F. and a rise or fall of temperature. But the whole E.M.F. in a wire due to this cause may be suitably called the Thomson E.M.F., as the reversible heat effect is called the Thomson effect; similarly to Peltier effect and E.M.F. Thus the Thomson E.M.F. in a wire whose terminal temperatures are t_1 and t_2, t_1 being the higher, is simply $\int_{t_2}^{t_1} \sigma dt$, acting when it is positive from the low to the high temperature. In a closed circuit of one metal, it vanishes; the terminals, which may now be anywhere, having the same temperature.

Regarding this, it was proved by Magnus that there was no current in a closed circuit of one metal, however the temperature varied, and also however the section varied. Though this has been shown later to be not always rigidly true in extreme cases, yet the departures are very small. Assuming its exact truth, it follows that in a circuit of one metal the E.M.F. from one point to another must be the same by either path; and generally, the integral E.M.F. between points at temperatures t_2 and t_1 must be independent of the intermediate temperatures, and therefore must equal $T_1 - T_2$, where T_1 and T_2 are the values at temperatures t_1 and t_2 of a function T of the temperature only (for a single metal), the E.M.F. per unit rise of temperature being therefore dT/dt. Therefore, if T_a and T_b are the values of T for two metals a and b, the complete E.M.F. in a circuit of two metals (which might be called the Seebeck E.M.F.) is

$$E_{ab} = [P_{ab}]_2^1 + [T_a - T_b]_2^1, \qquad (1)$$

where the square brackets indicate that the difference of the functions at temperatures t_1 and t_2 must be taken. The T's being reckoned from cold to hot, the Peltier E.M.F. from a to b, and t_1 being the higher temperature, when E_{ab} is positive the current sets in from a to b at the hot junction. In terms of σ we have

$$E_{ab} = [P_{ab}]_2^1 + \int_{t_2}^{t_1} (\sigma_a - \sigma_b) dt. \qquad \ldots\ldots\ldots\ldots\ldots\ldots\ldots\ldots (2)$$

How P and σ vary with the temperature there is nothing to show so far, except that we know that with copper and iron P falls to zero as the temperature rises from 0° to 275°, whilst, as Sir W. Thomson found, σ is positive in copper and negative in iron, or the E.M.F. is from cold to hot in copper and from hot to cold in iron. This was to the discoverer a most unexpected conclusion, he expecting to find σ either positive in both or negative in both, having a slight bias in favour of the former. Thus, both Thomson E.M.F.'s act the same way in the circuit as if from copper to iron at the hot junction.*

The phenomenon of thermo-electric inversion or reversal of direction of the current by raising the temperature of one junction, discovered by Cumming and Seebeck, is not peculiar to iron—though certainly especially noticeable when iron is one of the metals paired—but is an almost general property of any two metals or alloys, with sufficient range of temperature. It is so easily observed in the iron-copper couple that it is scarcely credible that many early observers should have been unable to verify it. An interesting way of observing it is with the telephone. Connect a low-resistance telephone with a copper wire, with an iron wire inserted, and make and break connection so rapidly by means of a rheotome that a weak current in the circuit will give a pure musical tone. Warming a junction with the hand will bring on the sound. Bringing it near a hot flame will be sufficient to raise the sound to its maximum, which is when the neutral point is reached. On further heating the sound falls steadily to a dead silence, and then returns. This is at about low red heat. Still further heating raises the sound steadily up to about twice the loudness at its former maximum, and at the highest temperatures the sound is about stationary. That is to say, on first removal from the source of heat, although the junction is cooling most rapidly, yet the sound is stationary for a moment, or falls only slowly, before the rapid fall sets in and the reverse phenomena occur. This shows that there is a second maximum; and, in fact, with a thin steel wire, there was a decided increase of sound on first removal from the source of heat, showing that the second maximum had been passed, which I could not observe with soft iron. The temperature of the junction at the second maximum could not be much less than the melting point of copper, 1,090° C., for the thin copper wire did partially melt sometimes. On the other hand, if the wires be thick, it is much more difficult to heat them sufficiently. The reason of the second maximum will appear later.

* [Thomson's researches are to be found collected in his "Mathematical and Physical Papers," vols. i. and ii.; his theory being in vol. i. and experiments mostly in vol. ii.]

In saying that there is no current in an unequally heated circuit of one metal, it is of course to be understood that the metal is all the same metal—not merely all copper, for instance, but the same copper. Different specimens may give strong thermo-electric currents. Thus, two copper wires of the same gauge and make gave, when one of their junctions was at bright red, a current nearly as strong as that in the copper-iron couple at the first maximum, which indicates an E.M.F. of about ·003 volt, and it was the same with several fresh joints, proving a great difference of quality. It sometimes happens, too, that the sudden insertion of a wire into a flame will bring on a faint sound, without any possibility of heating of junctions. This may be due to a difference of structure in parts of the same wire, either existing originally, or brought on by different parts of the wire having been subjected to different treatment before the experiment, and not having returned to identity.

Section IX*a*. The First and Second Laws of Thermodynamics.

The equation already given for the total E.M.F. in a thermo-electric circuit of two metals [equation (2) p. 314] in terms of the Peltier and Thomson forces, or the reversible heat effects per unit quantity of electricity at the junctions and in the unequally heated parts of the separate metals respectively, involves no hypothesis. If we could measure the values of PQ and of σQ (the Peltier and Thomson heats produced by the passage of any quantity Q of electricity) for all pairs of metals, and at a sufficient number of different temperatures, thus obtaining P and σ as functions of the temperature, we should have all the data required for the calculation of the thermo-electric force in any metallic circuit. But at this point of the inquiry a remarkable speculation of Sir W. Thomson comes in, furnishing an excellent example of the great aid to be derived from mathematical theorising in giving a direction to experimental inquiry and facilitating the practical completion of the theory. For the science of Thermodynamics, which at the time of its application to Thermo-electricity had only just been established on correct principles, furnishes information as to a relation between the Peltier and Thomson effects and the temperature, which, though not amounting to strict demonstration, is yet of great probability and suggestiveness.

There are in Thermodynamics two fundamental laws. The first expresses the equivalence of heat and work, an example of the principle of conservation of energy, and first applied to Thermodynamics by Clausius. Let a substance be made to go through a cycle of changes, at the end of which it returns exactly to its original state, having during the cycle been varying in pressure, temperature, and volume, and receiving or losing heat during its changes of temperature, and doing external work or having work done upon it by its pressure during its changes of volume. Let H_1, H_2, ..., be the amounts of heat successively given to the substance, heat given out being reckoned negatively, and let W be the whole external work performed by the substance during the cycle, then

$$\Sigma H = W$$

expresses the First Law of Thermodynamics. If ΣH be positive, or heat be given to the substance on the whole, it is accounted for by the external work done by it during the cycle, since the body is in the same state at the end as at the beginning, and contains the same amount of heat, whilst if ΣH be negative, there has been work done upon the substance. In the first case heat is transferred from higher to lower temperatures, in the latter from lower to higher. Any such arrangement is a heat engine, by means of which work may be obtained from heat, or conversely.

The mathematical expression of the Second Law is

$$\Sigma H/t = 0,$$

when the cycle of operations is completely reversible, t denoting the temperature at which the quantity H of heat is given to the substance. It is a consequence of Carnot's reasoning regarding reversible heat engines, when his erroneous assumption of the materiality of heat is discarded, and the equivalence of heat and work admitted in its place. As shown by Carnot, the test of perfection in a heat engine is its perfect reversibility. Thus, if it work between the temperatures t_1 and t_2 on any scale, the working substance receiving heat of amount H_1 from a hot body at the higher temperature t_1, and giving out an amount H_2 to a cold body at the temperature t_2, and performing therefore $W = H_1 - H_2$ of work in a cycle; and if the operations be completely reversible, so that by doing W of work on the substance, H_2 of heat is taken from the cold body and the greater amount $H_1 = W + H_2$ given to the hot body, such an engine is perfect; in this sense, that no other engine can possibly do a greater amount of work with the same supply of heat and working between the same temperatures. If there could be an engine B which should have a greater efficiency than a given reversible engine A, the two might be coupled, both working between the same temperatures, B working direct, receiving H_1 of heat from the hot body and driving A backward (since A is reversible) with an excess of useful work. For B would do more work in its direct action than would be needed to drive A backwards and give back to the hot body the heat H_1 which B received therefrom, and we could thus obtain a perpetual supply of work from heat at the lower temperature, *i.e.*, without letting heat down from a higher to a lower temperature, which is admittedly impossible. [This was made an Axiom by Clausius and Thomson, though in different forms.]

It follows that all heat engines working between the same temperatures have, if they are reversible, the same efficiency, or, in other words, the efficiency is independent of the nature of the working substance, and depends only upon the temperatures between which it works. *I.e.*, the ratio $W : H_1$ is a function of t_1 and t_2 only, or, more conveniently, the ratio $H_1 : H_2$ is a function of t_1 and t_2 only.

The nature of this function could therefore be found by experiment upon any substance.

Now, if the substance be air, and the scale of temperature be that of the air thermometer, whose zero is 273 °C. below the freezing point of water (air and other gases expanding nearly alike in the ratio 273 : 373 when raised from 0° to 100 °C. at constant pressure), then, by the known gaseous laws, and with the assistance of the experimentally known fact that a gas in expanding from one volume to another requires no heat (or nearly none) to be supplied to it to keep its temperature from falling (unless it be performing external work during the expansion, when an equivalent amount of heat must be supplied to the gas), it is easily shown that the ratio of the heat supplied to the substance at the higher to that given out by it at the lower temperature, when working reversibly, is simply the ratio of the temperatures. That is,

$$H_1 : H_2 = t_1 : t_2;$$

or

$$H_1/t_1 = H_2/t_2.$$

And, extending this to any reversible cycle, reckoning heats given out as negative, the corresponding equation is

$$H_1/t_1 + H_2/t_2 + H_3/t_3 + \dots = 0,$$

or simply

$$\Sigma H/t = 0,$$

each amount of heat being divided by the temperature at which it is supplied.

This, however, depends for its accuracy upon the degree of closeness with which air fulfils the gaseous laws, and also the above-mentioned property of requiring no supply of heat when expanding to keep its temperature constant. Some gases cool a little, others heat a little in expanding, and the gaseous laws are not perfectly fulfilled by any gas. Hence, although $\Sigma H/t = 0$ is very nearly true whatever be the nature of the working substance when the temperature is reckoned by the air thermometer, yet it is not exactly true. It would be exactly true with an imaginary perfect gas for thermometer. But, by actually *defining* the scale of temperature so as to make

$$H_1 : H_2 = t_1 : t_2,$$

when an engine works reversibly between any two temperatures t_1 and t_2, we have a scale which is independent of the properties of any particular substance, which is Sir W. Thomson's scale of absolute temperature. It is practically the same as that of the air thermometer.

In a thermo-electric circuit we have reversible heat effects resembling those taking place in the cycle of a reversible heat engine. Thus, in a circuit of two metals, if we make the temperature t_1 of one junction a little higher than t_2, that of the other, and if P_1 and P_2 are the corresponding Peltier forces, there is absorption of heat P_1Q and generation P_2Q at the hot and cold junctions respectively during the transfer of the quantity of electricity Q by the current set up, their difference $(P_1 - P_2)Q$ being expended in uniformly heating the circuit according to its resistance. Here P_1Q corresponds to H_1, the heat given at temperature t_1 to a substance working reversibly between t_1 and t_2, and P_2Q to H_2, the heat given out at the lower temperature; whilst $(P_1-P_2)Q$ corresponds to $H_1-H_2=W$, the work done by the substance. In both cases there is transference of heat from hot to cold, with performance of work. But the heat developed "frictionally" in the wire by the current is not a reversible operation. To make the correspondence complete, we must suppose that when the heat engine does W of work, it immediately converts it into heat again by working against frictional forces.

And, as we may work the engine backwards, transferring heat from cold to hot, taking H_2 from the cold body, and giving $H_1=H_2+W$ to the hot body, so we can, in the thermo-electric circuit, by sending a current through it in the opposite direction to and of greater strength than the natural current, carry heat from the cold to the hot junction, absorbing P_2Q at the cold and generating P_1Q of heat at the hot junction during the passage of Q.

There is also the ordinary conduction of heat going on, which is essentially of an irreversible nature, and it may be of far greater magnitude than the reversible effects. But this is completely ignored in applying the Second Law to the reversible effects, and is considered to have no influence on the phenomenon.

SECTION IX*b*. APPLICATION OF THE SECOND LAW TO THERMO-ELECTRICITY.

In the first place, if there were only the Peltier effects in a circuit of two metals, a and b, we should have

$$P_1Q/t_1=P_2Q/t_2,$$

where t_1 and t_2 are the absolute temperatures of the junctions, t_1 being the higher, P_1 and P_2 the Peltier forces, both reckoned from a to b, and Q the quantity of electricity passing whilst P_1Q of heat was absorbed at the hot and P_2Q generated at the cold junction; or

$$P_1/t_1=P_2/t_2.$$

And, this holding for any two temperatures, the quantity P/t should be the same for all temperatures, a constant depending upon the nature of the two metals concerned, $=p$ say. Hence

$$P=pt$$

would express the Peltier force at any temperature t, and the thermo-electric force of the circuit would be simply

$$E=P_1-P_2=p(t_1-t_2),$$

and therefore be proportional to the difference of temperatures of the junctions, whatever their absolute values might be.

This formula for E is nearly true for small differences of temperature when we keep t_2 constant and raise t_1. But the value of p must be taken differently according to what the mean temperature is. And the fact of inversion of current with many pairs of metals shows that the formula is quite wrong for wide ranges of temperature.

But now apply the Second Law to both the Peltier and the Thomson effects. Let σ_a and σ_b be the E.M.F.'s from cold to hot in a and in b per degree of rise of temperature. Then $\sigma_a Q$ and $\sigma_b Q$ are the heats absorbed when Q passes from cold to hot. Hence, taking Q as the unit of electricity for simplicity,

$$\frac{P_1}{t_1} - \frac{P_2}{t_2} + \int_{t_2}^{t_1} \frac{\sigma_a - \sigma_b}{t} dt = 0 \quad \text{.........................(3)}$$

expresses the Second Law. We have the − sign before P_2, because both P's are reckoned from a to b, and the − sign before σ_b, because both σ's are reckoned from cold to hot, whilst the current must go from a to b at one junction, and from b to a at the other, and from cold to hot in one metal, and from hot to cold in the other.

Equation (3) holding for any two temperatures, let the range be 1°, the mean temperature being t, and $t_1 = t + \frac{1}{2}°$, $t_2 = t_1 - \frac{1}{2}°$. We thus get, (or by simply differentiating (3)),

$$\frac{d}{dt}\left(\frac{P}{t}\right) + \frac{\sigma_a - \sigma_b}{t} = 0, \quad \text{.............................(4)}$$

which expresses the difference of "specific heats of electricity" in the two metals in terms of the variation of P/t with the temperature. Let b be a metal in which there is no reversible heat effect, *i.e.*, let $\sigma_b = 0$. Then by (4) we see that, if the Peltier force divided by the temperature varies with the temperature, its increase per degree equals the Thomson E.M.F. from hot to cold divided by the temperature.

By means of the relation in (4) we may eliminate the Thomson forces from the equation of E.M.F. This is,

$$E = P_1 - P_2 + \int_{t_2}^{t_1} (\sigma_a - \sigma_b) dt,$$

from a to b at the hot junction. Put for $\sigma_a - \sigma_b$ its value in terms of P, by (4), and integrate, and there results simply

$$E = \int_{t_2}^{t_1} \frac{P}{t} dt, \quad \text{.....................................(5)}$$

which is Sir W. Thomson's expression for the complete E.M.F. in a circuit of two metals in terms of the Peltier force only, at the temperatures intermediate between those of the junctions. In words, the thermoelectric force equals the product of the difference of temperature of the junctions into the mean value of P/t between t_1 and t_2.

Let the range be 1°, $t_1 = t + \frac{1}{2}$, $t_2 = t - \frac{1}{2}$. Then we get simply $E = P/t$. In fact, the Thomson effect decreases indefinitely in comparison with the Peltier effects at the junctions as the difference of temperature is made smaller and smaller. This all-important quantity P/t is called the thermo-electric Power. It depends upon the nature of both metals and on the temperature.

Thus, we do not need to measure the Peltier and Thomson effects to determine the E.M.F. in a circuit, but merely require to know how P/t varies with the temperature, and that may be found by observing the current in the circuit due to both P's with a small difference of temperature and varying the mean temperature.

It is convenient to refer all metals to a standard metal in which $\sigma=0$, whether such a metal really exist or not. Then, in a circuit made up of any number of different metals, the

$$\text{complete E.M.F.} = \Sigma\int\frac{P}{t}dt,$$

the summation including all the metals, each integral being taken between the terminal temperatures, for the metal concerned. P here is not now the real Peltier junction force from one metal to the next, but what it would be if the next metal were the standard. If all the junctions are at the same temperature, every integral vanishes, and there is no resultant E.M.F., the Peltier forces balancing, whilst the Thomson forces also balance, separately in each metal. And it follows that if any circuit be cut anywhere, and the ends joined through a series of metals whose junctions are all at the same temperature, the terminal conductors of the series being of the same metal, no change will be made in the E.M.F., *i.e.*, the E.M.F. of the new circuit is the same as that of the old. Of course there are additional reversible thermal effects introduced, but they exactly balance in the sum.

Further progress is facilitated by representing the changes in the thermo-electric powers P/t of different metals continuously in a diagram. Measuring temperatures from left to right, and the powers upwards, the power of any metal for all temperatures will be represented continuously by a line, which may be straight or curved for all we know without experiment. Sir W. Thomson's first thermo-electric diagram consisted of a number of lines which were all quite straight except that for brass, which shows some curvature. It has been confirmed by Professor Tait's experiments that the lines of thermo-electric power for all the metals are sensibly straight within wide ranges of temperature, with one or two exceptions. For generality we may take one straight and the other curved, arbitrarily.

Let the temperatures be measured along the horizontal base line, let the straight line a be the line of thermo-electric power for a metal a, and the line b, partly straight and partly curved, be that for a metal b, both referred to a standard in which $\sigma=0$. Pair a and the standard, and let the junctions be at temperatures t_1 and t_2. The verticals A_2t_2 and A_1t_1 present the power of a at those temperatures. Therefore the rectangles $A_2a_2Ot_2$ and $A_1a_1Ot_1$ represent the Peltier junction forces. Their difference is the excess of the area $A_1t_1t_2A_2$ over the area $A_1a_1a_2A_2$. But the mean value of the power of a multiplied by the difference of temperatures, which, by what was said before, is the total E.M.F., is represented by the area $A_1t_1t_2A_2$. Hence, since there is no Thomson E.M.F. in the standard metal, the area $A_1a_1a_2A_2$ represents the Thomson E.M.F. in the metal a. Similarly, by pairing b and the standard, the areas $B_1b_1Ot_1$ and $B_2b_2Ot_2$ represent the Peltier forces at the same temperatures t_1 and t_2, and the area $B_1b_1b_2B_2$ the Thomson integral force in the metal b.

Finally pairing a and b, the areas $A_1a_1b_1B_1$ and $A_2a_2b_2B_2$ show the Peltier forces, and of course the Thomson integral E.M.F.'s are as before. The complete E.M.F. in the circuit is the area $A_1A_2B_2B_1$. As regards the directions these four forces act in the circuit, it is only necessary to remember that they all act from places of higher to places of lower power, therefore from A_1 to B_1, and from A_2 to B_2 at the junctions, and from A_2 to A_1 and from B_1 to B_2 in the metals a and b respectively.

In the case illustrated, the Thomson forces predominate, and the current circulates in the direction $A_1B_1B_2A_2A_1$. When the higher temperature is raised to t_3, the lines cross, the metals are neutral to one another, the Peltier E.M.F. vanishes, and the current is a maximum. Further increase of temperature of the hot junction (keeping t_2 constant) reduces the current, because the Peltier force at the hot junction acts now from b to a. When the temperature of the hot junction reaches t_4 the current vanishes. Above t_4 it comes on again in the reverse direction, and gets stronger and stronger until t_5 is reached, where there is a second crossing of lines of power and maximum current (negative). From t_5 to t_6 the current gets weaker, and at t_6 reaches a minimum and begins to increase again. As the lines extend no further, we cannot say what will happen with further rise of temperature.

SECTION X. THE THERMO-ELECTRIC DIAGRAM AND ITS THEORY.

The thermo-electric diagram is a most valuable aid to an intelligent comprehension of the subject, and gives one rapidly a general view of the relative magnitudes and the directions of the E.M.F.'s in different parts of a linear circuit, whether of two or more metals, and especially so when the neutral temperature of two metals is within the range of temperature concerned.

Referring to the same diagram, in which the thermo-electric powers of two metals a and b are represented by the lines aA_2A_1 and bB_2B_1, of which the former is straight, and the latter, for the sake of generality, is supposed to be curved in part of its course, and also to change curvature so as to cross the line of a once in its straight portion and twice later at the higher temperatures t_5 and t_6, we have, when the two metals are paired to form a circuit, and their junctions kept at temperatures t_1 and t_2, the Peltier junction forces represented in magnitude by the rectangular areas $A_1B_1b_1a_1$ and $A_2B_2b_2a_2$, their directions in the circuit being respectively from A_1 to B_1 at the hot, and from A_2 to B_2 at the cold junction, *i.e.*, both downwards in the figure from the metal of higher to that of lower power. And, in addition, we have the Thomson E.M.F.'s in the separate conductors, owing to their ends being at different temperatures, whose integral amounts are represented by the areas $A_1A_2a_2a_1$ and $B_1B_2b_2b_1$, and whose directions are, as before for the Peltier forces, downwards in the figure, viz., from A_2 to A_1 in a (cold to hot), and from B_1 to B_2 in b

(hot to cold). And their resultant is the complete E.M.F. in the circuit, represented by the area $A_1B_1B_2A_2$, acting so that the current is from

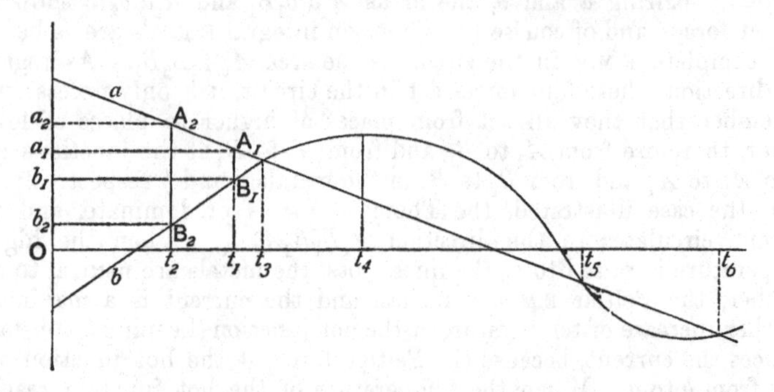

to b at the hot junction, and its direction round the circuit therefore similar to that of the motion of the hands of a watch.

These areas, by means of the relation, energy = E.M.F. × quantity (a relation whose meaning cannot be too carefully studied by those who think electricity is energy, and by those who conclude that E.M.F. is a form of energy, both of which views have their exponents), may also indicate the amounts of heat absorbed or generated in the circuit during the passage of the unit quantity of electricity, as the unit current for one second, or $1/n$ of the unit current for n seconds. At the hot junction, the current and the Peltier force there situated are similarly directed, and the area $A_1B_1b_1a_1$ shows the amount of heat absorbed there (difference of power of a and b multiplied by the temperature), *i.e.*, disappearing as heat and passing into the form of current energy, which latter is not to be confounded with the heat "frictionally" developed, which is energy dissipated. At the cold junction the current is against the Peltier E.M.F. there; hence the area $A_2B_2b_2a_2$ shows the amount of heat generated there passing from current energy into heat--not, however, in the frictional manner, irreversibly, but so that the generation becomes an absorption when the current is reversed. And in the substance of both conductors the current goes with the E.M.F., and hence the corresponding areas both represent heat absorbed.

But as regards these last E.M.F.'s, the diagram does not in any way indicate their distribution in the wires, but only their integral amounts, reckoned from end to end of the wires, depending on the terminal temperatures. Hence, the Thomson reversible heat effect, in a for example, may be very different in its distribution, according to the manner in which the temperature varies along the wire. If, for instance, the whole wire a and the whole wire b, except their portions close to the hot junction, be kept at or near the temperature of the cold junction, it is evident that the Thomson forces must be pretty nearly all collected near the hot junction, where the temperature of the wires falls rapidly. Or, if we greatly raised the temperature of the middle part only of one of the wires, we should introduce two opposed

E.M.F.'s of equal amounts, one each side of the place of maximum temperature, of greatest intensity where the temperature varied most rapidly. But, provided the terminal temperatures are the same, the integral E.M.F. and integral reversible heat effect in the wire are unaffected, and it is these alone of which the thermo-electric diagram takes cognisance, irrespective of the actual manner of variation of temperature between the given extremes.

The net result is that there is an absorption of heat per unit of electricity passing, whose amount is represented by the area $A_1B_1B_2A_2$ enclosed by the lines of the two metals, and by the vertical lines corresponding to their junction temperatures. It is the excess of the heat absorbed over that generated reversibly, and is accounted for by Joule's law as frictional heat in the whole circuit, distributed in proportion to the resistance of the different portions. And, of course, considering the areas to represent the E.M.F.'s, this area $A_1B_1B_2A_2$ is the complete thermo-electric force in the circuit, which, along with the resistance, determines the strength of current, provided there are no other E.M.F.'s acting.

So far as the four component E.M.F.'s are concerned, the above statements regarding reversible heats per unit of electricity do not require any alteration when the current passing is not that due to the thermo-electric forces alone, but also to some other intrinsic E.M.F., as of a galvanic cell in the circuit; with this reservation, that if the cell should reverse the current, the former absorptions of heat now become generations, and the former generations become absorptions, whilst the frictional heat per unit of electricity passing is now numerically equal to the excess of the cell's E.M.F. over that of the thermo-electric pair.

Dismissing now the borrowed galvanic cell, and considering thermo-electric forces alone, we may observe the effect produced by the powers of two metals becoming equal at a certain temperature. In the previous, both the junction temperatures were below t_3, the first neutral temperature. But, keeping the cold junction at the same temperature t_2, if we raise the temperature of the other above t_3, the Peltier force at the latter vanishes and reappears reversed, acting now from b to a. The complete E.M.F. reached its maximum at t_3, and is consequently thereafter reduced, and this will go on until the areas to right and left of the neutral point are equal, when the higher temperature is t_4. But we may accelerate this by warming the cold junction, and have zero current for any number of pairs of temperatures between t_2 and t_4, one below t_3 and the other above. Starting with both junctions at the neutral temperature itself, the effect of either heating or cooling one of them is to produce a current in the *same* direction. With both junctions above t_3 (or, more strictly, between t_3 and t_5, so as not to include a fresh crossing point), we again have only a single area to deal with, as when both were below t_3, though now the direction of current is reversed, from the line of the metal b being above instead of below that of a.

That one of the lines of power becomes curved (and it would be the same if both were curved) does not alter the method of reckoning the complete E.M.F. by the area enclosed between the lines of power and the verticals at the terminal temperatures. And obviously the Peltier forces, being always represented by rectangles, require no modification of treatment. But it is somewhat different with the Thomson force in b, the metal with curved line of power. For, draw a horizontal straight line from any point of the b line to the vertical through O, such as B_2b_2, and let its end B_2 travel along the b line, keeping always parallel to itself. The area it sweeps out between any two temperatures (as $B_2B_1b_1b_2$ between temperatures t_2 and t_1) measures the Thomson E.M.F. within that range in the metal b, so long as the horizontal line moves continuously either upward or downward; the E.M.F. being from hot to cold when it moves up, and from cold to hot when down. But when the line of power is curved, it may move up in one part of its course, and down in the rest; thus, if the junction temperatures are t_3 and t_5, it will move up to t_4, and thereafter down. In this case we must take the excess of the area swept out in the second part of the motion over that in the first to represent the Thomson E.M.F., which is from cold to hot. On the other hand, the Thomson E.M.F. in a between the same temperatures, also from cold to hot, requires no special treatment. As there are no Peltier forces, the complete E.M.F. equals the difference of the two Thomson E.M.F.'s, and this is simply the area bounded by the straight a line and the curved b line, the terminal verticals being non-existent. This example is curious from the complete absence of the junction forces and reversible heat effects there, such being confined to the interior of the metals. But a much more curious case was pointed out by Professor Tait, which may be readily understood from the diagram, viz., that in which a thermo-electric current is kept up in a circuit of two metals solely by the Thomson E.M.F. in one of them, there being thus absorption of heat in some parts with generation in other parts of the one, but no reversible effect at either junction or in the other conductor. Thus, pair the metal b and the standard metal whose line of power is the base line Ot_6, and keep their junctions at the temperatures at which the line b crosses the base line, the two neutral points of b with respect to the standard. Then we have no junction forces, and no force in the standard metal, but a large Thomson E.M.F. in the metal b.

When there are more than two metals in a thermo-electric circuit, the diagram naturally becomes more complex. The terminal temperatures of any wire being given, that portion of its line of power between these limits must be selected, and the same done for all the other wires. These detached lines must now have their ends joined by verticals in the proper order, thus making a closed circuit. The temperature and thermo-electric power are cyclic in the real electric circuit—*i.e.*, starting from any point, and going once round the circuit, they come back to their old values. Similarly in our closed diagram the power is cyclic, varying gradually with change of temperature in those portions made up of the lines of power of the different metals, and abruptly at the verticals, corresponding to the passage from one metal to another without change of temperature. Should there be no crossing of lines, the area bounded by our closed "curve" (in the general sense) gives us at once the complete E.M.F. If there are crossings, some areas will have to be reckoned negatively, and very curious complications may occur, as may be seen by inspection of a complete thermo-electric diagram for all the metals, with a considerable number of neutral temperatures of pairs of metals within a given range of temperature.

The method of reckoning the magnitude of the Thomson E.M.F. in a metal by the area swept out by a horizontal line, as above described, one end of which moves along the line of power, would be no longer correct if our horizontal base line were taken to represent the line of power of a metal in which the "specific heat of electricity" is not *nil*. Therefore, if no metal could be found in which there was no reversible heat effect, it would be highly convenient to imagine there to be one, and this imaginary standard metal would do quite as well as a real one. It is, however, satisfactory to know that there is a metal that fulfils the condition, or nearly so, viz., lead, in which the specific heat of electricity is *nil*, or very small, as found by Le Roux. Lead is therefore the appropriate standard metal whose line of power is the base line Ot_6.

No other metal has been found having this property, but some of Professor Tait's alloys of platinum and iridium answer the description. Their lines of power are therefore parallel to the base lead line, and the complete E.M.F. in a circuit formed of two of them, or of one of them and lead, is represented by a rectangle, being simply proportional to the difference of junction temperatures, current thus varying as difference of temperature. Such arrangements are most appropriate for measuring temperatures.

The lines of most of the metals are straight within wide ranges of temperature. This, which is remarkably convenient for the numerical calculation of thermo-electric forces, proves that the specific heat of electricity varies in general as the absolute temperature. For, the equation of a straight line being linear, let

$$p = r + st$$

express the power of any metal referred to lead in terms of the temperature t, where r and s are two constants. Then σ, the specific heat of electricity for that metal, being the E.M.F. from cold to hot per unit rise of temperature, must equal $-st$, or vary as the temperature. The constant s, depending on the nature of the metal only, that is, having no relation to what other metal it may be joined with to obtain a thermo-electric current, is positive for those metals whose specific heats of electricity are negative, and conversely. In fact $s =$ tangent of angle between the line of the metal referred to and the lead line, positive when it slopes up from left to right, whereas the Thomson force under the same circumstances is directed downwards. As for the constant r, it represents what the power referred to lead would be at the zero of absolute temperature, $-273°$ C., if the lines of lead and the metal concerned continued straight all the way to that limit. But when the line of a metal is curved, s is, of course, no longer independent of the temperature. Thus, in the figure, the value of s for the metal b changes from positive to negative at temperature t_4, and becomes positive again later. It is not likely that s keeps quite constant for any metal if sufficiently heated so as to appreciably change its structure (as by softening it), and it certainly varies greatly in iron and nickel, as regards the former of which Professor Tait found that its line of power became sinuous at high temperatures. But iron is anomalous in its behaviour in various other respects, as are the other magnetic metals, so we may expect anomalous thermo-electric behaviour.

The following are the formulæ for the component E.M.F.'s and their resultant in a circuit of two metals a and b. Let p_a and p_b be their powers with respect to lead at temperature t. Then

$$p_a = r_a + s_a t, \qquad p_b = r_b + s_b t,$$

where t is the absolute temperature, and r and s are constants. Put a in contact with lead, the junction E.M.F. is $p_a t$, the power multiplied by the temperature; thus

$$P_a = t(r_a + s_a t), \qquad P_b = t(r_b + s_b t),$$

P_a and P_b being the contact forces with lead. Put a and b in contact; their relative power is $p_a - p_b = p_{ab}$ say, or

$$p_{ab} = (r_a - r_b) + (s_a - s_b)t = (r_{ab} + s_{ab}t),$$

and therefore the contact force from a to b at temperature t is

$$P_{ab} = t(r_{ab} + s_{ab}t). \quad \text{.............................(1)}$$

In the circuit formed of a and b with junctions at temperatures t_1 and t_2 (t_1 being the higher) there are two contact forces whose values are found by giving t successively the values t_1 and t_2 in the last formula. Both being from a to b their difference constitutes the acting E.M.F. arising from contact force. This is,

$$r_{ab}(t_1 - t_2) + s_{ab}(t_1^2 - t_2^2). \quad \text{.........................(2)}$$

Next let T_a and T_b be the Thomson forces in a and b. Then since

$$T_a = \int_{t_2}^{t_1} \sigma_a dt \text{ from cold to hot, and } \sigma_a = -s_a t,$$

therefore

$$\left.\begin{aligned} T_a &= \int_{t_2}^{t_1} s_a t dt = \tfrac{1}{2}s_a(t_1^2 - t_2^2), \\ \text{and similarly} \quad T_b &= \tfrac{1}{2}s_b(t_1^2 - t_2^2). \end{aligned}\right\} \text{ both from hot to cold. ....(3)}$$

Adding together the two Peltier forces and the two Thomson forces, and attending to the signs so that when the sum is positive the current is from a to b at the hot junction, we find the complete E.M.F. to be

$$E_{ab} = (t_1 - t_2)\{r_{ab} + \tfrac{1}{2}(t_1 + t_2)s_{ab}\}, \quad \text{....................(4)}$$

which can, of course, be obtained directly from the formula

$$E_{ab} = \int_{t_2}^{t_1} p_{ab} dt, \quad \text{.... ....(5)}$$

which is Sir W. Thomson's general formula (before obtained), applicable whether the lines of power are straight or curved. Equation (4) may be written

$$E_{ab} = -s_{ab}(t_1 - t_2)(t_{ab} - t_0),$$

where t_0 is the mean temperature of the junctions, and t_{ab} the neutral temperature. But for numerical calculation of the E.M.F. of a thermo-electric pair (4) is most convenient. *I.e.*, multiply the difference of powers referred to lead at the mean temperature by the difference of temperature. A table of the values of the constants here called r and s is given in Everett's *Units and Physical Constants.*

SECTION XI. THE THERMO-ELECTRIC THEORY OF CLAUSIUS, AND OBJECTIONS THERETO.

Sir W. Thomson's thermo-electric theory was published in 1851, and, so far as linear conductors are concerned, in a complete form. His later papers (1854-6) contain a re-statement of the same in a somewhat simpler form, practical experimental results, and an extension of the theory to crystallised media, under which term are of course included isotropic conductors in a state of strain. In the meantime, under date 1853, the eminent German scientist, Professor Clausius, had also contributed to the theory. Apart from certain speculations as to the origin of the E.M.F.'s, Clausius's theory amounted to this. At the places of contact of different metals there are E.M.F.'s of thermal origin, which vary in strength with the temperature, and in such a manner as to be subject to the second law of Thermodynamics, *i.e.*, as regards the reversible heat-effects. Calling P the force from a to b, P is a function of the temperature. Hence there is no current set up in a circuit of a and b, if the junctions are equally hot, from the balancing of the E.M F.'s, which can only produce a static effect. Destroy the equality of junction temperature, and a current is set up due to the difference between the two contact forces; and since the second law requires that the contact force shall vary directly as the temperature, the complete E.M.F. in the circuit is

$$E = p(t_1 - t_2),$$

where t_1 and t_2 are the junction temperatures, and p is a constant for the two metals.

This state of things Clausius regarded as representing the regular phenomena, the normal behaviour of a thermo-electric circuit, and he considered the departures therefrom, especially when high temperatures occur, as due to changes of molecular condition or of structure produced by alteration of temperature, in support thereof instancing the differences between hard and soft steel depending on greater or less velocity of cooling, they behaving as different metals both mechanically and thermo-electrically. He also quotes an experiment of Seebeck, who found a ring of antimony to behave as if of two different materials, and which on rupture was found to be structurally different in different parts. Also Magnus's result that currents in circuits of one metal had their origin in want of homogeneity. (To which we may add that a strained wire is thermo-electrically different to the same wire unstrained. But it does not appear that a merely qualitative specification of a strain, *e.g.*, that a portion of a circuit of one metal is stretched, is sufficient; for the current may go one way across the hot junction of stretched and unstretched parts under small tension, and the other way with greater tension. Here, however, it may be mentioned that in the first case the strain is elastic, and in the second that there is a permanent elongation of the wires as well.)

Clausius thought that differences of structure might occur in a homogeneous metal (homogeneous when at one temperature throughout) when unequally heated, and, without developing this idea, proposed to explain the departures from the regular phenomenon by means of the new E.M.F.'s diffused through the unequally heated and therefore structurally different parts of the separate metals, acting together with the ordinary Peltier forces at the junctions.

Superficially regarded, this would seem to be simply the theory of the Thomson effect in a rudimentary form. But there are in reality two distinct ideas involved. The first is that the thermo-electric power of a metal (that is, p or P/t) may not be always constant, as in the supposed regular or normal phenomenon of Clausius, but may sometimes vary with the temperature, thus introducing abnormal reversible thermal effects in the interior of single metals. The second is a hypothesis as to the cause of the abnormal variation of p, viz., that it proceeds from change of structure. Some reasons against this are given below. In the meantime we may remark that in a circuit of copper and lead, for instance, the hypothesis that the copper, along which the temperature varies, may be considered as being made up of an immense number of pieces of different metals, though only differing very little from one piece to the next, at whose junctions there are small forces of the same nature as the large terminal forces at the junctions with the lead, and that all the reversible heat effects thereby resulting are subject to the thermodynamic laws, does lead to correct expressions for the Thomson E.M.F. and the total E.M.F. in the circuit, when we proceed to the limit, and hence make the number of divisions infinite, and the changes of material infinitely small.

As we remarked before, Clausius did not develop his idea, but (*Mech. Wärmetheorie*, Vol. II. Absch. VII.) he gives the development made by Budde in 1874. The supposed changes of structure are limited to be of a reversible character, so that heating a portion of a wire, and then cooling it back to its original temperature, is always accompanied by a restoration of the original structure; that is, the structure must depend upon the temperature only, and thus be the same in all parts of a wire which have the same temperature, and not be permanently altered by heating. This excludes the irreversible changes of structure sometimes produced, as when the sudden cooling of steel does not restore the structure it possessed before heating, but results in something notably different.

Now, instead of $p(t_1 - t_2)$ being the sum of the terminal Peltier forces, as in the regular phenomenon, we shall have $p_1t_1 - p_2t_2$, where p_1 and p_2 are the now unequal values of p at the junction temperatures t_1 and t_2.

And in the copper we shall have small forces, owing to the gradual change of p, the E.M.F. between two consecutive pieces whose junction temperature is t, and whose values of p differ by the amount dp, being tdp, so that $\Sigma\, tdp$ expresses the total Thomson E.M.F. Adding the terminal forces, attending to directions, their sum reduces to $\Sigma\, pdt$, where dt is the small difference of temperature between two consecutive pieces. Clausius's and Budde's theory leading precisely to the previously given results [Section IX*b*., p. 318], nothing more need be said of it, save as regards the hypothesis made that the variation of the thermo-electric power of a metal with the temperature results from a change of structure.

To examine this, perhaps the simplest way will be to interpret Clausius's views with the help of the thermo-electric diagram. Consider that when the so-called regular or normal phenomena are present the lines of power of all the metals are *parallel* straight lines, which, if we take the line of the standard metal horizontal, will be also horizontal, at different heights above or below the standard line. The E.M.F. in a circuit of two metals will be represented always by a rectangle, viz., the area enclosed by the two horizontal lines of the metals considered, and by the vertical lines corresponding to the junction temperatures. The rectangle will increase in area at a uniform rate as the difference of temperature increases, and if we keep the cold junction at a constant temperature and heat the other, if the current in the circuit be represented by a line, such line will be straight.

Clausius's hypothesis, then, would imply that if the lines of two metals should at some temperature lose their parallelism, and either diverge or converge, we should have evidence of change of structure. The mere existence of slope in the lines of power would therefore imply change of structure, and departure from the regular behaviour. But this is very different from the experimentally found state of things. No two metals have their lines parallel. The lines of power have all degrees of slope—some up, some down. The regular behaviour is conspicuous by its absence; departures from it are the rule, and so far from these departures only coming into play with wide differences of temperature, the constancy of the slopes, *i.e.*, the straightness of the lines, shows that the supposed departures are in force at all temperatures, and within the smallest range of temperature. And the lines of current, instead of being straight, are parabolas. (*See* equation (4), p. 326). Let t_2 be constant t_1, variable, E the ordinate, and t_1 the abscissa of the curve required.)

It would appear, then, that to regard the changes of thermo-electric power with the temperature, which are universal and not exceptional, as evidence of change of structure, is not justifiable. We should rather conclude that constancy (not absence) of slope implies constancy of structure; that if within a given range of temperature the line of a certain metal continues appreciably straight, whatever its slope may be, there is no appreciable change of structure, but that should it become curved, which is an exceptional phenomenon, we have real change of structure. This (curvature) we might anticipate to be the case at sufficiently high temperatures with the lines of all the metals, as it is well known that marked changes of physical state, as from solid to liquid, are not perfectly sudden. But this is quite a different thing from the change of structure going on at a uniform rate through hundreds of degrees that we shall have if we adopt Clausius's hypothesis, which, by the way, is wholly unnecessary for the mathematical development of the subject.

The peculiar manner in which the expression for the integral E.M.F. in a wire, say copper, viz, Σtdp, when added to the terminal junction forces with lead (taking lead for the metal paired with copper, in order not to have any E.M.F. except in the copper and at the junctions) becomes converted into Σpdt, which represents the total E.M.F., is worth illustrating explicitly. We may remark, in the first place, that we cannot get any resultant E.M.F. in our circuit by supposing that the diffused force in the copper wire is due merely to the variation of the contact force with respect to lead at different parts, although such might be suggested by considering that if a piece of lead were brought successively into contact with different parts of the copper wire, if insulated from one another, it would in each case take a definite potential with respect to the copper, the difference of potential being equal to P, the Peltier force, whose amount varies as the temperature of the copper varies. Let the copper wire be divided into four pieces, for instance (the process is the same for any number), and let the values of P for them be P_1, P_2, P_3, P_4. The terminal forces when the copper is in circuit with lead will be $-P_1$ and $+P_4$, and the intermediate forces at the three junctions of copper with copper would, on our supposition, be $P_1 - P_2$, $P_2 - P_3$, and $P_3 - P_4$. But their sum is $+P_1 - P_4$, which exactly cancels the junction forces with the lead, and leaves us no resultant E.M.F. In fact, we have a case of force derived from a potential function, of which a characteristic property is that the integral force round a circuit is zero.

Nor do we, for the same reason, get any assistance by putting pt for P in the above. But if, in the complete variation of pt, viz.,

$$d(pt) = tdp + pdt,$$

we neglect the latter part, and sum up the former, Σtdp, we get the integral force in the copper. And, neglecting the former part, and summing up the latter, we obtain Σpdt, which is the resultant E.M.F. Dividing the copper into four parts as before, let the temperatures of the five junctions be t_0, t_1, t_2, t_3, t_4; and the values of p be p_1, p_2, p_3, p_4 in the four pieces. Then the sum of all the forces is

$$-t_0 p_1 + (p_1 - p_2)t_1 + (p_2 - p_3)t_2 + (p_3 - p_4)t_3 + p_4 t_4,$$

where the first and last terms are the forces at the junctions with the lead, and the three intermediate represent Σtdp, the integral force in the copper. Now re-arrange the terms, pairing the first with the first part of the second, the second part of the second with the first part of the third, and so on, and we get

$$p_1(t_1 - t_0) + p_2(t_2 - t_1) + p_3(t_3 - t_2) + p_4(t_4 - t_3),$$

i.e., Σpdt, the expression for the complete E.M.F. in the circuit.

Reckoned per unit of length, the force in the copper is $t(dp/dx)$ in the direction of decrease of p. When we arrive at the junction with lead, we have an abrupt instead of a gradual change of p, so of course we cannot reckon the force per unit of length, but must take it in the lump, viz., tp or P, the Peltier contact force, the only proved contact force between metals which do not act chemically on one another.

Section XII. On Speculation and Explanation in Physical Questions.

Professor R. Clausius remarks (*Mech. Wärmetheorie*, Vol. II. Absch. XI., p. 337) concerning Sir W. Thomson's papers on thermo-electricity, after falling foul of the "specific heat of electricity," that the latter has not, so far as he is aware, given any explanation of the cause of thermo-electric currents. Now this is a very important remark, showing characteristics on both sides. There is a striking difference in the methods pursued by the two scientists in starting the subject, in laying the foundation for the mathematical development. It is true that Sir W. Thomson abstained from vain speculation and went straight to the point at once. There are reversible heat effects at the junctions of different metals when currents pass across them. There may be (although then unknown) similar effects in the metals themselves. There is no resultant E.M.F. in a circuit of one homogeneous metal, however it may be heated, and however its section may vary. The effects must be subject to the law of conservation of energy, that is, the First Law of Thermodynamics. They are very probably subject to the Second Law as well. Now, with these data, develop the laws governing the E.M.F.'s, without unnecessary hypotheses. Such is the method followed in Sir W. Thomson's papers (whatever may have been his private speculations), the truly scientific method in the strictest sense of the word, bearing in mind its derivation, and what science ought to mean—viz., knowledge, and discarding the vague extended meaning it has gradually acquired in the mouths of the unscientific.

On the other hand, Professor Clausius preludes his investigation, which, it may be remarked, has the same object and result, though applied only to the Peltier effect at junctions of different metals, by speculations on the causes of contact force in general, and of the thermo-electric force in particular, using hypotheses which appear based entirely on the materiality of electricity supposed to act directly at a distance on other electricity, and to be attracted differently by different kinds of matter. It is not easy to express these hypotheses in terms of less gross ideas of electricity without at the same time making them become mere ghosts, of no tangibility and of little utility. Such speculations should, in my opinion, be kept entirely apart from, and in particular should not precede, and so apparently form the groundwork of, a mere development of laws not in any way dependent on the hypotheses; so long as the object of inquiry is the laws, not the causes thereof. We shall endeavour, a little later, to put these speculations regarding contact force in a more modern form, by examining how far they are justified by facts, and dismissing the unessential parts. In the meantime, a few remarks on the general nature of explanations, and on the distinctions between the methods followed by the scientific and by the unscientific speculator, whose name is Legion (though usually spelt in other ways) may be useful to many who feel impelled, by natural instincts, to try and explain something.

It is human nature to speculate, and there will be always plenty of scope for speculation until everything is found out, which will not be for some few million years. We want to know the causes of things, why such and such things happen. Well, the first preliminary should be to find the laws of the phenomena. That is work for the scientific man, and usually difficult work, requiring scientific training and reasoning. When laws are established—which implies a very considerable knowledge of the facts, for otherwise the laws may not become evident —we may speculate on their causes. Or, since it would often be very tedious to wait until sufficient facts are known, we may speculate on the causes of phenomena without knowing anything about the laws governing them. Now this may be done by any one. Not that any one can find out a probable explanation of something strange, but any one can speculate. The more imaginative a man is the better for his speculative powers. Also, if he be unscientific, it is not desirable for him to know too much of the facts of the case, because facts are very unaccommodating, and form a great drawback to the free exercise of the speculative faculties.

The proper method for the unscientific speculator is to seize hold of one or two facts, let the imagination run riot, and develop their consequences without any regard to the quantitative relations which may be necessarily involved, and with complete ignoration of the thousands of other facts which might not fit into the hypothesis conveniently. For it is so easy to leave them out. Thus, there have been remarkable exhibitions of the afterglow in various parts of the world lately, and it is natural to speculate on the cause of such unusual occurrences. The influence of suspended particles in scattering and in absorbing light is well recognised, and it is very possible that suspended matter has something to do with it. Also, there was a great volcanic eruption in Java lately, which sent up an immense quantity of dust and smoke, which must take a long time to settle down again. Now let us speculate. It is known, by observations with the unprejudiced electrometer, that the earth is negatively electrified, and, being very large, must therefore contain an immense store of negative electricity. This is known to be self-repulsive, and since the law of force is that of the inverse square, when the particles of electricity are very close together, the repulsion is enormous. No wonder, then, that the thin crust of the earth is sometimes unable to withstand the pressure, but breaks down, when, of course, we have earthquakes and volcanic eruptions. In the latter case an immense quantity of the negative electricity is carried up with the dust and smoke, and is repelled to a great height; until, in fact, the force of gravitation on the matter which carries it balances the repulsion of the negative electricity left behind in or on the earth. Then, of course, the self-repulsive action of the electricity causes the smoke to diffuse itself all round the globe, if there be enough of it, and spread out in a great canopy. Evidently the smoke that rises from chimneys must also be negatively electrified, which explains the fact often observed of smoke rising up nearly vertically, and then spreading out horizontally. And many other curious phenomena may be readily explained by the same natural repulsion of the negative electricity.

But in the hands of the philosopher (not meaning metaphysicians who appropriate the title), with a proper attention to facts arranged in correct perspective, and in especial with a due attention to geometrical and quantitative relations in regard to space, time, motion, energy, etc., speculation becomes a very different thing from the above, and may be most usefully employed in forming hypotheses, which, though they may be themselves very improbable, may be provisionally of great utility, not merely to hang the facts together, but, on account of the inquiries they suggest, to serve as stepping-stones to a truer theory. Imagination is required no less than before, but it must be guided by strong sense and understanding.

Examples of useful scientific speculation are innumerable. They are usually gifted with importance by being termed theories, thus leading the uninitiated to take them for more than they are worth. Hypothesis would be a better name than theory, because its sound and associations suggest something supposititious and to be received with caution; whilst theory, on the other hand, has also the much more important meaning exemplified in Fourier's "Theory of Heat," Maxwell's "Theory of Electricity" (not the vortex hypothesis to be mentioned), or Rayleigh's "Theory of Sound," which have very little to do with speculations, but are mainly rigid developments of established laws. But it would certainly lead to a considerable loss of dignity were an investigator to speak of "my hypothesis" or "my speculations" on, for instance, the cause of magnetism, instead of the usual "my theory." For it is very well recognised that dignity, or the appearance thereof, has a very imposing effect on all, save those who take the trouble to look below the surface. Which is why lord mayors are dressed up in robes and chains, and the judges wear horsehair wigs.

As an example of the highest kind of scientific speculation we may mention a "theory" not very generally known, viz., Maxwell's hypothesis of molecular vortices to explain electricity and magnetism. That remarkable genius happily combined in his robust intellect great mathematical gifts, with immense powers of perception and mental realisation of consequences as a whole as well as in detail. Who but a man of the most vivid imagination could, as he did, frame a hypothesis to explain, by pure mechanism (not clockwork) obeying the dynamical laws, Electrostatics, and Electrokinetics, and Electromagnetism, all in one consistent scheme? Certainly imagination alone could not do it; a preliminary profound study of the facts and of the already formed theories and hypotheses was required, with the exceptional faculty of being able to digest all that learning, to assimilate the essential and reject the unnecessary parts; besides—it being pre-eminently a mathematical subject—the power of applying his mathematical knowledge. Mathematicians are as plentiful as mushrooms, but few have the applicative power. In fact, given the latter, a little mathematics will go a long way, and a knowledge extending no further than simple equations, if well applied, can be of immense utility to the practical electrician.

Regarding this hypothesis of molecular vortices, Maxwell was so modest as not even to give an account of the same in his great work on Electricity and Magnetism, merely referring to some of its results. This is, perhaps, greatly to be regretted. For it would be very useful in its suggestiveness to future electrical students; and, taken merely as a speculation, it appears to me to be of a far more useful kind than the speculations of the great German electricians who go out from the already three-parts extinct idea of the direct action of free electricity on itself at a distance, and develop their hypotheses in electrodynamics to suit, having to adopt strange devices in order to get rid of the electrostatic force as soon as ever the electricity is set moving. For physical theories, nothing can be more inharmonious with the modern spirit of physical science.

Maxwell's hypothesis is exceedingly unlikely to correctly represent the reality; the details are sufficient to show many improbabilities; but it proved that a dynamical explanation of electricity is possible, and that no actions at a distance are required. The final theory of electricity will most probably be strictly in terms of matter and motion, with, however, an intermediate medium of some kind, which cannot be done without. This will, perhaps, be very disagreeable to the imaginative unscientific, to whom dynamics is so odious that they must try and alter the fundamental notions, making, for example, force the square root of energy. As if, by any human possibility, the laws of motion could be capable of adjustment to suit individual eccentricities. They must be taken as they are. Other laws may prevail in other worlds, where spirits have their *habitat;* but in this special world of ours we must abide by whatsoever laws of motion we find working therein, or else receive a fearful punishment.

We are set down in space, to march with time, and have matter in motion everywhere around us. What space and time may be, the metaphysicians may decide, if they can; and, if they cannot, they might as well leave them alone to take care of themselves. It is enough for the scientist that they are—no matter what, and that Nature is not capricious, but orderly. In all speculations established elementary laws should be attended to, theories must be made to fit them, not the reverse, and the object should be to make a science as exact as possible, and bring everything under numerical relations. It may be objected that some branches of knowledge are so heterogeneous in their nature, and cover such an extent of ground, that they can never be brought under the rule of quantitative measurement. Yes, and for that very reason they must always remain speculative, and their theories consist mainly of imperfect generalisations, to be repeatedly altered, affording never-ending material for discussion and argument. Political Economy can never be a science in the same sense as Electricity, even if what takes place in electrical phenomena remain for ever unknown. But just in proportion as a branch of knowledge rises from being a heterogeneous collection of facts and imperfect laws to being a system, consistent in all its parts, so does it become scientific, and under the rule of exact relations. So long as there is uncertainty as to exactly how much a certain effect amounts to under given circumstances, it cannot be a finished science. Even Chemistry, that vast subject with so little coherence, is being made exact, now that chemical affinities are being measured, and the amounts of energy corresponding to the union of definite quantities of matter determined. Under the rule of numbers must all sciences come, to be worthy of the name. A few men may confound mathematics with metaphysics (strange delusion!), and vent their scorn upon the former—sour grapes to them. But it will not do. For them to ignore already established mathematical relations in their speculations is fatal to their accuracy. The mathematicians are very greatly to be thanked. Consider the present science of Electricity, with its various units, measuring instruments, and methods. Who have made this possible? The mathematicians. It would be very little use accumulating piles of facts without having the mathematicians to sift them, discover the numerical relations, test various theories with the mathematical touchstone, and gradually turn chaos into system, as they have done in Electricity.

Now, with regard to explanations, which it is the object of speculations to furnish. Some are more mysterious than the mystery to be explained, in this respect resembling the explanation of some of the Ancients as to how the world was supported, viz., on the back of an elephant, which in its turn stood on a tortoise's back, and goodness knows what the tortoise had to stand on. Such explanations find favour with the unscientific who cannot bear to have no explanation, who would rather worship a false god than none at all. Then there are poetical explanations of natural phenomena. As might be expected, these are *very* bad. The illustrious Goethe's explanation of colour should be a caution to poets to the end of time to keep to their poetry. He borrowed a prism, stuck it to his eye, and looked through it. He did not see Newton's spectrum, but something quite different, and hastily concluded that Newton's theory was all wrong, and set to work to write a book in which the whole thing was properly explained, and Newton's theory demolished. He, in his complete confidence in his theory, astonishing ignorance of the subject, contempt for Newton's theory, and hatred of the methods of the French mathematicians who had developed the laws of polarisation mathematically, displayed many of the characteristics of the unscientific explainer, whilst the complete ignoration of the great poet's theory by the scientists was no less characteristic of them.

That explanation was purely unscientific. But an explanation may be scientific without being real—the invention of fluids to explain electric and magnetic phenomena, for instance. These conceptions were, and are still, useful enough in their way, but we should avoid attaching any more reality to them than they deserve. Let them be servants, not masters. Again, light is popularly explained by the vibrations of a medium called the ether, and I understand Professor Tyndall to go so far as to believe it certain that the particles of ether really *vibrate* transversely to the direction of propagation. But here we should remember that what is known is that light propagated through space is a transverse periodic phenomenon of some kind, not that it consists of vibrations of the kind supposed, for many other transverse periodic arrangements may be imagined. A problem, the data of which are very imperfectly known, may have many solutions, instead of but one, and that a certain hypothesis seems to explain is no proof that it furnishes the correct explanation.

Also, it is common to attribute to explanations more virtue than they are entitled to. If we explain the electrification of two conductors by saying that the medium between them is polarised, which it may be, and very likely is, what we do is to electrify all the intermediate particles between the conductors, thus shifting the mystery without in any way explaining in what the polarisation consists. Or if, after Weber, we say that the molecules of iron are themselves magnets, with polarity and orientation, and that they may be rotated by external magnetic influence, so that by the consequent preponderance of molecules pointing their magnetic axes in certain directions rather than in others the iron may be made to show its magnetisation externally, although a very important step is made as regards the probable actual facts of magnetisation, do we thereby explain in any way, or even hint at an explanation of the Cause of Magnetism? Not in the least. The nature of the magnetism of a molecule is just as great a mystery as that of a collection of molecules forming a connected mass.

What, then, is a real explanation? Obviously nothing can be wholly explained, for that would require infinite discernment. But complexity may be resolved into simplicity, for one thing. This process of disentanglement, whereby by pure reasoning, without hypothesis, or with mere working hypotheses that may be thrown off, we may put a complex set of phenomena in order, as it were, and exhibit their mutual relations, may be all the explanation we can get, *i.e.*, the explanation of complex cases in terms of simpler ones of the same type. But, for another thing, some phenomena are so familiar to us and so universal that, although equally mysterious in themselves, they seem to less need explaining; so, taking them for granted, we seem to obtain a real explanation if we can resolve obscure facts into the familiar ones. This may be much more than the mere disentanglement above mentioned, being a change of type as well.

That remarkable triumph of hard-headed men, the kinetic theory of gases, is a case in point. The properties of gases are explained, and very completely up to a certain point, dynamically, with approximate estimates of the size and mass of molecules, on the ridiculously simple hypothesis that a gas consists of an immense number of small particles in motion, left entirely to themselves and the operation of the ordinary laws of motion and of collision of elastic spheres. This is something deserving the name of explanation, and is a real gain of knowledge. To matter and motion must electricity come before it can be said to be explained, though certainly it will be not quite so simple an affair as the above, the question being complicated with the ether as well.

After that, there is the nature of molecules, and of matter in general. And even if we resolve all matter into one kind, that one kind will need explaining. And so on, for ever and ever deeper down into the pit at whose bottom truth lies, without ever reaching it. For the pit is bottomless.

SECTION XIII. CHEMICAL CONTACT FORCE.

We now approach one of the most interesting subjects in the whole of electrical science, on which there has perhaps been more debate than on any other of its branches. He is a learned man who is fully acquainted with all the details in the history of the matter. But he may not be thereby made wise; on the contrary, he may easily become utterly confused in the attempt to reconcile the multitude of facts and hypotheses, especially as the observations are mostly only qualitative. He may wish to obliterate all that has been done, and start afresh in the unbiassed state of mind accompanying perfect ignorance.

Put any two metals in contact with one another, but otherwise insulated; they are said to acquire different potentials. That they are apparently at different potentials is made certain by the modern electrometric measurements, using no finger contacts or multiplying machines. Thus, zinc and copper in contact apparently differ in their potentials by about ·75 volt. Professors Ayrton and Perry found this to be so constant that they used it as a standard of comparison in their observations on the apparent differences of potential of other metals in contact. It is proved that when zinc and copper are put in contact, the zinc becomes positively, the copper negatively electrified, and that they act inductively on other conductors just as any two conductors similarly charged would.

If we join the zinc and copper by a wire of some other metal, say iron, instead of making immediate contact, just the same thing happens; the difference of potential is ·75 volt as before. This applies to all pairs of metals, whence follows the "summation law." If metals A and B in contact apparently differ in their potentials by x volts, and A and C by y volts, then B and C will differ by $x-y$ volts.

However, it is merely inferential that zinc and copper in contact really differ by ·75 volt. But assuming provisionally that such is the case, it follows that since in a state of electrical equilibrium the whole of the zinc is at one potential, and the whole of the copper also, there is an E.M.F. of ·75 volt acting at their junction from copper to zinc;

this being required to balance the supposed difference of potential. If so, if we pass an electric current from any source across the junction, there will be, by elementary principles, a continuous absorption of energy, when the current goes from copper to zinc, and evolution in the converse case, amounting per second to $\cdot 75 \times$ strength of current. Or, make a closed circuit of any number of metals and a battery; there will be similar absorptions and evolutions of energy at all the junctions, meaning by absorption that energy is taken in by the current from some source which—electrically speaking—may be called external, and by evolution that energy is given out by the current, or through its mediation.

But there is no evidence of any such relatively enormous conversions of energy going on at metallic junctions. The known thermo-electric forces are of such inferior strength as to be almost of a different order of magnitude. The source of energy is heat, *i.e.*, the energy of molecular agitations. There may be other small conversions of energy, but certainly none able to account for an E.M.F. of ·75 volt between copper and zinc, or ·6 volt between zinc and iron.

The thermo-electric forces being, then, so very small compared with the apparent contact forces now considered, we may neglect them altogether, in order to save continual reference to them and small corrections. Copper and zinc, then, when placed in contact, are necessarily at One potential.

It follows that if they were uncharged before being put in contact, and not in a field of electric force, they must have been at Different potentials. For, on contact, electricity passed from copper to zinc, reducing them to the same potential. But having been, as stated, uncharged in the first place, and not in a field of force (or, say simply, neither showing any signs of electrification), the air being then all at one potential, and the potentials of the copper and zinc differing from one another, must be different from that of the air, as thus defined:— Taking the air potential as zero, and that of the copper separately insulated as $(-x)$ volt, that of the zinc is $-(x+\cdot 75)$ volt. So far we do not know whether x is positive or negative, but we take it as positive here for convenience of statement. Thus a piece of uncharged zinc insulated in air has its potential $(x+\cdot 75)$ volt below that of the air, and a piece of uncharged copper insulated in air is also at a lower potential, but by a smaller amount, namely x volts. This requires that there shall be, over the whole zinc surface, an E.M.F. of strength $(x+\cdot 75)$ volt acting from zinc to air; and similarly over the whole copper surface an E.M.F. of strength x volt from copper to air.

Electricity, in conductors, is subject to the same law of continuity as an incompressible liquid. There cannot be current entering a certain space without there being at the same time an equal current flowing out of that space. At the surface of conductors electricity was once supposed to accumulate. Maxwell extended the law of continuity to the surrounding dielectric. There is great advantage in this view in facilitating conceptions. We may imagine an incompressible liquid filling *all* space, perfectly free to move by the slightest force in certain regions answering to pure conductors, with no tendency to return when displaced, but always meeting with resistance proportional to the velocity. Also perfectly free to move in the rest of space answering to a pure dielectric, and without frictional resistance, but now only elastically displaced, so that there is a force of reaction called into play proportional to the displacement, which will make the displacement subside when the force that produced it is removed.

Replace the material liquid by an imaginary something called electricity, filling all space (not the electricity of the mathematical definition, but capable of becoming it by displacement), let it be free to move in conductors when acted upon by electromotive force (answering to real force when the subject is a real fluid), but only capable of elastic displacement in the dielectric, and we may transfer results from one case to the other. We may remark, in passing, that the quasi-fluid cannot be really matter, because that would require electromotive force to be ordinary mechanical force.

If to the surface enclosing a portion of the material liquid in which there is no reactive force, but outside which there is, we apply uniform normal pressure or tension, the liquid is not moved, because the forces balance, but the pressure within is increased or decreased by an amount equal to the applied surface pressure or tension.

In the electric case, the uniform E.M.F. $(x+\cdot75)$, acting normally outward from a piece of zinc insulated in air, lowers its electric potential below that of the surrounding air by the amount $(x+\cdot75)$, but cannot displace electricity. Similarly the E.M.F. x, acting normally outward from the copper surface, lowers its potential below that of the air by the amount x. But the moment the copper and zinc are touched, we substitute at the place of contact metal for air; the force $(x+\cdot75)$ is removed from a portion of the zinc, and the force x from a portion of the copper surface; the differential force $\cdot75$ volt acts; there is a current from copper to zinc, from zinc to air, and from air to copper, which is stopped by the force of reaction of the electric displacement in the dielectric. The zinc and copper are reduced to the same potential; let this be y. Then, in the new state of equilibrium, the potential rises from y to $(y+x+\cdot75)$ in passing from the zinc to the air, then falls continuously along the lines of electric displacement in the air till the air outside the copper surface is reached, where it equals $(y+x)$, and then falls by the amount x in passing into the copper, where it is y, the common copper and zinc potential.

This may seem unnecessarily diffuse, but the importance of the subject, and the difference of the above from views in general acceptance demand a somewhat amplified statement.

The reason of the summation law readily follows. For let the zinc and copper, previously insulated, be joined by an iron wire. This, if insulated and free from charge, will have its potential lower than that of the air surrounding it by $(x+\cdot60)$ volt; or, $(x+\cdot60)$ is the E.M.F. from iron to air. In contact with the copper only, when their potentials equalised, the field of force in the air would show a difference of potential along any line of force of $\cdot60$ volt, and in contact with the zinc only, of 15 volt, the iron being positively electrified in the first case, and negatively in the second. But when the iron wire is intermediate between the zinc and copper, the force $(x + \cdot 60)$ from iron to air, since it can now draw electricity both ways, from the copper and from the zinc, can have no influence in altering the difference of potential between the air just outside the zinc and just outside the copper, although altering the actual potentials relative to the original potential of the air. If z is the final potential of the three metals, those just outside the copper, the iron and the zinc are $(z+x)$, $(z+x+\cdot 60)$, and $(z+x+\cdot 75)$, with a fall of $\cdot 75$ as before through the air from the zinc to the copper surfaces.

It may be remarked that the field of force is perfectly determinate with any number of metals in contact, between each of which and the air there is a given E.M.F. The bounding surface of the dielectric has then everywhere a given potential (+ a constant), and by Green's theorem this is sufficient to fully determine the distribution of force. Of course mathematical difficulties prevent the practical solution in general.

In the above we have, for simplicity, supposed the metals to be pure and homogeneous, and to have clean surfaces. Some little difference is made when there are surface impurities. The nature of the effect may be readily seen. Start, for example, with a piece of absolutely pure zinc, and put a small particle of iron on its surface. The iron and zinc are at once reduced to the same potential, with positive electrification of the zinc and negative of the iron, and a fall of potential of $\cdot 15$ volt through the air. Yet there will be no apparent electrification whatever, for the field of force can be only sensible quite close to the particle of iron, so that we cannot get at it. The air all round the zinc mass will be practically at one potential. If we enlarge the iron particle the field of force extends and becomes sensible at sensible distances, and so with further enlargement we can get sufficient separation of the parts of air at the extreme difference of potential to affect the electrometer inductively.

Similarly, when there are, as in commercial zinc, innumerable foreign particles exposed to the air, side by side with the zinc and in contact with it, there are innumerable local fields of force quite close to the surface set up by the unequal E.M.F.'s. But at a sensible distance from the surface there can be no appreciable force, the air potential will be there unaffected, and the zinc will appear uncharged.

Put this mass of impure zinc in contact with a mass of copper—it may be also impure—then, besides the complex local fields close to the surface there is the extended one which can influence the electrometer. The difference of potential cannot be so great as with perfectly pure zinc and copper, the impurities acting to reduce it.

Now change the medium. Let zinc and copper be in contact not in air, but in water, with a little acid to facilitate electrolysis; from being in a medium in which only elastic displacement can happen, let the zinc and copper be wholly immersed in an electrolyte. The surface E.M.F.'s are now probably not the same—it is very unlikely that they should be

—but there they are. Instead of their producing a mere momentary current, we now have a continuous current from zinc to liquid, liquid to copper, copper to zinc. The two metals are not now exactly at one potential, owing to the current, but practically all the fall of potential is in the liquid. The lines of force, which are of course also the lines of flow of current, are, when the sides of the vessel containing the liquid are sufficiently remote, distributed in the same manner as the lines of force in the corresponding case with air as the medium, though of course they become considerably altered if the vessel is small, the current being forced to be tangential at its sides.

The local superficial fields of force have now great importance, for there are naturally local currents to correspond between the zinc and its impurities with consequent waste of energy; waste in not being externally available. This is the same when the zinc is alone in the liquid. The purer the zinc the more slowly is it burnt in acid. Absolutely pure untarnished zinc would last for ever, owing to the balance of forces, but the least impurity getting on the surface would start galvanic action.

If a copper wire joins the zinc and copper, all being still wholly immersed, circumstances are not materially altered; the current goes from the zinc to the copper (say plates now), and also to the copper wire through the liquid and back through the wire; the current in the wire, however, is not everywhere of the same strength. But lift the wire out of the liquid together with that portion of the zinc plate to which it is attached, and the whole current (not counting the local currents) returns by the wire outside the vessel, and we have a full-blown galvanic cell.

The new E.M.F. introduced by the new contacts, viz., between the zinc and air, and the copper and air, do not in any way alter the integral E.M.F. in the circuit, nor can any difference of potential between the liquid and the air. The metals in connection may be nearly at one potential, or may differ by nearly the full E.M.F. of the cell, according to the resistance of the external wire. There is a large rise at the zinc-liquid surface, and a fall of much smaller amount at the liquid-copper surface, the excess of the rise at the zinc over the fall at the copper being equal to the available E.M.F. of the cell. But in other galvanic arrangements, as when there are two fluids, the E.M.F.'s and changes of potential become more complex.

The absorption of energy is at the zinc surface where the current goes with the E.M.F. there. The evolution is at the copper surface where the current goes against the E.M.F. there. The excess of the former over the latter becomes heat in the circuit.

At the zinc surface we know there is oxidation of zinc, and the supply of energy is readily accounted for. The heat which would have been produced locally if the zinc were burnt in oxygen now turns up in all parts of the circuit, through the intervention of the unknown electric agency, and the artificial disposition of conductors and insulators we have made.

The evolution of energy at the copper surface is more obscure. There is a local development of heat independent of the frictional heat in the circuit. The heating of galvanic batteries has not been fully investigated.

Regarding the cause of the E.M.F.'s, next to nothing is known. Separated zinc and oxygen have potential energy, they tend to unite, and in the act of union a store of energy is set free. At the same time there is E.M.F. from the zinc to the oxidising agent. But of the reason why zinc and oxygen should unite, or why E.M.F. should accompany the action, I have not come across any intelligible explanation. And I do not expect it.

But the known transformation of energy taking place at the zinc surface in our galvanic cell, together with the similarity of electrical conditions, enables us to conclude with a tolerable amount of certainty that the source of the electrostatic energy which is set up when zinc and copper are put in contact in air is oxidation of the zinc. The amount of oxidation is, of course, very small—infinitely unrecognisable. This will be evident on remembering what a large quantity of electricity must pass before any visible consumption of zinc takes place in the cell, or even before enough is consumed to be detectable by the most delicate chemical balance. In the air case the action is stopped in its very birth by the elastic reaction of the electric mechanism. The facts observed long ago by Sir W. Thomson confirm this conclusion regarding oxidation. The difference of potential is greatest when the zinc surface is clean—that is, in the best state for oxidation—and when the copper surface is already oxidised, and therefore in the worst state, amounting then to about 1·1 volt instead of only ·75 volt.

SECTION XIV. CONTACT FORCE AND HELMHOLTZ'S ELECTRIC LAYERS.

An important hypothesis regarding Contact Force was advanced by Helmholtz in his classical essay on the Conservation of Energy, of which a preliminary idea may be gained from the following sentences:—"It is evident that all phenomena in conductors of the first class (*i.e.*, those in which conduction of electricity takes place without electrolysis) follow from the assumption that different chemical elements have different attractions for the two electricities, which attractions act only through immeasurably small distances, whilst the electricities can act upon one another at greater distances also. The contact force will then consist in the difference of attraction which the particles of metal lying next the junction exert on the electricity at this place; and electrical equilibrium occurs when a particle of electricity which goes over from one to the other metal no longer either gains or loses kinetic energy."

This is part of Helmholtz's explanation of the phenomena which occur when two metals are put in contact, that they become oppositely charged and apparently at different potentials. To it we must add a brief statement of what is meant by the double electric layer. Assume that there is really an E.M.F. at the junction of two metals (we are not specially considering thermo-electricity), some tendency to produce a current from one to the other, to be measured by the amount of energy taken in per unit of electricity, and which must produce a current unless otherwise balanced—the place where the energy is taken in being the seat of the E.M.F. We must not suppose the E.M.F. to be confined strictly to a mathematical surface, but to extend through a small thickness t, so that if V be the E.M.F. from A to B, the impressed force per unit of length is V/t. If the conductors A and B are insulated there can be no continuous current, whether they are in contact at one or at any number of different places, provided in the latter case the junction E.M.F.'s are all equal, and all from A to B. Hence only one junction need be considered. Equilibrium of electricity requires that there shall be no electric force in any part of the conductors, including the junction. Therefore the impressed electric force V/t from A to B must be balanced by an equal force V/t acting from B to A through the thickness t at the junction, which will make the potential of A exceed that of B by the integral amount of this force, viz., V. Now, considering the force of reaction alone, we see that the field of force exactly resembles that between the plates of a charged condenser, for we have two opposed parallel surfaces with electric force acting normally to them through the intermediate space. The conclusion is that at the junction of A and B there is a double electric layer, a layer of positive electricity on the A side, and of negative on the B side, at the terminations of the lines of force, and that it is this double layer that is the cause of the electric equilibrium.

Before going further it will be well to distinctly separate four things.

(*a*). There is first the hypothesis that the contact force resides at the metal junction.

(*b*). Next, the hypothesis that it arises from different kinds of matter attracting electricity differently, though only at insensible distances.

(*c*). The statement that this E.M.F. is balanced by an equal, but oppositely directed force of reaction.

(*d*). The hypothesis that this force of reaction proceeds from a double electric layer.

Three hypotheses and one statement have to be considered, and we will take them in the order in which they are most easily disposed of. Commencing with the statement (*c*), there is very little to be said, because the statement is not open to any question. The most elementary notions regarding the balance of forces render the electrical example self-evident. If there is no current at any particular part of a conductor, there is no resultant electric force there; hence if there be any impressed force it must be balanced by a reaction.

(*a*). Now, regarding the first hypothesis, we have in Section XIII. given reasons against this view as respects the ordinarily observed differences of potential of metals in contact—the "contact force" of 75 volt between copper and zinc, for example—so its discussion here is rendered unnecessary. There is, however, a comparatively minute E.M.F of thermal origin undoubtedly existing at the contact place of different metals, and probably the same occurs at the junction of any two materials.

(*b*). The second hypothesis is, from its very material and speculative character, difficult either to grasp or to manipulate. It was intended to apply to the large differences of potential just now mentioned, and, speaking from memory of Helmholtz's Faraday Lecture delivered a few years ago, that scientist still maintains it, as well as the hypothesis of electric layers, to be later discussed. We may, however, apply (*b*) to the thermo-electric force at a metallic junction, or to air contacts, or by generalisation to any contact force we may choose to imagine. If copper and zinc attract electricity differently, though only at insensible distances, the differential force must certainly cause a momentary current across the junction when they are set in contact. And the same result would happen if the heat in the copper attracted electricity more than the heat in the zinc, or the same might be said of the ether, or if any other possible or impossible kind of differential attraction existed. Now, it is surely difficult enough to form a mental conception of what is happening when any kind of impressed E.M.F. is acting, with its corresponding transformation of energy, even when we know definitely that heat or chemical affinity is concerned. But contact force, with the supposititious attraction of matter for electricity, is rather harder to understand than without it. For the differential attraction being E.M.F., so are the separate attractions E.M.F.'s. Now, matter attracts matter, or, at any rate, things go on as if such were the case. But does matter attract electricity? Even on the material hypothesis of the direct action of electricity on electricity by attractive or repulsive forces, it is an enormous complication of the functions and properties of electricity to admit of attractive force between matter and electricity. The supposed attraction of mätter for matter is one kind of force, consistent in itself; that of electricity for electricity another kind, also self-consistent. But innumerable difficulties arise as soon as we admit the kind of cross-action supposed. We need not go into details; they will readily suggest themselves to any one acquainted with the theory of the dimensions of physical magnitudes. Apart from this side of the question, difficulties crop up on all sides when we pass from mere momentary currents to continuous currents, with continuous expenditure of energy in keeping them up, as in thermo-electric or voltaic circuits.

The following special argument is used by Clausius. He observes that whether observed differences of potential occur only through the differences in the attraction of metals for electricity may be left an open question, but he denies that all phenomena may be thus explained, that the hypothesis is, in fact, inadequate to explain thermo-electric currents. Thus, if two metals, A and B, form a closed circuit, and B attracts electricity more than A, B will take a positive and A a negative charge of sufficient amount to balance the difference of attraction, and there will be an end of it. If we heat one of the junctions we cannot get a current unless the attraction of A or B for electricity varies with the temperature, which Clausius considers very improbable.

Even allowing that such is the case, the result must be that every part of each metal will receive just as much electricity as it attracts, and there will be a state of equilibrium set up again. Thus Helmholtz's explanation is certainly incomplete. There is nothing to keep up the current.

Now Clausius, in other respects, agrees with Helmholtz's views, which, by the way, he observes are accepted by most other scientists (presumably continental, and as regards the contact layers mainly) But in order to bring them into harmony with the facts of thermo-electricity, he proposed the following addition :—"Heat itself is active in the creation and maintenance of the difference of potential at the junctions; for the molecular motion which we call heat tends to drive electricity from one to the other material, and this can only be stopped by the opposing force of the two electric layers thereby produced, when they have reached a certain thickness."

Having observed that Clausius, in his "Discussionen," pointedly called attention to the absence of any explanation of the origin of thermo-electric currents in Sir W. Thomson's papers on the subject, I was very curious to ascertain Clausius's explanation of the same in his chapters on Thermo-electricity. It is contained in brief in the last quoted sentence. But since the electric layers are surely not specially concerned with the Peltier contact force, but, if existent at all, are equally valid for any similarly situated impressed E.M.F., we find Clausius's explanation reduced to the statement regarding heat in the first part of the sentence, or any elaboration thereof not introducing any new hypothesis. This is rather disappointing. For what is contained therein beyond the recognition of the experimentally sufficiently obvious fact that heat is the source of the energy of the currents? Now, the reversible absorptions and evolutions of heat actually form the basis of Sir W. Thomson's theory, without explanation, of course. And also without any electric layers being brought into active co-operation.

An explanation of thermo-electricity will have to include not only the solution of the electrical problem in general, what actions take place necessarily during the existence of a current in a conductor or a dielectric however set up, but also how matter in its incessant agitations acts upon the electric system, using this term to indicate, very vaguely, an omnipresent agency of some kind, which must work according to dynamical laws. That electric currents are due to differential actions of matter on the system may be concluded from the various modes of electric excitement. It is, then, easily inferred that the action of matter on the system is always going on, that balanced states are of the kind exemplified in the theory of exchanges in the science of radiation, where bodies are continually emitting and receiving heat, and their molecules and the intermediate medium kept in a state of perpetual motion whilst every part remains in a stationary state on the average. All the various known actions of electricity will have to be included in one harmonious whole, consistent all round. It will have to be nothing short of a union of all the exact sciences with molecular science. That is for the future. For the present it would be a great step forward to know what relations the mysterious thermo-electric power of substances bears to their other physical properties. Here is a large field for investigation. And yet it is nothing. For we may be sure of this, that what is known is infinitely little compared with what is behind, and that the scientific investigator will never have reason to cease work for lack of matter for investigation, even keeping to terrestrial phenomena, let alone the study of the solar dermatology and other far away phenomena.

The remaining scientific hypothesis (*d*) is of a very different nature and of far greater importance, but it involves such extensive considerations of fundamental electrical laws that it must be separately dealt with

Section XV. Electric Layers do not Imply Electrification.

(*d*). Now, with respect to the hypothesis (*d*) of electric layers, widely believed in as realities, we shall endeavour to show that they are myths —that they cannot exist without violating principles where truth is as far as possible removed from being doubtful as anything can be in electrical science. We have already briefly described their supposed distribution in the case of E.M.F. acting at a surface, or through a stratum of small thickness. On one side of the stratum, that *to* which the force acts, there is supposed to be an accumulation of free positive electricity, and on the other side an accumulation of negative, which produce an electric field resembling that of a charged air-condenser, whose force wholly cancels the contact-force when there is equilibrium, and partly cancels it when there is current. The surface-density of the accumulations must depend upon the thickness of the stratum, being great when it is small, and conversely.

If t be the thickness of the stratum, and V the difference of potential, the electric force per unit length $= V/t$; hence, by the definition of the unit of electricity, applied to a surface distribution, the surface-density is $\sigma = V/4\pi t$.

These electric layers are brought into great prominence when the E.M.F. acts all over a closed surface, for example, when zinc is immersed in air. To make the force quite uniform we may imagine E.M.F. to be applied at those places where it is supported equal to that acting from the zinc to the air. The electric layers will now form a pair of closed surfaces, very close together, wholly surrounding the conductor, the positive layer outside, the negative within. This combination we may call a closed electric shell, from its obvious similarity to the closed magnetic shell which appears in the theory of magnetism. The electric force of the shell is wholly self-contained, that is, it is situated between the two layers of electricity, directed straight across the stratum from one to the other, with no electric force either within the inner or outside the outer layer. The potentials are uniform inside and outside the shell, but differ by the amount V, if V is the E.M.F. from the zinc to the air. For if a unit charge of electricity be carried from inside the conductor to the external air, it will travel against the electric force of the accumulations, and work must be done on the charge to the amount force × distance, or $V/t \times t = V$, which is therefore the excess of the outer above the inner potential. In this we consider the electric force due to the layers alone, for the resultant force being nothing, no work would really have to be done.

Comparing with a closed magnetic shell, if its positive side be the outer, the outer magnetic potential exceeds the inner by 4π × strength of the shell. This conforms to the above, remembering that the "strength" of a simple magnetic shell is defined to be the magnetic moment of unit of area, and is therefore = surface-density of magnetism × thickness of shell. We might similarly define the strength of the hypothetical electric shell, but it is not worth while doing so, as the amount of the difference of potential sufficiently settles it.

Now, these electric layers, if they existed, would be wholly independent of any real charge that we might communicate to the conductor. Say, for instance, we charge it by contact with some other metal. This will not alter the electric layers in any way. If they were there before, they are there still, for there is still the same difference of potential at the surface. The real charge, being connected by lines of force through the air with other conductors, is of course recognisable, but no tests can be applied to the associated layers. Perhaps it would be most reasonable, as it is simplest, to put the real charge outside the outer layer, rather than within the inner, or between them, if we must have the electric layers. But, although this extraordinary complication of the surface conditions by the presence of the layers may be used as an argument against their existence, still, such argument would be no proof that they do not or that they cannot exist. To obtain this in a plain form, it will be advantageous to both generalise and simplify. As often happens we lose nothing, and gain much by generalising, obtaining a broader view of a question freed from accidentals. The simplification of conditions again is desirable in order to separate the layers from the real charge, and exhibit them apart.

Let there be two wires of different metals and of the same gauge, whose ends are cut off straight across, and then their plane faces firmly pressed into one another to form a continuous wire. There is the Peltier contact force at this junction, and the electric layers corresponding must be very close together. There are also charges on the air surfaces of the two wires, due to the Peltier E.M.F., quite irrespective of the air E.M.F. and its layers. Now the layers do not depend on the special origin of the E.M.F. or upon the fact that the E.M.F. is at a junction of two metals. Let us therefore increase the thickness of the stratum in which the E.M.F. acts, stretching it out until from a stratum it becomes a cylinder or tube of E.M.F. of any length. Also do away with the other restriction, and let this tube of E.M.F. exist in any given length of a wire of one metal. If the force is uniformly distributed throughout the given space, then, just as in the case of the thin stratum, there will be the accumulations of free electricity at the plane ends of the lines of

E.M.F., whose surface-density will be $e/4\pi$ and $-e/4\pi$ respectively, e being the impressed E.M.F. per unit length. Next let the force be not uniformly distributed, but vary in intensity along the portion of wire considered, though still abruptly ceasing at the same places. We still have the terminal layers of course, but now in addition to them there must be a distribution of free electricity between them to correspond with the gradual variation of the force. The volume-density of this latter distribution is to be found by the same principle as the terminal surface-densities, by reckoning the excess of the number of lines of force leaving over those entering a small space. In mathematical language, the volume-density of the free electricity equals the convergence of the impressed force divided by 4π; this corresponds to the just given statement of the density of the terminal accumulations.

Thus, in general, when any impressed E.M.F. acts in a wire, we have not merely current, if the wire forms a closed circuit and the sum of the impressed forces taken round the circuit be not zero, and the surface electrification depending on the distribution of potential, to be recognised by the electric force in the surrounding air, but also, according to Helmholtz's hypothesis of electric layers, an internal distribution of free electricity depending on the manner of distribution of the impressed force.

Now it may be objected that, although when a distribution of electric force is given in air we can always find the free electricity to correspond by the convergence of the force, yet to apply the same method to the interior of a conductor is not justifiable, a good conductor, as a metal, being so very different in its properties from air, in which electrostatic observations are made. Again, why not apply the method to the actual E.M.F. in the wire? the resultant of the impressed force and that arising from difference of potential. This application is quite correct in a dielectric. If there is no force, or if the force has no convergence within a given space in air, there is no free electricity there. Similarly in the conductor; should there be equilibrium, there is no resultant force and no free electricity; whilst, on the other hand, should there be current, the resultant force, which is proportional to the current, has, like the current, no convergence, so there is still no free electricity.

But the matter is reduced to the simplest form by doing away with the surface charges altogether, by sufficiently enlarging the conductor. Theoretically we may consider an infinitely large conductor; practically, one whose boundary is far removed from the seat of the impressed E.M.F. Let the conductivity of this large conductor be uniform, and let there be a certain portion of it (not insulated from the remainder) in the form of a thin rod along which impressed E.M.F. acts in the direction of its length. The positive "pole" being defined to be that end to which the impressed force acts, the current will flow along this tube of impressed force from the negative to the positive pole, and issuing there, will spread out in all directions, at first in straight lines radiating from the pole, but sooner or later bending round to enter the tube at the negative pole in a similar manner.

Here the contact layers will be represented by an accumulation of positive electricity at the positive pole, and of negative at the negative pole, where the current leaves and enters respectively ; and since the boundary of the conductor is indefinitely removed, we have no surface charges to complicate the question. The only accumulations, if there be any, are those at the ends of the tube of impressed force, and the question arises, How did they get there ? In the first place, they were certainly not there before the impressed force began to act, so they must have been set up whilst the current was being established, which does not happen instantaneously. Now this establishment of the current is a very complex matter, but we do not need to enter upon it here at all, for this reason : It is admitted that currents in conductors always flow in closed circuits, that across any section of a tube of current (the whole system being divided into tubes) the same amount of current flows. It is clear, then, that no further accumulations than the hypothetical ones can take place after the current becomes steady. But not only that, whatever may have been the nature of the distribution of the induction currents before the steady state was reached, such currents must have formed closed tubes, and this excludes the possibility of any accumulations taking place at any time in the interior of the conductor.

The accumulations depending on the convergence of the impressed force are thus wholly mythical; and since the electric layers in the case of contact force are a special case of the same, they, too, are mythical. At a metallic junction there is no double layer accompanying the Peltier force, and at an air or liquid surface there is no electric shell accompanying the E.M.F. from the metal to the air or the liquid, but only the charge to be determined by the distribution of electric force in the air surrounding the conductor.

Impressed E.M.F. and Potential.

In a galvanic circuit with steady current, the impressed force is confined to a certain part of the circuit, namely, portions of the battery, the force in the remainder of the circuit being derived from difference of potential. But in other cases we have distributed E.M.F.'s, as in thermo-electricity, depending on the variation of temperature along a conductor, or in current induction; in both cases there is usually impressed force in every part of the circuit, as well as force derived from difference of potential, the actual force being their resultant.

To exhibit their general relation in the simplest manner, start with an infinitely extended conductor of uniform conductivity, in which there is a given distribution of impressed force. The most general definition of impressed force is that depending on energy, as, in fact, all forces, simple or generalised, are expressed by energy variations. If $\mathbf{e}$ be the impressed force within a certain unit volume of the conductor, and there be a current of density $\mathbf{C}$ at that place, and in the same direction as $\mathbf{e}$, then the amount of energy taken in per second within that unit volume by the current is represented by the product, $\mathbf{eC} = W$, say. Should $\mathbf{e}$ and $\mathbf{C}$ be in opposite directions, though still parallel, there is corresponding energy given out by the current. Should **e** and **C** be perpendicular, there is no energy either taken in or given out. When **e** and **C** are inclined to one another at any angle (which is almost sure to be the case in a large conductor, since **C** does not depend on **e** only at the spot, but upon the impressed force at other places as well), then to find the energy taken in by the current we must multiply the product of the sizes of **e** and **C** by the cosine of the angle between their directions, which result we may still represent by **eC**, using a notation which adapts itself conveniently to common algebra.

The frictional generation of heat is directly excluded. It is always a conversion *from* electrical energy *to* heat, and never the other way. Its amount is $\rho \mathbf{C}^2$ per unit volume, if ρ be the specific resistance. But the energy transformation concerned in impressed force is reversible with the current.

Given, then, a distribution of impressed force, **e**, throughout the conductor, we know that the whole amount of energy taken in per second is $\Sigma\, \mathbf{eC}$, summed up to include every place where **e** exists. We also know by Joule's law that the total frictional heat $= \Sigma\, \rho \mathbf{C}^2$. By conservation of energy these two sums must be equal. This implies some special relation between **e** and **C**. By Ohm's law $\rho \mathbf{C}$ is the actual resultant force corresponding to the current **C**; let this $= \mathbf{F} + \mathbf{e}$, so that **F** is the force that must be added to the impressed force at any place to make up the actual force. Then $\Sigma\, \mathbf{eC} = \Sigma\, \rho \mathbf{C}^2 = \Sigma\, \rho \mathbf{C} . \mathbf{C}$, by putting $\mathbf{F} + \mathbf{e}$ for $\rho \mathbf{C}$, becomes $\Sigma\, \mathbf{eC} = \Sigma\, \mathbf{eC} + \Sigma\, \mathbf{FC}$. Therefore $\Sigma\, \mathbf{FC}$ must vanish. Whatever be the impressed force, the supplementary force does no work upon the current on the whole. This characteristic will, with other considerations, serve to determine **F**.

There are two extreme cases of impressed force that mark themselves out distinctly. First, let the impressed force **e** be confined to a portion of the conductor forming a closed ring, the direction of **e** being along the ring, and its strength such that there is the same amount of **e** acting across every section of the ring. This is a closed tube of force. The current distribution follows immediately. If k be the specific conductivity, the reciprocal of ρ, then $\mathbf{C} = k\mathbf{e}$ simply. For Ohm's law is satisfied everywhere, and the continuity of the current is ensured by the continuity of the impressed force. Thus the current is entirely confined to the closed tube of force, and has no tendency to spread around, although there is no insulation of the tube from the remainder of the conductor. Not only is

$$\Sigma\, \mathbf{eC} = \Sigma\, \rho \mathbf{C}^2 \quad \text{or} \quad \Sigma\, k\mathbf{e}^2$$

on the whole, but the same is true for every part of the tube. The energy taken in anywhere is exactly equal to the frictional heat in the same space, so that there is apparently no transfer of energy. We have supplied just sufficient impressed force at each place to support the current, with no excess or lack. Examples occur in electromagnetic induction. The same applies to any number of closed tubes of impressed force, either wholly detached from one another or side by side in contact, and this includes the general case of any completely solenoidal distribution of impressed force, with the simple solution $\mathbf{C}=k\mathbf{e}$, and no apparent transfer of energy.

The other extreme case is any distribution of impressed force that can produce no current, although there is nowhere want of conductivity to prevent its doing so. For example, let the impressed force be distributed like the force due to a charge at a point, or a charge uniformly distributed over the surface of a sphere, that is, straight lines of force radiating from a centre. There can be no current, because there is no provision for supplying electricity *to* the centre, to satisfy the continuity of the current. Any number of such centres of force may coexist, producing a complex field of impressed force which cannot set up any current. The characteristic of such a field is that the work that would be done by the force on a charge, if it were carried round any closed path, would amount to nothing on completing a circuit; or, in other words, the work done from one point to another would be the same by any path; which implies that the impressed force is the space variation of a single valued function of position, say P; or $\mathbf{e}=\nabla P$. But since there is no current there is no resultant force, hence the impressed force $\mathbf{e}$ or ∇P must be cancelled by an equally great, oppositely directed force of reaction, $=-\nabla P$. This is the supplementary force $\mathbf{F}$, and it is derived from a potential, which is of course the electric potential. In the case of the closed tubes of force above, with current to match, there was no potential, nor any need of one. In the present case we have a potential, but no current.

The thermo-electric force arising from difference of temperature furnishes an example. In a single homogeneous conductor differences of temperature cannot keep up current. If p be this thermo-electric power, the force

$$\mathbf{e}=-t\nabla p \quad \text{per unit of length,}$$

t being the absolute temperature. Here ∇p is obviously derived from a "potential," viz., p, but it is not that which makes its product into t also be derived from a potential. The reason is that the thermo-electric power is a function of the temperature, and the direction in which the temperature falls fastest is also that in which the power varies fastest.

We have $$\mathbf{e}=-t\nabla p.$$

Let also $$\mathbf{e}=-\nabla P,$$

where P is another scalar, so that

$$\nabla P=t\nabla p.$$

Now $$\text{curl}\ \nabla P=0,$$

so also $$\text{curl}\ (t\nabla p)=0.$$

But $$\text{curl}\ \nabla p \text{ also }=0,$$

and there is only left the condition that

$$\nabla t \parallel \nabla p,$$

or p is a function of t.

Between these two extremes of no potential and no current lie all other possible varieties that may arise. By the properties of the space variation of vectors, any distribution of electric force may be considered as partly belonging to one extreme class and partly to the other. The given arbitrary distribution **e** considered to extend over all the boundless conductor (putting $\mathbf{e}=0$ where there is no impressed force) may be divided into two distinct distributions, say $\mathbf{e}_1$ and $\mathbf{e}_2$, of which the second is derived from a potential, say $\mathbf{e}_2=\nabla P$, and which, not being able to produce current by itself, may be left out altogether, whilst the first distribution $\mathbf{e}_1$ consists wholly of closed tubes of force, and is therefore suitable for a current system. The current will be $k\mathbf{e}_1$, and $\mathbf{e}_1$ will be made up of the impressed force **e** and of the supplementary force **F**, which is the same as $\mathbf{e}_2$ reversed, or $\mathbf{F}=-\nabla P$.

As regards the energy, we required to have $\Sigma\,\mathbf{FC}=0$. This is satisfied identically because **C** is wholly circuital, and **F** is derived from a potential. The line-integral of **F** round any one of the tubes of **C** vanishes, consequently the whole sum vanishes.

The following is from a slightly different point of view; though virtually equivalent, it may be thought less abstract. Start with a single tube of impressed force, with coincident current, and no potential. Next, remove impressed force from a portion of the ring, leaving a discontinuity. The impressed force alone is now obviously insufficient to maintain any current. Supplement it by two fields, consisting of lines of force radiating equally in all directions from the ends of the tube, which fields by their co-existence form a single system of tubes starting from one end and ending at the other end of the tube of impressed force, and by union therewith completing the circuit. The system of force is now complete, and the potential and current are known.

What we have to do in the general case is therefore this. Go over the field of impressed force. Wherever we find it continuous, as much force leaving as entering a small space, leave it alone. But if we come to a small space where more impressed force enters than leaves, or where there is convergence, supplement it by an equal amount of divergence, by letting the proper amount of force leave the space, and spread in all directions equally. Do this wherever the impressed force has convergence, and we make the system of force complete.

The auxiliary field **F** is thus made up entirely of central forces, varying as the inverse square of the distance, and is derived from a potential, or $\mathbf{F}=-\nabla P$. The "matter," self-repulsive, which, if it existed, would produce this force, would be distributed at the places of convergence of the impressed force; its volume-density would be σ, if σ stands for the convergence of $\mathbf{e}/4\pi$. The potential is therefore $P=\Sigma\,\sigma/r$, as usual.

We have endeavoured in the above to show the nature and necessity of an *electric* potential in the case of a conduction current, quite apart from any consideration of electro*static* potential, with which we have had nothing to do, having had no surface charges in question and no dielectric. Now, the potential function was originally invented to facilitate calculations relating to the attracting force of gravity, where real matter is concerned, situated at the centres of force, and its properties being equally applicable to electric and magnetic forces naturally led to support the idea that they were also due to attracting and repelling matter of some special kind situated at the centres of force. Now, we may idealise this entity, and let it be a mere symbol to represent the convergence of the force, which is almost what free electricity becomes in Maxwell's dielectric theory. But we have no justification to regard the σ above as representing free electricity according to old ideas, or even what free electricity becomes in Maxwell's theory, because there really is no convergence of force. The divergence represented by σ, that is, the divergence of the polar force, is merely introduced to cancel the convergence of the impressed force, which it does, producing continuity.

XXVIII.—THE INDUCTION OF CURRENTS IN CORES.

[*The Electrician*, 1884; § 1 to 10, May 3, p. 583; § 11 to 13, May 10, p. 605; § 14 to 16, May 31, p. 54; § 17, June 14, p. 103; §§ 18, 19, June 21, p. 133; § 20, July 12, p. 199; § 21 to 23, Aug. 23, p. 338; § 24 to 26, Aug. 30, p. 362; § 27 to 30, Sept. 6, p. 386; §§ 31, 32, Sept. 20, p. 430; § 33, Nov. 15, p. 7; § 34 to 36, Nov. 22, p. 28; §§ 37, 38, Nov. 29, p. 47; § 39, Dec. 20, 1884, p. 106; § 40, Jan. 3, 1885, p. 148.]

§ 1. The Ancients, who were no doubt sufficiently wise in their own generation, were wrong in concluding that the celestial bodies moved in circles because circular motion was perfect, without first ascertaining that they did move in circles; but this erroneous conclusion did not arise from any imperfection in the circle. For although the researches of the Moderns have established that a body, unacted upon by force, will move in a straight line, thus leading to the opinion that it is the straight line that is perfect, which again is corroborated by the decision of the powerful thinkers who have settled that simplicity is perfection, yet the ancient preference for the circle is defensible, since it is the simplest of all curved lines. Nothing can be rounder than a circle. And that there is a natural tendency for both the human body and understanding to move in circles is proved by the accounts of the doings of belated travellers in the wilds, and by the contents of a great mass of books. But to enter upon this subject would lead us too far afield; suffice it to say that in the following the circular motion of electricity, whether such be considered perfect or not, is involved from beginning to end, and that without it the mathematical difficulties to be surmounted would be far greater.

When a circular current changes its strength, the lines of induced electric force are also circles, in planes parallel to the first, and centred upon the same axis. The lines of electric force arising from variations in the strength of currents in any number of circular conductors in parallel planes, with a common axis, coincide in direction with the circuits themselves. The circular conductors, so far considered to be of infinitesimal sections, need not be all insulated from one another; so we see at once that the lines of flow of the currents induced in any conductor shaped as a solid of revolution whose axis is that of the inducing currents are also similar circles, and therefore the lines of electric force arising from the variation of these induced currents form similar circles, so that the circular characteristic is not lost. A circular cylinder, or straight hollow tube, forms a special and simple example. A further and remarkable simplification occurs when the circular conductors are packed closely and regularly together to form themselves a long hollow tube, within which a solid or hollow core is placed, their axes coinciding. The magnetic force due to equal current per unit length of the solenoid is nearly uniform in intensity throughout the space within, and is parallel to the axis. If infinitely long the force would be quite uniform, and there would be no force outside the coil. Then the induced currents in the core would be the same all along, and any thin slice cut out perpendicular to the axis would serve to illustrate the state of the whole core. With a practicable coil of finite length, the spreading out of the lines of force when nearing the ends of the coil naturally weaken the currents in the core there, but for obvious reasons we treat our coil and core in the following as forming a portion of an infinitely long similar coil and core, thus making the magnetic force due to the coil- and core-currents a function of the distance from the axis only, and not of distance along the axis also.

§ 2. *Geometrical and Electrical Data.*

Let l be the length of the coil, a its inner, and $a+b$ its outer radius, and c the radius of the core, which may be equal to or less than a. Also let, in the first place, the core be solid. Divide the core into an indefinitely great number of thin hollow tubes. Since, for reasons mentioned above, there is no motion of electricity from one tube to another, these tubes may be considered to be insulated from one another, so that the electrical system is reduced to an infinite number of known linear circuits. We know the paths of the currents, we have to find their strengths, and that of the magnetic force and other quantities concerned.

Let r be the radius of one of these tubes, h its small thickness, l its length, ρ its specific resistance, and R_1 the resistance of the tube, *i.e.* to the flow of electricity round it in the manner considered. The value of R_1 is obviously

$$R_1 = 2\pi r\rho/hl. \quad\quad (1)$$

Let the unit current flow round the tube, giving a current $1/l$ per unit length. The magnetic force it produces is of strength $4\pi/l$ within, parallel to the axis, and zero without. The section being πr^2, the surface-integral of the magnetic force is therefore $4\pi^2r^2/l$, and of the magnetic induction μ times as much, μ being the permeability of the matter within the tube. Hence the inductance* of the tube is

$$s = (2\pi r)^2\mu/l. \quad\quad (2)$$

* [Owing to the repeated occurrence of the old-fashioned circumlocutory phrase "coefficient of self-induction" in this paper, I have altered it to the modern equivalent "inductance." Of course "mutual inductance" means "coefficient of mutual inductance."]

The quotient of this by the resistance is the time-constant of the circuit, say α, where

$$\alpha = 2\pi rh\mu/\rho, \qquad (3)$$

and is the time a current left in the tube without other E.M.F. than that arising from its own self-induction would take to fall in strength from ϵ to 1 or 2·71 : 1; the E.M.F. of induction round the tube being equal to the rate of decrease of the amount of induction through it, which gives at once, by Ohm's law, if γ be the current at any moment,

$$-s\dot{\gamma} = R_1\gamma, \quad \text{or} \quad \gamma + \alpha\dot{\gamma} = 0,$$

whose solution is

$$\gamma = \Gamma\epsilon^{-t/\alpha},$$

if Γ is the current at starting, and t the time.

Thus the resistance of our tubes varies as the radius, their inductances as the square of the radius, and the time-constants as the radius. As regards their mutual inductances, the quantity s in equation (2) is plainly also the induction per unit current in the tube of radius r passing through any external tube; thus (2) gives us all the inductances, self and mutual, of the tubes. And as regards the coil, if there are N turns of wire in it per unit length of the core, the induction represented by s in equation (2) goes Nl times through the coil, whence

$$(2\pi r)^2 N\mu$$

is the mutual inductance of the tube and the coil.

The coil may also be treated as a system of concentric tubes, and its inductance found. If its depth b be very small compared with its radius a, the unit current in each wire of the coil, giving a current N per unit length, produces magnetic force of intensity $4\pi N$ within it, and induction $4\pi N\mu$, which goes Nl times through the coil over an area πa^2, making L_0, the inductance of the coil, be

$$L_0 = (2\pi Na)^2\mu l, \qquad (4)$$

provided the permeability is the same throughout the coil-opening. Practically, μ only differs from unity for iron; hence, if the core is of iron and does not fill the coil, (4) requires an obvious correction. To remember (4), observe that $2\pi Na$ is the length of wire per unit length of core,—the concentration of the wire, so to speak,—thus making the square of the concentration equal to the inductance per unit length of coil when the inner space is non-magnetic. L_0 is what the inductance of the coil would be if every turn had the same radius a.

§ 3. *Inductance of Coil-Circuit.*

Let L be the inductance of the coil-circuit, *i.e.*, the amount of induction through the circuit per unit current in it, when there are no other currents (as in the core, for example) to alter the magnetic force. This may be conveniently divided into four parts, say

$$L = L_1 + L_2 + L_3 + L_4, \qquad (5)$$

where L_1 is that part due to the induction in the core, L_2 to the induction in the space between the core and the coil, L_3 to the induction *in* the coil, and L_4 in the external circuit. L_4 is any arbitrary positive quantity. Let h be the magnetic force in the core due to unit current in the coil-circuit, that is, $h = 4\pi N$. The induction due to this is μh, going Nl times through the coil, over an area πc^2 (section of core), thus making

$$L_1 = \mu h . \pi c^2 . Nl = (2\pi Nc)^2 \mu l. \qquad (6)$$

Between the core and the coil the magnetic force is also h, the area is $\pi(a^2 - c^2)$, and it goes Nl times through the coil as before, hence

$$L_2 = (2\pi N)^2 (a^2 - c^2) l. \qquad (7)$$

In the thickness of the coil, h falls regularly from h to zero, because the current-density is constant; hence, if h_1 is the magnetic force at distance r from the axis, h_1 is given by

$$h_1 = h(a + b - r)/b.$$

At the same time the number of turns *outside* the circle of radius r as it increases from a to $a + b$ decreases regularly, and is therefore given by

$$N_1 l = Nl(a + b - r)/b.$$

Therefore we get

$$L_3 = \int_a^{a+b} h_1 . 2\pi r\, dr . N_1 l = 2(2\pi N)^2 \frac{l}{b^2} \int_a^{a+b} r(a + b - r)^2 dr,$$

which, worked out, gives

$$L_3 = \tfrac{1}{6}(2\pi N)^2 bl(4a + b). \qquad (8)$$

We need not stop to put the three parts of L for the coil together, as they will be more useful separately. We have supposed the wires to be of square section with insulating covering of no thickness. The correction is in general very small. The corresponding corrections for the resistance are, however, made in equation (9).

§ 4. *Resistance of Coil.*

The resistance of the coil-circuit is of course indefinite, but an expression for that of the coil alone will be useful, so we give it here, to get rid of preliminary matter at once. The volume occupied by the coil being

$$l\{\pi(a + b)^2 - \pi a^2\} = \pi bl(2a + b),$$

if there are m wires per unit length and depth, so that the former $N = m^2 b$, the whole length of the wire will be $m^2 \times$ above expression for volume, because in a unit of volume there are m^2 wires, each of length unity. And to get the resistance of the coil we must further multiply by the resistance of unit length of wire, which last equals its specific resistance divided by its section. The section $= 1/(m^2 c_1^2 4/\pi)$, where c_1 is the ratio of the radius of the covered to that of the bare wire; so if ρ_1 = specific resistance of the wire, and R the resistance of the coil,

$$R = \pi bl(2a + b) m^2 . \rho_1 . m^2 c_1^2 4/\pi = (2Nc_1)^2 \rho_1 l(2a + b)/b. \qquad (9)$$

§ 5. *Magnetic Force and Current in Core.*

Let H be the intensity of magnetic force at distance r from the axis. It is a function of r only. Let γ be the current-density at the same place, also a function of r. **H** is parallel to the axis, and γ is perpendicular to **H** and to **r**. Consider a tube of thickness dr, carrying therefore a current $\gamma\, dr$ per unit length of tube, producing magnetic force $4\pi\gamma\, dr$ within and none outside. $4\pi\gamma\, dr$ is therefore the amount by which H decreases in passing from r to $r+dr$, hence

$$\gamma = -\frac{1}{4\pi}\cdot\frac{dH}{dr} = -\frac{1}{4\pi}H' \quad\text{.......................(10)}$$

is the general relation between the magnetic force and the current in the core at any place, being a special case of

$$4\pi\boldsymbol{\gamma} = \text{curl } \mathbf{H},$$

which applies universally.

§ 6. *Electric Force and Current in Core.*

Let e be the electric force in the core at distance r from the axis. By Ohm's law, and by equation (10), we have

$$e = \rho\gamma = -\frac{\rho}{4\pi}H'. \quad\text{...........................(11)}$$

This electric force is derived entirely from magnetic induction, there being, from symmetry, no difference of potential in the core. (Should the core be placed excentrically in the coil, there would be a potential function to be taken into account by reason of the electric force of induction having a normal component producing displacement outwards.)

§ 7. *Electric Force and Magnetic Force in Core.*

The total induction passing through the tube of radius r being expressed by

$$\mu\int_0^r H.2\pi r.dr,$$

and the E.M.F. round the tube being, by the law of induction, the time-rate of decrease of this quantity, and also being $2\pi re$, we have

$$2\pi re = -\mu\int_0^r \dot{H}.2\pi r\, dr. \quad\text{........................(12)}$$

for any value of r. We employ, according to a common practice, much shortening expressions, a dot over a symbol to denote time-differentiation, and an accent as in (10) and (11) to denote differentiation with respect to the geometrical variable. Putting in (12) the value of e from (11), we have

$$rH' = \frac{4\pi}{\rho}\int_0^r \dot{H}\mu r\, dr, \quad\text{..........................(13)}$$

one form of the differential equation of the magnetic force. Differentiate

(13) with respect to r, and we obtain

$$\frac{1}{r}\frac{d}{dr}\left(rH'\right)=\frac{4\pi\mu}{\rho}\dot{H}, \qquad (14)$$

the partial differential equation, which is a special case of the general equation

$$\nabla^2\mathbf{H}=(4\pi\mu/\rho)\dot{\mathbf{H}},$$

holding universally in pure conductors.

§ 8. *Coil-Current and Core Magnetic Force.*

Whether the core fills the coil or not, the value of H at the boundary of the core fully settles the value of the coil-current, for the magnetic force in the intermediate space will be of the same strength from $r=c$ to $r=a$. Hence, if Γ is the current in the coil-circuit, and therefore in every wire of the coil, and H_c is the value of H at the boundary of the core, we have

$$H_c=4\pi N\Gamma \qquad (15)$$

§ 9. *E.M.F. in the Coil-Circuit.*

If R is the resistance of the coil-circuit complete, the total E.M.F. in it is, by Ohm's law, $R\Gamma$. Let E be the impressed force in the circuit, then $R\Gamma - E$ is the value of the E.M.F. arising from magnetic induction. Let this last be E_1. Then, by the induction law, E_1 = rate of decrease of induction through the whole circuit. Now, the induction through the circuit when there are no core-currents is simply $L\Gamma$, where L is the quantity in equation (5), whose parts are later defined. In addition to this, there is the induction through the coil due to the core-currents. Since when there are no core-currents the magnetic force in the core has the same value throughout, or $H=H_c$, the induction through the coil due to the core-currents only arises from the excess of the real magnetic force in the core over its boundary value, and is therefore given by

$$Nl\int_0^c \mu(H-H_c)2\pi r\,dr=2\pi Nl\mu\int Hr\,dr-L_1\Gamma, \qquad (16)$$

where L_1 is given in equation (6). The whole induction, say p, through the coil-circuit is therefore

$$p=(L-L_1)\Gamma+2\pi Nl\mu\int Hr\,dr, \qquad (17)$$

and we have

$$E-\dot{p}=R\Gamma,$$

or

$$E=R\Gamma+(L-L_1)\dot{\Gamma}+2\pi Nl\mu\int_0^c \dot{H}r\,dr, \qquad (18)$$

where it is desirable to get rid of the integration. Put $\dot{H}$ in terms of H' by (14), then

$$\int_0^c \dot{H}r\,dr=\frac{\rho}{4\pi\mu}\int_0^c \frac{d}{dr}(rH')dr=\frac{\rho c H_c'}{4\pi\mu},$$

by integration, which brings equation (18) to the form

$$\left.\begin{aligned} E &= R\Gamma + (L - L_1)\dot{\Gamma} + \tfrac{1}{2}Nl\rho c H'_c, \\ \text{or} \qquad E &= R\Gamma + (L - L_1)\dot{\Gamma} + \frac{L_1 \rho}{2\pi\mu c}\Gamma', \end{aligned}\right\} \quad \text{.......................(19)}$$

where in the last Γ is to be put in terms of H_c by (15).

Should the core fill the coil, we shall have $L_1 = L_0$, and if, further, the coil be of small depth compared with its radius, and there be no appreciable induction in the external circuit, then $L = L_1$, and the second term in (19) goes out altogether, leaving

$$E = R\Gamma + \frac{L_1 \rho}{2\pi c\mu}\Gamma' \text{.............................(20)}$$

as the equation of the E.M.F. in the coil. Its interpretation is easy. For now, with the limitations imposed, the E.M.F. per unit length of wire in the coil is the same throughout, and is the same as the value of e at the boundary of the core and coil, e being given by equation (11); consequently the whole E.M.F. in the coil arising from induction is $2\pi c . e . Nl$, which, by equations (11) and (15), is the same as the second term on the right-hand side of equation (20) with its sign changed. We may also note that the coefficient of $L\Gamma'$ in equation (20) is, by equation (3), the time-constant of a tube of unit depth with the mean radius c.

§ 10. *Oscillatory Currents.*

Let there be an oscillatory current in the coil-circuit, kept up by an impressed force $E \sin nt$, where E is constant, and n is proportional to the frequency, being $= 2\pi v$, if $v =$ number of complete waves per second. At starting, the current will not be simple-harmonic, but in a very short time, usually a small fraction of a second, the whole system of magnetic force and current will settle down to vary with the time according to the simple-harmonic law, though the coil-current will not be in coincident phase with the impressed force, nor with the core-currents. But the magnetic force H is a function of r, therefore put

$$H = H_1 \sin nt + H_2 \cos nt, \text{........................(21)}$$

where H_1 and H_2 are functions of r only. To find their form, insert the right side of equation (21) in equation (14), which H must satisfy. We get

$$\frac{1}{r}\frac{d}{dr}(rH'_1) = -xH_2, \qquad \frac{1}{r}\frac{d}{dr}(rH'_2) = +xH_1, \text{..............(22)}$$

where $x = 4\pi\mu n/\rho$; from which we find that both H_1 and H_2 are subject to

$$\frac{1}{r}\frac{d}{dr}r\frac{d}{dr}\frac{1}{r}\frac{d}{dr}r\frac{d}{dr}H_1 = -x^2 H_1. \text{.....................(23)}$$

To obtain a suitable series, put

$$H_1 = A_0 + A_1 r + A_2 r^2 + \dots,$$

insert the series in (23), perform the differentiations, and equate the coefficients of the different powers of r to 0 separately. We find that all the odd A's are zero, and that

$$H_1 = A_0 M + A_2 N,$$

where M and N are functions of r given by

$$\left.\begin{aligned} M &= 1 - \frac{x^2r^4}{2^24^2} + \frac{x^4r^8}{2^24^26^28^2} - \frac{x^6r^{12}}{2^2 \ldots 12^2} + \ldots, \\ N &= \frac{xr^2}{2^2} - \frac{x^3r^6}{2^24^26^2} + \frac{x^5r^{10}}{2^2 \ldots 10^2} - \ldots . \end{aligned}\right\} \qquad (24)$$

H_2 is of course the same in form as H_1, say with constants B_0 and B_2. But the equations (22) have to be separately satisfied. This requires that $A_0 = -B_2$ and $A_2 = -B_0$, which makes (21) take the form

$$H = (AM + BN) \sin nt + (AN - BM) \cos nt, \qquad (25)$$

where M and N are given in (24), and A and B are two arbitrary constants (the former A_0 and A_2). We should note that M and N are subject to

$$\left.\begin{aligned} \frac{1}{r}\frac{d}{dr}(rM') &= -xN \\ \frac{1}{r}\frac{d}{dr}(rN') &= +xM \end{aligned}\right\} \qquad (26)$$

which relations are very useful in transformations.

The functions M and N are of the oscillatory character. Oscillatory functions are always turning up in mathematical physics, even in questions having nothing to do with vibrations. M and N are related to Fourier's function for the cylinder, or Bessel's $J_0(z)$ function, where

$$J_0(z) = 1 - \frac{z^2}{2^2} + \frac{z^4}{2^24^2} - \ldots,$$

by the following equations :—

$$\left.\begin{aligned} J_0(r\sqrt{xi}) &= M - Ni, \\ J_0(r\sqrt{-xi}) &= M + Ni, \end{aligned}\right\} \qquad (27)$$

where i stands for $\sqrt{-1}$, from which we have

$$M^2 + N^2 = J_0(r\sqrt{xi})J_0(r\sqrt{-xi}),$$

$$M'^2 + N'^2 = J_0'(r\sqrt{xi})J_0'(r\sqrt{-xi}),$$

etc., and many other relations connected with energy properties. But it is easier to work with M and N than with the corresponding imaginary Bessel's functions.

In equation (25), where H is a known function of r and t, we have the solution for the magnetic force in the core, except as to two constants which fix the amplitude and the phase. By (15) the coil-current is known in terms of the core-boundary value of H, and by (10) the core-current is known.

The amplitude of H is the square root of the sum of the squares of the coefficients of the sine and cosine in equation (25). Denoting the amplitude by (H), we have therefore

$$(H)^2 = (A^2 + B^2)(M^2 + N^2) ; \qquad \text{(28)}$$

and if (Γ) be the amplitude of the coil-current,

$$(\Gamma)^2 = \frac{A^2 + B^2}{(4\pi N)^2}(M^2 + N^2), \qquad \text{(29)}$$

where the boundary values of M and N must be used.

§ 11. *Waves of Magnetic Force.*

From equations (25) and (24) we see that the magnetic induction due to the oscillatory coil-current travels into the core from its boundary in waves, decreasing rapidly in amplitude as they progress. It is not, however, a case of real [elastic] wave propagation, but of diffusion. To get an idea of the lengths of waves at given frequencies, let in (25) the constant $B = 0$, and $A = 1$, making

$$H = M \sin nt + N \cos nt.$$

At the moment $t = 0$, or at any succeeding moment making $\sin nt = 0$, we have $H = N$, where N is given in (24). N is zero at the axis, positive up to a certain distance, then negative, and so on until the boundary is reached, after which there are no more reversals of sign. The range of N on each side of zero increases fast with r. By inspection, the function N vanishes for the first time when its second term, disregarding its sign, is a little greater than the first, or xr^2 a little greater than 24, say 25. Now here $x = 4\pi\mu n/\rho$, so

$$4\pi\mu n r^2 = 25\rho$$

makes N vanish. Let the core be of copper, for which $\rho = 1700$ (*i.e.*, 1·7 microhm per c.c.) and $\mu = 1$. Also put $n = 2\pi v$, v being the number of waves per second. Then we have

$$4\pi . 2\pi v . r^2 = 25 . 1700, \quad \text{or} \quad vr^2 = 530.$$

Thus, if there are 530 waves per second, N vanishes at a distance of 1 cm. from the axis, and as it also vanishes at the axis, the length of the first semi-wave of magnetic force in the core is 1 cm. As it varies in general inversely as the square root of the frequency, we may get as many waves as we like into a core by increasing its diameter and the frequency. But we cannot get shorter waves at the same frequency than in copper in any other material except iron, and perhaps other magnetic metals, on account of ρ being least for copper.

But let the core be of iron. Its specific resistance is about 10,000, thus considerably reducing x, as compared with copper. On the other hand μ is usually a large number. Its value is so eminently variable that it is difficult to know what value to take. But $\mu = 100$ in good soft iron is probably not very far over or under a fair average. This makes

$$x = \frac{8\pi^2 . 100 . v}{10000} = \frac{8v}{10},$$

so

$$vr^2 = \frac{250}{8} = 31$$

makes $N=0$ for the first time, and thus a frequency of only 31 waves per second will make the length of the first semi-wave of magnetic force $=1$ cm. The speed $v=530$ mentioned in the case of a copper core above will reduce it to $\frac{1}{4}$ cm.

At the times making $\cos nt=0$, we have $H=M$; H is then unity at the axis, and 0 for the first time at such a distance as makes xr^2 a little over 8. This will give only half a semi-wave.

§ 12. *Amplitude of Magnetic Force.*

The maximum value of H at any point in the core varies as $(M^2+N^2)^{\frac{1}{2}}$. By squaring the two expressions in (24) and adding the results we find the following series for M^2+N^2:—

$$M^2+N^2=1+\frac{2y}{2^2 4^2}\Big(1+\frac{3y}{6^2 8^2}\Big(1+\frac{3\frac{1}{3}y}{10^2 12^2}\Big(1+\frac{3\frac{1}{2}y}{14^2 16^2}\Big(1+\frac{3\frac{3}{5}y}{18^2 20^2}(1+\ldots, \quad (30)$$

where we employ the brackets to show readily the degree of convergency of the series for a given value of y, which stands for x^2r^4, the quantity whose powers appear in the expressions for M and N. Every term being positive, it is verified that the amplitude increases continuously from the centre outward.

Compare the amplitudes of the magnetic force oscillations at the centre, at $\frac{1}{2}$ cm., and at 1 cm. First in a copper core, for which, as above, $y=25^2$ about for 1 cm. and $25^2/4^2$ for $\frac{1}{2}$ cm. With $y=24^2$ we find

$$M^2+N^2=1+18\Big(1+\frac{3}{4}\Big(1+\frac{2}{15}\Big(1+\frac{9}{224}\Big(1+\ldots=34{\cdot}37=(5{\cdot}8)^2.$$

And with $y=24^2/4^2$ we get

$$M^2+N^2=1+\frac{18}{16}\Big(1+\frac{3}{64}\Big(1+\frac{2}{240}\Big(1+\ldots=2{\cdot}18=(1{\cdot}47)^2.$$

Thus, we find the amplitudes at the centre, at $r=\frac{1}{2}$ cm., and $r=1$ cm., to stand in the ratios

$$1:1{\cdot}47:5{\cdot}86,$$

being almost exactly four times as great at the double distance. (We took 24^2 instead of 25^2 to simplify the fractions, thus making the distances a trifle smaller.)

In the iron core we shall of course have exactly the same proportions with the same values of y, but on account of the largeness of μ, whose square appears in y, the distances will be much less, or they may be the same with a lower frequency to suit. We see from the great rapidity of increase of the amplitude that in a large core, especially if it be of iron, the magnetic induction reaches the centre in comparatively small strength, quite insignificant in fact if the frequency be such as to cause there to be several waves in the core at once. The small inner magnetic force on the action-at-a-distance hypothesis is of course due to the magnetic force due to the coil-current being nearly cancelled by that due to the exterior core-currents. In the Maxwellian view of the matter the magnetic induction is a flux; it, or the molecular disturbance corresponding to it, travels in a conductor, diffusing itself according to the same laws as heat, with this remarkable difference, that the better the conductor the slower the diffusion. In the present case, the magnetic induction in the core being everywhere directed parallel to its axis, the diffusion takes place radially from the coil to the axis of the core and back.

§ 13. *Heat in Core and in Coil.*

Let W and w be the heats per second in the coil-circuit and in the core, when the oscillatory current passes in the coil. By Joule's law the rate of generation of heat in the coil-circuit is $R\Gamma^2$, but since Γ is a simple harmonic function of the time, this amounts to only

$$W = \tfrac{1}{2}R(\Gamma)^2 \text{ per second,} \qquad (31)$$

where (Γ) is the amplitude of Γ. Hence by equation (29),

$$W = \frac{\frac{1}{2}R}{(4\pi N)^2}(A^2+B^2)(M^2+N^2), \qquad (32)$$

where the boundary values of M and N are to be used.

In the core we have, similarly, $\rho\gamma^2$ = rate of generation of heat per unit volume, or $\frac{1}{2}\rho(\gamma)^2$ per second, (γ) being the amplitude of γ; and summing this up throughout the core, we get

$$w = \tfrac{1}{2}\rho l\int(\gamma)^2 2\pi r\,dr = \pi\rho l\int(\gamma)^2 r\,dr.$$

Here, by equations (10) and (25), we have

$$(\gamma)^2 = \frac{1}{(4\pi)^2}(H')^2 = \frac{1}{(4\pi)^2}(A^2+B^2)(M'^2+N'^2).$$

Hence

$$w = \frac{\pi\rho l}{(4\pi)^2}(A^2+B^2)\int(M'^2+N'^2)r\,dr. \qquad (33)$$

The integration is easily performed. Integrate one of the M''s and one of the N''s; thus,

$$\int(M'.rM'+N'.rN')\,dr = MrM'+NrN' - \int\left\{M\frac{d}{dr}(rM')+N\frac{d}{dr}(rN')\right\}dr.$$

Here the quantity under the integral sign on the right vanishes, as may be seen by equations (26); so taking limits $r=0$ and $r=c$ we bring (33) to

$$\left.\begin{aligned} w &= \frac{\pi\rho lc}{(4\pi)^2}(A^2+B^2)(MM'+NN') \\ &= c\pi\rho l(N\Gamma)^2\frac{MM'+NN'}{M^2+N^2}. \end{aligned}\right\}(r=c). \qquad (34)$$

In the last expression for the core-heat per second in terms of the coil-current, with A and B removed, we have a fraction to consider. The numerator is one half of the differential coefficient of the denominator, the expression for which is given in equation (30). Hence we find, by differentiating (30) with respect to r, the value of the numerator to be

$$MM'+NN'=\frac{y}{16r}\left(1+\frac{6y}{6^2 8^2}\left(1+\frac{5y}{10^2 12^2}\left(1+\frac{4\frac{2}{3}y}{14^2 16^2}\left(1+\frac{4\frac{1}{2}y}{18^2 20^2}\left(1+\ldots,\right.\right.\right.\right.\right. \quad (35)$$

where y stands for x^2r^4.

Now, when y is not too large, we may take only the first terms in this series, and in that for M^2+N^2, *i.e.*, take

$$M^2+N^2=1, \quad \text{and} \quad MM'+NN'=\frac{y}{16r}=\frac{x^2c^3}{16},$$

putting $r=c$. Further, put for x its full expression $8\pi^2 v\mu/\rho$, and (34) becomes

$$w=\frac{4\pi l}{\rho}(\pi^2c^2\mu vN)^2(\Gamma)^2. \quad \ldots\ldots\ldots\ldots\ldots\ldots\ldots (36)$$

From this we see that the core-heat, with the same coil-current per unit length of core (*i.e.*, $N\Gamma$), varies directly as the conductivity of the core, as its length of course, as the square of its section, as the square of its permeability, and as the square of the frequency. That it should vary as the square of the permeability will be understood on remembering that the inductive E.M.F. in the core, and therefore the current, varies as the time-rate of decrease of the induction, which is μ times the magnetic force, which makes the square of the current, and therefore the heating, vary as μ^2. Also we see at once how immensely greater the heating must be in iron than in copper, the high conductivity of the latter being little set off against the large value of μ^2. These conclusions require y to be small enough to make the second term in the expansion for M^2+N^2 small compared with unity. The full expression for the core-heat is got by multiplying the expression in (36) by Y, where

$$Y=\frac{1+\dfrac{6y}{6^2 8^2}+\dfrac{30y^2}{6^2\ldots 12^2}+\ldots}{1+\dfrac{2y}{2^2 4^2}+\dfrac{6y^2}{2^2\ldots 8^2}+\ldots},$$

by (35) and (30). Since the denominator increases faster than the numerator, the heat becomes less than is represented in equation (36), and, in fact, a small fraction thereof when y is made large.

We may easily make the core-heat as great as the coil-heat, even if the core is very small and non-magnetic, and no great frequency is required either, whilst with iron cores it may be hundreds of times as much. The heats in the core and in the coil being associated, it will be desirable to compare the core-heat, not with the heat in the whole coil-circuit, but in the coil only. Using, then, the equation (9) for the coil-resistance, viz.:—

$$R=(2Nc_1)^2\frac{\rho_1 l}{b}(2a+b), \quad \ldots\ldots\ldots\ldots\ldots\ldots\ldots (9) \textit{ bis.}$$

we find, by (36),

$$\frac{w}{W}=\frac{2\pi b(\pi^2c^2\mu v)^2}{c_1^2\rho\rho_1(2a+b)}. \quad \ldots\ldots\ldots\ldots\ldots\ldots\ldots (37)$$

§ 14. *Examples, and Remarks on Variable Permeability.*

Example 1.—Now let the coil-wire and the core be of copper, $\rho=\rho_1=1700$, and $\mu=1$, and take $c_1=1$, which is near enough. Also let the core fill the coil-opening, making $a=c$, and let this be 1 cm., and the depth $b=2$ cm. Then

$$w/W=\pi.100\,v^2/1700^2=v^2/9000 \text{ say.}$$

Thus, by the approximate formula it would require a frequency of 95 waves per second to make the core- and coil-heats equal. This value of the frequency, however, with the other data, makes y too large for Y to be altogether left out, and the core-heat is really about $\frac{3}{5}$ of the coil-heat.

Example 2.—With the same dimensions, let the core be of iron, with $\mu=100$ and $\rho=10{,}000$. Here $x=8v/10$ and $y=64\,v^2/100$. We find, by (37), $w/W=\pi v^2/17$. Therefore to get equal heats the approximate formula makes v as low as $(17/\pi)^{\frac{1}{2}}=2{\cdot}32$ waves per second.

Example 3.—The same iron core, but with $v=100$ per second. Here the approximate formula makes $w/W=1848$. But now y is so large that Y must be considered. Its value works out $Y=$about $\frac{1}{70}$. Consequently $w/W=1848/70=26{\cdot}4$, *i.e.*, 26·4 times as much heat in the core as in the coil. Increasing the size of the core also increases the ratio, with a corresponding increase in the size of the coil. In this last example put $\mu=1$, *i.e.*, core non-magnetic, but with the same conductivity, then $w/W=\pi/17$ only.

Some remarks * need be made concerning the heating of iron cores. In the first place, although the induction may be 100 or more times the magnetic force when the latter is weak, yet the ratio will get smaller and smaller as the force increases. Hence, as in the above

* [These remarks are not altogether wide of the mark. It was well known to me at the time, as a matter of common-place fact in the behaviour of iron in induction balances, that the magnetisation of iron exposed to weak magnetising forces, or rather to rapid variations thereof, was of the perfectly elastic type. That is, the permeability is a constant, usually a big number, as 100 or 200. Also that iron already strongly independently magnetised behaved similarly, that is, with elastic changes of induction under the influence of small variations in the magnetising force. Professor Hughes too, in his lecture "On the Cause of Evident Magnetism," asserted the perfectly elastic character of magnetisation within a limited range. On the other hand, Professor Ewing, of later hysteresial fame, not long after came to the singular conclusion that the initial permeability of iron was, or seemed to be, not big, but evanescent. Lord Rayleigh, however, disproved this in 1886 or 1887 by actual measurements with steady forces. There was no sign of μ vanishing, and the old view was completely confirmed.

The μ considered by magneticians of late years throughout very wide ranges of magnetising force is not elastic permeability at all, but is a cumulative matter, involving induced and intrinsic magnetisation simultaneously, and hysteresis. In spite of the recent remarkable extension of knowledge there is still much to be done. I do not know if observations have been made on the value of the elastic μ throughout the whole attainable range of the induction. Undoubtedly it becomes small when the induction is big, but whether it tends to unity, or can be lower, is unsettled. The possibility of μ being less than unity in vacuum, mentioned in the text, is entirely speculative, although we have no right to assert that the ether retains its properties unchanged when supporting extraordinarily great magnetic stress.]

calculations μ is treated as a constant, the results are only strictly applicable when the range of the oscillatory current is not so great as, by the magnetisation set up, to alter μ much. Should μ be much altered the strength of the induced current, and the consequent heating, will be correspondingly lowered.

Again, suppose that the oscillatory current in the coil-circuit is not the only current there, but consists of variations in the strength of a current of continuous sign. Here evidently the permanent current, by the permanent magnetisation it produces, lowers the value of μ, and the proper value as thus lowered must be taken in the above formulæ. In the extreme, when the permanent current is so strong as to practically saturate the iron, or produce nearly the greatest magnetisation it can legitimately receive under the circumstances, *i.e.*, without mechanical shocks to make the molecules settle into a state indicating a much greater magnetisation under the influence of the magnetising current than it would take unaided, then, unless the oscillatory current is so strong as to undo the work of the permanent current, the changes in the induction will be comparatively small. Let the oscillatory current be a small fraction of the permanent, then μ must be little greater than unity, so far as the variable magnetic force is concerned, and the induced currents and the heating little more than they would be in a non-magnetic core of the same conductivity, *i.e.*, considerably weaker than in a similar copper core.

Professor Hughes says he has discovered that air has a maximum capacity for magnetisation, and that it is equal to that of the purest soft iron. Here evidently, since for μ to be $\cdot 100$ or 50, or whatever it may be, is absurd, the undefined term capacity is used, so far as it is applied to air, to indicate something very different from μ, the magnetic permeability. Nor can it be κ, the ratio of the magnetisation to the magnetising force. As the question is one of considerable importance it is to be hoped that Professor Hughes's researches will cast much light upon this and many other little-understood parts of magnetic science. In the meantime we may here briefly point out one or two things in connection with saturation and conclusions from the mathematical theory. Let there be a coil, or long slender solenoid for simplicity, containing an iron core, and let a steady current be passing through the wire. The iron becoming magnetised, the action on a magnetic needle outside is increased above what it would be if the core were absent. Increase the current sufficiently, and the iron is said to become saturated, no further increase of magnetisation taking place. This is represented in the theory by supposing the value of μ to fall to unity. If this is practically reached, the core will then, so far as further increase of current goes, behave as if it were replaced by air. The action upon a needle at a distance should still go on increasing, and in proportion to the increase of current above the saturating value. But should the action on the needle stop increasing, or show signs of tending to a limit, not increasing as the current, the conclusion that the magnetisation of the core remains constant is erroneous. It must be decreasing. For our supposed observation shows that the magnetic induction tends to a limit, or at any rate does not increase as it should with $\mu = 1$. Every tube of induction goes through the core (with a correction for the space occupied by the coil), and in the core the induction is the resultant of the magnetic force and of 4π times the intensity of magnetisation, whilst the magnetic force is that due to the current in the coil by the ordinary formula plus the polar force of the magnetisation, which latter may be practically confined to the ends by lengthening the coil. The magnetic force then increasing in the same ratio as the current, and the induction in the air outside by our supposition not showing corresponding increase, the magnetisation of the core must be decreasing, or μ must have fallen below unity in the iron.

It is also obviously suggested that, the iron being removed, the mechanical force on a magnet outside, say on a needle at a sufficient distance to keep its magnetisation constant, may not continue to be proportional to the strength of current when the current is very strong. In this case, keeping to the relation $4\pi\gamma = \text{curl}\ \mathbf{H}$ between the current and the magnetic force, μ also becomes less than unity in the places of strongest force. And as we conclude from the behaviour of, say, copper in a strong field of force that the permeability of the copper is less by a trifle than that of the air enveloping it, if we have μ less than unity in the air it must be still more less than unity in the copper, and probably less than unity in vacuum also with a sufficiently strong current.

The published dynamo-formulæ giving relations between the current-strength and the E.M.F. are very much alike in this, that if you take away the iron they break down immediately. It is the fact of the magnetisation of iron not being proportional to the strength of current that is made use of to put a stop to the increase of the current when the speed is kept constant. But if μ falls in value in air or vacuum also, it will make the same or similar formulæ applicable to dynamos without any iron, in which present theory indicate indefinite increase of current when the speed is the least above a certain critical value, and the speed is kept constant, with therefore an indefinite supply of power. But we must return to the main subject from these parenthetical speculations.

§ 15. *Coil-Current in Terms of E.M.F.*

We require to know to what extent the current in the coil is altered by the presence of the conducting core, *i.e.*, given the impressed force $E \sin nt$, required Γ in terms of E. We have the equation of E.M.F. in the coil

$$E = R\Gamma + (L - L_1)\dot{\Gamma} + \frac{L_1\rho}{2\pi\mu c}\Gamma', \quad \text{.................(19) } bis$$

where, by (15), $4\pi N\Gamma = H_c$, the boundary value of H, and, by equation (25),

$$H = (AM + BN)\sin nt + (AN - BM)\cos nt. \quad \text{.........(25) } bis$$

Putting Γ in terms of H in equation (19) we get the two equations

$$4\pi NE = AP + BQ, \quad 0 = AQ - BP, \quad \text{.................(38)}$$

in which the new quantities P and Q are given by

$$\left.\begin{aligned} P &= RM - (L - L_1)nN + \frac{\rho L_1}{2\pi\mu c}M', \\ Q &= RN + (L - L_1)nM + \frac{\rho L_1}{2\pi\mu c}N'. \end{aligned}\right\} \quad \text{.................(39)}$$

From (38) we obtain

$$\left.\begin{aligned} (4\pi NE)^2 &= (A^2 + B^2)(P^2 + Q^2), \\ A = 4\pi N\frac{EP}{P^2 + Q^2}, &\quad B = 4\pi N\frac{EQ}{P^2 + Q^2}. \end{aligned}\right\} \quad \text{.............(40)}$$

A and B being thus known, the current and the magnetic force in the core and the coil-current become known at every moment. In equation (29), put for $A^2 + B^2$ its value given by the first of equations (40), and we get

$$(\Gamma)^2 = E^2\frac{M^2 + N^2}{P^2 + Q^2}, \quad \text{........................(41)}$$

giving the amplitude of the coil-current in terms of E and known functions, in which r is to be put $= c$.

The current at the time t is

$$\left.\begin{aligned} \Gamma &= (\Gamma) \cdot \sin(nt - \theta), \\ \text{where} \qquad \tan\theta &= \frac{QM - PN}{QN + PM}, \end{aligned}\right\}(r = c). \quad \text{.............(42)}$$

The angle θ determines the difference of phase between the current and the E.M.F.

The full expression for $P^2 + Q^2$ is, by (38),

$$P^2 + Q^2 = \{R^2 + (L - L_1)^2n^2\}(M^2 + N^2) + \left(\frac{\rho L_1}{2\pi\mu c}\right)^2(M'^2 + N'^2)$$
$$+ \frac{R\rho L_1}{\pi c\mu}(MM' + NN') + \frac{\rho L_1}{\pi c\mu}(L - L_1)n(MN' - NM'), \quad (43)$$

where, besides $(M^2 + N^2)$ and $(MM' + NN')$, for which the expansions have been given, equations (30) and (35), there are two other functions. If many terms are required we may derive the expansions of $M'^2 + N'^2$ and $MN' - NM'$ from $M^2 + N^2$ by differentiations. For we have

$$2(M'^2 + N'^2) = \frac{1}{r}\frac{d}{dr}r\frac{d}{dr}(M^2 + N^2),$$

as may be proved by equations (26). And

$$MN' - NM' = \frac{1}{2xr^2}\frac{d}{dr}r^2(M'^2 + N'^2),$$

as may also be proved by equation (26). We thus find

$$M'^2 + N'^2 = \frac{x^2r^2}{4}\left(1 + \frac{3y}{4^26^2} + \frac{10y^2}{4^2 \dots 10^2} + \frac{35y^3}{4^2 \dots 14^2} + \frac{126y^4}{4^2 \dots 18^2} + \dots\right), \quad (44)$$

$$MN' - M'N = \frac{xr}{2}\left(1 + \frac{6y}{4^2 6^2} + \frac{30y^2}{4^2 \dots 10^2} + \frac{140y^3}{4^2 \dots 14^2} + \frac{630y^4}{4^2 \dots 18^2} + \dots\right). \quad (45)$$

It may, however, according to circumstances, be easier to calculate the values of P and Q separately from equations (39). Of course we must then find the values of M, N, M', and N' separately, the first two by equations (24) and the last two from the expressions obtained by differentiating them.

§ 16. *First Approximation to Effect of Core-Currents in Altering Amplitude and Phase of Coil-Current.*

This we obtain by taking only the first terms of the different series, that is, we consider only the first power of $1/\rho$. How far this will approximate to the truth will depend on the size of y, as illustrated in previous examples. In equations (39) take

$$M = 1, \qquad N = \frac{xr^2}{2^2} = \frac{\pi\mu nc^2}{\rho}, \qquad M' = -\frac{4x^2r^3}{2^2 4^2} = -\left(\frac{\pi\mu nc^2}{\rho}\right)^2 \frac{1}{c},$$

$$N' = \frac{2xr}{2^2} = \frac{2\pi\mu nc}{\rho};$$

they then become

$$P = R - (L - \tfrac{1}{2}L_1)\frac{\pi\mu n^2c^2}{\rho}, \qquad Q = Ln + R\frac{\pi\mu nc^2}{\rho}, \quad \dots\dots\dots\dots(46)$$

so that the solution (42) reduces to

$$\Gamma = \frac{E}{(R^2 + L^2n^2 + RL_1\pi\mu n^2c^2/\rho)^{\frac{1}{2}}} \sin(nt - \theta) \quad \dots\dots\dots\dots(47)$$

and

$$\tan\theta = \frac{Ln}{R + \frac{1}{2}L_1\pi\mu n^2c^2/\rho}. \quad \dots\dots\dots\dots(48)$$

This solution, (47) and (48), shows that when induction of currents is permitted the retardation of phase is reduced, whilst the amplitude is also reduced. Put $\rho = \infty$ and we have the solution for the case where there is the same core inside the coil, but, by proper division, the currents cannot flow. The reduction in the strength of the coil-current from E/R to $E/(R^2 + L^2n^2)^{\frac{1}{2}}$ is then due to the self-induction, including that due to the magnetisation, and the retardation of phase is $\theta = \tan^{-1}Ln/R$. Now (47) and (48) show that, when the induced currents in the core are allowed to flow, and the speed is not so high, or the dimensions so great as to make it imperative to use more terms of the series, the effect on the amplitude and phase of the coil-current is the same as if the resistance of the coil-circuit were increased from R to

$$\{R^2 + RL_1\pi\mu n^2c^2/\rho\}^{\frac{1}{2}} = R + \tfrac{1}{2}L_1\pi\mu n^2c^2/\rho$$

approximately, thus reducing the time-constant, the strength of current, and the retardation of phase. Many phenomena which may be experimentally observed when rods are inserted in coils may be usefully explained in this manner.

§ 17. *Fuller Examination of Reaction of Core on the Coil.*

The core-currents may be allowed to flow or not, by, in the latter case, a suitable division of the core into insulated parts. We then have merely the insignificant currents which the dielectric will permit. We make no count of the heating from this cause, which will be considered later. Also, the alternate magnetisations and demagnetisations occurring in the core are supposed to be of the conservative character, involving no dissipation of energy at all when the currents are not allowed to flow. I do not know whether it has been definitely established that there is a dissipation of energy going on in iron cores apart from the heating as per Joule's law, although I conclude from indirect experiment that if there be any it is not great with moderate magnetising forces. On the above suppositions we may let the currents flow in the core, or stop them completely, as we please, without removing the core. If it be of a non-magnetic metal, with μ practically equal to unity, the effect on the coil-current of insulating the core is the same as removing it altogether. But if it be of iron its insulation and removal are of course not equivalent. Put $\rho=\infty$ to show insulation. Put $\mu=1$ as well to show removal, which will greatly alter the value of L, the inductance, when the core is of iron, but not at all if of copper.

Let $L=L_1$, *i.e.*, let the depth of coil be small compared with its radius, let the core fill the coil-opening, and let the external self-induction of the circuit be negligible. Allowing the core-currents to flow always diminishes the lag of the coil-current behind the impressed force, but the amplitude of the coil-current may be either reduced or increased, according to the frequency and other circumstances, especially the resistance of the coil-circuit. First, as regards the lag, that θ is always reduced may be thus seen. Put in equations (42) the values of P and Q given by equations (39); then

$$\tan\theta=\frac{S(MN'-NM')}{R(M^2+N^2)+S(MM'+NN')},$$

where S stands for $L_1\rho/2\pi\mu c=2L_1n/xc$. Now use in this the expressions lately given for the M and N functions in the brackets, and we shall find

$$\tan\theta=\frac{L_1n/R}{Y_1+(L_1n/R)(xr^2/8)Y_2}, \quad\text{.........................(48a)}$$

where Y_1 and Y_2 are two functions whose values are unity when $y=1$, and whose full expressions are

$$\left.\begin{aligned} Y_1&=\frac{1+\dfrac{2y}{2^24^2}\left(1+\dfrac{3y}{6^28^2}\left(1+\dfrac{3\frac{1}{3}y}{10^212^2}\left(1+\dfrac{3\frac{1}{2}y}{14^216^2}\left(1+\dfrac{3\frac{3}{5}y}{18^220^2}\right.\right.\right.\right.}{1+\dfrac{6y}{4^26^2}\left(1+\dfrac{5y}{8^210^2}\left(1+\dfrac{4\frac{2}{3}y}{12^214^2}\left(1+\dfrac{4\frac{1}{2}y}{16^218^2}\right.\right.\right.},\\ Y_2&=\frac{1+\dfrac{6y}{6^28^2}\left(1+\dfrac{5y}{10^212^2}\left(1+\dfrac{4\frac{2}{3}y}{14^216^2}\left(1+\dfrac{4\frac{1}{2}y}{18^220^2}\right.\right.\right.}{\text{Same denominator}}, \end{aligned}\right\} \quad\text{.......(49)}$$

the denominator in Y_2 being the same as in Y_1. Now it can be easily seen that Y_1 is greater than unity; this fact, since Y_2 is also positive, makes the denominator in (48) always greater than unity; and therefore $\tan\theta$ less than $L_1 n/R$, which is its greatest value, occurring when the core-currents are stopped.

But as regards the amplitude of the coil-current, if we call the impedance* of the circuit R_1, *i.e.*, what the resistance should be with the same impressed force to give the actual current-strength if there were no inductive E.M.F., then, by equation (41),

$$R_1^2 = \frac{P^2+Q^2}{M^2+N^2} = R^2 + S^2\frac{M'^2+N'^2}{M^2+N^2} + 2RS\frac{MM'+NN'}{M^2+N^2},$$

which, by the given expressions for the M, N functions, may be written

$$\left(\frac{R_1}{R}\right)^2 = 1 + \left(\frac{L_1 n}{R}\right)^2 Y_3 + \left(\frac{L_1 n}{R}\right)\frac{xr^2}{4}Y_4, \qquad (50)$$

in which Y_3 and Y_4 are given by

$$\left.\begin{aligned} Y_3 &= \frac{1+\frac{3y}{4^2 6^2}\left(1+\frac{3\frac{1}{3}y}{8^2 10^2}\left(1+\frac{3\frac{1}{2}y}{12^2 14^2}\left(1+\frac{3\frac{3}{5}y}{16^2 18^2}\right.\right.\right.}{1+\frac{2y}{2^2 4^2}\left(1+\frac{3y}{6^2 8^2}\left(1+\frac{3\frac{1}{3}y}{10^2 12^2}\left(1+\frac{3\frac{1}{2}y}{14^2 16^2}\left(1+\frac{3\frac{3}{5}y}{18^2 20^2}\right.\right.\right.\right.}, \\ Y_4 &= \frac{1+\frac{6y}{6^2 8^2}\left(1+\frac{5y}{10^2 12^2}\left(1+\frac{4\frac{2}{3}y}{14^2 16^2}\left(1+\frac{4\frac{1}{2}y}{18^2 20^2}\right.\right.\right.}{\text{Same denominator}}, \end{aligned}\right\} \qquad (51)$$

the denominator in both cases being the same. Now here Y_3 and Y_4 are less than unity, which is their value when $y=0$, and this permits R_1^2 to be either greater or less than $R^2+L^2n^2$, the value when the core-currents are stopped. In order that R_1 should be increased or decreased by permitting the core-currents to flow, we must have, by equation (50),

$$1 + \left(\frac{L_1 n}{R}\right)^2 Y_3 + \left(\frac{L_1 n}{R}\right)\frac{xr^2}{4}Y_4 > \text{ or } < 1 + \left(\frac{L_1 n}{R}\right)^2,$$

which is the same as

$$\frac{L_1}{R} < \text{ or } > \frac{\pi\mu r^2}{\rho}\,\frac{Y_4}{1-Y_3}. \qquad (52)$$

Here L_1/R is the time-constant of the coil-circuit. Its greatest value is when the coil is short-circuited, so we cannot increase it as we like; but by inserting resistance we may diminish it indefinitely. Thus L_1/R may be always made less than the quantity on the right side, and should it be greater when the coil is short-circuited, we may, according to the resistance we insert, cause the coil-current to be either increased or diminished by the reaction of the core-currents when the latter are permitted. Noting that Y_4 and Y_3 are fractions which decrease from unity as y increases, that is, as the frequency of the oscillations, and other

* ["Impedance" is here, and later, substituted for "apparent resistance." It is the ratio of the amplitude of the impressed force to that of the current when their variations are simple-harmonic.]

circumstances before mentioned, making $Y_4/(1-Y_3)$ large when y is small, and small when y is large, we see that in general at low frequencies the coil-current is likely to be weakened, and at high frequencies strengthened by the core-currents. The latter effect may be very considerable.

Example 1. *y small.*—Let $y=1$, or $(8\pi^2\mu v r^2)^2=\rho^2$. If the core is of copper, $\rho=1{,}700$, $\mu=1$. Let $c=r=1$ cm., then $8v=170$ gives the frequency required. We find by equations (51),

$$Y_3=\cdot 975, \qquad Y_4=\cdot 972,$$

so that the impedance R_1, by equation (52), is increased by allowing the core-currents to flow, if

$$L_1/R<(\pi/1{,}700)\div(972/25) \quad \text{or} \quad 1/15 \text{ sec. nearly.}$$

Now, under the circumstances, the time-constant must be less than this value. For, by equations (6) and (9), its greatest value, viz., when coil is short-circuited, is given by

$$L_1/R=\pi^2\mu b c^2/c_1{}^2\rho_1(2c+b),$$

in which we must for the present purpose put $\mu=1$. Also $\pi^2=10$, and $c_1{}^2\rho_1$ we may take $=2{,}000$, making

$$L_1/R=bc/400=b/400,$$

remembering that b must be small compared with c, and that $c=1$. Thus the amplitude of the coil-current is in this case reduced by the core-currents, whatever the resistance in circuit may be. With the time-constant $=1/4000$ we find the squares of the impedances as compared with the square of the real resistance to stand in the ratios

$$1 : 1\cdot 0011 : 1\cdot 0093,$$

the unity being with no core, the second number with self-induction only, the third with core-currents also, showing the effect of the latter cause in reducing the coil-current to be greater than that of the self-induction alone.

As regards the alteration in the amount of retardation, we find

$$\tan\theta=L_1 n/R=133\cdot 5/4000=\cdot 03337,$$

with self-induction only; and, by equation (48),

$$\tan\theta=\cdot 03337/(1\cdot 0206+\cdot 03337\times\cdot 9922/8)$$
$$=\cdot 03337/1\cdot 0247=\cdot 03256,$$

when the core-currents are permitted to flow. This difference may seem very insignificant, but when two coils are *balanced* against one another far smaller changes are experimentally observable. However, the example is not a favourable one for showing a large difference. If in this example we substitute an iron core, keeping other things the same, with the same value of $y=1$, with a lowered frequency to suit the altered permeability and conductivity, we simply multiply the former time-constant by the new value of μ, and we shall find, unless μ is exorbitantly large, that the small value $y=1$ still does not permit the coil-current to be increased by the core-currents. To get this result we must raise the now lowered frequency, give a larger value to y, and we shall find the coil-current much increased.

Example 2. $y=40^2$ Iron core.—Let $\mu=100$, $\rho=10{,}000$, $c=1$, and $v=50$. Then $x=40$, and $y=40^2$.

Here we find $M^2+N^2=198{\cdot}25$, showing that the amplitude of the magnetic force at the core's boundary is 14 times that at the axis. We also find

$$Y_1=4{\cdot}454,\quad Y_2={\cdot}176,\quad Y_3={\cdot}089,\quad Y_4={\cdot}039;$$

and these values inserted in equations (48*a*) and (50) give us

$$\left(\frac{R_1}{R}\right)^2=1+{\cdot}089\left(\frac{Ln}{R}\right)^2+{\cdot}396\left(\frac{Ln}{R}\right),\quad\text{.............(53)}$$

and
$$\tan\theta=\frac{Ln/R}{4{\cdot}454+{\cdot}881\,Ln/R},\quad\text{.......................(54)}$$

where Ln/R may be varied considerably.

By equation (52) the impedance of the coil-circuit is increased or reduced by permitting the core-currents to flow, according as

$$Ln/R<\text{ or }>\frac{{\cdot}396}{{\cdot}910}={\cdot}42.$$

By our initial supposition, we have $n=2\pi v=100\pi=314{\cdot}16$, therefore

$$L/R={\cdot}42/314{\cdot}16={\cdot}0013\text{ second}$$

is the critical value of the time-constant of the coil-circuit. Now, if the coil be short-circuited, the value of the time-constant can be far greater than this. For instance, by the formula for L/R given in the last example, L/R would be $\frac{1}{40}$ sec. if the depth of the coil were $\frac{2}{19}$ of its radius. It would then require a considerable resistance to be inserted in circuit with the coil to reduce the time-constant to ·0013 sec. If still more resistance be inserted, the core-currents will weaken the coil-current; if less, they will strengthen it. The amount of this strengthening when the coil has little external resistance in connection with it, we may see from equation (53) by taking therein $L/R=025$ and $n=314{\cdot}16$. We get

$$\left(\frac{R_1}{R}\right)^2=1+5{\cdot}524+3{\cdot}108=9{\cdot}632=(3{\cdot}10)^2,$$

the coil-current being therefore a little less than one-third of the strength it would have were there neither core-currents nor self-induction. Now stop the core-currents; we have

$$\left(\frac{R_1}{R}\right)^2=1+(7{\cdot}854)^2=(7{\cdot}86)^2\text{ say,}$$

which makes the coil-current a little over one-eighth of what it would be were there no self-induction as well. Thus, permitting the core-currents increases the strength of the current in the coil about $2\frac{1}{2}$ times. (The magnetisation of the core, on the other hand, is much weakened, except near its boundary. *At* the boundary it is made $2\frac{1}{2}$ times as strong, but as above mentioned it has only 1/14 part of the boundary-value at the centre; whereas, when the core-currents are stopped, the value is sensibly the same throughout the core at any moment.)

The retardation θ of the coil-currents behind the impressed force, which is given by

$$\tan\theta = Ln/R = 7{\cdot}854, \quad \text{or} \quad \theta = 82^\circ\ 40',$$

when the core-currents are stopped, is, by equation (54), brought down to

$$\tan\theta = 7{\cdot}854/11{\cdot}376, \quad \text{or} \quad \theta = 35^\circ,$$

by letting the core-currents flow.

We started with $y = 40^2$, and the values of Y_1, etc., corresponding. That is, $4\pi\mu nr^2/\rho = 40$. Consequently we may dig out from the same values of Y_1, etc., the results in a variety of other cases, varying the frequency and the permeability, resistance and radius of the core in any manner consistent with $y = 40^2$.

§ 18. *Induction in a Divided Core.*

To ascertain under what circumstances the heating, according to Joule's law, of a properly divided core might become sensible—general considerations telling us that it must be very small—let us, to bring the matter under mathematical treatment, specify a particular manner of division. This will be most shortly described by referring to a plane section of the core perpendicular to its axis. Divide the circular section by radii into sectors, and let the sectors be of two sets, one set having all an opening of say 1°, and the other set of say 9°. These are to alternate. The small sectors to be filled up with dielectric material, the large with metal. We thus stop the free flow of the circular induced currents in the core by the insulating barriers placed perpendicular to the lines of electric force. There are 36 condensers (the number is immaterial), joined up in sequence, which become charged and discharged during the passage of an oscillatory current in the coil-circuit, and according to old notions electricity accumulates on the bounding surfaces of the condensers. According to Maxwell's views, however, the currents flow in the same closed circuits as if the core were solid metal, but in the dielectric portions the electric elasticity brings a counter E.M.F. into play, thus preventing the passage of a continuous current, and weakening the strength of oscillatory currents.

Let ρ_1 and c_1 be the resistance and the electrostatic capacity per unit volume of the compound core, later more particularly defined. Let e, as before, be the impressed force per unit length arising from electromagnetic induction, γ the current-density, and now, in addition, D the electric displacement per unit area perpendicular to the lines of flow. Then, instead of the former equation, $e = \rho\gamma$, we shall have

$$e = \rho_1\gamma + D/c_1, \qquad (55)$$

with the additional relation

$$\gamma = \dot{D}. \qquad (56)$$

Differentiating equation (55) to t, and getting rid of D, by (56), we obtain

$$c_1\dot{e} = c_1\rho_1\dot{\gamma} + \gamma, \qquad (57)$$

which we must substitute for the old $e = \rho\gamma$.

Equation (10) holds as before, connecting the current and the magnetic force. So does equation (12), giving the E.M.F. of induction, if we give to μ therein a new value, to be given presently, on account of the core not being homogeneous. Eliminating e between equations (57) and (12) we obtain

$$\frac{1}{r}\frac{d}{dr}r\frac{d}{dr}(\dot{H}+H/\rho_1c_1)=\frac{4\pi\mu}{\rho_1}\ddot{H}, \quad\text{.................(14a)}$$

instead of equation (14), for the characteristic equation of magnetic force in the compound core. Its solution, suitable for oscillatory currents, requires the use of two functions of r, say H_1 and H_2, satisfying

$$\left.\begin{aligned}\frac{1}{r}\frac{d}{dr}r\frac{d}{dr}(H_1+H_2/\rho_1c_1)&=-x_1H_2,\\ \frac{1}{r}\frac{d}{dr}r\frac{d}{dr}(H_2+H_1/\rho_1c_1)&=+x_1H_1,\end{aligned}\right\}\quad\text{.................(22a)}$$

instead of equations (22). Here x_1 stands for $4\pi\mu/\rho_1$. And H_1 and H_2 may be found in series of ascending powers of r by

$$\left(\frac{1}{r}\frac{d}{dr}r\frac{d}{dr}+\frac{1}{\rho_1c_1}\right)^2H_1=-x_1^2H_1, \quad\text{.................(23a)}$$

which takes the place of (23).

Now as regards ρ_1, c_1, and μ. If the ratio of the angles of the wedges occupied by conducting and insulating material be n_1/n_2, and $n_1+n_2=1$, we shall have, if ρ be the specific resistance of the conductor,

$$\rho_1=n_1\rho,$$

the specific resistance of the compound core becoming somewhat reduced. And, if μ_1 and μ_2 be the magnetic permeabilities of the conducting and dielectric parts of the core, we have

$$\mu=n_1\mu_1+n_2\mu_2,$$

giving the value of μ to be used in equation (12) and in (14a) just given. Thus the permeability of the compound core, if the conducting part be of iron, is somewhat reduced.

Lastly, if c be the electrostatic capacity per unit volume of the dielectric only, that of the compound core is

$$c_1=c/n_2.$$

The time-constant ρ_1c_1 which appears in the characteristic equation (14a) is thus given by

$$\rho_1c_1=\rho cn_1/n_2,$$

where ρ and c belong to the metal and dielectric separately, ρ_1 and c_1 to the compound. The value of this time-constant is the same whether there are 20 or 20,000 wedges, alternately metal and dielectric, provided the ratio n_1/n_2 of the spaces they occupy, which appears in the last equation, is kept the same.

But this time-interval must in general be extremely small. For c, the electrostatic capacity of unit volume of the dielectric, is $K/4\pi$ in electrostatic measure, K being the dielectric constant, and, in the electromagnetic measure here required, we must further divide by V^2, where V is the velocity of light, $=3\times 10^{10}$. Thus we have

$$\rho_1 c_1 = \rho K n_1 / 4\pi V^2 n_2\,;$$

and K being a small number, and also $\rho = 10{,}000$ for iron, whilst V^2 is very large, $\rho_1 c_1$ must be very small, unless we make n_1/n_2 extravagantly large—*i.e*, reduce the thickness of the dielectric wedges greatly, compared with that of the conducting wedges, which of course has the effect of increasing the capacity of the condensers. If this impracticable thinness of dielectric could be carried out with proper insulation, we should, with an oscillatory current in the coil, make the induced currents in the core less by as little as we pleased from what they would be in a solid conducting core, and the heating of the compound core similarly approach that of the solid one.

The influence of increased speed may also be noticed. Consider a single closed circuit of conductors and condensers in series, such as we obtain by confining ourselves to the portion of core contained between the cylinders of radii r and $r+1$. If we further take only the unit length parallel to the axis, the section of the circuit has the unit area. R being the resistance of the circuit, viz., the sum of the resistances of the conducting portions; and C its capacity, viz., the reciprocal of the sum of the reciprocals of the capacities of the single condensers, we have

$$E = R\gamma + Q/C,$$

similarly to (55), E being the total impressed force in the circuit, γ the current, and Q the common charge of each condenser. Also $\gamma = \dot{Q}$, whence

$$C\dot{E} = RC\dot{\gamma} + \gamma.$$

Now let the E.M.F. be of the simple-harmonic type $E \sin nt$, where E is constant; the solution of the current is then

$$\gamma = \frac{nCE}{\{1+(nCR)^2\}^{\frac{1}{2}}}\cos(nt - \tan^{-1} nCR),$$

and the heat developed per second in the conducting part of the circuit is

$$\frac{\frac{1}{2}R(nCE)^2}{1+(nCR)^2} = \frac{1}{2}\frac{E^2}{R}\{1+(nCR)^{-2}\}^{-1}.$$

Here $\frac{1}{2}E^2/R$ is what the heating would be if the condensers were short-circuited; so, from the value of nCR we can easily see how much it is reduced by the insertion of the condensers. The value of RC is the same as that of $\rho_1 c_1$ before given; also $n = 2\pi v$, v being the wave-frequency. By sufficiently increasing the frequency we may make the heating approach as nearly as we like to $\frac{1}{2}E^2/R$ without necessarily making the dielectric portions of the circuit excessively thin; but it may be readily seen that for the heating in the closed circuit of condensers to be comparable with what it would be were they short-circuited, or the dielectric removed and conducting matter substituted, the frequency v must, on account of the presence of V^2 in the expression for RC, be itself comparable with the frequency of light vibrations. If, then, a properly divided iron core with oscillatory currents, or with such superimposed on a steady current passing in a coil surrounding it, should become sensibly heated, such heating cannot have arisen from the Joule effect of the induced currents.

§ 19. *Transmission of Energy into a Conducting Core.*

The magnetic energy per unit volume being $\mu H^2/8\pi$, where H is the magnetic force as before, if we suppose the magnetisation to be wholly induced, the magnetic energy per unit length of core of radius r is given by

$$\frac{\mu}{8\pi}\int 2\pi r\,dr\,.\,H^2 = \frac{\mu}{4}\int H^2 r\,dr. \quad\text{.....................(58)}$$

Also the rate of dissipation of energy in the same portion, being $\rho\gamma^2$ per unit volume, is given by

$$2\pi\rho\int\gamma^2 r\,dr = \frac{\rho}{8\pi}\int H'^2 r\,dr, \quad\text{.......................(59)}$$

since $\gamma = -H'/4\pi$. Now, energy can only enter the portion of core considered across its boundary, and, after having entered, is either stored up temporarily as magnetic energy, or is dissipated as heat through induced currents. Hence the rate of passage of energy into the space from outside equals the sum of the rate of increase of magnetic energy and of the rate of dissipation within the space. Thus, if by W we denote the amount of energy entering the core per second per unit area of its bounding surface, we obtain, by (58) and (59),

$$2\pi r W = \frac{\mu}{2}\int H\dot{H} r\,dr + \frac{\rho}{8\pi}\int H'^2 r\,dr.$$

Here, by equation (14),

$$\dot{H} = \frac{\rho}{4\pi\mu r}\frac{d}{dr}(rH').$$

Making the substitution and integrating, we get

$$W = \frac{\rho}{4\pi\mu}\frac{dT}{dr}, \quad\text{...........................(60)}$$

where $T = \mu H^2/8\pi$, the magnetic energy per unit volume. From this we see that the transmission of energy takes place from places of greater to places of less force irrespective of sign, and that the rate of transference per second is proportional to the rate of decrease of the density of the magnetic energy in the direction of transference.

This remarkably simple property, which applies to every part of the core at every moment, according to which the transmission of magnetic energy whilst induced currents are lasting is determined solely by the space-variation of its density, is not a general property of induction in conductors. If we inquire what the corresponding property is in general, by examining the rate at which energy is entering any given portion of a conductor through the imagined surface separating it from the rest, such being equal to the sum of the rate of increase of the magnetic energy and of the rate of dissipation as heat by induced currents within the portion considered, using the general relations

$$\text{curl } \mathbf{H} = 4\pi\gamma, \quad \text{and} \quad \nabla^2\mathbf{H} = 4\pi\mu\dot{\mathbf{H}}/\rho,$$

we find that W_s, the rate of passage of energy per unit area at any point across the surface whose normal has any direction **s**, or, briefly, the rate of transference in the direction of **s**, equals $\rho/4\pi$ times the component along **s** of the vector-product of the current-density and the magnetic force at the point. Or

$$\mathbf{W} = (\rho/4\pi)\mathrm{V}\gamma\mathbf{H},$$

in the brief vectorial form. The direction of maximum transference is therefore perpendicular to the plane containing the directions of the magnetic force and the current, and its amount per second proportional to the product of their strengths and to the sine of the angle between their directions.

Eliminating the current, and expressing the relation in terms of magnetic force only, we find

$$W_s = -\frac{\rho}{4\pi\mu}\frac{dT}{ds} + \frac{\rho}{(4\pi)^2}H\frac{dH_s}{dh}, \quad\quad (61)$$

where T is as before, the density of the magnetic energy, H_s the component of **H** in the direction of **s**, and h is measured along the direction of **H** itself.

Now, if we confine **H** to a constant direction, so that it cannot vary in strength in that direction, its component in any other direction **s** also does not vary with h. Then (61) reduces to

$$W_s = -\frac{\rho}{4\pi\mu}\frac{dT}{ds},$$

of which (60) is a special case.

§ 20. *Comparison of Induction in a Core with a Case of Fluid Motion.*

In order to obtain a full mental representation of the state of things in a physical problem of one kind, it may often be of some assistance if we can find one of another kind in which the quantities concerned are similarly connected. Thus, in the theory of the torsion of a solid elastic prism, to get a general idea of the warping that takes place when the section of the prism is not circular, and of the amount of its effect on the torsional rigidity, we may be assisted by the comparison with a hydrokinetic problem, in which the rotation about its axis of a box of the same shape as the prism, filled with incompressible perfect liquid, sets the liquid itself in motion when the section is not circular (Thomson and Tait, II., Art. 706). Suppose now the problem is one in which certain quantities subjected to given laws go through a series of complex changes in passing from one state to another. If we find another physical problem in which other quantities go through the same changes we have an interesting analogy, to say the least, even if the substituted problem be not more easily conceivable than the original.

But further, if in the substituted question the quantities concerned are everyday realities, so that their connections are readily grasped, we have not merely an interesting but a useful and valuable comparison. Again, though this is quite a separate matter and may or may not apply, we may possibly get some assistance in forming a physical theory of the unknown phenomena for which has been found a dynamical analogue.

Start the current in a circuit containing a coil with a conducting core by closing connection with a battery. In the transition from the initial state of no current or magnetic force to the final state of steady current in the coil-circuit alone and uniform magnetic force in the core, a series of complex changes, generally referred to as current-induction, is gone through by the magnetic force and the current, and something more than a superficial examination is needed to obtain a good grasp of the phenomena as a whole. We require, then, an analogous case in which we can readily see the course of events. There are various comparisons which may be made, but the substitutions are not usually sufficiently simple for the purpose. The only one I can find that is so, is the comparison of the magnetic force in the core with the motion of water in a pipe of the same shape. This requires some explanation before the correctness of the comparison can be appreciated.

Magnetic induction is mathematically subject to the law of continuity of an incompressible perfect liquid. This alone is of valuable assistance when we are regarding the nature of distribution of lines of force in a magnetic field. Further, if we ask what it is in the liquid motion that corresponds to the electric currents that accompany the magnetic force, the answer is that a current-line is represented by a vortex-line, a current-tube of infinitesimal section by a vortex-tube of infinitesimal section, the fluid within which is, at the moment, rotating with an angular velocity proportional to the strength of current in the current-tube, the axis of rotation being that of the tube. As current-tubes are closed upon themselves, so are vortex-tubes in a moving liquid. Any possible state of magnetic force with its corresponding electric current has its analogue in a similar state of liquid motion, lines of magnetic force being translated into lines of liquid velocity, and lines of electric current to vortex-lines.

The Newtonian equation of motion of a fluid particle, meaning thereby an extremely small portion of the fluid, expressing symbolically the definition of a force as the acceleration of momentum it would produce if it lasted for the unit of time, is

$$\mathbf{F} = \sigma \frac{D\mathbf{v}}{dt},$$

where σ is the density, $\mathbf{F}$ the force acting upon σ, or the force per unit volume, since σ is the mass per unit volume; $\sigma\mathbf{v}$ the momentum, and $D\mathbf{v}/dt$ the acceleration of σ's velocity. It is to be remembered that the force produces its full effect in the direction of its action, irrespective of what the actual velocity of the mass acted upon may be. $\mathbf{F}$ and $\mathbf{v}$ are vectors, and so is $D\mathbf{v}/dt$, which is a vector parallel to $\mathbf{F}$.

The force $\mathbf{F}$ is partly due to the stress on the particle arising from the matter around it, and partly to other causes, which we sum up under the name of external force, acting bodily on the fluid, by unknown agency. The force of gravity, for instance, which appears to act upon matter independently of the matter around it, is treated as an external force.

In any fluid at rest, the internal stress is a simple pressure, equal in amount in all directions about a point, but in general varying in amount from one point to another. In the ideal perfect fluid this is also the state of stress when the fluid is changing its shape. The variation of pressure constitutes a force tending to alter the motion of a particle. Calling the pressure p, and disregarding external force, we have $\mathbf{F} = -\nabla p$, the vector decrease of pressure, and

$$-\nabla p = \sigma \frac{D\mathbf{v}}{dt}$$

is the equation of motion of a fluid particle.

This insignificant-looking equation contains volumes of meaning, even without abstruse mathematical investigations to open them out. In fact, there is involved by it, and immediately visible when one looks at it through the proper glasses, a most remarkable property, that of the constancy of the circulation, with astonishing consequences.

The pressure p is a scalar—that is, it has a definite value at every point of the fluid, and requires no directional specification. Its variation $-\nabla p$ is, of course, a vector, having direction as well as magnitude, being in fact the force acting upon σ. Bearing this in mind, consider the state of the fluid at a fixed moment, and the alteration of velocity made by the force just after. Select a closed chain of particles, and travel once round it. However the pressure may vary along the circuit it comes back to its original value at the end. In mathematical language the line-integral of the force $-\nabla p$ round the closed chain is zero. By the above equation the same property must be true of the quantity on the right side of the equation, that is, of the acceleration of momentum. Let σ be constant. Then the line-integral of $D\mathbf{v}/dt$ is zero. Now, defining the "circulation" in a closed circuit to be the line-integral of the velocity, *i.e.*, $\int \mathbf{v}d\mathbf{s}$, where $d\mathbf{s}$ is a vector element of the circuit, its variation with the time is due, first, to the variation in the velocity, and, next, to the variation of form of the moving circuit; that is

$$\frac{D}{dt}\int \mathbf{v}d\mathbf{s} = \int \frac{D\mathbf{v}}{dt}d\mathbf{s} + \int \mathbf{v}\frac{Dd\mathbf{s}}{dt}.$$

We have already seen that the first integral on the right side vanishes. In the second, $\frac{Dd\mathbf{s}}{dt}$ is the rate at which the vector $d\mathbf{s}$ is changing, or the difference in the velocity at its ends, or $\frac{d\mathbf{v}}{ds}(ds)$,* making the quantity to be integrated $\mathbf{v}\frac{d\mathbf{v}}{ds}(ds)$, or $\frac{d}{ds}(\frac{1}{2}v^2)(ds)$; which, being the variation along

* [Here (ds) is the length of the vector element $d\mathbf{s}$.]

the circuit of a scalar, necessarily vanishes when summed up. Thus the circulation in the circuit does not change at all. The circulation along that circuit of particles, which may be any circuit in the fluid, whatever value it may have at one moment, will always have that value, and, what is more, always had it. Thus, vortex-lines move with the fluid, and vortex-tubes keep their strengths unaltered, however they may change their shapes. This results from the constancy of the circulation along a closed line of particles embracing a vortex-tube. The same reasoning applies when the density is not constant, but is a function of the pressure. Then $\nabla p/\sigma$ has the same property as that above mentioned for ∇p. Also, if the external force per unit mass be, like gravity, reducible to central forces, the same applies to it. From the constancy of the circulation in a perfect fluid we have the indestructibility and uncreatability of vortex motion, possible permanent differentiation of portions of the fluid from the rest, and Sir William Thomson's vortex atoms.

Although the equations of motion of a perfect fluid were formulated by Euler in 1755-9, and dozens of eminent men had been working at hydrokinetics later, a whole century elapsed before the property of the constancy of the circulation was discovered. No one had put on the right glasses, or had managed to focus them correctly, until Helmholtz in 1858 discovered the properties of vortex-motion in a perfect liquid, followed a few years later by Sir W. Thomson's extension of the same to compressible perfect fluids, by his theorem of constant circulation, from which, in fact, they follow by elementary reasoning.

Now, since every state of incompressible perfect liquid motion represents a system of magnetic force, as the liquid moves and carries its vortices along with it, its motion remains the representative of the magnetic field of a definite system of closed electric currents similarly moved. But obviously this system of currents does *not* correspond to what would happen in a conducting mass if we started with a given arrangement of magnetic force and then left it to itself. For the magnetic system subsides through the frictional generation of heat by the electric currents which accompany it, whilst the constancy of the circulation precludes the motion of the perfect liquid ceasing. We must introduce viscosity, or internal friction; in other words, give the liquid that property which all known fluids possess. That real fluids are viscous is known to the commonest observation. When at rest, equilibrium requires that there should be no tangential stress; when the parts of a fluid are changing shape there must be, because we can set fluids in motion by, purely tangential stress. The typical illustration is that of the circular bowl containing water set rotating about its vertical axis. If there were no tangential stress the water would remain at rest. In reality the tangential stress at its moving solid boundary pulls the outside layer of water round after the bowl, the outside layer pulls the next inner one, and so on up to the axis, where the water is the longest in getting up its motion. The final state is that the vessel and water rotate as one solid body. On stopping the bowl it drags back or retards the outer layer, then the next inside, and so on to the centre, where the liquid keeps up its motion the longest.

These tangential stresses are defined in amount by the coefficient of viscosity or of sliding friction. Let liquid be moving in horizontal layers in a definite direction, and let the velocity in any layer be proportional to its distance from the lowest layer, so that if x is the height of a layer, $v=ax$ is its velocity, a being constant. Let the motion be from left to right. Any layer is moving faster than the one below it, and slower than the one above it, and there is mutual stress of the frictional character between contiguous layers thus sliding, of amount ma per unit of area in contact, where a is the constant just mentioned —viz., the upward rate of increase of velocity—and m is the coefficient of viscosity.

Now in the equation of motion of a particle we have to take into account, in the expression for $\mathbf{F}$, of any force on σ arising from the sliding friction set up by the distortion it experiences in general as it moves. The result, by the analysis of stresses and strains, is, in the case of an incompressible liquid, to introduce a new force acting on the particle besides that arising from variation of pressure. Not to go into details which are not wanted here, this force is represented by $m\nabla^2\mathbf{v}$, which makes the equation of motion

$$\mathbf{F}-\nabla p+m\nabla^2\mathbf{v}=\sigma\frac{D\mathbf{v}}{dt},$$

where $\mathbf{F}$ is restricted to be the external force only. The constancy of the circulation is gone, fluid motion may be started by tangential stress alone, and, should there be no forces to keep it up, will cease by surface and internal friction, the energy of the motion producing heat. The rate of dissipation for the whole liquid by internal friction is *

$$4m\Sigma\,(\text{ang. vel.})^2 \text{ per second.}$$

But the special case to which our electrical problem corresponds is easily worked out. Let there be a long straight pipe of circular section, containing water or other practically incompressible viscous liquid, and let the motion be parallel to the length of the pipe, say from left to right, and be in cylindrical layers, *i.e.*, only varying in velocity from one layer to another. The cylindrical layers may slide over one another, but liquid must not move nearer to or further away from the axis of the pipe. Consider a layer of radius r and thickness dr moving with velocity v at r. The outward rate of increase of velocity being dv/dr, (the a above), the whole tangential force on the inner boundary of the layer per unit of length of pipe is

$$2\pi rm\frac{dv}{dr}, \text{ from right to left,}$$

* [This does not correctly distribute the dissipation, but gives the whole amount, when the fluid extends to infinity, or is at rest at its boundary. It is the practical formula to use, avoiding the lengthy calculations which arise when the correct distribution of waste is integrated. Take, for example, the case of Stoke's theory of a particle falling slowly through air, as given in Lamb's "Motion of Fluids."]

and on the outer boundary,

$$\text{the same, } +\frac{d}{dr}\left(2\pi rm\frac{dv}{dr}\right)dr, \quad \text{from left to right.}$$

The resultant moving force is therefore

$$2\pi m\frac{d}{dr}r\frac{dv}{dr}dr, \quad \text{from left to right,}$$

which must equal the acceleration of momentum

$$2\pi rdr\sigma\frac{dv}{dt},$$

if there be no other forces. Equating the last two expressions we find

$$\frac{1}{r}\frac{d}{dr}r\frac{dv}{dr}=\frac{\sigma}{m}\frac{dv}{dt} \quad \text{...............................(62)}$$

for the equation of motion of the layer of radius r. Comparing this with equation (14), for the magnetic force in the circular core inside a solenoidal coil, we see that they are of the same form. Magnetic force parallel to the axis is replaced by liquid velocity parallel to the axis, specific resistance by coefficient of viscosity, and magnetic permeability by liquid density. (Of course there is also the silly 4π in the electrical case, arising from the faulty definition of the strength of a pole.)

There may besides be, in the liquid, force arising from variation of pressure; but we do not require it. Let the ends of the long pipe be joined together, or immersed in the ocean, and let the liquid be set in motion by uniform tangential force applied to its boundary, acting parallel to the length of the pipe, of strength X per unit of surface. The surface-equation of the liquid is of the same form as the boundary-equation of magnetic force in the core, *i.e.*, the equation of E.M.F. in the coil surrounding it. For, if m_1 be the coefficient of sliding friction between the liquid and the solid pipe, the frictional retarding force on the outer layer of liquid is m_1v per unit area, making $X-m_1v$ the actual force from left to right on the outer side of the boundary-layer per unit area. On the other side of the layer, of infinitesimal thickness dr, there is the frictional stress acting in the opposite direction, of amount $m\,dv/dr$ per unit area, so the equation of motion of the layer is

$$(X-m_1v)-m\frac{dv}{dr}=\sigma dr\frac{dv}{dt},$$

the right-hand member of which vanishes with dr, thus giving

$$X=m_1v+m\frac{dv}{dr} \quad \text{........ .... (63)}$$

for the boundary-equation. Comparing this with equation (20), which we may write, using (6) and (15),

$$\frac{E}{l_1}=\frac{R}{4\pi Nl_1}H+\frac{\rho}{4\pi}\frac{dH}{dr}, \quad \text{...................... (20a)}$$

where l_1 is the whole length of wire in the coil, we see that the boundary-equations are of the same form when the depth of the coil is small compared with its radius, and the core fills the interior space. (A modification can be made in the liquid problem to make the boundary-equations agree when the coil is not of small depth.)

Having thus a perfect correspondence of mathematical conditions, we may, in considering the nature of induction in the core, dismiss altogether the complicated and cumbrous imagery of a set of currents acting and reacting upon one another at a distance, which it is only possible to manage in simple cases, and substitute the following method:—Let an E.M.F. act in the coil-circuit, variable in any arbitrary manner. The magnetic force set up in the core will, at any moment and at every place, correspond in direction and intensity with the fluid velocity set up in a pipe similar to the core, filled with incompressible viscous liquid, if it be acted upon by superficially applied tangential force, uniform in amount per unit area, acting parallel to the axis, such force to vary in intensity in the same manner as the applied E.M.F. in the former case. Of course the viscosity, surface-friction and density of the liquid must be properly chosen to suit the electrical data, as we may see on comparing (62) with (14), and (63) with (20a).

The wave-like propagation of magnetic force into the core from its boundary when an oscillatory E.M.F. acts in the coil-circuit, and the rapid decrease of amplitude in going inward, and the insignificance of the magnetic force except near the boundary when the oscillations are rapid, are made perfectly easy to follow by the fluid analogue, wherein, with a similar to-and-fro tangential force on the liquid boundary, the motion is propagated inward by means of sliding friction.

In the core the current-density is $-(4\pi)^{-1}(dH/dr)$, and the current-lines are circles in planes perpendicular to the axis. In the pipe the angular velocity of instantaneous rotation is $-\frac{1}{2}(dv/dr)$, and the vortex-lines are similar circles. That the fluid moving in straight lines can be rotating will be seen by considering that if we impress upon every part of a small mass of the fluid in which the velocity varies a velocity equal to that of its central portion, but in the reverse direction—that is, do away with its bodily translational velocity—there is left only the relative motion of its parts, and that consists of a shear combined with an equal similar rotation. There is current in the core at any point only when the magnetic force varies in the neighbourhood, and in the liquid there is differential rotation only when there is sliding of layers. The heat of the induced currents corresponds to the frictional heat developed by the sliding, and the coil-heat to the heat of friction against the pipe. The case of steady E.M.F. in the coil will be next considered, and graphically illustrated.

§ 21. *Normal or Harmonic Distributions of Magnetic Force.*

The problem of determining the manner in which the magnetic force and induced current in a core inserted within a long solenoidal coil vary, when the coil-circuit containing a battery or other source of steady E.M.F. is suddenly closed, is somewhat more simply managed by reversing it, and finding the manner in which the magnetic force subsides when the E.M.F., after having set up the permanent state, is suddenly removed, without interrupting the circuit or altering its resistance. And this is a special case of the more general, but theoretically quite as easily managed, problem of starting at a given moment with *any* distribution of magnetic force in the core, subject only to the condition of being directed parallel to the axis, and only varying in intensity with distance from the axis, and, leaving the system to itself, determining the subsequent state of things, the coil-circuit being open or closed, but without other E.M.F. than that arising from the subsidence oi the core's induction. The initial intensity of magnetic force may vary from layer to layer in a perfectly arbitrary manner, either continuously or abruptly.

As regards the arbitrariness, we may obviously, by means of a previously acting arbitrarily variable E.M.F. in the coil-circuit, set up an infinite number of different states of magnetic force. But they will not be arbitrary, because the magnetic force will, in all such cases, vary continuously in intensity from layer to layer. That abrupt changes of intensity are admissible in the initial state may be thus shown. Considering the core as made up of a great though finite number of thin concentric tubular shells, we have a set of linear electric circuits. From their coefficients of self and mutual induction, and their resistances, we have all the data required for determining what will happen when we start with given currents in these circuits, and leave the system to itself. The circuits being independent, the initial strengths of current in them may have any values we please; and since there is no breach of continuity in passing from a finite number of circuits to an infinite number, making up a solid core, it follows that in the mathematical treatment of the subject the initial current, and therefore also the magnetic force, may vary abruptly in passing from the axis outward, although of course special means would be required to set up the discontinuities.

We may also, from the theory of linear circuits, see at once what the form of our solutions must be in the case of a continuous core. For, if there are n circuits, there are n distinct rates of subsidence, and the currents at time t after the moment of leaving the system to itself are given by n equations of n terms each, of the form,

$$\left.\begin{aligned} \gamma_1 &= A_1\epsilon^{D_1 t} + A_2\epsilon^{D_2 t} + A_3\epsilon^{D_3 t} + \dots, \\ \gamma_2 &= B_1\epsilon^{D_1 t} + B_2\epsilon^{D_2 t} + B_3\epsilon^{D_3 t} + \dots, \\ \gamma_3 &= C_1\epsilon^{D_1 t} + C_2\epsilon^{D_2 t} + C_3\epsilon^{D_3 t} + \dots, \\ &\dots\dots\dots\dots\dots\dots, \end{aligned}\right\} \qquad (64)$$

$\gamma_1, \gamma_1, \dots$ being the currents at the time t in the first, second, etc., circuits. Here there are n constants $D_1, D_2, \dots$, which are the same for every current, and n^2 constants $A, B, C, \dots$, but the *ratios* of the constants in any column are fixed by the electrical data (as are also D_1, D_2, etc.), leaving only one constant in each column arbitrary, making n altogether, whose values may be found from the given initial strengths of the n currents, thus completing the solution.

Now, these circuits being concentric shells to start with, in passing to the case of a continuous core, making the number of shells infinite, we see first that the form of the solution must remain unaltered. Next, that the number of constants D_1, D_2, etc., becomes infinite, whilst they are the same for all parts of the core. Thirdly, that, supposing $\gamma_1, \gamma_2, \ldots$ are the currents in consecutive shells passing from the axis outward, the coefficients $A, B, C, \ldots$ in any one column of (64) become the successive values of a continuous function of r, whose magnitude alone is left arbitrary, so that we may write (66) in the form of a single equation

$$\gamma = A_1 u_1 \epsilon^{D_1 t} + A_2 u_2 \epsilon^{D_2 t} + A_3 u_3 \epsilon^{D_3 t} + \ldots, \quad \text{.............(65)}$$

γ signifying the current at distance r from the axis, $u_1, u_2, \ldots$ being functions of r determined solely by the electrical data, and $A_1, A_2, \ldots$ constants to settle the absolute magnitude of each term.

And lastly, we see that, since at the time $t = 0$, we have

$$\gamma_0 = A_1 u_1 + A_2 u_2 + A_3 u_3 + \ldots,$$

and since the initial current γ_0 is arbitrary, it must be possible to expand any function of r in a series of u's, by properly determining the magnitude of the A's.

Special proofs of the possibility of the expansion of any function in series of a definite kind are usually of a singularly obscure and unsatisfactory nature, quite apart from their speciality or want of applicability to other forms of series. But if, as in the above, we pass from the solution of a set of linear differential equations with one variable, the time, to that of a partial differential equation, we see the absolute necessity of the possibility of the expansion. It would be a miracle were the expansion impossible. Have we got the right *form* of function in the first place, and next, have we got *all* of them, to satisfy the conditions of a physical problem? If so, the possibility of the expansion requires no proof.

By (65), and the relation $4\pi\gamma = -dH/dr$ between the current and magnetic force, the general solution for the magnetic force is of the same form, viz.:

$$H = A_1 u_1 \epsilon^{D_1 t} + A_2 u_2 \epsilon^{D_2 t} + \ldots, \quad \text{..................(66)}$$

where u_1, u_2, etc., are functions of r to be found by putting the elementary solution $u\epsilon^{Dt}$ for H in the characteristic equation of H, viz.:—

$$\frac{1}{r}\frac{d}{dr} r \frac{dH}{dr} = \frac{4\pi\mu}{\rho}\frac{dH}{dt}. \quad \text{....................(14) } bis$$

Since $\frac{d}{dt}\epsilon^{Dt} = D\epsilon^{Dt}$, we obtain

$$\frac{1}{r}\frac{d}{dr} r \frac{du}{dr} = \frac{4\pi\mu}{\rho} Du, \quad \text{.........................(67)}$$

whose solution gives the function u. Being of the second order, it has two distinct solutions, say v and w, so that

$$u = av + bw \quad \text{................................(68)}$$

is the complete solution. One of v and w may be found at once by assuming

$$u = a_0 + a_1 r + a_2 r^2 + \ldots,$$

inserting in (67), and making it true for every power of r. This gives

$$v=1+\frac{4\pi\mu}{\rho}\frac{r^2D}{2^2}+\left(\frac{4\pi\mu}{\rho}\right)^2\frac{r^4D^2}{2^24^2}+\left(\frac{4\pi\mu}{\rho}\right)^3\frac{r^6D^3}{2^24^26^2}+\dots \quad \dots(69)$$

Put $4\pi\mu D/\rho=-n^2$, for subsequent convenience, and we have

$$v=1-\frac{n^2r^2}{2^2}+\frac{n^4r^4}{2^24^2}-\frac{n^6r^6}{2^24^26^2}+\dots=J_0(nr) \quad \dots\dots(70)$$

for one solution of (67). This function is usually denoted by $J_0(nr)$, and was first employed by Fourier. Whether he invented it or discovered it is a doubtful point; the question is raised whether mathematical truths lie within the human mind alone, or whether the infinite body of known and unknown mathematics could exist in a dead universe. But this is metaphysics, which is all vanity and vexation of spirit.

The other solution may be shown to be

$$w=\frac{2}{\pi}\left\{v\log nr+\frac{n^2r^2}{2^2}-(1+\tfrac{1}{2})\frac{n^4r^4}{2^24^2}+(1+\tfrac{1}{2}+\tfrac{1}{3})\frac{n^6r^6}{2^24^26^2}-\dots\right\}\dots(71)$$

For proof it is sufficient to test that it satisfies (67), and is not the same as (70).

The general form of the u's in (66) is thus completely known. But we do not want the second solution, w, at all at present in dealing with a core solid to its centre, because w becomes infinite when $r=0$, on account of the logarithm. Its coefficients, therefore, require to be zero to make the magnetic force finite at the axis of core, *i.e.*, w does not come in at all. It will occur later. At present we have

$$H=A_1v_1\epsilon^{D_1t}+A_2v_2\epsilon^{D_2t}+\dots, \quad \dots\dots(72)$$

in which the v's only differ from one another in having a different value of n or D.

The nature of the function v is shown in Fig. 1. Distance from the axis of the core is measured along the base line to the right, and the

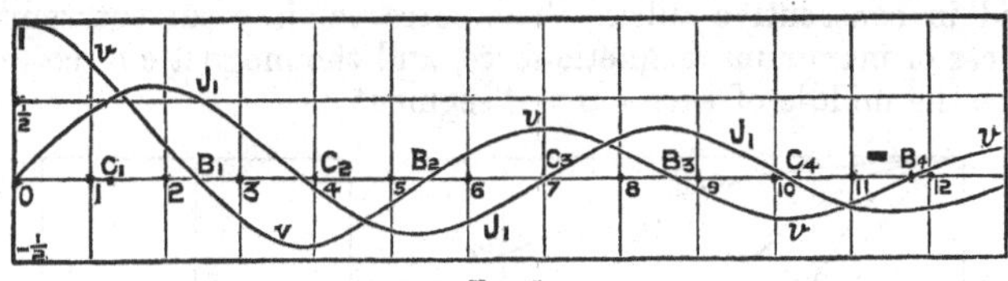

FIG. 1.

value of v upward. The curve v crosses the base line an infinite number of times. After a few fluctuations it becomes very nearly a sinusoidal curve, but with diminishing amplitude, varying inversely as the square root of the distance from the axis.

The magnetic interpretation is, if the intensity of magnetic force in a core at different distances from its axis be represented by the curve v, the magnetic force will subside everywhere at the same rate. When it has fallen to $1/m$ of its initial intensity at any point, it has fallen to $1/m$ of its intensity at every other point, so that the curve representing the force at any moment remains similar to itself.

The other curve in Fig. 1, marked J_1, shows in a similar manner the strength of the current accompanying the magnetic force; its equation is

$$J_1(nr) = -\frac{d}{d(nr)}J_0(nr), \qquad (73)$$

or, by (70),

$$J_1(nr) = \frac{nr}{2}\left(1 - \tfrac{1}{2}\frac{n^2r^2}{2^2} + \tfrac{1}{3}\frac{n^4r^4}{2^24^2} - \ldots\right). \qquad (74)$$

The curve v may be drawn to any vertical scale, since this amounts merely to fixing the absolute intensity of magnetic force. But as to the horizontal scale, this depends upon what the boundary conditions are. By means of a suitable boundary condition we may arrange to have the boundary at any distance along the base in Fig. 1, where of course the curve must stop.

§ 22. *Example I.—Coil-Circuit Interrupted.*

If the coil be absent, or its circuit broken, the boundary-value of the magnetic force is compelled to be zero. Now in Fig. 1, $v=0$ at B_1, B_2, etc., and at any of these places we may imagine the core to terminate and the curve stop. This gives a definite series of values to n, of which the first four and their squares are given by

$$\left.\begin{aligned} an_1 &= 2{\cdot}405, & an_2 &= 5{\cdot}520, & an_3 &= 8{\cdot}654, & an_4 &= 11{\cdot}791; \\ a^2n_1^2 &= 5{\cdot}783, & a^2n_2^2 &= 30{\cdot}471, & a^2n_3^2 &= 74{\cdot}888, & a^2n_4^2 &= 139{\cdot}037. \end{aligned}\right\} \qquad (75)$$

Now, in Fig. 2, the base line, OP, represents a, the radius of the core, and the four curves, the first four normal distributions of magnetic force, being the curve v in Fig. 1, drawn upon four different horizontal scales, so as to reach the boundary, P, at the first, second, third, and fourth roots of $v=0$. There are an infinite number of other normal systems, corresponding to the higher roots. The normal system v_m divides the core into a solid central cylinder, surrounded by $m-1$ concentric tubes, at whose boundaries the magnetic force vanishes, being oppositely directed in consecutive tubes. The corresponding current vanishes at the places of maximum magnetic force, and the magnetic force vanishes at about the middle of each current-segment.

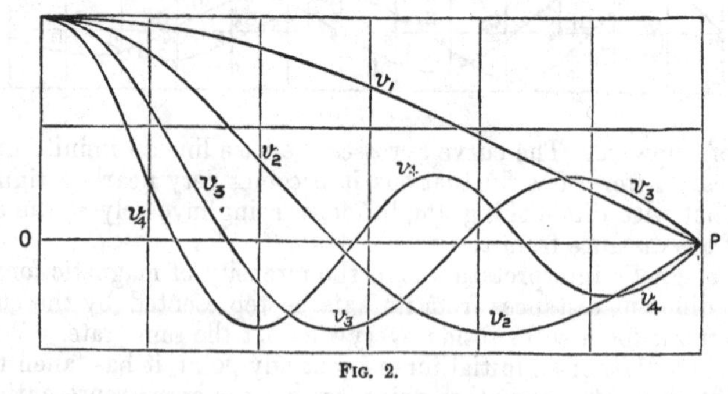

FIG. 2.

The time-constant measuring the slowness of subsidence of a normal system is

$$-1/D = 4\pi\mu/\rho n^2,$$

thus being proportional to the magnetic permeability, to the conductivity, to the sectional area of the core, and inversely to the square of na. Of all metals except iron, copper is the one in which any normal system subsides most slowly; on the other hand, the great permeability of soft iron overbalances its comparatively low conductivity and makes it the metal of slowest subsidence.

Copper.—$\rho = 1{,}700$, $\mu = 1$. These make $4\pi\mu/\rho = \cdot 0074$, and the time-constants of the first four normal systems to be, by (75),

$$\cdot 0013a^2, \quad \cdot 00024a^2, \quad \cdot 000099a^2, \quad 000054a^2, \quad \text{seconds,}$$

where a is the radius of the core in centimetres.

Iron.—$\rho = 10{,}000$, $\mu = 100$ say. These make $4\pi\mu/\rho = \cdot 1256$, and the time-constants

$$\cdot 0217a^2, \quad \cdot 0041a^2, \quad \cdot 0017a^2, \quad \cdot 0009a^2,$$

which are 17 times as large as for a copper core of the same radius. In an iron core of 10 cm. radius (5 in. diameter), the time taken by the first and most important normal system of magnetic force, inducing a current only in the core itself, to fall in strength from 2·718 to 1 would be 2·17 seconds. If of 1 m. radius it would take 217 seconds.

§ 23. *Note on Earth-Currents.*

The remarkable slowness of subsidence of currents in large masses of metal, or, equivalently, in proportionately larger masses of badly-conducting material, is vaguely suggestive in regard to earth-currents. Without intending any strict comparison, it may be remarked, first, that although the "earth" may be poorly conducting, yet there is a good deal of it. And next, that the "earth-currents" observed on long lines of telegraph during magnetic storms are, neglecting the minor fluctuations, remarkably alike in their behaviour in some respects. Although a big "wave" may set in one way, and continue of great strength, for some considerable time, even minutes, to be followed by another wave reversing the current, yet the transition from one to the other never takes place suddenly; even the most rapid reversal of a big wave takes several seconds to accomplish. A similar sluggishness may be observed in all large changes of current-strength; and it is suggested, without any hypothesis as to the cause of these earth-currents, what keeps up such powerful currents for a long time in one direction at one place in at least the superficial portion of the earth (of course in closed circuits) of which portions find their way into telegraph lines, that the characteristic sluggishness is due to the magnetic retardation as the currents in the earth change their strength and distribution.

§ 24. *Determination of Constants—Conjugate Property.*

Referring to Fig. 2, by superimposing any number of normal systems, of which the first four only are shown, taken of any absolute magnitudes, we may produce an immense variety of distributions of magnetic force. They will all, by the manner of their construction, be decomposable into the normal systems from which they arose. Now, by the reasoning previously stated, when the whole series of normal systems is taken, we may, by properly choosing the values of the A's in the series $A_1v_1 + A_2v_2 + \ldots$, make it represent *any* chosen function of r; of course single-valued. The decomposition of the given function into normal systems is most easily effected by making use of the conjugate property $\int v_1v_2r\,dr = 0$, possessed by every pair of different normal functions, the limits of integration being 0 and a, the axis and the boundary of the core.

The proof is easy. For u_1 and u_2, by (67), satisfy

$$\frac{1}{r}\frac{d}{dr}r\frac{dv_1}{dr} + n_1^2v_1 = 0, \quad \text{and} \quad \frac{1}{r}\frac{d}{dr}r\frac{dv_2}{dr} + n_2^2v_2 = 0.$$

Multiply the first of these by v_2r, the second by v_1r, subtract the second result from the first, and then integrate with respect to r from $r=0$ to $r=a$. We find immediately,

$$(n_1^2 - n_2^2)\int v_1v_2r\,dr = \Big[\,r(v_1v_2' - v_2v_1')\,\Big]_0^a, \quad \ldots\ldots\ldots\ldots(76)$$

the accents denoting differentiation to r. The right-hand member vanishes at both limits, because v_1 and v_2 vanish at $r=a$, v_1' and v_2' being then finite, whilst both factors of rv_1' and rv_2' vanish with r, v_1 and v_2 being then finite. Hence, if n_1 and n_2 are different, we have

$$\int v_1v_2r\,dr = 0; \quad \ldots\ldots\ldots\ldots\ldots\ldots\ldots\ldots(77)$$

the conjugate property to suit the present case. Now, given H_0 as a function of r, the initial distribution, to expand it in v's, thus

$$H_0 = A_1v_1 + A_2v_2 + A_3v_3 + \ldots,$$

multiply both sides by rv_1dr and integrate between limits 0 and a. We get

$$\int H_0rv_1\,dr = A_1\int v_1^2r\,dr;$$

since, by (77), all the rest vanishes. This gives the value of A_1, and similarly for the other coefficients, any one being given by

$$A = \int H_0vr\,dr \Big/ \int v^2r\,dr, \quad \ldots\ldots\ldots\ldots\ldots\ldots(78)$$

using the particular u concerned. The denominator may be evaluated by (76). For, if in (76) we make $n_1 = n_2$, making $v_1 = v_2$ and $v_1' = v_2'$, it assumes the form $0 \times \int v^2r\,dr = 0 - 0$, and the integral, being the sum of squares, and therefore not vanishing, is to be found by the ordinary process. Differentiate (76) with respect to n_1^2, and then make $n_1^2 = n_2^2 = n^2$; this gives

$$\int v^2r\,dr = \left[\,r\left(\frac{dv}{dn^2}v' - v\frac{dv'}{dn^2}\right)\right]_0^a \quad \text{generally}, \ \ldots\ldots\ldots(79)$$

and $$= av'(dv/dn^2), \ (r=a), \quad \text{in present case.}$$

Now $\frac{dv}{dn^2}=\frac{1}{2n}\frac{dv}{dn}$, and, by (70), $r\frac{dv}{dr}=n\frac{dv}{dn}$; therefore

$$\int v^2 r\,dr=\frac{a^2}{2n^2}\left(\frac{dv}{da}\right)^2=\frac{a^2}{2}\{J_1(na)\}^2, \quad\quad (80)$$

dv/da being the value of dv/dr at $r=a$. Hence, by (78) and (80),

$$A=\frac{2n^2}{a^2}\frac{\int H_0 vr\,dr}{\left(\frac{dv}{da}\right)^2}=\frac{2}{a^2}\frac{\int H_0 vr\,dr}{\{J_1(na)\}^2}. \quad\quad (81)$$

Thus, when H_0 is given, the expansion is effected, and the magnetic force is known at any subsequent time.

§ 25. *Special Case.* $H_0 =$ *constant.*

Put a battery in the coil-circuit. It sets up ultimately $H_0=$ constant in the core. Break the circuit, or make its resistance very high, removing the E.M.F. Never mind the slight loss of energy in sparking, as, the core being free for induced currents, nearly all the energy will be dissipated therein. We can now evaluate the numerator in (81).

$$\int H_0 rv\,dr=H_0\int rv\,dr=-\frac{a}{n^2}\frac{dv}{da}H_0=+\frac{a}{n}J_1(na)H_0,$$

by (70) first, integrating, and then comparing with (74). This makes (81) become

$$A=2H_0/naJ\,(na),$$

and the complete solution is therefore

$$H=2H_0\sum\frac{J_0(nr)}{J_1(na)}\cdot\frac{\epsilon^{Dt}}{na}, \quad\quad (82)$$

the summations to include all values of n.

The current density is got from this by differentiation; thus

$$\gamma=-\frac{1}{4\pi}H'=+\frac{H_0}{2\pi}\sum\frac{J_1(nr)}{J_1(na)}\cdot\frac{\epsilon^{Dt}}{a}. \quad\quad (83)$$

In establishing the solution (82) of the most simple case that presents itself (curve to be given in Fig. 3), we have entered more into detail than at first intended; as a refresher to the memory of readers who are acquainted with the methods employed, for the education of readers who may be only learning them, and because in the more complex cases to follow much of the reasoning will be exactly similar, so that the investigations may be given more briefly without loss of intelligibility. At the same time there are numerous details and side matters of interest that must be omitted. Otherwise, we should "go on for ever," like the brook in the poem.

§ 26. *Magnetic Energy and Dissipation.*

The magnetic energy per unit volume being $\mu H^2/8\pi$, the amount T in unit length of core, by (82) and the conjugate property (77) making products vanish, is

$$T = H_0^2 \frac{4\mu}{8\pi} \int_0^a 2\pi r\, dr \sum \frac{J_0^2(nr)}{J_1^2(na)} \cdot \frac{\epsilon^{2Dt}}{n^2 a^2} = \frac{\mu}{2} H_0^2 \cdot \sum \frac{\epsilon^{2Dt}}{n^2}, \quad \text{.........(84)}$$

by integrating and using (80). (In these series it is of course only necessary to operate on one term, since all terms are alike in their properties.)

This, of course, vanishes ultimately, the D's being negative. At starting, $t = 0$, we know, from H_0 being constant, that $T = \pi a^2 \mu H_0^2 / 8\pi = \mu a^2 H_0^2 / 8$; hence we must have, by (84),

$$\tfrac{1}{8} \mu a^2 H_0^2 = \tfrac{1}{2} \mu H_0^2 \Sigma n^{-2}; \quad \text{or,} \quad \tfrac{1}{4} = \Sigma (na)^{-2},$$

the values of na being those in (75), and their companions. This is true, for, examining the form of the expansion (70) of $J_0(nr)$, the sum of the squares of the reciprocals of the roots of $J_0(nr) = 0$ = coefficient of the second term with the sign changed, *i.e.*, $\frac{1}{4}$.

To test that the formulæ make all the energy be dissipated in the core according to Joule's law; the heat per second per unit volume being $\rho\gamma^2$, the total per unit length of core, from the beginning to the end of the discharge, by (83), is

$$\frac{\rho H_0^2}{2\pi a^2} \int_0^\infty dt \int_0^a r\, dr \sum \frac{J_1^2(nr)}{J_1^2(na)} \epsilon^{2Dt}; \quad \text{.....................(85)}$$

where again we omit products, on account of the conjugate property of the normal systems of current, thus proved :—

$$\int J_1(n_1 r) J_1(n_2 r) r\, dr = \frac{1}{n_1 n_2} \int \frac{dv_1}{dr} \frac{dv_2}{dr} r\, dr = \frac{r}{n_1 n} \cdot \frac{n_1^2 v_1 v_2' - n_2^2 v_2 v_1'}{n_1^2 - n_2^2}, \quad \text{....(86)}$$

by integrating and using (79). This vanishes when the limits are 0 and a.

Evaluating by differentiation for the case $n_1 = n_2 = n$, we get

$$\int_0^a J_1^2(nr) r\, dr = \frac{a^2}{2} J_1^2(na), \quad \text{.........................(87)}$$

and using this in (85), we obtain, after integrating to t as well,

$$-\sum \frac{\rho H_0^2}{8\pi D}, \quad \text{or} \quad \frac{\mu H_0^2 a^2}{2} \sum \frac{1}{(na)^2} = T, \ (t = 0),$$

as before found.

§ 27. This equivalence of total heat to initial magnetic energy can of course be predicted beforehand, as a certain consequence of the elementary laws underlying the structure of our solutions for the magnetic force and the current, and therefore its verification merely serves to show that we have not got upon a wrong track in the pursuit of a mathematical wild goose, which may happen unless proper tests are occasionally applied.

§ 28. *Remarks on Normal Systems.*

In (82) the magnetic force (and in (83) the current) is expressed as the sum of a number of terms each of which is the product of a function of r, the distance from the axis, into a function of t, the latter being such as to show a decrease in the magnitude of the term with the time in the same manner as the charge of a condenser decreases when discharged through a conductor of insensible self-induction; or as the current decreases in a linear circuit of insensible electrostatic capacity and subject only to its own self-induction; or as the momentum of a body decreases when resisted by a force proportional at any moment to its momentum.

Now, as may be seen from the values of the first four A's given below, the series is very slowly convergent at the start ($t=0$), and a large number of terms would then have to be taken to make the sum come to H_0 within, say, a millionth part. But, owing to the different rates of subsidence of the normal systems, the series soon becomes rapidly convergent. The higher ones subside so rapidly that in a short time, before the first has sensibly altered, only it and a few of the following are of any importance, and as time progresses those left drop out of practical existence one by one till at length, before the magnetic force has fallen to half strength at the centre of the core, the first normal system is left alone. The distribution of magnetic force in the core from the axis outward is then represented simply by the curve v_1 in Fig. 2, which continues to be the distribution during the remainder of the discharge, only falling in strength according to the exponential law. Thus, by considering the first normal system only, calculating the values of A and of D belonging to it, we obtain important information, for we know thereby the solution except for small values of the time. But the neglected terms completely alter the character of the subsidence at the commencement of the discharge, as we may see from the value of A_1, viz., $1{\cdot}667\,H_0$, which is, at the axis, $\frac{2}{3}$ greater that H_0, the real initial strength, and of A_2, which (negatively) is also greater than H_0. This, however, will be made fully evident from the curves of the magnetic force and current at the axis and boundary. The values of the first four A's are

$$A_1=1{\cdot}667\,H_0,\quad A_2=-1{\cdot}065\,H_0,\quad A_3={\cdot}856\,H_0,\quad A_4=-{\cdot}737\,H_0.$$

The above remarks concerning the great relative importance of the first normal system of slowest subsidence may, to a certain extent, be generalised. Given any arbitrary distribution to start with, the system of slowest subsidence will soon be left decaying alone, all its companions having faded and gone, unless it should happen that one or more of them be of such great initial strength that their influence continues sensible, masking that of the first system. Or, the initial distribution may not contain the one of slowest subsidence at all, or any number of systems may be absent. But in the various practical cases that arise of initial distributions set up by a battery, the system of slowest subsidence has the greatest amplitude, and the amplitude of the others decreases with their rapidity of subsidence. It is easy, however, by suitably-arranged boundary conditions, though they may not always be practicable, to introduce various anomalous peculiarities, quite altering the character of the subsidence, or reversing it, making the force increase.

In the above case the roots na are all real, giving real D's. This is, however, not necessary. The n's may be, some or all, imaginary, when, of course, the corresponding normal systems and their rates of subsidence are imaginary. By pairing two connected imaginary terms, the unreal parts mutually cancel, leaving a real dual system which subsides with oscillations instead of in the former manner. In all cases the finding of the proper normal systems is of primary importance, and in all cases their amplitudes, to make up a given distribution, may be found by the magical process of selection of coefficients by the conjugate property of the normal systems, either as above exemplified, or in a similar, though more general manner.

§ 29. *Example* 2. *Coil-Circuit Closed. Coil of Negligible Depth.*

Let the core be charged as before in the first place, and then be discharged by removing the E.M.F. from the coil-circuit, but leaving it closed. There will be a current in the coil (the extra-current) in the same direction as the original current, and as the induced current in the core. Its effect, we see at once, is to retard the rapidity of discharge, and to keep up the core's magnetisation longer, if it be magnetisable, in the ordinary sense. The theory makes no distinction between iron and any other metal, except in the value of the coefficient of magnetic permeability. In the last example, the boundary condition which settled the values of n and D was $H_a = 0$, because there was no current outside the core, the coil-current being stopped. (We disregard dielectric currents, to be later considered, as they are of utterly insignificant magnitude in comparison with the conduction currents in coil or core.) Now, there being a current in the coil, the boundary magnetic force is not zero, but stands in a constant ratio to the coil-current. The coil-current, again, is proportional to the boundary E.M.F. of induction, and therefore, by Ohm's law, to the boundary core-current. Hence the boundary magnetic force and the density of the boundary core-current are constrained to preserve their ratio constant during the whole period of the discharge, and the same is true for every normal system of magnetic force with its accompanying current. Determining the value of the constant by (15) and (20), taking $a =$ radius of core filling the coil, and the depth of coil small compared with a, and putting $E = 0$, since there is no externally impressed force, we have

$$0 = H + sH', \qquad \ldots\ldots(88)$$

where $s = L\rho/2\pi\mu aR$, L being the inductance, R the resistance of the coil-circuit, ρ the specific resistance, and μ the permeability of the core.

Put $J_0(nr)$ for H in (88), and we obtain

$$0 = J_0(nr) + s\frac{d}{dr}J_0(nr), \quad \text{at } r = a,$$

or,

$$J_0(na)/J_1(na) = sn, \qquad \ldots\ldots(89)$$

by (73). In Fig. 1, giving the curves of J_0 and J_1, we must find the values of na satisfying (89), instead of the places where $J_0 = 0$. The effect is to shift the position of all the roots to the left, through different distances, say, as in a calculated case, from B_1 to C_1; B_2 to C_2, etc.

The value of n_1a is reduced in a far greater ratio than n_2a, etc., and since the time-constants of the normal systems vary inversely as $(na)^2$, the effect is to make the first system of still greater relative importance than before.

In the case to be illustrated, I have taken $s=1$ and $a=1$, which corresponds to not unpractical coil-data. The determinantal equation (89) is then $J_0(n)=nJ_1(n)$; or, in full,

$$0=1-\frac{3n^2}{2^2}+\frac{5n^4}{2^24^2}-\frac{7n^6}{2^24^26^2}+\dots.$$

The first five roots and their squares are

$$\left.\begin{aligned} n_1&=1\cdot256, & n_2&=4\cdot079, & n_3&=7\cdot155, & n_4&=10\cdot271, & n_5&=13\cdot48,\\ n_1^2&=1\cdot577, & n_2^2&=16\cdot637, & n_3^2&=51\cdot203, & n_4^2&=105\cdot492, & n_5^2&=181\cdot71,\end{aligned}\right\} \quad (90)$$

which may be compared with (75) to see the relative and absolute changes in the values of the time-constants.

In the first normal system the magnetic force falls only about 40 per cent. from the axis to the boundary, instead of to zero, whilst it subsides nearly four times as slowly. The amount of shifting of the roots by closing the coil-circuit depends mainly upon the time-constant of the circuit, which varies with the external resistance. Starting with infinite resistance, the roots are at B_1, B_2, etc., corresponding to $J_0(na)=0$. Reduce the resistance; they move to the left, and of course stop when the coil is short-circuited. Now, if we could reduce the coil-resistance indefinitely without altering the number of wires, say by increasing the specific conductivity, the roots would be given by $J_1(na)=0$, viz., zero, and the values of na at the other places where the curve J_1 in Fig. 1 crosses the axis of abscissæ.

From this we may, without calculation, derive some interesting information. The first normal system $A_1J_0(n_1r)\epsilon^{D_1t}$ becomes, with $n=0$, and consequently $D=0$, simply A_1. If then the initial distribution H_0 was constant, we have $H_0=A_1$, and $A_2=A_3=\dots=0$. The core's induction does not subside at all, being wholly represented by the first normal function, which has become a constant. But should the initial distribution vary in intensity from layer to layer, the value of A_1 will be the *mean* strength taken over the section of the core, whilst the departure from the mean intensity will be represented by the other normal systems $A_2J_0(n_2r)$, etc. These, having finite time-constants, will subside, leaving the mean intensity of force, which will remain steady, with no current in the core, but with a current in the coil of the necessary strength to cause the magnetic force, *i.e.*, as given by $H=4\pi N\Gamma$, N being the number of turns of wire per unit length of core, and Γ the current in each of them. (Of course we should rather say that the magnetic force causes the current, but this is a mere question of words. The essential idea, whichever we like to consider as causing the other, is that the magnetic force and the current are inseparably bound.) This is an illustration of Maxwell's theory of the impermeability of a perfect conductor to magnetic induction. Our core is bounded by a perfectly conducting shell, and whatever magnetic induction was in it cannot get out of it; it can merely settle down to a steady state, in case it was previously non-uniformly distributed.

From this we see that with a core of high magnetic permeability, bounded by a coil of high conductivity short-circuited, any irregularities in the initial distribution of induction will quickly vanish, leaving the mean strength to subside at the slow rate of the first normal system, being during subsidence only a trifle less strong at the boundary than at the axis.

There is another case in which we arrive at precisely the same normal systems given by the roots of $J_1(na)=0$, as when the coil-resistance becomes infinitely small, but with entirely different resulting phenomena. For, by continuously increasing the specific resistance of the core we shall make the roots pass continuously from those of $J_0(na)=0$ with $\rho=0$, to those of $J_1(na)=0$, with $\rho=\infty$. The last is of course practicable, as it merely means a nonconducting core. But the relation between D and n being

$$4\pi\mu D = -n^2\rho,$$

we have, for all the finite values of n, $D=\infty$, *i.e.*, any irregularities in the initial distribution disappear instantaneously, not gradually as in the last case. But for the value $n_0=0$, with $\rho=\infty$, we have D_0 finite. In fact, the determinantal equation, expressed in terms of D instead of n^2, reduces to

$$0=1+LD/R$$

simply, the coefficients of the higher powers of D vanishing. Hence $D_0=-R/L$, and the solution is

$$H=H_0\epsilon^{-Rt/L},$$

as we know it should be, there being no possibility of induced currents in a nonconducting core to alter the character of the subsidence. The higher normal systems have become ghosts, and gone to that region where exist all the roots save one of an equation of the first degree.

Returning to the general subject of this section, very little modification of the investigation in Example 1 is needed to complete the solution. For, by (89), $v_1/v_1'=v_2/v_2'=$ constant for every normal system. Thus the right-hand member of (76) still vanishes, giving us the same conjugate property (77) as before, and the same expression (78) for A, whose further development is, however, different. Thus, to find its denominator, use (79), remembering that

$$r\frac{dv}{dr}=n\frac{dv}{dn}, \quad \text{and} \quad \frac{dv}{d(n^2)}=\frac{1}{2n}\frac{dv}{dn};$$

then

$$\int v^2r\,dr=\frac{1}{2n^2}\left[r^2(v')^2-v\frac{d}{dr}(rv')\right]=\frac{1}{2n^2}[r^2(v')^2+n^2r^2v^2], \quad \text{by (67)},$$

$$=\tfrac{1}{2}a^2[\{J_0(na)\}^2+\{J_1(na)\}^2], \quad \text{by (73)},$$

$$=\tfrac{1}{2}a^2(1+s^2n^2)\{J_1(na)\}^2, \quad \text{by (89)}, \quad \ldots\ldots\ldots\ldots\ldots\ldots(91)$$

which we see returns to (80), when $R=\infty$, making $s=0$. This completes the solution when the initial H_0 is unstated. But H_0 is to be constant; hence, evaluating the numerator of (78), we have the same expression, $H_0aJ_1(na)/n$, as before, and we obtain, generally,

$$A = 2H_0 \div \{naJ_1(na)(1+s^2n^2)\}, \qquad (92)$$

which, in the calculated case, becomes, with $s=1$, and $a=1$,

$$A = 2H_0 \div \{n(1+n^2)J_1(na)\}, \qquad (93)$$

which, taken in the formula

$$H = \Sigma AJ_0(nr)\epsilon^{Dt},$$

gives us the complete solution at time t, the current being got by differentiation. The magnetic energy per unit length of core at the start is

$$\tfrac{1}{8}\mu a^2 H_0^2 = T = \sum \frac{\mu}{8\pi}\int_0^a 2\pi r\,dr\{AJ_0(nr)\}^2 = \tfrac{1}{2}\mu H_0^2 \Sigma\, n^{-2}(1+s^2n^2)^{-1}, \qquad (94)$$

by (89) and (91), no products being required. At time t the corresponding value of T is got by multiplying each term in the summation by ϵ^{2Dt} with the proper value of D.

With $s=1$, as in the calculated case, the proportions of the initial energy going to the first few normal systems we find to be given by

$$\tfrac{1}{8} = \cdot 1230 + \cdot 0017 + \cdot 00018 + \cdot 00003 + \ldots .$$

Thus, although the second and higher normal systems are very important for a short time from the commencement of subsidence, yet the energy of the first system is no less than $\frac{123}{125}$ of the total. The proportion is not so great when the circuit is open (Example 1) being then only $\frac{88}{125}$ of the whole.

§ 30. *Description of Fig. 3. Subsidence of Induction in Core.*

This is to show the manner of subsidence of the magnetic force at the axis and at the boundary, and of the coil-current in three cases.

Time is measured from left to right, from $t=0$ to $t=1\cdot 4$ second, as it happens. Strength of magnetic force (or of current) is measured upwards. The time-constant of the coil is $\frac{1}{2}$ sec. and its radius 1 cm. That is,

$$s=1, \quad a=1, \quad D=-n^2, \quad L/R=\tfrac{1}{2}.$$

As the coil is of very small depth, there must be a great number of turns per cm., or else a soft iron core, to make the time-constant so large.

In the first place, suppose that the core is properly divided, to stop the flow of induced currents, and that we start with a steady current in the coil-circuit, and, removing the E.M.F., leave it to itself. The curve h_1h_1 shows the manner of its subsidence. It is the ordinary exponential curve, and is given by $h_1 = \epsilon^{-2t}$, if the value at the start be taken as unity. The characteristic property is that if the current fall from strength 1 to $\frac{1}{2}$ in the time t, it falls from $\frac{1}{2}$ to $\frac{1}{4}$ in the next interval of time t, from $\frac{1}{4}$ to $\frac{1}{8}$ in the third, and so on. The magnetic force, induction, and the magnetisation, if of soft iron, subside in the same manner in all parts of the core on the usual hypothesis of no retentiveness.

Now undo the insulation of the core, so that an induced current can flow as the coil-current subsides, other things being the same. The subsidence of the coil-current is now shown by the curve H_aH_a. Comparing this with h_1h_1, there is seen to be a general resemblance, with two notable differences: First, and most important, there is a very sudden drop in the strength of the coil-current at the commencement, the curve being nearly vertical. This is the first effect of allowing the core-current to flow, it being of considerable density, but practically confined to near the boundary. The current is apparently transferred from the coil to the core. But later on, as the core-current travels into the core, the coil-current subsides less quickly, and finally we have the second difference, a slower subsidence than when the core was divided.

These two effects are characteristic of induction in general between a primary and a secondary, a sudden drop in the strength of the primary, sometimes of very large amount, accompanying the simultaneous appearance of the secondary current in the same direction, and a later more slow subsidence of what is left of the primary current than if the secondary circuit were interrupted.

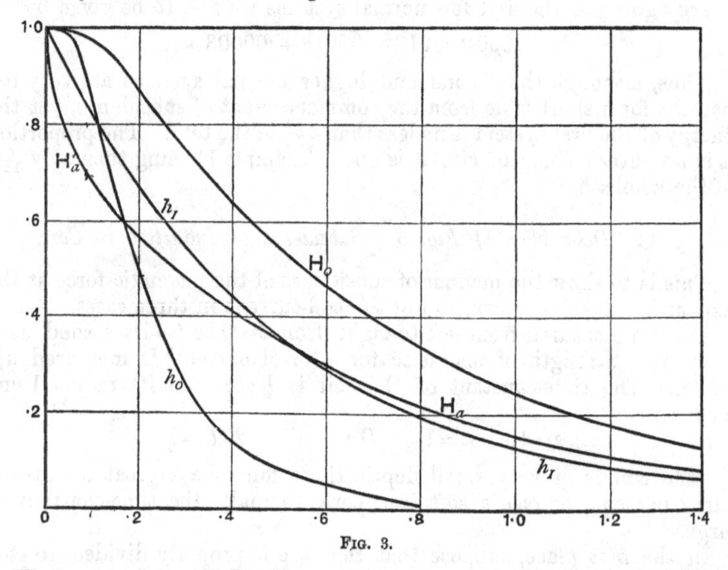

FIG. 3.

The same curve H_aH_a shows the manner of subsidence of the magnetic force at the boundary of the core, and of the density of the induced current close to the coil, for they both keep pace with the coil-current just outside.

Under the same circumstances the fall in the strength of the magnetic force at the axis of the core is shown by the curve H_0H_0, whose principal characteristics are the preliminary retardation and the opposite curvature of the first portion of the curve from the remainder. Although, according to the formula, the axial force commences to subside instantaneously, yet for a certain time the fall is so excessively small as to be quite insensible. The terms of the series are alternately positive and negative, and, with small values of t, a very large number of them and careful calculation is required to obtain evidence of any fall at all. But, this dead period got over, the rapidity of subsidence increases fast. Ultimately the curve settles down to be the curve of subsidence of the first normal system.

Between H_aH_a and H_0H_0 lie the curves of subsidence of magnetic force at points between the boundary and the axis; as we pass outward from the axis the length of the dead period decreases, and ceases entirely at the boundary.

So far with the coil-circuit closed. Now, starting as before, with the steady current and magnetic force, make the resistance of the coil-circuit very great (the E.M.F. to be removed). The current in it disappears at once, and the subsidence of the magnetic force in the core is greatly accelerated, the curve being shown by h_0h_0 for points on the axis, corresponding to H_0H_0 with circuit closed.

If we wish to see how the current and force rise to their final strengths when a steady E.M.F. is put in the coil-circuit, initially free from current (as also the core), all we have to do is to turn the diagram (Fig. 3) upside down, and view its reflection in a mirror.

Regarding the dead interval, there is this curious property. At the time $t=0$, the rate of decrease of H at the axis is obviously zero. Not only that, but each of the whole series of successive differential coefficients of H with respect to t is then zero, except the ∞th, and that is infinitely great. This statement requires interpretation, of course, because the number of differential coefficients is infinitely great. The interpretation may be got by considering the core to consist of a finite though very great number, m, of concentric cylinders. Then all the differential coefficients $=0$ except the mth, which is very large. Increase m indefinitely, and we approximate as nearly as we please to the solid core, with the above result.

§ 31. *Telegraph Cable Analogue.*

Let a cable be constructed according to the following simple specification:—Its electrostatic capacity to vary in simple proportion to the distance from one end O. Its conductance* to vary in simple proportion to the distance from O. That is all, except that its self-induction must be negligible. Let $r=$ distance measured from O, then rc and r/k are the capacity and conductance per unit length, c and k being constants. Let v be the potential at distance r from O, then the current there will be $-(r/k)(dv/dr)$, and the charge per unit length will be rcv, making the differential equation of the potential be

$$rc\dot{v} = -\frac{d}{dr}\left(-\frac{r}{k}\frac{dv}{dr}\right), \quad \text{or} \quad ck\dot{v} = \frac{1}{r}\frac{d}{dr}r\frac{dv}{dr}.$$

This being of the same form as the equation of H in the core, all the

* ["Conductance" is here substituted for "conductivity." It means the reciprocal of the resistance.]

previous solutions may be translated into the solutions of problems connected with signalling through the cable.

The strength of force at distance r from the axis of the core corresponds to the potential of the conductor at distance r from the end O; the strength of the circular current in the core to the strength of current in the cable divided by the distance from O. Let the other end of the cable be at P. This corresponds to the boundary of the core.

The cable being constructed, submerged, and a station opened at the end P, let an intelligent operator make connection between the cable and earth through a battery. The electrostatic potential will be propagated through the cable in the same manner as the magnetic force is sent into the core when the coil-circuit current is similarly operated upon. There will be no current at the end O, because the conductance is zero there, so it will be of no use to put a recorder there, though as the potential varies, an electrometer might be made use of. But, by cutting a small piece off the end, we may let some current pass, and then use a recorder, without practically altering the nature of the propagation of the potential.

An artificial line approximately fulfilling the conditions would perhaps serve as well, in case it should not be found convenient to have the cable constructed. Or, it may be left wholly to the imagination.

§ 32. *Example 3. Coil of Any Depth.*

When the depth of the coil is not small compared with the radius of the core, an appreciable fraction, or it may be a large fraction, of the magnetic energy is contained between the inner and outer boundary of the coil, *i.e.*, in the wire itself, its covering, and the air-spaces, and requires to be taken into consideration, as it modifies somewhat the boundary condition, and the solution. Putting $E=0$ in (19), and $H=4\pi N\Gamma$, we have

$$0=RH+(L-L_1)\dot{H}+\frac{L_1\rho}{2\pi\mu c}H'$$

for the boundary condition. Put $H=J_0(nr)\epsilon^{Dt}$, and it becomes

$$0=\{R+(L-L_1)D\}J_0(nc)-\frac{L_1\rho n}{2\pi\mu c}J_1(nc).$$

Here, for generality, we let the boundary be at $r=c$, the space from $r=c$ to $r=a$ being air. Putting D in terms of n, we get

$$J_0(nc)/J_1(nc)=(L_1n\rho/2\pi\mu c)\{R-(L-L_1)n^2\rho/4\pi\mu\}^{-1}, \quad \text{......(95)}$$

the determinantal equation, whose roots give us the admissible values of n, and therefore also of D.

In finding the coefficients A in the solution $H=\Sigma AJ_0(nr)\epsilon^{Dt}$, we must remember that now $J_0(nr)$ from $r=0$ to $r=c$, *i.e.*, from axis to boundary of the core, is not a complete normal system; to complete it, we have $J_0(nc)$ from $r=c$ to $r=a$, and $J_0(nc)\{1-(r-a)/b\}$ from $r=a$ to $r=a+b$, if b be the depth of the coil. The three taken as a single set constitute a complete normal system to which the conjugate property $\Sigma v_1v_2r\mu=0$

applies. Also remember that the permeability is μ in the core, and unity without it.

Thus,

$$0=\mu\int_0^c J_0(n_1r)J_0(n_2r)r\,dr+\int_c^a J_0(n_1c)J_0(n_2c)r\,dr+\int_a^{a+b} J_0(n_1c)J_0(n_2c)\left(1-\frac{r-a}{b}\right)^2 r\,dr \qquad \ldots(96)$$

is the expression of the conjugate property in its physically most meaningful form, *i.e.*, expressing directly that the mutual energy of two normal systems is *nil*.

Otherwise, use equation (76), putting $J_1(nc)$ in terms of $J_0(nc)$ by (95). We shall obtain, after reductions,

$$\frac{\mu l}{2}\int_0^c J_0(n_1r)J_0(n_2r)r\,dr+(L-L_1)\frac{J_0(n_1c)}{4\pi N}\,\frac{J_0(n_2c)}{4\pi N}=0, \qquad \ldots\ldots\ldots(97)$$

expressing the same truth as (96), but now in terms of the normal systems of current in the coil (instead of magnetic force from $r=c$ to $r=a+b$), and the inductances.

Here $J_0(n_1c)/4\pi N$ and the similar expression with n_2 instead of n_1 are the proper coil-currents corresponding to the normal core magnetic force systems $J_0(n_1r)$ and $J_0(n_2r)$. Calling the latter H_1 and H_2, and the former Γ_1 and Γ_2, we may write

$$0=\frac{\mu l}{4\pi}\int_0^c H_1H_2 2\pi r\,dr+(L-L_1)\Gamma_1\Gamma_2, \qquad \ldots\ldots\ldots\ldots(98)$$

expressing, in the simplest form, that the mutual energy of a pair of normal systems is zero, a principle to be considered in a later section. If, then, H_0 and Γ_0 be the complete initial core magnetic force and the coil-current, the value of any coefficient A_m in the proper expansion is

$$A_m=\frac{\dfrac{\mu l}{4\pi}\displaystyle\int H_0H_m 2\pi r\,dr+(L-L_1)\Gamma_0\Gamma_m}{\dfrac{\mu l}{4\pi}\displaystyle\int H_m^2 2\pi r\,dr+(L-L_1)\Gamma_m^2}. \qquad \ldots\ldots\ldots\ldots(99)$$

The values of the numerator in case $H_0=$ constant, and of the denominator in any case, are

$$\left.\begin{aligned}\text{Numr.}&=J_1(nc)\left\{\frac{\mu lc}{2n}+\frac{L-L_1}{4\pi N}\,\frac{\Gamma_0 sn}{1+(L-L_1)D/R}\right\},\\ \text{Denr.}&=\frac{\mu lc^2}{4}\{J_1(nc)\}^2\left[1+\frac{\{1+(L-L_1)/(4\pi N)^2\}s^2n^2}{\{1+(L-L_1)D/R\}^2}\right].\end{aligned}\right\}\ldots(100)$$

It is useful to use Γ_0 and the inductances, for these reasons. First, not only may H_0 be arbitrary (a function of r), but Γ_0 may have any value we please, quite independent of H_0. Thus we might have it given that $H_0=0$; then we use only the second part of the numerator. Γ_0 having any stated value implies a certain distribution of force outside the core, constant between the core and the coil, and varying in the depth of the coil in the manner stated in the earlier part of this section. If there be no force in the core at the same time, there is a discontinuity at its boundary, implying a surface-current in the proper direction, and of strength sufficient to cancel the magnetic force in the core due to the coil-current. This boundary-current is of course purely imaginary, having infinite volume-density. But the moment the subsidence commences the discontinuity is rounded off, the surface-current spreading into the core, becoming the real induced current accompanying the passage of magnetic force into the core from outside, where it previously existed.

Next, whilst in Example 2 (circuit closed, coil of small depth) we supposed there to be no self-induction in the external circuit, we may now remove this restriction, as well as that relating to the depth of the coil. There may be any amount of self-induction in the external part of the circuit, provided it be unaccompanied by induced currents in metal in the neighbourhood; *e.g.*, there may be other coils in the circuit, either without cores, or with cores divided to stop induced currents. This self-induction must be included in the value of L above, no other alteration being required, except in equation (96), which does not allow for the external induction.

§ 33. *Two Coils, with Cores, in Sequence.*

From the already-obtained solutions in the cases of open and closed coil-circuit, we may deduce the solutions in many other cases of interest. Thus any number of similar coils with similar cores, having their cores charged in any manner to begin with, may be joined in sequence, and the resulting phenomena completely determined in terms of two solutions only. We start with a pair of similar coils. Let H_1 and H_2 be the strengths of magnetic force in the first and second core at distances r_1 and r_2 from their axes. When the coils are connected in circuit, the current must be the same in both; hence

$$H_1 = H_2 = 4\pi N\Gamma \quad \text{..........................(101)}$$

is the continuity condition, the boundary values of H_1 and H_2 being used, whilst Γ is the common coil-current.

Let L and R be the inductance and resistance of the circuit, both complete. The circuit may contain, besides the two coils mentioned, any number of other coils, either without cores, or with cores in which no current is permitted. The self-induction arising from their presence is included in L, so they may be dismissed altogether from consideration, and when we speak of the coils, we refer only to those with cores in which induced currents can flow. Let M be that part of L due to a single core—viz., $M = (2\pi Nc)^2 l\mu$ (equation (6), putting M for L_1 for subsequent convenience). Then $L - 2M$ is the inductance not counting the cores, and $-(L - 2M)\dot{\Gamma}$ the induced E.M.F. to correspond. Also, the induced E.M.F. arising from the core-induction is given by the third term with sign changed of either of equations (19), and is, owing to the coils and cores being alike, of the same form for both cores; so we get, when there is no impressed force in the circuit,

$$0 = RH_1 + (L - 2M)\dot{H}_1 + \frac{M\rho}{2\pi\mu c}(H_1' + H_2'),$$

for the equation of E.M.F. in the coil-circuit. Or, putting $s = M\rho/2\pi\mu cR$,

$$0 = H_1 + \frac{L-2M}{R}\dot{H}_1 + s(H_1' + H_2'), \quad \text{.............(102)}$$

A normal system is of the form

$$h_1 = AJ_0(nr_1)\epsilon^{Dt}, \qquad h_2 = BJ_0(nr_2)\epsilon^{Dt};$$

D being the same in both to make h_1 and h_2 subside at the same rate, whilst n is the same in both because the coils are alike. By (101), we therefore have (at the boundaries, $r_1 = r_2 = c$),

$$AJ_0(nc) = BJ_0(nc), \quad \text{........................(103)}$$

which gives $A = B$, provided $J_0(nc)$ does not vanish. Putting the expressions for h_1 and h_2 in (102) we get

$$A\left\{J_0(nc)\left(1 + \frac{L-2M}{R}D\right) - snJ_1(nc)\right\} = snBJ_1(nc); \quad \text{......(104)}$$

and, eliminating A and B from (103) and (104) by cross-multiplication, we arrive at the determinantal equation

$$J_0(nc)\left\{\left(1 + \frac{L-2M}{R}D\right)J_0(nc) - 2snJ_1(nc)\right\} = 0, \quad \text{.......(105)}$$

from which we see that there are two series of roots, viz., those of $J_0(nc) = 0$ and those of $\{\ldots\} = 0$. These, for distinction, we shall call the first and the second set of roots. The first set is the same as we have when the coil-circuit is interrupted, and the magnetic energy of the core is converted into heat in the core itself. The second set is the same as we have with a single coil and core, when the coil-circuit is closed. (Examples 1 and 3.) With the second set of roots we have $A = B$, as shown above; with the first set, by using (104), since (103) fails, $A = -B$. In this last case there is no coil-current accompanying the subsidence of the normal system of magnetic force. Thus,

$$\left.\begin{aligned} H_1 &= \Sigma_2 AJ_0(nr_1)\epsilon^{Dt} + \Sigma_1 BJ_0(nr_1)\epsilon^{Dt}, \\ H_2 &= \Sigma_2 AJ_0(nr_2)\epsilon^{Dt} - \Sigma_1 BJ_0(nr_2)\epsilon^{Dt}, \end{aligned}\right\} \quad \text{............(106)}$$

where the 1 and 2 following the Σ's indicate that the first or the second set of roots is employed.

Hence, by addition and subtraction, using the same value of r in both cases,

$$\tfrac{1}{2}(H_1 + H_2) = \Sigma_2 AJ_0(nr)\epsilon^{Dt}, \quad \text{....................(107)}$$

$$\tfrac{1}{2}(H_1 - H_2) = \Sigma_1 BJ_0(nr)\epsilon^{Dt}. \quad \text{....................(108)}$$

The expressions for A and B have been already given, in terms of the initial distributions, so we need not repeat them. Thus (81) gives B, when for H_0 there we write the initial value of $\frac{1}{2}(H_1 - H_2)$, and this is a function of r; and (99) gives A, with corresponding alterations to suit the initial value of $\frac{1}{2}(H_1 + H_2)$. Thus the theoretical solution is complete.

We may now consider the physical significance of the above, and how to practically apply it. In the first place, suppose the cores are charged to the same strength, and are then joined up so that they *add* their inductive E.M.F.'s in the coil-circuit, the impressed forces being removed. Or, more simply, put them in series with a battery, and then remove the latter without breaking the circuit. From symmetry it is evident that the magnetic force subsides in the same manner in both cores, contributing equally to the circuit E.M.F., and that there is a current in the circuit. But as with the first set of roots there can be no coil-current, it follows that the second set of roots, with the corresponding normal systems, must be alone in operation.

Next, starting with cores charged to equal strengths, let the coils be joined so as to produce opposed E.M.F.'s in the coil-circuit. These E.M.F.'s being equal, there can be no coil-current. The boundary magnetic forces are zero, and the normal systems are those of the first set of roots only, the magnetic force in both cores subsiding in the same manner as if the circuit were interrupted.

Thirdly, let there be initially different strengths of magnetic force in the two cores. Decompose them thus :—

$$\left.\begin{aligned} H_1 &= \tfrac{1}{2}(H_1 + H_2) + \tfrac{1}{2}(H_1 - H_2), \\ H_2 &= \tfrac{1}{2}(H_1 + H_2) - \tfrac{1}{2}(H_1 - H_2). \end{aligned}\right\} \qquad (109)$$

The portion $\frac{1}{2}(H_1 + H_2)$ being the same in both cores, has the second set of normal systems only, thus giving us the solution (107); whilst the portion $\frac{1}{2}(H_1 - H_2)$ being opposite in the two cores (as regards the direction of induced E.M.F. in the circuit) produces no current in the circuit, and has therefore the first set of normal systems, thus giving rise to the solution (108). Combining the thus obtained (107) and (108) we obtain the solution for each core by itself, expressed in equations (106).

We have, in this deduction of (106), supposed H_1 and H_2 initially constant, but the reasoning plainly holds good when H_1 and H_2 are functions of r_1 and r_2, and are decomposed as in (109) into like and unlike auxiliary systems.

Now we have, in Fig. 3, given the curve of subsidence of magnetic force, initially steady, at the axis of a core, corresponding to the first set of roots of (105), and also the curves of subsidence at the axis and at the boundary for a particular case of the second set, viz. : coil of small depth, and a certain value of its time-constant. Therefore from curves h_0h_0 and H_0H_0 (Fig. 3) (the last being with $2s = 1$ in (105), and $a = c$), or a similar curve instead of H_0H_0 when $2s$ is not $= 1$, we can draw the curves to suit the present problem.

If, at starting, the cores are equally charged, and the coils are joined so as to add the E.M.F.'s, the subsidence of magnetic force takes place similarly in both cores, being represented by H_aH_a at their boundaries, (which also shows the subsidence of the coil-current), and by H_0H_0 at their axes.

But if they are connected so that the E.M.F.'s are opposed in the circuit, there is no coil-current, and the subsidence in the cores is accelerated, being, at their axes, represented by h_0h_0.

Next, let one core only be charged, to begin with, and its coil be then connected with the second coil with uncharged core, the battery being removed. The subsidence (axial) in the first core is got by taking the mean of H_0H_0 and h_0h_0 in Fig. 3; *i.e.*, construct the curve whose ordinate for any value of the time is one half the sum of the ordinates of H_0H_0 and h_0h_0 at the same time. And the curve of magnetic force at the axis of the second core is got by constructing the curve whose ordinate is one half the difference in the ordinates of H_0H_0 and h_0h_0. The resulting curve, starting from zero, shows initial retardation, rises rapidly to a maximum, and falls thereafter slowly to zero again, as a portion of the energy originally in the first core is transmitted into the second, there to be dissipated.

If both cores are charged to start with, but to different strengths, we must construct our curves by taking the ordinates of H_0H_0 and h_0h_0 in the proper proportions shown by the decomposition in (109), adding them for one core, subtracting for the other, to obtain the ordinates of the required curves.

The boundary-force and the coil-current curves may be got from the curve H_aH_a alone by merely altering its scale vertically, there being no coil-current to correspond to h_0h_0. That is, directly the coils are connected, the current alters its strength to suit the mean initial force $\frac{1}{2}(H_1+H_2)$. For example, if the second core is uncharged the current drops suddenly to half-strength.

§ 34. *Three Similar Coils and Cores in Sequence.*

With the same notation, the equation of E.M.F. in the circuit is, by an obvious extension of (102),

$$0 = H_1 + \frac{L-3M}{R}\dot{H} + s(H_1' + H_2' + H'), \quad \ldots\ldots\ldots\ldots(110)$$

and if $\quad AJ_0(nr_1)\epsilon^{Dt}, \quad BJ_0(nr_2)\epsilon^{Dt}, \quad CJ_0(nr_3)\epsilon^{Dt}$

be a normal system, continuity of current in the circuit requires that

$$AJ_0(nc) = BJ_0(nc) = CJ_0(nc);$$

giving $A = B = C$, when $J_0(nc)$ does not vanish. Corresponding to (105) we have the determinantal equation

$$\{J_0(nc)\}^2\left\{\left(1+\frac{L-3M}{R}D\right)J_0(nc) - 3snJ_1(nc)\right\} = 0; \quad \ldots.(111)$$

with two sets of roots, as in the last case, the first set corresponding to circuit broken, the second set to a single coil with circuit closed.

Regarding the ratios $A/B/C$ when the roots are those of the first set, we may be guided by physical considerations. Decompose the initial magnetic forces H_1, H_2, H_3 into

$$\begin{aligned} H_1 &= \tfrac{1}{3}(H_1+H_2+H_3) + \tfrac{1}{3}(H_1-H_2) + \tfrac{1}{3}(H_1-H_3), \\ H_2 &= \qquad \text{ditto} \qquad + \tfrac{1}{3}(H_2-H_1) + \tfrac{1}{3}(H_2-H_3), \\ H_3 &= \qquad \text{ditto} \qquad + \tfrac{1}{3}(H_3-H_1) + \tfrac{1}{3}(H_3-H_2). \end{aligned}$$

The portion $\frac{1}{3}\Sigma H$ common to all three obviously requires the use of the second set of roots. The other six terms cancel one another as regards E.M.F. in the coil-circuit. Thus, $\frac{1}{3}(H_1-H_2)$ in first core and

$\frac{1}{3}(H_2 - H_1)$ in the second cancel, and similarly for the other two pairs. They therefore require us to use the first set of roots only. We may write the decomposition thus,

$$\left.\begin{array}{l} H_1 = \frac{1}{3}\Sigma H + (H_1 - \frac{1}{3}\Sigma H), \\ H_2 = \frac{1}{3}\Sigma H + (H_2 - \frac{1}{3}\Sigma H), \\ H_3 = \frac{1}{3}\Sigma H + (H_3 - \frac{1}{3}\Sigma H). \end{array}\right\} \qquad (112)$$

Consequently the solution for the three cores is to be found in a similar manner to before. Starting with any arbitrary distributions, substitute for them, first a distribution $\frac{1}{3}\Sigma H$ in every core, and let them subside according to the second set of roots, adding their effects in the coil-circuit. And let simultaneously the complementary distributions $H_1 - \frac{1}{3}\Sigma H$, etc., subside, each in its own core independently, according to the first system. The resultants will give the actual state of things. The curves of axial subsidence, when the initial distributions are constant, may be found from curve h_0h_0 (Fig. 3), and either H_0H_0 or a similar one, by constructing three new curves whose ordinates are made up of those of H_0H_0 and h_0h_0 taken in the proper proportions expressed in (112).

§ 35. *Any Number of Coils in Sequence.*

The extension to any number m of equal coils, with equal cores, in sequence, is plain. The determinantal equation is

$$\{J_0(nc)\}^{m-1}\left\{\left(1 + \frac{L - mM}{R}D\right)J_0(nc) - msnJ_1(nc)\right\} = 0, \qquad (113)$$

with the two sets of roots as before, the repetition of the roots of $J_0(nc) = 0$ only affecting the ratios of the constants A, B, C, etc., in a normal system. The decomposition is

$$\left.\begin{array}{l} H_1 = \frac{1}{m}\Sigma H + \left(H_1 - \frac{1}{m}\Sigma H\right), \\ \cdots\cdots\cdots\cdots \\ H_m = \frac{1}{m}\Sigma H + \left(H_m - \frac{1}{m}\Sigma H\right). \end{array}\right\} \qquad (114)$$

The mean distribution $m^{-1}\Sigma H$ is what makes the coil-current. This distribution must be imagined to exist in all the cores, each with its proper complementary distribution; the mean distribution to subside one way, using the second set of roots, the complementary to subside as if the coil-circuit were interrupted. It is unnecessary to write out the developments, such being merely repetitions of the solutions in Examples 1 and 2 or 3. But the general case of any number of dissimilar coils and cores joined in sequence, of which the above is a special case, does not admit of the reduction to two simple solutions, so its theory will form the subject of a later section.

§ 36. *Equal Coils, with Cores, in Parallel.*

Two coils in parallel being the same as two in sequence, nothing more need be said about it. We may consider the theory of three coils in parallel, and its natural extension to any number. Let P and Q be the common terminals of the three branch conductors containing the coils; every branch to have the same resistance and self-induction, and to contain a coil with a core, the three coils being alike, and also the three cores. Let Γ_1, Γ_2, Γ_3 be the three currents, reckoned from P to Q, and V the fall of potential from P to Q. This being the same for the three branches, we have the equation of E.M.F.

$$V = R\Gamma_1 + (L - M)\dot{\Gamma}_1 + S\Gamma_1' \quad \dots\dots\dots\dots\dots\dots (115)$$

in the first branch, and two similar equations with Γ_2 and Γ_3 written for Γ_1. Here $S = M\rho/2\pi\mu c$, and L and M are the inductance of a branch, and that part of it due to the core. We have also the equation of continuity

$$\Gamma_1 + \Gamma_2 + \Gamma_3 = 0; \quad \dots\dots\dots\dots\dots\dots\dots\dots (116)$$

the sum of the currents meeting at P or at Q being zero.

A normal system of magnetic force in the cores being

$$A_1 J_0(nr_1)\epsilon^{Dt}, \qquad A_2 J_0(nr_2)\epsilon^{Dt}, \qquad A_3 J_0(nr_3)\epsilon^{Dt},$$

with the currents in the coils

$$A_1 J_0(nc)\epsilon^{Dt}/4\pi N, \quad A_2 J_0(nc)\epsilon^{Dt}/4\pi N, \quad A_3 J_0(nc)\epsilon^{Dt}/4\pi N$$

to correspond, substitution of the latter for Γ_1, Γ_2, Γ_3 in (115) and (116) gives, by eliminating V and the A's, the determinantal equation

$$J_0(nc)[\{R + (L - M)D\}J_0(nc) - snJ_1(nc)]^2 = 0, \quad \dots\dots\dots (117)$$

or say $xy^2 = 0$; giving the two sets of roots belonging to $x = 0$, and $y = 0$. The equation $x = 0$ means that there is no current in any of the coils, and $y = 0$ that the terminals P and Q are at the same potential, as if the three branches were put on short-circuit.

There being two manners of subsidence of initially given distributions of magnetic force, which we shall refer to here and later as the first manner (no coil-current, or circuit broken), and the second manner (branch containing a coil on short-circuit), it only remains to properly effect the division of the given distributions into two distributions, such that one will give no current in the coils, and the other will give currents subject to the condition of continuity (116). Now, if we start with a distribution H_1 in the first core, and none in the others, and substitute the distributions $\frac{1}{3}H_1$, $\frac{1}{3}H_1$, $\frac{1}{3}H_1$ in the first, second, and third cores, these will subside in the first manner, owing to the balanced E.M.F.'s; whilst the distributions $\frac{2}{3}H_1$, $-\frac{1}{3}H_1$, $-\frac{1}{3}H_1$, which, with the former, make up the given real distributions H_1, 0, 0, will give currents complying with (116); *i.e.*, a current in the first branch from P to Q dividing equally between the second and third branches, in which it is directed from Q to P. This is the proper division so far as H_1 is concerned, and those for H_2 and H_3 may be similarly constructed. Putting all together, we find the proper division when H_1, H_2, H_3, are all finite, is

$$\left.\begin{aligned} H_1 &= \tfrac{1}{3}\Sigma H + (H_1 - \tfrac{1}{3}\Sigma H), \\ H_2 &= \tfrac{1}{3}\Sigma H + (H_2 - \tfrac{1}{3}\Sigma H), \\ H_3 &= \tfrac{1}{3}\Sigma H + (H_3 - \tfrac{1}{3}\Sigma H), \end{aligned}\right\} \quad \dots\dots\dots\dots\dots\dots (118)$$

the common mean distribution $\frac{1}{3}\Sigma H$ in each core to subside in the first manner, and the remainders in the second manner, furnishing the coil-currents.

Contrast with the case of three coils in sequence, the division being represented in equations (112). These are the same as (118), but whereas in the latter case the mean distribution furnishes no coil-current, in the former it is the mean distribution that supplies the coil-current.

The extension to any number m of equal coils in parallel may be shortly stated. The determinantal equation is

$$xy^{m-1}=0, \qquad (119)$$

which may be compared with (113), the corresponding equation when the coils are in sequence, which, written similarly to (119), is $yx^{m-1}=0$.

The proper division of the initially given distributions of magnetic force in the cores is given by equations (114), only noting that the mean distribution $m^{-1}\Sigma H$ must now subside in the first manner, instead of, as in the case to which (114) relates, in the second manner, with the corresponding change as regards the complementary distributions.

§ 37. m_1 *Coils in Sequence with* m_2 *Coils in Parallel.*

It may be inferred from the preceding, that if we join up equal coils in any manner to form a linear system of conductors, the determinantal equation of the system will be merely $x^a y^b=0$, a and b being integers. This will be evident from the section to follow on unequal coils, so at present we take it for granted, to avoid unnecessary repetition. The general case of a linear system, with equal coils, will be considerably lightened by first taking a simple case combining coils in sequence and coils in parallel, say two coils in sequence with three in parallel, as in the figure.

The numerals showing the position of the coils, and the lines their connections, given at a certain moment there to be distributions of magnetic force H_1, H_2, etc., in the five cores, and that there are no external impressed forces, the manner of subsidence is required, knowing that it is the resultant of the two manners corresponding to $x=0$ and $y=0$.

Let there be, in the first place, a distribution of magnetic force H_1 in the first core only, the rest being uncharged. Whatever current is set

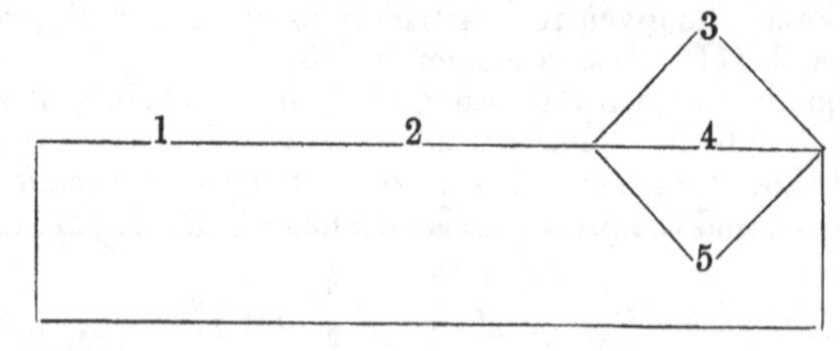

up in the 1st coil, we must, by continuity, have the same current in the 2nd, and by continuity and symmetry one-third of this current in the

3rd, 4th, and 5th coils, all similarly directed, say from left to right. The proportions are 1, 1, $\frac{1}{3}$, $\frac{1}{3}$, $\frac{1}{3}$, and these must be the relative magnitudes of the magnetic forces at the boundaries of the five cores, after the subsidence has commenced. Also, at the first moment, we have to cancel the magnetic force in the 2nd, 3rd, 4th, and 5th cores. Thus we must have another set, z, -1, $-\frac{1}{3}$, $-\frac{1}{3}$, $-\frac{1}{3}$, where z, for the 1st coil, is yet unsettled. To find it, by the principle that this set must set up no current in the coils, the common $-\frac{1}{3}$ in the 3rd, 4th, and 5th branches must be balanced by $+\frac{1}{3}$ in the branch formed by the 1st and 2nd taken together. This gives $z=\frac{4}{3}$, and $1+z=\frac{7}{3}$, the complete boundary-force in the first core on the same scale. So, multiplying by $\frac{3}{7}H_1$ throughout, we find

$$h_1=\tfrac{4}{7}H_1, \quad h_2=-\tfrac{3}{7}H_1, \quad h_3=h_4=h_5=-\tfrac{1}{7}H_1, \qquad \text{.........(120)}$$

as the five distributions which subside in the first manner, without current in coils. A similar set of distributions corresponds to H_2, since the second coil is in sequence with the first, giving

$$h_1=-\tfrac{3}{7}H_2, \quad h_2=\tfrac{4}{7}H_2, \quad h_3=h_4=h_5=-\tfrac{1}{7}H_2, \qquad \text{.........(121)}$$

to subside in the first manner.

With H_3 the distributions are different. The current of strength 1 in the third coil, due to E.M.F. in itself only, divides thus,

$$\tfrac{1}{5}, \quad \tfrac{1}{5}, \quad 1, \quad -\tfrac{2}{5}, \quad -\tfrac{2}{5},$$

in the five coils, and these are the proportions of the boundary magnetic forces. To cancel them at the first moment in all cores except the 3rd, we require the supplementary set

$$-\tfrac{1}{5}, \quad -\tfrac{1}{5}, \quad z, \quad \tfrac{2}{5}, \quad \tfrac{2}{5};$$

where z must evidently be $\frac{2}{5}$. This gives us, multiplying by $\frac{5}{7}H_3$,

$$h_1=h_2=-\tfrac{1}{7}H_3, \qquad h_3=h_4=h_5=\tfrac{2}{7}H_3, \qquad \text{............(122)}$$

to subside in the first manner. The corresponding distributions for H_4 and H_5 are got by writing first H_4 and then H_5 for H_3 in (122). Putting together (120), (121), (122), and the two equations similar to the last, we find h_1, h_2, ..., the complete distributions, to subside in the first manner, to be

$$\left.\begin{aligned} h_1 &= \tfrac{4}{7}H_1-\tfrac{3}{7}H_2-\tfrac{1}{7}(H_3+H_4+H_5),\\ h_2 &= -\tfrac{3}{7}H_1+\tfrac{4}{7}H_2-\tfrac{1}{7}(\ldots\ldots\ldots\ldots\ldots),\\ h_3=h_4=h_5 &= -\tfrac{1}{7}(H_1+H_2)+\tfrac{2}{7}(\ldots\ldots\ldots\ldots\ldots). \end{aligned}\right\} \qquad \text{.........(123)}$$

The complementary distributions H_1-h_1, H_2-h_2, etc., are to subside in the second manner, as if the coils were short-circuited.

In all the preceding relating to coils in sequence and in parallel, we have, for facility of description, referred to the H's as the initial distributions in the *cores*. But should the coils be not of very small depth, we must, along with the core-distribution H, consider the coil-distribution of magnetic force; and, again, if there be external self-induction associated with the coil, there is a certain unknown distribution, though known amount, of magnetic induction, depending on the strength of the coil-current. These supplements to the H's have to be treated in the same manner as the H's, and may be imagined to be included in them.

§ 38. *Any Combination of Equal Coils, with Cores.*

It may have been observed by the attentive reader (I hope I have at least one), especially in a study of the contents of the last section relating to five coils, that there is a general principle underlying the method of finding the two sets of distributions, a knowledge of which is essential in finding how the subsidence takes place. The principle, like most principles, shows itself more prominently by putting on one side all the insignificant details of special solutions (although the study of special solutions is, if not indispensable, at least very desirable), and viewing the matter more generally. First, to define what is meant by "any combination ..." in the title of this section. It is any combination as hereinafter set forth and described. The unit, which is the element of the combination, is a coil with a core, joined in sequence with an external conductor. Any number of such units, all alike as regards main coils, cores, and external conductors, may be connected in sequence to form a branch; and, finally, any number of branches, containing different numbers of units, may be joined together to form a network, a linear system of conductors, so far as the wire of the coils and external conductors is concerned.

Given at any moment the state of magnetic force (subject to the previous restrictions as to direction and symmetry) in all the cores, with, if necessary, the strength of current in each coil at the same time, what happens when the system is left to itself without impressed E.M.F., that is, left under the influence of the E.M.F.'s of induction only? We may, for instance, imagine the separate units, previously separately charged, to be instantaneously connected together to form the aforesaid combination.

The complete solution may be expressed in terms of the already-obtained two solutions. The one is that which expresses what happens when a unit, as above specified, previously charged, is left to itself with the coil-circuit interrupted; the other expresses what happens under the same circumstances except that the coil-circuit is closed. In the first case we have induced currents in the core only; in the second case in the coil also, and with a different and slower manner of subsidence of the core's magnetisation, if there be any, or in general, of the magnetic induction. The determinantal equation of the combination is of the form $x^a y^b = 0$, and we have $x = 0$ for the first manner and $y = 0$ for the second manner of subsidence. We only require to decompose the initially given states into two sets, such that in one of them we shall have $x = 0$, and subsidence as if all the wires were disconnected, and in the other $y = 0$, or subsidence as if every unit formed an independent closed circuit.

This may be done generally, as done in the last section with a simple combination, by making use of the following property connecting steady impressed forces and currents in a linear system of conductors. Let the resistances of the different branches be R_1, R_2, ..., the impressed forces acting in them E_1, E_2, ..., and let the currents be C_1, C_2.... These currents are what result from the simultaneous action of all the impressed forces. They may be found by the application of Ohm's law to every conductor; either symmetrically, or, when the system is not very complex, by the easier ways that present themselves.

Now remove the impressed forces E_1, E_2, ..., and substitute others F_1, F_2, ..., of such strengths that

$$F_1 = E_1 - R_1 C_1, \qquad F_2 = E_2 - R_2 C_2, \quad \text{etc.}$$

There will now be no current in any of the conductors.

This may be very easily seen. For, if in the original distribution of current, the rises of potential in the directions of the currents C_1, etc., in the conductors of resistances R_1, etc., be ΔP_1, ΔP_2, etc., the complete E.M.F.'s in the branches are $E_1 - \Delta P_1$, $E_2 - \Delta P_2$, etc. Hence, by Ohm's law,

$$E_1 - \Delta P_1 = R_1 C_1, \qquad E_2 - \Delta P_2 = R_2 C_2, \quad \text{etc.}$$

Thus, our defined second set of impressed forces to take the place of E_1, etc., are given by

$$F_1 = \Delta P_1, \qquad F_2 = \Delta P_2, \quad \text{etc.}$$

and, consequently, by the elementary potential property, there is no resultant impressed force in any of the closed circuits that can be made up in the linear system, and therefore no current in any of the branches.

(In tri-dimensional steady flow, ρ being the specific resistance, $\mathbf{E}$ the impressed force per unit length, γ the current-density, and P the potential, we have

$$\mathbf{E} - \nabla P = \rho \gamma$$

in the first distribution; and the second distribution of impressed force is $\mathbf{F} = \nabla P$, producing no current.)

We have now only to apply this property to our combination. Let an impressed force of unit strength act steadily in one of the coils, say the first, and find the distribution of current set up. Let it be e_1/r, e_2/r, e_3/r, etc., where r is the common resistance of every unit in the combination. Thus e_1, e_2, e_3, ... are a set of proper fractions, and the set of impressed forces of strengths $1 - e_1$, $-e_2$, $-e_3$, etc., in the first, second, etc., coils would, by the preceding, set up no current in the coils. Let H_1 be the initial magnetic force in the first core, a function of the distance from the axis, and let the other cores be uncharged. The distributions $e_1 H_1$, $e_2 H_1$, $e_3 H_1$, etc., if they existed alone would subside in the second manner, as if every unit in the combination formed a separate closed circuit. The supplementary distributions $(1 - e_1)H_1$, $-e_2 H_1$, $-e_3 H_1$, etc., if they existed alone, would subside in the first manner, without current in the coils. Superimpose these two distributions, and we have the actual initial distribution, and at any subsequent moment the state of the system will be the resultant of what these distributions would have then become. Similarly we may treat the initial magnetic forces in the other cores. If the second coil be in sequence with the first the same set of ratios e_1, e_2, e_3, ..., is to be used. But in passing to another branch in the combination, we must, in general, find another set of ratios. Having gone over the whole combination in this way, it is a question of simple addition or subtraction to obtain the final resultant solutions.

§ 39. *Dissimilar Coils. Characteristic Function of a Linear System of Conductors, and Derivation of the Differential Equation.*

As might be expected, when, in the combination as specified in the last section, the coils and cores are made dissimilar in their dimensions and materials, it is no longer possible to express the general solution in terms of two comparatively simple solutions, although with special relations amongst the coil-data it may sometimes be expressed in terms of a limited number of simpler solutions. Now, so far as the mere algebra goes we can easily arrive at the general solution when the coils are dissimilar, and in fact more easily than when there are particular relations introducing indeterminateness in the general solution, requiring special attention; but, owing to difficulties of interpretation, the general solution has very limited utility, except in showing how we might, if we gave the time to it, completely solve the problem numerically. But there are, on the journey to the general solution, some views by the way which make it more interesting and instructive than it would otherwise be, and it is to exhibit them, rather than a mere mathematical complication, that the present section is written, before leaving coils in combination and proceeding to other parts of the subject of induction, in cores and elsewhere.

Starting, in the first place, with a linear system of conductors, self-contained, that is, having no connection with other conductors, let there be q points connected by $\frac{1}{2}q(q+1)$ conductors, the least number that will join every point with all the rest. Let their potentials be $P_1, P_2, \ldots$. To distinguish the different conductors, use double suffixes. Thus, let K_{12} be the conductance (reciprocal of resistance) of the conductor joining the points 1 and 2, and Γ_{12} the current from 1 to 2, with a similar notation for the rest.

Continuity of current requires that the sum of the currents leaving any point shall vanish. Thus, for point 1, we have

$$\Gamma_{12}+\Gamma_{13}+\Gamma_{14}+\ldots+\Gamma_{1q}=0. \qquad \ldots\ldots\ldots\ldots\ldots\ldots(124)$$

There are $q-1$ other equations of this kind, got by changing the first suffixes.

Next, let E_{12} be the impressed force from 1 to 2, and similarly for the other conductors; then, by Ohm's law,

$$\Gamma_{12}=K_{12}(E_{12}+P_1-P_2) \qquad \ldots\ldots\ldots\ldots\ldots\ldots(125)$$

expresses the current in the branch joining the points 1 and 2 in terms of the impressed force in that branch and the potentials at its ends. There are $\frac{1}{2}q(q+1)$ equations of this kind. Insert in (124), and we have

$$(K_{12}E_{12}+K_{13}E_{13}+\ldots+K_{1q}E_{1q})+(K_{12}+\ldots+K_{1q})P_1$$
$$=(K_{12}P_2+K_{13}P_3+\ldots+K_{1q}P_q), \ldots(126)$$

with $q-1$ other equations of the same kind. Here the quantity in the first ( ) is the current that would leave the point 1 by all the conductors meeting there if the impressed forces were the only forces; call it C_1, and similarly for the other points. Also, put

$$-K_{11}=K_{12}+K_{13}+\ldots+K_{1q}, \qquad \ldots\ldots\ldots\ldots\ldots\ldots(127)$$

and we get

$$\left.\begin{aligned} C_1 &= K_{11}P_1 + K_{12}P_2 + K_{13}P_3 + \dots + K_{1q}P_q, \\ C_2 &= K_{21}P_1 + K_{22}P_2 + K_{23}P_3 + \dots + K_{2q}P_q, \\ &\dots\dots\dots\dots\dots\dots\dots\dots\dots\dots \\ C_q &= K_{q1}P_1 + K_{q2}P_2 + \dots \qquad + K_{qq}P_q. \end{aligned}\right\} \quad \dots\dots\dots\dots(128)$$

Here, as we have q linear equations and q potentials, it looks as if we could find them. Thus, if K is the determinant of the K's, and Z_{rs} the coefficient of K_{rs} in this determinant, so that

$$K = K_{11}Z_{11} + K_{12}Z_{12} + \dots + K_{1q}Z_{1q}, \quad \dots\dots\dots\dots(129)$$

with similar expressions, we should have

$$KP_1 = C_1Z_{11} + C_2Z_{21} + \dots + C_qZ_{q1}, \quad \dots\dots\dots\dots(130)$$

and similar expressions for the other potentials. But K_{11}, K_{22}, ..., are not independent of the other real K's, being defined by (127). This structure of the diagonal K's makes all the Z's identical, and the sum of the K's in any one row or column of (128) vanish, so that if Z is the common value of the Z's, (129) becomes

$$K = (K_{11} + K_{12} + \dots + K_{1q})Z = 0 \times Z,$$

by (127), and (130) becomes

$$0 \times P_1 = (\Sigma C)Z = 0 \times Z,$$

since the sum of the C's also vanishes. We cannot find the q potentials, but only $q-1$ of them, supposing the remaining one to be given.

Maxwell's investigation, Vol. I., Art. 280, differs from the above in the system not being self-contained, having external connections at the q points, at which there are given currents Q_1, Q_2, ..., entering the system. This only requires us to write $C_1 - Q_1$ instead of C_1, $C_2 - Q_2$ for C_2, etc., in (128). But regarding the fact that only $q-1$ equations are independent, Maxwell ascribes it to the necessary condition $\Sigma Q = 0$. This reason is surely erroneous, for in the above there are no Q's to consider, and the reason fails.

We may therefore reject one of equations (128), say the last, leaving p equations, $(p = q - 1)$; thus

$$\left.\begin{aligned} C_1 - K_{1q}P_q &= K_{11}P_1 + K_{12}P_2 + \dots + K_{1p}P_p, \\ C_2 - K_{2q}P_q &= K_{21}P_1 + K_{22}P_2 + \dots + K_{2p}P_p, \\ &\dots\dots\dots\dots\dots\dots\dots\dots\dots\dots \\ C_p - K_{pq}P_q &= K_{p1}P_1 + K_{p2}P_2 + \dots + K_{pp}P_p, \end{aligned}\right\} \quad \dots\dots\dots(131)$$

which gives the p potentials in terms of the C's and the rejected P, viz., P_q. This last we may put $=0$; as, if it is not to vanish, the effect is merely to add a common potential all round.

The determinant of K's in this last system of equations is the same as the before mentioned, Z, except as regards sign, being negative when p is even and positive when p is odd. Its full expression may be written down mechanically. It may be called the characteristic function of the linear system. It contains in itself the expressions for the resistance external to every branch in the system, which expression can be got by inspection. Let there be only one impressed force E_{12}, and find Γ_{12} the current in the same conductor. If G_{rs} is the coefficient of K_{rs} in Z, we find

$$\Gamma_{12}=K_{12}E_{12}\left\{1+\frac{K_{12}}{Z}(G_{11}+G_{22}-2G_{12})\right\}. \quad \text{............(132)}$$

Now, Z containing only first powers of the K's, we may write, by picking out the terms containing K_{12}, in the full expression for Z from which the diagonal K's have been eliminated,

$$Z=K_{12}X+Y, \quad \text{..........................(133)}$$

where X and Y do not contain K_{12}. Further, we may show that

$$-X=G_{11}+G_{22}-2G_{12};$$

which, by (132), gives

$$\Gamma_{12}=K_{12}E_{12}(1-K_{12}X/Z)=K_{12}E_{12}Y/Z, \quad \text{by (133),}$$
$$=E_{12}(K_{12}^{-1}+X/Y)^{-1}, \quad \text{by rearrangement. ....(134)}$$

Consequently, the resistance external to K_{12} is X/Y. Similarly, by isolating any particular conductance in the expression for Z, as in (133), we obtain the resistance external to the conductor isolated.

Thus, let there be 4 points joined by 6 conductors (the well known "Bridge"). The characteristic function is, if $s_1=a+b+c+d$,

$$Z=(a+b)(c+d)e+(a+c)(b+d)f+s_1ef+ab(c+d)+cd(a+b). \quad (135)$$

The letters here stand for the conductances of the branches as shown in the diagram, it being supposed that the four other branches g, h, i, j, are non-existent.

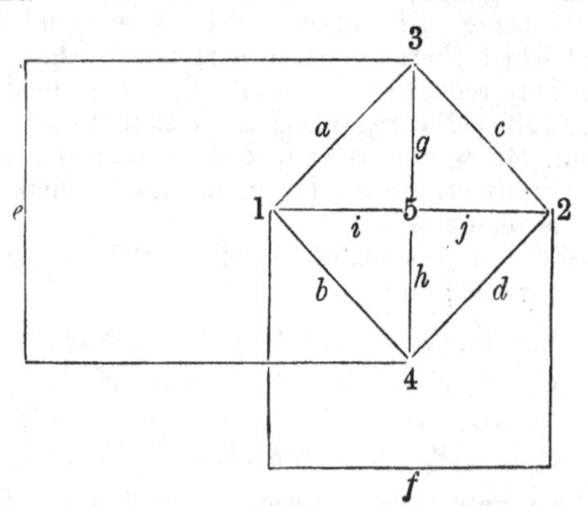

As arranged in (135), it is suitable for showing the resistance external to either e or f, those symbols being isolated.

Next, 5 points joined by 10 conductors. In the figure there is an additional junction 5, and four new conductors g, h, i, j connecting it with the former four points. The characteristic function is now the determinant of the K's from K_{11} to K_{44}, and is, the conductances being expressed by the letters in the diagram, expressed thus, using the additional symbol $s_2=g+h+i+j$,

$$\begin{aligned}Z=e\{&s_2(a+b)(c+d)+s_1ij+(g+h)[i(c+d)+j(a+b)+ij]\}\\&+f\{s_2(a+c)(b+d)+s_1gh+(i+j)[g(b+d)+h(a+c)+gh]\}\\&+ef\{s_1s_2+(g+h)(i+j)\}+ij\{(a+c)(b+d)+g(b+d)+h(a+c)\}\\&+s_2\{ab(c+d)+cd(a+b)\}+gh\{(a+b)(c+d)+i(c+d)+j(a+b)\}\\&+gi(bc+bd+cd)+gj(ab+ad+bd)+hj(ab+ac+bc)\\&+hi(ac+ad+cd)+ghij. \qquad (136)\end{aligned}$$

As before, it is arranged suitably for giving the resistance external to either e or f; the complete coefficient of e, divided by the remaining terms, being the resistance external to e. Similarly for f. To get the resistance external to another conductor, say a, this symbol must be isolated. Or, more easily, rearrange the diagram, changing the lettering suitably.

The characteristic function also gives us the differential equation of the system when the self-induction in some or all the branches is taken into account. Thus, if R_{12} is the resistance from 1 to 2, and L_{12} the inductance, write $R_{12}+L_{12}D$ instead of R_{12} (or $1/K_{12}$) in the characteristic function, and similarly for the others; this being done,

$$Z=0 \qquad (137)$$

is the differential equation of the system, if D stands for d/dt, D^2 for d^2/dt^2, etc. And, D being algebraic, (137) is the determinantal equation, whose roots give the admissible time-constants of subsidence, these being the negative reciprocals of the roots D_1, D_2, etc.

§ 39. Equation (137) is also the determinantal equation when there are coils with cores in the different branches, but with a different meaning attached to the coefficients K. Let there be a coil with a core in every branch. To distinguish them, it suffices to number the principal symbols before used for a single coil. These are, Γ the coil-current, R the resistance of the branch containing the coil, having N windings per unit length l; L the complete inductance of the branch, and M that part of it due to the core; H the magnetic force in the core of length l, and radius c, at distance r from the axis; ρ and μ the specific resistance and permeability of the core; $S=M\rho/2\pi\mu c$. Then

$$P_1-P_2=R_{12}\Gamma_{12}+(L_{12}-M_{12})\dot{\Gamma}_{12}+S_{12}\Gamma'_{12}, \qquad (138)$$

as in (115), is the equation of E.M.F., which takes the place of (125), whilst (124) remains unaltered.

§ 40. Now let a normal system of magnetic force in the cores be

$$Am_{12}J_0(n_{12}r_{12})\epsilon^{Dt}, \quad Am_{13}J_0(n_{13}r_{13})\epsilon^{Dt}, \quad \text{etc.}, \qquad (139)$$

merely changing the subscripts in passing from one branch to another. Here D is the same in all. That makes it a normal system. Again, the n's are known in terms of the D, by

$$4\pi\mu_{12}D=-n_{12}^2P_{12}, \quad \text{etc.}$$

Finally, the m's in (139) are a set of ratios to be found, and A is a common factor, fixing the actual size of the system. We have also

$$\Gamma_{12}=Am_{12}J_0(n_{12}c_{12})\epsilon^{Dt}/4\pi N_{12}, \quad \text{etc.} \qquad (139a)$$

Inserting this expression for Γ_{12} in (138), we get, with $t=0$,

$$P_1 - P_2 = \Gamma_{12} y_{12}/x_{12}, \quad \text{.......................(140)}$$

where $x_{12} = J_0(n_{12}c_{12})$, and

$$y_{12} = [R_{12} + (L_{12} - M_{12})D]J_0(n_{12}c_{12}) - S_{12}n_{12}J_1(n_{12}c_{12}), \quad \text{......(141)}$$

as in (117), § 36. Here y_{12} and x_{12} are constants; also (140) would be the expression of Ohm's law if y_{12}/x_{12} were the resistance of the branch 1, 2, and there were no impressed force. So all we have to do is to put $C_1=0$, $C_2=0$, etc., $P_q=0$, in equations (131), and write x_{12}/y_{12} instead of K_{12}, to get the potential equations. But, the left members being now zero, we have

$$Z=0 \quad \text{.......................(142)}$$

necessarily. This is the determinantal equation of D, when in the full expression for Z we write x/y for K. It should be cleared of fractions to avoid missing roots. Thus, the expressions (136) equated to zero, with x/y, as defined in (141), put for every conductance, and cleared of fractions, will be the determinantal equation when every branch in the diagram contains a coil with a core. The ratios m in (139) next have to be dug out of the potential and continuity equations. That done, there is only left the size of A to be settled. Given the state of magnetic force h in the cores, and the coil-currents γ, at the moment $t=0$, A is to be found by the property of the vanishing of the mutual energy of a pair of normal systems, which makes the mutual energy of the given system with respect to a normal system, as defined in (139) and (139a), with $t=0$, equal to twice the energy of the normal system itself, and gives

$$A = \frac{\Sigma \tfrac{1}{2}\mu l \int hmJ_0(nr)rdr + \Sigma(L-M)\gamma mJ_0(nc)/4\pi N}{\Sigma \tfrac{1}{2}\mu l \int m^2\{J_0(nr)\}^2 rdr + \Sigma(L-M)\{mJ_0(nc)/4\pi N\}^2}. \quad \text{...(143)}$$

Here the Σ's indicate inclusion of all the coils, so no suffixes are written, and the integration is from the axis to the boundary in every core. With this, we leave the combination.

XXIX. REMARKS ON THE VOLTA-FORCE, ETC.

[*Journal Soc. Tel. Eng.*, March, 1885.]

Professor Lodge has done me the honour of mentioning me, in the course of his learned memoir on the "Seat of the Electromotive Forces in the Voltaic Circuit,"* as having published some similar statements to some of his. It may be of interest to the members of this Society to learn what they were. The views to which Professor Lodge's researches have conducted him are so very similar to mine, except on some purely

* [Read at the Montreal meeting of the Brit. Assn., Sept., 1884.]

speculative points, that it might be merely necessary to point out these points. But as my previous remarks on the subject do not extend to any great length, I may as well quote the article containing them. This will be advantageous, because, so far as the Volta-force is concerned, they are disconnected from the historical matter which forms so interesting a part of Professor Lodge's memoir, by publishing which, together with the valuable data he has been at so much trouble to collect, he has conferred a great benefit on the present generation of electricians. Being thus isolated, as well as from the subject of thermo-electricity, they may enable some readers who have not given much attention to the matter to more easily understand the plain course of the argument apart from the speculative points which are incapable of present verification. I shall add some additional remarks on the general subject, and wind up with a statement of what I consider to be the correct mathematical representation of the relations of impressed force and potential in condenser circuits, when there are impressed forces in the dielectrics as well as in the conductors—relations which are perhaps not yet fully recognised, if I may judge from some remarks of Professor Lodge's on the total unlikeness between the Volta-force conditions and an ordinary condenser investigation. But there is nothing revolutionary in the case; and when we make use of Maxwell's most admirable theory of dielectric currents and displacement, the case of impressed force in the dielectric becomes as easy to follow as when it is in the conducting part of a circuit. The theory of current in the dielectric I consider to be certainly true, however difficult its experimental verification may be (though far more difficult to disprove), and to be the natural outcome of Faraday's way of looking at things electrical.

I may premise, in the first place, that, like many others, I had been for many years profoundly dissatisfied with the paradoxical state into which electrical theory, in other respects so consistent all round, was thrown by Sir W. Thomson's conclusion, from his experiments with his wonderful electrometers, that old Volta was right, and that there could be no doubt the whole thing was simply chemical action at a distance. He proved decisively that the setting of two metals in contact did charge them like a condenser, and concluded that the E.M.F. was situated at the junction of the metals, founding upon this a notable calculation of the probable size of atoms by the manufacture of brass. Now, if it had been merely a question of explaining the Volta-force phenomenon, which, however interesting a matter, is not one of such paramount importance as to render other considerations secondary, it would have mattered little whether the localisation of the impressed force at the metallic junction were correct or not. But it involved the whole question of the relations of impressed force and potential, and the doing of work on or by the current. Sir W. Thomson's "Electrostatics and Magnetism," Maxwell's Treatise, and Professor Jenkin's "Electricity and Magnetism," all came out at about the same time. In the first, we had limited utterances on the subject, but Sir W. Thomson's paper on the "Size of Atoms" made his position clear. In the second, a remarkably complete theory of electricity was made indistinct in many places by the reservations concerning the potential of conductors, which was all the more surprising when it is remembered that Maxwell was no believer in the existence of the metallic-junction force. In the third was an account of previously unpublished experiments of Sir W. Thomson's, and a most decided statement of the theory, which asked us to believe that the situation of the E.M.F. of a voltaic cell is not in the cell at all, but outside it, although admittedly the energy, keeping up the current when the circuit is closed, is derived from the chemical actions going on in the cell itself. This paradox must have intensely puzzled all readers, save those who brought a strong faith into operation in the course of their scientific studies.

Now, if we ignore the Volta-force experiments altogether, the general theory of impressed force, potential, and the taking in or giving out of energy by the current, is clear and explicit, contains no paradoxes, and is in harmony with general dynamical principles. Was it really worth while to upset the theory because some very curious experiments were difficult of explanation? Certainly theory must ultimately be made to agree with facts; but when some few facts do not apparently fit into a theory which suits a much greater number of other facts, it becomes a question of balance of advantages whether it would be better to alter theoretical notions, or to leave the facts unexplained for the time, waiting for further information, or for new light on the question of fitting the facts into the theory. I think, in the present instance, considering the extraordinary character of the alteration of theory, that the best course would have been to let the experiments wait for an explanation. All the more so on account of the long time it takes for views taught by the leaders of scientific thought, and accepted by their followers, to be eliminated should they turn out to be erroneous. Besides that, it was not exactly a case of altering a theory to suit facts, but rather to suit certain conclusions from facts, which might or might not be correct. The fact is, that the air outside zinc is at a different potential from that in the air outside copper when the two metals are in contact. The conclusion, quite distinct from the fact, is that the difference of potential is produced by an impressed force at the metallic junction, which makes the zinc and copper be at different potentials, with the further result that the E.M.F. of a battery is outside it. Nor can we clear up the matter by defining the potential of a conductor as the potential somewhere else (itself a paradox, which I learn, with great surprise, has always been taught by Sir W. Thomson), namely, in the air outside it. This makes the zinc and copper be at different potentials, *because* the air potentials are different, and necessitates an impressed force at the junction of the metals. It is the same case, slightly differently expressed.

Such was the extent of my respect, almost amounting to veneration, for Sir W. Thomson's opinions, on account of his invaluable labours in science, inexhaustible fertility, and immense go, that I made the most strenuous efforts to understand the incomprehensible, impelled thereto also by a feeling that it might be prejudice on my own part that made it incomprehensible. But, failing to understand it, I finally gave it up, and evolved the views explained in the article hereafter quoted out of my own inner consciousness, and of course felt immensely relieved in my mind at once. That the application of the well-known heats of combination method to find what the differences of potential should be for different pairs of metals gave figures very poorly agreeing with the observed differences of potential, I did not attach much importance to, considering how the state of the surfaces might alter results, and the unknown value of the thermo-electric force, which may not be so small at a metal-gas contact as at a metal-metal junction, and the unknown influence of the nitrogen in the air, which may not be wholly inert in the matter. That there was some sort of a general kind of agreement was quite as much as could be expected. Moreover, if there were absolute disagreement, it would not, in my opinion, shift the seat of the E.M.F.'s from the air surfaces, but merely alter our views as to their cause.

I am inclined to confidently believe that the mere statement that the E.M.F. of a voltaic cell is *not* at the place where the energy transformation which keeps up the current occurs, is in itself sufficient, when rightly understood, to fully discredit any theory which necessitates that statement, when the matter is viewed generally from the modern dynamical stand-point. All the physical sciences are bound to become branches of dynamics in course of time, and anything contradicting the principles of dynamics should be unhesitatingly rejected. Without having made an exhaustive study of dynamics, I have yet managed to come to the conclusion that a force cannot act where it is not—meaning by acting, the doing of work. If the doing of work at one place involves the doing of work at another, the force doing the work at the second place is there, not at the first place. Of course there must be some connection between the two places whereby energy is transmitted between them, and wherever there is a transfer of energy going on there is force. All working forces involve transfer of energy, and the measure of a force, whether simple or generalised, is the amount transferred per unit change of the variable to which the force corresponds. In an isolated conservative system, all transfers of energy are internal, and its total energy remains constant. If such a system receives energy, this involves impressed force, and a system communicating the energy. What to consider as impressed force depends upon how large we make our system, and is therefore considerably a matter of choice. Thus, if there be two systems, each conservative in the absence of the other, but with a transfer of energy between them, and therefore impressed force on either system when it is considered by itself, and if we include the two in a single system, their mutual forces cease to be impressed, becoming internal. These reminders are merely to illustrate force considered as impressed. Add to the above, that forces which always involve loss of energy from the system, never a gain—namely forces of the frictional character depending on the velocities of the moving parts—are conveniently not reckoned as impressed, but as dissipative forces, reserving the term "impressed" for reversible actions, and we have a brief outline of the nature of the dynamical system which is represented, in a skeleton form, by the electromagnetic equations.

In the absence of conducting matter, the system would be conservative and keep its energy unchanged in amount, or only lose it by setting bodies in motion. But disregard this, and let there be no relative motions of masses permitted. The presence of conductors introduces dissipative forces proportional to velocities (strength of current being a generalised velocity), and the energy tends to become used up through the Joule heat of currents in conductors; excepting that part may be locked up, as it were, as when insulated conductors are electrified, or when there is intrinsic magnetisation. It does not then waste itself; otherwise there is continual waste, which must be compensated by impressed forces if the electric and magnetic energies are to be kept up. Their seat is in the ether. The actual constitution of the ether is unknown. It never can be *known*; but a constitution may be invented for it which shall admit of propagating heat and light and electromagnetic disturbances to produce observed results. If the ether is made to propagate light (say, by vibrations), and will not propagate electromagnetic disturbances, it cannot be of the right construction, and another must be found. The transfer of energy in any conductor (isotropic) takes place not with the current, but perpendicular thereto, as I showed in *The Electrician* for June 21, 1884,* thus being delivered into a wire from the dielectric without. This does not hold good in the dielectric itself, where it is perpendicular to the electric force, or nearly parallel to the wire, into which it is continuously wasting itself, keeping up the conduction current. Both cases are included in the statement that the transfer of energy takes place perpendicular to the electric and to the magnetic *forces* of the system. (See *The Electrician*,† January 10, 1885, *et seq.*) That this is the correct statement is verified by its holding good in cases of strain or crystalline structure when the current or displacement is not parallel to the electric force, nor the magnetic induction to the magnetic force.

I have no doubt that some day a tolerably simple constitution of the ether will be invented, making it do anything reasonable that is required of it, which may in course of time come to be believed in as a reality, as light is now believed to be propagated by transverse displacements. But taking things as they are, with an unknown constitution of the ether, the electromagnetic equations indicating a dynamical system, all actions on the system not included in the internal forces expressed by the equations must be impressed, and impressed somewhere. And impressed forces require to be very freely introduced, because, however complex the relations may appear to be in the electromagnetic equations on first acquaintance, the essential parts are fundamentally very simple, and the infinitely numerous minor actions which are not accounted for can only be made to show themselves, electrically speaking, by means of the use of impressed forces on the one hand, and auxiliary investigations on the other. Now, although when to simplify matters we ignore details (as

* ["Induction in Cores," § 19, p. 377 *ante*.]

† ["Electromagnetic Induction and its Propagation," Art. XXX. later, Sections ii. and iv.]

when we consider a linear electric circuit as a whole, its state being defined simply by the strength of current), we lose sight of the real energy transformations, and see only their result in the total, and cannot definitely say where energy is taken in or given out reversibly; yet if we, as we are justified in doing, take every portion of space by itself, we are bound at the same time to consider every impressed force to act upon the electromagnetic system at the very place where it is situated; and the sole conclusive direct evidence of there being an impressed force at a certain spot is in there being a reversible transformation of energy taking place there on the passage of an electric current; whilst we have indirect evidence, which may be equally conclusive, if we can show that there is no such action anywhere else.

If the Volta-force experiments were twenty times as difficult to explain as they have been considered to be, I do not see that there would be any sound reason for not concluding, or rather taking it for granted, quite apart from the Volta-force phenomena, that in a voltaic circuit, where we know that there is a transformation of energy going on, which accounts for the Joule heat in the circuit, the impressed force is exactly where an ignorant man would suppose it to be, namely, in the cell itself, although the exact distribution therein may be difficult to ascertain, owing to the complex nature of the actions. If it be not in the cell that energy is taken in by the current (to use an expression which should not be understood literally), but at an external junction, where there is no appreciable change occurring, it would follow that the energy of the chemical combination taking place in the cell did not result in an impressed force there, but first passed out of the battery to the junction, and was there taken in by the current. It must go to the junction first, to account for no change occurring there, and in the passage it must not act on the electromagnetic medium, for that would mean impressed force in the cell. But no one would wish to believe in this roundabout process.

Maxwell's formulæ for the distribution of electric and magnetic energy in space may not be correct. But if others were substituted, giving the same results in the sum, and consistent with the laws of induction, we should still have to put the impressed force in the cell, unless we assume action at a distance, without the intervention of a medium, or save appearances by having two mediums.

I now interpolate my remarks on the Volta-force and the voltaic cell, which appeared in *The Electrician* for February 2nd, 1884. It is necessary to say, to account for the very short manner in which the energy definition of impressed force is considered in the first part of the article, that it was one of a series, succeeding some on the subject of thermo-electricity, following therein Sir W. Thomson's beautiful thermodynamic theory, as applied to linear circuits. In this theory we have abundant illustration, in both the Peltier and the Thomson effects, of energy being taken in or given out by the current in any part of a circuit according as the current goes with or against the impressed force.

[Here followed the paper referred to, already given, Section xiii., Art. XXVI., p. 337 *ante*.]

After making the above statements I proceeded to make some more in the two following numbers of the series (March 1 and 31, 1884), on matters connected with contact force, in connection with the layers of electricity supposed, originally by Helmholtz, to accompany impressed forces, as well as on the relations of potential and impressed force. They are, together, too long for quoting here, but I think their perusal might be useful to some who have not already made up their minds on these matters. But I may quote an extract immediately illustrating the application of contact layers to the Volta-force experiments.

[Extracts from Section xv., Art. XXVI., p. 346 *ante*, followed here.]

I may add to this extract that I have no faith whatever in the existence of these layers of electricity, but must refer to the articles for particulars. I am sorry not to have Professor Lodge with me in this part of the matter, as the wide publicity he gives to his views adds force to his powerful advocacy. I may add that layers of electricity in connection with impressed force are apparently widely believed in, even by followers of Maxwell; possibly in the last case because he was not always true to his principles, putting, for example, free electricity in the interior of conductors, in defiance of his law of continuity of the current. There may be free electricity in conducting matter if it be dielectric as well, and heterogeneous (and all conductors may support elastic displacement to some extent), but its distribution will not by any means be the same as that supposed.

Practically these things matter very little, but theoretically they matter a great deal, as it is important to have theory as definite as possible, as well as consistent with itself.

To return to the Volta-force. Although we are agreed on all essential points, I cannot very well follow Professor Lodge's straining atoms, nor see their utility in the argument. This is because I know nothing about atoms. I cannot think that he knows much more. But, on the other hand, we do know something, however little, about the law of the electric current, that it "flows" in closed circuits, as if it were an incompressible liquid, and that in consequence there can be no current leaving a conductor, or displacement, if the impressed force act equally all round it, owing to the balance of the E.M.F.'s. The real interpretation of the quasi-incompressibility we do not know, but admitting it, there is no difficulty. We also know that chemical affinity, or tendency to chemical combination, is measurable in terms of E.M.F., so that, as there is chemical affinity between oxygen and zinc, and air contains oxygen, there must (irrespective of the argument based upon the absence of reversible effects at metal junctions) be an outward acting E.M.F. all over zinc alone in air, and therefore no current or displacement, as above. Electrically expressed, it is intelligible.

But when we, after Professor Lodge, put the electrical conditions in the background, and consider the oxygen atoms round a piece of zinc all straining at and trying to combine with it, we may well ask, why don't they do it? All the more should they do it if they are straining *all* round, unless they get wedged together, so that it is necessary to remove some of them to let the rest go nearer, possibly quite close enough to combine with the zinc. No, without such an irrational supposition we must fall back upon the electrical argument, and then we do not want the imagery of straining atoms, for it is intelligible by itself. If it were thermo-electric force, for example, it would be necessary to recast the imagery, whilst the electrical argument remains the same.

I am not objecting to the use of the imagination. That would be absurd; for most scientific progress is accomplished by the free use of the imagination (though not after the manner of professional poets and artists when they touch upon scientific questions). But when one, by the use of the imagination, has got to a definite result, and then sees a stricter way of getting it, it is perhaps as well to shift the ladder, if not to kick it down. For I find that practically, in reading scientific papers, in which fanciful arguments are much used, it gives one great trouble to eliminate the fancy and get at the real argument. Nothing is more useful than to be able to distinctly separate what one knows from what one only supposes.

To return again to the Volta-force. When we desire to go further and inquire what is the exact nature of the energy transformation that takes place when the E.M.F. is removed from a part of the zinc surface by contact, say, with copper, thus causing a current to pass in the circuit copper-zinc-air-copper, setting up a state of electric displacement in the dielectric, with a certain definite amount of potential energy depending on the capacity of the condenser which the arrangement forms, we enter a very difficult and speculative matter, whose solution, one way or another, will not alter the preceding. Take two flat plates of zinc and copper, for example, and put them close together, not touching. The moment they are connected the condenser becomes charged. Let U be the potential energy of displacement. This equals $\frac{1}{2}EQ$, if E be the difference of potential and Q the charge, which again $=ES$, if S be the capacity of the condenser. Now, at the same time as the potential energy, U, was set up, an equal amount of Joule heat was generated (an example of a general law, concerning which those interested in the matter may be referred to an article by me, now awaiting space in a forthcoming number of *The Electrician**). Thus we have $2U$ of work done. Besides this, if there be a copper-air force of strength x, so that $x+E$ is the zinc-air force, and if this force was present when the current passed, there must have been xQ of work done at the copper surface. Thus

$$\begin{aligned}(E+x)Q &= \text{work done by the zinc-air force,}\\ &= \tfrac{1}{2}EQ, \quad \text{electrostatic energy,}\\ &+ \tfrac{1}{2}EQ, \quad \text{Joule heat,}\\ &+ xQ, \quad \text{work done at the copper-air surface.}\end{aligned}$$

Now, as there is a loss of energy $\frac{1}{2}EQ$ necessarily, it cannot be a case

* [April 25, 1885. "Electromagnetic Induction," Art. XXX., Section vii., later.]

of mere pulling backwards and forwards of atoms, with conservation of energy.

But irrespective of this loss, I am strongly of opinion that there is not a mere yielding to a tendency to chemical combination, but an actual combination when the current passes—an actual minute amount of oxidation of the zinc, as expressed in the article above quoted.

To illustrate, start with a conductively closed voltaic circuit. It is admitted that the steady generation of heat in the circuit is derived from the energy of chemical combination in the cell. For we know that there is combination, and that the heat of combination is of the right amount. Now, suppose we suddenly insert a very large condenser in the circuit, so that the current is still kept up, although the circuit is interrupted, in the common language, the energy being now delivered into the dielectric of the condenser, as well as into the conductor, though at a decreasing rate, owing to the elastic displacement set up putting a gradual stop to it, till finally the current becomes insensible. Will it not be granted that during the whole time the decreasing current passed, the energy was derived from chemical combination, even down to the last dregs of current, including the weak current of apparent absorption? and that the same applies when we decrease the capacity of the condenser till at last we come down to a voltaic cell with two bits of copper wire attached to the plates? If not, where shall we draw the line between chemical combination on the passage of a current, the E.M.F. being measurable by the heat of combination, and merely a yielding to the tendency, with its necessary indefiniteness? And why should we draw any line?

Or, in another form, let the copper be in the cell first, then put in the zinc with attached wire, thus passing a minute quantity of electricity. Is it not due to chemical combination? If not, we are placed in the gratuitously difficult position that we must pass a certain quantity of electricity before any combination occurs at all. Now, I do not want to assert that electrical laws, holding good on a large scale, necessarily continue true in the same form on all smaller scales, however small. This would bring us to fractional parts of an atom at last. But in the experiment just mentioned, however feeble the effects may be, they must be still far elevated above atomic fractions, and I therefore see no reason for drawing the line.

Nor do I see any reason for drawing the line in the corresponding air experiment with copper and zinc. The air tends to oxidise the zinc, but cannot when the zinc is alone, for reasons intelligible when electrically expressed, otherwise indefinite, as before mentioned, there being a balance of E.M.F.'s. But destroy this balance, removing the cause that prevented combination occurring, by putting copper in contact with the zinc. I can see no reason why it should not occur, lasting till there is again electrical equilibrium. That the air battery and the voltaic cell are not exactly alike, air not being an electrolyte transporting ions delivering up their imaginary charges, I do not consider any objection. We do not need transport of matter. The dielectric carries the current, according to Maxwell's theory, and that is what is required. The action at the copper-air surface is, of course, very obscure, necessarily more so than in the case of the voltaic cell, where it is obscure enough.

If we examine according to the law of induction what occurs when zinc and copper are connected, we find that the disturbance commences at the place of contact. This is, however, merely an example of the general principle that when we alter the electrical conditions anywhere, causing a previously steady state to be upset, the disturbance commences at the place where the alteration was made. Thus, if a battery be on at one end of a submarine cable, with its farther end insulated, and we then put the latter to earth, the "signal" will travel *to* the battery. In the Volta-force case, suppose that, on making contact, we instantaneously remove the air from a circular patch of zinc, the copper touching it all over the patch. Then the disturbance will start from the circle bounding the patch, the zinc-copper-air line. This line is the first line of magnetic force. As it comes on, electric current flows round this line, infinitely close at first, partly through the metal and partly through the air, and the magnetic force spreads laterally, and with it the electric current. But practically we could not instantaneously make contact over such a patch, but would commence contact nearly at a point, which would be the first origin of the disturbance.

Now we know that some pressure is required to make a good contact. When it is light, it is microphonic, has considerable resistance, which is variable with the strength of current (when in a closed circuit with a battery), and there is really air between the supposed touching surfaces, through which the current passes. Have any experiments been made to ascertain the influence of such microphonic contacts on the magnitude of the Volta-force difference of potential? Does the difference of potential come on gradually with increasing pressure, or does it come on all at once at a certain stage in the operation? Further, I would ask, has mechanical work done in making the contact any concern in the matter, as in squeezing out air?

I conclude, for the reason given at the commencement of this paper, with the theory of impressed force in condenser circuits. Let there be three condensers, which is enough for generality.

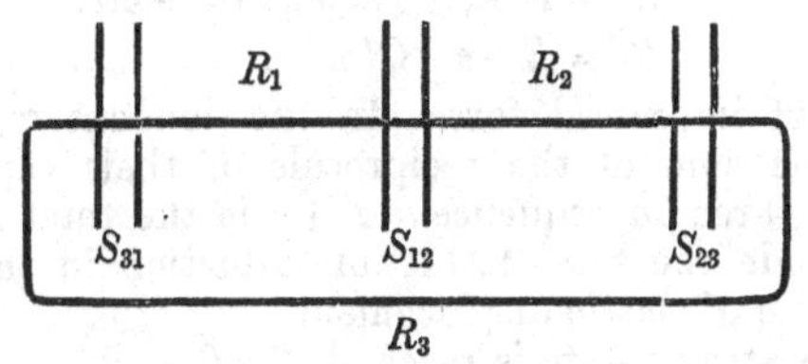

Let the positive direction in the circuit be clockwise, right to left below, left to right above. The three conductors have resistances R_1, R_2, R_3, and the capacities of the condensers are S_{31}, S_{12}, S_{23}, the numbers showing between which conductors they are placed.

Let the potentials on the left sides of the condensers be P_{31}, P_{12}, P_{23}, and the falls of potential in passing through them be p_{31}, p_{12}, p_{23}. Let

E_1, E_2, E_3 be the impressed forces in the conductors, and e_{31}, e_{12}, e_{23} those in the condensers. E_1 is the total impressed force in R_1, anyhow distributed; strictly it is the total of the impressed force resolved parallel to the length of the wire.

Let C be the current in the circuit and Q the common charge, supposing that we start with the condensers uncharged, so that Q is the integral displacement. Then, by Ohm's law applied to each conductor, we have

$$\left.\begin{array}{l} P_{31}-p_{31}-P_{12}+E_1=R_1C, \\ P_{12}-p_{12}-P_{23}+E_2=R_2C, \\ P_{23}-p_{23}-P_{31}+E_3=R_3C. \end{array}\right\} \quad \text{.....................(1)}$$

Adding these, we find

$$E-\Sigma p=RC, \quad \text{..............................(2)}$$

where R is the sum of the resistances, Σp the sum of the falls of potential through the dielectrics, and E the sum of the impressed forces in the conductors. From this we see that when the steady state of no current is reached, and $C=0$,

$$E=\Sigma p,$$

or it is the impressed forces in the conductors only, irrespective of what the forces may be in the condensers, that is opposed by the sum of the condenser differences of potential. If no total impressed force in the conductors, the sum of the falls of potential is zero. I have forgotten to mention that the E's and e's include the E.M.F.'s of induction, though, strictly speaking, we should confine the term "impressed force" to a force which is neither that of induction nor derived from difference of potential.

Now, also,

$$Q=S_{31}(e_{31}+p_{31})=S_{12}(e_{12}+p_{12})=S_{23}(e_{23}+p_{23}) \quad \text{..........(3)}$$

is the common charge, and

$$C=\dot{Q}=S_{31}(\dot{e}_{31}+\dot{p}_{31})=\text{etc.}; \quad \text{..................(4)}$$

or, when the impressed condenser forces are steady,

$$C=\dot{Q}=S_{31}\dot{p}_{31}=\text{etc.}$$

Put (3) in (2); then

$$RC=E-(Q-S_{31}e_{31})/S_{31}-\text{etc.}$$

Or,

$$RC=E+e-Q/S, \quad \text{..................................(5)}$$

where $e=$ sum of impressed forces in the condensers, and S is the reciprocal of the sum of the reciprocals of their capacities, or the capacity of the three in sequence. $E+e$ is the total impressed force (this must include the total E.M.F. of induction in the circuit), and Q/S the back force of elastic displacement.

And when the steady state is reached, $C=0$, and

$$Q=S(E+e),$$

or the common charge = total capacity × total impressed force (which now of course contains no E.M.F. of induction).

Q being known, by (3),

$$p_{12}=Q/S_{12}-e_{12}, \quad \text{etc.,}$$

give the falls of potential in the condensers, and (1) with $C=0$ gives the falls in the conductors. To go into the manner of fall, the exact distribution of impressed force is required to be known. It must exactly cancel the impressed force in the conductors; and in the condensers the impressed force per unit length + the fall of potential per unit length = $(4\pi/c) \times$ displacement per unit area, c being the specific capacity. The displacement per unit area is of course $Q \div$ area of the condenser considered.

The absolute value of the potential anywhere is left arbitrary, and it has no absolute value, not signifying any physical state, which is signified by the electric force. The potential is a quantity that by its variations gives an auxiliary distribution of force, which together with the impressed force, makes up a complete system of force to suit the continuity of electricity, Ohm's law and Maxwell's law to match. There is no electrification *in* the conductors, or *in* the dielectrics, however the impressed force varies in distribution. The only electrification is at the boundaries between the dielectrics where the displacement is elastic, and the conductors, where it is not. But we need not be misled by the term displacement to think there is anything displaced in the direction of displacement.

If a condenser contain no impressed force, its fall of potential is proportional to its elastance (reciprocal of capacity); if it contain impressed force, it is the sum of the impressed force and the difference of potential that is proportional to the elastance.

We may write

$$p_{12} = S\{(E+e)/S_{12} - e_{12}/S\};$$

so, if

$$S : S_{12} = e_{12} : e,$$

or the impressed force in any condenser is the same fraction of the total impressed force in all the condensers as the elastance of the condenser is of the total elastance, we have

$$p_{12} = ES/S_{12},$$

the fall of potential now depending on the impressed force in the conductors, being the same fraction as before mentioned of the total conductor impressed force. And if $E=0$, then $p_{12}=0$; that is, if the impressed forces be in the condensers only, and proportional in each to its elastance, there is no difference of potential in any part of the system if they be uniformly spread in each condenser; otherwise the conductors are all at the same potential, and the condensers appear uncharged, but there are variations in the condensers themselves.

By (2) and (3) the differential equation of the current is

$$C/S + R\dot{C} = \dot{E} + \dot{e},$$

or, adding the E.M.F. of induction separately, which is $-L\dot{C}$, if L be the inductance of the circuit,

$$C/S + R\dot{C} + L\ddot{C} = \dot{E} + \dot{e},$$

which, save in the presence of e, does not differ from the ordinary case of a coil and condenser. It is in the potential details that the cases of impressed force in the conductors and in the dielectrics require to be distinguished.*

In the Volta-force experiment (say, copper and zinc), we have a feeble thermo-electric force at the metallic junction, feeble thermo-electric forces in the zinc and copper if temperature varies, and two big forces, one of $x + \cdot 8$ volt, say, in a thin layer of the dielectric next the zinc surface, and one of x volt similarly at the other end of the dielectric. On charging, the rise of potential is very nearly as much at the zinc layer, with very nearly as much fall as before at the copper layer.

The theory of impressed force and potential in a dielectric is curiously illustrated by the phenomenon of absorption. The electric elasticity is not perfect; under the action of the stress the dielectric slowly yields, and with it a part of the displacement set up by an impressed force outside ceases to be of the elastic character, becoming intrinsic, and the difference of potential falls, requiring more current to enter to keep up the difference of potential.

The first discharge, like the first charge, is of elastic displacement. What is left, which shows no signs of being there at all, was elastic, but is no longer so. We may regard it as being kept up by uniformly distributed impressed force in the dielectric itself, arising from altered state of the dielectric produced by loss of elasticity. In time it recovers itself, or the impressed force is taken off, when the residual charge shows itself by the difference of potential it can now produce.

To really discharge the condenser at once, we must apply, after the first discharge, an opposite impressed force of the right amount, of course apparently charging the condenser oppositely to before. Leave it to itself, disconnected, and the apparent charge will gradually disappear.†

Residual magnetisation in soft iron is somewhat analogous, but the effect is of far greater magnitude, and there is permanent set as well, which becomes predominant in intrinsic steel magnets. But we can set up permanent set of displacement also in a dielectric, as by passing a current through warm glass and then cooling it. It is then like a permanent magnet.

If we had conductors for magnetic induction (analogous to electric conductors), we, by magnetising a plate of iron, setting up residual magnetisation, could apparently discharge it so as to show no force outside. It would then be like the charged condenser (in which "absorption" has occurred) after its apparent discharge.

* [The above investigation may be compared with that on p. 376 *ante*, relating to condensers in sequence, subjected to a simple-harmonic impressed force.]

† [See Sections x. and xii. of the next Art., XXX.]

XXX. ELECTROMAGNETIC INDUCTION AND ITS PROPAGATION.

[*The Electrician*, 1885-6-7. Section I., Jan. 3, 1885, p. 148; II., Jan. 10, p. 178; III., Jan. 24, p. 219; IV., Feb. 21, p. 306; V., March 14, p. 366; VI., April 4, p. 430; VII., April 25, p. 490; VIII., May 15, p. 6 (vol. 15); IX., June 12, p. 73; X., July 3, p. 134; XI., July 17, p. 170; XII., August 7, p. 230; XIII., August 21, p. 270; XIV., August 28, p. 290; XV., September 4, p. 301; XVI., October 9, p. 408; XVII., Nov. 13, p. 6 (vol. 16); XVIII., Nov. 27, p. 46; XIX., Dec. 11, p. 86; XX., Dec. 18, 1885, p. 106; XXI., Jan. 1, 1886, p. 146; XXII., Jan. 15, p. 186; XXIII., Jan. 22, p. 206; XXIV., March 26, p. 386. The second half of this article is in Vol. II., with the references thereto.]

SECTION I. ROUGH SKETCH OF MAXWELL'S THEORY.

Conductivity, Capacity, and Permeability.

In the electromagnetic scheme of Maxwell there are recognised to be three distinct properties of a body considered with reference to electric force and magnetic force, viz., conductivity, electrostatic capacity, and magnetic permeability. The body may support a conduction current, it may support electric displacement, and it may support magnetic induction. These three phenomena may, and in general do, coexist at any one point. Quantitatively considered, they are all vector magnitudes, having definite directions as well as strengths, which are reckoned per unit area perpendicular to their directions in terms of chosen units.

The facility of supporting the three states of conduction current, electric displacement and magnetic induction varying with the nature of the medium for equal amounts of energy concerned, brings in three coefficients, the electric conductivity k, the electric capacity c, and the magnetic permeability μ. At first sight it might appear as if three other vector magnitudes related to the former by these coefficients were involved; but in reality there are but two, the electric force and the magnetic force, the former being connected with both the conduction current and the displacement.

First we have Ohm's law. **C** being the conduction current-density, **E** the electric force, and k the specific conductivity,

$$\mathbf{C} = k\mathbf{E}. \quad \text{.........(Conduction current)} \quad (1)$$

Far more is known about conductivity than about capacity or permeability. In an unstrained isotropic metal, k appears to depend on the temperature only, and not to vary rapidly with it. That is, k is practically a constant, which simplicity is of great utility. Within wide limits k is independent of the current or the electric force.

The range of conductivity in different media is very great. From the conductivity of copper to that of cold glass is such an enormous range as to compare with astronomical ratios, and it speaks well for electrical science that it can compare definitely such widely differing magnitudes.

Dry air in its ordinary state appears to have no conductivity. But it is a vacuum that is the perfect non-conductor in Maxwell's theory. Where there is no matter, in the ordinary sense, there is no dissipation of energy; and ether, whatever it be, is perfectly conservative and non-dissipative, dynamically considered. Dissipation of energy is a necessary accompaniment of a conduction current, so far as is known; though of course a perfect conductor can be imagined in which a continuous current developed no heat. But ether cannot be this perfect conductor consistently with the propagation of magnetic disturbances, for none can be propagated in a perfect conductor. Grant that they are propagated in pure ether (space from which all "matter" has been removed) without loss of energy in the medium, and it follows that ether is the perfect non-conductor. This however, somewhat anticipates electric displacement and magnetic induction.

Equation (1) is a vector equation. In an isotropic medium k is a scalar constant. We may symbolise **E**, **C**, or other physical vector magnitudes by geometrical vectors, lines drawn of the proper lengths and in the proper directions. Thus **E** is one vector, **C** is another, and when, as ordinarily, k is a scalar constant, (1) says simply that **C** and **E** are parallel, and that **C** is k times as long as **E**. Vector quantities are compounded like velocities; in a vector equation containing n vectors, separated by + or − signs, the n vectors form the n sides of a polygon. But two straight lines cannot enclose a space, so, in equation (1), **C** and k**E** are parallel and equal.

But in a body eolotropic as regards conductivity, **C** and **E** are only exceptionally parallel. Using the same equation (1) to represent the relation between them, k, from being a scalar constant, becomes a linear operator; k**E** must be regarded as a single symbol, being **E** operated upon by k in a certain manner, turning it into a new vector k**E**. The operation is a little complex when expressed in Cartesian co-ordinates referred to any axes, so it is better to define once for all the meaning to be attached to k when eolotropy is to be included, and then use equation (1), rather than be repeating the Cartesian operations over and over again. The following defines the operation k, and the same will serve for c and μ later. First let there be no rotatory power. Then, in three directions, mutually perpendicular, fixed in a body at the point considered, depending on its structure there, Ohm's law, as ordinarily considered, is obeyed. That is to say, if electric forces $\mathbf{E}_1$, $\mathbf{E}_2$, $\mathbf{E}_3$ act successively parallel to the above mentioned directions of the principal axes of conductivity, and $\mathbf{C}_1$, $\mathbf{C}_2$, $\mathbf{C}_3$ be the corresponding currents, $\mathbf{C}_1$ will be parallel to $\mathbf{E}_1$, $\mathbf{C}_2$ to $\mathbf{E}_2$, and $\mathbf{C}_3$ to $\mathbf{E}_3$, and we shall have

$$\mathbf{C}_1 = k_1\mathbf{E}_1, \quad \mathbf{C}_2 = k_2\mathbf{E}_2, \quad \mathbf{C}_3 = k_3\mathbf{E}_3,$$

and

$$C_1 = k_1E_1, \quad C_2 = k_2E_2, \quad C_3 = k_3E_3, \quad \text{.................}(2)$$

where k_1, k_2, k_3 are scalar constants, being the principal conductivities, and C_1 is the tensor or magnitude of $\mathbf{C}_1$, C_2 of $\mathbf{C}_2$, etc. From these we may find the current when the force acts in any other direction than parallel to one of the principal axes. For if **E** be the force, let its components parallel to the axes be $\mathbf{E}_1$, $\mathbf{E}_2$, $\mathbf{E}_3$; the components of the current will then be $\mathbf{C}_1$, $\mathbf{C}_2$, $\mathbf{C}_3$, as defined by (2). Compounding them, we get $\mathbf{C}$. Thus the relation of $\mathbf{C}$ to $\mathbf{E}$ requires a knowledge of the principal conductivities and the directions of the principal axes.

But should the body possess rotatory power, the above process is incomplete. Let ϵ be a vector, directed parallel to the conductivity axis of rotation, and of length properly chosen; then, to the current as found by the above process must be added another current expressed by

$$\mathrm{V}\epsilon\mathbf{E}, \quad \text{..............(Vector product)} \quad (3)$$

which stands for a vector whose direction is perpendicular to the plane containing ϵ and $\mathbf{E}$, and whose length equals the product of the length of ϵ into that of $\mathbf{E}$, into the sine of the angle between their directions. This also defines the prefix V before two vectors. The + direction is defined thus. Let ϵ and $\mathbf{E}$ be the short and the long hands of a watch. Let ϵ point to XII. and $\mathbf{E}$ anywhere else. The angle between ϵ and $\mathbf{E}$ is measured positive in the usual direction of motion of the hands, and the direction of $\mathrm{V}\epsilon\mathbf{E}$ when positive is from the face to the back.

It is possible, consistent with the linear principle, for k_1, k_2, k_3 to be all zero, and ϵ not zero. Then

$$\mathbf{C} = \mathrm{V}\epsilon\mathbf{E}$$

simply; the current is always perpendicular to the force, of maximum strength when ϵ and $\mathbf{E}$ are perpendicular, and vanishing when they are parallel.

Returning to equation (1), multiply it by $\mathbf{E}$. Then

$$\mathbf{EC} = \mathbf{E}k\mathbf{E} = Q, \text{ say. } \quad \text{......(Dissipativity)} \quad (4)$$

Q is the dissipativity per unit volume. It is, in the first place, the rate of working of the force $\mathbf{E}$, and next, by the experimental law of Joule, the rate of generation of heat per unit volume. (4) is scalar equation. All our equations will be either wholly scalar or wholly vector. In case of isotropy, with k a scalar constant, we may write

$$Q = kE^2;$$

or, since

$$E = k^{-1}C,$$

$$Q = k^{-1}C^2,$$

where k^{-1} is the specific resistance, a more familiar form of Joule's law. But in general, when k is a linear operator, we must not take $\mathbf{E}k\mathbf{E} = kE^2$, unless $\mathbf{E}$ act parallel to one of the principal axes, when we may do so, with the appropriate value of k for that axis. When $\mathbf{E}$ and $\mathbf{C}$ are not parallel, the product $\mathbf{EC}$ means the strength of $\mathbf{E}$ multiplied by that of $\mathbf{C}$, and by the cosine of the angle between their directions; which of course includes the common algebraic meaning of $\mathbf{EC}$, since when $\mathbf{E}$ and $\mathbf{C}$ *are* parallel, $\cos 0° = 1$. Referred to three rectangular axes, if E_1, E_2, E_3 are the scalar components of $\mathbf{E}$, and C_1, C_2, C_3 those of $\mathbf{C}$, then

$$\mathbf{EC} = E_1C_1 + E_2C_2 + E_3C_3, \quad \text{.....(Scalar product)} \quad (5)$$

which is an equivalent definition of $\mathbf{EC}$.

Coming next to specific capacity, although there are media, as air, which appear to have no conductivity, yet, by the continuity of the electric current, they can support current; not steady, but transient, and stopped elastically. By an obvious mechanical analogy the integral current is termed the electric displacement. Let this be **D**, and let **E**, as before, be the electric force. We have then

$$\mathbf{D} = c\mathbf{E}/4\pi. \quad \text{.....(Electric displacement)} \quad (6)$$

The excrescence 4π is a mere question of units, and need not be discussed here. The 4π's are particularly obnoxious and misleading in the theory of magnetism. Privately I use units which get rid of them completely, and then, for publication, liberally season with 4π's to suit the taste of B.A. unit-fed readers. Of course, if it comes to numerical comparisons we should have to consider the ratios of units in the ordinary to what I may call the rational system. Sometimes it is $\sqrt{4\pi}$, sometimes $(4\pi)^{-\frac{1}{2}}$, sometimes 4π, sometimes unity, but in the mere algebra it is simply a matter of putting in 4π's here and there in translating from rational to ordinary units. [*See* pp. 199, 262.]

In a dielectric medium, the force and the displacement are simultaneous, like the force and the current in a conductor. Time does not appear in the equations. In an isotropic dielectric, c is simply a scalar constant; in an eolotropic dielectric it is, as described above for k, a linear vector operator, with this difference, however, that there is no rotatory vector ϵ, so that the relation of **D** to **E** is settled by the values of the principal capacities, and their axes.

Multiply (6) by $\frac{1}{2}\mathbf{E}$; then

$$\tfrac{1}{2}\mathbf{ED} = \mathbf{E}c\mathbf{E}/8\pi = U, \text{ say.} \quad \text{(Electric energy)} \quad (7)$$

U is the electric energy per unit volume, the work done by the force on the displacement as they rise from 0 to their final values, or the final displacement multiplied by the mean force which produced it. This energy is stored, and is recoverable in work like the energy of a perfectly elastic strained spring. It is unnecessary to assume that there is any real displacement of anything in the direction of the electric displacement. All the electric and magnetic quantities are more or less abstractions, measurable abstractions, whose real signification is as yet unknown.

Far less is known of c than of k, and it is not so agreeably definite as k. Solid dielectrics appear to have imperfect electric elasticity, as they have imperfect mechanical elasticity. The bent spring, with the applied force removed, and brought quietly to rest, is not exactly in its equilibrium position. A small part of the displacement remains, and slowly disappears. This is easily shown when not visible to the eye by using a microphonic contact; though, by the way, the variability of the contact itself makes it a bad method. Most likely there is no such thing as a perfect return even with small displacements; we cannot draw a hard and fast line to mark the limit of perfect elasticity.

All non-conductors are dielectrics. Bad conductors are also dielectrics. Good conductors, even the best, may be dielectrics as well, so that with a force **E** we shall have a conduction current $k\mathbf{E}$ and a displacement $(4\pi)^{-1}c\mathbf{E}$ co-existing. But in such case, as well as in the case of known dielectric power of bad conductors, $k\mathbf{E}$ is not the complete or true current, unless the displacement remains steady. The time-variation of the displacement is itself an electric current, and the true current is the sum of the conduction current and of the rate of increase of the displacement. Let Γ be the true current; we then have, in a conducting dielectric, or dielectric conductor,

$$\mathbf{C} = k\mathbf{E}, \qquad \mathbf{D} = c\mathbf{E}/4\pi,$$

$$\Gamma = \mathbf{C} + \dot{\mathbf{D}} = k\mathbf{E} + c\dot{\mathbf{E}}/4\pi. \quad \text{......(True current)} \quad (8)$$

Put $c=0$ in a pure conductor, and $k=0$ in a pure dielectric. It is the true current that is "the current" when we come to induction and variable states.

In the equation $\Gamma = \mathbf{C} + \dot{\mathbf{D}}$ we have three vectors. They form the three sides of a triangle, unless $\dot{\mathbf{D}}$ should be parallel to $\mathbf{C}$. But $\dot{\mathbf{D}}$ may not be parallel to $\mathbf{C}$, nor need it be parallel to $\mathbf{D}$. If we charge a condenser formed of two large flat opposed conductors very close together, the displacement current, when setting up the displacement, is, by general reasoning, parallel to the displacement—at least away from the edges. But this is not invariable. When charged conductors are discharged, the displacement current does not in general follow the tubes of displacement. To do so would require instantaneous propagation of the disturbances to infinite distances. The displacement current may be perpendicular to the displacement—viz., when the displacement at a certain place changes its direction without changing in amount.

Multiply (8) by $\mathbf{E}$; then

$$\mathbf{E}\Gamma = \mathbf{E}k\mathbf{E} + \mathbf{E}c\dot{\mathbf{E}}/4\pi = Q + \dot{U}. \quad \text{....................}(9)$$

The rate of working of the force is accounted for partly in heating (Q per second), and partly in the increase in the energy U of the displacement. (Equations (4) and (7)). The first is lost from the system, the latter is stored.

Whilst conductivity depends on the presence of matter, the existence of capacity is independent of matter, though modified in amount by its presence. That is, capacity is a function of the ether, which is the standard dielectric medium of least capacity. Ether is a very wonderful thing. It may exist only in the imaginations of the wise, being invented and endowed with properties to suit their hypotheses; but we cannot do without it. How is energy to be transmitted through space without a medium? Yet, on the other hand, gravity appears to be independent of time. Perhaps this is an illusion. But admitting the ether to propagate gravity instantaneously, it must have wonderful properties, unlike anything we know.

Coming next to permeability, all bodies sustain magnetic induction, and most of them to nearly the same degree. $\mathbf{H}$ being the magnetic force, $\mathbf{B}$ the induction, and μ the permeability,

$$\mathbf{B} = \mu\mathbf{H}. \quad \text{.........(Magnetic induction)} \quad (10)$$

μ is taken as unity in ether (in the "electromagnetic" system of units), and is either a little greater or a little less in most bodies. But in some bodies it, very singularly, runs up to large numbers. Iron is the principal offender; then come nickel and cobalt, minor magnetics, but far removed from the crowd of almost unmagnetisable substances. Fe = 56, Ni and Co about 58·5. What can it be?

The linear connection between **H** and **B** is very unsatisfactory. Not merely does μ vary with the temperature, and enormously from one piece of iron to another, being, with moderate strength of magnetic force, largest in the softest iron and smallest in hard steel, but it varies with the magnetic force, first increasing with the force, and then, more importantly, decreasing greatly; how far down is unknown. To make matters worse, part of the induction produced by applied magnetising force becomes fixed, for the time, remaining after the removal of the force. Thus the linear connection between **H** and **B** must be taken with salt. But within moderate limits, and excluding permanent magnetisation, which requires separate consideration, μ in equation (10) may be taken to be, like k and c before, a scalar constant in case of isotropy, and a linear vector operator in eolotropic media, being then, like c, self-conjugate, or without the rotatory power.

μ in soft iron is said to run up to 5,000 or 10,000 (Rowland's experiments. I forget the exact figures). But in general it is very far lower than these tremendous figures. From experiments on the retardation of coils made some years ago, including straight solenoids, I concluded that μ = from 50 to 200 was safe, [for small forces].

Not **B**, but $\mathbf{B}/4\pi$ should be the magnetic induction to compare with **D**, the electric induction, or displacement. So, dividing (11) by 4π, and then multiplying by $\frac{1}{2}\mathbf{H}$, we have

$$\tfrac{1}{2}\mathbf{HB}/4\pi = \mathbf{H}\mu\mathbf{H}/8\pi = T, \text{ say. (Magnetic energy)} \quad (11)$$

T is the energy of the magnetic induction per unit volume, when wholly induced, and acting conservatively, [within the elastic limits].

SECTION II. ON THE TRANSMISSION OF ENERGY THROUGH WIRES BY THE ELECTRIC CURRENT.

Consider the electric current, how it flows. From London to Manchester, Edinburgh, Glasgow, and hundreds of other places, day and night, are sent with great velocity, in rapid succession, backwards and forwards, electric currents, to effect mechanical motions at a distance, and thus serve the material interests of man.

By the way, is there such a thing as an electric current? Not that it is intended to cast any doubt upon the existence of a phenomenon so called; but is it a current—that is, something moving through a wire? Now, although nothing but very careful inculcation at a tender age, continued unremittingly up to maturity, of the doctrine of the materiality of electricity, and its motion from place to place, would have made me believe it, still, there is so much in electric phenomena to support the idea of electricity being a distinct entity, and the force of habit is so great, that it is not easy to get rid of the idea when once it has been formed. In the historical development of the science, static phenomena came first. In them the apparent individuality of electricity, in the form of charges upon conductors, is most distinctly indicated. The fluids may be childish notions, appropriate to the infancy of science; but still electric charges are easily imaginable to be quantities of a something, though not matter, which can be carried about from place to place. In the most natural manner possible, when dynamic electricity came under investigation, the static ideas were transferred to the electric current, which became the actual motion of electricity through a wire. This has reached its fullest development in the hands of the German philosophers, from Weber to Clausius, resulting in ingenious explanations of electric phenomena based upon forces acting at a distance between moving or fixed individual elements of electricity. It so happened that my first acquaintance with electricity was with the dynamic phenomena, and after I had read with absorbed interest that instructive book, Tyndall's "Heat as a Mode of Motion." This may explain why, when it came later to book-learning regarding electricity, I had the greatest possible repugnance to all the explanations, and could not accept the electric current to be the motion of electricity (static) through a wire, but thought it something quite different. I simply did not believe, except so far as mere statements of experimental facts were concerned. This had its disadvantages; one can get on faster if one has sufficient faith—which we know moves mountains—to accept a certain hypothesis unhesitatingly as a fact, and work out its consequences undoubtingly, regardless of the danger of fixing one's ideas prematurely.

As Maxwell remarked, we know nothing about the velocity of electricity; it may be an inch in a year or a million miles in a second. Following this up, it may be nothing at all. In fact, it is only on the hypothesis that the electric current is something moving, a definite quantity in a given space, that we can entertain the idea of its possessing velocity. Then, the product of its hypothetical density into its velocity is the measure of the current; but, being a mere hypothesis, unless we chose to accept it, to talk of the velocity of electricity in the electric current becomes meaningless. On the other hand, when we apply the ideas of abstract dynamics to electricity, and compare the electric current to a velocity, it is not the above supposititious velocity of electricity that is referred to in any way. It has no meaning now. It is the supposed velocity of electricity in the electric current; whereas, in the dynamical theory, it is the electric current itself that is a velocity, in the generalized sense, with the electromotive force as the generalized force; so that force × velocity = activity. In only one sense do I think we can speak of the velocity of electricity, consistent with Maxwell's theory, viz., by the hypothesis that the electric current in a wire is the continuous discharge of contiguous charged molecules, when plainly we can call the velocity of motion of a molecule the velocity of the charge it carries. As between the molecules we have the electric medium the ether, this view of the conduction current ultimately resolves itself into "displacement" currents in a dielectric.

But is there not the fact that we can send a current into a long circuit, and that it plainly travels along the wire, taking some time to arrive at the other end? Does that not show that electricity travels through the wire? To this I should have answered formerly, when filled with "Heat as a Mode of Motion," that it is a fact that there is a transformation of energy in the battery, and that this energy is transmitted through the wire, there suffering another transformation, viz., into heat; that when the current is set up steadily, the heat is generated uniformly; that the electric current in the wire is therefore some kind of stationary motion of the particles of the wire, not exactly like heat, but having some peculiarity of a directional nature making the difference between a positive and a negative current; but that there was no evidence in the closed circuit of any motion of electricity through the wire, but only of a transfer of energy through the wire.

However, leaving personal details of no importance to anyone but myself, let us consider the transmission of energy through a wire. To fix ideas, let our circuit be an insulated suspended wire from London to Edinburgh, and that we transmit energy to Edinburgh from a battery in London, the circuit being completed through the earth. Let the current be kept on. In the first place the phenomenon is steady. It does not change with the time. Next we find that the magnetic force about the wire is the same everywhere at the same distance, or the wire is in the same condition as regards the magnetic induction outside it, and when we apply our knowledge to the interior of the wire, regarded as a bundle of smaller wires, we find that the magnetic force *in* the wire does not vary along its length. Again, heat is being generated within the wire at a uniform rate (a part of the steadiness above mentioned), and next, this phenomenon is also the same all along the wire. Heat is undoubtedly a kinetic phenomenon, hence the electric current is also, at least in part, a kinetic phenomenon. The electric current is not itself heat; but as its existence in the wire involves the continued production of heat, we conclude that some kind of motion is necessarily involved in the electric current apart from the heat produced, and from the uniformity of effect in different parts of the wire, that it is a kind of stationary motion. Again, the electric force is the same all through the wire. There seems no difference between one part and another. Outside the wire, in the dielectric, however, there is a difference, for the electric force varies not only at different distances from the wire but also at the same distance outside different parts of the wire. (We disregard here all irregularities due to other conductors and currents.)

Passing to the battery, the complexity of conditions makes it more difficult to follow, though the state of electric force and magnetic force and heat generation is reducible to the same, and may be made identically the same as in the wire by properly choosing its shape, etc. But in the battery there is a very remarkable thing taking place, namely, the loss of chemical energy at a steady rate; and in the system generally, a still more remarkable thing, an exactly equivalent steady gain of heat. Heat that might have been produced on the spot by the chemical action, otherwise conducted, appears all over the circuit. How does it get there? The natural answer is, through the wire. But to get to the further parts of the wire it must go through the nearer, hence there must be what we may call an energy-current, which, in the wire, at a given place, would be the rate of transfer of energy through a cross section there. Now, which way is the energy-current directed? It would seem only fair to let it go both ways equally from the battery. Let it be so first. Then there is an energy-current entering the wire, equal to one-half the dissipativity, which falls in strength regularly up to the middle of the wire, where it is zero. It falls in strength on account of the heat generation. Similarly the other energy-current goes through the earth to Edinburgh almost unabated in strength, and is then directed from Edinburgh to the middle of the wire, where its strength also falls to nothing. This seems absurd. Then let the energy-current be directed one way only, say with the positive current. If the positive pole of the battery is to line, we have an energy-current in one direction all round the circuit, London to Edinburgh, and back through earth. If of maximum strength at the battery it falls nearly to nothing at the distant end, and quite to nothing through the earth up to the other pole of the battery. But we have no data whatever to fix whereabouts the place of maximum energy-current is. It requires a second assumption. The reader may similarly consider the effect of reversing the battery, or of making the energy-current be directed with the negative current. There is no getting at anything definite, except that the energy-current must vary very widely, though regularly, in strength, whilst there is nothing to fix which way it is directed, or where the maximum strength is. Again, the energy-current is a kinetic phenomenon, and as it varies so widely in different parts, we might expect different parts of the wire itself to be in different electrical states, which is exactly what we do not do; for though its potential varies, yet potential is not a physical state, but a mere scientific concept.

Had we not better give up the idea that energy is transmitted through the wire altogether? That is the plain course. The energy from the battery neither goes through the wire one way nor the other. Nor is it standing still. The transmission takes place entirely through the dielectric. What, then, is the wire? It is the sink into which the energy is poured from the dielectric and there wasted, passing from the electrical system altogether. All [the above mentioned] difficulties now disappear.

That the energy of the battery passes into heat immediately would require its instantaneous transmission to all parts of the wire, which cannot be entertained. There must be an intermediate state or states, after leaving the battery and before becoming heat. And there must be a definite amount of energy in transit at a given moment; in the steady state this must be of constant amount, just as the total rate of transmission is of constant amount. We must not, however, individualize particular elements of energy, and follow their motions, but regard the matter quantitatively only. The energy in transit may be compared to the energy of a machine which is transmitting motion; if done at a steady rate, it remains constant and definite, and the rate of transmission is definite.

Now, in Maxwell's theory there is the potential energy of the displacement produced in the dielectric parts by the electric force, and there is the kinetic or magnetic energy of the magnetic induction due to the magnetic force in all parts of the field, including the conducting parts. They are supposed to be set up by the current in the wire. We reverse this; the current in the wire is set up by the energy transmitted through the medium around it. The sum of the electric and magnetic energies is the energy of the electric machinery which is transmitting energy from the battery to the wire. It is definite in amount, and the rate of transmission of energy (total) is also definite in amount.

It becomes important to find the paths along which the energy is being transmitted. First define the energy-current at a point to be the amount of energy transferred in unit time across unit area perpendicular to the direction of transmission. As the present section is argumentative and descriptive only, we cannot enter into mathematical details further than to say that if **H** be the vector magnetic force, and **E** the vector electric force, not counting impressed forces, the energy-current, as above defined, is $\mathbf{VEH}/4\pi$ (see equation (3) for definition of V). This is true universally, irrespective of the nature of the medium as to conductivity, capacity, and permeability, or as to eolotropy or isotropy, and true in transient as well as in steady states. A line of energy-current is perpendicular to the electric and the magnetic force, and is a line of pressure. We here give a few general notions.

Return to our wire from London to Edinburgh with a steady current from the battery in London. The energy is poured out of the battery *sideways* into the dielectric at a steady rate. Divide into tubes bounded by lines of energy-current. They pursue in general solenoidal paths in the dielectric, and terminate in the conductor. The amount of energy entering a given length of the conductor is the same wherever that length may be situated. The lines of energy-current are the intersections of the magnetic and electric equipotential surfaces. Most of the energy is transmitted parallel to the wire nearly, with a slight slant towards the wire in the direction of propagation; thus the lines of energy-current meet the wire very obliquely. But some of the outer tubes go out into space to an immense distance, especially those which terminate on the further end of the wire. Others pass between the wire and the earth, but none in the earth itself from London to Edinburgh, or *vice versâ*, although there is a small amount of energy entering the earth straight downwards, especially at the earth "plates." If there is an instrument in circuit at Edinburgh, it is worked by energy that has travelled wholly through the dielectric, then finding its way into the instrument, where it enters the coil and is there dissipated, or else used up by the visible motions it effects in moving parts of the instrument; which, however, is a different kind of affair from dissipation, as it involves impressed force.

Now, go into the line-wire. A tube of energy-current arriving at the surface of the wire by a long slant, at once turns round and goes straight to the axis. In passing from the battery to the wire through the dielectric the energy-current is continuous, the state being steady (or the ether machinery frictionless); but directly it reaches the conducting matter of the wire dissipation commences and the current begins to fall in strength, and on reaching the axis has fallen to nothing. Not a fraction of an erg is transmitted along the wire. Some small part of the energy leaving the battery may enter it again, but most of the dissipation in the battery itself is accounted for by the weakening of strength in tubes which are on their way to leave the battery.

Put the battery in the middle of the line; earth at both ends. Now, one half of the energy-current tubes leaving the battery sideways turn round to one section of the line, the other half to the other section. Otherwise the case is similar to the last.

When we have a double wire looped without earth, and battery at one end, most of the energy is transmitted between the wires.

In a circular circuit, with the battery at one end of a diameter, its other end is the neutral point; the lines of energy-current are distributed symmetrically with respect to the diameter.

On closing the battery circuit there is an immediate rush of energy into the dielectric, and, at the first moment, into all bodies in the neighbourhood of the battery, and wasted there in induced currents according to their conductivity. In the variable state the tubes of energy-current are themselves in motion. It takes some time to set the electric machinery going steadily. Also the energy-current is not continuous in the dielectric, for the potential energy of displacement and the magnetic energy have to be supplied at every place. But, in the end, the energy-current becomes continuous in the dielectric, goes round an external conductor instead of entering it, as it would do in the transient state, and finally reaches the conductor to which the battery is connected, penetrating which it terminates.

If we neglect the magnetic energy, as in Sir W. Thomson's original telegraph theory, against the energy of electric displacement, we can easily get a general idea of the setting up of the permanent state in a long suspended wire; a submarine cable is more complex on account of the sheath. The energy reaches the beginning of the wire first, and only reaches the end, save insignificantly, later on. But the theory indicates instantaneous setting up of current at the far end, though not in recognisable amount. This result follows from the neglect of the magnetic energy. In a dielectric medium the velocity of undisturbed propagation is $(c\mu)^{-\frac{1}{2}}$; where c is the capacity, and μ the permeability; that the magnetic energy $=0$ is equivalent to assuming $\mu=0$ everywhere, whence instantaneous transmission. The "retardation," however, arises from the setting up of the potential energy of displacement. But, strictly speaking, we must not neglect μ. It is, then, not so easy to follow the transient state without simplifications. There is an oscillatory phenomenon in the dielectric, a to-and-fro transmission of energy and pressure parallel to the wire all round it with a velocity whose possible maximum is that of undisturbed transmission. This is modified as it progresses by dissipation in the wire, and so gets wiped out. This usually occurs so rapidly that the waves are of importance only at the battery end of a long wire. The electric machinery must have mass, as well as elasticity, by reason of this phenomenon, since there is reason to believe (from Maxwell's theory of light) that it is not the air, but something between the air molecules that is the electromagnetic medium, the air merely modifying the phenomena somewhat.

In the state of steady current through a submarine cable, with an iron sheath outside the dielectric, the energy is transmitted wholly through the gutta percha or other suitable insulator (neglecting the small amount going to earth), thus going nearly parallel to the wire, practically quite parallel, except as regards the lines near the wire itself, as they all eventually meet the wire. There is no transmission in the sheath lengthwise, though there is dissipation there if it should contain, as it does sometimes, part of the return current. In the transient state there is, of course, always dissipation in the sheath more or less, besides the loss of energy to magnetise it.

Now to speak more generally. In the steady state of current due to any impressed forces, the tubes of energy-current start sideways from the places of impressed force, where energy is supplied to the electric system, and travel through definite paths, without loss in dielectric, with loss in conducting parts, to terminate finally in conducting matter; or else they may go from one place of impressed force to another with or without dissipation on the way when the current is with the impressed force at one source, and against it at the other. But with special arrangements (solenoidal) of impressed force, there is no transmission of energy in the steady state.

Since on starting a current the energy reaches the wire from the medium without, it may be expected that the electric current in the wire is first set up in the outer part, and takes time to penetrate to the middle. This I have verified by investigating some special cases.

Increase the conductivity of a wire enormously, still keeping it finite, however. Let it, for instance, take minutes to set up current at the axis. Then ordinary rapid signalling "through the wire" would be accompanied by a surface-current only, penetrating to but a small depth. The disturbance is then propagated parallel to the wire in the manner of waves, with reflection at the end, and hardly any tailing off. With infinite conductivity there can be no current set up in the wire at all. There is no dissipation; wave propagation in the medium is perfect. The wire-current is wholly superficial—an abstraction—yet it is nearly the same with very high conductivity. This illustrates the impenetrability of a perfect conductor to magnetic induction (and similarly to electric current), applied by Maxwell to the molecular theory of magnetism. Whatever state of magnetic induction and of current there may be in a perfect conductor is a fixture. If we move the conductor about in a magnetic field, superficial currents are instantaneously induced, whose only function is to ward off external induction and keep the interior state unchanged.

In a thermo-electric circuit of two metals, with one junction a little hotter than the other, there is a transmission of energy from one junction to the other through the dielectric, with a trifling amount of loss in the circuit generally. Here the source of the electric energy is heat, and the final result is heat. One junction is cooled, the other is heated, reversibly. Now, heat is the energy of molecular agitation, and at first sight the only difference is that the agitation is a little more brisk at one junction than at the other. Again, all parts of the circuit are agitating the ether. It would appear, then, that the ordinary molecular agitations set up no electric manifestations on account of their irregularity; although the electric machinery may be influenced vigorously, yet it must be done in some regularly symmetrical manner to constitute an impressed electric force. At the junctions there is a change of material, the molecules are different, and at their contact some directed quality is given to the agitations. This is very vague, no doubt, but is merely to point out that the impressed force is a symmetrical kind of radiation.

After these general remarks the temporarily interrupted mathematical treatment will be resumed.

SECTION III. RESUMPTION OF ROUGH SKETCH. EXTENSIONS.

Real transient, and suggested dissipative Magnetic Current.

As the rate of increase of the displacement in a non-conducting dielectric is the electric current, so the rate of increase of $\mathbf{B}/4\pi$ may be called the magnetic current. Let it be $\mathbf{G}$. Then

$$\mathbf{G} = \dot{\mathbf{B}}/4\pi = \mu\dot{\mathbf{H}}/4\pi. \qquad \text{(Magnetic current)} \quad (12)$$

Like electric displacement currents, magnetic currents are transient only, *i.e.*, they cannot continue indefinitely in one direction, like an electric conduction current. Also, like electric currents in a dielectric, they are unaccompanied by heat generation. In ether, the electric current and the magnetic current are of equal significance.

There is probably no such thing as a magnetic conduction current, with dissipation of energy. If there be such, analogous to an electric conduction current, then let

$$\mathbf{G} = g\mathbf{H} + \mu\dot{\mathbf{H}}/4\pi. \qquad (13)$$

Here $g\mathbf{H}$ is the magnetic conduction current, which, added to the undoubted magnetic current as in (12), gives $\mathbf{G}$ the true magnetic current. g may be scalar, or similar to k, with rotatory ϵ. Multiply (13) by $\mathbf{H}$. Then, using (11),

$$\mathbf{HG} = \mathbf{H}g\mathbf{H} + \dot{T}. \qquad (14)$$

Here $\mathbf{H}g\mathbf{H}$ is the rate of dissipation. Compare with (9).

Effect of g in a Closed Iron Ring.

The permanency of state of a steel magnet makes it improbable that

g has any existence at all, so that the conduction magnetic current is quite imaginary. But we may inquire what would happen in a closed ring of iron under magnetising force, on the supposition that g exists. Let the ring be uniformly lapped with wire, through which we pass a current from a voltaic battery.

If the radius of the ring be large compared with its section, the core may be treated as straight, and the manner in which the current would rise in the coil and the accompanying core phenomena may be easily worked out by a slight modification of the corresponding case with $g=0$ [Art. xxvii., § 29, Example 2, p. 394]. Let a be the radius of the core, also of the coil of negligible depth surrounding it, having N windings per unit length of core. Let k and μ be the conductivity and permeability of the core, and $\mathbf{H}$ (parallel to the axis) the magnetic force at distance r from the axis. The differential equation of H will be

$$\frac{1}{r}\frac{d}{dr}r\frac{dH}{dr}=(4\pi)^2 gkH+4\pi k\mu\dot{H};$$

whence $J_0(nr)\epsilon^{mt}$ is a normal system of magnetic force, if

$$-m=\frac{n^2}{4\pi k\mu}+\frac{4\pi g}{\mu}.$$

Thus the effect of g is to increase the reciprocal of the time-constant of every normal system by the same quantity $4\pi g/\mu$; in this respect resembling the effect of uniform leakage along a telegraph line, and having a similar result, viz., to accelerate the establishment of the permanent state. When this is reached, we do not have uniform strength of magnetic force in the core; but, if H_0 is the strength at the axis, that at distance r therefrom is

$$H=H_0\left(1+\frac{xr^2}{2^2}+\frac{x^2r^4}{2^2 4^2}+\dots\right),$$

where $x=(4\pi)^2 gk$. This is accompanied by core-currents parallel to the coil-current, of density

$$-\frac{1}{4\pi}H_0\frac{xr}{2}\left(1+\tfrac{1}{2}\frac{xr^2}{2^2}+\dots\right).$$

The coil-current will be a little less strong than if $g=0$; for the work of the battery is spent not merely in supporting the coil-current, but in heating the core, both by reason of the weak electric current in the core and the supposed weak magnetic current $g\mathbf{H}$. The back E.M.F. in the coil will be of strength

$$-F\left\{1+\frac{R}{4\pi Lg}\,\frac{1+\frac{xa^2}{2^2}+\dots}{1+\frac{1}{2}\frac{xa^2}{2^2}+\dots}\right\}^{-1},$$

where $F=$ E.M.F. of battery, R resistance of coil-circuit, and L its inductance without the core—*i.e.*, with air replacing it. Or, since g is to be small,

$$-F(1+R/4\pi Lg)^{-1}.$$

If $L/R = \cdot 01$ second, $g = 1/4\pi$ would make the back force $= 1/101$ of the battery force, so g, if existent, must be very small.

In the following, $g = 0$, so that equation (12) is the equation of the magnetic current.

First Cross Connection of Magnetic and Electric Force.

In the foregoing we have been dealing with the direct connection of the electric force and its consequences, electric conduction current and displacement, and of the magnetic force and magnetic induction. We have also brought in the displacement current in a dielectric, and the true current in a conducting dielectric. Also, to balance the displacement current, we have introduced the magnetic current. But, so far, we have no relations whatever between the electric and the magnetic quantities, which we must have, in order to make a consistent system.

The first cross connection is expressed by

$$\text{curl}\,\mathbf{H} = 4\pi\Gamma, \quad \text{..............................(15)}$$

H being the magnetic force and Γ the true current. Here "curl" is, like sin and cos, the symbol of an operation. It is so recurrent in electromagnetism that it might be termed *the* electromagnetic operator. It may be defined with reference to Cartesian coordinates thus: If H_1, H_2, H_3, are the three rectangular components of **H**, those of curl **H** are

$$\frac{dH_3}{dy} - \frac{dH_2}{dz}, \quad \frac{dH_1}{dz} - \frac{dH_3}{dx}, \quad \frac{dH_2}{dx} - \frac{dH_1}{dy}. \quad \text{............(16)}$$

But the most useful definition is that which is virtually contained in the fundamental Theorem of Version:—The line-integral of a vector **H** round any closed curve or circuit (or the "circulation" of **H**) equals the surface-integral of another vector, viz., curl **H**, over any surface bounded by the circuit. Apply this to small squares in planes perpendicular to **x**, **y**, and **z** successively, and the three expressions given in (16) for the components of curl **H** follow at once. Apply the theorem to suitably chosen infinitely small areas in any system of coordinates and we obtain the proper expressions in, usually, a far simpler manner than by laborious transformations of differential coefficients. Whilst the expressions for the components vary according to the system of coordinates chosen as most suitable for a special problem, the theorem, on the other hand, is universal, and gives us the inner meaning of the operation. It is far the best in general investigations not to employ any system of coordinates, but to emancipate one's self from their complexity by employing symbols which only relate to the intrinsic meaning of the operations; besides which, there is a great gain in the ease of manipulation. In the present paper the meanings of all forms of expression likely to be unfamiliar are briefly stated, and we shall avoid occupying valuable space by lengthy formulæ.

The operator "curl" is connected with rotation thus: if **H** be the instantaneous velocity at a point in a moving fluid, curl **H** is a vector whose direction is that of the axis of instantaneous rotation of the fluid surrounding the point, and whose length equals twice the angular velocity of rotation.

Notice that (15) contains no physical constants. It is therefore, in a sense, a purely geometrical equation. Given a system of magnetic force **H**, mentally represented by lines or tubes of force mapping out space in one way, by the operator "curl" we find another system of lines or tubes mapping out space in another way, viz., the lines and tubes of current. Whether **H** be wholly continuous or not, the derived Γ is necessarily continuous [that is, circuital]. The curl of a vector can have no divergence anywhere, which we express by

$$\operatorname{div} \Gamma = 0; \quad \text{or,} \quad \frac{d\Gamma_1}{dx} + \frac{d\Gamma_2}{dy} + \frac{d\Gamma_3}{dz} = 0, \quad \text{..............(17)}$$

which defines "divergence" with reference to Cartesian coordinates. The divergence of Γ is the amount of Γ leaving a point, reckoned per unit volume. When Γ, as here, signifies electric current, it is continuous; as much current leaves as enters any volume, or the integral amount leaving it, reckoning that entering it as negative, is zero. That (17) is involved in (15) is tested by differentiating the three components in (16) to x, y, and z respectively and adding them, when (17) results.

Given **H**, we have Γ, by (15), perfectly definite. But given Γ (necessarily continuous), **H** is not definitely fixed by (15). For, on finding one function **H** satisfying (15) with Γ given, we may add to **H** any function **I** such that curl **I** = 0, without disturbing the relation (15). The nature of **I** is given by

$$\mathbf{I} = -\nabla\Omega; \quad \text{or,} \quad I_1 = -d\Omega/dx, \quad I_2 = -d\Omega/dy, \quad I_3 = -d\Omega/dz, \quad (18)$$

where Ω is a *scalar* function of position, a scalar potential in fact. We require some other condition than (15) to find **H** completely when Γ is given; this is, that the magnetic induction $\mathbf{B} = \mu\mathbf{H}$, (equation (10),) is continuous, or div **B** = 0. **H** is now perfectly definite. If μ = constant, or all space is equally magnetisable isotropically, then **B** is the same multiple of **H** everywhere, hence div **H** = 0, so that the proper solution of (15) is that function **H** satisfying (15) which is continuous, like Γ. But **H** is not continuous when μ varies from one part of the field to another.

Having now defined "curl," "divergence," and ∇ applied to a scalar function, consider (15) from a less abstract point of view, in the light of the Version Theorem. Let there be any closed circuit in space,—whether passing through conducting or dielectric matter is immaterial. The amount of current passing through the circuit in the positive direction (that passing the other way being counted negatively) equals the circulation of **H** round the circuit $\div 4\pi$. The actual distribution of Γ is got by taking the circuit infinitely small and applying it to all parts of the field. Let us, whilst considering a finite circuit, yet take it sufficiently small to make the current pass all one way through it. Then, setting up current through the circuit, we set up magnetic force round it.

But there is another way of setting up magnetic force round the circuit, viz., by motion of the circuit itself in a previously undisturbed electric field. Thus, let there be a steady field of electric force, say in air, with therefore steady electric displacement, and no electric current. Let the closed circuit be a thin wire. When at rest in the field there is no current through it, and no magnetic force round it. But if we move the circuit so that the amount of electric displacement through it varies, there is electric current through the circuit, to be measured by the rate of increase of the amount of displacement through it at any moment; or, in another form, by the number of tubes of displacement added to the circuit per second by the motion of the circuit across them. Hence there will be magnetic force round the circuit, and if it be a thin iron wire, it will become magnetised by the motion in the electric field. In general, the motion of matter in an electric field sets up magnetic force.

As an example, fix a thin circular iron ring in air. Call the line through its centre perpendicular to its plane the axis. Let there be no current or magnetic force in the first place. Now shoot a small bullet, having an electrical charge, through the ring, along its axis. The electric displacement due to the charge will be continually changing; thus, there is a system of electric current in the air accompanying the motion of the bullet. The velocity of propagation of disturbances in air is so great that, unless the velocity of the bullet be *not* a very small fraction of the velocity of propagation, we may neglect the disturbance in the field of force due to the latter velocity not being infinite, and suppose that the bullet carries with it in its motion its normal field of force (radiating straight lines) unchanged. The distribution of displacement current about the moving bullet is then the same as that of the lines of magnetic force that would come from it if it were uniformly magnetised parallel to the axis, or line of actual motion in the real case, and the lines of magnetic force accompanying the displacement currents are circles centred upon the axis, in planes perpendicular thereto, the strength of magnetic force in the air being inversely proportional to the square of the distance from the centre of the bullet, and directly proportional to the cosine of the latitude; the equator being the circle on the bullet's surface in the plane perpendicular to the axis passing through the centre of the bullet. (With very high velocity this distribution of displacement current and magnetic force is departed from.) The fixed ring coincides with the lines of magnetic force during the whole motion of the bullet, and is therefore solenoidally magnetised thereby, most strongly when the magnetic force is strongest there, *i.e.*, when the bullet has just reached the centre of the ring, and the current through the ring is a maximum. The current through the ring may be measured either by the displacement current through a surface bounded by the ring, or by the rate at which the ring cuts the lines of electric force (supposed undisturbed) of the bullet.

Next, fix the charged bullet and move the ring instead, so that their relative motion shall be as before. There is exactly the same amount of electric displacement through the circuit added per second as before, in corresponding positions of the bullet and ring, with, therefore, the same magnetic force in the ring and the same magnetisation. Otherwise, however, there is a great difference in the two experiments. In the first case, changing electric displacement or electric current all through the dielectric, the greatest strength of current being at the poles of the bullet; whilst in the latter case the field is practically undisturbed except near the moving ring itself. Compare with the induction of electric force in a ring in a magnetic field, first when the field is moving, and next when the ring is moved in the field.

The induced magnetic force per unit length in a wire moved perpendicularly across the lines of force in an electric field equals the amount $\times 4\pi$ of electric displacement of the field crossed by the unit length of wire per second, and is perpendicular to the electric displacement and to the direction of motion. In general,

$$\mathbf{h} = \mathrm{V}\mathbf{D}\mathbf{v} \times 4\pi, \qquad \text{(18}a\text{)}$$

where **D** is the displacement of the field, **v** the velocity, **h** the induced magnetic force, and V is as in equation (3). There are, of course, corrections due to the reactions set up, due to the wire not being infinitely thin, and to finite length.*

In electromagnetic units, c in air $= (v_1)^{-2}$, if $v_1 =$ velocity of propagation $= 3 \times 10^{10}$ cm. per sec. Therefore, in the case of motion of a thin wire perpendicularly across the lines of force in a uniform electric field of strength E,

$$h = Ev(v_1)^{-2} = Ev/(9 \times 10^{20}).$$

Let $E = 10^{12}$ c.g.s., or 10^4 volts per cm., which is less than the disruption force in air in its ordinary state, then

$$h = v/(9 \times 10^8).$$

To get magnetic force of strength 10^{-5} c.g.s., v must equal 90 metres or 300 feet per second.

Magnetic Energy of Moving Charged Spheres.

In passing, I may remark that J. J. Thomson (*Phil. Mag.*, April, 1881) found the magnetic energy ΣT due to a sphere of radius a with an electric charge q moving with velocity v in a medium of permeability μ to be

$$\Sigma T = \tfrac{2}{15}\mu q^2 v^2/a.$$

I find that the $\frac{2}{15}$ should be $\frac{1}{3}$. Also, he found the mutual magnetic energy ΣT_{12} of two infinitely small spheres at distance r with charges q_1 and q_2, moving with velocities defined by the rectangular components u_1, u_2, u_3, and v_1, v_2, v_3, with u_1 and v_1 the velocities parallel to the line joining the spheres, to be

$$\Sigma T_{12} = (\mu q_1 q_2/3r)\,(u_1 v_1 + u_2 v_2 + u_3 v_3),$$

against which I find it to be

$$\Sigma T_{12} = (\mu q_1 q_2/2r)\,(2u_1 v_1 + u_2 v_2 + u_3 v_3).$$

I do not know what corrections, if any, have been published, and should be glad to receive information on the point, whether in corroboration of my results or otherwise.

* [The force defined by (18*a*) I now term the motional magnetic force, and its companion (21*a*) below, the motional electric force. Examples of their use will occur in later papers.]

SECTION IV. COMPLETION OF ROUGH SKETCH.

Second Connection between Electric Force and Magnetic Force.

The equation (15), $\text{curl}\,\mathbf{H} = 4\pi\Gamma$, expressing a relation, independent of physical constants, between the magnetic force and the electric current, is an extension of Ampere's results for linear circuits. By Γ must be understood Maxwell's true current—that is, the sum of the conduction current and the displacement current if the body considered be both a dielectric and a conductor, or the conduction current alone or the displacement current alone if the body have no dielectric capacity or no conductivity respectively. All bodies are either conducting or dielectric, or both, and ether is dielectric, so that electric current may exist everywhere. Putting Γ in terms of $\mathbf{E}$ by the equation of true current (8), [p. 443], we get

$$\text{curl}\,\mathbf{H} = 4\pi k\mathbf{E} + c\dot{\mathbf{E}}, \qquad \text{.........................(19)}$$

which is one connection between $\mathbf{E}$ and $\mathbf{H}$.

The second connection may be obtained by translating Faraday's law of induced electric force in a linear circuit into a mathematical form. It is remarkable that the ideas of Faraday, who was no mathematician, should admit of immediate translation into mathematical language; a fact due to his dispensing with the direct action-at-a-distance hypothesis, and employing the intermediate mechanism of lines or tubes of force. In popular language, the total E.M.F. of induction round a linear circuit is measured by the number of lines of force taken out of the circuit per second. Here the conventional connection between the assumed positive direction of translation through a circuit, and the assumed positive direction of motion in the circuit must be remembered. Selecting either direction through a circuit as the positive direction of translation, look through the circuit in this direction. Then the positive direction of rotation is right-handed, or with the hands of a watch whose front faces the spectator. Thus, increasing the number of lines of force through a circuit sets up negative E.M.F. round it.

So far in a medium of unit permeability. But when we make allowances for differences of magnetic permeability, it is not the variation of the magnetic force $\mathbf{H}$, but of the magnetic induction $\mathbf{B} = \mu\mathbf{H}$, which determines the induced E.M.F. The amended statement is that the total E.M.F. of induction round a circuit equals the rate of decrease of the amount of magnetic induction through the circuit. Now, since we have here a line-integral, viz., of the electric force of induction round a circuit, and a surface-integral, viz., of $-\mu\dot{\mathbf{H}}$ or $-\dot{\mathbf{B}}$ over any surface bounded by the circuit, we may at once apply the Version Theorem before referred to [p. 443] and deduce

$$\text{curl}\,\mathbf{E} = -\dot{\mathbf{B}} = -\mu\dot{\mathbf{H}}, \qquad \text{.......................(20)}$$

which is one form of the second relation between $\mathbf{E}$ and $\mathbf{H}$.

The following method is also instructive. Since the rate of increase of the magnetic induction at a point equals $4\pi\mathbf{G}$, where $\mathbf{G}$ is the magnetic current, as defined by equation (12), we may state the law of induced electric force thus:—The total E.M.F. of induction round a circuit in the negative direction equals 4π times the total magnetic current through the circuit in the positive direction. Now compare this statement with the statement regarding equation (15) [p. 443], viz., that the total magnetic force round a circuit equals 4π times the total electric current through the circuit, and change this so as to produce the statement in the last sentence. We must change magnetic force to electric force taken negatively, and electric current to magnetic current. Hence

$$\text{curl}\,\mathbf{H} = 4\pi\Gamma \quad \text{.........................(15)}\,bis$$

becomes

$$-\text{curl}\,\mathbf{E} = 4\pi\mathbf{G}, \quad \text{..........................(21)}$$

which is equivalent to (20).

We have, in order to simplify the establishment of (20) or (21), avoided mentioning the E.M.F. induced in a linear circuit by its motion in the field, which may or may not be varying independently. The amount of induction added to or taken out of a circuit from this cause may be obviously represented by a line-integral, as it depends upon the rate at which the different elements of the circuit cross the lines of induction. If the induction were of the same strength at all the moving parts of the circuit, and they all moved at right angles to their lengths and also perpendicularly across the lines of induction in the same sense, the total E.M.F. would be of strength $= B \times$ rate of increase of area of circuit. But when B varies, and likewise the velocity of the different elements across the lines of B, each element must be considered separately. The amount contributed to the total E.M.F. by an element of unit length equals the component parallel to its length of

$$\text{V}\mathbf{v}\mathbf{B}, \quad \text{...............................(21}a\text{)}$$

if $\mathbf{v}$ be the vector velocity. But, if there be current induced, this brings in working mechanical forces, and should therefore be separately considered. At present we return to the case of c, k, and μ constant with respect to the time, and no parts moveable.

In equation (21), $\mathbf{E}$ is the electric force of induction only, not the actual electric force. There may be in addition electrostatic force, and also impressed electric force. But the electrostatic force is polar; it is derived from a scalar potential. If this be P, the force is $-\nabla P$. But $\text{curl}\,\nabla P = 0$, as was before remarked [p. 444] with reference to Ω, consequently the polar force may be included in $\mathbf{E}$ in equation (21). Similarly any polar force may be included in $\mathbf{H}$ in the previous equation (15). Now in all our equations, from (1) up to (14), not containing any relations between $\mathbf{E}$ and $\mathbf{H}$, those symbols mean the actual resultant electric and magnetic force from all causes. Hence, in order that the two equations (15) and (21) may harmonise with the preliminary equations (1) to (14), not only in space where there is no impressed force, but at the places where such exist as well, we must, whilst still using $\mathbf{E}$ and $\mathbf{H}$ to denote the actual forces, deduct from them the impressed forces in using the relations (15) and (21). So, let **e** be the impressed electric force, and **h** the impressed magnetic force. Our two connections between **E** and **H** are then

$$\text{curl}\,(\mathbf{H}-\mathbf{h}) = 4\pi\Gamma = 4\pi k\mathbf{E} + c\dot{\mathbf{E}}, \quad \text{.................(22)}$$

$$\text{curl}\,(\mathbf{e}-\mathbf{E}) = 4\pi\mathbf{G} = 4\pi g\mathbf{H} + \mu\dot{\mathbf{H}}, \quad \text{.................(23)}$$

where the coefficient g of magnetic conductivity is introduced to show the symmetry, and may be put $=0$. We have now a dynamically complete system.

The subject of impressed force will be considered in a following section more fully, especially as règards impressed magnetic force, and its interpretation in terms of magnetisation. In the meantime we may define impressed electric force thus. If **e** be the impressed electric force at a point, and Γ the electric current there, **e**Γ of energy is taken into the electromagnetic system there per unit volume per second. Similarly, we may define the impressed magnetic force **h** at a point, by saying that if there be a magnetic current **G** there, **hG** of energy is taken in per second per unit volume by the electromagnetic system there. In general, **e**Γ and **hG** are scalar products [*see* equation (5)], having the ordinary signification when **e** is parallel to Γ, or **h** to **G**; in other cases to be multiplied by the cosine of the angle between **e** and Γ or between **h** and **G**.

The Equation of Energy and its Transfer.

We must find the rate of working of the impressed forces, and compare with the dissipativity and with the changes taking place in the energy of displacement and the magnetic energy. Multiply (22) by (**e** − **E**), and (23) by (**h** − **H**), and add the results. We get

$$4\pi\{(\mathbf{e}-\mathbf{E})\Gamma + (\mathbf{h}-\mathbf{H})\mathbf{G}\} = (\mathbf{e}-\mathbf{E})\,\text{curl}\,(\mathbf{H}-\mathbf{h}) + (\mathbf{h}-\mathbf{H})\,\text{curl}\,(\mathbf{e}-\mathbf{E}).$$

Or

$$\mathbf{e}\Gamma + \mathbf{hG} = \mathbf{E}\Gamma + \mathbf{HG} + \{(\mathbf{H}-\mathbf{h})\,\text{curl}(\mathbf{E}-\mathbf{e}) - (\mathbf{E}-\mathbf{e})\,\text{curl}(\mathbf{H}-\mathbf{h})\}/4\pi, \quad (24)$$

by rearrangement. **E**Γ and **HG** here occurring have been already expressed in terms of the dissipativity Q, the electric energy of displacement U, and the magnetic energy T; *see* equations (9) and (14). Thus

$$\mathbf{E}\Gamma + \mathbf{HG} = Q + \dot{U} + \dot{T}. \quad \text{.........................(25)}$$

On the left side we have the rate of working per unit volume of the actual forces **E** and **H** on the currents Γ and **G**; on the right side the dissipativity, or rate at which energy is being lost from the system irreversibly, producing heat according to Joule's law, and the rate of increase of the electric and magnetic energies, all per unit volume.

Now looking at (24), the left side expresses the rate at which energy is being taking in (reversibly) per unit volume, in virtue of the impressed forces **e** and **h**. Therefore the excess of (**e**Γ + **hG**) over (**E**Γ + **HG**) must be the energy leaving the unit volume per second through its sides. Now, **X** and **Y** being any two vectors,

$$\mathbf{Y}\,\text{curl}\,\mathbf{X} - \mathbf{X}\,\text{curl}\,\mathbf{Y} = \text{div}\,\mathbf{VXY}; \quad \text{.................(26)}$$

or in full, by (5), (16), (17), and (3),

$$Y_1(dX_3/dy - dX_2/dz) + Y_2(dX_1/dz - dX_3/dx) + Y_3(dX_2/dx - dX_1/dy)$$
$$- X_1(dY_3/dy - dY_2/dz) - X_2(dY_1/dz - dY_3/dx) - X_3(dY_2/dx - dY_1/dy)$$
$$= (d/dx)(X_2Y_3 - X_3Y_2) + (d/dy)(X_3Y_1 - X_1Y_3) + (d/dz)(X_1Y_2 - Y_1X_2);$$

using the numbers 1, 2, 3, to denote the x, y, and z components. Let

then $$\mathbf{W} = \mathrm{V}(\mathbf{E} - \mathbf{e})(\mathbf{H} - \mathbf{h})/4\pi, \quad \text{......................(27)}$$

then by (26), with $\mathbf{X} = \mathbf{E} - \mathbf{e}$, and $\mathbf{Y} = \mathbf{H} - \mathbf{h}$, equation (24) becomes

$$\left.\begin{aligned} \mathbf{e}\Gamma + \mathbf{hG} &= \mathbf{E}\Gamma + \mathbf{HG} + \operatorname{div} \mathbf{W} \\ &= Q + \dot{U} + \dot{T} + \operatorname{div} \mathbf{W}, \end{aligned}\right\} \quad \text{....................(28)}$$

the equation of energy put in its most significant form. Summing up through all space, **W** goes out; or the total work per second of the impressed forces equals the total dissipativity plus the rate of increase of the total electric and magnetic energy.

W is the vector rate of transfer of energy, or what we before [p. 438] termed the energy-current, a vector whose direction is that of the transfer of energy, and whose magnitude equals the amount transferred per second across unit area of a plane perpendicular to that of transfer. Note [p. 438] that impressed forces were said to be not counted; hence as **E** and **H** are the actual forces now, the impressed forces are deducted, as shown in (27). The magnitude of **W** is the product of the strengths of the two forces and the sine of the angle between their directions, and the direction of **W** is perpendicular to both forces, with the before-stated convention regarding positive directions.

The general nature of the energy-current was described in Section II. "On the transmission of energy through wires by the electric current" [p. 434], where, however, only impressed electric force was considered. The same general results apply to impressed magnetic force; energy proceeding from places where such exists, to be dissipated as heat in conducting matter, or to increase the electric and magnetic energies, or to go to other places of impressed magnetic force. But there are great practical differences between impressed electric and magnetic force, owing to the transient nature of magnetic currents and other causes.

Differential Equations of **E** *and* **H**.

By eliminating **E** or **H** between (22) and (23) we obtain the characteristic equation of **E** or of **H**. Put $g = 0$, and eliminate **H**. Then, $$\operatorname{curl} \mu^{-1} \operatorname{curl}(\mathbf{e} - \mathbf{E}) = \operatorname{curl} \dot{\mathbf{h}} + 4\pi k\dot{\mathbf{E}} + c\ddot{\mathbf{E}}, \quad \text{............(29)}$$

which is the equation of **E**. Here **e** and **h**, being impressed, must be supposed to be given. μ^{-1} is the operator inverse to μ, that is, in the general case of eolotropy μ^{-1} is defined by the three principal axes and the values $1/\mu_1$, $1/\mu_2$, $1/\mu_3$, along them, as was explained [p. 430] in speaking of k. Similar remarks apply to k^{-1} and c^{-1} should they occur.

In space where there is no impressed electric force, and no, or else constant, impressed magnetic force,

$$\text{curl}\,\mu^{-1}\,\text{curl}\,\mathbf{E} + 4\pi k\dot{\mathbf{E}} + c\ddot{\mathbf{E}} = 0. \qquad (30)$$

In a non-dielectric conductor,

$$\left.\begin{aligned} &\text{curl}\,\mu^{-1}\,\text{curl}\,\mathbf{E} + 4\pi k\dot{\mathbf{E}} = 0, \\ \text{and} \quad &\text{curl}\,k^{-1}\,\text{curl}\,\mathbf{H} + 4\pi\mu\dot{\mathbf{H}} = 0. \end{aligned}\right\} \qquad (31)$$

Here propagation of **E** and of **H** is by diffusion. And in a non-conducting dielectric,

$$\left.\begin{aligned} &\text{curl}\,\mu^{-1}\,\text{curl}\,\mathbf{E} + c\ddot{\mathbf{E}} = 0, \\ &\text{curl}\,c^{-1}\,\text{curl}\,\mathbf{H} + \mu\ddot{\mathbf{H}} = 0. \end{aligned}\right\} \qquad (32)$$

Here propagation is by waves, *i.e.*, propagation of **E** or **H**, not of energy.

c and μ self-conjugate; k not necessarily so.

We should note that

$$\dot{T} = (d/dt)(\mathbf{H}\mu\mathbf{H}/8\pi) = \mathbf{H}\mu\dot{\mathbf{H}}/8\pi + \dot{\mathbf{H}}\mu\mathbf{H}/8\pi,$$

whilst

$$\mathbf{HG} = \mathbf{H}\mu\dot{\mathbf{H}}/4\pi;$$

so for **HG** to equal $\dot{T}$ we require

$$\mathbf{H}\mu\dot{\mathbf{H}} = \dot{\mathbf{H}}\mu\mathbf{H},$$

i.e., μ must be self-conjugate, or contain no rotatory ϵ [equation (3), p. 431]. Similarly, for $\dot{U}$ to equal $\mathbf{E}\dot{\mathbf{D}}$, c must be self-conjugate. But there is no such limitation thrown upon k the electric conductivity operator, nor would there be upon g the magnetic conductivity operator, did such exist. There are other proofs of these conclusions, but the above are very short. There is, however, an objection to be raised against the rotatory conductivity vector ϵ, which want of space does not permit to be mentioned at present.

SECTION V. IMPRESSED MAGNETIC FORCE. INTRINSIC MAGNETISATION.

The energy definition of impressed electric force, due originally, if not explicitly, at least substantially, to Sir W. Thomson, has long been well recognised by most writers on electrical subjects, especially since the practical introduction of dynamo machines, accumulators, etc., which raised the energy transformations concerned in electrical phenomena from being matters of almost purely scientific interest to matters of the extremest practical commercial importance.

But in our last we gave an energy definition of impressed magnetic force, precisely similar to that of impressed electric force. Thus, if **h** be the impressed magnetic force at a point, and **G** the magnetic current there, the rate of working is **hG** per unit volume, and this amount of energy is taken in per second by the electromagnetic system at the place, and is employed in increasing the energy of electric displacement and the magnetic energy, or wasted in the heat of conduction currents. Also it may be used in effecting bodily motions when there is yielding to the mechanical forces, or in chemical work, etc. Should there be no impressed force, electric or magnetic, except an impressed force **h** in a single unit volume, we have

$$\mathbf{hG} = \Sigma(Q + \dot{U} + \dot{T}),$$

the summation extending through all space, where Q is the dissipativity, U the energy of displacement, and T the magnetic energy, all per unit volume. In order to identify the quantity thus defined, and show the relation it bears to the quantity termed intensity of magnetisation, let there be no electric force in the field, and its state be steady. The second equation of induction (23), goes out, since $\mathbf{G} = 0$, and the first, (22), is reduced to

$$\text{curl}\,(\mathbf{H} - \mathbf{h}) = 0. \qquad (33)$$

Integrating once, we have

$$\mathbf{H} = \mathbf{h} + \mathbf{F}, \quad \text{where} \quad \text{curl}\,\mathbf{F} = 0, \quad \text{or} \quad \mathbf{F} = -\nabla\Omega. \qquad (34)$$

Thus the actual magnetic force **H** differs from the impressed force **h** by a polar force **F**, a force which, when analysed, is found to be made up by the superposition of radial forces proceeding from points. Ω is the potential, a scalar, variable from point to point.

But, so far, there is nothing to settle what particular distribution of polar force **F** must be. A second condition is wanted. Now we know from equation (23) that the magnetic current, like the electric current, is always circuital, *i.e.*,

$$\text{div}\,\mathbf{G} = 0, \quad \text{therefore} \quad \text{div}\,\dot{\mathbf{B}} = 0, \qquad (35)$$

and, if we take the time-integral, we find

$$\text{div}\,\mathbf{B} = f(x, y, z),$$

any scalar function of position, independent of the time. If, then, the magnetic induction **B** were not also circuital, its divergence would continue unaltered at any place, however the field might otherwise vary. It could only be altered by convection, shifting the arrangement of matter. It would then, by a suitable arrangement of matter, be possible to have a unipolar magnet, a quantity of matter round which the magnetic force was everywhere directed outward, or everywhere inward. This being contradicted by universal experience, we must conclude that

$$\text{div}\,\mathbf{B} = 0, \quad \text{as well as} \quad \text{div}\,\mathbf{G} = 0. \qquad (36)$$

The second equation of induction (23), if we use the full expression for **G** there given, is too general, requiring the limitation $g = 0$, from the absence of magnetic conductivity. The now added limitation $\text{div}\,\mathbf{B} = 0$, as it does not contradict the second equation of induction, must be considered as an auxiliary condition. Though not necessarily dependent upon the first limitation $g = 0$, it is yet intimately connected with it.

Our two equations are, therefore, (33) and (36), or,

$$\text{curl}\,(\mathbf{H}-\mathbf{h})=0, \quad \text{and} \quad \text{div}\,\mu\mathbf{H}=0, \quad \text{.................}(37)$$

and the question is, given μ the permeability, and $\mathbf{h}$ the impressed force, find $\mathbf{H}$ the real force. Or, using (34), we have

$$\text{div}\,\mu(\mathbf{h}+\mathbf{F})=0, \qquad \text{curl}\,\mathbf{F}=0\,; \quad \text{.................}(38)$$

and now, with the same data, we have to find $\mathbf{F}$. By the second of (38), $\mathbf{F}$ is restricted to be a polar force; by the first it is still further restricted to have, when multiplied by μ, a given amount of divergence, thus,

$$\text{div}\,\mu\mathbf{F} = -\,\text{div}\,\mu\mathbf{h}.$$

The two conditions together make $\mathbf{F}$ perfectly definite, and therefore $\mathbf{H}$ and $\mathbf{B}$ definite through all space. Of this a variety of proofs may be given (the first was given by Sir W. Thomson in 1848, relating to similar equations, μ being isotropic). The following perhaps puts the matter in as simple a form as it can be put, and is best adapted to the present circumstances.

In the first place there cannot be two solutions of (38) for $\mathbf{F}$. For if (38) are satisfied by $\mathbf{F}$ and also by $\mathbf{F}+\mathbf{f}$, we have, by subtraction,

$$\text{div}\,\mu\mathbf{f}=0, \qquad \text{curl}\,\mathbf{f}=0, \quad \text{.....................}(39)$$

as the equations that $\mathbf{f}$ must satisfy. But consider the quantity $\Sigma\,\mathbf{f}\mu\mathbf{f}$ integrated through all space. It cannot be negative, for every element of it is positive or else zero. Thus

$$\mathbf{f}\mu\mathbf{f}=\mu_1\mathrm{f}_1^{\,2}+\mu_2\mathrm{f}_2^{\,2}+\mu_3\mathrm{f}_3^{\,2},$$

if μ_1, μ_2, μ_3 be the principal permeabilities, and f_1, f_2, f_3 the corresponding components of $\mathbf{f}$. We suppose the permeability always positive (to deny this would lead to absurdities), hence $\mathbf{f}\mu\mathbf{f}$ is always positive, or else zero, viz., when $\mathbf{f}=0$. But, by the second of (39), $\mathbf{f}=-\nabla p$, if p is the potential of $\mathbf{f}$, and therefore, by a potential property,

$$\Sigma\,\mathbf{f}\mu\mathbf{f}=\Sigma\,p\,\text{div}\,\mu\mathbf{f}, \quad \text{..........................}(40)$$

and therefore vanishes, by the first of (39). Hence $\mathbf{f}=0$, making the supposed two solutions identical.

Next, to show that there is one solution, consider $\Sigma\,\mathbf{H}\mu\mathbf{H}$ through all space. This is also positive, or zero, whatever $\mathbf{H}$ may be, by the same reasoning, so has necessarily a minimum value. If $\mathbf{H}$ be quite arbitrary, the minimum is zero, when $\mathbf{H}=0$ everywhere. But $\mathbf{H}=\mathbf{h}+\mathbf{F}$, and here $\mathbf{h}$ is constant. Let $\mathbf{F}$ vary. The corresponding variation in $\Sigma\,\mathbf{H}\mu\mathbf{H}$ is

$$\delta\,\Sigma\,\mathbf{H}\mu\mathbf{H}=\Sigma\,\delta\mathbf{F}\mu(\mathbf{h}+\mathbf{F})+\Sigma\,(\mathbf{h}+\mathbf{F})\delta\mu\mathbf{F}=\Sigma\,2\delta\mathbf{F}\mu(\mathbf{h}+\mathbf{F}). \quad \text{......}(41)$$

Now subject $\mathbf{F}$ to either of the two equations (38), say the second, so that $\mathbf{F}=-\nabla\Omega$. Then, similarly to (40),

$$\delta\,\Sigma\,\mathbf{H}\mu\mathbf{H}=\Sigma\,2\delta\Omega\,\text{div}\,\mu(\mathbf{h}+\mathbf{F}). \quad \text{..................}(42)$$

Hence, to make $\Sigma\,\mathbf{H}\mu\mathbf{H}$ a minimum requires

$$\text{div}\,\mu(\mathbf{h}+\mathbf{F})=0,$$

which is the first equation (38). Thus, when $\mathbf{F}$ satisfies both equations, it makes $\Sigma\,\mathbf{H}\mu\mathbf{H}$ a minimum. But this quantity *has* a minimum, therefore there *is* a solution of the equations (38), or $\mathbf{F}$ is a definite vector for every point of space.

If we assume that μ contains a rotatory vector ϵ, so that

$$\left.\begin{aligned}\mu\mathbf{H} &= \mu_0\mathbf{H} + \mathrm{V}\epsilon\mathbf{H},\\ \text{let}\qquad \mu'\mathbf{H} &= \mu_0\mathbf{H} - \mathrm{V}\epsilon\mathbf{H},\end{aligned}\right\}\qquad(43)$$

then μ' is conjugate to μ, and

$$\tfrac{1}{2}(\mu + \mu')\mathbf{H} = \mu_0\mathbf{H},$$

where μ_0 is self-conjugate or non-rotatory. Instead of (41) we shall have

$$\delta\,\Sigma\,\mathbf{H}\mu\mathbf{H} = \Sigma\,2\delta\mathbf{F}\mu_0(\mathbf{h} + \mathbf{F}),\qquad(44)$$

and the minimum is given by

$$\operatorname{div}\mu_0(\mathbf{h} + \mathbf{F}) = 0,$$

which is not the proper condition.

There is a similar failure in the mathematically analogous problem of conduction current kept up by impressed electric force, when the conductivity k is rotatory.

In connection with the above, we may notice two special solutions of (37). First, if $\operatorname{curl}\mathbf{h} = 0$, then $\operatorname{curl}\mathbf{H} = 0$, which, with $\operatorname{div}\mu\mathbf{H} = 0$, requires $\mathbf{H} = 0$, and therefore $\mathbf{B} = 0$. That is, if the impressed force be wholly polar, there is no induction. The simplest example is a closed magnetic shell of uniform strength, and any thickness, an assemblage of magnets put together side by side in such a manner that there is no induction anywhere.

Secondly, if $\operatorname{div}\mu\mathbf{h} = 0$, then $\operatorname{div}\mu\mathbf{F} = 0$, which, with $\operatorname{curl}\mathbf{F} = 0$, requires $\mathbf{F} = 0$, therefore $\mathbf{H} = \mathbf{h}$, and $\mathbf{B} = \mu\mathbf{h}$. Here the impressed force everywhere produces the full induction, and there is no polar force.

Comparing our equations with those occurring in the problem of magnetisation, we find that, if $\mathbf{I}$ be the intensity of intrinsic magnetisation, it is related to $\mathbf{h}$ thus :—

$$\mathbf{I} = \mu\mathbf{h}/4\pi.\qquad(45)$$

$\mathbf{h}$ may therefore be called the intrinsic magnetic force, if we like. The real magnetisation is the sum of the intrinsic and the "induced," which we shall call $\mathbf{i}$, and the ordinary form of the magnetic induction equation is equivalent to

$$\mathbf{B} = \mathbf{F} + 4\pi(\mathbf{I} + \mathbf{i}),\qquad(46)$$

where $\mathbf{F}$ is the polar force due to both the intrinsic and the induced magnetisation. It is the same as $\mathbf{F}$ above. And, to identify (46) with the equation $\mathbf{B} = \mu\mathbf{H}$ we use always, we have first

$$\mathbf{i} = \kappa\mathbf{F},$$

giving the induced magnetisation in terms of the polar force, κ being the coefficient of induced magnetisation, next the equation (45), and lastly,

$$\mu = 1 + 4\pi\kappa\,;$$

so that (46) becomes

$$\mathbf{B} = \mathbf{F} + \mu\mathbf{h} + 4\pi\kappa\mathbf{F} = \mu(\mathbf{h} + \mathbf{F}) = \mu\mathbf{H}.\qquad(47)$$

It will be seen that this separation of magnetisation into intrinsic and induced is a roundabout way of treating the subject, and that there is considerable simplification obtained by always using the equation $\mathbf{B} = \mu\mathbf{H}$, both in an intrinsic magnet and without it, ever employing the two ideas of magnetic force and magnetic induction, and letting magnetisations alone. The quantity **B** is the magnetic induction as ordinarily understood; **H** is also the magnetic force as ordinarily under stood everywhere, except where there is intrinsic magnetisation. There, however, we add the intrinsic or impressed force **h**, making

$$\mathbf{H} = \mathbf{h} + \mathbf{F}$$

in general, just as in a conduction current system we add the impressed force, where there is any, to the polar electric force and that of induction to obtain the resultant efficient electric force. **H** is thus always the resultant magnetic force from all causes, and although in the above we have considered no electric currents to exist, yet we may add that should there be any,

$$\mathbf{H} = \mathbf{h} + \mathbf{F} + \mathbf{F}_1,$$

where $\mathbf{F}_1$ is the magnetic force of the current; and $\mathbf{B} = \mu\mathbf{H}$ always.

The term "intrinsic," as applied to magnetisation, is used by Sir W. Thomson, but not by Maxwell, though he gives the same theory of induced magnetisation as the former. It is not always clear in Maxwell's treatise whether by magnetisation he refers to the intrinsic only or to the actual, including the induced. Maxwell's equations of disturbances, also, break down when they are applied to the interior of an intrinsic magnet, owing to his use of the equation

$$\text{curl}\,\mathbf{H} = 4\pi\Gamma,$$

as the relation between the magnetic force and the current, whereas it must be—equation (32)—

$$\text{curl}\,(\mathbf{H} - \mathbf{h}) = 4\pi\Gamma,$$

where there is intrinsic magnetisation, if we are to obtain consistent results.

Although, in identifying **I** with $\mu\mathbf{h}/4\pi$, we have, by means of the two equations of induction, (32) and (33), and the equation of energy deduced therefrom, obtained a justification of the energy definition of **h**, similar to that for impressed electric force, yet, as the consequences are of some importance in variable states, they may be advantageously followed up from the impressed force point of view.

Magnetic Energy. Double Work of Magnet.

Let there be a distribution of impressed magnetic force **h** in a medium of given variable permeability μ—for example, a magnet in air containing soft iron and any other substances. Imagine the impressed forces to be put on suddenly. We know, by the above, that a certain definite distribution of magnetic induction is set up, which is steady when the arrangement of matter is fixed. During the transient state there is magnetic current everywhere unless $\mu = 0$ somewhere, which we must believe to be impossible, since μ is very little less than unity for any known substance. The magnetic currents are wholly closed. They are accompanied by electric currents, also closed, in all space, in general. When they have ceased, there is no electric force anywhere, and the magnetic induction at any point is the time-integral of the magnetic current $\times 4\pi$. Compare the work done by the impressed forces with that done by the actual forces through all space. The latter is, per unit volume,

$$\int \mathbf{HG}dt = \int \mathbf{H}\mu\dot{\mathbf{H}}/4\pi . dt = \int \frac{d}{dt}(\mathbf{H}\mu\mathbf{H}/8\pi)dt = \mathbf{H}\mu\mathbf{H}/8\pi,$$

where, in the last expression, **H** is the final value of the magnetic force. In the integrals, **H** is the variable value at time t. Let T be the whole work thus done in all space, then

$$T = \Sigma\, \mathbf{H}\mu\mathbf{H}/8\pi. \qquad (48)$$

On the other hand, the work done by the impressed force per unit volume is

$$\int \mathbf{hG}dt = \int \mathbf{h}\mu\dot{\mathbf{H}}/4\pi . dt = \mathbf{h}\mu\mathbf{H}/4\pi;$$

so, if T_1 is the whole work done by the impressed forces,

$$T_1 = \Sigma\, \mathbf{h}\mu\mathbf{H}/4\pi, \qquad (49)$$

where the summation may also extend through all space, since where there is no **h** nothing is contributed to the sum. Now $\mathbf{H} = \mathbf{h} + \mathbf{F}$, and

$$\Sigma\, \mathbf{F}\mu\mathbf{H} = \Sigma\, \Omega \operatorname{div} \mu\mathbf{H} = 0, \qquad (50)$$

because **F** is polar and μ**H** circuital. (Similar to (40).) So in (49) we may add **F** to **h**, making

$$T_1 = \Sigma\, (\mathbf{h} + \mathbf{F})\mu\mathbf{H}/4\pi = \Sigma\, \mathbf{H}\mu\mathbf{H}/4\pi = 2T, \qquad (51)$$

by (48). The impressed forces therefore do double the work of the actual magnetic forces during the transient state. The excess is done by the electric forces. For, integrating (28), to the time, with $\mathbf{e} = 0$, and also through all space, to get rid of W,

$$\Sigma \int \mathbf{hG}dt = \Sigma \int \mathbf{E}\Gamma dt + \Sigma \int \mathbf{HG}dt;$$

and, since finally $U = 0$, we have

$$T_1 = \int Q dt + T. \qquad (52)$$

Hence, by (51), $\qquad T = \int Q dt,$

the total heat in conductors arising from induced currents.

One half the work done by the impressed forces is wasted in heat of induced currents, the other half is the magnetic energy set up, expressed by (48). Now, suddenly remove the impressed forces; there will be a similar inverse transient state, during which, as the magnetic induction subsides, the whole of the energy T will find its way to the conducting parts to be there wasted as heat. The intrinsic magnet itself, it should be remembered, would come in for a large share of this in general, or, in special cases, the whole. T is thus the whole amount of work that can be got out of the magnetic system by removing the intrinsic forces, which is why we term it the magnetic energy. [In the above investigation it is assumed that there is a steady state. When there cannot be, there are exceptional peculiarities, treated of in later papers.]

It may be objected that the above is unrealisable, that we cannot put on, or suddenly remove, the retentiveness of a magnet, and so obtain the whole magnetic energy existing in a given configuration in work. Admitting this, it may be remarked that we can do something equivalent, or at least approximate thereto, in the following manner. By means of a properly chosen distribution of impressed electric force we may set up electric current that shall exactly neutralise the field of the magnet, producing a state of no induction. Now suddenly cut off the impressed electric forces that kept up the currents. We then start with no induction, and terminate with the proper distribution of induction due to the magnet, and the impressed electric forces do no work, being cut out. Hence the effect is the same as suddenly putting on the impressed magnetic forces, doing $2T$ of work, one half magnetically, the other half being expended in the heat of induced currents. Some simple examples to illustrate this will be worked out later.

This extreme case will serve to illustrate the meaning of the distinction between the work done by $\mathbf{h}$ and by $\mathbf{H}$, also the meaning to be attached to magnetic energy. There are other ways, of course, of using up the energy T above said to be wasted in heat. Thus, if we alter the configuration of the matter in our magnetic system, we usually alter T. Let it be increased from T to $T+\delta T$. Then the impressed forces $\mathbf{h}$ will do $2\delta T$ of work, one half magnetically, in increasing T, the other half mechanically, by the mutual stresses assisting the motion, and this latter half will be partly wasted in heat at once by the induced current accompanying the motion, and all may be ultimately thus wasted.

This naturally brings us to the subject of the mechanical forces.

SECTION VI. THE MECHANICAL FORCES AND THEIR POTENTIAL ENERGY.

The expression for the magnetic energy T may be conveniently put into another remarkable form, thus: By (50),

$$\Sigma\,\mathbf{F}\mu\mathbf{h} = -\Sigma\,\mathbf{F}\mu\mathbf{F};$$

then, by (48) and (51), we obtain

$$T = \Sigma\,\mathbf{h}\mu\mathbf{h}/8\pi - \Sigma\,\mathbf{F}\mu\mathbf{F}/8\pi, = T_0 - M, \quad \text{say} \qquad (53)$$

Here the magnetic energy T is expressed as the difference of two quantities T_0 and M, of which the first is constant, being the maximum value of T. Short-circuit the magnet by an infinitely permeable skin; there will be then no induction outside the skin, and T will be the greatest possible consistent with not altering the interior permeability μ and impressed force $\mathbf{h}$. But it will not in general be as great as T_0. This requires there to be no polar force, so that **h** at any place can produce the full induction μ**h**. But if we imagine every element of volume surrounded by an infinitely permeable skin, T becomes T_0. Should the magnet be uniformly magnetised, and of uniform permeability, it is sufficient to coat the outer surface with an infinitely permeable skin to increase T to T_0. In some other cases no skin is needed.

For the same reason as before given for T, T_0 and M are both necessarily positive. The quantity M is the same as what Sir W. Thomson calls the "mechanical value" of the magnetic system, or the amount of work that would have to be done against repulsions to build up the intrinsic magnet if it were given in the state of infinitely slender filaments, magnetised parallel to their lengths, placed infinitely widely apart. Or, reversing the operations, imagine the intrinsic magnet to be divided into infinitely slender filaments parallel to the lines of intrinsic magnetisation, and the filaments cut up into short straight pieces (though infinitely long compared with their diameters). Then if the elementary parts thus defined be infinitely widely separated from one another, and from all matter susceptible to induced magnetisation, the work done during the separation by the polar forces would amount to M. But I am unable to verify the statement that M may be either positive or negative ("Electrostatics and Magnetism," Art. 731, end of p. 565). The form $\Sigma\,\mathbf{F}\mu\mathbf{F}/8\pi$ (not given in the paper quoted) shows that it must be always positive. The following are the principal forms :—

$$\left.\begin{aligned} M &= T_0 - T = \Sigma\,\mathbf{h}\mu\mathbf{h}/8\pi - \Sigma\,\mathbf{H}\mu\mathbf{H}/8\pi = \Sigma\,\mathbf{F}\mu\mathbf{F}/8\pi \\ &= \tfrac{1}{2}\Sigma\,\Omega\rho_1 = \tfrac{1}{2}\Sigma\,\Omega_1\rho = \Sigma\,\mathbf{F}\mathbf{F}_1/8\pi = -\tfrac{1}{2}\Sigma\,\mathbf{I}\mathbf{F}_1 = -\tfrac{1}{2}\Sigma\,\mathbf{I}_1\mathbf{F}\,; \end{aligned}\right\} \quad \ldots\ldots\ldots\ldots(54)$$

to understand which, it is necessary to say that $\mathbf{I}_1 = \mu\mathbf{h}/4\pi$ is the intensity of intrinsic magnetisation, and **I** the actual, the sum of the intrinsic and the induced; $\mathbf{F}_1$ the polar force of the intrinsic, and **F** the actual polar force; ρ_1 the density of free intrinsic magnetism, and ρ the actual density; Ω_1 the potential of $\mathbf{F}_1$, and Ω of **F**; and, lastly, $\mathbf{H} = \mathbf{h} + \mathbf{F}$. For the intrinsic magnetisation, we have

$$\mathbf{F}_1 = -\nabla\Omega_1, \qquad 4\pi\rho_1 = \operatorname{div}\mathbf{F}_1 = -4\pi\operatorname{div}\mathbf{I}_1.$$

Similarly, if the number $_2$ refer to the induced magnetisation, we have

$$\mathbf{F}_2 = -\nabla\Omega_2, \qquad 4\pi\rho_2 = \operatorname{div}\mathbf{F}_2 = -4\pi\operatorname{div}\mathbf{I}_2.$$

Lastly, $\quad \mathbf{F} = \mathbf{F}_1 + \mathbf{F}_2, \qquad \Omega = \Omega_1 + \Omega_2, \qquad \rho = \rho_1 + \rho_2, \qquad \mathbf{I} = \mathbf{I}_1 + \mathbf{I}_2.$

The connection between the induced and intrinsic magnetisations is

$$\mathbf{I}_2 = (\mu - 1)(\mathbf{F}_1 + \mathbf{F}_2)/4\pi = (\mu - 1)\mathbf{F}/4\pi,$$

which makes the induction **B** be

$$\mathbf{B} = \mathbf{F} + 4\pi\mathbf{I}_1 + 4\pi\mathbf{I}_2 = \mu(\mathbf{F} + 4\pi\mathbf{I}_1/\mu) = \mu(\mathbf{F} + \mathbf{h}) = \mu\mathbf{H}.$$

The various forms in (54) are got by application of the elementary potential property

$$\Sigma\,\mathbf{f}\mu\mathbf{f} = \Sigma\,p\operatorname{div}\mu\mathbf{f}$$

through all space, **f** being any "polar" force, whose potential is p, with the assistance of the various relations following (54). The forms in the first line of (54), in terms of forces, are the most important.

Now return to the subject from the impressed force point of view. (Our language may be suggestive of our believing magnetic induction to be a purely static state, but such a conclusion is not meant to be conveyed). We suppose there to be impressed magnetic force in the intrinsic magnet, of strength h at any place, which is always present. The impressed forces try to do as much work as they can. They have in any configuration of the system (referring to the external arrangement of magnetisable matter) already done the amount T magnetically, in setting up the state of induction **B**, which $= \mu$**H** everywhere, **H** being the magnetic force as ordinarily understood outside the intrinsic magnet, including inductively magnetised matter, and the same with the impressed force **h** added, where there is **h**, that is, in the intrinsic magnet, which is of course inductively magnetised as well, unless its permeability should be unity. The impressed forces take advantage of all displacements of the system to do more work, if possible. If parts of the system be free to move, move they will, in such a manner as to let the impressed forces do more work, and increase T. The generalised "force," assisting a displacement dx, is expressed by

$$dT/dx, \quad \text{or,} \quad -dM/dx;$$

since, T_0 being constant, any increase in T is accompanied by an equal decrease in M.

Any increase of permeability increases the induction and T, unless there be counteracting decrease elsewhere. A sphere of soft iron has no tendency to move anyway when placed in a perfectly uniform field of magnetic force. T is the same for any position of the sphere. But if the field be not uniform it will move so as to increase T. Any small piece of matter inductively magnetised will move in the direction of fastest increase in the square of the force of the field if its permeability be greater than that of the medium in which it moves; and in the direction of fastest decrease when its permeability is less. This is irrespective of the direction of the force. Thus iron moves to, and bismuth from, either magnet pole, and in certain positions they may move straight across the lines of force. This also happens when a wire conveying an electric current attracts iron, the motion being across the lines of force. (This is not a case of intrinsic magnetic force, but the principle is the same.)

Imagine a uniformly intrinsically magnetised magnet to be wholly surrounded by imaginary impermeable matter to begin with. There is no induction anywhere, and $T = 0$. Let outside the magnet there be matter of all degrees of permeability with no retentiveness, and divisible as much as we please, all floating in the standard medium of unit permeability. If we remove some of the impermeable matter from the surface of the magnet, the impressed forces immediately act, and some induction comes out into the surrounding space, and with it there are mechanical stresses set up, which, if yielded to by the matter, assisted by suitable guidance if required, will have the effect of bringing the most permeable matter to and driving the least permeable matter away from the magnet. All the while, the impressed forces **h** are working, increasing T the magnetic energy, and equivalently reducing M the "mechanical value," which was T_0, its greatest value, at starting; all the while doing an equal amount of work mechanically, viz., on the matter set in motion, which we may conveniently dispose of by frictional resistance. In the end, supposing we have infinitely permeable matter to surround the magnet with, the whole work done by the impressed forces will be $2T_0$, half magnetically, half mechanically. The final magnetic energy is T_0, the mechanical value or potential energy, *nil*. The distribution of T as it rises in value from *nil* to T_0 becomes ultimately confined to the magnet alone. For in the final state, the magnet is short-circuited by the infinitely permeable skin. It is only necessary for the ends of the lines of impressed force to be connected by infinitely permeable matter, and this is most simply done by the skin. To obtain the magnetic energy T_0 that is left locked up in the magnet, the impressed force must be removed; then, the equations of induction show that T_0 of heat is generated by the induced currents accompanying the subsidence of the induction.

Regarding the before-given definition of the mechanical value, notice that the more slender a filament (longitudinally magnetised) is, the less important is the effect of the polar force on the induction inside, which differs little from $\mu\mathbf{h}$, except near the ends, $\mu\mathbf{h}$ being the maximum induction $\mathbf{h}$ can produce itself. Thus by slitting up the magnet into filaments as described, and separating them infinitely, we have a final state in which the impressed forces have done infinitely nearly the full amount of work they can do, the same amount as if the magnet, without any slitting and separation, were short-circuited, if it be uniformly magnetised.

When work is done by external agency against the mechanical forces, as in drawing soft iron away from a magnet, we reduce T by the same amount. There is, during the motion, magnetic current in the magnet opposed to the impressed force, and the work done against $\mathbf{h}$ is twice the decrease in T. Half of this is accounted for by the magnetic energy returned to the magnet (becoming latent, as it were), the other half by the work done mechanically in drawing away the soft iron.

$2M$ might be called the potential energy of the impressed forces in any configuration, being at a maximum $2T_0$ when the forces are prevented from working by an impermeable skin, and zero when short-circuited (with necessary modifications for irregular distributions of $\mathbf{h}$). There are so many senses in which the energy of a magnet may be understood that it is necessary to be precise in stating one's meaning. Therefore, I repeat that by the magnetic energy I always mean the quantity T, which has the value $\mathbf{H}\mu\mathbf{H}/8\pi$ (or $\frac{1}{2}$ force $\times$ induction $/4\pi$) per unit volume, both in the magnet and without (and also when there are electric currents, only then it will not be the magnetic energy of the magnet alone), the intrinsic force, where there is any, to be included in the reckoning of $\mathbf{H}$, the magnetic force. We are thereby enabled to make use of electromagnetic ideas without bringing in the hypothetical Ampèrean currents. The inclusion of $\mathbf{h}$ in the magnetic force, making $\mathbf{B}=\mu\mathbf{H}$ always, is specially useful in simplifying both ideas and formulæ, without loss of generality. The theory of magnetism is quite difficult enough already (owing to imperfect retentiveness and variable permeability) without additional gratuitous difficulties.

To return to the mechanical forces set up by a magnet. There is a very important reservation to be made when considering the forces and the variation in the value of T as they are yielded to. The motion must be sufficiently slow to not appreciably alter the force by electric currents set up. That is, the variation of T the magnetic energy (of the magnet) does not ever give exactly the value of the generalised force, and in rapid motions it may be something very different.

For distinctness, consider a round bar magnet (intrinsic) and a round cylinder of soft iron moving in a line with its axis. As the soft iron moves, the induction increases or decreases, both in the magnet and the soft iron, according as it approaches or recedes from the magnet. From the symmetry, the lines of induced E.M.F. are circles about their common axis, and as they are conductors there are currents set up in both (there are also similar circular currents in the dielectric, but not involving waste of energy); their directions are such as to retard the increase of induction on approach, and retard the decrease on recession; hence the attraction of the magnet and soft iron is reduced as they approach, and increased as they recede from one another. More work must be done externally against the attraction in drawing away the soft iron than is done by the magnet in the reverse motion. The difference is accounted for fully, according to the laws of induction, by the heat of the induced currents. This will be greatest in the magnet itself, less in the soft iron, and a very little in surrounding conductors. This effect has nothing to do with any lagging or retardation of magnetisation in the soft iron, which, if there be any, requires separate reckoning, but is of the same nature as the resistance to the motion of a (practically) unmagnetisable conductor in the magnetic field. Substitute a cylinder of copper for the soft iron; there is no appreciable force now when the cylinder is very slowly moved, as there was no appreciable departure from the normal attraction of the magnet for the soft iron when it was very slowly moved; in both cases the currents set up by rapid motions waste energy, as heat in the magnet and in the soft iron or copper respectively, and this waste must be externally accounted for. Or, supposing the magnet to draw the soft iron from rest at a certain distance, the kinetic energy communicated to the soft iron mass after it has moved a certain length will be less than the increase of T during the motion, the deficit being wasted by the heat of induced currents.

The following brings into a strong light the connection between intrinsic force and intrinsic magnetisation. Suppose we double the permeability in every part of a magnetic system, how will it affect the magnetic energy? That depends on whether we keep the intrinsic force constant or the intrinsic magnetisation. If we keep the intrinsic force constant, we double the magnetic energy, since we keep the actual force unchanged as well, whilst we double the induction. On the other hand, if we keep the intrinsic magnetisation constant, we halve the magnetic energy; for we halve the force everywhere, whilst keeping the induction unchanged.

Similarly, any increase of permeability outside a magnet increases T and the induction. Also any increase of permeability inside the magnet increases the induction and T provided the intrinsic force at the place is unaltered. T_0 is also increased. But if, whilst increasing the permeability at a place inside a magnet, we keep the intrinsic magnetisation constant, we reduce T.

The conduction current analogue will make this plain. In any system of conduction current kept up by impressed E.M.F., any increase of conductivity outside the seat of impressed force will increase the current, and also the heat generation; the same is true if we increase the conductivity at a place where there is impressed force, if we do not alter the impressed force. But if, whilst, say, doubling the conductivity at a certain place where there is impressed force, we halve the strength of the latter, we decrease the current and the heat generation.

Section VII. Work done by Impressed Forces during Transient States.

When we charge a condenser by means of a voltaic battery a transient current is set up in the circuit, which is quickly stopped by the elastic reaction of the electric displacement in the dielectric. There is then a certain amount of electrostatic energy set up in the condenser, say U, and, during the charge, a certain amount of heat was generated in the conductor. That its value is also U (expressed as energy, to save the perfectly useless introduction of the mechanical equivalent of heat), may be seen at once on remembering that when we discharge the condenser through the same resistance (without impressed force), the current passes through the same series of values at corresponding times as during the charge, and must therefore generate the same heat, which, being now derived from the potential energy of the condenser, must amount to U. And we further see that whether the discharge circuit has or has not the same resistance as the charge circuit, the heat during the charge and discharge are equal, namely U. Thus, in charging the condenser, the battery does $2U$ of work, half of which is accounted for by the Joule-heat during the charge, and the other half by the energy of displacement in the condenser.

This property is wholly irrespective of the manner in which the charge takes place, if no other work be finally done than in heating and in setting up electrostatic energy. Thus a coil may be inserted in the circuit, which may materially alter the manner of the charge, and render it oscillatory; still, the heat will amount to U as before. And if we put another coil near the first, so that there is a current during the charge in it as well as in the main circuit, the heat will still be exactly U, provided we include the heat in the secondary coil as well.

Similarly, in charging a submarine cable, the distant end being insulated or only connected to earth inductively through condensers, so that the final state is one of no current (practically), the total heat during the charge exactly equals in value that of the electrostatic energy set up in the dielectric of the cable, condensers, etc., when we count the Joule-heat in the conductor, sheath, and wherever else there may be conduction current during the charge.

The general law, of which the above are examples, is as follows:—If, in any arrangement of matter, conducting (metallically), or dielectric, or both, originally uncharged and free from current, we cause any steady impressed forces to suddenly commence to act, and we keep them on, whose distribution is such that the final state is one of no current, the Joule-heat generated in the conducting parts during the transient state will exactly equal in amount the value of the final electrostatic energy set up. The impressed forces may be either in the dielectric or the conducting matter. If in the former, it does not matter how they are distributed, for the final state will be one of no current; but if in conducting matter the distribution of impressed force ceases to be permissibly arbitrary. In a linear conducting circuit, for example, their sum must be zero round the circuit. If partly conductive, partly inductive, this ceases to be necessary, but the impressed forces must act equally over the whole cross-section of the linear conductor, otherwise the final state will not be one of no current.

Suppose, however, other things being the same, the distribution of impressed force is left perfectly arbitrary in the conductors as well as in the dielectrics, with the result that the final state is a certain distribution of steady electric current in conductors, of electrostatic energy in dielectrics, and of magnetic energy in both, all to be definitely known from the given data, the distribution of impressed forces, of conductivity, capacity, and permeability. This we may conveniently divide into three cases; first, the final magnetic energy negligible in comparison with the electrostatic; next, the electrostatic energy negligible in comparison with the magnetic; and last, the real case, both being counted.

In the first case we have a transient state, during which the actual electric force anywhere is that due to the impressed force on the spot and the changing electrostatic force, (though "static" is rather misapplied), with electric current, both in conductors and dielectrics, leading to a final state in which the current is confined to conductors. Now, if there had been no electrostatic capacity, the final state of current would have been set up instantaneously, the activity of the impressed forces would be $\Sigma \mathbf{eC}_0$, $\mathbf{e}$ being the impressed force anywhere and $\mathbf{C}_0$ the current, the summation to include all places where $\mathbf{e}$ exists. This activity would have existed from the first moment, so that at the time t after putting on the impressed forces, the whole work done by them would have been $\Sigma \mathbf{eC}_0 t$, wholly accounted for in Joule-heating. In reality, when there is electrostatic energy set up as well, the whole work done by the impressed forces up to the time t, to include the transient state, exceeds the amount $\Sigma \mathbf{eC}_0 t$, which would have been done had there been no electrostatic energy to set up, by the amount $2U$, if U is the final energy of electric displacement. Besides doing an additional amount of work U in setting up the energy of electric displacement, the battery does an equal additional amount, which is accounted for as extra Joule-heating.

Thus, in charging a shunted condenser, or a cable whose further end is to earth (if we do not count the electromagnetic induction), U being the value of the final electrostatic energy, the battery will do $2U$ more work than if the electrostatic capacity were *nil*, half in extra heating, half in setting up the electrostatic energy.

In the second case we have a great difference. Here electromagnetic induction is predominant. The law is now (other things being the same) that the impressed forces do $2T$ *less* work than they would have done had there been no magnetic energy to set up, T being the value of the final magnetic energy. In the electrostatic case the work done was

$$\Sigma \mathbf{e}\mathbf{C}_0 t + 2U;$$

it is now

$$\Sigma \mathbf{e}\mathbf{C}_0 t - 2T,$$

up to any time t including the transient state. Hence the Joule-heat (instead of being U more) is now $3T$ less than if no magnetic work had been done.

In both cases, U and T, the electric and magnetic energy, are recoverable, appearing as Joule-heat in the conductors when the impressed forces are removed; but the doing of electrostatic work makes the impressed forces work faster, and of magnetic work slower. The one is potential energy, the other kinetic. The one is connected with elasticity, the other with inertia.

Lastly, coming to what is more usually the case, both electric and magnetic energy set up. During the transient state the coexistence of the two inductions causes a singularly complex state of affairs, by no means the mere resultant of the two taken separately. Yet the law, which we might guess from the preceding, is that the additional work done by the impressed forces above the amount $\Sigma \mathbf{e}\mathbf{C}_0 t$, that they would have done had there been no electric and magnetic energy to set up, amounts to $2U - 2T$, being $2U$ more on account of the electric energy U, and $2T$ less on account of the magnetic energy T; whilst the Joule-heat is increased by $U - 3T$.

This includes, of course, all the preceding special cases. Thus, in the case of charging a condenser, the final current $C_0 = 0$, and $T = 0$. The additional work $2(U - T)$ may of course be either positive or negative, according to the values of U and T.

The following proof covers the whole, the impressed forces being arbitrarily distributed, and the matter having any conductivity, capacity, and permeability. Also, eolotropy in these three respects is included.

Let $\mathbf{e}$ be the steady impressed force at any place, put on at the time $t = 0$, and kept on. Let $\mathbf{E}$, $\mathbf{H}$, Γ, $\mathbf{G}$, be the electric force, magnetic force, electric current, and magnetic current, at the time t after the commencement, and $\mathbf{E}_0$, $\mathbf{H}_0$, Γ_0, their final values, $\mathbf{G}_0$ being zero.

The activity of $\mathbf{e}$ is $\mathbf{e}\Gamma$ at any moment, and the total activity is $\Sigma \mathbf{e}\Gamma$ through all space, or wherever $\mathbf{e}$ exists. Let $\mathbf{F}$ be any "polar" electric force, then

$$\Sigma \mathbf{F}\Gamma = 0, \qquad (55)$$

because $\mathbf{F}$ has no curl, and Γ no divergence; a well-known theorem that is made visibly true by considering that the tubes of Γ are closed, whilst the line-integral of $\mathbf{F}$ in any circuit is zero.

Now, in the transient state, the equation of induction is

$$\text{curl}\,(\mathbf{e}-\mathbf{E})=4\pi\mathbf{G}\,; \qquad \text{..........(56)}$$

which becomes, in the final state,

$$\text{curl}\,(\mathbf{e}-\mathbf{E}_0)=0\,;$$

whence

$$\mathbf{E}_0=\mathbf{e}+\mathbf{F}_0, \qquad \text{..........(57)}$$

where $\mathbf{F}_0$ is polar, or $\mathbf{F}_0=-\nabla P$; P being a scalar potential, the electric potential. Choose then $\mathbf{F}=\mathbf{F}_0$; then, by (55) and (57),

$$\Sigma\,\mathbf{e}\Gamma=\Sigma\,\mathbf{e}\Gamma+\Sigma\,\mathbf{F}_0\Gamma=\Sigma\,\mathbf{E}_0\Gamma\,; \qquad \text{..........(58)}$$

or the activity is the same on putting the final real electric force for the impressed force.

Now

$$\Gamma=\mathbf{C}+\dot{\mathbf{D}}, \qquad \text{..........(59)}$$

$\mathbf{C}$ being the conduction and $\dot{\mathbf{D}}$ the displacement current, $\mathbf{D}$ being the displacement (elastic). Therefore, by (58) and (59),

$$\begin{aligned}\Sigma\,\mathbf{e}\Gamma&=\Sigma\,\mathbf{E}_0\mathbf{C}+\Sigma\,\mathbf{E}_0\dot{\mathbf{D}},\\ &=\Sigma\,\mathbf{E}_0k\mathbf{E}+\Sigma\,\mathbf{E}_0\dot{\mathbf{D}}, \quad \text{because } \mathbf{C}=k\mathbf{E}\,;\\ &=\Sigma\,\mathbf{C}_0\mathbf{E}+\Sigma\,\mathbf{E}_0\dot{\mathbf{D}}, \qquad \text{..........(60)}\end{aligned}$$

k being the conductivity, and $\mathbf{C}_0=k\mathbf{E}_0$ the final conduction current, $=\Gamma_0$.

But

$$\text{curl}\,\mathbf{H}_0=4\pi\mathbf{C}_0\,; \qquad \text{..........(61)}$$

therefore

$$\Sigma\,\mathbf{C}_0\mathbf{E}=\Sigma\,\frac{\mathbf{C}_0}{\text{curl}}\,\text{curl}\,\mathbf{E}=\Sigma\,\mathbf{H}_0\,(\text{curl}\,\mathbf{e}/4\pi-\mathbf{G}),$$

by (61) and (56). Therefore

$$\Sigma\,\mathbf{C}_0\mathbf{E}=\Sigma\,\mathbf{e}\,\text{curl}\,\mathbf{H}_0/4\pi-\Sigma\,\mathbf{H}_0\mathbf{G}=\Sigma\,\mathbf{e}\mathbf{C}_0-\Sigma\,\mathbf{H}_0\mathbf{G}.$$

Putting this in (60), we get

$$\Sigma\,\mathbf{e}\Gamma=\Sigma\,\mathbf{e}\mathbf{C}_0+\Sigma\,\mathbf{E}_0\dot{\mathbf{D}}-\Sigma\,\mathbf{H}_0\mathbf{G}. \qquad \text{..........(62)}$$

This is true at every moment. Now integrate (62) to the time, from 0 to t, to include the transient state (t must mathematically be infinity), and we get

$$\Sigma\,\mathbf{e}\int\Gamma dt=\Sigma\,\mathbf{e}\mathbf{C}_0t+\Sigma\,\mathbf{E}_0\mathbf{D}_0-\Sigma\,\mathbf{H}_0\mathbf{B}_0/4\pi, \qquad \text{..........(63)}$$

$\mathbf{D}_0$ being the final displacement, $\mathbf{B}_0$ the final magnetic induction. But

$$U=\Sigma\,\tfrac{1}{2}\mathbf{E}_0\mathbf{D}_0, \qquad T=\Sigma\,\tfrac{1}{2}\mathbf{H}_0\mathbf{B}_0/4\pi$$

are the values of the final electric and magnetic energies. So (63) becomes

$$\Sigma\,\mathbf{e}\int\Gamma dt=\Sigma\,\mathbf{e}\mathbf{C}_0t+2U-2T, \qquad \text{..........(64)}$$

which is the required result, showing that the work done by the impressed forces is increased by $2U$ on account of the electric energy, and reduced by $2T$ on account of the magnetic energy, the Joule-heat being increased by $U-3T$.

I have written out the above rather fully. Without the explanations, it goes simply thus,

$$\Sigma \mathbf{e}\Gamma = \Sigma(\mathbf{e}+\mathbf{F}_0)\Gamma = \Sigma \mathbf{E}_0\Gamma = \Sigma \mathbf{E}_0\mathbf{C} + \Sigma \mathbf{E}_0\dot{\mathbf{D}}$$
$$= \Sigma \mathbf{C}_0\mathbf{E} + \Sigma \mathbf{E}_0\dot{\mathbf{D}} = \Sigma \mathbf{H}_0 (\text{curl } \mathbf{e}/4\pi - \mathbf{G}) + \Sigma \mathbf{E}_0\dot{\mathbf{D}}$$
$$= \Sigma \mathbf{e}\mathbf{C}_0 + \Sigma \mathbf{E}_0\dot{\mathbf{D}} - \Sigma \mathbf{H}_0\mathbf{G},$$

whose time-integral is

$$\Sigma \mathbf{e}\int \Gamma dt = \Sigma \mathbf{e}\mathbf{C}_0 t + 2U - 2T.$$

This shows how much may be put in a small compass.

We should remark that it is Γ the true current that is circuital, in general, not $\mathbf{C}$ or $\dot{\mathbf{D}}$ separately, except in the final state, when the current is wholly conductive. Also, that we twice make use of the theorem

$$\Sigma \mathbf{A} \text{ curl } \mathbf{B} = \Sigma \mathbf{B} \text{ curl } \mathbf{A} \qquad (65)$$

through all space, $\mathbf{A}$ and $\mathbf{B}$ being any vector functions; of this (55) above is a special case. Giving proofs of all the potential properties made use of is out of the question. I entered fully into these matters in former articles. It is customary in mathematical investigations in electromagnetism to virtually prove this and similar theorems over and over again in the course of working out results, instead of merely quoting them; like proving a proposition in "Euclid" *ab initio*, from the axioms and definitions. It would, however, be very desirable to have special names for the various useful vector theorems connected with the ∇ operator. This is sometimes done by quoting a man's name, and leads to confusion, if two theorems are called, for instance, Laplace's theorem. I think the three fundamental theorems of Slope, Version, and Divergence would be recognisable by these names by anyone acquainted with the theorems, though not previously with these names for them. From them follow a number of others of the greatest utility, of which (65) is an example.

SECTION VIII. ELECTRIC ENERGY. CIRCUITAL DISPLACEMENT.

In the theory of electrostatics a tube of displacement has a beginning and an end, at its beginning there being positive, at its end negative electrification. The terminations of the tubes are usually upon conducting surfaces; there may, however, be interior electrification in the dielectric, if so, it has got there by convection, or by disruption. Impressed force in the dielectric is not considered in the theory of electrostatics. But should there be any, there will usually be closed tubes of displacement without electrification, as well as terminated tubes, due to the presence of conductors; and should there be no conductors, the displacement set up by impressed force is wholly circuital. This will be briefly considered later. At present we take the case of circuital displacement in a dielectric arising from electromagnetic induction. Conductors are temporarily excluded for simplicity. Let there be any state of electric displacement **D** and magnetic induction **B** in an infinitely extended dielectric, without impressed forces. A possible state of induction and displacement is meant, of course. For instance, set up any state of displacement by impressed force, which then remove, leaving the system to itself. **E** and **H** being the electric and magnetic forces at any moment, we have

$$\mathbf{B} = \mu\mathbf{H}, \qquad \mathbf{D} = c\mathbf{E}/4\pi; \quad \text{.........................(66)}$$

μ being the permeability and c the specific capacity. The tubes of **B** are always closed. Those of **D** are also closed, if there be no bodily electrification. We cannot, therefore, express the electric energy in terms of the scalar electric potential and the electrification. The appropriate form is in terms of the magnetic current and its vector-potential.

Let **Z** be such that

$$-\operatorname{curl}\mathbf{Z} = c\mathbf{E}, \quad \text{.................................(67)}$$

which is possible because this is the general integral of $\operatorname{div} c\mathbf{E} = 0$, expressing that the displacement is closed.* **Z** is the vector-potential of the magnetic current. It is given by

$$-\mathbf{Z} = \Sigma\frac{\operatorname{curl} c\mathbf{E}}{4\pi r}, \quad \text{or} \quad = \operatorname{curl}\Sigma\frac{c\mathbf{E}}{4\pi r}, \quad \text{...............(68)}$$

by potential properties. Here r is the distance from the point where **Z** is reckoned, of the element of the quantity summed up through all space. To **Z**, as given by (68), any polar term may be added without affecting (67); **Z**, after (68), being circuital, like **B** and **D**. Since the second equation of induction is

$$-\operatorname{curl}\mathbf{E} = \mu\dot{\mathbf{H}} = 4\pi\mathbf{G}, \quad \text{.......................(69)}$$

G being the magnetic current, the equation of **Z**, by (67) and (69), is

$$\operatorname{curl} c^{-1}\operatorname{curl}\mathbf{Z} = 4\pi\mathbf{G}; \quad \text{.......................(70)}$$

from which we see that when c is constant, we have

$$\mathbf{Z} = c\,\Sigma\,\mathbf{G}/r, \quad \text{..............................(71)}$$

verifying that **Z** is the vector-potential of the magnetic current.

If U be the electric energy (energy of electric displacement), we have

$$U = \Sigma\,\tfrac{1}{2}\mathbf{ED} = \Sigma\,\tfrac{1}{2}\frac{\mathbf{D}}{\operatorname{curl}}\operatorname{curl}\mathbf{E} = \Sigma\,\tfrac{1}{2}\mathbf{ZG}, \quad \text{...............(72)}$$

by potential properties, and (67) and (69).

These may be instructively compared with the corresponding magnetic equations. If **A** be Maxwell's vector-potential of the electric current, we have

$$\operatorname{curl}\mathbf{A} = \mathbf{B} = \mu\mathbf{H}, \quad \text{..........................(67}a)$$

* [The first use (not then, but now) known to me of the function **Z** in a dielectric, to give the displacement by curling, is in Professor Fitzgerald's paper "On the Electromagnetic Theory of the Reflection and Refraction of Light," *Phil. Trans.*, 1880.]

resulting from integrating $\operatorname{div} \mathbf{B} = 0$. And the first equation of induction is

$$\operatorname{curl} \mathbf{H} = c\dot{\mathbf{E}} = 4\pi\Gamma, \qquad \text{.........................(69a)}$$

Γ being the electric current. So, T being the magnetic energy, we have

$$T = \Sigma \tfrac{1}{2}\mathbf{HB}/4\pi = \Sigma \tfrac{1}{2}\frac{\mathbf{B}}{\operatorname{curl}}\operatorname{curl} \mathbf{H}/4\pi = \Sigma \tfrac{1}{2}\mathbf{A}\Gamma, \qquad \text{.........(72a)}$$

by (67a) and (69a). And, if μ be constant, we have

$$\mathbf{A} = \mu \Sigma \Gamma/r. \qquad \text{...............................(71a)}$$

These four equations marked a correspond to the former equations with the same numbers. We have also

$$\mathbf{E} = -\dot{\mathbf{A}} - \nabla P, \qquad \text{...........................(73a)}$$

and, to correspond thereto, $\quad \mathbf{H} = -\dot{\mathbf{Z}} - \nabla\Omega, \qquad \text{.......................(73)}$

P being the scalar single-valued electric potential, and Ω the scalar single-valued magnetic potential. (73a) is Maxwell's equation. To prove (73), we may merely remark that by (69a) the magnetic force round a closed curve equals 4π times the electric current through the curve; and by (67) the same relation holds between $-\mathbf{Z}$ and 4π times the displacement. Or, differentiate (67) to the time, and compare with (69a).

So far as the energy expressions in (72) and (72a) go, it does not matter whether Ω and P are counted or not, though they usually exist, especially if there are variations of permeability or capacity from place to place.

Other forms of U and T:—When $\mathbf{A}$ and $\mathbf{Z}$ are wholly circuital,

$$U = \Sigma \dot{\mathbf{A}}c\dot{\mathbf{A}}/8\pi - \Sigma \tfrac{1}{2}P\sigma, \qquad T = \Sigma \dot{\mathbf{Z}}\mu\dot{\mathbf{Z}}/8\pi - \Sigma \tfrac{1}{2}\Omega\rho,$$

if $\qquad -4\pi\rho = \operatorname{div} \mu\dot{\mathbf{Z}} \quad$ and $\quad -4\pi\sigma = \operatorname{div} c\dot{\mathbf{A}};$

(73) and (73a) holding good. Here σ is imaginary electrification, and ρ imaginary magnetic matter, the first being where c and the second where μ varies.

Simple Example of Closed Displacement.

In this example there is a conductor in the field, but as, from symmetry, it will be obvious that the displacement is wholly closed, it will not matter. If an intrinsic magnet be at rest in a dielectric, there is no electric force, but merely a state of magnetic force. But if it be set in motion there is immediately a field of electric force set up as well, and of displacement and electric current due to changing displacement. Whether the displacement does not or does cause electrification will depend upon whether it is, at the surface of a conductor, wholly tangential or not, for it is the normal component that introduces surface electrification. Now, if a straight bar magnet of circular section be carried through the air parallel to its length, the lines of electric force are clearly circles about the line of motion, so that the displacement in the air is wholly circuital. Or, let a uniformly intrinsically magnetised sphere move through a dielectric in the direction of its axis of magnetisation with constant velocity v small compared with that of the propagation of light, so that we may regard the sphere's field of magnetic force to be rigidly attached to it and move with it without change. A certain state of electric force in circles about the line of motion, the continuation both ways of the sphere's axis, is set up, changing at any *fixed* point continuously. But if we travel through the air with the sphere, the electric field is stationary. We may thus regard the magnet as carrying with it, in rigid connection, a certain constant electric field as well as its magnetic field. The first approximation to the solution, by far the most important part, is readily found.

Let M be the magnetic moment of the sphere, of radius a; z distance measured along the line of motion, from a fixed origin, of the centre of the sphere at time t; r the distance of any point, P, from the centre of the sphere, and θ the angle $\mathbf{r}$ makes with $\mathbf{z}$. The magnetic potential at P is

$$\Omega = (M/r^2)\cos\theta.$$

Let v be the velocity of the sphere, then

$$z = z_1 - vt,$$

t being the time and z_1 constant.

This gives $$d\Omega/dt = -v\,d\Omega/dz;$$

so that the magnetic current at the point P is

$$\mathbf{G} = \mu\dot{\mathbf{H}}/4\pi = -\nabla\dot{\Omega}/4\pi = (v/4\pi)\nabla(d\Omega/dz);$$

or, using the above value of Ω,

$$\mathbf{G} = (vM/4\pi)\nabla\{(1-3\cos^2\theta)/r^3\}, \qquad\text{(74)}$$

the differentiation ∇ being conducted at P.

Calculate the total of $\mathbf{G}$ through the circle r, θ. This is, if $\omega = \cos\theta$,

$$\frac{vM}{4\pi}\iint r^2 d\phi\, d\omega \frac{d}{dr}\,\frac{1-3\cos^2\theta}{r^3} = \frac{3vM}{2r^2}\omega(1-\omega^2). \qquad\text{(75)}$$

The circle r, θ is a line of electric force $\mathbf{E}$. So, by the relation (69),

$$-2\pi rE\sin\theta = 6\pi vM\omega(1-\omega^2)/r^2, \quad\text{or}\quad E = -\tfrac{3}{2}vMr^{-3}\sin 2\theta.$$

This could be got more simply, but we wanted an expression for $\mathbf{G}$, and having it, made use of the relation (69) applied to a closed curve, or the Theorem of Version.

The last equation gives the strength of electric force at the point r, θ referred to the sphere's centre and axis of motion. The potential energy U of the displacement is

$$U = \Sigma\, cE^2/8\pi = \iiint \frac{c}{8\pi}\left(\frac{3vM}{r^3}\right)^2 \omega^2(1-\omega^2) r^2 dr\, d\omega\, d\phi.$$

The limits for r are a and ∞; for ϕ, 0 and 2π; for ω, -1 and $+1$. This gives

$$U = cv^2M^2/5a^3 = v^2M^2/5v_1^2a^3, \qquad\text{(76)}$$

if v_1 is the velocity of light, the electromagnetic value of c being the reciprocal of v_1^2.

U is only a small fraction of T the magnetic energy of the sphere. This, by the formula

$$T=\Sigma\,\mu H^2/8\pi=\Sigma\,B^2/8\pi\mu,$$

is easily shown to be $T=3M^2/a^3\mu(\mu+2)$,

if μ be the permeability of the sphere, the outside value being unity. Thus, if $\mu=1$ in the sphere also,

$$U/T=\tfrac{1}{5}v^2/v_1^2.$$

Travelling with the sphere, we have a steady electric field, and no current. But at a fixed point, the current, or rate of increase of displacement, is

$$c\dot{E}/4\pi=-(cv/4\pi)(dE/dz)=3v^2M\sin\theta(1-3\cos^2\theta)/4\pi r^4v_1^2.$$

As before remarked, the solution is the first approximation. In a complete theory there would be no discontinuity in the electric force at the surface of the magnet, as the above supposes, and there would be electric current in the magnet, with waste of energy by the Joule heating, thus requiring a continuously applied mechanical force to keep up the motion. Whilst, therefore, we should have disturbance inside the magnet, the solution outside would be not exactly that given. In fact, we see that the calculated dielectric current itself has its magnetic field, thus slightly altering the assumed magnetic field, that of the magnet at rest; and to the motion of this new magnetic field (very weak) there corresponds a new electric field, and so on. However insignificant these corrections may be in point of magnitude, they are yet required to make up a complete system satisfying the laws of induction. Taking, however, the above field of electric displacement by itself, we may close the magnetic currents in the appropriate manner on the surface of the magnet itself. The surface value of E is, by (75),

$$-\tfrac{3}{2}Mva^{-3}\sin 2\theta, \qquad\qquad(77)$$

and it is tangential. Hence the same expression divided by 4π, and taken positively, is, by the surface interpretation of (69), the strength of the complementary surface magnetic current, directed at right angles to the electric force—that is, along the meridional lines, if the poles be those points of the sphere cut by the line of motion through the centre. This system, with the former, makes a closed system of magnetic current, whose vector-potential may be taken to be given by

$$Z=cMvr^{-2}\cos\theta,\quad\text{parallel to }\mathbf{z}. \qquad\qquad(78)$$

For this satisfies (67). This is literally the vector-potential of a surface magnetic current of strength $\sigma\mathbf{v}$, parallel to $\mathbf{z}$, σ being the surface-density of free magnetism; but it is unnecessary to calculate the part due to the complementary current required to close it.

We can now check the value of U by the formula (72).

$$U=\Sigma\,\tfrac{1}{2}\mathbf{ZG}=\tfrac{1}{2}\frac{3}{4\pi}\frac{Mv}{a^3}\frac{cMv}{a^2}\iint\omega^2(1-\omega^2)a^2d\omega\,d\phi. \qquad\qquad(79)$$

Here we have to integrate the scalar product of $\mathbf{Z}$ and $\mathbf{G}$ through all space, $\mathbf{Z}$ being given by (78) and $\mathbf{G}$ by (74) outside the magnet, and by

(77) (when divided by -4π) on its surface. But the volume-integral outside the magnet vanishes, because **Z** contains Q_1 and **G** contains Q_3 as a factor (zonal harmonics). Hence there is simply the surface integral left, expressed by the right side of (79), the z-component of the surface **G** alone counting. It gives the value (76) again.

The reason why (78) suffices for use in the formula $\Sigma \frac{1}{2}\mathbf{ZG} = U$ is because **G** is circuital. If, on the other hand, we employed the full formula for **Z**, we would not need to close **G**. A surface-current $\sigma\mathbf{v}$ parallel to **z** would then suffice. Thus, if $\mathbf{Z} = \mathbf{Z}_1 + \mathbf{Z}_2$ and $\mathbf{G} = \mathbf{G}_1 + \mathbf{G}_2$, wherein **Z** and **G** are both circuital, whilst the parts $\mathbf{Z}_2$ and $\mathbf{G}_2$ are polar, we have

$$\Sigma\,\mathbf{ZG} = \Sigma(\mathbf{Z}_1 + \mathbf{Z}_2)\mathbf{G}_1 = \Sigma\,\mathbf{Z}_1(\mathbf{G}_1 + \mathbf{G}_2),$$

since

$$\Sigma\,\mathbf{ZG}_2 = 0 \quad \text{and} \quad \Sigma\,\mathbf{Z}_2\mathbf{G} = 0.$$

SECTION IX. IMPRESSED ELECTRIC FORCE IN DIELECTRICS.

A comparison is often made between distributions of magnetic induction and of electric current. There is, however, a far more satisfactory analogy between magnetic induction and electric displacement in a dielectric, which may be pushed much further before correspondence ceases. So far as mere distributions in space go, of the three phenomena of conduction current, electric displacement in a dielectric, and magnetic induction, we may conveniently compare them simultaneously.

First, let there be a distribution of impressed electric force $\mathbf{e}_1$, in a conductor of conductivity k (infinitely extended in the general case, with the conductivity different at different places), setting up a steady state of electric force $\mathbf{E}_1$, and conduction current **C**. We have the three conditions

$$\mathbf{C} = k\mathbf{E}_1, \qquad \operatorname{div}\mathbf{C} = 0, \qquad \operatorname{curl}(\mathbf{e}_1 - \mathbf{E}_1) = 0. \quad \text{..........(80}a\text{)}$$

Secondly, in a non-conducting unelectrified dielectric of capacity c, in which a distribution of impressed electric force $\mathbf{e}_2$, sets up a steady state of electric force $\mathbf{E}_2$, and displacement **D**, we have

$$\mathbf{D} = c\mathbf{E}_2/4\pi, \qquad \operatorname{div}\mathbf{D} = 0, \qquad \operatorname{curl}(\mathbf{e}_2 - \mathbf{E}_2) = 0. \quad \text{........(80}b\text{)}$$

Thirdly, in a medium of permeability μ, in which a distribution of impressed magnetic force **h**, sets up a steady state of magnetic force **H**, and induction **B**, we have

$$\mathbf{B} = \mu\mathbf{H}, \qquad \operatorname{div}\mathbf{B} = 0, \qquad \operatorname{curl}(\mathbf{H} - \mathbf{h}) = 0. \quad \text{............(80}c\text{)}$$

These three sets of conditions are exactly similar. We have in each case a "force" and a "flux." The first condition is the linear relation between the force and the flux, *i.e.*, Ohm's law, etc. The second condition is that of continuity of the flux, asserting that its divergence or convergence is zero everywhere, or that the flux is circuital. The third is the force equation, what the equation of induction becomes when the state is steady; the third conditions in (80a) and (80b) being examples of

$$\operatorname{curl}(\mathbf{e} - \mathbf{E}) = \mu\dot{\mathbf{H}},$$

with $\dot{\mathbf{H}}=0$, and that in (80c) arising from

$$\mathrm{curl}\,(\mathbf{H}-\mathbf{h})=4\pi\dot{\Gamma},$$

with $\Gamma=0$. The difference between the actual and the impressed force, or the natural force of the field itself, has no rotation, or its line-integral round any closed line whatever is zero. Or,

$$\mathbf{E}_1=\mathbf{e}_1+\mathbf{F}_1, \qquad \mathbf{E}_2=\mathbf{e}_2+\mathbf{F}_2, \qquad \mathbf{H}=\mathbf{h}+\mathbf{F}_3,$$

where $\mathbf{F}_1$, $\mathbf{F}_2$, $\mathbf{F}_3$ are "polar" forces entirely, derived from single-valued scalar potentials, whose space-variations give the polar forces; thus

$$\mathbf{F}_1=-\nabla P_1, \qquad \mathbf{F}_2=-\nabla P_2, \qquad \mathbf{F}_3=-\nabla\Omega,$$

where P_1 and P_2 are electric potentials and Ω magnetic potential.

The three conditions serve to determine unambiguously the complete solution, so far as the force and the flux are concerned, when the impressed force and the distribution of conductivity, etc., are given. To the impressed force we require to add a polar force to make up a complete system of force satisfying the continuity of the flux. At the poles, or places where the polar force converges, or diverges, we may, if we like, put imaginary matter, electric or magnetic, as the case may be, repelling according to the inverse-square law, and regard the potentials as the potentials of the matter. Each distribution of impressed force requires a particular polar force to supplement it; except when the impressed force is so distributed that it can by itself satisfy the continuity of the flux, and the linear relation between the flux and the force. Thus, when

$$\mathrm{div}\,k\mathbf{e}_1=0, \quad \text{or} \quad \mathrm{div}\,c\mathbf{e}_2=0, \quad \text{or} \quad \mathrm{div}\,\mu\mathbf{h}=0,$$

no polar force is needed, and there is none, or the potential does not vary. We have then

$$\mathbf{C}=k\mathbf{e}_1, \qquad \mathbf{D}=c\mathbf{e}_2/4\pi, \qquad \text{and} \qquad \mathbf{B}=\mu\mathbf{h}$$

respectively. On the other hand, should the impressed force be itself polar in its distribution, there is no flux produced. We then have

$$\mathbf{e}_1=-\mathbf{F}_1, \ \text{etc.}, \qquad \text{and} \qquad \mathbf{E}_1=0, \ \text{etc.}$$

Now, if in the above three problems, the distributions of impressed force—electric or magnetic, as the case may be—are identical, and also the distributions of conductivity, etc., in space, then also the three fluxes have identical distributions. Practically, as neither the permeability nor the capacity (in non-conductors, at any rate) can vanish, we must not let the conductivity be zero anywhere in the conduction current problem; *i.e.*, all space must be conducting, more or less, to get identical distributions of current to those of induction and displacement respectively. We cannot confine either of the last to definite closed channels, as we do electric currents, by arrangement of matter, although we can do so by proper distributions of the impressed force, viz., in the above mentioned cases of no polar force.

The distributions of Joule-heat per second (or dissipativity), of electric energy (or energy of displacement), and of magnetic energy are also similar, being $\mathbf{E}_1 k \mathbf{E}_1$, $\mathbf{E}_2 c \mathbf{E}_2/8\pi$, and $\mathbf{H}\mu\mathbf{H}/8\pi$ respectively per unit volume; and if their totals through all space be Q, U, and T respectively, we have

$$Q = \Sigma\,\mathbf{E}_1 k \mathbf{E}_1 = \Sigma\,\mathbf{e}_1 k \mathbf{E}_1 = \Sigma\,\mathbf{e}_1 k \mathbf{e}_1 - \Sigma\,\mathbf{F}_1 k \mathbf{F}_1, \quad \text{.........(81}a)$$

$$8\pi U = \Sigma\,\mathbf{E}_2 c \mathbf{E}_2 = \Sigma\,\mathbf{e}_2 c \mathbf{E}_2 = \Sigma\,\mathbf{e}_2 c \mathbf{e}_2 - \Sigma\,\mathbf{F}_2 c \mathbf{F}_2, \quad \text{.........(81}b)$$

$$8\pi T = \Sigma\,\mathbf{H}\mu\mathbf{H} = \Sigma\,\mathbf{h}\mu\mathbf{H} = \Sigma\,\mathbf{h}\mu\mathbf{h} - \Sigma\,\mathbf{F}_3 \mu \mathbf{F}_3. \quad \text{.........(81}c)$$

In the first form of expression, as $\Sigma\,\mathbf{E}_1 k \mathbf{E}_1$, the quantity summed is the actual amount in the unit volume. In the second form the summation extends only where there is impressed force, being of the scalar product of the impressed force and the actual flux.

In the third form, there are two summations, both necessarily of positive amounts, whose difference gives the dissipativity, etc. The first extends only where there is impressed force, being the greatest value of Q, etc. The second extends over all space where there is polar force, and vanishes when there is none. The parts depending on the polar forces may also be expressed in terms of the potential and of imaginary matter. Thus—

$$\Sigma\,\mathbf{F}_1 k \mathbf{F}_1 = \Sigma\,P_1 \rho_1, \quad \text{if} \quad \rho_1 = \operatorname{conv} k \mathbf{e}_1, \quad \text{............(82}a)$$

$$\Sigma\,\mathbf{F}_2 c \mathbf{F}_2/8\pi = \Sigma\,\tfrac{1}{2} P_2 \rho_2, \quad \text{if} \quad \rho_2 = \operatorname{conv} c \mathbf{e}_2/4\pi, \quad \text{.........(82}b)$$

$$\Sigma\,\mathbf{F}_3 \mu \mathbf{F}_3/8\pi = \Sigma\,\tfrac{1}{2}\Omega\sigma, \quad \text{if} \quad \sigma = \operatorname{conv} \mu \mathbf{h}/4\pi. \quad \text{.........(82}c)$$

Note here that the distribution of imaginary matter is not the same (in general) as that before mentioned, measured by the divergence of the polar force. Here the matters are distributed where the impressed force, multiplied by the conductivity, etc., has convergence. In (82c), σ is the density of imaginary magnetic matter on the ends of a magnet. But in (82a) and (82b), ρ_1 and ρ_2 are not distributions of electrification, for there is none in either case. The electrification is measured by the divergence of the displacement, which is zero under the stated conditions.

We may also employ vector-potentials in all three cases. Thus

$$Q = \Sigma\,\mathbf{E}_1 \mathbf{C} = \Sigma\,\mathbf{H}_1 \mathbf{G}_1, \quad \text{....................(83}a)$$

$$U = \Sigma\,\tfrac{1}{2}\mathbf{E}_2 \mathbf{D} = \Sigma\,\tfrac{1}{2}\mathbf{Z}\mathbf{G}_2, \quad \text{....................(83}b)$$

$$T = \Sigma\,\tfrac{1}{2}\mathbf{H}\mathbf{B}/4\pi = \Sigma\,\tfrac{1}{2}\mathbf{A}\Gamma. \quad \text{....................(83}c)$$

The relations of these new quantities are

$$\operatorname{curl} \mathbf{H}_1 = 4\pi\mathbf{C}, \qquad \operatorname{curl} \mathbf{e}_1 = 4\pi\mathbf{G}_1, \quad \text{....................(84}a)$$

$$\operatorname{curl} \mathbf{Z} = c\mathbf{E}_2, \qquad \operatorname{curl} \mathbf{e}_2 = 4\pi\mathbf{G}_2, \quad \text{....................(84}b)$$

$$\operatorname{curl} \mathbf{A} = \mu\mathbf{H}, \qquad \operatorname{curl} \mathbf{h} = 4\pi\Gamma. \quad \text{....................(84}c)$$

In the conduction current case (83a) and (84a), $\mathbf{H}_1$ is the magnetic force of the current, and $\mathbf{G}_1$ is an imaginary distribution of magnetic current, viz., where the impressed force has rotation, or varies laterally. In (83b) and (84b), $\mathbf{G}_2$ is also an imaginary magnetic current, similarly related to the impressed force, whilst in the magnetic case (83c) and (84c), Γ is similarly related to the impressed magnetic force, and is the well-known imaginary electric current which would (if it were a real current) correspond to the same state of magnetic induction as the impressed force **h** sets up. The summations do not extend over the whole region of impressed force, but to portions only, perhaps round a single line. This we will illustrate in the conduction current case. The formula $Q = \Sigma \mathbf{H}_1\mathbf{G}_1$ has some suggestiveness in connection with the transfer of energy, but turns out to have no very important application. As we remarked before, if the impressed force be polar, there is no flux. For there to be any flux at all, the impressed force must have curl somewhere, and $Q = \Sigma \mathbf{H}_1\mathbf{G}_1$ shows the exact dependence of the activity on the situation and amount of this curl. Take a simple voltaic circuit, copper and zinc in acid, and the copper continued by a copper wire to the zinc outside the liquid. Suppose the impressed forces are entirely confined to the metal-acid surfaces, and are of uniform strength over each metal. Then the places of summation are the two wind-and-water lines. One place gives the amount of energy leaving the zinc per second, the other the amount arriving at the copper, and their difference is the amount of Joule-heat in the circuit. Next suppose that the zinc-air force equals the zinc-acid force, and that the copper-air force equals the copper-acid force. There is now only one place of curl of impressed force, viz., the air-boundary of the copper-zinc junction, and the summation round that line only gives the value of Q. But not only may we thus shift the places of summation outside the battery, but we may locate them altogether away from the circuit if we like by suitable dispositions of impressed force outside the circuit, which will not in any way disturb the state of magnetic force and current provided we keep the same impressed forces in the battery as before, although they will alter the paths of the transfer of energy. In problems (*b*) and (*c*) there is, of course, no transfer of energy at all after the steady states, which are alone considered, have been set up.

In problems (*b*) and (*c*) we may also compare the mechanical stresses. They are such as to increase U or T respectively when allowed to act, by letting the impressed forces do more work. Two bar magnets repel with like poles, and attract with unlike poles approached, whilst either pole of either will attract soft iron and repel bismuth. Similarly, take two bars of a dielectric and put in them impressed electric force parallel to their lengths. Like ends will repel and unlike attract one another, whilst either pole will attract a piece of solid dielectric of greater capacity than air (in which all are immersed), in which there is no impressed force, or repel it if its capacity be less than that of the air, or other surrounding medium. The same will happen if there be intrinsic displacement in the solid bars, such as arises from so-called "absorption." (If in the conduction current problem there were a corresponding tendency for Q to increase, then a copper ball in mercury conveying a current would be attracted by either electrode, where the current has greater density.)

Dismissing now the problems (*a*) and (*b*), consider something quite peculiar to (*c*), that of dielectric displacement. Both the conduction current and the magnetic induction are continuous, as expressed by the second condition in each case. We have also written down the same condition for the electric displacement. No arrangement of impressed electric force can *create* any discontinuity in the displacement, provided there be no conduction anywhere. Should there be any discontinuity, its proper measure is the divergence of the displacement. That is, ρ the volume-density of the electrification—for it is nothing else than discontinuity in the displacement—is given by

$$\rho = \text{div}\,\mathbf{D}.$$

Supposing there to be any such discontinuity, it will remain a fixture, or only be varied in distribution by motion of matter carrying the electrification with it—of matter, because it appears most probable that there cannot be any electrification without the presence of matter; or, the displacement in the electromagnetic medium itself, the ether, is always continuous. To get rid of the electrification there must be conduction. Conversely, to create electrification there must be conduction. The two statements go together.

Now the essential property of a conduction current is dissipation of energy by the heating effect produced. Destruction of electrification by disruption of the dielectric is therefore of the nature of a conduction current, at least partly, although the simple metallic conduction law is not followed, and, in fact, it is not known definitely what is the exact course of events, even when considered merely electrically. But we should never, in attempting to explain something, go from the complex and ill-understood to the comparatively simple and mathematically expressible, but pursue the other course. We may say then, with tolerable certainty, that to create electrification there must be (1), the presence of matter; (2), impressed electric force, *i.e.*, a cause in operation which tends to produce an electric current (if we like we may say which tends to produce magnetic force, and regard the current as an affection of the magnetic force), which is neither the polar electric force nor the electric force of induction, these being the only two naturally belonging to the electromagnetic medium; (3), conduction, with dissipation of energy.

Nothing is easier than to create electrification experimentally or involuntarily by the contact or friction of bodies. Imaginative explanations may be readily made up, and are likely to be of very little value. It is, then, necessary to be somewhat general, or vague, in order to keep on the straight and narrow path. We may create electrification by the contact of different conductors, either by means of the small known contact-force of thermal origin, or of the much larger air-surface contact-forces of chemical origin, and, so far, it is easy to recognise the presence of the matter, of the impressed force, and of the conduction. Similarly, by the contact of a conductor and a dielectric, if there be similar impressed forces present. No friction is absolutely necessary. But if we set two dielectrics in contact, without any connection with conducting matter, no impressed electric forces can, without a change of conditions, set up any electrification. Some friction is needed, and with it there is conduction, or equivalently disruption. The actual original impressed electric forces need not be great, and are probably small. It is the act of separating mechanically the opposite electrifications produced by the friction, at first extremely close together, that is the main cause of the high difference of potential observed after separation, together with the similar separation taking place during the friction. The electric field is, for the most part, set up during the separation, and derives its energy in the main from the mechanical work then done.

Now, having got electrification—ρ per unit volume, say—the field due to it is settled by the three conditions,

$$\mathbf{D} = c\mathbf{E}/4\pi, \qquad \operatorname{div}\mathbf{D} = \rho, \qquad \operatorname{curl}\mathbf{E} = 0, \quad \ldots\ldots\ldots\ldots\ldots(85)$$

differing from the former, equations (80*b*), in the second and third, in the absence of impressed force and the existence of ρ. The field is definitely determined by (85); the force is polar completely, and the energy is

$$\Sigma\,\mathbf{E}c\mathbf{E}/8\pi = \Sigma\,\tfrac{1}{2}P\rho,$$

if P be the potential.

Should there be also impressed forces, the actual field will be the sum, in the vector sense, of the two fields, due separately to the electrification and to the impressed forces. Not only that, but the total energy of displacement will be the sum of the amounts in the separate fields; or, the mutual energy of the two fields is zero. This is true because the displacement in the field of the impressed force is everywhere continuous, and the force of the other field wholly polar.

Electrification has no magnetic analogue, the magnetic induction being always continuous. It is important that the distinction between electrification, as above considered, and *imaginary* free electricity (as when the force is discontinuous, though not the displacement) should be clearly recognised. The latter is the analogue of free magnetism.

Similarly, there is no electrification in a conductor supporting a current, provided it be not dielectric as well. In the latter case it is the true current, the sum of the conduction and displacement current, that is continuous, making them sometimes separately discontinuous, thus

$$-\operatorname{div}\mathbf{C} = +\operatorname{div}\dot{\mathbf{D}} = \dot{\rho}; \qquad \operatorname{div}\Gamma = 0;$$

$\mathbf{C}$ being the conduction, $\dot{\mathbf{D}}$ the displacement current, and their sum Γ. It is usually only at the surface of a conductor that there is electrification; should, however, the specific capacity of the conducting matter itself be variable from place to place, there will generally be interior electrification as well, during the existence of a conduction current.

Section X. Dielectric Displacement and Absorption.

The most remarkable and distinguishing feature of Maxwell's theory of electromagnetism is his dielectric current, whose introduction into the theory gives us a dynamically complete system, with propagation of disturbances in time through the medium surrounding and between conductors, doing away with the mathematically expressible but practically unimaginable, instantaneous actions of currents upon one another at a distance, and similar, though more simply expressible, instantaneous forces between imaginary accumulations of the "electric fluid." Some might say that the distinguishing feature is his dielectric displacement. But it is scarcely that, strictly. For that a dielectric medium was put into a state of polarisation by electric force was Faraday's idea, and this polarisation is only another name for electric displacement. Again, Sir W. Thomson had given mathematical expression to the idea of polarisation in a dielectric; the statement $\mathbf{D} = c\mathbf{E}/4\pi$, expressing the displacement or polarisation in terms of the electric force, had been given in a somewhat similar form by him. Whether we call it displacement or polarisation does not matter; the important step made by Maxwell was the recognition that changes in the displacement constitute a real electric current (though without dissipation of energy in Joule-heating), and that the electric current, whether conductive or not, is always continuous. Accumulations are done away with altogether, and with their abolition the fluid or fluids become meaningless.

As regards the term displacement, though it may be objected to as misleading, suggesting a real displacement in a certain direction, whereas the phenomenon, though undoubtedly having some directional peculiarity, is probably not of the nature of a simple displacement, just as the conductive current is not the motion of a fluid through a wire, yet on the other hand, it is to be remarked that the term current is firmly fixed in use, and, to accompany it, there could be no better term than displacement. The time-integral of the current, from the zero configuration, in a dielectric, is the displacement; conversely, the current is the time-variation of the displacement. It would therefore be a pity to abolish the term displacement, unless we simultaneously abolish current; and it would be hard to find two words which fit together so well.

If we define the displacement as the time-integral of the current anywhere, not merely in a non-conducting dielectric, the current is the time-variation of the displacement anywhere. The displacement, reckoned from a proper zero at a certain time (any time when there was no dielectric displacement) is then, like Maxwell's true current, a vector magnitude of no convergence at every moment and everywhere. In some special investigations this is useful. We may then speak of conductive displacement and elastic displacement, but I believe in general it is best to confine the term displacement to elastic displacement (in a dielectric) only, unless we are careful to qualify the word by a preliminary adjective. In the absence of any qualification displacement in a dielectric is meant, and, should it be also conducting, the conductive displacement must be separately reckoned.

The difference between Maxwell's and older views regarding electricity and the electric current is instructively brought into prominence by making a small change in Maxwell's system, or rather in taking an imaginably possible, though really untrue, special case of the same. Put $c = 0$ everywhere. That is, stop all elastic displacement. Make no other change. The electric current is left continuous, and is now necessarily confined to conductors only, and at their surfaces must always be tangential. But there will still be magnetic force and induction, mutual forces between circuits, induction of one circuit on another, and there will still be transfer of energy through the medium from sources to sinks of energy. We have simply done away with the "elastic yielding of the connecting mechanism." The velocity of propagation of waves, as the specific capacity is imagined to be reduced to nothing, becomes infinitely great, instead of being only that of light. The theory of the induction of linear circuits becomes that given by Maxwell, wherein dielectric displacement is ignored. No effect is produced upon the distribution of steady current set up by steady impressed forces, but in the first transient states there are great changes. In a linear circuit, for example, the current is now absolutely constrained to keep in the conductor, so that there can be no surface charges or static retardation. Though there is no electrification, there is still electric potential. But we cannot charge a condenser. Therein lies the difference from reality. To be able to do this, without employing Maxwell's dielectric current, we are necessitated to suppose that the current is not continuous, but that it is the real motion of something that can accumulate in places. We are also obliged to change the relation between magnetic force and current, as it implies continuity of the current everywhere. Many other changes are also required to make a consistent system, for one change necessitates another, and we shall ultimately come to something extremely different from Maxwell's system.

In view of the extreme relative simplicity of Maxwell's views, and their completeness without any artificial contrivances to save appearances, and in their modernness, referring to modern views regarding action at a distance, one is almost constrained to believe that the dielectric current, the really essential part of Maxwell's theory, is not merely an invention but a reality, and that Maxwell's theory, or something very like it, is *the* theory of electricity, all others being makeshifts, and that it is the basis upon which all future additions will have to rest, if they are to have any claims to permanency.

Electric displacement is primarily a phenomenon of the ether. Ether is perfectly elastic. It must be so if there be no absorption of the energy of radiation during transmission through space. This conclusion is of course independent of Maxwell's view of light being itself an electromagnetic disturbance. If the energy of displacement be potential energy, the displacement, whatever it really be, is of a perfectly elastic character, in the absence of ordinary matter.

But when electric displacement occurs in a solid dielectric, if there be, as there must be, mutual influence between the ether and matter, we may expect the elastic properties of the matter to be communicated, apparently, to the ether. Thus, no solid is perfectly elastic, and, consequently, electric displacement in a solid dielectric is not perfectly elastic, as it is assumed to be in the formula $\mathbf{D} = c\mathbf{E}/4\pi$, with c invariable at any place, the linear relation between the displacement and the electric force.

A solid, under the influence of externally-applied force, is strained. If the strain be under a certain magnitude it is assumed to be perfectly elastic, so that whilst the stress remains steady the strain does not change; and on the removal of all stress the strain entirely ceases. But in reality, if a certain applied force deform an elastic solid, and the force be kept on, the deformation slowly increases, as if the elasticity slowly decreased. And on the removal of the applied force the original zero configuration is not immediately recovered. The residual deformation will then slowly subside. If, after the first approximate return to the original state, the body be, by constraint, prevented from changing shape, the solid, which at first did not react against the constraint, will gradually do so, so that on the removal of the constraint, a second return, in a lump, towards the original configuration, will take place.

Remembering that a dielectric under electric stress is in a state of strain, we may expect there to be a corresponding electrical phenomenon. The displacement produced by a given constant impressed force should be, at first, appreciably quite elastic, but should thereafter slowly increase. On the removal of the impressed force the original displacement should at once subside, leaving a small residual displacement which should subside very slowly, as it came on. Or, if the residual displacement be fixed, so that it cannot subside, it should appear to gradually come back into existence, as elastic displacement referred to the original zero, so that on the removal of the constraint a second sudden subsidence should take place.

This is what happens in the phenomenon of electric absorption in a perfectly insulating dielectric. Whether it be really true or not that a part of the electric displacement becomes intrinsic by reason of some *quasi*-rigid connection with a similar phenomenon taking place in the strained solid, there is no doubt that there is a remarkable resemblance in the details of the two cases. That the presence of the matter causes the displacement to be increased from what it would be with the same electric force in ether is a separate matter, as this refers to elastic displacement ($c > 1$ in all dielectrics, if unity in vacuo).

An elastic spring is therefore the most correct analogue to a condenser when we wish to make up a mechanical illustration of the electro-elastic properties of a solid dielectric. We may, for instance, take a flat spring, clamp it firmly at one end, and apply pressure or pull to the free end in a direction perpendicular to its flat sides. Consider the applied force to represent the E.M.F. of a battery joined to a condenser, and the displacement of the free end of the spring to represent the electric displacement in the condenser. The displacement will be proportional to the force, in both cases, approximately; and if by any suitable means we magnify, mechanically or optically, the motions of the spring under varied circumstances, we shall see a corresponding set of phenomena to those occurring under similarly varied circumstances in the case of the electric displacement.

But an illustration, which, though less exact, is more easily followed by the mind's eye, when we cannot render visible the absorption properties of a spring, is that which occurred to me when first making acquaintance with Maxwell's mechanical illustration, the parallel vertical tubes, containing water. This being rather complex, I substituted the following:—Electric displacement in a condenser is represented by the actual displacement of a piston in a cylinder from its position of natural equilibrium in the middle thereof, when the cylinder is perfectly air-tight and contains an equal amount of air on each side. To give motion to the piston, rods may be attached to it, passing through holes in the closed ends of the cylinder. Call the ends of these rods, outside the cylinder, *a* and *b*, and the two corresponding air spaces in the cylinder A and B, *a* and A being to the left, *b* and B to the right. If pressure be applied to *a* or a pull to *b*, or both, the piston will be displaced from left to right, and small displacements will be proportional to the corresponding applied forces.

The displacement of the piston corresponds to the total electric displacement in a condenser; the applied force to the E.M.F. of the battery on the condenser; the back pressure of the rod *a* to the difference of potential of the condenser plates; the displacement of *a* inward to the positive charge, and of *b* outward to the equal negative charge. Insulation of the condenser is represented by fixing *a* or *b* so that the displacement cannot change.

If we like to carry the illustration further, we may cause the rods *a* and *b* to meet with frictional resistance when moving, proportional to the speed of their motion. This speed will correspond to the strength of current, and the coefficient of friction to the resistance of the conductor joining the poles of the condenser. We may go further, and suppose the mass of the piston and rods to represent the inductance of the electric circuit, thus obtaining an illustration of the oscillatory discharge which occurs with suitable values of the resistance, capacity and inductance. Although this analogy, which is well known in one form or another, is very close, and therefore educationally valuable, it should be remembered that it suggests that the momentum of an electric current is that of matter moving with the current, or of the current itself, if it be the motion of matter, having therefore necessarily momentum. So far it is apt to mislead; for electricity has no momentum itself, or kinetic energy. The momentum is that of the magnetic induction, or is proportional thereto, and it, and the (nominal) energy of the current, exist wherever there is magnetic induction, not merely in the wire.

To imitate absorption, make the piston very slightly leaky. Then if *a* be pushed in by a steady pressure, the first displacement of the piston is elastic with reference to the proper zero, the middle of the cylinder. But air then leaks slowly from B to A on account of the increased pressure in B, which causes the back pressure of *a* to decrease, and allows the same applied force to slowly increase the displacement. (This corresponds to the slow continuous increase of electric displacement in a condenser when a constant battery is kept on). Or, if *a* be fixed, the back pressure will slowly fall whilst the displacement remains constant. (If the condenser be insulated, its difference of potential will fall whilst its charge remains constant.) Unfix *a*; the first return equals the first displacement, approximately, but there is left a small displacement which will slowly subside if the piston-rods be free; or, if a be fixed, a pressure at a will be gradually brought on by the slow leakage from A to B, so that a second sudden motion of the piston back to its proper zero can be got on unfixing a. (Remove the battery, but close the circuit still; the first discharge will approximately equal the first charge, but there is left a small charge which will slowly subside if the circuit remain closed; or, if it be insulated, a difference of potential of the same kind as before will gradually come on, so that a second discharge can be got on again closing the circuit.)

To instantly remove the displacement of the piston after leakage has occurred, we must apply an opposite force, a pressure at b or pull at a of the right amount to bring the piston to the middle. If we then fix a and leave the thing to itself the pressure of the rod at b will gradually subside. Similarly if we, after absorption has occurred in a condenser, and the first discharge has been taken, charge it oppositely by a reverse E.M.F. of the right amount to make the real displacement zero, and leave the condenser insulated, the apparent opposite charge will gradually disappear.

Since the first discharge equals the first charge we may regard the capacity of the condenser as being constant, and the displacement at any moment to consist of two parts, first that due to the battery E.M.F., which can be got rid of at any moment, and next a temporary intrinsic displacement which is kept up by impressed electric force in the dielectric itself arising from its altered structure, or the changed zero of its elastic deformation. The displacement due to a certain total E.M.F. in the circuit is the same however it be distributed, whether in the conductor or dielectric, if, over any cross-section, it be evenly distributed. But only the part in the conductor causes difference of potential in a steady state between the ends of the conductor, so that the intrinsic displacement due to internal impressed force gives no external sign of its existence until, by the removal of the impressed force, difference of potential is developed between the terminals of the condenser.

In the case of the piston in the cylinder, the intrinsic force keeping up displacement of the piston after the rods have been set free is the pressure of air in A against the piston from left to right, which, had it not been for the leakage, would have been in B, and have pressed the piston the other way.

Another effect may be mentioned. If the dielectric of the condenser be of a such a nature that its capacity increases or decreases with the temperature, then, on suddenly charging it, there will be a cooling or a heating effect produced in the dielectric. This has also its parallels in the metal spring and "spring of the air" illustrations.

Intrinsic displacement in a dielectric, without conducting matter surrounding it, to render the displacement wholly latent, has also some interesting features, which will be considered later.

SECTION XI. THE PRINCIPLE OF THERMAL RESISTANCE.

Suppose we, with a sure faith in the truth of the principle of Conservation of Energy, and a knowledge of the equivalence of work and heat, observe that the motion of a magnet in the neighbourhood of a closed circuit generates heat in it, that this ceases when the motion ceases, and that we satisfy ourselves that this heat is the only final result of the motion, so far as energy is concerned. This heat must be the equivalent of work done. As we have no reason to suppose that the magnet is in a different state at the end from what it was in at the beginning of the motion, we cannot attribute the heat to a loss of potential energy by the magnet. Hence we may conclude that the heat is the equivalent of work done against resistance to the motion of the magnet. That is to say, however the magnet move, its motion is resisted. Here we have Lenz's law, without any reference to the direction of the current induced in the circuit, but merely as regards conservation. In fact we made no mention of current. If now we bring in other knowledge, that there is a current induced in the circuit, the heat being proportional to the square of its strength, we are still left without means of determining its direction in a given case. Only finally, when we utilise Ampère's determination of the mutual forces between magnets and currents, can we exactly say in which direction the induced current will be in a given case, for it must always be such as to resist the motion.

Lenz's law is not, however, the subject of the present section, but is used to point out the distinction between the above kind of resistance to motion and the kind involved in the Principle of Thermal Resistance, to which we now proceed. It is such a large subject, and there are so many ways of treating it, that it is difficult to know how to begin; after consideration I adopt a method which I have not met with, and which is therefore, if not novel, at least unusual; believing that, whether it be better or worse than other methods, there is advantage in viewing a truth from all possible sides, to allow another law, that of the survival of the fittest, which, like that of thermal resistance, results from averages, to have a chance of operating.

We cannot, in general, alter the configuration of a body without doing work upon it, or letting the body do work. Considering any solid elastic body, for example, a straight wire: we cannot twist it without doing work. The motion is therefore resisted. On the other hand, if we let the wire untwist, it can do work itself against external resistance. Here, of course, we must have conservation of energy when all actions are taken into account, and nothing novel, so far, is presented. But there is, during the changes of configuration of an elastic body, another kind of resistance brought into play, depending upon the rate of change of configuration, at least, usually. Suppose we twist a wire slowly, and at every stage of the process note exactly the amount of the applied force, or do the same by small instalments. By summation we know the total work done in producing a given twist. Similarly, if we let the wire untwist slowly by small instalments, the force will be the same in the same configuration as during the twisting, and the same amount of work will be done by the body. But if we twist the wire *suddenly* to the same extent as before, *more* work will have to be done, or there will be an additional resistance to the motion at every moment; and if the wire suddenly untwist, it will do *less* work, or again, there is a resistance to the motion which did not exist before.

We may put it this way: If we measure statically the forces in different configurations we do not by summation get the total work done. There is always an additional resistance to the motion, opposing the change of configuration, never assisting it. I call this thermal resistance, because, whenever it occurs, there is a thermal effect also produced. It is not, however, necessarily a heating, as one might expect, but may be either a heating or a cooling. But, whichever it be, the change of configuration is always resisted. That is the cardinal fact that must be remembered.

If, during the change of configuration, the body be allowed to part with or to receive heat so as to neutralise the thermal effect if it be a heating or a cooling respectively (as by conducting the operation very slowly, and not thermally insulating the body), the thermal resistance itself is evanescent. On the other hand, if the body be thermally insulated, so that heat cannot leave or enter it, it will be of the full amount. This will serve to elucidate the effect above mentioned of *suddenly* changing the configuration. We then give no time for heat to escape or be taken in appreciably. Now given this law, or principle of thermal resistance, and a statement of one effect upon a body, deduce another. I give a few examples.

(1). We observe that heat lengthens and cold shortens a bar. What should be the thermal effect of suddenly stretching it? It must be such as to oppose the stretching; that is, to shorten the bar, and is therefore a cooling.

(2). What should be the effect of removing the stretching force? To oppose the return to the unstretched length, therefore to lengthen the bar; therefore a heating.

(3). Water above 4° C. expands by heating, what is the effect of compression? Such as to oppose the compression; therefore a heating.

(Notice there are two effects—the thermal resistance and the heating or cooling effect.)

(4). Water below 4° C. contracts by heating. The effect of compression is, therefore, to cool it, as the compression is thereby resisted.

(5). Water expands in vaporising. Hence pressure raises the boiling point.

(6). Water expands in freezing. Hence pressure lowers the freezing point.

(7). An india-rubber band at ordinary temperatures, stretched by a weight, lifts it when heated. What should result from suddenly stretching the band? The motion must be resisted; that is the invariable fact. The thermal effect is therefore a heating, for that, by the previous, lifts a weight.

(8). A twisted wire is suddenly twisted further. Is it a heating or a cooling effect that is produced? Whichever it be, it must increase the torsional rigidity, so as to oppose the twisting. If, then, heat lessens the torsional rigidity, it is a cooling effect, and conversely.

(9). Compressing a gas heats it. Hence heat applied to a gas increases its pressure, for this is the only way to make the compression be additionally resisted. This property of a gas is so well known that it is more difficult to recognise the principle. It is hard to imagine the possibility of a gas being cooled by compression. Yet the principle involved is identically the same as in the less obvious illustrations.

Other mechanical illustrations may be multiplied indefinitely, but the above will be sufficient. When we wish to go further, and make applications to electricity or magnetism, it is necessary to be very particular that the necessary conditions are complied with. The thermal resisting force is always opposed to the motion, and so far resembles a frictional force; but the thermal effect, unlike that of friction, is reversible with the direction of motion, and the motion produced by heat is reversed if cold be applied. Thus, during the stretching of a spring, the pull of the spring is $F+f$, if F be what it would be if there were no thermal effect, and f is the small increase produced by, or accompanying it. On the other hand, when the spring shortens, the strength of force is $F-f$ at a corresponding stage, and at the same temperature, f being the same quantity as before. Regarding the forces as vectors, $\mathbf{F}$ is constant in direction, $\mathbf{f}$ changes, being with $\mathbf{F}$ in the stretching, and against it in the unstretching. That is, it is always against the direction of motion. The heat effect, if it be a cooling during the stretching, is a heating during the unstretching. And as regards the effect of heat, if heating increases, cooling must decrease the elasticity. If in some peculiar state, both heating and cooling produced the same effect on the elasticity, or if either stretching or unstretching produced the same thermal effect, we could not immediately apply the principle without reservation. Further investigation would be needed.

(10). Given a circuit of two metals A and B, at one temperature initially. We observe that slightly heating one junction causes a current from A to B, and cooling it causes a current from B to A. Here is a perfectly reversible thermal effect, although accompanied by other strictly irreversible effects in the circuit. In accordance with the principle, what should be the thermal effect at the same junction on passing a current from an external source from A to B? The current must be made weaker than it would otherwise be, hence the current due to the thermal effect is from B to A, hence it is a cooling, by the previous knowledge. At the other junction it must be a heating, for the current is there from A to B. Thus there is a transfer of heat from the first junction to the second. Notice the peculiarity that the thermoelectric force in the circuit, due to both junctions, weakens the main current; that is the first conclusion. But the E.M.F. at both junctions is from A to B, whence it follows that at the first junction the E.M.F. is weaker than at the second. Further inquiries would lead us to the full theory of thermoelectricity. [*See* pp. 309 to 327.]

(11). Given that the capacity of a condenser is increased by heating and decreased by cooling the dielectric, what should be the thermal effect of suddenly charging it? The charging must be opposed, hence a decrease of capacity or a cooling effect. Similarly, suddenly discharging the condenser should produce a heating effect, which we may conclude thus:—The discharge must be resisted. Now, the rate of discharge depends upon the difference of potential of the ends of the discharge wire. This must be reduced, therefore. But a reduction of potential with the same charge means an increase of capacity, for which there must be, by the above, a heating effect. Observe here that these thermal effects are independent of the absorption phenomenon.

(12). A curious result of thermal resistance is that a perfectly elastic solid, if such really existed, vibrating, would come to rest without any external or internal friction to cause it. Thus in a vibrating spring, thermal resistance opposes its motion whether it be moving to or from the equilibrium position. Now, if the spring could give out and take in heat instantaneously, so as to keep its temperature constant, the effect would vanish, and, without friction, the spring go on for ever. Similarly, if the spring could be thermally insulated, the cooling and heating effects produced when moving to or from the equilibrium position would balance, and the spring could go on for ever. But practically, neither one nor the other condition can be complied with, and the spring must be brought to rest without friction entirely by the thermal effect being practically radiated or conducted away in one motion, without an exactly counterbalancing receipt of heat in the opposite motion.

There is another way of looking at the principle which is useful, viz., to direct attention to the flow of heat into or out of the body when it is being strained, supposing that the body is, in the first place, in thermal equilibrium with its environment, and that it can receive or lose heat instantaneously. Thus, in example (9), compressing a gas drives heat out of it. Now, go to the other end of the operation. Take heat out of a gas. It compresses itself. Or take example (1). Stretching a bar draws heat into it, and sending heat into a bar makes it stretch itself. From this point of view we regard two events as being invariably connected—motion of matter of a certain type, and a flow of heat in a certain way. But the method is less general than the preceding, assuming, as it does, that the flow of heat is always permitted, likewise the motion.

In all cases whatever in which the principle of thermal resistance has been experimentally tested, it has been found to be correctly followed by Nature. Can we then assert its invariable truth, and apply the principle unhesitatingly to hitherto unverified cases, possibly unverifiable? Is it possible to give a rigid demonstration of its truth? The first question may [within certain limits] be answered in the affirmative, the latter not. We cannot prove it even as we prove the truth of conservation of energy, viz., by seeing that it is a necessary truth in pure dynamics, and extending our notions to all the operations of nature; observing that it is experimentally true in a great many cases, and convincing ourselves of its universal application with the assistance of a little faith. For we cannot deduce the principle of thermal resistance from the laws of dynamics. It would be no breach of conservation of energy were it to be exactly reversed, were strains to be assisted by thermal effects. All we can really do is to convince ourselves that, being true in all observed cases, and its negation leading to extraordinary consequences which are *not* observed, though not dynamically impossible *à priori*, it must, by faith, be generally true.

First, put the principle in a mathematical form. Let an elastic body be strained from one configuration to another, keeping it always at one temperature t, to allow which a quantity of heat H leaves the body, whilst W is the work done upon the body. Then the principle asserts simply that $dW/dt \div H$ is positive, or that dW/dt and H are either both positive or both negative, which is easily verified by the above examples. More work is caused to be done by the thermal effect if it be not than if it be allowed to escape.

Next, to see the consequences of its negation, follow the example of the founder of thermodynamics, Sadi Carnot, when, in the most masterly and scientific manner, he set to work to find the cause of the motive power of heat. After the above strain, bring the body back exactly to its original state through the same series of intermediate configurations, but at a slightly different temperature $t + dt$. This requires there to be two other operations (2nd and 4th), viz., to raise the temperature by dt in the second configuration, and to lower it by dt when it has got back to the first. Now the body does work $W + (dW/dt)dt$ at the higher temperature. Hence, in the complete cycle, the body does work $(dW/dt)dt$. What else happens is that an amount of heat H is lowered in temperature by the amount dt. Now, without any experience to guide us, H and dW/dt may be algebraically positive or negative, and not of the same sign necessarily. But if they could be of opposite signs, work would be obtained through a substance *raising* the temperature of heat. If then we take it as axiomatic that it is impossible by conveying heat from a cold to a hot body to obtain mechanical effect, then we prove that the law of thermal resistance is universally true, at least for bodies in mass, and inanimate.

Now this axiom, so called, is really the principle Carnot was led to, viz., work is derived from heat by *lowering* its temperature. Carnot's principle is thus a consequence of the principle of thermal resistance. In Carnot's cycle for a gas engine heat is taken in at a higher and given out at a lower temperature, and the pressure is greater at the higher than at the lower, so that the gas does more work in expanding than is done in compressing it back to its original state, and thus we have finally heat lowered in temperature, and work done by the gas. In his water and vapour cycle it is similar, so he concluded that the necessary condition of obtaining work by thermal agency was the lowering the temperature of heat. He was only wrong as regards the *quantity* of heat lowered in temperature.

Now this being a consequence of the principle of thermal resistance, and the various examples above given being commonly regarded as consequences of the Second Law of Thermodynamics, with which Carnot was most assuredly not acquainted, how is it that Carnot was acquainted with *them?* I do not say he thought of them all. But he knew some of them, and, had he been asked what would be the result in a given case, according to his principle, he would have given the correct answer. Only his reason for the answer would have been erroneous. For the only legitimate reason that Carnot could have for applying his principle universally would be that he had proved that all reversible engines had the same efficiency between the same temperatures, so that his principle, known to be true for one body, must be true for all.

But this proof was vitiated by his erroneous assumption of the materiality of heat, then universally believed in, and was only put right long after, by Clausius and Thomson, on the basis of the equivalence of heat and work, fully established experimentally by Joule. Carnot founded his proof on a perfectly unexceptionable axiom (on his view of the nature of heat). Work could be obtained without any thermal agency if all reversible engines had not the same efficiency between the same temperatures. The substituted axioms of Clausius and Thomson are by no means so satisfying, considered as axioms. They express truths, they involve Carnot's principle, they involve the principle from which Carnot's is derived, but they are not axioms, unless a law of Nature, only to be learnt by experience, is an axiom.

Clausius said that heat will not pass from cold to hot by itself, or without compensation. True enough, by definition of cold and hot, if the cold and hot bodies be in contact. Otherwise not self-evidently true, though a law of Nature. Thomson said we cannot get work out of a body by cooling it below the lowest temperature of surrounding objects. This I also admit to be true, knowing that it involves Carnot's principle, and believing that; but it is not self-evidently true. A third axiom that can be used is that we cannot convert heat into work without lowering the temperature of heat. This I believe to be the best of all, being the simplest expression of the truth of Carnot's principle that work is got by lowering the temperature of heat.

Of course Carnot was wrong in his quantitative estimate of the efficiency of reversible engines, but his principle remains unaltered. When, with the principle of thermal resistance, we combine that of the equivalence of work and heat, and use Carnot's criterion of the perfectness of an engine, that it must be reversible, we arrive at an absolute measure of temperature independent of any particular substance, and using it, give quantitative expression to the principle of thermal resistance, viz. :—

$$dW/dt = H/t,$$

where t is temperature according to the scale of equal dilatations of an imaginary perfect gas under constant pressure, whose energy, at constant temperature, is independent of its volume. It now becomes a form of the Second Law of Thermodynamics.

There is a reason for everything, and therefore (if that be an axiom) for the axioms of thermodynamics, better called laws. The reason of conservation of energy the student of dynamics can understand, if he can grant that the laws of matter in motion observed in masses are true for the smallest parts of bodies. The reason of thermal resistance is also, by the aid of the kinetic theory of gases, becoming evident, and will no doubt some day be established for all bodies. It arises from the irregular nature of the motions we call heat. We cannot control single molecules. Could we do so, down would go the law of thermal resistance and heat could be converted straight into work, molecular irregular motions to an equivalent amount of motion of a mass, the irregularity disappearing, without at the same time having to lower the temperature of another quantity of heat, as at present required. But what we cannot do with inanimate matter may be going on always to a certain extent in living matter; not because living matter is exempt from any law of Nature, but that we do not yet know all its laws. The perpetual running down of the available energy of the universe is a matter that must be cleared up. It is incredible that it can always have been going on, and dismal in its final result if uninterrupted. It is therefore the duty of every thermodynamician to look out for a way of escape.

Section XII. Electrization and Electrification. Natural Electrets.

A dielectric, not including ether, the universal medium, when under the influence of electric force, may be said to become electrized. We must not say electrified, as that refers to something entirely different from electrization, namely, electrification, or discontinuity in the displacement. The proper measure of the intensity of electrization will appear presently; in the meantime we need only observe that electrization is the analogue of magnetization, and, like it, requires the presence of matter.

Electrization may be approximately perfectly elastic with reference to the standard zero state, as in a dielectric in which absorption does not occur, so as to disappear on the removal of the exciting cause. This must be an impressed electric force somewhere, except in a transient inductive state, when there may be none, unless we choose to consider the electric force of induction as impressed; or else there must be electrification somewhere. There may also be residual electrization, namely, when absorption occurs. This may tend naturally to wholly subside, or a part of it may remain. The body is then permanently electrized. But, apart from artificial production of permanent electrization, it exists naturally in pyro-electric crystals, if nowhere else.

A word is evidently wanted to describe a body which is naturally permanently electrized by internal causes. Noticing that "magnet" is got from "magnetism" by curtailment at the third joint from the end, it is suggested that we may get what we want by performing the same operation upon electricity. An "electric," which is what results, would be a very good name for an intrinsically electrized body, but for two reasons. First, it was once used to signify what we should now call a dielectric or an insulator; and secondly, electric is now used as an adjective, or, equivalently, electrical. The former of these objections is of hardly any weight, that use of the word as a substantive being wholly obsolete. The latter is heavier, but still of no great importance. Another word that suggests itself is electret, against which there is nothing to be said except that it sounds strange. That is, however, a mere question of habit. Choosing, at least provisionally, the second word suggested, to avoid collision with the adjective, we may then say that certain crystals, if no other bodies, are natural electrets; that solid insulating substances may be made electrets artificially, with a greater or less amount of permanency; that liquid insulators can only be electrets for a minute interval of time, if at all; whilst gases, whose particles are always vigorously wandering about, are never electrets. But all insulators can be electrized; they are only electrets when the electrization, or part of it, is intrinsic.

Let **J** be the intensity of intrinsic electrization anywhere, and let

$$\mathbf{J} = c\mathbf{e}/4\pi,$$

c being the specific capacity at the place. Then **e** is the intrinsic electric force, and the displacement **D** and actual electric force **E** are determinable fully by the conditions

$$\mathbf{D} = c\mathbf{E}/4\pi, \qquad \text{div}\,\mathbf{D} = \rho, \qquad \text{curl}\,(\mathbf{e} - \mathbf{E}) = 0\,;$$

supposing that there is no conductivity. Here ρ is the volume-density of electrification, if there be any. Should there be conductors present, charged or uncharged, their effect is determinable by the additional conditions of there being no force in any conductor (unless there be impressed force therein), and every separate conductor to have a given total charge, zero or finite, according as the conductor is not or is electrified when placed by itself in a space previously free from electric force.

The physical explanation of electrization is of course matter of speculation, as it depends not only on the nature of molecules, but also on their relation to the ether. Weber's theory of induced magnetization obviously suggests its analogue. Accordingly, we might say that the molecules of a body are always permanent electrets; but should there be, in a very small portion of it, but still large enough to contain a very great number of molecules, an average uniformity of distribution in all directions of their axes of electrization, there will be no external signs of its being electrized, or the intensity of electrization is zero. Next, the molecules admit of rotation, so that under the influence of externally caused electric force the axes of electrization are turned to a greater or less extent towards coincidence with the direction of the electrizing force. This preponderance of electrization in one direction of the molecules in our small volume naturally causes there to be a finite intensity of evident electrization, or of electric moment. Should this angular displacement of molecules be only elastically resisted, the evident electrization will wholly disappear on the removal of the electrizing force, or the body was only inductively electrized. Should there be a slow slipping, the phenomenon of absorption will occur, there being no longer a complete return of the molecules to their original state on the removal of the electrizing force. The body is then made an electret. Should the slip be permanent, the body is an artificially made permanent electret.

Such a theory is of course very empirical, and admits of considerable variation according to the hypothesis we adopt regarding the angular displacement and its tendency to subside. There is of course a maximum intensity of electrization which cannot be exceeded, viz., when all molecules are turned the same way. It is only one way of accounting for evident electrization, and may be utterly unlike the reality. Weber's theory of rotation of molecules has a great many recommendations, but I do not think they go further than to make it anything better than a working hypothesis, even with the support of Professor Hughes's experiments.

But, admitting that molecules are electrets, we may go a step further, and, in a manner, account for it. In the first place, the state of electric force and displacement determined by the above equations would be identically given in another manner. Do away with $\mathbf{e}$, that is, do away with $\mathbf{J}$, the intrinsic electrization, and substitute for it an arrangement of circuital magnetic current of density

$$-\operatorname{curl} \mathbf{e}/4\pi = \mathbf{G}, \quad \text{say.}$$

The magnetic current $\mathbf{G}$ and the intrinsic electric force $\mathbf{e}$ are equivalent, so far as the distribution of displacement is concerned. In the case of a bar uniformly electrized longitudinally, the current $\mathbf{G}$ will be entirely superficial, going round the bar. But by Ampère's device, we may substitute a network of currents for a current bounding the network. Thus we may get down to the molecules, and ascribe their electrization to molecular magnetic currents whose moments equal their electric moments. But I do not put this forward as being at all a physical explanation of why molecules are electrets, if they be. There is merely a mathematical equivalence.

Nor do I, in the parallel case of explaining the magnetism of a molecule by means of a molecular electric current, admit that any step towards a physical explanation of magnetism has been made. Not that I attach much importance to the common objection originating from the fact that an electric current in a conductor requires a continual supply of energy to keep it up, owing to the Joule-heating that goes on. For we observe the effects of heat in mass, and ascribe heat to the kinetic energy of agitation of the molecules. Not of the parts of molecules, but of their wholes. Now, considering a molecule as an atom, for simplicity, if an atomic current generated heat as a current in a wire does, the heat would be the energy of agitation of the parts of the atom, which, though indivisible, may yet not be of unchangeable form. This communication of energy would surely alter the nature of the atom. In fact, it is assuming an atom to be like a body in the ordinary sense. In brief, conduction as we know it is an affair in which molecules or atoms as wholes are involved, so far as the heat is concerned, and it would be far more wonderful if atoms had internal resistance than, as we are obliged to suppose, that atoms are perfect conductors, if we choose to have atomic currents.

My reason for not considering a molecular electric current to be a physical explanation of magnetism is, that although the more closely we look at the matter, apart from old-fashioned notions of an electric current as something going round and round, the more we are led to the conclusion that the magnetized molecule with its field of force and the Ampèrean current with its field of force are really one and the same thing, yet we are brought no nearer to an understanding of either, so that explaining the former by the latter is futile, regarded physically. But let us return, for the present, to electrization.

Dismissing altogether all ideas concerning the possible electrization of single molecules, keep to evident electrization. To illustrate it with sufficient comprehensiveness, without any great complication, or difficult calculations, let the electrized body be of spherical shape, and let us perform a series of simple operations upon it.

(1). Let there be a uniform field of electric force of strength **E**, and displacement **D**, in air of specific capacity c, practically equal to unity. Then bring into the field a perfectly neutral solid spherical dielectric of radius a and specific capacity c_1. Call those points on its surface where it is cut by a straight line through its centre parallel to the force of the undisturbed field the poles, the negative pole being that where the line of force enters, and the positive where it leaves the sphere. This line is the axis. Supposing c_1 greater than c, the lines of force will be drawn in by the sphere symmetrically with respect to the axis. The displacement is continuous, and is therefore made greater within the sphere than in the field previously. On the other hand the electric force within the sphere is weakened. Both are parallel to the axis and uniform within the sphere. Let D_1 and E_1 be the displacement and force, then

$$D_1 = c_1 E_1/4\pi, \qquad \text{and} \qquad D_1 = 3Dc_1/(2c + c_1).$$

Thus it is not possible to make the displacement more than three times as great as before, and that is when $c_1 = \infty$. It is twice as great when $c_1 = 4c$. There is no real electrification, but the sphere appears electrified, to surface-density D_1 at the positive pole, and elsewhere proportional to the sine of the latitude. The potential, not counting that of the undisturbed field, which increases uniformly parallel to the axis, is

$$\sigma r \cos\theta \text{ within}, \qquad \text{and} \qquad \sigma a^3 \cos\theta/r^2 \text{ without}$$

the sphere, where $\sigma = E(c_1 - c)/(2c + c_1)$,

at a point distant r from its centre, θ being the co-latitude. If the sphere be taken out of the field, its apparent electrification will disappear, unless absorption or conduction have occurred.

(2). Whilst in the field, as above, let absorption occur, and go on until the displacement has become increased by D_2, such an amount as would be caused by an impressed force e_2 uniformly distributed parallel to the axis. We must not write $D_2 = c_1 e_2/4\pi$; D_2 is less than this amount. For continuity in the displacement requires that when the displacement in the sphere increases from some cause in itself, there must be a corresponding change in the external displacement. The effect of the intrinsic electrization of intensity

$$J_2 = c_1 e_2/4\pi$$

is to make $$D_2 = 2J_2 c/(2c + c_1)$$

within the sphere. The apparent charge is increased in density to $(D_1 + D_2)\cos\theta$.

(3). Next, remove the sphere from the original field of force. There is left the field due to the intrinsic electrization. The displacement within the sphere is D_2, and the apparent charge is of density $D_2 \cos\theta$. The potential is

$$\rho r \cos\theta \text{ within and } (\rho a^3/r^2)\cos\theta \text{ without}$$

the sphere, where $\rho = c_1 e_2/(2c + c_1)$.

In fact, the field is exactly similar to the magnetic field of a uniformly magnetised sphere. If the absorption gradually disappear, so will the external field.

(4). But whilst the intrinsic electrization remains appreciably steady, let us cover the sphere with a metal coating, or in any other way produce surface-conduction. There will be a current through the sphere from the negative to the positive side, and oppositely in the conducting coating. Thus the displacement within the sphere must be increased by the short-circuiting; it becomes

$$D_3 = c_1 e_2/4\pi = J_2,$$

the greatest displacement possible without external aid. The external field is done away with altogether. There is now a real distribution of electrification on the surface of the sphere, of density $-J_2 \cos\theta$; *i.e.*, negative on the positive side and positive on the negative side. The potential is zero inside and outside. The electret is now in the state of the dielectric of a condenser in which absorption has occurred, and a first discharge taken, the remaining charge not being accompanied by difference of potential between the metal plates.

(5). On removing the metal coating, the surface-charge gives no signs of its presence.

(6). Keeping the sphere insulated, let the intrinsic electrization subside. As it does so, an external field of the opposite character to before appears, the real surface-electrification remaining constant, whilst the displacement it exactly neutralized becomes less. When the intrinsic electrization has gone, supposing it to wholly disappear, the displacement, inside and outside, is merely that due to the surface-charge. Inside, the displacement is

$$+c_1 J_2/(2c + c_1),$$

and the potential is $\quad -4\pi J_2 r \cos\theta/(2c + c_1)$.

Outside, the potential is $\quad -4\pi J_2 a^3 \cos\theta/r^2(2c + c_1)$.

The apparent electrification has density

$$-2J_2 c \cos\theta/(2c + c_1).$$

(7). Finally we may get rid of the surface-charge by again putting on the conducting coating, after which the sphere will be in its original neutral state.

The study of the theory of electrization is in some respects more important than of magnetization; on account of its greater generality it is more instructive. We have conductors and dielectrics, real and apparent electrification. The magnetic problems are less general on account of the absence of magnetic conductors, with corresponding absence of any magnetic representative of real electrification, as in (4) above, the magnetic matter or free magnetism being, except as regards a constant factor, the representative of the apparent or imaginary electricity, as in (3), where there was no real electrification.

As regards natural electrets, Sir W. Thomson's theory of pyro-electricity is (so far as is known to me) contained in a short article in "Nicol's Cyclopædia," reprinted in Vol. I. of "Mathematical and Physical Papers," Art. 48, p. 315. Being only a few lines in length, I can scarcely be quite certain that, when fully developed, it would be exactly as I state it—that is to say, when details are gone into, similarly to the above—though, generally speaking, there is no room for ambiguity. A pyro-electric crystal is a natural electret. In its neutral state, however, its external field of force has been done away with by surface-conduction and convection. Disregarding eolotropy, it is then in the condition of the sphere in (5) above, the surface being charged so that its density equals the divergence of the internal displacement. Warming or cooling the electret, by altering the internal stresses, alters the intensity of electrization, whereby the surface-charge no longer exactly balances it. If, for example, warming decreases the intensity of electrization, the positive end appears to acquire a negative charge, the negative end a positive, like the sphere in (6) above. If it be kept at the higher temperature, and surface-conduction occur, its surface-charge will readjust itself to again balance the internal displacement. Evidently we cannot get rid of the surface-charge as in (7), unless we can make the intensity of electrization zero.

The electret may, however, without change of temperature, produce external force by breaking it across its axis of electrization, when, evidently, the two pieces will not be neutral, unless the act of fracture should cause exactly the right amount of electrification on the fractured surfaces, which is highly improbable. But by surface-conduction again the two pieces will each become neutral.

Thermodynamic principles have also been applied to the case by Sir W. Thomson, under date 1879. This is the easiest part of the matter, considered qualitatively, if we know exactly the influence of heat. Applying the principle of thermal resistance discussed in the last section, we see that moving a natural electret about in an electric field must cause thermal effects. If, for example, as above, heat decrease the intensity of electrization, by decreasing the capacity, if we suddenly put the electric in an electric field so as to increase the displacement, the increase must be resisted, and this requires the electret to be heated. If now, we suddenly invert it, so as to decrease the displacement, this decrease must be resisted, hence a cooling effect. Should e vary as well as c, the application will be more complex. The phenomenon is no doubt very insignificant, but is very curious when we consider that it must occur in neutral crystals showing no external signs of their electrization.

SECTION XIII. SIMULTANEOUS CONDUCTION CURRENT AND ELASTIC DISPLACEMENT.

In the ordinary intercourse of man with his environment he is more or less accustomed to overlook, ignore, or treat as non-existent all phenomena whose recognition is not of immediate practical utility to him. For instance, very few people are even aware, until their attention is forcibly called to it, of the multitude of most singular optical phenomena that occur in the everyday use of the eyes, some of them very difficult of explanation. Their perception would not be of immediate utility to the average man. They are therefore left unnoticed, as if they were not, until even their recognition becomes difficult. The phenomena occur, and are seen by the eye mechanically, but the mind's eye is blind to them.

Most electricians, in a somewhat similar manner, are accustomed to confine their attention to only one part of the wonderful phenomena occurring during the existence of a conduction current. They may think of the wire, of its resistance, the current in it, and the E.M.F. causing it. Or possibly, in a more advanced stage, they may think also of the heating of the wire, and of external work done, as in driving a motor, with the necessary consideration of conservation of energy as regards the amount of work done. Still, however, it is the current in the wire that is the first object of attention, and what goes on outside it is ignored. This is perfectly natural, for what we call the strength of the current is the one thing that is, in the steady state, practically altogether independent of the external conditions; whilst, again, the steady state is of great importance, and may be brought under calculation with comparative ease. We are aware of the existence of an external magnetic field, and also of an electric field, or of an electromagnetic field having two sides, the electric and the magnetic. They vary under different external circumstances. But as the steady conduction current is independent of them they are ignored. This independence, which is really a fact of an extraordinary character, is, by habit, taken for granted, and ceases to have anything remarkable about it.

If we have a closed circuit with a steady impressed force in it, there results a steady current in the circuit, whose strength is calculated from the data relating to the wire itself, and is independent of the distribution of matter around, provided we do not disturb the insulation at the immediate surface of the conductor. If we bring a mass of iron near the circuit, or in general alter the distribution of external matter as regards its permeability to magnetic induction, there results merely a momentary disturbance of the conduction current. When this has ceased, the current is just as it was before, although the external state has been considerably altered, there being in particular a different magnetic field, and a change in the distribution of the rays of energy converging to the wire. This, with the similar transient disturbance of current by motion of a magnet or another closed current, comprises what is ordinarily understood by the induction of currents. Only the magnetic energy is concerned, in the main.

There are also the inductive effects due to alterations, not of the magnetic permeability, but of the electric capacity outside the wire, or by altering the external electric field. The current induced, or the alteration made in the previously steady current, is, as in the case of magnetic induction, transient only, and no disturbance is produced finally in the distribution of current. (Except indirectly, as by altering the structure of the wire itself, and its conductivity.)

The effect of surrounding a wire supporting a current with soft iron is to decrease the current temporarily. On the other hand, the effect of surrounding it with gutta percha is to increase the current on the whole temporarily, it being increased in that part of the circuit next the source more than it is decreased in the other part. If in the former case T be the increase made in the magnetic energy, the battery does $2T$ less work than if the iron had not been brought to the wire. In the latter case, if U be the increase made in the electric energy, the battery does $2U$ more work than if the gutta percha had not been brought to the wire.

Both the inductions, electric and magnetic, are in simultaneous action always. In a large class of cases, however, especially with condensers, the magnetic induction is of comparative insignificance. In another large class, as of coils containing iron, and circuits closed, the electric induction is negligible. In intermediate cases, when neither is negligible, we have very complex effects. But in all cases there is no permanent effect on the distribution of conduction current, however greatly the electric and magnetic fields outside the circuit are permanently altered, and likewise the manner of transit of energy from the source to the parts of the circuit away from the source. In brief, when the impressed forces are given, the conductivity conditions alone determine the steady state of the current. And, although it is not at once seen, the same is true whatever electric capacity or permeability the conductor itself may possess, so that there is interior electric or magnetic energy. The latter is recognised. Also, the electric capacity of very bad conductors is known, but there is little or no information regarding the capacity of good conductors, though for various reasons, I believe in its existence. The coefficient of permeability μ is known to never vanish. That of conductivity k has an enormous range, and may perhaps also vanish altogether—for instance in planetary space, if nowhere else. Similarly, the coefficient of electric capacity c may have a larger range than it is at present supposed to have in bad conductors. We tacitly assume $c=0$ in good conductors in general.

It will be understood that in thus speaking of the electric capacity (specific) of a conductor, we do not in any way refer to the Capacity of a Conductor of electrostatics, which is really the capacity of the surrounding dielectric, by all analogy with usage in the corresponding conduction problems. We refer to the specific capacity of the material of the conductor for elastic displacement in itself. This has, of course, nothing to do with the nonconducting dielectric outside. It is sometimes said that the specific capacity of a conductor is infinite. This is a mischievous delusion. It is tolerable, for mere mathematical purposes, sometimes to assume $c = \infty$ in a conductor, and ignore its conductivity altogether; *i.e.*, do away with the conductor, and substitute for it a dielectric of infinitely great specific capacity. We may, similarly, sometimes conveniently replace a voltaic cell (mathematically) by a condenser of infinite capacity and constant difference of potential. Or, we may consider a modern accumulator as a condenser, for limited purposes, in spite of the obvious radical difference in their physical nature. But the absurdity of considering a conductor as of infinite specific capacity is readily seen. For although we may, by increasing greatly the specific capacity of a dielectric, imitate, in some respects only, a conductor, yet as specific capacity refers to elastic displacement, such displacement must afterwards subside when allowed to, which is wholly different from the behaviour of conductive displacement, which has no tendency to return. The proper condition is not $c = \infty$, but $c = 0$ (unless we know there is elastic displacement in the conductor, as we do when it is badly conducting), with the auxiliary condition, k finite. Conductive and elastic displacement are wholly distinct things, and may be combined in any proportion. The former dissipates energy, the latter stores it.

In the following is an investigation of the general problem of steady impressed electric force and electrification in space with any distributions of electric capacity and conductivity. So far as I know, it has not hitherto been fully treated, but only piecemeal, and under limitations. For instance, Maxwell's treatise does not consider impressed force in the dielectric at all, and the internal capacity of conductors but imperfectly. Nor, again, is the determination of the state of the dielectric due to currents in conductors treated, which presents some curious peculiarities. Also, it is of advantage to deal comprehensively with the matter, since in practice we really have this simultaneous conduction and elastic displacement, either at the same or at different places, but connected.

Given k and c the electric conductivity and capacity, for conduction current $\mathbf{C}$ and elastic displacement $\mathbf{D}$, at every point of space, and also the distribution of impressed force $\mathbf{e}$; find the steady state, or, rather, show how it is determinate. $\mathbf{E}$ being the electric force, we have

$$\mathbf{C} = k\mathbf{E}, \qquad \operatorname{div} \mathbf{C} = 0, \qquad \operatorname{curl} (\mathbf{e} - \mathbf{E}) = 0, \quad \text{...........(86)}$$

$$\mathbf{D} = c\mathbf{E}/4\pi, \qquad \operatorname{div} \mathbf{D} = \rho, \qquad \operatorname{curl} (\mathbf{e} - \mathbf{E}) = 0, \quad \text{...........(87)}$$

for the full statement of conditions at a point. First, the linear relations, viz., Ohm's law and Faraday's (or Thomson's or Maxwell's). The second of (86) expresses that the conduction current is continuous everywhere, and the second of (87) defines the electrification density ρ in terms of the displacement, viz., its divergence, the proper measure of the discontinuity of the elastic displacement. The third conditions are identical. They express that the actual force $\mathbf{E}$ is the sum of impressed $\mathbf{e}$ and of a polar force $\mathbf{F}$, such that $\operatorname{curl} \mathbf{F} = 0$. If $c = 0$ at a certain place we have only (86) to consider; if $k = 0$, then only (87); if neither vanishes, then both. Also, there will usually be surface conditions, which will come later.

First imagine $c = 0$ everywhere, and k finite everywhere, that is, an infinite conductor whose conductivity nowhere quite vanishes, and of no capacity. Then we may write either

$$\operatorname{div} k(\mathbf{e}+\mathbf{F})=0, \qquad \operatorname{curl}\mathbf{F}=0, \qquad \text{.................(88)}$$

and determine **F**; or else

$$\operatorname{div}\mathbf{C}=0, \qquad \operatorname{curl}(\mathbf{C}/k-\mathbf{e})=0, \qquad \text{.................(89)}$$

and determine **C**. It is obvious that the first of (88) has any number of solutions for **F**; we must then show that one (and only one) of them satisfies the second of (88), thus fixing **F**. Or, it is obvious that the second of (89) has any number of solutions for **C**; we must then show that one (and only one) of them satisfies the first of (89), thus fixing **C**. Thus, since the first of (88) expresses that **C** is circuital, choosing any circuital **C** settles **F**. Or,

$$k\mathbf{F}=\operatorname{curl}\mathbf{A}-k\mathbf{e}$$

is the general solution, wherein **A** is any vector whatever. Similarly the general solution of the second of (89) is

$$\mathbf{C}/k=\mathbf{e}-\nabla P,$$

wherein P is a perfectly arbitrary scalar.

Now, whatever **F** and **f** may be, if $\mathbf{F}_1=\mathbf{F}+\mathbf{f}$, then

$$\Sigma\,\mathbf{F}_1k\mathbf{F}_1=\Sigma\,(\mathbf{F}+\mathbf{f})k(\mathbf{F}+\mathbf{f})=\Sigma\,\mathbf{F}k\mathbf{F}+\Sigma\,\mathbf{f}k\mathbf{f}+\Sigma\,2\mathbf{F}k\mathbf{f},$$

the Σ indicating summation through all space. But if **F** satisfies both of (88) whilst $\mathbf{F}_1$ satisfies only the first, the third summation on the right vanishes. For **F** is polar, and $k\mathbf{f}$ is circuital, so that, by the elementary property of the polar force, $\Sigma\,\mathbf{F}k\mathbf{f}=0$ for any one infinitely slender tube of $k\mathbf{f}$, and therefore for all. Thus we are reduced to the first and second summations on the right side. The first is fixed, the second may vary, but is necessarily positive, every element of it being positive. Hence $\mathbf{f}=0$ makes the summation $\Sigma\,\mathbf{F}_1k\mathbf{F}_1$ a minimum. But $\mathbf{F}_1$ is any solution of the first of (88), whilst **F** is a solution of both, so we prove that any solution $\mathbf{F}_1$ of the first, which also satisfies the second of (88), makes $\Sigma\,\mathbf{F}_1k\mathbf{F}_1$ a minimum. But this quantity has a minimum, for it is necessarily positive, unless $\mathbf{F}_1=0$ everywhere. Hence there is a solution of (88), viz., **F**, when $\mathbf{f}=0$. There cannot be two solutions. For, if **F** and $\mathbf{F}+\mathbf{f}$ be both solutions of (88), we must have

$$\operatorname{curl}\mathbf{f}=0 \qquad \text{and} \qquad \operatorname{div} k\mathbf{f}=0,$$

and therefore $\Sigma\,\mathbf{f}k\mathbf{f}=0$, which can only be by $\mathbf{f}=0$, making **F** the sole solution.

We may treat (89) similarly. Let **C** satisfy both conditions, whilst $\mathbf{C}+\gamma$ satisfies only the second, so that γ satisfies $\operatorname{curl}\gamma/k=0$. Then the total Joule-heat per second is

$$\Sigma\,(\mathbf{C}+\gamma)k^{-1}(\mathbf{C}+\gamma)=\Sigma\,\mathbf{C}k^{-1}\mathbf{C}+\Sigma\,\gamma k^{-1}\gamma+\Sigma\,2\mathbf{C}k^{-1}\gamma.$$

Here the third summation on the right vanishes, because **C** is circuital and γ/k polar. The second summation is necessarily positive, therefore $\gamma=0$ makes the heat a minimum. This closes the current, making $\mathbf{C}+\gamma$ become **C**, satisfying both conditions. As the minimum is necessary, the necessity of a solution of (89) follows, and that it is unique is shown by $\Sigma \gamma k^{-1}\gamma$ vanishing when γ is circuital and $k^{-1}\gamma$ polar, requiring that $\gamma = 0$, if we assume that $\mathbf{C}$ and $\mathbf{C}+\gamma$ are both solutions of (89).

As regards (88), we may state the result thus. There is one distribution of polar force $\mathbf{F}$, a linear function of which, namely, $k\mathbf{F}$, has a given distribution of convergence, viz.,

$$\text{conv}\, k\mathbf{F} = \text{div}\, k\mathbf{e}.$$

Or, there is one distribution of force $\mathbf{E}$ differing from $\mathbf{e}$ by a polar force alone, a linear function of which, namely, $k\mathbf{E}$ or $\mathbf{C}$, is continuous everywhere. As regards (89), we may similarly state that there is only one circuital current $\mathbf{C}$, a linear function of which, namely, $\mathbf{C}/k$ or $\mathbf{E}$, has a given distribution of curl, viz.,

$$\text{curl}\, \mathbf{C}/k = \text{curl}\, \mathbf{e}.$$

Or, assuming that Ohm's law and the continuity of the current are always obeyed, let $\mathbf{C}$ vary. Then $\Sigma \mathbf{F}k\mathbf{F}$ is made a minimum by that one distribution of $\mathbf{C}$ which makes $\mathbf{F}$ polar. And, assuming that Ohm's law is obeyed, and also that $\mathbf{F}$ is always polar, let $\mathbf{F}$ vary. Then $\Sigma \mathbf{EC}$, the heat, is made a minimum by that one distribution of the current that is completely circuital.

The gist of the above, and also of the more abstruse and complex demonstrations that may be given, is expressed by the theorem $\Sigma \mathbf{FC} = 0$, or, a polar force, $\mathbf{F}$, does no work on a circuital flux, $\mathbf{C}$, arising from the property of a polar force that its total round any closed curve is zero. Given first Ohm's law, the linear connection between the actual electric force and the current-density, and also that the current must be circuital. If the impressed force be so distributed that it alone is sufficient to satisfy Ohm's law and continuity (that is, when $\text{div}\, k\mathbf{e} = 0$), the force $\mathbf{e}$ is the actual force, and the current is $k\mathbf{e}$. But should this condition not be complied with by the impressed force, it is clear that an auxiliary or complementary force, $\mathbf{F}$, is required which, together with $\mathbf{e}$, shall make up the actual force $\mathbf{E}$ to satisfy continuity and Ohm's law. The question, then, is, how is this complementary force to be found? Any number of $\mathbf{F}$'s may be made up, but only one of them is completely polar. We then have $\Sigma \mathbf{FC} = 0$, and the total heat expressed by

$$\Sigma \mathbf{EC} = \Sigma \mathbf{eC} = \Sigma \mathbf{e}k\mathbf{e} - \Sigma \mathbf{F}k\mathbf{F}.$$

It is naturally suggested by this making of the total heat equal to the activity of the impressed forces, that we examine the effect of subjecting the complementary force, not to being polar, but to satisfying $\Sigma \mathbf{FC} = 0$, or $\Sigma \mathbf{eC} = \Sigma \mathbf{EC}$, which is possible in other ways than by a polar force. Thus, knowing that $\mathbf{F}$ uniquely determined polar satisfies this condition, and supposing that $\mathbf{E}$, $\mathbf{C}$, and $\mathbf{F}$ represents the real solution, let us alter the current in any manner, keeping it circuital, also obeying Ohm's law, and finally keeping the heat per second always equal to the activity of the impressed forces. Let $\mathbf{f}$ be the alteration in the complementary force, then $k\mathbf{f}$ is the alteration in the current, and is circuital. Also the alteration of $\Sigma \mathbf{eC}$ is $\Sigma \mathbf{e}k\mathbf{f}$. But, by our final restriction above,

$$\Sigma (\mathbf{F}+\mathbf{f})(\mathbf{C}+k\mathbf{f}) = 0, \qquad \text{or} \qquad \Sigma \mathbf{f}(\mathbf{C}+k\mathbf{f}) = 0,$$

which reduces to $\quad \Sigma \mathbf{e}k\mathbf{f} = -\Sigma \mathbf{f}k\mathbf{f}$

on putting $\mathbf{C}=k(\mathbf{e}+\mathbf{F})$, and then $\Sigma\mathbf{F}k\mathbf{f}=0$, $\mathbf{F}$ being polar and $k\mathbf{f}$ circuital. But $\Sigma\mathbf{f}k\mathbf{f}$ is positive, hence $\Sigma\mathbf{e}k\mathbf{f}$ is negative. Hence $\Sigma\mathbf{eC}$ is reduced. That is to say, any change made in $\mathbf{C}$ from the real current, subject to Ohm's law, continuity and conservation, decreases the heat and the (equal) activity of the impressed forces. Hence, in the real case, the impressed forces work as fast as ever they can, and this is by having the auxiliary force polar.

Taking a simple linear circuit, to exemplify, we find that there are only two solutions, the real, and $\mathbf{C}=0$. Both satisfy conservation (meaning $\Sigma\mathbf{eC}=\Sigma\mathbf{EC}$), but no other current will. But we need not be misled by $\mathbf{C}=0$ being, in this case, the only other solution than the real. For if we introduce a shunt, making three wires joining two points, and have an impressed force in only one of the wires, we may create additional solutions without number. For two simple circuits can be made containing the impressed force, furnishing two extreme solutions, in which no change from the real current is made in the current in one or the other branch not containing the impressed force. Between these extremes, by changing the current in all three branches suitably, we get any number of other solutions, in all of which continuity, Ohm's law, and conservation are satisfied. The heat varies from zero up to a certain maximum, which occurs in the real case; then the auxiliary force is completely polar, or has a potential. Of course, did we ever find a seeming departure from the proper distribution of force related to the known impressed force, we should naturally ascribe it to *other* impressed forces, and so come back to potential again.

SECTION XIV. CONDUCTION AND DISPLACEMENT (*Continued*).

Supposing now that we have, by the previous, satisfied ourselves that the distribution of current in the steady state is uniquely determinate by the subjection of the current to Ohm's law and continuity, and of the complementary force to being wholly polar, which last condition makes the activity of the impressed forces the greatest possible subject to conservation. The question then arises, how is this affected by the conductor being a dielectric as well, so that there must be elastic displacement in it? If we compare the dielectric conditions (87) with those for the conduction current (86), we see that they are of exactly similar form, except in the presence of ρ, the volume-density of electrification. If, then, we start with $\rho=0$, and proceed to find the displacement produced by the impressed force on the supposition that there is no conductivity anywhere (just as before, in the conduction problem, we assumed the specific capacity to be zero everywhere), we know that the displacement is uniquely determinate, as was the current before. Like the current, it is circuital, and is so distributed as to make (not the heat, but) the potential energy of displacement a maximum. The total work done by the impressed force is twice this. The other half is not accounted for. In the transient state the energy is partly electric, partly magnetic; in the absolutely steady state it is wholly electric; practically there is always conductivity somewhere, so that half of the work is spent in the Joule-heat of induced conduction currents, and is not radiated away, and, so to speak, lost. But with that we have no concern at present.

Now this distribution of displacement will not in general be consistent with that of electric current determined by the conductivity conditions. To be consistent, c the electric capacity, and k the electric conductivity, must be everywhere in the same ratio, or capacity × resistance (both specific) must be a constant. Hence, in general, either only one of the solutions is correct, or else neither. A little consideration will show that it is the conductivity solution that is correct, and that we must, after finding the distribution of electric force **E** from the conductivity conditions without any reference to those of capacity, determine the displacement and electrification to correspond to the thus-found electric force by the first two conditions of (87), viz.,

$$\mathbf{D} = c\mathbf{E}/4\pi, \qquad \rho = \text{div } \mathbf{D}.$$

For we thus satisfy all the conditions, which we could not do by starting with the dielectric problem, finding **E** to suit, and from it, finally, the distribution of current. Why we are able to solve the problem by the conductivity conditions alone is mathematically accounted for by the presence of ρ in the continuity condition relating to the displacement. The distribution of displacement **D**, and the distribution of electrification ρ, are such as to have no polar force when existing together; or the polar force of the displacement alone and that of the electrification alone are equal and opposite. Thus the electric field is not in any way disturbed, and the current is therefore also undisturbed and independent of the existence of elastic displacement.

On this point it should be remembered that whereas in considering a magnet as a collection of very small magnets, or any very small portion of the magnet as being polarised, the positive end of a polarised particle is that end to which the vector polarisation points, and is the end upon which positive magnetic matter may be assumed to be collected; yet on the other hand, if we do the same with electric displacement, it is the negative electrification that is on the positive side of a particle, that to which the vector displacement points. We cannot, in the magnetic problem, imitate the above state of electric displacement and electrification having no polar force. For the magnetic analogue of **D**, which is $\mathbf{B}/4\pi$, **B** being the magnetic induction, is always continuous.

Now, to go further. At the commencement we assumed that the conductivity nowhere quite vanished, so that we have been, so far, considering the current and displacement produced by given impressed force in an infinite conductor, in which k is finite everywhere, whilst c is quite arbitrary, and may vanish in certain portions if we please; it being the conductivity conditions alone that determine the field of force. But, practically, we must regard certain parts of space as being wholly nonconducting. This, though apparently included in the preceding, viz., by taking the special value $k=0$ in certain spaces, really, to a great extent, necessitates a changed mode of treatment. It is not sufficient to find **E** from the conductivity conditions, putting $k=0$ in nonconducting space. For this will not give the proper field in the nonconductors, but only in the conductors, unless at the same time $c=0$ in the former, or, in special cases, $c=$ constant.

Also, it may happen that when the conductivity is finite, though very small, in certain parts of space, it will be practically necessary to suppose it to be quite zero, on account of there being a first approximately steady state, and then, a long time after, a wholly different, quite steady state; which last, though it is not what is practically wanted, is what the conductivity conditions give. Consider, for example, a submarine cable whose ends are earthed through condensers. Put a battery in circuit at one end. The practical steady state is reached quickly, in a few seconds (of course not counting disturbances or absorption). But as the dielectric is slightly conducting, if we keep on the battery, the "charge of the cable" will in time—minutes or hours, according to the insulation—nearly disappear; theoretically, in an infinite time. This is the real final state, but the first approximately steady state is what is practically wanted, given, very closely, by assuming $k=0$ exactly in the dielectric.

Now divide all space into two sets of regions, the conducting and the wholly nonconducting, including in the latter the dielectric in such a case as just mentioned. All conducting matter which is continuously connected must be regarded as a single conductor. Thus all the wires on a line of poles, if they are earthed, together with the earth, and all that is in conducting connection with it, form strictly a single conductor. But if we loop two of the wires, removing earth, they form a separate conductor. (Leakage ignored.) There may thus be any number of distinct conductors, each self-continuous, but wholly separated from all the rest by nonconducting matter, or else unbounded partly; though the last is, as regards conductivity, practically unrealisable. Similarly the rest of space, the nonconducting space, forms a number of self-continuous regions, each either wholly bounded by conducting matter, or else partly unbounded; which last is practically realised.

Now, selecting any one of the conducting regions, let the impressed force **e** be given in it. It is readily shown, by the simplest modification of the conduction problem for all space, that **C** (and of course **E**) are definitely fixed by the distribution of **e** and of conductivity. The modification consists in applying the space-integrals to the finite space occupied by the conductor, at the same time introducing the surface condition that the current is confined to the conductor, or is tangential at its boundary, which condition is expressed by $\mathbf{CN}=0$, if **C** be the current-density and **N** the unit vector normal to the surface. But this modification may be wholly avoided by integrating through all space as before, thus including surface terms, on the assumption $k=0$ outside the conductor. The solution obtained applies to the conductor only; the part for the external space must be wholly rejected. There is an exception, namely, if $c=0$ as well as $k=0$ in the whole external space, when the external solution for the electric force will be the correct one, as will also be the case sometimes if $c=$ constant.

Thus the internal electric state of any conductor depends solely on the conductivity and the impressed force in it, and is independent of all external conditions. (This is not true for the magnetic state of the conductor, which will be influenced by the external conditions not only as regards permeability, but as regards current and magnetisation; but at present it is only the electric state that is in question.) We therefore settle the electric state of the whole of the conducting parts of space, one conductor at a time. Not only that, but in the same manner as before done in the case of the infinite conductor, we settle the displacement and electrification in every conductor, one at a time, from the known electric force in them, making no alteration whatever in the electric field.

There remain, finally, the nonconducting regions, and now the matter gets rather more complex. So far, our knowledge ceases at the boundaries between the conducting and the nonconducting regions. But at every point on these boundaries, on the nonconducting side, the tangential component of the electric force is fixed. For the second equation of induction is

$$\mathrm{curl}\,(\mathbf{e} - \mathbf{E}) = 4\pi\mathbf{G} = \mu\dot{\mathbf{H}}, \quad \ldots\ldots\ldots\ldots\ldots\ldots(23)\ bis$$

where $\mathbf{G}$ is the magnetic current, $\mathbf{H}$ the magnetic force, and μ the permeability. Here $\mathbf{E} - \mathbf{e} = \mathbf{F}_1$, the electric force of the field by itself, not counting impressed force. Applying the Version Theorem to this, interpreting it for a mere surface, we see that the tangential component of $\mathbf{F}_1$ must be the same on both sides of the surface. Otherwise the surface would be the location of a magnetic current-sheet, of strength $\mathbf{G}$ per unit area. This continuity of the tangential component of the electric (similarly of the magnetic) force of the field is true whether the state be steady or not. When not steady, $\mathbf{F}_1$ is not polar. At present $\mathbf{F}_1 = \mathbf{F}$, and is polar. We know, then, the tangential component of $\mathbf{F}$ (including its direction) at every part of all the boundaries. In terms of potential, we know its surface-variation over every boundary, but not its absolute magnitude. The last is unknown, not merely as regards a constant, but as regards n constants, one for each conductor, which, if we choose to employ potentials, must be mutually reconciled, so as to leave only one of them arbitrary. Now we cannot do this without knowing the state of the dielectric nonconductors. The potential must be found last of all, even in the conductors, whose electric state is independent of one another. Obviously, even were there no other reason to give, this would be a powerful argument against potential representing any physical state, in the same manner as electric force, current, etc., do.

As we took the conductors one at a time before to completely find their electric state independently, so now we may take the nonconducting regions one at a time, and find the electric state of each independently of the others. Selecting, then, a single nonconducting region, the conditions to be satisfied within it are,

$$\mathbf{D} = c\mathbf{E}/4\pi, \qquad \mathrm{div}\,\mathbf{D} = \rho, \qquad \mathrm{curl}\,(\mathbf{e} - \mathbf{E}) = 0, \quad \ldots\ldots(87)\ bis$$

where now, of course, $\mathbf{e}$ is the impressed force at the point considered. Or, eliminating $\mathbf{D}$, and putting $\mathbf{E} = \mathbf{e} + \mathbf{F}$, $\mathbf{F}$ being the polar force,

$$\mathrm{div}\,c(\mathbf{e} + \mathbf{F}) = \rho, \qquad \mathrm{curl}\,\mathbf{F} = 0. \quad \ldots\ldots\ldots\ldots\ \ldots\ldots(90)$$

The electrification density ρ is now not to be *found*, as in the conductors previously, but must be *given*. Thus $\mathbf{e}$, ρ, and c are given at each point, and $\mathbf{D}$, $\mathbf{E}$, $\mathbf{F}$ have to be found, or simply $\mathbf{F}$, since that fixes the others.

There is also the before-stated boundary condition; $\mathbf{FN}$ being the normal component of $\mathbf{F}$, the vector normal component is $(\mathbf{FN})\mathbf{N}$, so that it is

$$\mathbf{F} - (\mathbf{FN})\mathbf{N} = \mathrm{V}\mathbf{N}\mathrm{V}\mathbf{FN} \qquad \ldots\ldots\ldots\ldots\ldots\ldots\ldots\ldots(91)$$

that is given over the bounding surface or surfaces. Or, simply $\mathrm{V}\mathbf{FN}$, the tangential $\mathbf{F}$ turned through a right angle on the surface, may be regarded as given.

A preliminary examination, as regards energy, of the three conditions (90) and (91), shows that they are insufficient to determine $\mathbf{F}$. We need to know, also, the boundary representatives of ρ, the electrification. But it is not the electrification at every point of the boundaries, but only their totals, that is needed, *i.e.* the charges. Let q_1, q_2, etc., be the charges of the separate surfaces, then

$$q_1 = \Sigma\,\mathbf{DN}_1, \qquad q_2 = \Sigma\,\mathbf{DN}_2, \quad \text{etc.} \quad \ldots\ldots\ldots\ldots\ldots(92)$$

must be given, $\mathbf{D}$ being the displacement at a surface, and $\mathbf{N}_1$, $\mathbf{N}_2$, etc., unit normals from the conductors to the nonconductor.

For it is clear, by the continuity of the electric current, that if a conductor, wholly surrounded by a nonconductor, had a charge before the impressed force in the conductor began to act, such charge will not be altered in amount, though its distribution may be changed, when the impressed force acts. Similarly, impressed force in the dielectric cannot alter its amount, but only its distribution, nor can the introduction of external electrification. The same applies partly when the conducting surface wholly surrounds a dielectric; neither $\mathbf{e}$ in the conductor nor in the dielectric can charge it. But, on the other hand, the introduction of any interior electrification, by continuity of the current, requires the surface to have a total charge equal to the whole interior electrification taken negatively. In any case, then, q_1, q_2, etc., must be known, independently of external electrification, in the one case, or depending on interior electrification in the other.

It will be convenient to break up the problem into two, thus:—

(*a*). Given $\mathbf{e}$, ρ, and q_1, q_2, etc., and that $\mathrm{V}\mathbf{FN} = 0$. Find $\mathbf{F}$.

(*b*). Given $\mathrm{V}\mathbf{FN}$, and that $\mathbf{e} = 0$, $\rho = 0$, $q_1 = 0$, etc. Find $\mathbf{F}$.

We take (*b*) first, as it is the simpler, and is also directly connected with the conduction current. That is to say, we suppose that before the impressed forces in the conductors began, the whole system was free from electrification or impressed force in the nonconductors. We have

$$\operatorname{div} c\mathbf{F} = 0, \qquad \operatorname{curl}\mathbf{F} = 0, \qquad \Sigma\,\mathbf{DN} = 0, \qquad \mathrm{V}\mathbf{FN} = 4\pi\mathbf{g}, \quad \ldots(93)$$

$\mathbf{F}$ being the force in the region. The first two apply to every point in the region, the third to any conducting bounding surfaces as a whole, whilst in the last $\mathbf{g}$ is given over the whole boundary. We have

$$\Sigma\,\mathbf{F}_1 c\mathbf{F}_1 = \Sigma\,\mathbf{F}c\mathbf{F} + \Sigma\,\mathbf{f}c\mathbf{f} + \Sigma\,2\mathbf{F}c\mathbf{f}, \quad \ldots\ldots\ldots\ldots\ldots(94)$$

if $\mathbf{F}_1 = \mathbf{F} + \mathbf{f}$, whatever they be. Suppose now $\mathbf{F}$ to be a solution of (93), and $\mathbf{F} + \mathbf{f}$ to satisfy the second and fourth conditions, so that $\mathbf{F}_1$ is a solution of the second and fourth, but is otherwise unrestricted. By the first, $\mathbf{D}$ is continuous within the region, and, by the third, at its boundaries as much leaves as enters the region. We may therefore close $\mathbf{D}$ completely outside the region without disturbing $\mathbf{D}$ within it. $\mathbf{D}$, thus extended, is a circuital flux, and therefore may be represented by

$$4\pi\mathbf{D} = \text{curl}\,\mathbf{Z},$$

where $\mathbf{Z}$ is determinate in various ways. This makes

$$\Sigma\,\mathbf{F}c\mathbf{f} = \Sigma\,\mathbf{f}\,\text{curl}\,\mathbf{Z}/4\pi = \Sigma\,\text{conv}\,\text{V}\mathbf{f}\mathbf{Z}/4\pi, \quad \ldots\ldots\ldots\ldots\ldots(95)$$

This summation extends through the region. By the Convergence Theorem, the last form is at once expressible as a surface-integral over the boundary, viz.,

$$\Sigma\,\mathbf{N}\text{V}\mathbf{f}\mathbf{Z}/4\pi = \Sigma\,\mathbf{Z}\text{V}\mathbf{N}\mathbf{f}/4\pi = 0,$$

the vanishing taking place because $\text{V}\mathbf{f}\mathbf{N} = 0$. Thus $\Sigma\,\mathbf{F}c\mathbf{f} = 0$. Therefore, by inspection of (94), $\Sigma\,\mathbf{F}_1 c\mathbf{F}_1$ is made a minimum by $\mathbf{f} = 0$, making $\mathbf{F}_1 = \mathbf{F}$, and proving $\mathbf{F}$ to exist. That it is unique follows by the same reasoning as in the conductivity problem.

The physical interpretation of the above, as regards $\mathbf{Z}$, which is best taken as the magnetic impulse, or time-integral of the transient inductive magnetic force, is interesting, but must be left over. In the above, however, $\mathbf{Z}$ may be any vector whose curl is $\mathbf{D}$ in the nonconductor, its curl in the conductors being arbitrary. Thus, if we close $\mathbf{D}$ in any manner through the conductors, and call the complete system $\mathbf{D}_1$, then

$$\mathbf{Z} = \text{curl}\,\Sigma\,\mathbf{D}_1/4\pi r + \text{any polar vector}$$

satisfies the requirement. Or, instead of the curl of the vector-potential of $\mathbf{D}_1$, we may form the vector-potential of its curl [p. 206, *ante*].

As regards the transformation used in (95), if $\mathbf{f}$ and $\mathbf{Z}$ are *any* vectors, we have

$$\text{conv}\,\text{V}\mathbf{f}\mathbf{Z} = \mathbf{f}\,\text{curl}\,\mathbf{Z} - \mathbf{Z}\,\text{curl}\,\mathbf{f};$$

and (95) follows because $\mathbf{f}$ has no curl.

Lastly, the assumption made in the above demonstration that $\mathbf{F}_1$ is possible, to satisfy the second and fourth of (93), though nearly obvious, is made evident by constructing any scalar potential whose tangential variation at the boundaries is of the required amount, its value elsewhere being arbitrary, and letting $\mathbf{F}_1$ be its slope.

SECTION XV. CONDUCTION AND DISPLACEMENT (*Conclusion*).

As regards other causes influencing the electric field in the space external to the conductors, there remains the problem (*a*) of the last section. If, in it, we put $\mathbf{e} = 0$, we reduce it to the common electrostatic problem :—Given the volume electrification within, and also the charges upon equipotential surfaces bounding the nonconducting region (at least when they require to be independently stated, apart from the volume electrification), show that the field of force is determinate. We might take this for granted, and confine our attention to impressed force only; but as it makes scarcely any difference to include electrification, we shall do so. We have then the conditions

$$\operatorname{div} c(\mathbf{e}+\mathbf{F}) = 4\pi\rho, \qquad \operatorname{curl} \mathbf{F} = 0,$$

to be satisfied by $\mathbf{F}$ within the nonconducting region; and at any bounding surface,

$$\Sigma\, \mathbf{DN} = q, \qquad \mathrm{V}\mathbf{FN} = 0,$$

the first of these expressing that the given charge q on the surface is the total normal displacement from it into the nonconductor, and the second that the force has no tangential component at the surface, or is normal thereto, or that the surface is equipotential.

Assuming $\mathbf{F}$ to satisfy the above, alter it to $\mathbf{F}_1 = \mathbf{F} + \mathbf{f}$ in any way that does not alter the electrification or the charges. That is, subject $\mathbf{f}$ to

$$\operatorname{div} c\mathbf{f} = 0, \qquad \Sigma\, \mathbf{N}c\mathbf{f} = 0$$

within the region, and at a surface, respectively. We have merely to show that this change increases the electric energy $\Sigma\, \mathbf{F}c\mathbf{F}/8\pi$, a quantity that must have a minimum, to show that $\mathbf{F}$ is determinate, by reasoning used before, which need not be repeated; and, since

$$\Sigma\, \mathbf{F}_1 c \mathbf{F}_1 = \Sigma\, \mathbf{F}c\mathbf{F} + \Sigma\, \mathbf{f}c\mathbf{f} + \Sigma\, 2\mathbf{F}c\mathbf{f},$$

wherein the second sum on the right is positive, we have only to show that $\Sigma\, \mathbf{F}c\mathbf{f}$, the third sum, vanishes. Now, by the first condition for $\mathbf{f}$, we have $c\mathbf{f}$ perfectly continuous within the region; and, by the second, as much enters as leaves any bounding surface. That is, $c\mathbf{f}$ is any circuital flux whatever, either closed entirely within the region, or else partly or wholly through the rest of space. Therefore

$$c\mathbf{f} = \operatorname{curl} \mathbf{Z},$$

$\mathbf{Z}$ being any function whose curl is $c\mathbf{f}$, and

$$\Sigma\, \mathbf{F}c\mathbf{f} = \Sigma\, \mathbf{Z}\mathrm{V}\mathbf{NF} = 0;$$

the first summation extending through the region, the second over the bounding surfaces, and vanishing because $\mathrm{V}\mathbf{NF} = 0$ at every point thereof.

This finishes the question of the determinateness of $\mathbf{F}$. In any conductor it is settled by the distribution of impressed force and of conductivity therein, and the interior displacement and electrification follow. Then, knowing the tangential force set up at the boundaries of the nonconductors due to impressed force in the conductors, this is sufficient to settle the force all through the nonconductors when the distribution of capacity is given, provided there be no electrification in them, or impressed force, or boundary charges. No electrification or charges will be produced, as the displacement in the nonconductors can be closed through the conductors. Should there be electrification, etc., in the nonconductors, these have no effect in the conductors, produce no boundary tangential force, and their effect in the nonconductors may with convenience be separately considered. Knowing $\mathbf{F}$ to be determinate through all space, we may construct a potential to suit it, fixing its value arbitrarily at any one point. Should we, however, make the potential the direct object of attention in our investigations, instead of the force, we should greatly complicate certain parts of them. For example, to show that **e** in the conductors fixes **F** in the nonconductors when there is no electrification, etc. If there are n conductors, n fields of force (namely in them) are found by the conductivity conditions. The potential is therefore known over the surface of all the conductors, except as regards a constant for each. If we assume these constants to be known, we may prove that the thus-fixed boundary potential fixes the potential throughout all the nonconducting regions. But the resulting fields of force in them will be (except by extraordinary accident) entirely wrong. For the potential thus determined gives charges to the bounding surfaces, whereas there should be none. We must, therefore, to get matters right, communicate to the bounding surfaces exactly equal charges of the opposite kind, and distribute them in such a manner as not to disturb the already correctly settled tangential force, *i.e.*, distribute them equipotentially. The field of force will then get right, and the potential will be the sum of the wrong potential and that due to the charges added to cancel those given by the wrong potential.

As regards **e** in the nonconductors, it may be due to thermal or chemical causes, or be any other impressed electric force. Intrinsic electrisation is also included under **e**; thus, if **J** be its intensity, $\mathbf{J}=c\mathbf{e}/4\pi$ gives us the value of **e** to correspond. [*See* p. 489.] So far as producing polar force is concerned, a distribution of **e** acts in precisely the same manner as would a distribution of electrification of volume-density $\sigma = \text{conv}\, c\mathbf{e}/4\pi$. Here, however, the similarity ceases. It is quite different as regards the displacement and the energy, as **e** contributes to the displacement equally with polar force, thus, $\mathbf{D}=c(\mathbf{e}+\mathbf{F})/4\pi$. Sometimes **F** acts to assist **e**, but more frequently it is the other way. We should also remark that this *false* electrification σ may, like ρ, the real electrification, have surface distributions over conductors; whenever in fact, $c\mathbf{e}$ is not tangential. The surface-density is then $-\mathrm{N}c\mathbf{e}/4\pi$. But, unlike a real charge, which can be varied in distribution by influence, the false charge is fixed under the same circumstances. It (σ) should never be referred to as electricity or electrification without the prefix *false*, or some other qualification to distinguish it from Maxwell's electrification or free electricity, which is always discontinuity produced in the elastic displacement, only to be got rid of by conduction, with dissipation of energy in producing heat. In fact, σ is as false electrification as the distribution of electric current round a bar magnet, which would correspond to the same magnetic field, is a false electric current.

Various Expressions for the Electric Energy.

There being any steady state, with impressed forces both in the conductors and in the nonconductors, also electrification and surface charges, let us obtain expressions for the electric energy. We may consider the whole field of displacement as made up of four fields, viz., that in the nonconductors due to impressed force in the conductors; ditto, due to impressed force in the nonconductors; ditto, due to electrification and surface-charges; and lastly, that in the conductors. The total electric energy will be the sum of the energies of the separate fields, together with their mutual energies, if they be not conjugate. We shall denote the four fields by the suffixes 1, 2, 3, 4; in all other respects employing the same notation.

1. First, let U_1 be the electric energy in the nonconductors due to $\mathbf{e}_4$ in the conductors. Then,

$$U_1 = \Sigma \tfrac{1}{2}\mathbf{E}_1\mathbf{D}_1 = \Sigma \tfrac{1}{2}\mathbf{F}_1 c\mathbf{F}_1/4\pi = \Sigma \tfrac{1}{2}P_1\mathbf{D}_1\mathbf{N} = \Sigma \tfrac{1}{2}\mathbf{Z}_1\mathbf{g}_1. \quad \text{......}(96)_1$$

Here the first form of U_1 is the form common in all cases, the energy per unit volume being half the scalar product of the force and the displacement. The second form is virtually the same, only putting the displacement in terms of the force, which is polar. These summations of course extend throughout the whole nonconducting regions. The third form is a surface-summation over the whole boundary, of half the product of the scalar potential and the normal displacement. The fourth form is also a summation over all the boundary, of half the scalar product of the vector-potential $\mathbf{Z}_1$ and the false magnetic current $\mathbf{g}_1$. We have

$$-\operatorname{curl} \mathbf{Z}_1 = 4\pi\mathbf{D}_1, \qquad \mathrm{V}\mathbf{F}_1\mathbf{N} = 4\pi\mathbf{g}_1.$$

$\mathbf{Z}_1$ is the magnetic impulse, or time-integral of the transient magnetic force during the variable states which would occur if the impressed force were suddenly cancelled, ending in the removal of the displacement $\mathbf{D}_1$ in the nonconductors. Or, without altering the value of the summation, we may take $\mathbf{Z}_1$ to be the time-integral of the actual magnetic force, taken negatively, arising from putting on the impressed force, though this is less convenient in general. Should there be no electric current finally, $\mathbf{e}_4 + \mathbf{F}_4 = 0$, if $\mathbf{F}_4$ is the polar force in the conductor, and therefore $4\pi\mathbf{g}_1 = \mathrm{V}\mathbf{N}\mathbf{e}_4$. In this case also,

$$U_1 = \Sigma \tfrac{1}{2}\mathbf{e}_4 \int \Gamma \, dt,$$

Γ being the true current at time t after starting $\mathbf{e}_4$, so that $\int \Gamma dt$ is the total displacement, elastic and conductive. The other half of the work done by $\mathbf{e}_4$ is spent in heat. Observe that it is necessary for the impressed force in the conductor to reach to its boundary, and to have a tangential component there, for it to be able to set up displacement outside without at the same time producing current in the conductor.

2. Next, let U_2 be the electric energy in the non-conductors due to impressed force in them only. Then,

$$\left.\begin{aligned} U_2 &= \Sigma \tfrac{1}{2}\mathbf{E}_2\mathbf{D}_2 = \Sigma \tfrac{1}{2}\mathbf{e}_2\mathbf{D}_2 = \Sigma \tfrac{1}{2}\mathbf{Z}_2\mathbf{g}_2 \\ &= \Sigma \tfrac{1}{2}\mathbf{e}_2 c\mathbf{e}_2/4\pi - \Sigma \tfrac{1}{2}\mathbf{F}_2 c\mathbf{F}_2/4\pi = \Sigma \tfrac{1}{2}\mathbf{e}_2 c\mathbf{e}_2/4\pi - \Sigma \tfrac{1}{2}P_2\sigma, \end{aligned}\right\} \quad (96)_2$$

The first form requires no remark. The second results because $\mathbf{E}_2 = \mathbf{e}_2 + \mathbf{F}_2$, and $\Sigma \mathbf{F}_2\mathbf{D}_2 = 0$. In the third form $\mathbf{Z}_2$ is the magnetic impulse (of $\mathbf{e}_2$ now, of course), and

$$-\operatorname{curl} \mathbf{Z}_2 = 4\pi\mathbf{D}_2, \qquad -\operatorname{curl} \mathbf{e}_2 = 4\pi\mathbf{g}_2;$$

the object of introducing 4π's being to harmonise as well as possible with the usual electromagnetic formulæ. In the fourth form we have the difference of two sums, the first a constant, the second the energy of the field of the polar force by itself, as if due to electrification. In the fifth form the energy of the polar force is expressed in terms of its potential and the false electrification σ. All the summations are taken throughout the nonconductors. There may possibly be surface summations as well, if $\mathbf{e}$ exists at the boundaries. All the formulæ in $(96)_2$ are the same as if there were no conductors present.

3. Let U_3 be the common electric energy due to electrification ρ and surface charges q_1, q_2, etc.

$$U_3 = \Sigma\,\tfrac{1}{2}\mathbf{E}_3\mathbf{D}_3 = \Sigma\,\tfrac{1}{2}\mathbf{F}_3\mathbf{D}_3 = \Sigma\,\tfrac{1}{2}P_3\rho + \Sigma\,\tfrac{1}{2}P_3 q. \quad \text{..........}(96)_3$$

The first and second forms are identical, the force being polar. In the third form the energy is expressed in terms of the potential and electrification. We may also here employ the magnetic impulse, by a device, but with no utility.

4. Let U_4 be the electric energy in the conductors due to $\mathbf{e}_4$ in them. Then

$$U_4 = \Sigma\,\tfrac{1}{2}\mathbf{E}_4\mathbf{D}_4 = \Sigma\,\tfrac{1}{2}(\mathbf{e}_4 + \mathbf{F}_4)\mathbf{D}_4 = \Sigma\,\tfrac{1}{2}\mathbf{e}_4\mathbf{D}_4 + \Sigma\,\tfrac{1}{2}P_4\rho_4, \quad \text{.....}(96)_4$$

Here $\mathbf{F}_4$ is the polar force in the conductors due to $\mathbf{e}_4$, and P_4 its potential; ρ_4 is the interior electrification, if any. Should there be none, $\mathbf{D}_4$ is closed within the conductor, and we may write

$$U_4 = \Sigma\,\tfrac{1}{2}\mathbf{Z}_4\mathbf{g}_4,$$

if $\quad \operatorname{curl}\mathbf{Z}_4 = -4\pi\mathbf{D}_4, \quad$ and $\quad \operatorname{curl}\mathbf{e}_4 = -4\pi\mathbf{g}_4.$

It is clear that a similar form exists for the energy in any volume in which there is no electrification, $\mathbf{g}$ being represented by the tangential boundary force turned through a right angle, besides the part depending on $\mathbf{e}_4$, if it exist in the volume. (Here $\mathbf{e}_4$ is the same as in the previous concerning U_1.)

5. Lastly, there are the mutual energies. We get rid of U_{14}, U_{24}, and U_{34} at once by observing that the fourth field is in a different place from the first, second, and third, so that these quantities are zero. We can also get rid of U_{13} and U_{23}; the first and third fields, though coexistent, are conjugate, and so are the second and third. Thus,

$$U_{13} = \Sigma\,\mathbf{D}_1\mathbf{F}_3 = \Sigma\,\mathbf{Z}_1\mathrm{V}\mathbf{N}\mathbf{F}_3 = 0,$$

$\mathbf{Z}_1$ being the same quantity as before. The second summation extends over the boundaries, and vanishes because $\mathbf{F}_3$ is normal to them. Similarly show that U_{23} vanishes. There remains only U_{12}, which does not necessarily vanish.

$$\left.\begin{aligned} U_{12} &= \Sigma\,\mathbf{E}_1\mathbf{D}_2 = \Sigma\,\mathbf{E}_2\mathbf{D}_1 = \Sigma\,P_1\mathbf{D}_2\mathbf{N} \\ &= \Sigma\,\mathbf{Z}_1\mathbf{g}_2 = \Sigma\,\mathbf{Z}_2\mathbf{g}_1 = \Sigma\,\mathbf{e}_2\mathbf{D}_1. \end{aligned}\right\} \quad \text{..................}(96)_5$$

Here the quantities are the same as before used in treating U_1 and U_2. Finally, U being the complete electric energy in all space,

$$U = U_1 + U_2 + U_{12} + U_3 + U_4.$$

Here the first three components of U may be taken together conveniently. Let the impressed forces in the conductors and the nonconductors start at the same moment, and $\mathbf{Z}$ be the magnetic impulse, $=$ former $\mathbf{Z}_1 + \mathbf{Z}_2$, then

$$U_1 + U_2 + U_{12} = \Sigma \tfrac{1}{2}\mathbf{Zg},$$

where $\mathbf{g} =$ former $\mathbf{g}_1 + \mathbf{g}_2$, that is,

$$4\pi\mathbf{g} = -\text{curl}\,(\mathbf{E}_1 + \mathbf{E}_2),$$

with the equivalent boundary representative.

Finally, we may, by means of the various conjugate properties, put together the potential results. Let P be the resultant potential from all causes; ρ the electrification density whether in conducting or nonconducting matter, defined by $\rho = \text{div}\,\mathbf{D}$, including in it surface electrification; σ the density of false electrification, defined by $\sigma = \text{conv}\, c\mathbf{e}/4\pi = \text{conv}\,\mathbf{J}$, if $\mathbf{J}$ be the (equivalent) intensity of electrisation, *i.e.*, $\mathbf{J} = c\mathbf{e}/4\pi$, whether in the conductors or not; then

$$U = \Sigma \tfrac{1}{2}\mathbf{eJ} + \Sigma \tfrac{1}{2}P(\rho - \sigma) \quad \text{.......................(97)}$$

expresses the total electric energy, the summations being through all space. Notice the minus sign before the false electrification. (97) may also be easily proved directly.

Here, and in any of the summations containing a scalar potential, any constant may be added to the potential, owing to the sum of the electrification being zero, ditto of false. There is an apparent exception when, as may be imagined, tubes of displacement go out to infinity, though this cannot really occur. In such a case, we must choose P to vanish at infinity. But we may get over this by regarding the end of space as an electrified surface. The surface-density will be infinitely small, but the charge on it finite, being the exact negative of all the other electrification. Counting it, the constant in P again becomes arbitrary.

We may extend the results to apply to transient states by including in $\mathbf{e}$ the electric force of induction, $= -\dot{\mathbf{A}}$, if $\mathbf{A}$ be such that $\text{curl}\,\mathbf{A} = \mathbf{B}$, the magnetic induction. Considering U_1 during the transient state, we find

$$U_1 = \Sigma \tfrac{1}{2}\mathbf{Z}_1\mathbf{G}_1 + \Sigma \tfrac{1}{2}\mathbf{Z}_1\mathbf{g}_1$$

at any moment, $\mathbf{Z}_1$ being the magnetic impulse that would arise from removing $\mathbf{e}$ at that moment, $\mathbf{G}_1$ the *real* magnetic current $\dot{\mathbf{B}}/4\pi$ at the moment, and $\mathbf{g}_1$, as before, the false magnetic current at the boundary. The first summation extends through the nonconductors. When the steady state is reached, $\mathbf{G}_1 = 0$, and $\mathbf{Z}_1$ and $\mathbf{g}_1$ arrive at their before-used steady values. Similarly for U_2, U_4, and U_{12} during their transient state.

Section XVI. Magnetic and Electric Comparisons.

(*a*). Let two large masses of iron be united by a thin iron wire. For simplicity of exposition let the two masses and the wire be homogeneous throughout, and only "elastically" magnetisable. Call the masses P and Q, and the wire w. Let air surround them to a great distance, to get rid of foreign influences. The masses may be conveniently imagined brought near one another, and the wire not to join them straight, but to be led round through the air, although these assumptions are immaterial.

Now let a portion of the wire, which we shall call x, be intrinsically magnetised to intensity I parallel to its length. Here, again, we may conveniently imagine x to be very short, to be a mere disc, in fact, perpendicular to the axis of the wire. The magnetisation I is the same as an impressed magnetic force **h** similarly distributed, if $\mathrm{I} = \mu\mathbf{h}/4\pi$, and μ is the permeability of the iron. Instead of this impressed force in x, we may equivalently have an electric current round x, say in a coil of very small depth. Or, instead of this real current, we may refer to a false electric current round x.

The magnetic induction passing through x due to **h** (or its equivalents) is both altered in amount and in distribution from what it would be were x alone in the air. Its amount is increased by the permeance of the magnetic circuit being increased. And its distribution is altered, on account of the permeability being changed unequally in different places. It is, to a certain extent, led along the wire to the mass P (of course with a large amount of leakage on the way), through the air to Q, and through the wire back to x, completing the circuit. But if we increase the permeability of the iron we increase the induction, and at the same time cause relatively more of it to pass through P and Q, with relatively less loss from the wire (out on one side of x and in on the other). Indefinitely increase the permeability. The total induction in the circuit reaches its greatest value. The leakage becomes relatively insignificant, and by making the wire thinner may be reduced indefinitely. Then nearly all the induction is led through the wire to P, through the air to Q, and back to x through the wire. The total induction depends on the form and relative position of P and Q (with which, to be exact, we must count the two portions of w on opposite sides of x). In this final state the magnetic force in the air is everywhere perpendicular to the surface of P, Q, and w, except at x, where it is partly tangential.

(b). Replace the iron by a solid dielectric nonconductor of high electric capacity or low electric elasticity. Replace the impressed magnetic force **h** by impressed electric force **e**; or by intrinsic electrisation of intensity $\mathbf{J} = c\mathbf{e}/4\pi$, if c is the electric capacity; or, finally, by a magnetic current round x, similar to the before-used electric current, except that it must circulate the reverse way. The flux is now elastic electric displacement. Its distribution is similar to that of magnetic induction in (a). On indefinitely increasing c, or doing away with the electric elasticity in the solid dielectric, the greatest displacement is obtained. It then passes through the air mainly from P to Q. The electric force in the air is normal to the surface of the solid dielectric, except at x again.

(c). If in the case (b), before making the capacity infinite, and when, consequently, the final state (b) is not assumed, we communicate to the solid dielectric throughout its whole substance any finite degree of electric conductivity, we shall cause the displacement in the air to immediately (practically) assume the final state (*b*). The conductivity may vary anyhow, and likewise the electric capacity, which last may in fact be zero now. This is the common case of charging a condenser, except that it is usual to consider the conductor as having no electric capacity.

(*d*). If, in the case (*a*), of magnetic induction, we could communicate to the iron any finite degree of what, by analogy, we may call magnetic conductivity, rendering it impossible for the iron to support magnetic force without a magnetic conduction current, with dissipation of energy, it would be unnecessary to make the permeability infinite to obtain the final distribution in the air there mentioned. The permeability might be finite or zero, and the magnetic conductivity have any distribution in the iron. The final state would be as in (*c*), with magnetic induction instead of electric displacement. The examples (*a*), (*b*), and (*c*) are real, except as regards the assumption of infinite values of μ and c in (*a*) and (*b*) respectively, which may be conveniently imagined to become very great, though not really infinite, thus letting us approximate to the required results without using impossible conceptions. The example (*d*) is unreal. It is introduced to show the analogy.

In examples (*b*) and (*c*) the displacements out from P to the air, and in from the air to Q are identical. (The final state (*b*) is referred to.) But there is no electrification in case (*b*), whilst there is in case (*c*). If in case (*b*) we remove the wire from P and Q to a distance, we shall at the same time do away with the former apparent charges of P and Q. There will be left merely the small displacement from the wire on one side to that on the other side of x where the impressed e is, some small part of which will of course go through P and Q *viâ* the air. But if, in case (*c*), we remove the wire, P and Q will be left charged as before; besides that we have an insignificant wire-charge. Herein lies the difference between the elastic and conductive displacement.

Similarly, if in the example (*a*) we remove the wire from P and Q, we at the same time do away with their apparent magnetic charges. Whilst, if we could realise the case (*d*), on removal of the wire the bodies P and Q would be left really charged magnetically. They would apparently be unipolar magnets, though without any interior polarisation, it being a matter of the terminal induction of the medium between them. In the final states of (*a*) and (*b*) there is finite interior induction or displacement, being the exact complements of those outside in the air, although there is no interior magnetic or electric force, and consequently no interior magnetic or electric energy. But in examples (*d*) and (*c*), having the same external induction or displacement as in (*a*) and (*b*), we have no interior induction or displacement, as well as no magnetic or electric force and no interior energy.

(*cc*). Going back to (*c*) with an impressed electric force in the wire, let us join P to Q by a second conducting wire, attaching it to them anywhere, but for distinctness, away from the first wire. We shall now have a steady conduction current in the closed circuit made, as well as elastic displacement in the air, having a different distribution to the previous; the amount of change depending materially upon the resistance of the second wire compared with that of the rest of the circuit, being small when it is great, and conversely. The electric force is no longer normal to the conductors at any part of their boundary. There may now be any amount of internal electric energy, existing independently of the external, depending upon the internal capacity of the conductor, in which there is electric force, and consequently elastic displacement if there be capacity for it.

(*bb*). If, in the final state (*b*), we unite P to Q by a second dielectric wire also of infinite capacity, we make a closed dielectric circuit which has no elastance (or elastic resistance to displacement), and in which infinite displacement corresponds to finite impressed force, and likewise infinite electric energy. Clearly this is an impossible example, and we should never reach the final state. Let us therefore modify the conditions somewhat. Do away with the impressed force **e** at x, and substitute a bodily distribution of impressed force similar to that of the polar electric force (reversed) in the conductor in example (*cc*). Or, substitute equivalent intrinsic electrisation. Or, finally, a distribution of magnetic current on the boundary, given by $\mathbf{g} = \mathrm{V}\mathbf{FN}/4\pi$, if **F** is the external electric force in example (*cc*), and **N** a unit vector normal from the solid to the air. This magnetic current goes round the wire at x in the same direction as in (*b*), and oppositely round all the rest of the circuit. It is strong at x and weak elsewhere, of such strength, that total at x equals that elsewhere.

With this changed distribution of impressed force or equivalents, the external electric field will be identically that in (*cc*). The impossible infinite internal displacement is abolished, the total impressed force round the circuit (of the solid dielectric) is zero; there is a finite internal displacement which exactly closes the external; and there is no internal energy, because there is no electric force. (That is to say, as c is increased the internal energy becomes less and less without limit, the electric force going down to zero, the displacement assuming a finite value.)

(*aa*). If in the final state (*a*) we unite P to Q by a second iron wire, also of infinite permeability, we make a circuit of infinite permeance in which is a finite impressed force, so that we have infinite internal induction and magnetic energy to correspond. Modify as in (*bb*). Do away with the impressed magnetic force **h** at x, and substitute a bodily distribution of impressed magnetic force (or intrinsic magnetisation) similar to that of impressed electric force in (*bb*) or of the polar auxiliary force (reversed) in (*cc*). Or, substitute a distribution of electric current on the boundary given by $\gamma = \mathrm{V}\mathbf{NH}/4\pi$, if **H** is the external magnetic force, when similarly distributed to the external electric force in (*cc*) and (*bb*). This boundary false electric current is similarly arranged to that of magnetic current in (*bb*), except in reversal of direction of circulation everywhere.

Under the changed circumstances, there will be no internal magnetic force, and no magnetic energy, but finite magnetic induction, which will exactly close the external induction, which, again, has the same distribution as the displacement in the air in cases (*cc*) and (*bb*).

(*dd*). We should cause the same external induction as just arrived at, without modifying the distribution of impressed force, if we could impart magnetic conductivity to the iron-circuit. That is to say, in example (*d*) unite P to Q by a second magnetically conductive iron wire. If the distribution of magnetic conductivity be the same as that of electric conductivity in (*cc*), we shall have the external magnetic induction distributed like the electric induction in (*cc*) and (*bb*), and like the magnetic induction in (*aa*). There would be a steady magnetic conduction current in the iron-circuit, similar to that of electric current in (*cc*), and the internal magnetic energy (like the internal electric energy in (*cc*)) would be arbitrary, depending upon the permeability of the iron. (In this replacement of electric conduction current by magnetic conduction current, the electric conductivity of the iron-circuit is ignored. Similarly in (*ddd*), later).

Any magnetic field is accompanied either by true electric current Γ or by impressed magnetic force **h**. For the latter we may substitute a distribution of false electric current γ, such that if it were real the induction would be the same. Similarly, in a nonconducting dielectric in which there is no electrification, the elastic displacement is accompanied either by a true magnetic current $\mathbf{G} = \dot{\mathbf{B}}/4\pi$, if **B** is the induction, or by impressed electric force **e** (or equivalent electrisation). For the latter we may substitute a false magnetic current **g**. Also, U being the whole electric and T the whole magnetic energy, we have

$$4\pi\gamma = \text{curl}\,\mathbf{h}, \qquad 4\pi\mathbf{g} = -\text{curl}\,\mathbf{e}, \qquad T = \Sigma\,\tfrac{1}{2}\mathbf{A}(\Gamma + \gamma), \qquad U = \Sigma\,\tfrac{1}{2}\mathbf{Z}(\mathbf{G} + \mathbf{g}).$$

Here **A** and **Z** are the electric and the magnetic impulses that would arise on the sudden removal of **h** and of **e** respectively (also partly due to Γ or to **G**), when the induction or the displacement would wholly subside and spread.

Now, if we confine ourselves to a limited region, we shall still have the electric energy within it expressed in the same manner, provided we include in the false magnetic current **g** a boundary magnetic current given by $\mathbf{g} = \mathrm{V}\mathbf{F}\mathbf{N}/4\pi$, if **F** be the electric force (not counting impressed **e**) and **N** a unit normal from the boundary to the region; and further, provided that it be possible to close the displacement outside the region, thus not interfering with that within it. This is not always possible. And the magnetic energy is expressible in the same manner, if we include in γ a boundary false electric current given by $\gamma = \mathrm{V}\mathbf{N}\mathbf{H}/4\pi$, if **H** is the magnetic force, not counting impressed **h**; with no reservation as to the possibility of closing the induction outside the region, because it is always possible.

But in neither case is it generally true that these distributions of current (including those on the boundary), would, if they existed alone, set up the identical displacement in the one case and induction in the other, independent of what may be beyond the region. (To illustrate this, we need merely take a piece of a round tube of induction, with plane ends cutting the tube perpendicularly. If the induction is uniform and steady, the only electric current is the false current round the round part of the tube, and this solenoid will clearly not produce uniform induction, but induction that spreads out in passing from the middle to either end of the tube.) But we may very easily produce independence of the external state, by short-circuiting the unclosed displacement or induction, as the case may be, by making either the electric capacity or the permeability infinitely great beyond the region. Or, merely by making c or μ infinite in a mere boundary-skin, or over enough of it to completely close the displacement or the induction. When this is done, the currents, magnetic or electric as the case may be, produce the exact given displacement or induction within the region, without external aid.

In examples (*b*) and (*c*) we have identical displacement outside the solid dielectric or the conductor. Since the states are steady, $\mathbf{G}=0$. And, as there is no impressed force in the air, which is our region of electric energy, $\mathbf{g}$ is reduced to the surface-current $\mathbf{g}=\mathrm{V}\mathbf{FN}/4\pi$. This is zero except round x (because $\mathbf{F}$ is normal elsewhere), and there it is $\mathbf{g}=\mathrm{V}\mathbf{Ne}/4\pi$, which is the value of $\mathbf{g}$ in the expression $U=\Sigma\frac{1}{2}\mathbf{Zg}$ for the electric energy in the air.

In examples (*a*) and (*d*) the total magnetic energy in the air in the identical distributions of induction is $T=\Sigma\frac{1}{2}\mathbf{A}\gamma$, since there is no true electric current. And the false electric current (or it might be a true current in a thin coil) is confined to round x, being given by $\gamma=\mathrm{V}\mathbf{hN}/4\pi$.

But in example (*cc*), with electric current in the wire, and in (*bb*) as modified, the magnetic current is no longer confined to be round x, but extends over the whole conductor in (*cc*), and solid dielectric in (*bb*); being $\mathbf{g}=\mathrm{V}\mathbf{FN}/4\pi$, wherein $\mathbf{F}$ is no longer normal to most of the surface. This $\mathbf{g}$, used in $U=\Sigma\frac{1}{2}\mathbf{Zg}$, gives us the electric energy in the air in both cases. But there is a perfectly arbitrary amount in the conductor in example (*cc*), and none in example (*bb*). In order, therefore, that in (*bb*) the magnetic current should exactly correspond to the external displacement, we must, as before explained, short-circuit the latter beyond the air; *i.e.*, make $c=\infty$ outside the air-region. We then get the case (*bb*) as modified, and see the reason of the modification.

Similarly, in examples (*aa*) and (*dd*) the energy of the identical external magnetic inductions is $T=\Sigma\frac{1}{2}\mathbf{A}\gamma$, wherein γ is the boundary electric current given by $\gamma=\mathrm{V}\mathbf{NH}/4\pi$. But in (*dd*) there is also an arbitrary amount of internal magnetic energy, so that the boundary-current γ would not generally, existing alone, produce the actual external magnetic induction. It can be made to do so by abolishing the impressed force $\mathbf{h}$ at x in (*dd*), and making $\mu=\infty$ throughout the iron. Then we obtain (*aa*) as modified.

(*ccc*). In example (*cc*), with dissipation of energy by conduction, there is also magnetic energy, not before mentioned, as it would have introduced some confusion. For all space its amount is $T=\Sigma\frac{1}{2}\mathbf{AC}$, if $\mathbf{A}$ is the electric impulse and $\mathbf{C}$ the conduction current in the wire. That part of it in the air is $\Sigma\frac{1}{2}\mathbf{A}\gamma$ where $\gamma=\mathrm{V}\mathbf{NH}/4\pi$. This is a boundary-current, not round the wire, but along it, of the same total amount as the real current in it, as if it were all pressed to the surface. It would not alone set up the external field, but would do so if we short-circuit as much of the induction of the field as is unclosed (at the boundary).

(*ddd*). Similarly, in the unreal example (*dd*) with dissipation by the magnetic conduction current in the iron-circuit, there is, besides the magnetic energy, also electric energy. In all space, the amount is $U = \Sigma \frac{1}{2}\mathbf{ZG}$, if $\mathbf{G}$ be the magnetic current in the wire. That in the air alone amounts to $\Sigma \frac{1}{2}\mathbf{Zg}$, where $\mathbf{g}$ is a surface magnetic current given by $\mathbf{g} = \mathrm{V}\mathbf{FN}/4\pi$. It is along the wire, like $\mathbf{G}$ inside, and of the same total amount, as if $\mathbf{G}$ were pressed to the boundary. It would exactly correspond to the external displacement, if the latter were short-circuited at the boundary.

The unreal (*d*), (*dd*), and (*ddd*) are merely brought in because they, to a certain extent, assist the other comparisons.

SECTION XVII. THE MAGNETIC FIELD DUE TO IMPRESSED E.M.F.

Sections XIII., XIV., and XV. were principally devoted to the consideration of the electric field set up by impressed electric force, also as modified by previously existing electrification. There is also simultaneously a magnetic field, if there be electric current. This will depend on how the impressed forces are distributed, which question we need not return to further than to say that should they be wholly in nonconducting regions there can be no steady current, but merely a transient one producing elastic displacement; and that if there be impressed force in a conducting region, which is the first condition for there to be a steady current, it must not be polar in its distribution therein, and with its lines perpendicular to the boundary, or there can be no current again. (The term polar force is borrowed from magnetism to signify any force whose distribution is such that its integral round any closed curve is zero. This is the most useful property by which to identify a polar force. The lines of force start from certain places and terminate at others; these places are the poles, in an extended sense; any pole, positive or negative, may be conceived to send out a definite amount of "force," uniformly in all directions, *i.e.*, according to the inverse square law. The mathematical expression is $\mathrm{curl}\,\mathbf{F} = 0$, if $\mathbf{F}$ be polar; the boundary representation of $\mathrm{curl}\,\mathbf{F}$ being the tangential component of $\mathbf{F}$ turned through a right angle on the surface. That the lines of $\mathbf{F}$ must be perpendicular to a series of surfaces does not sufficiently identify a given force with a polar force, as this is not inconsistent with the integral force in a circuit being a finite quantity, and so giving rise to the corresponding flux.) Thus to have current in a conductor the impressed force must have a finite value in at least one closed path entirely within the conductor. The current thus depends upon the "curl" of the impressed force. This is of great importance in the theory of the Volta-force or other boundary forces. The curl of any force is always arranged in closed lines, *e.g.*, the closed line at the common meeting-place of zinc and copper (in contact) and a medium surrounding them. [As stated, the current produced by impressed force depends upon its curl, but this does not necessitate that the impressed force should be *in* the conductor. If the curl is the same, the force may equally well be outside it, and yet produce the same fluxes. Of this, more later.]

Supposing now the arrangement is such that there is current, the determination of the state of the magnetic field from it is, in comparison with the determination of the state of the electric field anywhere, a comparatively simple matter, the former being a reduced and greatly simplified form of the latter, with changed meanings of the quantities concerned. The flux is the magnetic induction. That has no divergence, to begin with—one simplification as compared with electric displacement. Next, there is no magnetic current, as it ceases when the state becomes steady. And, finally, the ratio of the flux to the force (magnetic), or the permeability, is everywhere finite, so that there is no division of space into permeable and impermeable regions. In brief, the full statement of the conditions is contained in

$$\mathbf{B} = \mu\mathbf{H}, \qquad \text{div}\,\mathbf{B} = 0, \qquad \text{curl}\,(\mathbf{H} - \mathbf{h}) = 4\pi\mathbf{C}\,; \quad \ldots\ldots\ldots(98)$$

the first being the linear connection between the flux and the force, the second expressing the continuity of the flux, and the third the relation between the magnetic force $\mathbf{H}$ and the current $\mathbf{C}$, in which $\mathbf{h}$ is the impressed force of intrinsic magnetisation $\mathbf{I} = \mu\mathbf{h}/4\pi$. As, by the linearity of the equations, the field due to $\mathbf{C}$ and to $\mathbf{h}$ is simply the sum of their separate fields, we may put $\mathbf{h} = 0$ at once, and therefore deal entirely with induced magnetisation (*quasi*-elastic).

If we integrate the third equation (98) on this understanding, we see that $\mathbf{H}$ may be any vector whose curl is $4\pi\mathbf{C}$, and is therefore indeterminate as regards a polar force, $\mathbf{F}$. We have then $\mathbf{H} = \mathbf{h}_1 + \mathbf{F}$, wherein, on account of the presence of $\mathbf{F}$, we may choose $\mathbf{h}_1$ to have no divergence. It is then definitely given by

$$\mathbf{h}_1 = \text{curl of vector potential of } \mathbf{C} = \text{curl}\,\Sigma\,\mathbf{C}/r,$$

if r be the distance from an element of $\mathbf{C}$ to the place where $\mathbf{h}_1$ is reckoned.

As thus defined, $\mathbf{h}_1$ is what is usually called the magnetic force of the current. It *is* the magnetic force of the current if there be no variation of permeability anywhere; otherwise it is what the magnetic force would be if there were no such variation. The absolute value of the permeability, provided it be the same everywhere, is a matter of indifference, so far as $\mathbf{h}_1$ is concerned. The polar force $\mathbf{F}$ therefore represents the change made in the magnetic force by variations of permeability, due, of course, principally to the presence of iron.

For limited purposes, $\mathbf{h}_1$ may be regarded as the impressed magnetic force due to a distribution of intrinsic magnetisation $\mathbf{I}_1 = \mu\mathbf{h}_1/4\pi$. That is, if we abolish the electric current and substitute $\mathbf{I}_1$, the magnetic force and the induction would be unchanged. But whilst $\mathbf{h}$ of real intrinsic magnetisation may be arbitrary, and is usually in very limited portions of matter, the lines of $\mathbf{h}_1$, as we have chosen it, are closed, and extend over all space in general, though by particular arrangements of current they may be shut out from certain spaces.

Let T be the total magnetic energy due to $\mathbf{C}$. The two most noteworthy forms for T are

$$T = \Sigma \tfrac{1}{2}\mathbf{HB}/4\pi = \Sigma \tfrac{1}{2}\mathbf{AC}, \quad \ldots\ldots\ldots\ldots\ldots\ldots\ldots(99)_1$$

the first summation being of the scalar product of the magnetic force and induction ($\div 8\pi$), and probably representing the real distribution of the magnetic energy; the second of half the scalar product of the current and the electric impulse $\mathbf{A}$, which is excessively unlikely to be anything near the real distribution. In $(99)_1$ $\mathbf{A}$ may be any vector whose curl is $\mathbf{B}$; but to give it the most physical significance, it is best to take it to be the electric impulse arising from inertia, or the time-integral of the electric force of induction on sudden removal of the impressed force $\mathbf{e}$ keeping up $\mathbf{C}$. It is a scientific concept which does not express any physical state or condition. The electric impulse $\mathbf{A}$ at a given place does not depend upon the magnetic state there, but upon its condition everywhere; as, on removal of $\mathbf{e}$, disturbances are propagated to the place, these determine the electric force of induction whose time-integral is $\mathbf{A}$.

Other forms are useful in showing the influence of variable permeability. Thus

$$\left.\begin{aligned} T &= \Sigma \tfrac{1}{2}\mathbf{h}_1\mathbf{B}/4\pi \qquad (\text{because } \Sigma\,\mathbf{FB} = 0), \\ &= \Sigma \tfrac{1}{2}\mathbf{h}_1\mathbf{I}_1 - \Sigma \tfrac{1}{2}\mathbf{F}\mu\mathbf{F}/4\pi = \Sigma \tfrac{1}{2}\mathbf{h}_1\mathbf{I}_1 - \Sigma \tfrac{1}{2}\Omega\rho. \end{aligned}\right\} \ldots\ldots\ldots\ldots(99)_2$$

Here $\mathbf{h}_1$ and $\mathbf{I}_1$ are definitely known, by the preceding. The new quantities Ω and ρ are the magnetic potential and the volume-density of magnetic matter; the polar force being the slope of Ω, or $\mathbf{F} = -\nabla\Omega$, whilst ρ is given by

$$\rho = \operatorname{conv} \mathbf{I}_1 = -(4\pi)^{-1}\mathbf{h}_1\nabla\mu, \qquad (\text{when } \mu \text{ is scalar}).$$

Thus ρ exists only at places where the permeability varies, therefore mainly at the bounding surfaces of different kinds of matter, or, disregarding perfectly abrupt changes, in thin layers at the bounding surfaces.

If $\nabla\mu$ be perpendicular to $\mathbf{h}_1$, we have $\rho = 0$. Hence, starting with μ constant, when $\mathbf{h}_1$ is the real magnetic force, if we select a complete tube of force, or any region bounded wholly by lines of force, and alter its permeability to any other value (constant throughout the region), the magnetic force will be unaltered, whilst the induction will be altered within the region in the same ratio as the permeability. If now we choose to ignore the changed permeability, we may ascribe the altered induction to an additional electric current, on the surface of the region, perpendicular to the magnetic force and of the proper strength to produce the increased induction. In $(99)_1$, $\mathbf{A}$ will be altered by a quantity depending upon this false current. By adding more and more tubes of force to the region, we finally include all space, or alter the permeability everywhere in the same proportion. Then, as might be expected, the false current to account for the increased induction on ignoring the altered permeability, occupies the same situation as the real, in fact increasing its strength in the same ratio as the permeability was increased.

There is another form for T which is very curious, related to the electric energy due to the same impressed electric force $\mathbf{e}$. Let U be the total electric energy, then $U = \Sigma\frac{1}{2}\mathbf{ED}$, if $\mathbf{E}$ be the electric force and $\mathbf{D}$ the elastic displacement. Now this last is the time-integral of $\dot{\mathbf{D}}$, the transient displacement current during the charge. Let $\dot{\mathbf{D}}_1$ be the simultaneous transient conductive current, so that their sum is the transient current of induction during the charge, a circuital system of current which finally ceases, and which, when added to the final current $\mathbf{C}$, makes up the actual current Γ at any moment during the charge. Thus,

$$\Gamma = \mathbf{C} + \dot{\mathbf{D}} + \dot{\mathbf{D}}_1.$$

The time-integral of $\dot{\mathbf{D}}_1$ is $\mathbf{D}_1$, the complement of $\mathbf{D}$; the two together being circuital. Then, to match $U = \Sigma\frac{1}{2}\mathbf{ED}$, we shall have

$$T = -\Sigma\tfrac{1}{2}\mathbf{ED}_1. \qquad \text{......................}(99)_3$$

This we may verify by a former equation. Assuming it to be true, we have

$$\left.\begin{aligned} U - T &= \Sigma\tfrac{1}{2}\mathbf{E}(\mathbf{D} + \mathbf{D}_1) = \Sigma\tfrac{1}{2}\mathbf{e}(\mathbf{D} + \mathbf{D}_1), \\ \text{or,} \qquad 2(U - T) &= \Sigma\mathbf{E}\int(\Gamma - \mathbf{C})dt = \Sigma\mathbf{e}\int(\Gamma - \mathbf{C})dt, \end{aligned}\right\} \text{............}(99)_4$$

by definition of $\mathbf{D}$ and $\mathbf{D}_1$, which expresses that the work done by $\mathbf{e}$ during the transient state is $2(U - T)$ more than if the current started instantly everywhere in its final distribution. This proof of $(99)_3$ therefore rests upon equation (64). But it is easily proved directly, by using the electric impulse $\mathbf{A}$. Thus, by $(99)_1$,

$$T = \Sigma\tfrac{1}{2}\mathbf{AC} = \Sigma\tfrac{1}{2}\mathbf{A}k\mathbf{E}, \quad \text{if } k = \text{conductivity},$$

$$= \Sigma\tfrac{1}{2}\mathbf{E}\int k\dot{\mathbf{A}}dt.$$

Now, let $\mathbf{E}_1$ be the electric force in the transient state, *i.e.*

$$\mathbf{E}_1 = \mathbf{e} + \mathbf{f} - \dot{\mathbf{A}}, \qquad \text{where } \mathbf{f} \text{ is polar.}$$

Then
$$k\dot{\mathbf{A}} = k(\mathbf{e} + \mathbf{f}) - (\Gamma - \dot{\mathbf{D}}),$$

therefore,
$$T = \Sigma\tfrac{1}{2}\mathbf{E}\int\{k(\mathbf{e} + \mathbf{f}) - (\mathbf{C} + \dot{\mathbf{D}}_1)\}dt,$$

by definition of $\mathbf{D}_1$; and, since $\Sigma k\mathbf{eE} = \Sigma\mathbf{eC} = \Sigma\mathbf{EC}$, and $\Sigma\mathbf{fC} = 0$, we are reduced to $(99)_3$, as required.

The equation $(99)_4$ may be transformed to

$$U - T = \Sigma\tfrac{1}{2}\mathbf{Zg}, \qquad \text{......................}(99)_5$$

if we have

$$-\operatorname{curl}\mathbf{Z} = 4\pi(\mathbf{D} + \mathbf{D}_1), \qquad \text{and} \qquad -\operatorname{curl}\mathbf{e} = 4\pi\mathbf{g}.$$

Here $\mathbf{g}$ is the false magnetic current corresponding to the impressed force, $\mathbf{e}$, going round the lines of $\mathbf{e}$, roughly speaking. And $\mathbf{Z}$ may be taken to be the magnetic impulse at a point, the analogue of $\mathbf{A}$, as we may thus verify. By the just-given equation of $\mathbf{Z}$, differentiating to the time, we get

$$-\operatorname{curl}\dot{\mathbf{Z}} = 4\pi(\dot{\mathbf{D}} + \dot{\mathbf{D}}_1) = 4\pi(\Gamma - \mathbf{C}) = \operatorname{curl}(\mathbf{H}_1 - \mathbf{H}),$$

if $\mathbf{H}_1$ be the magnetic force at any moment during the variable period, $\mathbf{H}$ being its final value. Therefore

$$-\dot{\mathbf{Z}} = \mathbf{H}_1 - \mathbf{H} + \mathbf{F}_1,$$

where $\mathbf{F}_1$ is polar, showing that $-\dot{\mathbf{Z}}$ may be taken as the magnetic force of induction. Or thus: remove $\mathbf{e}$ suddenly, the time-integral of the magnetic force in the variable period following at any point will be $\mathbf{Z}$. The definition of $\mathbf{Z}$ may be extended so that its time-variation shall be the actual magnetic force; but it is simplest, and harmonises best with the electric impulse, to make it refer to the variable period only.

Take $\mathbf{e}$ to exist only across a single thin slice of a wire, the most elementary case. Then $\mathbf{g}$ will be round the boundary. $\mathbf{D}+\mathbf{D}_1$ is then the integral current through $\mathbf{g}$ during the charge (not counting the final current, if any); or, reversed, it is the integral current through $\mathbf{g}$ when $\mathbf{e}$ is removed. If this current be oscillatory, it may amount to nothing in the total. If so, the potential and kinetic energies were equal. The value of $\mathbf{Z}$ at the place of $\mathbf{g}$ is also zero, of course, by $(99)_5$, the magnetic force there reversing with the current. (The place of $\mathbf{g}$ is where energy leaves the seat of $\mathbf{e}$ when it is working.)

Another pair of allied expressions for the parts of U and T *outside* the conductors is expressed by

$$T = \tfrac{1}{2}\int dt\, \Sigma \operatorname{conv} \mathbf{VE}_1\mathbf{H}/4\pi, \qquad U = \tfrac{1}{2}\int dt\, \Sigma \operatorname{conv} \mathbf{VEH}_1/4\pi. \quad \ldots(99)_6$$

Here $\mathbf{E}_1$ and $\mathbf{H}_1$ are the values of the electric and magnetic forces at time t after $\mathbf{e}$ was started, and $\mathbf{E}$, $\mathbf{H}$, their final values; the time-integral to include the variable period, and the Σ being summation through the whole space outside the conductors.

By means of the convergence theorem we may at once transform these expressions to surface-integrals over the boundaries of the conductors. Thus, let $\mathbf{N}$ be the unit normal out from conductor to non-conductor, and

$$\mathbf{VE}_1\mathbf{N} = 4\pi\mathbf{g}_1, \qquad \mathbf{VNH}_1 = 4\pi\gamma_1, \qquad \mathbf{VEN} = 4\pi\mathbf{g}, \qquad \mathbf{VNH} = 4\pi\gamma;$$

γ and γ_1 being therefore boundary (false) electric, and $\mathbf{g}$, $\mathbf{g}_1$ boundary (false) magnetic currents. Then

$$\left.\begin{aligned} T &= -\Sigma\tfrac{1}{2}\mathbf{H}\int \mathbf{g}_1 dt = -\Sigma\tfrac{1}{2}\gamma\int \mathbf{E}_1 dt = \Sigma\tfrac{1}{2}\mathbf{A}\gamma, \\ U &= -\Sigma\tfrac{1}{2}\mathbf{E}\int \gamma_1 dt = -\Sigma\tfrac{1}{2}\mathbf{g}\int \mathbf{H}_1 dt = \Sigma\tfrac{1}{2}\mathbf{Zg}, \end{aligned}\right\} \ldots\ldots\ldots\ldots(99)_7$$

wherein of course the Σ's are mere surface-integrals, since on the surface only are the quantities $\mathbf{g}$ and γ. In the last forms, which we had occasion to employ in the last section, $\mathbf{A}$ and $\mathbf{Z}$ are, as there and in the former part of the present section, the electric and magnetic impulses. Notice, however, that the $\mathbf{g}$ in $(99)_7$ is not the same as the $\mathbf{g}$ in $(99)_5$, and there defined; the present $\mathbf{g}$ extending over all the boundary in general; although, should $T=0$ (no final current), we see by $(99)_5$ and $(99)_7$ that they are then identical, situated round $\mathbf{e}$, which must reach the boundary.

Here we may notice a peculiarity of interest in connection with the difference of treatment of vectors according as they are circuital or polar. Suppose we have two conducting bodies in air, and charge them oppositely by an impressed force in a connecting wire. The electric energy set up is $U = \Sigma \frac{1}{2}\mathbf{Zg}$, $\mathbf{g}$ being round $\mathbf{e}$ only. But now suppose we disconnect the wire from the bodies and take it, with $\mathbf{e}$, away to a distance. We have altered the field very slightly, so that $\mathbf{Z}$ has nearly the same value anywhere as before, and U also. But $\mathbf{g}$ has gone altogether. How then does $\Sigma \frac{1}{2}\mathbf{Zg}$ apply, the force being normal to the conductors?

Notice that, as stated in the last section, this formula only applies when it is possible to close the displacement outside the region in which U exists, so as not to pass through it and alter the field. Thus the formula applies to all the space between the two oppositely charged conductors provided we leave out a little piece joining them, along which to let the displacement return. This little piece may be reduced to a mere line, thus infinitesimally altering the field. The formula then reduces to a line-integral, which will be found to become

$$\tfrac{1}{2} \text{ charge} \times \text{difference of potentials},$$

the common electrostatic formula.

On the other hand, if we join the conductors by a wire through which they will discharge (unless balanced by impressed force in it), the formula acquires reality at once; $\mathbf{g}$ at first moment being at place of contact (round the spark), and thereafter over the conducting surface generally. During the discharge there will be a real magnetic current $\mathbf{G}$ in space, and the value of U at any moment will be $\Sigma \frac{1}{2}\mathbf{Z}(\mathbf{G}+\mathbf{g})$. [See last section, p. 513]

SECTION XVIII. NORMAL ELECTROMAGNETIC SYSTEMS. ENERGY CONJUGATE PROPERTIES.

The specification of the complete state of the electromagnetic field at a given moment requires a knowledge of seven quantities. We must, in the first place, know the electric capacity and conductivity, and the magnetic permeability, c, k, and μ. Next, we require to know the electric and the magnetic force, $\mathbf{E}$ and $\mathbf{H}$. From these five data we know, by the linear relations, the conduction current $\mathbf{C} = k\mathbf{E}$, the elastic displacement $\mathbf{D} = c\mathbf{E}/4\pi$, and the magnetic induction $\mathbf{B} = \mu\mathbf{H}$. We also know the electric energy $\frac{1}{2}\mathbf{ED}$, the magnetic energy $\frac{1}{2}\mathbf{HB}/4\pi$, and the dissipativity $\mathbf{EC}$; all per unit volume. But, in addition, we require to have given the impressed electric and magnetic forces, $\mathbf{e}$ and $\mathbf{h}$. Then, by the two equations of induction,

$$\left.\begin{aligned} \text{curl}\,(\mathbf{H}-\mathbf{h}) &= 4\pi\Gamma = 4\pi(\mathbf{C}+\dot{\mathbf{D}}), \\ \text{curl}\,(\mathbf{e}-\mathbf{E}) &= 4\pi\mathbf{G} = \dot{\mathbf{B}}, \end{aligned}\right\}$$

we know the true electric current Γ, and therefore the displacement current $\dot{\mathbf{D}}$; and also the magnetic current $\mathbf{G}$. As for the electrification, it is known because $\mathbf{D}$ is known, of which it is the divergence. The seven data may be otherwise stated; for instance, instead of **E** and **H**, we may have **D** and **B**, or **C** and **B**. As regards the number of distinct numbers on which these seven quantities depend, if we take any three rectangular axes of reference, the four vectors **E**, **H**, **e**, and **h** require three numbers each, making twelve, and the three operators c, k, and μ require six numbers each (if there be no rotatory k), making eighteen. Thus altogether thirty numbers are concerned; or thirty-three, if there be rotatory conductivity. In case of isotropy, the number is fifteen, owing to c, k, and μ being then simply scalars.

We not only know the complete state of the system, but the rate at which it is changing; for $\dot{\mathbf{E}}$ and $\dot{\mathbf{H}}$ are known, and therefore, if the impressed forces be given at every moment, we can find the changes it goes through under their influence; or, if they be absent, in settling down to a state of equilibrium.

The equation of activity at any moment is

$$\Sigma\,\mathbf{e}\Gamma + \Sigma\,\mathbf{h}\mathbf{G} = Q + \dot{U} + \dot{T};$$

the left member being the total activity of the impressed forces in all space, and the right its equivalent, the sum of the dissipativity Q (Joule-heat per second) and the rates of increase of the electric and magnetic energies.

Now, suppose no relative motion of masses is permitted, thus making c, k, μ functions of position only, and excluding the impressed forces brought into play by such motions. If now **e** and **h** be constant with respect to the time, the system will settle down to a steady state, in which $\Sigma\,\mathbf{eC} = Q$ simply. If, further, there be no **e** in conductors, or, more generally, only such distributions as may cause elastic displacement, but no steady conduction current, the final field is simply that due to **h** and to the **e** left, and to electrification and its surface equivalent, the charges of conductors.

But, by the linearity of equations, the inductive phenomena during the subsidence to the final state under the influence of steady **e** and **h** may be got by superimposition. We may therefore, in investigating subsidence, take $\mathbf{e} = 0$, $\mathbf{h} = 0$, and no electrification or charges on conductors; so that the subsidence is to a final state of no **E** or **H** anywhere.

We then have, at every moment after removal of impressed forces,

$$Q + \dot{U} + \dot{T} = 0; \qquad \text{(100)}$$

the rate of decrease of the sum of the electric and magnetic energies being equal to the dissipativity. Q, U, and T are all necessarily positive, being sums of squares, or else of positive scalar products. (For instance $\mathbf{EC} = \mathbf{E}k\mathbf{E}$; if **C** is parallel to **E**, it is a square, $k\mathbf{E}^2$; if not parallel, their mutual angle must be always acute.) This necessary positivity is of the greatest importance, as it excludes the possibility of indefinite increase of normal systems of force left to themselves, making them always subside, either without or with oscillations.

Let, next, there be two systems of electric and magnetic force distinguished by the suffixes $_1$ and $_2$, so that their equations of induction are

$$\text{curl}\,\mathbf{H}_1 = 4\pi\Gamma_1, \quad -\text{curl}\,\mathbf{E}_1 = 4\pi\mathbf{G}_1,$$
$$\text{curl}\,\mathbf{H}_2 = 4\pi\Gamma_2, \quad -\text{curl}\,\mathbf{E}_2 = 4\pi\mathbf{G}_2.$$

Using the third and second of these, we find by space-integration,

$$\Sigma\,\mathbf{E}_1\Gamma_2 = \Sigma\,\mathbf{E}_1\,\text{curl}\,\mathbf{H}_2/4\pi = \Sigma\,\mathbf{H}_2\,\text{curl}\,\mathbf{E}_1/4\pi = -\Sigma\,\mathbf{H}_2\mathbf{G}_1.$$

Similarly, by the first and fourth, we shall get $\Sigma\,\mathbf{E}_2\Gamma_1 = -\Sigma\,\mathbf{H}_1\mathbf{G}_2$; so that we have

$$\Sigma(\mathbf{E}_1\Gamma_2 + \mathbf{H}_2\mathbf{G}_1) = 0, \quad \text{and} \quad \Sigma(\mathbf{E}_2\Gamma_1 + \mathbf{H}_1\mathbf{G}_2) = 0;$$

or,

$$\left.\begin{aligned}\Sigma\,\mathbf{E}_1\mathbf{C}_2 + \Sigma\,\mathbf{E}_1\dot{\mathbf{D}}_2 + \Sigma\,\mathbf{H}_2\dot{\mathbf{B}}_1/4\pi = 0,\\ \Sigma\,\mathbf{E}_2\mathbf{C}_1 + \Sigma\,\mathbf{E}_2\dot{\mathbf{D}}_1 + \Sigma\,\mathbf{H}_1\dot{\mathbf{B}}_2/4\pi = 0.\end{aligned}\right\} \quad \text{.................(101)}$$

If we add these, we shall obtain

$$Q_{12} + \dot{U}_{12} + \dot{T}_{12} = 0, \quad \text{.......................(102)}$$

the equation of mutual activity, U_{12} and T_{12} are the mutual electric and magnetic energies, and Q_{12} the mutual dissipativity, or the excess of the total dissipativity when the two fields co-exist over the sum of the separate dissipativities.

Let, next, the arrangement of $\mathbf{E}$ and $\mathbf{H}$ be such that in subsiding they change in magnitude only, not in distribution. Let $\mathbf{E}_0$ and $\mathbf{H}_0$ be the distributions at time $t=0$, and, at time t,

$$\mathbf{E} = \mathbf{E}_0\epsilon^{nt}, \quad \mathbf{H} = \mathbf{H}_0\epsilon^{nt}. \quad \text{.....................(103)}$$

The constant n is the reciprocal of a time, and is of course negative, if the subsidence be real. The larger n, the more rapid the subsidence. $\mathbf{E}$ and $\mathbf{H}$ thus defined constitute a normal system of electric and magnetic force.

Now,

$$\left.\begin{aligned} Q &= \Sigma\,\mathbf{E}k\mathbf{E}, & U &= \Sigma\,\mathbf{E}c\mathbf{E}/8\pi, & T &= \Sigma\,\mathbf{H}\mu\mathbf{H}/8\pi,\\ &= Q_0\epsilon^{2nt}, & &= U_0\epsilon^{2nt}, & &= T_0\epsilon^{2nt},\end{aligned}\right\}$$

by (103), if Q_0, U_0, T_0, be the initial values. From this, $\dot{T} = 2nT$, and $\dot{U} = 2nU$, which makes the equation of activity (100) become

$$Q + 2n(U+T) = 0;$$

or the ratio of the energy left at any moment to its rate of leaving is constant, $= -(2n)^{-1}$.

If then the two systems to which equations (101) refer be both normal, with rates of subsidence n_1 and n_2, we shall have

$$\left.\begin{aligned} Q_1 + 2n_1(U_1 + T_1) = 0,\\ \text{and} \qquad Q_2 + 2n_2(U_2 + T_2) = 0,\end{aligned}\right\} \quad \text{.......................(104)}$$

when they exist separately; and, in addition, when they co-exist, the equations (101), in which $d/dt = n_1$ or n_2, according to whether it operates on the first or second system. Now

$$\mathbf{E}_1\mathbf{D}_2 = \mathbf{E}_2\mathbf{D}_1, \qquad \mathbf{H}_1\mathbf{B}_2 = \mathbf{H}_2\mathbf{B}_1, \qquad \text{and} \qquad \mathbf{E}_1\mathbf{C}_2 = \mathbf{E}_2\mathbf{C}_1,$$

if there be no rotatory conductivity coefficient; so that (101) becomes

$$\left.\begin{aligned}\tfrac{1}{2}Q_{12}+n_2U_{12}+n_1T_{12}=0,\\ \tfrac{1}{2}Q_{12}+n_1U_{12}+n_2T_{12}=0,\end{aligned}\right\}\text{.....................}(105)$$

where

$$Q_{12}=2\Sigma\,\mathbf{E}_1\mathbf{C}_2,\qquad U_{12}=\Sigma\,\mathbf{E}_1\mathbf{D}_2,\qquad T_{12}=\Sigma\,\mathbf{H}_1\mathbf{B}_2/4\pi.$$

Between the equations (105) we may eliminate in succession either the Q, or U, or T. Thus we get, if we leave out the common factor (n_1-n_2),

$$\left.\begin{aligned}0&=U_{12}-T_{12},\\ 0&=\tfrac{1}{2}Q_{12}+(n_1+n_2)T_{12},\\ 0&=\tfrac{1}{2}Q_{12}+(n_1+n_2)U_{12},\end{aligned}\right\}\text{.....................}(106)$$

which are the universal conjugate properties of normal systems.

The first tells us that the mutual potential and kinetic energies of two normal systems are equal. This being true at every moment during the subsidence, it follows, by (102), that the mutual dissipativity is derived from them equally. This is the interpretation of the second and third of (106), the second saying that the mutual dissipativity equals twice the rate of decrease of the mutual kinetic energy, and the third that it equals twice the rate of decrease of the mutual potential energy.

Any one of (106) enables us to decompose a given initial state of electric and magnetic force into the sum of normal distributions, when the nature of the latter has been found, and hence determine the manner of subsidence. Thus, let $\mathbf{E}_0$, $\mathbf{H}_0$, be the initial state, and $\mathbf{E}_1$, $\mathbf{H}_1$, $\mathbf{E}_2$, $\mathbf{H}_2$, $\mathbf{E}_3$, $\mathbf{H}_3$, etc., the various normal distributions, with constants n_1, n_2, etc. Let

$$\left.\begin{aligned}\mathbf{E}_0&=A_1\mathbf{E}_1+A_2\mathbf{E}_2+A_3\mathbf{E}_3+\ldots,\\ \mathbf{H}_0&=A_1\mathbf{H}_1+A_2\mathbf{H}_2+A_3\mathbf{H}_3+\ldots,\end{aligned}\right\}$$

where the A's are coefficients fixing the absolute magnitudes of the normal solutions. If the A's can be found, then, at time t later, we shall have

$$\mathbf{E}=A_1\mathbf{E}_1\epsilon^{n_1t}+A_2\mathbf{E}_2\epsilon^{n_2t}+\ldots,$$

$\mathbf{E}$ being what $\mathbf{E}_0$ then becomes; and a similar equation for $\mathbf{H}$.

The A's are found thus:—$\mathbf{E}_0$ and $\mathbf{H}_0$ being given, and the rth coefficient A_r, being required, calculate the mutual potential and the mutual kinetic energy of the given state with respect to the rth normal distribution. Let them be U_{or} and T_{or}. Their values are

$$U_{or}=A_1U_{1r}+A_2U_{2r}+A_3U_{3r}+\ldots+A_rU_{rr}+\ldots,$$
$$T_{or}=A_1T_{1r}+A_2T_{2r}+A_3T_{3r}+\ldots+A_rT_{rr}+\ldots.$$

Subtract the second line from the first, and there results

$$U_{or}-T_{or}=A_r(U_{rr}-T_{rr});\qquad\text{or,}\qquad A_r=(U_{or}-T_{or})/(U_{rr}-T_{rr}),\quad(107)_1$$

since, by the first of (106), all the remaining terms cancel. Thus, the rth coefficient equals the excess of the mutual potential over the mutual kinetic energy of the given state and the rth normal state, divided by twice the excess of the potential over the kinetic energy of the normal state itself.

The second and third of (106) give two equivalent relations, by widely different processes, viz.,

$$A_r = (Q_{or} + 4nT_{or})/(Q_{rr} + 4nT_{rr}), \quad \text{..............}(107)_2$$

and

$$A_r = (Q_{or} + 4nU_{or})/(Q_{rr} + 4nU_{rr}). \quad \text{..............}(107)_3$$

As the first of (106) is the easiest to remember, so it is in general the most readily applied, giving $(107)_1$. But some special cases should be noticed.

If there be no potential energy, but only kinetic energy and dissipativity; that is, in all problems in which dielectric displacement is not taken into account, as, for instance, any combination of conductors between which there is electromagnetic induction, but with no condensers, we have

$$c = 0, \quad T_{12} = 0, \quad Q_{12} = 0; \quad \text{and} \quad A_r = T_{or}/T_{rr} = Q_{or}/Q_{rr}. \quad \text{...}(108)_1$$

If there be no kinetic energy, but only potential energy and dissipativity; that is, in all cases in which electromagnetic induction is ignorable, as in any combination of conductors and condensers, but not coils, we have

$$\mu = 0, \quad U_{12} = 0, \quad Q_{12} = 0; \quad \text{and} \quad A_r = U_{or}/U_{rr} = Q_{or}/Q_{rr}. \quad (108)_2$$

If there be potential and kinetic energy, but no dissipation,

$$k = 0, \quad U_{12} = T_{12}; \quad \text{and} \quad A_r = (U_{or} - T_{or})/(U_{rr} - T_{rr}). \quad \text{...}(108)_3$$

In this last case conductors are excluded. We have a strictly conservative system, from which all radical friction is excluded. It goes on oscillating for ever, but never does any useful work. We must therefore abolish it. A peculiarity connected with $(108)_3$ will be noticed in the next section. Also that the general properties (106) are true whether the rates of subsidence of the two systems be unequal or equal, although in the latter case special procedure is required.

The nature of the normal distributions themselves depends upon the distribution of c, k, and μ throughout space. We have

$$\text{curl}\,\mathbf{H} = (4\pi k + cn)\mathbf{E}, \qquad -\text{curl}\,\mathbf{E} = \mu n\mathbf{H}, \quad \text{...........}(109)$$

in a normal arrangement. Hence, either $\mathbf{E}$ or $\mathbf{H}$ being found, the other follows. Thus, eliminate $\mathbf{H}$ to get the equation of $\mathbf{E}$,

$$\text{curl}\,\mu^{-1}\,\text{curl}\,\mathbf{E} + n(4\pi k + n)\mathbf{E} = 0. \quad \text{...............}(110)$$

Any solution of this is a normal $\mathbf{E}$, and the corresponding $\mathbf{H}$ is definitely fixed by the second of (109). Not counting the simple cases of linear circuits and similar problems, (110) has been solved in three dimensions in a very few cases.

Presuming we have obtained the normal solutions, the question arises, what values of n shall we take? We must take all that satisfy all the conditions of the problem. One form of the determinantal equation, whose roots give all the admissible values of n, is the equation of activity itself,

$$Q + \dot{U} + \dot{T} = 0; \qquad \text{or,} \qquad Q + 2n(U + T) = 0,$$

applied to the normal solutions. It is an equation in n only, with various constants, but independent of x, y, z, and t. That is, if in some special case of (110), we know the normal solution, we can find the equation of n by writing out the equation of activity extended over the whole system. But the equation of n is usually to be obtained from the boundary conditions, when the normal functions are known through bounded spaces. [This is the proper way. The other way may be accidentally, but is not generally true.]

If we extend our calculation of the excess of the mutual potential over the mutual kinetic energy of two normal systems through a bounded space, instead of all space, we shall obtain, not $U_{12} - T_{12} = 0$, but

$$4\pi(n_1 - n_2)(U_{12} - T_{12}) = \Sigma\, \mathbf{N}(\mathrm{V}\mathbf{E}_2\mathbf{H}_1 - \mathrm{V}\mathbf{E}_1\mathbf{H}_2), \quad \text{........(111)}$$

the summation being over the boundary, $\mathbf{N}$ being the unit normal drawn inward. Hence $U - T$ for a single normal system n is given by

$$\left.\begin{aligned} 8\pi(U - T) &= \Sigma\,(\mathbf{E}c\mathbf{E} - \mathbf{H}\mu\mathbf{H}) \\ &= \Sigma\, \mathbf{N}(\mathrm{V}\mathbf{E}\mathbf{H}' - \mathrm{V}\mathbf{E}'\mathbf{H}), \end{aligned}\right\} \quad \text{..................(112)}$$

where the accent means differentiation with respect to n. The first is a volume-, and the second a surface-summation.

There are, of course, corresponding boundary forms for the second and third of (106).

The general properties of normal systems (100), (102), (104), (106), (107), and (108), are not peculiar to the special dynamical connections which are involved in the electromagnetic equations, but belong to any dynamical system in which forces of reaction are proportional to displacements, and resistances to velocities, with reciprocal relations amongst the coefficients which are equivalent (in the electromagnetic case) to the three linear relations between forces and fluxes being of a symmetrical nature; or c, k, and μ self-conjugate, with no rotatory power. Conservation of energy requires this to be true for c and μ; and (106) are not true unless k be also self-conjugate.

Section XIX. Remarks on Normal Electromagnetic Systems. Conditions of Possibility of Oscillatory Subsidence. Equal Roots, and their Effects.

In the last section I omitted to define the three symbols, U_{rr}, T_{rr}, and Q_{rr}, except by implication. They express the *doubles* of the potential energy, kinetic energy, and dissipativity of the rth normal system $\mathbf{E}_r$, $\mathbf{H}_r$; being quantities formed in the same manner as Q_{12}, U_{12}, and T_{12}, defined just after equation (105).

As a preparation for what follows it will be useful to bear in mind the general character of the subsidence to equilibrium of a displaced elastic body, which, for our purpose, may be simply a stretched elastic thin wire fixed at its ends. Let it be bent into the form of the arc of a bow, or, more accurately, into the form of the sinusoidal curve, and then be left to itself. If there be no resistance the wire will go on vibrating for ever with uniform frequency, always preserving the sinusoidal form. But if there be resistance to its motion, proportional at every moment to its speed, its amplitude of vibration will continuously decrease, although the frequency (lowered) will be still uniform. By a sufficient increase in the coefficient of resistance (say, by motion in a viscous fluid), we shall ultimately stop the vibrations, the displaced wire returning to, but never crossing its equilibrium position. The displacement at time t in the original frictionless vibration was represented by

$$(a_1 \sin + b_1 \cos)c_1 t.$$

Friction makes it $\epsilon^{-c_2 t}(a_2 \sin + b_2 \cos)c_3 t,$

showing the oscillatory subsidence to equilibrium. When, by sufficient increase of resistance, the oscillations are just stopped, it is

$$\epsilon^{-c_4 t}(a_3 + b_3 t);$$

and finally, further increase makes it

$$a_4 \epsilon^{-c_5 t} + b_4 \epsilon^{-c_6 t},$$

the sum of two independent non-oscillatory subsidences.

In general, if inertia be altogether negligible, but not elastic yielding, or the friction, there can be no oscillations. Similarly, if the elastic yielding be negligible, but not inertia and friction, there can be no oscillations. To have oscillations we require both inertia and elastic yielding; besides that the resistance must not be too great.

Coming now to the electromagnetic applications, we shall expect the subsidence of normal systems to come under these four types. If there be no elastic displacement, and therefore no potential energy, the subsidence of a normal system must be non-oscillatory; and it must be real subsidence, not indefinite increase according to the same law. Similarly, if there be no inertia ($\mu = 0$, no magnetic induction, no magnetic or kinetic energy), the subsidence must also be real and non-oscillatory. But if neither elastic displacement nor inertia be negligible, there will be either non-oscillatory or oscillatory real subsidence, according to the relative importance of the resistance. In these three cases there is supposed to be always resistance. But if there be none but only elasticity and inertia to consider, the normal systems will be simple harmonic with respect to the time, and go on vibrating for ever. Cases in which two of the three quantities c, k, and μ are non-existent, scarcely belong to the present subject. And the fourth case above (vibrations in dielectric media, with *no* dissipation) does not occur in ordinary problems, as it requires unrealisable conditions.

Now the equation of activity of a normal system is

$$2n(U+T) + Q = 0,$$

where $U = \Sigma \mathbf{E}c\mathbf{E}/8\pi, \qquad T = \Sigma \mathbf{H}\mu\mathbf{H}/8\pi, \qquad Q = \Sigma \mathbf{E}k\mathbf{E}.$

Here $\mathbf{E}$ and $\mathbf{H}$ constitute a normal system, $\mathbf{E}$ being a solution of (110), and $\mathbf{H}$ derived from it by the second of (109); or else $\mathbf{H}$ being a solution of the $\mathbf{H}$ equation corresponding to (110), and $\mathbf{E}$ derived from it by the first of (109). (If there be no inertia, the electric force is polar. Then the single scalar, the electric potential, will serve for variable.)

If n be real, the normal functions $\mathbf{E}$ and $\mathbf{H}$ are real, or may be so chosen as to be real. Then also Q, U, and T are real. Further, it is necessary in the electromagnetic applications that they cannot be negative. This is secured by the angle between a force and a flux being less than 90° at the most.

($c=0$.) First, ignore elastic displacement. Then $c=0$, U is non-existent, and

$$2nT+Q=0.$$

We see at once that if n be real, it must be negative. If, then, we show that it cannot be imaginary, we prove that all the n's are real and negative, when c vanishes, but not μ and k. Thus, if i stand for $(-1)^{\frac{1}{2}}$, let

$$n_1=a+bi, \qquad n_2=a-bi,$$

be a pair of imaginaries. They turn $\mathbf{H}$ into $\mathbf{L}+\mathbf{M}i$ and $\mathbf{L}-\mathbf{M}i$ respectively, and $\mathbf{E}$ into $\mathbf{L}_1+\mathbf{M}_1 i$ and $\mathbf{L}_1-\mathbf{M}_1 i$ respectively, $\mathbf{L}$, $\mathbf{M}$, etc., being real. Using n_1, the expressions for T and Q become

$$\left.\begin{aligned} T&=\Sigma(\mathbf{L}\mu\mathbf{L}-\mathbf{M}\mu\mathbf{M}+2i\mathbf{L}\mu\mathbf{M})/8\pi, \\ Q&=\Sigma(\mathbf{L}_1k\mathbf{L}_1-\mathbf{M}_1k\mathbf{M}_1+2i\mathbf{L}_1k\mathbf{M}_1). \end{aligned}\right\}$$

Using these in the equation of activity, with $n=n_1$, and separating the real from the imaginary parts, we get

$$0=\Sigma\{2a(\mathbf{L}\mu\mathbf{L}-\mathbf{M}\mu\mathbf{M})-4b\mathbf{L}\mu\mathbf{M}+8\pi(\mathbf{L}_1k\mathbf{L}_1-\mathbf{M}_1k\mathbf{M}_1)\},$$
$$0=\Sigma\{2b(\mathbf{L}\mu\mathbf{L}-\mathbf{M}\mu\mathbf{M})+4a\mathbf{L}\mu\mathbf{M}+16\pi\mathbf{L}_1k\mathbf{M}_1\}.$$

But also, by the conjugate properties in $(108)_1$, we have the mutual T and Q of the two systems n_1 and n_2 both zero; or

$$\Sigma(\mathbf{L}\mu\mathbf{L}-\mathbf{M}\mu\mathbf{M})=0, \qquad \Sigma(\mathbf{L}_1k\mathbf{L}_1-\mathbf{M}_1k\mathbf{M}_1)=0;$$

the imaginary parts cancelling. These bring the previous equations to

$$b\Sigma\,\mathbf{L}\mu\mathbf{M}=0, \qquad a\Sigma\,\mathbf{L}\mu\mathbf{M}/4\pi+\Sigma\,\mathbf{L}_1k\mathbf{M}_1=0.$$

From the first of these we conclude that $b=0$, unless $\mathbf{L}$ and $\mathbf{M}$ are the magnetic forces of two normal systems, which is not the case here. The imaginary parts are therefore non-existent, which brings us to

$$n_1=a, \qquad n_2=a, \qquad T=0, \qquad Q=0.$$

What we wanted to show was that imaginaries could not exist. In addition, we show that if there be a pair of *equal* n's, they will make the kinetic energy and the dissipativity of the (equal) normal systems both zero. The only way this can happen, T and Q being the sum of quantities that cannot be negative, is for each of their elements to vanish, and, therefore, $\mathbf{E}=0$, $\mathbf{H}=0$. That is, if n be double (or repeated any number of times), that value of n will make the normal functions vanish over all space.

($\mu=0$.) Next, ignore magnetic induction. Then $\mu=0$, T is non-existent, and

$$2nU+Q=0.$$

We can show that the n's are all real and negative, excluding oscillatory subsidence, and that the first conditions of a repeated n are

$$U=0, \qquad Q=0,$$

which necessitate the vanishing of the normal functions for that value of n. But, owing to the peculiarities arising from the division of space into conducting and non-conducting regions, the matter cannot be shortly treated, and will be returned to.

(c, k and μ.) Take next the general case of T, U, and Q all existent. Write the activity equation thus,

$$2n^2(U+T)+nQ=0,$$

and solve as a quadratic. Then,

$$n=\{-Q\pm(Q^2-4n^2UT)^{\frac{1}{2}}\}/4U,$$
$$n=\{-Q\mp(Q^2-4n^2UT)^{\frac{1}{2}}\}/4T.$$

Remembering that if n is real, Q, U, and T are all positive (if not zero), we see that $Q^2>4n^2UT$, or $Q^2>\dot{U}\dot{T}$, or the dissipativity greater than the geometrical mean of the rates of decrease of the potential and kinetic energies, must be true. And n is negative. The limit of reality is reached when $Q^2=4n^2UT$; or

$$U=T, \qquad 4nU+Q=0, \qquad 4nT+Q=0.$$

Thus $U_{12}=T_{12}$, the general conjugate property of two normal systems (n_1 and n_2) when they are unequal, is also true when they are equal, *i.e.*, when n is a double root of the determinantal equation of n. This includes the previous special cases of either $c=0$, or else $\mu=0$. Further information regarding imaginary n's may be obtained by separating the real from the imaginary parts in the above.

($k=0$.) When we take $k=0$, in the equation of the normal **E** functions, we have

$$\text{curl}\,\mu^{-1}\text{curl}\,\mathbf{E}+cn^2\mathbf{E}=0, \qquad \mu n\mathbf{H}=-\text{curl}\,\mathbf{E}. \quad \text{.......(113)}$$

If, on the other hand, we take $c=0$, we have an equation for **E** of the same form, but containing n instead of n^2. Hence the same normal functions serve in both cases, if $4\pi kn$ and cn^2 be exchanged. Former conclusions regarding n in the case $c=0$ are therefore now true of n^2. That is, every n^2 is real and negative, making the n's pairs of oppositely signed equal imaginaries, as

$$n_1=ai, \qquad n_2=-ai, \qquad n_3=bi, \qquad n_4=-bi, \qquad \text{etc.,} \quad \text{...(114)}$$

where a, b, etc., are real, indicating simple harmonic oscillations without subsidence.

The property $U_{12}=T_{12}$ is true for any two roots, whether naturally associated or not; *i.e.*, for n_1 with respect to all the rest, including its companion n_2, whose square is the same. But also, the second and third conjugate properties (106), keeping in the (n_1-n_2) factor there omitted, are

$$(n_1^2-n_2^2)U_{12}=0, \qquad (n_1^2-n_2^2)T_{12}=0,$$

hence $U_{12}=0=T_{12}$, except if $n_1^2=n_2^2$; that is, in case of the naturally paired n's. Also, the equation of activity becomes

$$U+T=0,$$

for every single root n.

As the first of (113) contains n^2, if we take the $\mathbf{E}$ normal functions from it, they will be identical in pairs, $\mathbf{E}_1=\mathbf{E}_2$, $\mathbf{E}_3=\mathbf{E}_4$, etc., for the roots (114).

But then the second equation (113) shows that the corresponding $\mathbf{H}$ functions are the negatives of one another in pairs, thus $\mathbf{H}_1=-\mathbf{H}_2$, $\mathbf{H}_3=-\mathbf{H}_4$, etc. Thus the expansions of $\mathbf{E}_0$ and $\mathbf{H}_0$, the initial states of electric and magnetic force, become

$$\mathbf{E}_0=(A_1+A_2)\mathbf{E}_1+(A_3+A_4)\mathbf{E}_3+\dots,$$
$$\mathbf{H}_0=(A_1-A_2)\mathbf{H}_1+(A_3-A_4)\mathbf{H}_3+\dots.$$

The mutual potential energy of any two double normal systems is zero, and the same is true of the mutual kinetic energy. We therefore have

$$\Sigma\,\mathbf{E}_0c\mathbf{E}_1=(A_1+A_2)\Sigma\,\mathbf{E}_1c\mathbf{E}_1,$$
$$\Sigma\,\mathbf{H}_0\mu\mathbf{H}_1=(A_1-A_2)\Sigma\,\mathbf{H}_1\mu\mathbf{H}_1\,;$$

giving A_1 and A_2 in terms of the initial state $\mathbf{E}_0$, $\mathbf{H}_0$. Using these, and putting the solutions in the appropriate real form, taking $\mathbf{H}_1i=\mathbf{M}_1$, $\mathbf{H}_3i=\mathbf{M}_3$, etc., we find that

$$\left.\begin{aligned}\mathbf{E}_0&=\frac{\mathbf{E}_1}{\Sigma\,\mathbf{E}_1c\mathbf{E}_1}(\Sigma\,\mathbf{E}_0c\mathbf{E}_1\cos-\Sigma\,\mathbf{H}_0\mu\mathbf{M}_1\sin)at+\dots,\\ \mathbf{H}_0&=\frac{\mathbf{M}_1}{\Sigma\,\mathbf{E}_1c\mathbf{E}_1}(\Sigma\,\mathbf{E}_0c\mathbf{E}_1\sin+\Sigma\,\mathbf{H}_0\mu\mathbf{M}_1\cos)at+\dots\end{aligned}\right\}$$

express the values $\mathbf{E}_0$ and $\mathbf{H}_0$ reach at time t later.

The proof that there cannot be any imaginary n^2's requires some modification from the proof of absence of imaginary n's in the case $c=0$, owing to the changed conjugate properties. It also shows that a repeated n^2 makes the normal functions vanish. (*See* Thomson and Tait on Cycloidal Motion with no Dissipative Forces, "Natural Philosophy," vol. I., part 2).

(*Equal roots.*) This remarkable property of the vanishing (with equal roots) of the normal functions in case any one of the three c, k, and μ is zero, is closely connected with another property, viz., that of shutting out the $t\epsilon^{nt}$ term from the solutions. Looking to the formulæ at the commencement of this section, we see that on the boundary between oscillatory and non-oscillatory subsidence we have, instead of the form $a_1\epsilon^{n_1t}+a_2\epsilon^{n_2t}$, that of $(a+bt)\epsilon^{nt}$. Also, when by a change in the value of some constant, two roots are made to approach one another, and then again diverge imaginary, between the two states we have a pair of equal roots. If, then, the oscillatory form of solution is possible we have the $t\epsilon^{nt}$ term on the very verge of oscillation. Now, in certain cases we know that oscillations are impossible; they require both kinetic and potential energy to be concerned; so, if either be absent, something must happen to prevent solutions taking the oscillatory form. That something is the vanishing of the normal functions, thus excluding the $t\epsilon^{nt}$ terms, and making the solution in case of a double root take the form $a\epsilon^{nt}$, the same as if the root were not repeated.

Let n_1 and n_2 be a pair of n's, and write down the corresponding terms of the $\mathbf{E}_0$ expansion. They are

$$\frac{\Sigma(\mathbf{E}_0 c\mathbf{E}_1 - \mathbf{H}_0\mu\mathbf{H}_1)}{\Sigma(\mathbf{E}_1 c\mathbf{E}_1 - \mathbf{H}_1\mu\mathbf{H}_1)}\mathbf{E}_1\epsilon^{n_1t} + \frac{\Sigma(\mathbf{E}_0 c\mathbf{E}_2 - \mathbf{H}_0\mu\mathbf{H}_2)}{\Sigma(\mathbf{E}_2 c\mathbf{E}_2 - \mathbf{H}_2\mu\mathbf{H}_2)}\mathbf{E}_2\epsilon^{n_2t}. \quad \ldots\ldots(115)$$

Here there are no restrictions put upon c, k and μ; $\mathbf{E}_1$, $\mathbf{H}_1$, and $\mathbf{E}_2$, $\mathbf{H}_2$, are the normal functions corresponding to n_1 and n_2, and the $U_{12} = T_{12}$ formula has been employed to find A_1 and A_2, the coefficients of $\mathbf{E}_1$ and $\mathbf{E}_2$ (the fractions).

At first glance it might appear that if, by some change in the value of some electrical constant concerned, or generally, by a changed distribution of c, k, μ, the roots n_1 and n_2 are made to approach and finally reach equality, making $\mathbf{E}_1$ and $\mathbf{H}_1$ also approach to and finally be the same as $\mathbf{E}_2$ and $\mathbf{H}_2$, their coefficients A_1 and A_2 will also approach and ultimately be equal. But, in general, nothing could be further from the truth, and instead of equality, we shall have infinite inequality, on account of the denominators approaching zero from opposite sides, sending one A up to positive and the other down to negative infinity. For if $n_1 = n + h$, and $n_2 = n - h$, where h is very small, we shall have

$$\mathbf{E}_1 = \mathbf{E} + h\mathbf{E}', \qquad \mathbf{H}_1 = \mathbf{H} + h\mathbf{H}',$$
$$\mathbf{E}_2 = \mathbf{E} - h\mathbf{E}', \qquad \mathbf{H}_2 = \mathbf{H} - h\mathbf{H}',$$

if the accent denote differentiation to n, provided the functions $\mathbf{E}$, $\mathbf{H}$, and their differential coefficients do not vanish. These make

$$\Sigma(\mathbf{E}_1 c\mathbf{E}_1 - \mathbf{H}_1\mu\mathbf{H}_1) = 2h\Sigma(\mathbf{E}c\mathbf{E}' - \mathbf{H}\mu\mathbf{H}').$$

This is the value of the denominator of A_1 in (115), and that of A_2 is the same taken negatively, thus showing the infinite divergence of the A's. [There is an example worked out on p. 90.]

The two terms, when united, and h made to vanish, give rise to a solution of the form

$$C_1\mathbf{F}\epsilon^{nt} + C_2\mathbf{E}t\epsilon^{nt},$$

where C_1 and C_2 are new constants, $\mathbf{E}$ is the old $\mathbf{E}$, and $\mathbf{F}$ is a new function derived from $\mathbf{E}$. From this we see that when the repeated n makes the normal functions vanish, as when any one of c, k, μ is zero, the second term goes out altogether.

The double-root solution in these cases of vanishing $\mathbf{E}$ and $\mathbf{H}$ is

$$\frac{\Sigma(\mathbf{E}_0 c\mathbf{E}' - \mathbf{H}_0\mu\mathbf{H}')}{\Sigma(\mathbf{E}'c\mathbf{E}' - \mathbf{H}'\mu\mathbf{H}')}\mathbf{E}'\epsilon^{nt}, \quad \ldots\ldots\ldots\ldots\ldots\ldots\ldots(116)$$

differing from the original form only in this, that instead of the normal functions $\mathbf{E}$ and $\mathbf{H}$, we take their differential coefficients with respect to n. This single term takes the place of the former two terms.

If the root n be triple, $\mathbf{E}'$ and $\mathbf{H}'$ will also vanish. Then take $\mathbf{E}''$ and $\mathbf{H}''$ instead; and similarly go on to further differentiations in case of further repetitions of n.

If $N = 0$ be the determinantal equation of the n's, the function $U - T$ of a normal system contains N', the differential coefficient of N with respect to n, as a factor. $N' = 0$, in addition to $N = 0$, is the condition that n is a double root. Similarly, the function $U - T$, not of **E** and **H**, but of **E**$'$ and **H**$'$, contains N'', the second differential coefficient of N, as a factor, and so on.

Before leaving this curious subject of the effect of equal rates of subsidence, we should notice that when the duplicity of an n, making the conjugate properties of unequal n's hold good for two of the same value, necessitates the simultaneous vanishing of the normal functions, it does so in virtue of the positivity of Q, U, T, as before mentioned. But should they be allowed to be negative, although, for example, in the case $c=0$, we still have $T=0$, $Q=0$, when an n is double, there is no longer any necessity for **E** and **H** to simultaneously vanish. Then we have the $t\epsilon^{nt}$ term, and the $t^2\epsilon^{nt}$ term if a triple root, and so on. The vanishing of T will then depend on its expression containing N, for the special value of n, as a factor. As our expressions for Q, U, and T are in the form of the sum of scalar products, we can only make any one of them negative by allowing that the force and the flux, in some parts of space at least, can make an obtuse angle with one another; that is, be opposed, which is a contradiction to common sense. In special applications, involving only a limited number of degrees of freedom, the positivity of U, T, and Q will require that certain functions of the electrical constants, usually determinants, cannot be negative for any values of the electric variables.

Section XX. Some Cases of Subsidence of Displacement.

In the electromagnetic scheme we have the equations of a dynamical system, involving the potential energy of elastic displacement (or of electric polarisation, if that very vague term be preferred; *any* vector function may be made up of polarised elements, whether it be circuital or polar, so it is as well not to attach too much importance to the idea of polarisation), the kinetic energy (or magnetic energy), probably of a rotational motion, and dissipation of energy by forces analogous to frictions proportional to velocities, when the electric current in a conductor is taken as a generalised velocity. There is nothing peculiarly electrical until we specify the connections of the different magnitudes. It is one out of the infinite number of dynamical systems subject to

$$Q + \dot{U} + \dot{T} = 0,$$

the general equation of activity when energy is neither communicated to the system nor allowed to be withdrawn except through the irreversible frictional forces.

The three qualities to which c, μ, and k refer, relate to the potential energy, the kinetic energy, and the dissipativity. In order to render practically simple the theory of special cases, it is necessary to place restrictions upon their values, restrictions that we may know to be untrue. This is perfectly legitimate, as it is the common-sense procedure in all matters of reasoning to simplify as far as possible. But it becomes necessary to be careful in the interpretation of the extreme results of a limited theory.

Consider, for example, the discharge of a condenser through a wire. The first approximation to its theory is got by ignoring inertia. If q_0 be its initial charge, that left at time t later is $q_0\epsilon^{-t/t_1}$, where t_1 is the time-constant, the product of the capacity of the condenser and the resistance of the wire. An appropriate mechanical illustration is the restoration to equilibrium of a bent spring of negligible mass in a viscous fluid.

But if we push this to extremes, by shortening the discharge-wire indefinitely, this theory says that the discharge will always be of the same character, though finally instantaneous. This is entirely wrong. The influence of inertia may be negligible when the resistance is great, but is not when it is small. We allow for inertia by introducing the inductance of the circuit, bringing in an electromotive force proportional to the rate of decrease of the current. Then we find that when the resistance of the wire is below a certain value the discharge becomes oscillatory. This is quite correct, and the theory as amended is then true within a far wider range than before. But it, again, must not be pushed to extremes. It shows that if the resistance be reduced to nothing, whilst the inductance of the circuit is finite, as it must be, the oscillations continue for ever undiminished in strength, with frequency $(2\pi)^{-1}(sp)^{-\frac{1}{2}}$, if s be the inductance and p the capacity of the condenser. *I.e.*, short-circuiting a condenser would never get rid of its charge, except momentarily, when the energy is all kinetic. Here, of course, the objection is that we cannot indefinitely reduce the resistance in circuit, on account of the resistance of the metallic coatings, previously neglected, when the external resistance was great in comparison. Allowing for that, we still have oscillatory subsidence. But when we consider further that a short-circuited condenser can scarcely be treated as a *linear* circuit, and that we have ignored the dissipation of energy by the oscillatory phenomenon in the magnetic field producing vibratory electric currents in neighbouring conductors, we see that the complete theory of a short-circuited condenser may be only roughly represented by taking into account three constants, the capacity of the condenser, the inductance, and the resistance. What is true enough within certain limits (or uncertain, because no definite line can be drawn between the true and the false) may be wholly untrue beyond them, owing to circumstances of the minutest previous significance becoming then of (relatively) paramount importance.

Retardation in a medium in which $\mu = 0$, $c/k =$ *constant.*

This is a very singular case, and of considerable interest. In general, we have

$$\operatorname{div}\left(4\pi k + c\frac{d}{dt}\right)\mathbf{E} = 0,$$

to express the continuity of the true electric current, $\mathbf{E}$ being the electric force, k the conductivity, and c the specific capacity.

It will be convenient to put $c/4\pi = p$. The new quantity p is the capacity per unit volume considered as a condenser. We may avoid ambiguity by using the word "specific" in connection with c, in the absence of a better nomenclature.

In any normal system $d/dt = n$, a constant, so that the above becomes

$$\text{div}\,(k+pn)\mathbf{E} = 0, \qquad \text{.........................(117)}$$

or, since the displacement is $\mathbf{D} = p\mathbf{E}$,

$$\text{div}\,(k/p+n)\mathbf{D} = 0. \qquad \text{.........................(118)}$$

Now, if there be no inertia, or $\mu = 0$, and no impressed forces, we shall also have

$$\text{curl}\,\mathbf{E} = 0, \qquad \text{or} \qquad \text{curl}\,\mathbf{D}/p = 0. \qquad \text{.................(119)}$$

In (117), $(k+pn)$ is a function of position, k and p being variable from place to place, whilst n is constant. Apply Sir W. Thomson's theorem of determinancy. If $(k+pn)$ be everywhere positive, the only solution of (117), subject to the first of (119), is $\mathbf{E} = 0$. Similarly, if $(k+pn)$ be everywhere negative, the only solution is $\mathbf{E} = 0$. In both cases the point of the demonstration is that $\Sigma\,(k+pn)\mathbf{E}^2$ is necessarily positive if $(k+pn)$ be everywhere positive, and negative if it be everywhere negative. This quantity $\Sigma\,(k+pn)\mathbf{E}^2$ is $Q + \dot{U}$, and is therefore zero. It follows that, in any normal system, $(k+pn)$ must be positive in some parts of space and negative in others (unless it be zero everywhere). Therefore, if k_1/p_1 be the least, and k_2/p_2 the greatest value of k/p, it follows that (k_1/p_1+n) is negative and (k_2/p_2+n) is positive. Hence the values of n for all the normal systems lie between $-k_1/p_1$ and $-k_2/p_2$. Or, their time-constants all lie between the greatest and least values of p/k. If then k/p is the same everywhere, there is only one rate of subsidence for any initial state, given by $(k/p+n) = 0$. (To show that there cannot be imaginary n's, make use of $U_{12} = 0$, $Q_{12} = 0$, applying them to the solutions corresponding to a supposed pair of imaginaries. It follows that the unreal part of the roots is zero, and that the normal functions vanish in the case of equal roots.)

Given that the initial displacement is $\mathbf{D}_0$ in a medium in which p/k is constant, and $\mu = 0$, and that it is left without impressed force, we therefore obtain the subsequent state in the following manner. Let

$$\text{div}\,\mathbf{D}_0 = \rho_0,$$

so that ρ_0 is the initial electrification. Find $\mathbf{D}_1$, such that

$$\text{div}\,\mathbf{D}_1 = \rho_0, \qquad \text{curl}\,\mathbf{D}_1/p = 0.$$

$\mathbf{D}_1$ is uniquely determinable by these conditions. Then $\mathbf{D}_0 = \mathbf{D}_1 + (\mathbf{D}_0 - \mathbf{D}_1)$, where $(\mathbf{D}_0 - \mathbf{D}_1)$ is a system of circuital displacement. It will subside instantaneously, leaving $\mathbf{D}_1$, which will then subside so that the displacement $\mathbf{D}$ at time t later is given by

$$\mathbf{D} = \mathbf{D}_1 \epsilon^{-kt/p}. \qquad \text{.............................(120)}$$

The conduction current is $k\mathbf{D}/p$, and the displacement current the negative of the same, so that the true current is zero. It is not a case of propagation at all, every elementary condenser discharging through its own resistance. It is the instantaneous vanishing of the circuital displacement that is connected with propagation, it being what would happen if $k=0$ with the same distribution of p. First the displacement readjusts itself to make the electric force polar with the same electrification; and then, what is left subsides everywhere at the same rate, according to (120).

Now, any distribution of impressed force sets up a corresponding distribution of circuital conduction current, and, therefore, since k and p are everywhere in the same ratio, of circuital displacement, without electrification. But it is only displacement with electrification that has a finite rate of subsidence. Hence there is no retardation whatever in connection with impressed force. However it vary with the time, the corresponding displacement will vary with it instantaneously. Evidently this is a case in which inertia is not negligible. Maxwell (Vol. I., chap. x.) treats of the case $p=$ constant, $k=$ constant. The extension to $p/k=$ constant allows us to distribute capacity as we please, and so obtain immediately the solutions of various problems connected with shunted condensers.

Now let there be inertia. Although (119) is no longer true, yet (118) is; and, since $(k/p+n)$ is constant, it may be written

$$(k/p+n)\,\mathrm{div}\,\mathbf{D}=0.$$

There is, therefore, no electrification in any normal system, unless $(k/p+n)=0$. It follows that if there be electrification initially, the above process of dividing $\mathbf{D}_0$ into $\mathbf{D}_1$ and $\mathbf{D}_0-\mathbf{D}_1$ is applicable to give us the part of the subsequent state depending on electrification. Thus (120) is true whether there be magnetic induction or not, the left member, however, being not the complete displacement, but only that depending upon the initial electrification. The other part of the initial displacement, $\mathbf{D}_0-\mathbf{D}_1$, will subside, not as before, instantaneously, but according to the nature of the normal distributions other than the $(k/p+n)=0$ solution, depending upon the distribution of k and μ in space, and also upon the initial state of magnetic induction.

Why (120) is true in spite of inertia, is because there is no true current, the force being polar, and therefore no magnetic induction in connection with the electrification solution. As before, no electrification can be produced by any impressed forces, so that the $(k/p+n)=0$ solution may be left out of account. The retardation in connection with the effects of impressed force will depend wholly upon the other n's.

As a simple application of the preceding, let $k=0$ everywhere except in a single wire, forming a closed circuit. It is a perfectly insulated dielectric wire whose conductivity may vary as we please along it, provided its capacity vary in the same ratio. Let now k and p signify the conductance and capacity per unit length of wire, and D the total displacement across its section. Then D/p is the electric force per unit length, and

$$\frac{d}{dx}\left(k+p\frac{d}{dt}\right)\frac{D}{p}=0$$

is the equation of continuity, if x be distance measured along the wire. Or,

$$(k/p+n)\,dD/dx=0,$$

in a normal system. If then D_0 be the initially given displacement, divide it into D_1 and D_0-D_1, such that

$$dD_1/dx=dD_0/dx=\rho_0,$$

the initial electrification of a cross-section, and such that the E.M.F. round the circuit is zero. At time t later, the displacement is

$$D=D_1\epsilon^{-kt/p}, \qquad \text{.........................(121)}$$

the part D_0-D_1, which is circuital, and is the mean initial displacement all round the circuit, instantly vanishing (if there be no inertia). As before, there is no true current during the subsequent subsidence. By the "mean" displacement is meant the quotient of the total initial E.M.F. round the circuit by the total elastance, that is, $(\Sigma\, D/p)\div(\Sigma\, 1/p)$, the summation extending round the circuit.

To corroborate, insert a conductor in the circuit having no capacity. There will now be two normal rates of subsidence, one of which is the previous. If n_1 be the new n, it is given by

$$0=K+K_1+Sn_1,$$

where K and K_1 are the conductances of the old and of the new wire, and S is the total capacity of the old wire, *i.e.*, $S=(\Sigma p^{-1})^{-1}$, the reciprocal of the sum of the elastances round the circuit. The solution will now be

$$D=(D_0-D_1)\epsilon^{n_1 t}+D_1\epsilon^{-kt/p}. \qquad \text{..................(122)}$$

As the auxiliary wire is shortened, n_1 goes out to negative infinity. Then we return to the former instantaneous subsidence of the mean displacement when the whole wire has capacity, (122) becoming (121). When K_1 is finite, there is necessarily electrification somewhere; if not in the old wire itself, then at its ends, where it joins on to the new one. If the circuit be open instead of closed, there will be no instantaneous subsidence in any case; the solution is then (121) with D_0 instead of D_1 on the right side.

Any impressed force in the circuit will only suffer retardation in its effects as regards the n_1 term. It would be very convenient, as well as wonderful, if some ingenious inventor could construct a telegraph cable whose electrostatic capacity should be in the conductor instead of outside it. Having it there, however, a first approximation towards lessening the retardation is to give greater conductivity to the insulating covering. Even a leakage-fault raises the speed of working considerably. Nothing is worse for rapid signalling (when pushed to limiting speeds) than the most perfect insulation. The lower it can be made (natural high conductivity, not due to faults which by getting too bad would stop communication) consistent with getting enough current at the receiving end, the better, and much better, it is for the signalling. Of course there are other considerations, but we must return to the immediate subject.

We can easily obtain the effect of inertia in modifying the solution (121). Let s be the inductance per unit length of wire, constant for purposes of calculation, and really so if the wire be circular. Let also k and p be constant. We have only to examine how the circuital displacement, $(D_0 - D_1)$, subsides, in which alone magnetic induction is concerned. Let Γ be the true current over the cross-section, like D. The electric force to correspond is that of inertia, viz., $-s\dot{\Gamma}$. Hence

$$\Gamma = \left(k + p\frac{d}{dt}\right)(-s\dot{\Gamma}), \qquad \text{or} \qquad \Gamma = (k+pn)(-sn\Gamma),$$

in a normal system; and the determinantal equation is

$$spn^2 + skn + 1 = 0,$$

giving the two n's,

$$n_1 \text{ or } n_2 = -k/2p \pm (k^2/4p^2 - 1/ps)^{\frac{1}{2}}.$$

Besides the circuital displacement, the initial current may be arbitrary. Let it be Γ_0. Then at time t later,

$$\left.\begin{aligned} \Gamma &= \frac{\Gamma_0 + n_1(D_0 - D_1)}{2 + ksn_1}\epsilon^{n_1 t} + \frac{\Gamma_0 + n_2(D_0 - D_1)}{2 + ksn_2}\epsilon^{n_2 t}, \\ D &= -spn_1 \times \text{ditto} \qquad -spn_2 \times \text{ditto} \qquad + D_1\epsilon^{-kt/p}. \end{aligned}\right\}$$

The current is oscillatory if

$$k/p < 2/(ps)^{\frac{1}{2}},$$

and non-oscillatory if it be greater. This differs completely from the condenser and coil theory; for now we get oscillations by reducing the inductance, whereas in the other case, it is by reducing the inductance that we get rid of oscillations.

Although in this solution we take into account the magnetic field, yet we only regard that part of the electric field that is within the conductor, so that the specific capacity c in the wire must be much greater than in the surrounding air to render the latter negligible.

SECTION XXI. A NETWORK OF LINEAR DIELECTRIC CONDUCTORS, OR OF SHUNTED CONDENSERS.

Let any number of points be connected by linear conductors, thus forming a network of any degree of complexity. They will be referred to as Branches. Let each branch consist of any number of conductors in sequence, to be called the Shunts. Let every shunt have its ends joined to the poles of a condenser by wires whose resistances we do not count. This makes the combination complete. We have a linear combination of inductive branches exactly similar to the conductive; they are side by side, as it were, and in connection at certain points. We may regard the conductors as shunts to the condensers, or the other way, as we please, but the former plan is perhaps the best.

Instead of thus shunting the condensers by external conductors, we may do away with the shunts, giving instead equal conductances to the condensers themselves, thus making a combination of unshunted leaky condensers.

Or, we may abolish the external condensers, and give equal dielectric capacity (uniformly distributed *in* the wires) to the former shunts themselves, thus making a network of dielectric conductors. The theory is in all cases the same, with certain exceptions as regards the effects of impressed force and of magnetic induction. In many respects the theory is most simply expressed by having the two qualities, conductivity and dielectric capacity, coincident, as in the dielectric conductors, instead of side by side, as in the case of the shunted condensers.

If the shunted condensers were all disconnected from one another so as to form independent circuits, partly conductive and partly dielectric, any charges they might have would discharge through their shunts, each charge at its proper rate depending upon the time-constant of the condenser concerned, which is rp, if r be the resistance of the shunt and p the capacity of the condenser. (Inertia is ignored.) When they are in connection, as above described, the time-constants of the normal systems will all lie between the greatest and the least of the time-constants of the separate shunted condensers. If these be all equal, there is but one time-constant, viz., the common value of rp. In this case, if the condensers be charged in any manner and then be left in connection without impressed force anywhere in the system, the charges will at once readjust themselves to a new distribution, to be found by the two considerations that the E.M.F. in any circuit in the new state is zero, and that the charges that disappear form a system of circuital displacement in the combination. This new state will then subside uniformly everywhere, each condenser discharging through its own shunt. If an impressed force be introduced at the junction of two shunted condensers, say in an infinitely short wire joining one shunt to the next, it sets up the appropriate state of conduction current in the branches and of charge in the condensers instantly; these charges are equal in all the condensers in one branch, and in different branches are simply proportional to the currents in the branches. The same will be true if the impressed force be in a shunt, if there be an equal and similarly directed impressed force in the corresponding condenser. (In the case of the dielectric wires there is no need for this reservation.) But if the impressed force be in a shunt only, the charge of its condenser will be opposite to that of the others in the same branch, and there will be retardation according to the common time-constant. Thus, if the condensers be charged in any manner which could be produced by impressed forces in any of or all the branches (equally in shunts and condensers), and be left to themselves, the subsidence is instantaneous. There is only retardation in connection with those parts of the system of charges which could not be produced in the described manner. And, considering electromagnetic induction, if it operate equally on conductor and dielectric, as in the case of the dielectric wires, it will only affect the discharge of the parts that before subsided instantly, the subsidence being no longer immediate; whilst the other parts will subside just as before, independently of inertia;

for as the conductive and dielectric currents are equal and opposite, there is no true current and no magnetic induction in connection therewith.

The above are conclusions from the general theory in the last section. As regards the proof of the limits between which the time-constants must lie, let E be the E.M.F. in any condenser, p its capacity, and k the conductance of its shunt. Then, in any normal system, if Q be the dissipativity and U the potential energy,

$$\Sigma\,(k+pn)E^2 = Q + 2nU = 0,$$

the Σ to include all the shunted condensers. If n do not lie between the greatest and least values of $-k/p$, the summation cannot vanish, as it must; therefore every n does lie between these limits.

The differential equation of the combination, and the determinantal equation of the rates of subsidence, are most directly found by the method used in the paper on Induction in Cores, when treating of combinations of coils [p. 415]. The sum of the steps of potential in any circuit must be zero; get, then, the expression for the step of potential between any two points in terms of the currents, and we have one equation for every circuit. Eliminate the currents by their conditions of continuity, and the result is the differential equation, or the determinantal equation, according as we treat d/dt as the differentiating operator or as a constant.

In the present case, if Γ be the sum of the currents in a shunt and in its condenser, reckoned the same way in both (or the true current in the dielectric conductor), we have

$$\Gamma = (k+pn)E,$$

if n stand for d/dt. Every condenser has an equation of this form. Here E is the fall of potential through the shunt and through the condenser. Since Γ is the same along the whole of any one branch, the fall of potential between its ends is

$$\Gamma\Sigma\,(k+pn)^{-1},$$

the Σ to include all the condensers in the branch. Hence, if the combination consist of only one closed circuit,

$$\Sigma\,(k+pn)^{-1} = 0,$$

when cleared of fractions, is the differential, or the determinantal equation, according as n is d/dt or algebraical. That is, we equate the (generalised) resistance of the circuit to zero. Thus, if there be three condensers, and y stand for $k+pn$, the equation is

$$(y_1)^{-1} + (y_2)^{-1} + (y_3)^{-1} = 0; \qquad \text{or,} \qquad y_1y_2 + y_2y_3 + y_3y_1 = 0.$$

The determinantal equation of m condensers in one circuit is of the $(m-1)^{\text{th}}$ degree; one freedom is lost. The missing root is the negative infinity root of the instantaneous subsidence. To bring it to finiteness, put in the circuit a conductor without condenser. Then *its* y is its real conductance, say k_0, and the equation is

$$k_0{}^{-1} + \Sigma\, y^{-1} = 0.$$

This has the full number m of roots.

Let q_1, q_2, ..., be the charges, and E_1, E_2, ..., be the E.M.F.'s of the condensers (of capacity p_1, p_2, ..., whose shunts have conductances k_1, k_2, ...,) going round the circuit in the + direction of Γ. A charge is + when the displacement is in the + direction. Then we have

$$q_1 = p_1 E_1, \qquad q_2 = p_2 E_2, \quad \text{etc.,}$$

or,

$$q_1 = p_1 \Gamma/(k_1 + p_1 n), \qquad q_2 = p_2 \Gamma/(k_2 + p_2 n), \quad \text{etc.}$$

Therefore, in a normal system, the ratios of the charges are

$$q_1 : q_2 : \ldots = (k_1/p_1 + n)^{-1} : (k_2/p_2 + n)^{-1} : \ldots = p_1/y_1 : p_2/y_2 : \ldots,$$

and thus we have m sets of ratios, by giving to n its m values in succession. To determine the absolute size of a particular normal system, use $U_{12} = 0$, taking p_1/y_1, ... as the normal functions. If A be their common multiplier, we get

$$A = (\Sigma D y^{-1}) \div (\Sigma p y^{-2}),$$

if D_1, D_2, etc., be the initial charges.

When, by shortening the wire k_0, we send the root depending upon its presence to $-\infty$, the above ratios become 1 : 1 : 1 : ... in the normal system of this root. Taking, then, 1, 1, 1, etc., as the normal functions,

$$A = (\Sigma D p^{-1}) \div \Sigma p^{-1}$$

gives the common charge of all the condensers that instantly disappears. It is the charge due to the initial E.M.F. in the circuit. Its disappearance makes the electric force polar.

If all the time-constants p/k of the separate condensers are equal we have $y = 0$ repeated $(m-1)$ times. The charges at time t are therefore

$$q_1 = (D_1 - A)\epsilon_0, \qquad q_2 = (D_2 - A)\epsilon_0, \quad \text{etc.,}$$

where $\epsilon_0 = \epsilon^{-kt/p}$, and A is given by the previous equation.

So far relating to a single closed circuit, the next simplest case is that of any number of branches uniting two points. Here the sum of the currents leaving either point is zero. If Γ_1, Γ_2, ..., be the currents in the branches, all reckoned parallel, p_{11}, p_{12}, ..., the capacities in the first branch, p_{21}, p_{22}, ..., in the second, with a similar notation for the other quantities, we have

$$\Gamma_1 = (k_{11}/p_{11} + n)q_{11} = (k_{12}/p_{12} + n)q_{12} = \ldots,$$

$$\Gamma_2 = (k_{21}/p_{21} + n)q_{21} = (k_{22}/p_{22} + n)q_{22} = \ldots,$$

etc., and the sum of the Γ's is zero. These give the determinantal equation

$$(\Sigma y_1^{-1})^{-1} + (\Sigma y_2^{-1})^{-1} + (\Sigma y_3^{-1})^{-1} + \ldots = 0,$$

each summation to include all the y's in one branch only. That is, the sum of the generalised conductances of the branches in parallel is zero.

The number of missing roots is one less than the number of branches. The full number, equal to the number of condensers, may be got by inserting condenserless conductors in all the branches except one; if in that one also, it makes no difference in their number, though altering their magnitude.

If there be no inserted condenserless conductors, it will be necessary to determine what part of each charge instantly disappears. We have to make the electric force polar, and therefore equalise the E.M.F.'s in the different branches reckoned the same way between the two points, and do it by making equal changes in the initial charges in any one branch. The charges q at time t, after they were given D, are, if k/p is the same for every condenser, given by

$$q_{11} = (D_{11} - A_1)\epsilon_0, \qquad q_{21} = (D_{21} - A_2)\epsilon_0, \quad \text{etc.},$$
$$q_{12} = (D_{12} - A_1)\epsilon_0, \qquad q_{22} = (D_{22} - A_2)\epsilon_0, \quad \text{etc.},$$
$$\text{etc.} \qquad\qquad \text{etc.}$$

where ϵ_0 is, as before, the time-function of the repeated root, and there is one A to be found in each branch. It will not be necessary to take up space by describing how they are got in this special case, or in writing them out, as the following method, applicable to any combination, will apply.

In any network of linear conductors there is a certain number of degrees of freedom, *i.e.*, the number of branches in which the currents must be given in order that they may be known in all the rest. Thus, in the common "Bridge," the currents in three branches being given, those in the rest follow.

(If m points be joined by $\frac{1}{2}m(m-1)$ conductive branches, the number of current-freedoms is $\frac{1}{2}(m-1)(m-2)$. This is $(m-1)$ less than the number of branches.)

This number of current-freedoms is just the number of the missing $(-\infty)$ roots in the determinantal equation when the branches have condensers connected along them as described at the beginning. As for the equation itself, if the characteristic function of the conductive combination be known, it may be got by turning every k into $k+pn$ in it, and equating the result to zero.

(The characteristic function is of the degree $(m-1)$ in terms of the conductances (one less than the number of points); hence, when for k we put $(k+pn)$, the determinantal equation is of the degree $(m-1)$ in n, so that the roots are fewer in number than the branches by the number of current-freedoms in the conductive network, if there be but one condenser in each branch, and fewer in number than the condensers by the number of current-freedoms in the conductive network if there be many condensers in each branch. If, on the other hand, there are no condensers, but we take account of the self-induction of every branch, we get the determinantal equation by turning k into $(k^{-1}+sn)^{-1}$, if s be the inductance of a branch. There will now be $(m-1)$ fewer roots than the number of branches.)

Now, suppose k/p is the same for every condenser, and we want to know how the initial charges subside. Let us number the branches 1, 2, 3, etc., and choose (arbitrarily) a certain direction in each for the positive direction in which the current, E.M.F., and charges (displacements) are reckoned. Let every capacity have two suffixes, the first to denote which branch is referred to, the second to show its position in the branch; and do the same with the charges and the conductances. The currents Γ only want one suffix, to show which branch is referred to. We have, then, in the case of the infinity roots, if A_{11}, etc., are the charges that disappear,

$$A_{11}=A_{12}=A_{13}=\ldots=A_1, \quad \text{say,}$$
$$A_{21}=A_{22}=A_{23}=\ldots=A_2, \quad \text{say,}$$

etc.; that is, the same portion of the charge of every condenser in one branch disappears instantly. Besides that, A_1, A_2, ..., in the different branches, are connected together by the same conditions of continuity as the currents in the different branches. That is, A_1, A_2, ..., form a system of circuital displacement.

The solution is therefore of the same form as in the previous equations, and we have only to find one A for each branch to complete the solution. Initially, we have

$$D_{11}=A_1+(D_{11}-A_1), \qquad D_{21}=A_2+(D_{21}-A_2), \quad \text{etc.,}$$
$$D_{12}=A_1+(D_{12}-A_1), \qquad D_{22}=A_2+(D_{22}-A_2), \quad \text{etc.,}$$
$$D_{13}=A_1+(D_{13}-A_1), \qquad D_{23}=A_2+(D_{23}-A_2), \quad \text{etc.,}$$
$$\text{etc.} \qquad\qquad \text{etc.}$$

The system of the A's is circuital. That of the $(D-A)$'s is such that its electric force is polar. The mutual energy of the latter and any circuital displacement is therefore zero. The mutual energy of the D's and any circuital displacement is therefore equal to that of the A's and the same. Let this "any circuital displacement" be the charges set up in the system by a unit impressed force in any branch (equally in shunt and condenser). For instance, let d_{11}, d_{12}, d_{13}, etc., be the charges of the condensers in branches 1, 2, 3, etc., due to unit impressed force in branch 1. Then the mutual energy of the D's and d's equal that of the A's and d's. But the latter equals twice the product of the impressed force of the d's into the displacement of the A's; or, since the impressed force is unity and is in the branch 1 only, it equals $2A_1$ itself. Hence A_1 is one half the mutual energy of the D's and d's. Or,

$$A_1=d_{11}\Sigma D_1/p_1+d_{12}\Sigma D_2/p_2+d_{13}\Sigma D_3/p_3+\ldots,$$

where the first Σ relates to branch 1, the second to branch 2, and so on.

Similarly, if d_{21}, d_{22}, d_{23}, ..., are the charges in 1, 2, 3, ..., due to unit impressed force in 2, we have

$$A_2=d_{21}\Sigma D_1/p_1+d_{22}\Sigma D_2/p_2+d_{23}\Sigma D_{23}/p_3+\ldots.$$

Thus the A's are known in terms of initial charges and of the d's. The latter may be found in precisely the same manner as the current in the branches due to the unit impressed forces. In fact, instead of mutual energy, we may employ the idea of mutual dissipativity and activity. Let γ_{11}, γ_{12}, γ_{13}, etc., be the currents in 1, 2, 3, ... due to unit impressed force in 1. Then

$$A_1=\gamma_{11}\Sigma D_1/k_1+\gamma_{12}\Sigma D_2/k_2+\gamma_{13}\Sigma D_3/k_3+\ldots$$

is an alternative form of A_1. In the reasoning we should now imagine the *D*'s to be currents (not closed), and the *A*'s closed currents, and speak of mutual dissipativity or of activity. The matter is therefore reduced to the problem of finding how current due to impressed force in any branch divides through the conductive system.

(When the time-constants of the condensers (shunted) are not equal, the charges that are left after the first readjustment require to be decomposed into their proper normal systems, to be done by the $U_{12}=0$ property. This does not present anything unusual.)

The following is the tridimensional representative of the above method of finding the *A*'s. Referring now everything to the unit volume, let $\mathbf{D}_0$ be the initial displacement, and p the capacity. Divide $\mathbf{D}_0$ into two parts, of which one is circuital, whilst the electric force of the other is polar. That is, let $\mathbf{A}$ be the circuital displacement, so that

$$\operatorname{div}\mathbf{A}=0, \qquad \operatorname{curl}(\mathbf{D}_0-\mathbf{A})/p=0;$$

find $\mathbf{A}$. The mutual energy of $\mathbf{D}_0$ and any circuital displacement equals that of $\mathbf{A}$ and the same, because the force $(\mathbf{D}_0-\mathbf{A})/p$ is polar. Let the any circuital displacement be that due to unit $\mathbf{e}$ at any point, and call it $\mathbf{d}$. Then

$$\Sigma\,\mathbf{d}\mathbf{D}_0/p=\Sigma\,\mathbf{d}\mathbf{A}/p=\Sigma\,(\mathbf{e}+\mathbf{f})\mathbf{A},$$

if $\mathbf{f}$ is the polar force of $\mathbf{e}$,

$$=\Sigma\,\mathbf{e}\mathbf{A}=\text{tensor of }\mathbf{A}\text{ at the place of }\mathbf{e},$$

if $\mathbf{e}$ be parallel to $\mathbf{A}$.

Thus we know the distribution of $\mathbf{A}$ as soon as we know the displacement due to impressed force.

Section XXII. The Mechanical Forces and Stresses. Preliminary. The Simple Maxwellian Stress.

As this is not a treatise upon the theory of Elasticity, it will be only necessary to say so much on the subject of stresses in general as will serve to introduce us to the principal formulæ connecting stresses with the corresponding mechanical forces, which we may find useful hereafter. This can be done very briefly.

A simple stress is either a tension or a pressure acting in a certain line. It implies the existence of mutual force between contiguous parts of the substance in which it resides, and of a corresponding state of strain, with storage of energy in the potential form, *i.e.*, depending upon configuration, though perhaps ultimately resolvable into kinetic energy. Thus if we fasten a cord to a beam, and hang a weight to its free end, the cord is slightly stretched, the work done by gravity during the stretching is somehow stored in the altered configuration, and the cord is put into a state of tension. At its lower end the tension in the cord is equal to the weight attached, at its upper end to the same plus the weight of the cord. The state of strain of course extends to the beam, and to the beam's attachments, and so round to the earth, to which we ascribe the gravitational force, which is somehow stressed across the air to the weight, and from the weight to the earth.

If the stretched cord be in motion in its own line, as when a horse tugs a barge along a canal, there is, besides the transfer of energy through space by the onward motion of the horse, rope, barge, and dragged water, carrying their kinetic energy with them, a transfer of energy through the rope from the horse to the barge, and through the strained barge to the water, where it is wasted in friction. The rate of transfer per second equals the product of the tension of the rope into its speed, and the direction of transfer is against the direction of motion.

If motion be transmitted from one machine to another by means of a horizontal endless band, the transfer of energy is through the stretched half of the band, and is again proportional to its speed (and against its motion), and to the difference of tensions of the two sections; a uniform tension meaning continuously stored potential energy.

A pressure is a negative tension. If the tension or pressure in a cord or rod be not uniform in amount across every section, we see at once that any small piece of the cord is pulled in opposite directions by forces of different amounts. Their difference is the mechanical force on the small piece considered, and measures its rate of acceleration of momentum. Thus, if P be the tension across a section at distance x from one end, dP/dx is the mechanical force at that place, per unit length of cord, acting in the direction of $\mathbf{x}$ positive. Unless otherwise balanced, it increases the momentum of the cord. Thus $-dP/dx$ is the force that must be applied to keep the stress-difference from working. Examples :—(1). A vertically hanging cord; unequal tension; applied force gravity. (2). In the endless horizontal band moving with uniform speed, there is no force resulting from its tension except where it changes in intensity, for example where it passes over pulleys. At a pulley where the band gives out energy, the force and velocity product is positive, and where it receives energy, negative. These forces of the stress-variation are the negatives of the forces the pulleys exert on the band.

The most general stress considered in the common theory of elasticity consists of three simple stresses (pressures or tensions) acting in three lines at right angles to one another in a substance.

When, as is necessary in general, the axes of reference are not the lines of action of the mutually perpendicular simple stresses, the following notation is the most convenient. Although a tension or a pressure is not a vector in the usual sense, since it, although acting in a certain line, acts both ways, yet we may consider only one side of a stress at a time, and so represent the stress on any plane by a vector. On this understanding, let $\mathbf{P}_1$, $\mathbf{P}_2$, $\mathbf{P}_3$, be the vector stresses per unit area on planes whose normals are $\mathbf{x}$, $\mathbf{y}$, $\mathbf{z}$ respectively. These are the forces exerted by the matter on the positive side on that on the negative side of the three planes, and, being forces, are vectors. Let the scalar components of $\mathbf{P}_1$ be P_{11}, P_{12}, P_{13}, etc.; and $\mathbf{i}$, $\mathbf{j}$, $\mathbf{k}$ be unit vectors parallel to $\mathbf{x}$, $\mathbf{y}$, $\mathbf{z}$. Then

$$\left.\begin{aligned}\mathbf{P}_1&=\mathbf{i}P_{11}+\mathbf{j}P_{12}+\mathbf{k}P_{13},\\ \mathbf{P}_2&=\mathbf{i}P_{21}+\mathbf{j}P_{22}+\mathbf{k}P_{23},\\ \mathbf{P}_3&=\mathbf{i}P_{31}+\mathbf{j}P_{32}+\mathbf{k}P_{33}.\end{aligned}\right\}\qquad(1a)$$

(The first of a double suffix fixes the plane, and the second the direction of the force.) Here there are nine components in a general stress. But examination of the force on a unit cube arising from this stress shows at once that the transverse stresses must be equal in pairs, $P_{12} = P_{21}$, etc., if the force is to be purely translational, thus reducing the number to six. Then, the translational force due to the stress is

$$\mathbf{i}\,\mathrm{div}\,\mathbf{P}_1 + \mathbf{j}\,\mathrm{div}\,\mathbf{P}_2 + \mathbf{k}\,\mathrm{div}\,\mathbf{P}_3\,; \quad \text{..................(2a)}$$

i.e., the x-component is div $\mathbf{P}_1$, etc.

Should, however, we admit the possibility of nine coefficients (as we may do, at least on paper, in some kinds of magnetic and electric stresses), the x-component of the translational force is not the divergence of $\mathbf{P}_1$, but of its conjugate; thus

$$x\text{-component is not} \quad = dP_{11}/dx + dP_{12}/dy + dP_{13}/dz,$$

$$\text{but is} \quad = dP_{11}/dx + dP_{21}/dy + dP_{31}/dz\,; \quad \text{........(3a)}$$

a distinction which disappears when $P_{12} = P_{21}$, etc. Besides this, there is rotational force arising from the stress, whose vector moment per unit volume is [the torque per unit volume]

$$\mathbf{i}(P_{23} - P_{32}) + \mathbf{j}(P_{31} - P_{13}) + \mathbf{k}(P_{12} - P_{21}), \quad \text{.............(4a)}$$

which also vanishes when $P_{21} = P_{12}$, etc. Should this be the case, the negative of (2*a*) is the applied force required for equilibrium. If not, then the negative of (3*a*) is the x-component of the applied force required to balance the translational force, and the negative of (4*a*) is required to balance the torque.

There are three applications of this theory of stress. The first is in the dynamical theory of elastic bodies; the second is, after Faraday and Maxwell, in the explanation of forces of unknown origin by means of stress in a medium; and the third application consists in the use of the stresses, not for explanation, but for purposes of investigation.

Thus, as from a given state of stress we derive the corresponding mechanical forces by differentiations, so we may obtain a state of stress that will produce a given distribution of force of any origin by integrations. The former is an exact process; the latter is to a certain extent indefinite; for we may clearly add to the state of stress that gives rise to certain forces any state of stress that gives rise to no forces. We should naturally choose the simplest forms that present themselves, unless there should be reasons against this. We have a choice of formulæ for yet another reason, viz., when it is not the exact distribution of force that is known, but only its resultant effect on a solid body, of which examples will occur later. We need not bind ourselves to the hypothesis that a certain state of stress really exists in a certain case, but merely use the stress-vectors as auxiliary functions to assist the reasoning, if the investigations should be assisted thereby.

The gravitational application made by Maxwell requires a pressure along a line of gravitational force combined with an equal tension in all directions perpendicular to it. But the intensity of the stress is something stupendous, being at the earth's surface

$$=\frac{1}{8\pi}\left(\frac{614\times 10^{25}}{3928.(637\times 10^{6})^{2}}\right)^{2}=59\times 10^{10} \text{ dynes per sq. cm.,}$$

or 3770 tons per square inch. (Maxwell made it ten times as much, so I give the above figures, in which I do not see any error; the unit mass is that of 3928 grams, 614×10^{25} that of the earth, and 637×10^{6} cm. its radius.) In the earth, on the supposition of uniform density, it would be proportional to the square of the distance from the centre. But a very severe mental tension is caused by an endeavour to imagine this stress to really exist. Yet the action of gravitation must be transmitted somehow.

First Electromagnetic Application.

No objection on the score of enormously great stresses being required applies when the electric and magnetic mechanical forces are in question. In fact the method seems peculiarly fitted for their explanation. The cause of this would seem to be twofold. Gravitational matter is all attractive, and is collected in great lumps. The electric and magnetic "matters," on the other hand, are comparatively superficial affairs, and are always in equal amounts of opposite kinds. If, as some suppose, the earth is full of electricity, it might as well not be there, for all the good it does.

As a first simple application, let us confine ourselves to a portion of space in which there are no impressed electric or magnetic forces, and the dielectric capacity, the conductivity, and the magnetic permeability are all constants, *i.e.*, a homogeneous isotropic medium. There are three mechanical forces to be accounted for by a state of stress.

(1). The mechanical force on electrification. This is, per unit volume,

$$\mathbf{E}\rho=\mathbf{E}\operatorname{div}\mathbf{D}=p\mathbf{E}\operatorname{div}\mathbf{E}, \quad \text{......................(5a)}$$

if $\mathbf{E}$ and $\mathbf{D}$ are the electric force and displacement, and $p=c/4\pi$ the condenser capacity per unit volume. It acts parallel to $\mathbf{E}$, which is the force per unit density.

(2). The mechanical force called by Maxwell the Electromagnetic force. This is,

$$\mathrm{V}\Gamma\mathbf{B}=\mathrm{V}(k\mathbf{E}+p\dot{\mathbf{E}})\mu\mathbf{H}, \quad \text{........................(6a)}$$

k being the conductivity, μ the permeability, $\mathbf{H}$ and $\mathbf{B}$ the magnetic force and induction, and Γ the true current, the sum of the conduction current and that of elastic displacement. It is perpendicular to both the current and the induction, and is in strength equal to the product of their tensors into the sine of the angle between their directions. Its existence in a dielectric is speculative, but it is difficult to do without it.

(3). A mechanical force that we may call the Magnetoelectric Force. It is

$$4\pi\mathrm{V}\mathbf{D}\mathbf{G}=\mathrm{V}\mathbf{D}\dot{\mathbf{B}}=p\mu\mathrm{V}\mathbf{E}\dot{\mathbf{H}}, \quad \text{......................(7a)}$$

where **G** is the magnetic current, or the time-variation of the magnetic induction $\div 4\pi$. Its existence anywhere is speculative, but it is absolutely needed as a companion to the last. It is perpendicular to the electric displacement and to the magnetic current. If **v** be the velocity, the activity of this force is

$$4\pi\mathbf{v}\mathrm{V}\mathbf{DG} = 4\pi\mathbf{G}\mathrm{V}\mathbf{vD}\,; \quad \ldots\ldots\ldots\ldots\ldots\ldots\ldots\ldots(8a)$$

hence $4\pi\mathrm{V}\mathbf{vD}$ is the magnetic force "of induction," due to the motion, the second form of (8*a*) expressing *its* activity. The existence of this magnetic force due to motion in an electric field was concluded before by general reasoning. [*See* p. 446. Notice that the impressed force required to balance the magnetoelectric force is the negative of (7*a*); so that the motional magnetic force, regarded as impressed, is the negative of (8*a*). Similarly as regards the electromagnetic force and the motional electric force, p. 448. Some worked out examples will follow.]

The magnetoelectric force can only exist in transient states. The electromagnetic force exists in steady states as well, but then there must be dissipation of energy going on. The force on electrification is independent of whether the state is steady or transient.

The forces (1) and (3) are explained by a simple Maxwellian stress, electric; whilst (2) is explained by a similar magnetic stress. A simple Maxwellian stress consists of a tension along a certain line combined with an equal lateral pressure. Let U_1 and T_1 be the electric and magnetic energies per unit volume, or $\frac{1}{2}p\mathbf{E}^2$ and $\frac{1}{2}\mu\mathbf{H}^2/4\pi$. Then U_1 is the intensity of the electric stress, and T_1 that of the magnetic stress. The tension is parallel to the electric force in the one case, and to the magnetic force in the other, the pressures being perpendicular to their directions.

The electric stress on any plane, defined by its unit vector normal **N**, is

$$(\mathbf{EN})\mathbf{D} - U_1\mathbf{N}, \quad \ldots\ldots\ldots\ldots\ldots\ldots\ldots\ldots(9a)$$

that is, a force parallel to **D** of intensity $\mathbf{EN} \times$ tensor of **D**, combined with a normal pressure of intensity U_1. Similarly the magnetic stress on the plane is

$$(\mathbf{HN})\mathbf{B}/4\pi - T_1\mathbf{N}, \quad \ldots\ldots\ldots\ldots\ldots\ldots\ldots(10a)$$

i.e., a force parallel to **B** of intensity $\mathbf{HN} \times$ tensor of $\mathbf{B}/4\pi$, combined with a normal pressure of intensity T_1. By taking $\mathbf{N} = \mathbf{i}, \mathbf{j}, \mathbf{k}$, in succession, we may obtain the corresponding three stress vectors on their planes. But the simple Maxwellian stress is fully defined by the single expression (9*a*) or (10*a*), according as it is electric or magnetic, **N** being in any direction we please.

To prove that these stresses give the required forces, it is sufficient to differentiate them. The divergence of the **N**-plane stress-vector is the **N**-component of the mechanical force due to the stress. Thus the divergence of (9*a*) gives the **N**-component of the forces (1) and (3), whilst that of (10*a*) gives the force (2). As the transformations will occur later in a more general manner, space will not be occupied by them here. Whether these stresses be realities or not (physically), there can be no doubt as to their appropriateness.

The medium is in equilibrium in all places where there is no electrification, or electric or magnetic current. Thus, in the region outside a wire supporting a steady conduction current, the stress-vectors have no divergence, and there is no mechanical force arising from the stresses. On the plane containing both the electric and magnetic forces, the stress is a normal pressure, of intensity $U_1 + T_1$, acting in the line of transfer of energy. If, further, the electric and magnetic forces are perpendicular, as when the circuit lies in one plane, we have a tension $U_1 - T_1$ in the line of the electric, and a tension $T_1 - U_1$ in the line of the magnetic force. Lastly, if also U_1 and T_1 are equal, we have left only the previously mentioned simple pressure.

Generally, let the normal **N** to any plane make an angle θ with **E** and ϕ with **H**; then the force on the **N**-plane is compounded of a normal tension

$$U_1 \cos 2\theta + T_1 \cos 2\phi,$$

and two tangential forces

$$U_1 \sin 2\theta, \qquad \text{and} \qquad T_1 \sin 2\phi,$$

the first being in the plane of **E** and **N**, the second in that of **H** and **N**.

There is another important case (not a steady state) in which the stress reduces to a pressure in one line, viz., in the propagation of a plane wave through a homogenous isotropic nonconducting medium. Let **z** be measured in the direction of propagation, **x** and **y** at right angles to **z**; then if **E** is parallel to **x**, **H** is parallel to **y**. If we look along **z** in the + direction, and **E** be + upward, **H** will be + to the right. They keep time together in all their variations of intensity at any place, and are of such relative magnitude that U_1 and T_1 are equal. Thus,

$$\mathbf{E} = \mu v \mathrm{V}\mathbf{HN}, \qquad \mathbf{H} = cv\mathrm{V}\mathbf{NE}, \qquad v^2 = (\mu c)^{-1}, \quad \text{.........}(11a)$$

if v be the speed of the wave and **N** a unit vector parallel to **z**. Or if E_0 and H_0 are the tensors (magnitudes, apart from direction) of **E** and **H**,

$$E_0 = \mu v H_0, \qquad H_0 = cvE_0.$$

Here, **E** and **H** being perpendicular and such that $U_1 = T_1$, the stress is a simple pressure $P = 2U_1 = 2T_1$ in the line of **z**. The only mechanical force arising therefrom is one parallel to **z**, due to the variation of P along **z**. This force is the sum of the electromagnetic and magneto-electric forces, which are equal, and parallel to **z**, each represented by $-\frac{1}{2}dP/dz$, per unit volume.

Since the medium is not in equilibrium under the stress P, there is translatory motion in the line of **z**. This requires the medium to be compressible. Thus a wave of compression travels with the electromagnetic wave. The compression is, however, only an effect of, not the electromagnetic disturbance itself. Thus, in the case of a simple harmonic wave, there is a translatory to-and-fro motion of the parts of the medium in the line of propagation, accompanying the wave; having double its frequency, as there are two maxima of pressure in a wave-length. Assuming a light-ray to be an electromagnetic wave of this kind, and taking the amplitude of **H** to be ·02 c.g.s. in strong sunlight, requiring the amplitude of **E** to be 6×10^8, or 6 volts per cm., with an electric current-density of 240 c.g.s., the maximum translational force, $-dP/dz$ parallel to **z**, is about 5 dynes per cubic cm. It is here supposed that the disturbance is simple-harmonic, and that $v\lambda/\pi = 4 \times 10^5$, if λ is the wave-length. The translational momentum parallel to **z** is, in general, $2U_1/v +$ a constant independent of the time.

This motion of the medium parallel to **z**, not to be confounded with the internal motions of the disturbance, must react upon the electromagnetic wave. For, if $\mathbf{v}_1$ be the **z**-velocity (vector), the electric force induced by the motion is $\mu \mathrm{V}\mathbf{H}\mathbf{v}_1$, and the magnetic force induced by the motion is $c\mathrm{V}\mathbf{v}_1\mathbf{E}$; to a first approximation **E** and **H** are altered to these extents. If we compare these expressions with (11*a*) above, we see that $v\mathbf{N} = \mathbf{v}_1$ makes the electric and magnetic forces induced by the motion equal to the original electric and magnetic forces. This result of a uniform speed of motion of the medium in the direction of propagation does not, however, mean more than the expression of the fact that if we travel with a wave, and at the same speed, the wave will appear stationary.

The size of $\mathbf{v}_1$ depends upon the density of the medium, varying inversely with it. But $\mathbf{v}_1$ is not likely to be anything but a very minute fraction of the velocity of propagation, and therefore negligible, unless we artificially increase the electromagnetic or magnetoelectric forces by passing a ray of light through a strong magnetic or electric field. Noticing that H = ·02 is quite small, we can greatly multiply the electromagnetic force by sending a ray across the lines of force of a strong magnetic field, whilst keeping its direction the same (along the ray). If, on the other hand, we send a ray parallel to the lines of force of the field, there is transverse electromagnetic force, and transverse motion produced, far exceeding the original in amount. Under such circumstances we might expect the effect of $\mathbf{v}_1$ to be not negligible. These remarks, it will be noticed, rest upon the existence of the supposed stress.

SECTION XXIII. THE MECHANICAL ACTION BETWEEN TWO REGIONS.

Summary of some results of Vector Analysis.

In order to keep the present section within limits, it will be desirable to first give a short summary of certain general relations, discussed at length in previous articles.

Let there be a distribution throughout space of a vector **H**, to be mentally realised by drawing lines following its direction, packing them closely where H is strong, loosely where it is weak. **H** may be the intensity of electric or magnetic force, the mechanical force on the unit of the corresponding matter. The field of **H** is decomposable into two fields of very different natures, say, **H** = **F** + **K**, such that **F** is a polar force, its line-integral round any circuit being zero, and **K** has no divergence, or is circuital. From this property it follows that $\Sigma\,\mathbf{FK}$ through all space is zero, making

$$\Sigma\,\mathbf{H}^2 = \Sigma\,\mathbf{F}^2 + \Sigma\,\mathbf{K}^2.$$

The divergence of $\mathbf{H}$ is that of $\mathbf{F}$ only; the curl of $\mathbf{H}$ is that of $\mathbf{K}$ only. Let

$$\text{div}\,\mathbf{H} = 4\pi\rho, \qquad \text{curl}\,\mathbf{H} = 4\pi\Gamma; \quad \text{..............(12}a\text{)}$$

then ρ is the density of the "matter" of $\mathbf{F}$, and Γ is the density of the "current" of $\mathbf{K}$. (Say magnetic matter and electric current, or electric matter and negative magnetic current; but, so far as the present section is concerned, the matter and current are simply defined by (12a). Let P and $\mathbf{A}$ be the potentials at any point, Q, of the matter and the current, according to

$$P = \Sigma\,\rho/r, \qquad \mathbf{A} = \Sigma\,\Gamma/r, \quad \text{....................(13}a\text{)}$$

r being the distance from ρ or Γ to the point Q where P or $\mathbf{A}$ is reckoned. Then

$$\mathbf{F} = -\nabla P, \qquad \mathbf{K} = \text{curl}\,\mathbf{A}, \quad \text{...................(14}a\text{)}$$

show the derivation of $\mathbf{F}$ and $\mathbf{K}$ from the corresponding potentials. Also

$$\mathbf{F} = -\Sigma\,\rho\mathbf{f}, \qquad \mathbf{K} = -\Sigma\,\mathrm{V}\Gamma\mathbf{f}, \quad \text{................(15}a\text{)}$$

if $\mathbf{f}$ be the vector force at the place of ρ or Γ, due to unit matter at Q, where $\mathbf{F}$ or $\mathbf{K}$ is reckoned. If we call the quantity $\Sigma\,\mathbf{H}^2/8\pi$ the energy of $\mathbf{H}$, the total energies of $\mathbf{F}$ and $\mathbf{K}$ are

$$\Sigma\,\tfrac{1}{2}\mathbf{F}^2/4\pi = \Sigma\,\tfrac{1}{2}P\rho, \qquad \Sigma\,\tfrac{1}{2}\mathbf{K}^2/4\pi = \Sigma\,\tfrac{1}{2}\mathbf{A}\Gamma. \quad \text{.........(16}a\text{)}$$

The mechanical force per unit volume is

$$\rho\mathbf{H} + \mathrm{V}\Gamma\mathbf{H}. \quad \text{............................(17}a\text{)}$$

Limitation to a bounded region.

The above referring to all space, in order to apply the results to a bounded region we must suppose $\mathbf{H} = 0$ outside it, for our temporary purpose, but without altering $\mathbf{H}$ within the region. This makes ρ and Γ zero in the outer region, keeps them unaltered in the inner region, and, owing to the sudden cessation of $\mathbf{H}$ at the boundary, introduces new matter and current there. Let σ_1 and γ_1 be the surface representatives of ρ and Γ. They are given by

$$\mathbf{N}_1\mathbf{H} = 4\pi\sigma_1, \qquad \mathrm{V}\mathbf{N}_1\mathbf{H} = 4\pi\gamma_1, \quad \text{...............(18}a\text{)}$$

corresponding to (12a), $\mathbf{N}_1$ being the unit normal from the boundary to the inner region. If, now, we include this surface matter and current in the former ρ and Γ, the results are applicable to the inner region outside which $\mathbf{H}$ has been abolished, or to all space if we like to keep the outer $\mathbf{H}$ zero. P must be the potential of ρ and σ_1, and $\mathbf{A}$ the potential of Γ and γ_1.

As $\mathbf{H}$ in the inner region is the same as before $\mathbf{H}$ in the outer region was abolished, it follows that the force in the inner region due to the surface matter and current is identically the same as that due to the abolished ρ and Γ in the outer region. And, as there is now no force in the outer region, it follows that the force in the outer region due to the surface matter and current is the negative of that due to the matter and current in the inner region. If, for example, there was originally no ρ and no Γ in the outer region, the force in the inner region due to the surface matter is the negative of that due to the surface current, so that together they produce no force in the inner region, whilst in the outer region their joint effect is the negative of that of the ρ and Γ in the inner region.

Similarly we may treat the outer region as self-contained, by making $\mathbf{H}=0$ in the inner region, and introducing surface matter and current given by

$$\mathbf{N}_2\mathbf{H}=4\pi\sigma_2, \qquad \mathrm{V}\mathbf{N}_2\mathbf{H}=4\pi\gamma_2,$$

$\mathbf{N}_2$ being the unit normal from the boundary to the outer region. Since $\mathbf{N}_2=-\mathbf{N}_1$, σ_2 and γ_2 are the negatives of the former σ_1 and γ_1. The force in the inner region due to σ_2 and γ_2 is the negative of that due to the ρ and Γ in the outer region, whilst in the outer region it is the same as that due to the abolished ρ and Γ in the inner region. It will be convenient to have a fixed way of reckoning the normal and the surface matter and current. Let the normal be always + from the boundary to the outer region, and be called $\mathbf{N}$, the same as the former $\mathbf{N}_2$. Similarly call the surface matter and current σ and γ, corresponding to $\mathbf{N}$, being the former σ_2 and γ_2, so that $\sigma_1=-\sigma$, and $\gamma_1=-\gamma$.

Internal and External Energies.

Let ρ_1, Γ_1, and ρ_2, Γ_2 be the matter and current densities in the inner and the outer regions, σ and γ the surface matter and current (as just defined), P_1, $\mathbf{A}_1$, P_2, $\mathbf{A}_2$, and P_0, $\mathbf{A}_0$, the potentials of the inner, the outer, and the boundary matter and current, after (13a). Then the energies in the internal and the external regions are

$$\left.\begin{aligned}&\Sigma\tfrac{1}{2}(\mathbf{A}_1-\mathbf{A}_0)(\Gamma_1-\gamma)+\Sigma\tfrac{1}{2}(P_1-P_0)(\rho_1-\sigma),\\ \text{and}\quad&\Sigma\tfrac{1}{2}(\mathbf{A}_2+\mathbf{A}_0)(\Gamma_2+\gamma)+\Sigma\tfrac{1}{2}(P_2+P_0)(\rho_2+\sigma),\end{aligned}\right\}\quad\ldots\ldots\ldots(19a)$$

respectively, by (16a). But also, by disregarding the boundary altogether, the total energy in both regions is

$$\Sigma\tfrac{1}{2}(\mathbf{A}_1+\mathbf{A}_2)(\Gamma_1+\Gamma_2)+\Sigma\tfrac{1}{2}(P_1+P_2)(\rho_1+\rho_2). \quad \ldots\ldots\ldots(20a)$$

So the sum of the two expressions in (19a) equated to that in (20a) gives us the necessary relation

$$\Sigma(\mathbf{A}_1\Gamma_2+P_1\rho_2)=\Sigma\{\gamma(\mathbf{A}_2-\mathbf{A}_1)+\sigma(P_2-P_1)+\mathbf{A}_0\gamma+P_0\sigma\};$$

both members being expressions for the mutual energy of the matter and current in the two regions. If there is no ρ or Γ in the outer region, this is equivalent to

$$0=\Sigma(\mathbf{A}\gamma+P\sigma-\mathbf{A}_0\gamma-P_0\sigma),$$

and the external and the internal energies are

$$\Sigma\tfrac{1}{2}(\mathbf{A}_0\gamma+P_0\sigma), \qquad \text{and} \qquad \Sigma\tfrac{1}{2}(\mathbf{A}\Gamma+P\rho)-\Sigma\tfrac{1}{2}(\mathbf{A}_0\gamma+P_0\sigma),$$

respectively. The first of these we may also write as $\Sigma \frac{1}{2} P\sigma$, if $\mathbf{H} = -\nabla P$, as it can be expressed; or else as $\Sigma \frac{1}{2}\mathbf{A}\gamma$, if $\text{curl}\,\mathbf{A} = \mathbf{H}$, which is possible when $\Sigma\sigma = 0$.

Mechanical Force between the Regions.

There being matter and current in either or both regions, the resultant force on the inner region is

$$\Sigma(\rho_1 \mathbf{H} + \mathrm{V}\Gamma_1 \mathbf{H}), \quad \ldots\ldots\ldots\ldots\ldots\ldots\ldots\ldots (21a)$$

the summation extending throughout the region. It is zero on any region in which there is no matter or current. Similarly

$$\Sigma(\rho_2 \mathbf{H} + \mathrm{V}\Gamma_2 \mathbf{H}) \quad \ldots\ldots\ldots\ldots\ldots\ldots\ldots\ldots (22a)$$

is the resultant force on the outer region, the summation extending through *it*. Since the summations in (21*a*) and (22*a*) together include all space, the one sum is the negative of the other. Or, as (21*a*) is the resultant force *of* the outer *on* the inner region, and (22*a*) is that *of* the inner *on* the outer region, the one is the negative of the other. That is to say, action and reaction are equal and opposite; or, stress is mutual; or, a complete dynamical system cannot set itself moving, when taken as a whole. Both (21*a*) and (22*a*) are expressed by

$$\pm \Sigma \tfrac{1}{2}(\sigma \mathbf{H} + \mathrm{V}\gamma \mathbf{H}), \quad \ldots\ldots\ldots\ldots\ldots\ldots\ldots\ldots (23a)$$

taken over the boundary, using the + sign to express the force of the outer on the inner region, and the − sign to express that of the inner region on the outer. Comparing (23*a*) with (22*a*) and (21*a*), we see that boundary matter and current take the place of the matter and current in the inner or the outer region as the case may be. We may verify the equivalence of (23*a*) to the others by differentiation, applying the perennially useful and labour-saving Theorem of Convergence to either region with the common boundary; but the reason of (23*a*) and its necessity may be more simply seen thus. When we abolish the external field, and put $-\sigma$ and $-\gamma$ on the boundary, we make the inner region, with the boundary matter and current, a complete system, on which there is no resultant force. The resultant force on the surface matter and current $-\sigma$ and $-\gamma$ is therefore the negative of that on the internal ρ and Γ. But the surface matter and current are, so far as the mechanical force between the regions is concerned, equivalent to the external matter and current. Hence the resultant force on σ and γ is the same as that of the ρ and Γ in the outer region on the ρ and Γ in the inner. Hence, in (21*a*) we may put σ and γ for ρ_1 and Γ_1. But then, since by this we turn the volume-integral into a surface-integral, we must take the mean value of $\mathbf{H}$ through the infinitely thin layer of the surface matter and current. This is $\frac{1}{2}\mathbf{H}$, since $\mathbf{H} = 0$ outside when the surface matter and current are taken instead of the external ρ_2 and Γ_2. Hence the presence of the $\frac{1}{2}$ in (23*a*).

The vector in that expression, viz.,

$$\tfrac{1}{2}(\sigma \mathbf{H} + \mathrm{V}\gamma \mathbf{H}), \quad \ldots\ldots\ldots\ldots\ldots\ldots\ldots\ldots (24a)$$

is the vector stress, according to the last section, regarding one side of it only, the force of the outer on the inner region per unit area of the boundary. Putting $\sigma = \mathbf{NH}/4\pi$, $\gamma = \mathrm{V}\mathbf{NH}/4\pi$, it takes the form

$$\mathbf{H}(\mathbf{HN}/4\pi) - \mathbf{N}(\mathrm{H}^2/8\pi). \quad \ldots\ldots\ldots\ldots\ldots\ldots\ldots(25a)$$

Hence it is a simple Maxwellian stress of intensity $\mathrm{H}^2/8\pi$.

It is not necessary, in reckoning the resultant mutual force between the regions, to take $\mathbf{H}$ in the formulae, *i.e.*, to take the intensity of the force due to the matter and current in both regions. Thus, $\mathbf{H}_1$ being the force-intensity due to ρ_1 and Γ_1, and $\mathbf{H}_2$ that due to ρ_2 and Γ_2, the resultant forces on the inner and outer regions are

$$\Sigma(\rho_1\mathbf{H}_2 + \mathrm{V}\Gamma_1\mathbf{H}_2) \quad \text{and} \quad \Sigma(\rho_2\mathbf{H}_1 + \mathrm{V}\Gamma_2\mathbf{H}_1) \quad \ldots\ldots(26a)$$

respectively. Comparing with (21a) and (22a) we see that this is equivalent to saying that

$$0 = \Sigma(\rho_1\mathbf{H}_1 + \mathrm{V}\Gamma_1\mathbf{H}_1), \qquad 0 = \Sigma(\rho_2\mathbf{H}_2 + \mathrm{V}\Gamma_2\mathbf{H}_2). \quad \ldots\ldots(27a)$$

Remember that in the inner region $\mathbf{H}_2$ is the same as the force-intensity due to $-\sigma$ and $-\gamma$; whilst in the outer region $\mathbf{H}_1$ is the same as that due to $+\sigma$ and $+\gamma$. The (26a) expressions are equivalent to the boundary summations

$$\Sigma(\sigma\mathbf{H}_2 + \mathrm{V}\gamma\mathbf{H}_2) \quad \text{and} \quad -\Sigma(\sigma\mathbf{H}_1 + \mathrm{V}\gamma\mathbf{H}_1);$$

or, in terms of the surface $\mathbf{H}_1$ and $\mathbf{H}_2$ only, to

$$\pm\Sigma\{\mathbf{H}_1(\mathbf{NH}_2) + \mathbf{H}_2(\mathbf{NH}_1) - \mathbf{N}(\mathbf{H}_1\mathbf{H}_2)\}/4\pi. \quad \ldots\ \ldots\ldots(28a)$$

H *and P in either region due to ρ and Γ in the other.*

For distinctness, let there be no ρ or Γ in the outer region. $\mathbf{H}$ is then given by

$$\mathbf{H} = -\Sigma(\rho\mathbf{f} + \mathrm{V}\Gamma\mathbf{f}) = -\nabla\Sigma p\rho + \mathrm{curl}\,\Sigma p\Gamma, \quad \ldots\ldots\ldots(29a)$$

$\mathbf{H}$ being reckoned at a fixed point Q, and p and $\mathbf{f}$ being the potential and force-intensity at the place of ρ and Γ due to unit matter at Q. That is,

$$p = 1/r, \qquad \mathbf{f} = \mathbf{r}_1/r^2,$$

if r be the distance from Q to the place of ρ or Γ, and $\mathbf{r}_1$ a unit vector along $\mathbf{r}$ from Q. For ρ and Γ we may substitute the surface matter and current, when the point Q is in the outer region; thus,

$$\mathbf{H} = -\Sigma(\sigma\mathbf{f} + \mathrm{V}\gamma\mathbf{f}) = -\nabla\Sigma p\sigma + \mathrm{curl}\,\Sigma p\gamma. \quad \ldots\ldots\ldots(30a)$$

Then, since σ and γ depend only upon the boundary $\mathbf{H}$, if $\mathbf{H}$ be given only over a closed surface, we know $\mathbf{H}$ through the whole external space, so far as it depends upon ρ and Γ within the surface, and therefore definitely throughout the external space when there is ρ and Γ only within the surface (or, in the extreme, upon it). As we change the form of the boundary, the distributions of the surface σ and γ change. There may be matter only, viz., when $\mathbf{H}$ is normal everywhere, or the boundary is an equipotential surface (say due to an electrically charged conductor). There would be current only if $\mathbf{H}$ could be everywhere tangential, but this is not possible if $\mathbf{H}$ be magnetic force, at least without having current in the outer region. In general, we have both matter and current. Although the external field is definitely fixed by the surface **H**, we can get no information from it as to the internal field, except that $\Sigma\sigma = \Sigma\rho$, or the matter on the surface is the same in amount as that within it. If this be not zero, there is matter of amount $-\Sigma\sigma$ or $-\Sigma\rho$ on the surface at infinity, if we stop the extension of space. A closed current is equivalent to equal amounts of + and − matter; so is a magnet.

In the second form of (30*a*), the force **H** is derived from the scalar potential of σ and the vector potential of γ. We can, however, derive it from a scalar potential only, thus:—Since the surface **H**, say $\mathbf{H}_0$, is given, the surface potential can be found, except as regards a constant, by a line-integration on the surface from a fixed point, whose potential is taken as zero. Thus, P_0 being the surface potential,

$$P_0 = -\int \mathbf{H}_0 d\mathbf{s}, \quad \text{.........................(31}a\text{)}$$

$\mathbf{H}_0 d\mathbf{s}$ being the scalar product of $\mathbf{H}_0$ and $d\mathbf{s}$, the vector element of the line of integration. Then, if P is the potential at the external point Q, we shall have

$$P = \Sigma p\sigma - \Sigma(\mathbf{fN})P_0/4\pi, \quad \text{.....................(32}a\text{)}$$

and $\mathbf{H} = -\nabla P$. The first Σ gives that part of the external potential due to the surface matter, $\sigma = \mathbf{H}_0\mathbf{N}/4\pi$. The second part is the scalar potential of the current; that is, if it be electric current, it is its magnetic potential. Or, it is the external magnetic potential of a closed magnetic shell, normally magnetised to strength $P_0/4\pi$. For, if **I** be the vector magnetic moment of a small magnet, its potential at Q is $-\mathbf{If}$; in the present case $\mathbf{I} = \mathbf{N}P_0/4\pi$, and $P_0/4\pi$ is the moment per unit area. It is the same as Maxwell's "current-function" of a current-sheet. Test (32*a*) by the Convergence Theorem. The indeterminateness of P_0 as per (31*a*) as regards a constant does not affect (32*a*), the external potential of a closed magnetic shell of uniform strength being zero.

The external P in terms of the Surface P_0.

Let it be the surface P_0 that is given, not $\mathbf{H}_0$ the surface force. A part of P is known, viz., the second term on the right of (32*a*). But the first term being in terms of the force, through σ, must be got rid of. Let x denote the operator that finds the external potential of the magnetic shell; that is, $x = -\Sigma(\mathbf{fN})/4\pi$ operating upon P_0, finds xP_0 the external potential of the shell of strength $P_0/4\pi$; this potential xP_0 is not the same as P. Let x_0P_0 denote the surface value of xP_0, not the same as P_0. Then we may denote by x_0^{-1} the operator which, acting upon P_0, finds $x_0^{-1}P_0$, the strength of shell whose potential *is* P_0 at the surface. Then, at an external point,

$$P = xx_0^{-1}P_0.$$

To render this process intelligible, expand x_0^{-1} thus:—

$$x_0^{-1} = \{1-(1-x_0)\}^{-1} = 1 + (1-x_0) + (1-x_0)^2 + \dots .$$

Then we have

$$P = x\{P_0 + (1-x_0)P_0 + (1-x_0)^2P_0 + \dots\}, \quad \dots\dots\dots\dots\dots\dots(33a)$$

giving the external P in terms of the surface P_0 by direct operations, when for x we put its rational equivalent $-\Sigma(\mathbf{fN})/4\pi$.

The first approximation is xP_0. The next is $x(P_0 - x_0P_0)$, and so on. It is a process of exhaustion. But it only works when $\Sigma\rho = 0$. (33a) therefore solves the problem of finding the potential throughout a region bounded by a closed surface (taking $\mathbf{N}$ as the inward normal) in terms of the surface potential, the ρ and Γ being on the other side. And, when ρ and Γ are inside, it finds the external potential if $\Sigma\rho = 0$.

Annihilation of the Surface Current.

To get rid of γ and substitute matter giving the same external potential. Thus, given $\mathbf{H}_0$. The first of (30a) gives $\mathbf{H}$ in terms of $\mathbf{H}_0$ through σ and γ. The first distribution of matter is $\sigma = \mathbf{NH}_0/4\pi$, and the force due to it is $-\Sigma\mathbf{f}(\mathbf{NH}_0)/4\pi$. Call this $y\mathbf{H}_0$, and let its surface value be $y_0\mathbf{H}_0$. Then

$$\mathbf{H} = yy_0^{-1}\mathbf{H}_0 = y\{\mathbf{H}_0 + (1-y_0)\mathbf{H}_0 + (1-y_0)^2\mathbf{H}_0 + \dots\}, \quad \dots\dots(34a)$$

in direct operations.

Or thus, $\mathbf{H}_0$ gives a first σ and γ. Find the field due to the first γ. It has a normal and a tangential component, and therefore gives a second σ and γ. Find the field due to the second γ, which gives a third σ and γ; and so on. As we proceed, the γ left gets smaller and smaller, and is finally annihilated, leaving a distribution of matter, the sum of all the σ's, say σ_0. Then $\mathbf{H} = -\Sigma\mathbf{f}\sigma_0$, where σ_0 is given by

$$\sigma_0 = \mathbf{N}(1-z)^{-1}\mathbf{H}_0/4\pi, \quad \text{where} \quad z = \Sigma\mathbf{fVN}/4\pi,$$

or,

$$\sigma_0 = \mathbf{N}(\mathbf{H}_0 + z\mathbf{H}_0 + z^2\mathbf{H}_0 + \dots)/4\pi. \quad \dots\dots\dots\dots\dots\dots\dots\dots(35a)$$

Annihilation of the Surface Matter, when possible.

Starting with $\mathbf{F}_0$, it gives a first σ and γ, and the $\mathbf{H}$ due to the latter is

$$\Sigma\mathbf{Vf}\gamma = \Sigma\mathbf{VfVNH}_0/4\pi = w\mathbf{H}_0 \quad \text{say}.$$

Let $w_0\mathbf{F}_0$ be the surface value. Then

$$\mathbf{H} = ww_0^{-1}\mathbf{H}_0 = w\{\mathbf{H}_0 + (1-w_0)\mathbf{H}_0 + (1-w_0)^2\mathbf{H}_0 + \dots\}. \quad \dots(36a)$$

Or thus; in the manner (35a) was got, but annihilating σ instead of γ, we shall have $\mathbf{H} = \Sigma\mathbf{Vf}\gamma_0$, where γ_0 is the finally-arrived-at current given by

$$\gamma_0 = \mathbf{VN}(1-y)^{-1}\mathbf{H}_0/4\pi; \quad \text{where} \quad y = -\Sigma\mathbf{f}.(\mathbf{N}/4\pi).$$

Or

$$\gamma_0 = \mathbf{VfVN}(\mathbf{H}_0 + y\mathbf{H}_0 + y^2\mathbf{H}_0 + \dots)/4\pi. \quad \dots\dots\dots\dots(37a)$$

Here y is the same as in (34a).

P in Terms of the Surface $\mathbf{H}_0$.

When the ρ and Γ are in the inner region, the first approximation to the value of P at the external point Q is $\Sigma p\mathbf{NH}_0/4\pi$, say $u\mathbf{H}_0$; and the force due to this is $-\nabla u\mathbf{H}_0$. The complete external potential is

$$P = u(-\nabla u_0)^{-1}\mathbf{H}_0 = u\{1 + (1 + \nabla u_0) + \dots\}\mathbf{H}_0, \quad \dots\dots\dots(38a).$$

But if the point Q be inside, and the ρ and Γ outside, this formula does not give the internal P (with **N** as the inward normal), but leaves it indeterminate as regards a constant.

A *in Terms of the Surface* $\mathbf{H}_0$.

The first approximation to **A** (such that $\mathbf{H} = \text{curl}\,\mathbf{A}$) being $\Sigma p\mathrm{V}\mathbf{N}\mathbf{H}_0/4\pi$, say $q\mathbf{H}_0$, and the force due to this being curl $q\mathbf{H}_0$, the complete external **A** at Q, when ρ and Γ are inside, and $\Sigma\rho = 0$, is

$$\mathbf{A} = q(\text{curl}.q_0)^{-1}\mathbf{H}_0 = q\{1 + (1 - \text{curl}.q_0) + \dots\}H_0. \quad \dots\dots\dots(39a)$$

But if $\Sigma\rho$ or $\Sigma\sigma$ is not zero, we shall have

$$\mathbf{H} = \text{curl}\,\mathbf{A} - \nabla P_1,$$

where **A** is given by (39*a*), and P_1 is the potential due to $\Sigma\rho$ distributed equipotentially. The external energy will now be expressed by

$$\Sigma\tfrac{1}{2}\mathbf{A}\gamma + \tfrac{1}{2}P_1\Sigma\rho$$

where $\gamma = \mathrm{V}\mathbf{N}\mathbf{H}_0/4\pi$; the mutual energy of the fields of **A** and P_1 being zero.

Remarks on these formulæ.

The process indicated by the right member of (33*a*) consists in the substitution of magnetic shells for matter, and finding their potentials. Let the final result be P_1, and P_{10} be its surface value. Then $P_{10} = P_0$ if $\Sigma\rho = 0$. Otherwise, their difference must be a constant, say $P_0 - P_{10} = P_{20}$. For xP_{20} must be zero, therefore P_{20} is constant. It is the potential of the surface when the quantity of matter $\Sigma\rho = \Sigma\sigma$ is distributed over it equipotentially.

Similarly, in the (36*a*), (37*a*) annihilation of the surface matter, so far as is possible, we arrive at the force which differs from the real force by that due to the matter $\Sigma\sigma$ which is left, and which is distributed equipotentially.

It will be observed that whilst the first σ and γ together produce no field on one side of the surface, the annihilation of either completely alters this. The complete σ_0, for example, produces a field on both sides, although it is the same on one side as that due to σ and γ, *i.e.*, on the side where **H** was under investigation.

There is always a distinction between the external and the internal regions as regards the determination of P from the surface *force*. It fixes the external force, when ρ and Γ are inside, and it also fixes the potential, so as to vanish at an infinite distance. It also fixes the internal force when the ρ and Γ are outside, but cannot then fix the potential. (*E.g.*, no surface force, yet a constant internal potential, depending upon external matter.) This puts a difficulty in the way of the estimation of the external P in terms of the surface P_0 when the ρ and Γ are inside, and $\Sigma\rho$ is not zero, even when we apply Green's method. We do not arrive at the proper Greenian distribution of matter, but at another, giving a surface potential differing from what is wanted by a known constant, so that we have to find another distribution, to give this constant potential.

SECTION XXIV. ACTION BETWEEN A MAGNET AND A MAGNET, OR BETWEEN A MAGNET AND A CONDUCTOR SUPPORTING AN ELECTRIC CURRENT. THE CLOSURE OF THE ELECTRIC CURRENT. ITS NECESSITY.

The section before the last being preliminary to the subject of the stresses, and the last section being of a perfectly abstract nature, the one to follow this will be on the magnetic stress in general, as modified by differences of permeability and other causes. The present section is of an intermediate nature. Though dealing with the magnetic stress outside magnets, its principal object is to direct attention to the vexed question of the closure of the electric current; which I endeavour, as far as I can, to bring down to a question of definition.

Let us forget, if possible, for a time, all knowledge of the electric current—or rather, let us make no use of it. Suppose that we are fully acquainted with the mechanical actions of rigid magnets upon one another. That there is probably no such thing as a perfectly rigid magnet (that is, in the larger theory, an unmagnetisable magnet, whose permeability is unity, or the same as that of the enveloping medium) is immaterial. We suppose there is. Except that the argument would be more complex, it would not alter our general conclusions to take a magnetisable magnet.

We may put any collection of little rigid magnets together to form a complex magnet, having any distribution of magnetisation. Its external field of force is to be got by observing the mechanical force it exerts upon one pole of an exceedingly weak and slender magnetised filament, uniformly and longitudinally magnetised, so as to localise its poles strictly at its ends. Upon the basis of the definition of a unit pole, that it repels a similar unit pole at unit distance with unit force, we can map out the external field of force. Let $\mathbf{F}$ be the magnetic force intensity, that is, the force on a unit pole placed in the field. $\mathbf{F}$ is subject to the conditions

$$\operatorname{div} \mathbf{F} = 0, \qquad \operatorname{curl} \mathbf{F} = 0,$$

outside the magnet. The lines of $\mathbf{F}$ start from the surface of the magnet, proceed in curved paths through the air, and end upon its surface again at other places.

If we describe a closed surface in the air, completely enclosing the magnets, of whose position within the surface we might be ignorant, we should be entirely unable from our examination of the field outside the surface, to determine the interior distribution of magnetisation, or even to determine the situation of its poles, that is, the distribution of the imaginary magnetic matter, to which, if we ascribe self-repulsive force according to the inverse-square law, we may attribute the external force. We might ascribe $\mathbf{F}$ to a distribution of matter σ over the surface itself, of total quantity zero; or to a distribution of a vector quantity γ over the surface in closed lines, producing external **F** according to another law, viz., $\mathrm{V}\mathbf{f}\gamma$ = force at a point Q due to the element γ; **f** being the force at γ due to unit matter at Q; or to combinations of both σ and γ, one of which is unique, inasmuch as it produces no internal field. Contract the surface until it reaches the magnet itself. Then we may ascribe the external field to σ, or to γ, or to combinations, definitely, on the surface of the magnet, or in its interior in various way, but not definitely. The distribution of magnetisation is, *a fortiori*, still more undiscoverable, for there may be distributions corresponding to no magnetic matter, giving no external force.

But, making use of our knowledge of the effect of building up a magnet from smaller ones, we may suppose that the magnetisation is known. Let the vector magnetisation be **I**, and its convergence be ρ, the density of the magnetic matter. Then

$$\operatorname{div}\mathbf{F} = 4\pi\rho, \qquad \operatorname{curl}\mathbf{F} = 0$$

fully determine **F**. It is of considerable practical utility in theoretical reasoning not to treat the surface- and the volume-densities separately, but to include them both in ρ, the volume-density. Thus, if a magnet be quite uniformly magnetised, there is, strictly speaking, no ρ. The convergence of **I** is on the surface. But, by supposing **I** not to cease abruptly on reaching the surface, but gradually, however rapidly, through a thin surface-layer, we make the matter have a space distribution, of the same total amount per unit of surface as the surface distribution it represents. We can always derive the surface expressions from those of the volume with great ease, when we want them; whilst our work is much simpler without them when we do not want them.

The resultant force between any two magnets may now be easily represented. $\Sigma\mathbf{F}_2\rho_1$ is the resultant force on a magnet whose matter density is ρ_1, due to another magnet whose polar force is $\mathbf{F}_2$. And $\Sigma\mathbf{F}_1\rho_2$ is the force of the first magnet on the second. (We need not trouble about the forces of rotation at present, which are nearly as simply represented.) These are equal and opposite. Also $\Sigma\mathbf{F}_1\rho_1 = 0$ and $\Sigma\mathbf{F}_2\rho_2 = 0$; that is, a magnet cannot translate *itself* (or rotate itself either). So if $\mathbf{F} = \mathbf{F}_1 + \mathbf{F}_2$, making **F** the actual force of the field, the forces on the magnets are $\Sigma\mathbf{F}\rho_1$ and $\Sigma\mathbf{F}\rho_2$ respectively.

Describe any closed surface separating one magnet from the other; let the first magnet be inside. The force on it may, as in the last section, be represented by the surface-integral

$$\Sigma\,\tfrac{1}{2}(\sigma\mathbf{F} + \mathrm{V}\gamma\mathbf{F}),$$

if $\sigma = \mathbf{NF}/4\pi$, $\gamma = \mathrm{V}\mathbf{NF}/4\pi$, **N** being a unit normal from the inner to the outer region. Or, which is the same thing, by

$$\Sigma\{\mathbf{F}(\mathbf{NF}) - \mathbf{N}(\tfrac{1}{2}\mathbf{F}^2)\}/4\pi,$$

in terms of the Maxwellian stress.

The following is the process of showing that this stress gives rise to the required mechanical forces. The quantity summed up is the vector stress on the plane whose normal is **N**. If we fix the direction of **N**, its divergence is the **N**-component of the force per unit volume, by the principles of varying stress.

Now $$\text{div}\,\{\mathbf{F}(\mathbf{NF})\} = (\mathbf{NF})\,\text{div}\,\mathbf{F} + \mathbf{F}\nabla(\mathbf{NF}),$$

and $$\text{div}\,\{\mathbf{N}(\tfrac{1}{2}\mathbf{F}^2)\} = \frac{d}{dn}\left(\tfrac{1}{2}\mathbf{F}^2\right) = \mathbf{F}\frac{d\mathbf{F}}{dn},$$

if n be length measured along **N**.

Also, $$\mathbf{F}\left\{\frac{d\mathbf{F}}{dn} - \dot{\nabla}(\mathbf{N}\dot{\mathbf{F}})\right\} = \mathbf{N}\mathrm{V}\mathbf{F}\,\text{curl}\,\mathbf{F},$$

so, **S** being the vector stress on the **N** plane,

$$\begin{aligned} 4\pi\,\text{div}\,\mathbf{S} &= (\mathbf{NF})\,\text{div}\,\mathbf{F} - \mathbf{N}\mathrm{V}\mathbf{F}\,\text{curl}\,\mathbf{F}, \\ &= \mathbf{N}\{\mathbf{F}\,\text{div}\,\mathbf{F} - \mathrm{V}\mathbf{F}\,\text{curl}\,\mathbf{F}\}, \\ \text{div}\,\mathbf{S} &= \mathbf{N}\{\mathbf{F}\rho + \mathrm{V}\Gamma\mathbf{F}\}, \end{aligned}$$

if $\Gamma = \text{curl}\,\mathbf{F}/4\pi$. This being the **N**-component of the force, the force itself is

$$\mathbf{F}\rho + \mathrm{V}\Gamma\mathbf{F}$$

per unit volume; or, since $\Gamma = 0$, the magnetic force of the magnets being polar, simply $\mathbf{F}\rho$ per unit volume.

The length of this process depends upon our wishing to develop the term $\mathrm{V}\Gamma\mathbf{F}$. If it were not for that, we would see at once that

$$\nabla(\mathbf{NF}) = \frac{d\mathbf{F}}{dn}$$

when **F** is polar, **N** being in any direction, and so get the force $\mathbf{F}\rho$ immediately.

The question now asks itself (remembering that we are ignorant of the electric current), what is this Γ, whose vanishing cuts the work short at the beginning, in our case of **F** being polar. Can it really represent any physical magnitude?

It is defined by $\text{curl}\,\mathbf{F} = 4\pi\Gamma$, and is necessarily zero in the case of magnets. It indicates *closed* lines of **F**, which are impossible with a strictly polar force in all space. Γ is a vector which is necessarily circuital. This is a mathematical consequence of its definition.

Furthermore, whilst **F**, as a polar force, with $\Gamma = 0$, is a special kind of distribution of a vector, if we allow Γ to be not zero **F** becomes of the most general type possible, *any* distribution of force, or any field of force, without the polar limitation. Given the divergence, and the curl of a vector, the vector itself is fixed, if it is to vanish at infinity.

Supposing, then, we allow that Γ can exist, we can predict what the mutual force between a magnet and it will be. Let there be Γ in a certain region, outside of which there is only ρ, therefore only magnets. The resultant force on this region is equivalent to $\mathrm{V}\Gamma\mathbf{F}$ per unit volume. We cannot localise the force—we can only know its total amount. It is necessarily $\Sigma\,\mathrm{V}\Gamma\mathbf{F}$. The reason we cannot say that $\mathrm{V}\Gamma\mathbf{F}$ is the force on unit volume is that Γ is necessarily circuital, and so we cannot work down to a unit volume without having current in the external region as well, which is against our previous knowledge.

So far, Γ has a merely speculative existence. It is got by making the assumption that there can be circuital magnetic force. Admit that there can be, the laws of Γ follow. Γ is necessarily circuital. The magnetic force it produces at Q is $\Sigma \mathbf{f}\Gamma$, if $\mathbf{f}$ be the force at Γ due to unit matter at Q. The force between one Γ and another, and between Γ and ρ follow, viz., that the resultant force on any region containing ρ and closed Γ is

$$\Sigma(\rho \mathbf{H} + \mathrm{V}\Gamma\mathbf{H}),$$

H being the actual intensity of the field due to all the ρ *and* Γ.

Now, we do know what Γ is, under certain circumstances. By the researches of Ampère, the father of electrodynamics, we know that Γ measures the density of current in a conductor when it is steady. His researches were conducted in a very different manner, and are indeed very difficult to follow, like most novel researches, but the results are exactly these, without his hypotheses as to the action between different elements of a current. We virtually measure the strength of current in a conductor, when we use a galvanometer, by the line-integral of **H** round a current, and that is the amount of the quantity $\Gamma = \mathrm{curl}\,\mathbf{H}/4\pi$, passing through the line of integration.

Also, steady currents are closed. So far, then, we identify Γ with the conduction current.

But *our* Γ is closed under any circumstances. We know also that conduction currents are not always closed; for instance, when we charge a condenser. We still measure the conduction current in its transient state by Γ. We do so by the continuously changing instantaneous magnetic force if the charge or discharge be slow enough; otherwise, by the ballistic method, which is virtually the same. Γ being then unclosed in the conductor, has necessarily its exact complement, to close it, outside the conductor, *i.e.*, our Γ has, though it may be only called current when in the conductor.

But our Γ, being identified with current when in a conductor, both in steady states when the current is closed in the conductor, and in transient states when it is closed through the dielectric, and this Γ in the dielectric being related to the magnetic force in the same way as if it were conduction current, why should we not call it electric current also? As it demonstrably exists, we see that the closure of the current is reduced to a question of a name. It would be positively illogical not to call it electric current.

To sum up:—

1. From magnetic knowledge only, there should be no circuital magnetic force in the space outside magnets.

2. If we admit the existence of circuital magnetic force, or, say generally, if we admit that the line-integral of the force in a circuit in air can be finite, we arrive at a vector quantity Γ having also the property of being circuital, and we can find the mechanical force between it and magnets or other Γ, and can thus definitely measure Γ.

3. This Γ we know to be conduction current, when steady.

4. But in transient states, conduction currents are not always circuital.

5. But a part of Γ still measures the conduction current.

6. The other part, the complement of the conduction current, is outside the conductor, continuous with the conduction current, and closing it.

7. Then why not call it electric current?

We see that it is not a question for experiment, for no amount of experimenting could alter this reasoning, but of definition, an agreement to call a certain function of the magnetic force always by one name, viz., the electric current, which, if in a conductor, heats it and wastes energy, whilst in a nonconductor does not, energy being stored potentially. It is, of course, needless to add that this current in a nonconductor is Maxwell's current of displacement, **D**, the rate of increase of the displacement, whilst $\mathbf{E}\dot{\mathbf{D}}$ is the activity of the electric force **E** to match, and $\frac{1}{2}\mathbf{ED}$ the stored potential energy of displacement.

NOTE on equation (32*a*) [p. 553].—For mnemonical purposes, the following is a concise form of this equation. The potential being given $=P_0$ over a closed surface, due to matter or current within or on it, the potential P at any external point Q is

$$P = -\Sigma p^2 \frac{d}{dn}(P_0/p), \quad \text{..................(32}a\text{) } bis.$$

if p be the potential due to unit matter at Q, and n be length measured along the normal outward.

[The second half, Sections 25 to 47, of this Article, is in vol. 2].

END OF VOL. I.

ROBERT MACLEHOSE, PRINTER TO THE UNIVERSITY, GLASGOW.

Printed in the United States
By Bookmasters